Research Perspectives on Functional Micro- and Nanoscale Coatings

Ana Zuzuarregui
CIC nanoGUNE Consolider, Spain

Maria Carmen Morant-Miñana
CIC nanoGUNE Consolider, Spain

A volume in the Advances in Chemical and Materials Engineering (ACME) Book Series

Information Science REFERENCE
An Imprint of IGI Global

Published in the United States of America by
Information Science Reference (an imprint of IGI Global)
701 E. Chocolate Avenue
Hershey PA, USA 17033
Tel: 717-533-8845
Fax: 717-533-8661
E-mail: cust@igi-global.com
Web site: http://www.igi-global.com

Library of Congress Cataloging-in-Publication Data

Library of Congress Cataloging-in-Publication Data

Names: Zuzuarregui, Ana, 1984- editor. | Morant-Minana, Maria Carmen, 1979- editor.
Title: Research perspectives on functional micro- and nanoscale coatings / Ana Zuzuarregui and Maria Carmen Morant-Minana, editors.
Description: Hershey, PA : Information Science Reference, [2016] | Includes bibliographical references and index.
Identifiers: LCCN 2015051301| ISBN 9781522500667 (hardcover) | ISBN 9781522500674 (ebook)
Subjects: LCSH: Protective coatings--Materials. | Nanocomposites (Materials) | Thin films.
Classification: LCC TA418.76 .R47 2016 | DDC 667/.9--dc23 LC record available at http://lccn.loc.gov/2015051301

This book is published in the IGI Global book series Advances in Chemical and Materials Engineering (ACME) (ISSN: 2327-5448; eISSN: 2327-5456)

British Cataloguing in Publication Data
A Cataloguing in Publication record for this book is available from the British Library.

For electronic access to this publication, please contact: eresources@igi-global.com.

Advances in Chemical and Materials Engineering (ACME) Book Series

J. Paulo Davim
University of Aveiro, Portugal

ISSN: 2327-5448
EISSN: 2327-5456

Mission

The cross disciplinary approach of chemical and materials engineering is rapidly growing as it applies to the study of educational, scientific and industrial research activities by solving complex chemical problems using computational techniques and statistical methods.

The **Advances in Chemical and Materials Engineering (ACME) Book Series** provides research on the recent advances throughout computational and statistical methods of analysis and modeling. This series brings together collaboration between chemists, engineers, statisticians, and computer scientists and offers a wealth of knowledge and useful tools to academics, practitioners, and professionals through high quality publications.

Coverage

- Fatigue and Creep
- Chemical Engineering
- Ceramics and glasses
- Computational methods
- Composites
- Sustainable Materials
- Electrochemical and Corrosion
- Biomaterials
- Thermo-Chemical Treatments
- Multifuncional and Smart Materials

IGI Global is currently accepting manuscripts for publication within this series. To submit a proposal for a volume in this series, please contact our Acquisition Editors at Acquisitions@igi-global.com or visit: http://www.igi-global.com/publish/.

The Advances in Chemical and Materials Engineering (ACME) Book Series (ISSN 2327-5448) is published by IGI Global, 701 E. Chocolate Avenue, Hershey, PA 17033-1240, USA, www.igi-global.com. This series is composed of titles available for purchase individually; each title is edited to be contextually exclusive from any other title within the series. For pricing and ordering information please visit http://www.igi-global.com/book-series/advances-chemical-materials-engineering/73687. Postmaster: Send all address changes to above address.

Titles in this Series

For a list of additional titles in this series, please visit: www.igi-global.com

Emerging Research on Bioinspired Materials Engineering
Mohamed Bououdina (University of Bahrain, Bahrain)
Engineering Science Reference • copyright 2016 • 441pp • H/C (ISBN: 9781466698116) • US $215.00 (our price)

Petrochemical Catalyst Materials, Processes, and Emerging Technologies
Hamid Al-Megren (King Abdulaziz City for Science and Technology, Saudi Arabia) and Tiancun Xiao (Oxford University, UK)
Engineering Science Reference • copyright 2016 • 539pp • H/C (ISBN: 9781466699755) • US $225.00 (our price)

Applying Nanotechnology to the Desulfurization Process in Petroleum Engineering
Tawfik A. Saleh (Department of Chemistry, King Fahd University of Petroleum and Minerals, Saudi Arabia)
Engineering Science Reference • copyright 2016 • 555pp • H/C (ISBN: 9781466695450) • US $235.00 (our price)

Quantitative Structure-Activity Relationships in Drug Design, Predictive Toxicology, and Risk Assessment
Kunal Roy (Jadavpur University, India)
Medical Information Science Reference • copyright 2015 • 703pp • H/C (ISBN: 9781466681361) • US $295.00 (our price)

Handbook of Research on Recent Developments in Materials Science and Corrosion Engineering Education
Hwee Ling Lim (The Petroleum Institute, UAE)
Engineering Science Reference • copyright 2015 • 493pp • H/C (ISBN: 9781466681835) • US $325.00 (our price)

Processing Techniques and Tribological Behavior of Composite Materials
Rajnesh Tyagi (Indian Institute of Technology (BHU), Varanasi, India) and J. Paulo Davim (University of Aveiro, Portugal)
Engineering Science Reference • copyright 2015 • 464pp • H/C (ISBN: 9781466675308) • US $225.00 (our price)

Handbook of Research on Diverse Applications of Nanotechnology in Biomedicine, Chemistry, and Engineering
Shivani Soni (Alabama State University, USA) Amandeep Salhotra (City of Hope National Medical Center, USA) and Mrutyunjay Suar (KIIT University, India)
Engineering Science Reference • copyright 2015 • 820pp • H/C (ISBN: 9781466663633) • US $395.00 (our price)

Innovative Applications and Developments of Micro-Pattern Gaseous Detectors
Tom Francke (Myon, Sweden) and Vladimir Peskov (CERN, Switzerland)
Engineering Science Reference • copyright 2014 • 303pp • H/C (ISBN: 9781466660144) • US $215.00 (our price)

IGI GLOBAL
DISSEMINATOR OF KNOWLEDGE
www.igi-global.com

701 E. Chocolate Ave., Hershey, PA 17033
Order online at www.igi-global.com or call 717-533-8845 x100
To place a standing order for titles released in this series, contact: cust@igi-global.com
Mon-Fri 8:00 am - 5:00 pm (est) or fax 24 hours a day 717-533-8661

Editorial Advisory Board

Table of Contents

Detailed Table of Contents

Photoactive nanoparticles are smart systems that exhibit unique optical properties. In general, their intrinsic properties are size dependent. The degree and type of response to size are both related to their composition. Nanoparticles usually require to be capped with organic ligands in order to be dispersible in an aqueous or organic media, thus leading to nanoparticle colloidal dispersions and enhancing the processability of the material. The organic ligand also plays a key role in their preparation. In addition, the high surface-to-volume ratio of the nanoparticles combined with the affinity of the ligands for the nanoparticle surface can be used to place a large number of functional molecules at their periphery. The purpose of this chapter is to understand the synergism between nanoparticles and organic ligands with regard to their preparation, performance, and applicability.

Functional polymeric coatings comprise a diverse array of organic/inorganic functional groups, properties, and applications. This chapter attempts to provide a comprehensive review on the science and engineering of functional polymer coatings for both the novice and experienced chemist/chemical engineer. Functional polymer coatings are coatings that are designed through careful selection of organic/inorganic functional groups and controlled synthesis on the substrate's surface. The design leads to changes in the surface properties (e.g., adhesion, wettability, corrosion resistance, wear resistance, anti-fouling, non-stick, and anti-scratch) of a particular substrate (metal, glass, wood, or plastic). Functional polymeric coatings represent the "next generation" of coatings. These coatings dynamically respond to their environment, as opposed to a static coating that is used as a simple barrier, or for decorative purposes. This review is divided into three sections: (1) functionalization of polymers, (2) polymer properties and surface analytical techniques, and (3) applications.

This chapter focuses on resins based on non-hydrolyzed, monomeric and polymeric alkoxysilanes. As alternative to classical sol-gel processing, the resins are applied to a surface without a preceding hydrolysis step. Only after application, hydrolysis and condensation of the alkoxysilyl groups occur by means of atmospheric moisture to result cross-linked organic-inorganic hybrid coatings. While the use of non-hydrolyzed silanes is well established, for example by applying polyethyl silicate as binder for zinc-rich anti-corrosive primers, this chapter describes the chemical structures of various novel organic-inorganic hybrid precursors that have significantly extended the area of application to adhesives and scratch-resistant, repellent, or anti-fouling coatings. At present, individual resins are produced and applied at industrial scale in the fields of protective coatings and automotive topcoats.

Research topics related to the production of nanocomposites are the most important directions of development of new semiconductor engineering, ensuring high nanocomposites obtaining useful properties in the scope of biophysical characteristics, biomedical and piezoelectric applications. We present two case studies as Hydroxyapatite are in medical applications and aluminum nitride as acoustic wave sensor. Hydroxyapatite, is the main inorganic structure of the tooth enamel and bone and is a biomaterial that is commonly used in biomedical applications that involve bone substitution, drug delivery and bone regeneration because of its excellent biocompatibility, high bioactivity and good osseoconductivity. Since the past decade. Aluminum nitride (AlN), an electrical insulating ceramic with a wide band gap of 6.3 eV, is a potentially useful dielectric material very important in fields such as optoelectronic and micro electronics.

This chapter describes the functionalization of the surface of the glass through thin film inorganic coatings. Such filters called dichroic filters, following the physical principle of interference of light using the rules of optical physics. The design of the optical multilayer materials with high and low refractive index and optimizing the predetermined optical responses. The AFM quantify the real thickness and, calibrate the deposition plant, using as an input data for the simulation to evaluate the dispersion index refraction and absorption. The dichroic filters separate the incident radiation into two or more predetermined optical bands. The materials used are deposited using PVD Reactive Magnetron Sputtering, allows to increase the deposition rate, to obtain good homogeneity range of the surface deposited. The set point of deposition which allows to obtain stoichiometric oxides is analyzed with the technique RBS. These optical filters, also called dichroic are applied in the research field of the splitting photovoltaic concentration.

This chapter focuses on the plasmonic effects that appear in the ellipsometric functions and the pseudodielectric function when metal thin films and nanoparticles are analyzed by spectroscopic ellipsometry in the visible, near infrared and ultraviolet regions of the electromagnetic spectrum. The chapter is structured in two large sections. The first section reviews the basics of total internal reflection ellipsometry (TIRE), based on the excitation of surface polaritons in metal thin films. The conditions required to excite polaritons in TIRE systems are analyzed along with the main characteristics of those electromagnetic waves. The second section of the chapter is devoted to study the optical properties of plasmonic resonances in nanostructures and the characteristics introduced in the dielectric functions. The treatment of optical anisotropies and Fano resonances in the ellipsometric models is discussed. The last section of the chapter reviews the state of the art of the technique in biosensing applications.

This chapter outlines the application of coatings for implants and/or scaffolds used for bone tissue engineering (TE). Previously orthopaedic implants and/or scaffolds were designed considering mainly their mechanical aspects. However, due to the criticality of osteointegration with the surrounding tissues after implanting, the biological aspects of implants and/or scaffolds are becoming crucial. Recent trend is to use functional coatings like anti-infective, Ca-P and biomolecules coatings onto implants and/or scaffolds to improve anti-infective, osteoconductive and osteoinductive properties, respectively; thereby improving the osteointegration of the implant and/or scaffold with the surrounding tissue. Here, the application of different types of coatings on implants and/or scaffolds for bone TE will be described. The use of coatings as a drug and gene delivery carrier will also be covered in brief. As many of the coatings are still in preclinical testing stage, challenges associated with successful clinical use will be discussed as well.

The possibility of using photo-stimulus to control flow in microfluidics devices is very appealing as light can provide contactless stimulation, is biocompatible and can be applied in a non-invasive and highly precise manner. One of the most popular ways to achieve photo-control flow in microfluidic channels is throughout the use of photo-responsive molecules. We review here the different principles and strategies of using photo-responsive molecules to induce or control liquid motion using light, which include the use of photo-controlled polymeric actuators, photo-sensitive coatings, or photo-sensitive surfactants. We further analyse the capability of these approaches to induce flow control throughout the photo-operation of valves, photo-control of electro-osmotic flows or photo-manipulation of discrete microliter-sized droplets.

In the proposed chapter, the authors will present an overview about the role of the main transparent conductive oxides as functional coatings and its incorporation into emerging large area flexible photovoltaic technologies. In particular, the authors will describe the state of the art in this area, with the highlight of establishing the importance of using transparent conductive oxides as an essential part of an industrial chain. In this sense, it will be emphasized the main requirements of transparent conductive oxides, the main deposition techniques employed for its fabrication, the different material alternatives that emerge as substitutes of conventional ones and, finally, the evaluation of the main risks found when the coatings will be incorporated in an industrial production chain of flexible devices.

Surfaces of aerospace components are subjected to heat, water, ice, electromagnetic waves, corrosion, abrasion, wear, thermal shock and/or fire during their lifetime. Materials science and engineering have introduced the use of coatings to protect the surface of the materials under these operating conditions. Materials such as ceramics, polymers, and metals and their combinations either as composites or hybrid materials, as well as functionalized versions of them, have been proposed as different design alternatives. Six different types of coatings can be identified: anti-corrosion, abrasion and wear-resistant, thermal barrier and flame-retardant, conductive, anti-icing, and superhydrophobic. This chapter will review the most important advances in functional nanocoatings, which are being researched and those already used in aerospace applications, as well as the profits that these materials can provide to the surfaces of the components where them are used.

The need to create new enhanced corrosion resistant coatings has grown because of the financial burden corrosion places on certain industries such as oil industry. Organic inhibitors have been extensively reported to protect metals and metallic alloys from corrosion although they have a negative impact on environment. Development of better corrosion resistant coatings such as metal alloys, metal-ceramics, polymers, and hydrophobic films are examples of corrosion resistant coatings for metals and alloys. Recently, superhydrophobic coatings have been widely implemented into many fields including anti-fogging transparent materials, self-cleaning surfaces, biomedical and corrosion applications. In the proposed chapter, a comprehensive review will be dedicated for the fundamentals and developments of superhydrophobic materials including theoretical background; superhydrophobicity in nature; preparation techniques; and recent attempts to develop superhydrophobic surfaces.

Electrodeposition of only molybdenum onto substrates is difficult, therefore molybdenum is typically deposited with iron-based alloys such as nickel. The deposition of such alloys is known as an induced codeposition mechanism. The electrodeposition of nickel-molybdenum alloys using alkaline plating solutions is covered in this chapter. The mechanism for deposition of nickel-molybdenum is reviewed, as well as the influence of the plating parameters on the coatings. Characterization of the coatings by scanning electron microscopy and x-ray diffraction is discussed and how deposition parameters affect morphology, composition, and crystallite size. Nickel-molybdenum alloys offer enhanced corrosion protection and mechanical properties as coatings onto various substrates. A survey of the resulting hardness and Young's modulus is presented for several research studies. Corrosion parameters for several studies are also compared and show the percentage of molybdenum in the coatings affects these values.

In conventional thermal barrier coatings (TBCs) the major problem is the spalling of a ceramic coating. This could be due to the large thermal stresses which are generated during thermal cycling on the oxidizing environment. One of the ways to improve the life span and overcome mentioned problems is introducing the concept of functionally graded materials (FGM) into the TBC. Functionally graded materials are referred to as a class of advanced materials that are distinguished by variation in their properties with varying their dimensions. Through employing a functionally graded thermal barrier coating (FG-TBC), an intermediate layer with a gradual compositional variation is embedded between the top and the bond coats. This layer(s), composed of ceramic and metal in various ratios, can achieve a gradual composition variation, thereby leading to gradual changes in microstructures and better mechanical and physical properties.

Foreword

The principle of coating has always been an effective way towards adding a desired functionality to a material which otherwise would not be present. First attempts to perform even area-selective coatings can be seen in many caves as parietal art some of which reach 40.000 years of age. The paintings serve as an ancient and eternal hard disk that records stories and transmits them to later generations. Starting from there, coating has been considered a very efficient procedure for protecting or preserving goods and therefore has been continuously developed and optimized.

Certainly, if we talk about coatings in the current time, our understanding is more technical. It is habitual to use the term "functional coatings", although technically seen all coatings are inherently functional and there would be no necessity for explicitly mentioning it. Nevertheless, the terminology was largely adopted with the aim to emphasize that the described coating is vital part of a specific technology for which this particular coating was developed. Thus, the term shall stress the difference between what we consider being a trivial coating and a coating developed specifically for a technological application.

During the past centuries our society has witnessed a great development of coatings in terms of both methodologies and materials and for a variety of purposes. One of the main drivers of the intense development has clearly been the electronics industry that strongly relies on a plethora of coating technologies in order to improve their processor chips, memories, etc. However, corrosion protection, biomedical applications, energy conversion or storage and further applications play meanwhile an equally important role for the development of a variety of coating strategies.

In the meantime we can find coating technologies suitable for most of the goals we want to achieve. There are physical methods such as thermal evaporation, sputtering, pulsed laser deposition, etc., gas phase chemical deposition methods that include mainly chemical vapor deposition (CVD) or its subset atomic layer deposition (ALD), and wet-chemical strategies including sol-gel techniques, electroplating, polymerizations, and so on. Each of those techniques has advantages and disadvantages over the other, which makes the overall field being a colorful toolbox with a plethora of complementary tools.

Coatings can be approached with three different points of view. First, there is the technological aspect where plenty of researchers or engineers invest considerable effort in order to improve technologies and/or instrumentation for coating processes. The work of those people was and still is of critical importance, since only their work made coating technologies easily available for everybody, including industry. This part of the world of coatings is not covered with the present book and the interested reader may find plenty specialized literature elsewhere.

The second and third point of view refer to the chemical aspect, that is, the development of chemicals that enable the deposition of a functional material at all and the view from the functionality perspective that is needed for a specific application, respectively. This book is focused on the later mentioned two aspects.

The reader will find examples of exciting applications that are enabled by various kinds of coatings and materials. Those span various emerging application fields including biomedical applications, optics and plasmonics, photovoltaics, corrosion protection, microfluidics, and so on. The routes to obtain the desired coatings vary between physical and chemical or electrochemical methods, which gives additional value to the book as it is not focused on a specific deposition technology. It rather shows that a wise choice of the pathway is necessary for obtaining the best possible results. Eventually the final application or device will benefit.

The involved chemistries are utmost important for many of the procedures. Therefore some chapters address this issue with examples from metalorganic or polymer chemistry, which guides the reader into strategic considerations from the chemical point of view.

Putting together all those aspects, this book is a truly multidisciplinary reference for the reader interested in functionalizing materials by coatings. It gives some insight into the planning needed for obtaining the desired results. Likewise it may act as source of inspiration for the reader as it describes a variety of concepts that may be adapted to further application areas and in this way enable exciting applications of functional coatings.

Still, there is plenty of room for innovative research and development. Much remains to be done and some of the existing concepts may serve as a starting point for newcomers to the field who eventually will further improve the development and implementation of coating materials and methodologies in the forthcoming decades. History has shown that coating follows our civil development for a long time and therefore it is obvious that the future of our technologies will also be dependent on functional coatings.

Mato Knez
San Sebastián, October 2015

Preface

A coating is defined as a material that is applied to cover the surface of an object, usually known as substrate, with a defined purpose. When the purpose of the coating is devoted to the change of one or more properties of the treated material for techonological or scientific applications, it is renamed as functional coating. In our age-old culture functional coatings are the ideal solution to solve problems arising in daily life situations and tools. In the same way as chemistry is a part of our everyday life, functional coatings can be found in almost every object, gadget or device you can see or touch. However, in the last 20 years the advances made in the preparation of different functional coatings with diverse composition have allowed the development of the so-called smart coatings. In this period of time, the functional coatings field has evolved from just decorative and protective films to coatings with multiple functions and applications. These applications consist of thin or thick films that can be the fundamental unit for next generation devices. To fulfill this objective, it is necessary to develop new methods or adapt and optimize the existing ones to fabricate novel coatings with unique properties.

The methodology employed to obtain the described functional materials and their subsequent modification depends on the advanced properties we want to imbue to the new materials. Different techniques (both well established and novel ones) have enabled the deposition of these materials from gas, solution or solid phase and the transference of their unique properties to the substrate. Vapor deposition techniques (such as atomic layer deposition, chemical vapour deposition, physical vapor deposition) have been used widely in microelectronics and microsystems. Chemical, electrochemical and spin or dip coating techniques stand out as the prefferred ones to implement functional coatings from liquid sources with unique and advanced features. Roll-to-roll coating processes are employed to transfer the standard protocols and functional coatings to flexible substrates in the industrial network. Although solid phase fabrication methods are almost nonexistent in the academic areas, they are highly used in industry and commercial fields. Laser micro- nanomachining and thermal spray coatings can yield surfaces with antibacterial behaviour, conductive, protective and superhydrophobic properties. These techniques provide a great number of possibilities to transform the selected substrates from basic materials to multifunctional materials which are capable of changing their properties after an appropiate stimuly in a controlled way and then become an useful, active and smart component of the targeted device.

The nature of the selected material for the coating implementation ranges from metal or semiconductor nanoparticles or layers to inorganic, organic or hybrid films and polymers. Functional coatings are materials or composites with specified functional groups or features, which show specific physical, chemical, or biological properties or capabilities depending on the nature of the film or particle and the interaction with the sensitive material or surface treated. In the last years, much effort has been devoted to the implementation of metallic, inorganic and organic coatings both in industrial and academic fields.

The configuration or structure of the coating is designed depending on the application. Optical coatings usually consist of multilayers that change de reflection properties to obtain surfaces with high selectivity, antirreflective coatings or erosion protection. Structured coatings generally present superhydrofobic properties and can be employed in the production of self-cleaning surfaces and in devices intented to work in oily environments. Resistance to high temperatures can be accomplished using multilayers or hybrid coatings that can be even employed as fire retardant elements if they are composed fo specific materials or compounds. Aditionaly, hybrid nanocomposites can be used for many applications. Photoresponsive, photoluminiscent or conductive nanoparticles can be dispersed in an organogel or polymer matrix in order to obtain films with self-erasable, optical and electrical outstanding properties respectively. Self-assembled monolayers can be used to fabricate antiffogging surfaces for instance. Melted coatings are usually employed in the area of non-volatile memories to confere dual properties resulting of the mixture of well defined interfaces. Coatings formed enterely by nanoparticles can be used for numerous applications such as drug-delivery, bactericide or gas sensors.

This book covers key aspects in the preparation methods, modification, organization and utilization of the functional coatings in micro, nano and biotechnology. The book can be divided in two sections. The first one (chapter 1–6) is related to general aspects of synthesis of coatings and characterization of inorganic precursors, polymers and hybrid components. The second part (chapter 7-15) is related to specific applications of the materials developed by many research groups all over the world.

The objective of chapter 1 is to compile the advances in the field of functional coatings in order to obtain a comprehensive view of the different approaches that can be carried out when functionalizing surfaces or materials with layers or nanoparticles from different components and with particular applications or purposes. Metal–alkoxo complexes constitute a very important group of precursors to fabricate oxide materials that are in the foundation of modern electronics. Moreover, capping semiconductors or metallic nanoparticles with organic ligands results in an interesting approach to achieve hybrid photoactive compounds with novel properties that can be applied in the form of coating. A summary of these techniques and some of these applications are described in chapter 2 and chapter 3, respectively. Chapter 4 describes the preparation, functionalization, analysis and usages of the next generation of polymeric functional coatings. Chapter 5 describes the chemical structures of various novel organic-inorganic hybrid precursors that can be applied to fabricate coatings with many more functionalities that the ones employed nowadays in the industry. The last chapter of the first section is dedicated to nanocomposites. In chapter 6 the use of hydroxyapatite for medical applications and the implementation of acoustic wave sensors using aluminum nitride as dielectric material are presented.

Coming to the applications part of the book, some optical functionalities of coatings are presented. Chapter 7 describes the fabrication of dichroic filters on glass surfaces through thin film inorganic multilayer coatings implemented by PVD reactive magnetron sputtering, whereas chapter 8 focuses on the plasmonic effects found when analyzing metal thin films and nanoparticles with spectroscopic ellipsometry for biomedical applications. As a matter of fact, biomedicine seems to be one of the most demanding fields in the fabrication of micro and nanoscale functional coatings, as is evidenced in chapter 9 where the application of biocompatible coatings for implants and/or scaffolds used in bone tissue engineering is outlined. Chapter 10 presents the bridge between both optical and biomedical applications of functional coatings, describing the use of photo-stimulus to control flow in microfluidic devices.

The last part of the second section of the book is devoted to some of the technological applications of smart coatings. Chapter 11 provides an overview about the role of the main transparent conductive oxides as functional coatings and its incorporation into emerging large area flexible photovoltaic technologies. Functional coatings are widely used in aerospace components (chapter 12), which are subjected to extremely harsh conditions that demand properties as hydrophobicity and abrasion resistance, among others. However, the use of superhydrophobic coatings has not been limited to aerospace materials as it can be inferred from chapter 13. The application of superhydrophobic surfaces especially in anti-corrosion elements is one of the key topics of the chapter. Chapter 14 goes deeper in this research area, providing an accurate description of the state of the art of the implementation of coatings for corrosion protection, the ones based on Ni-Mo alloys in particular. Finally chapter 15 discusses the use of functionally graded materials in thermal barrier coatings and remarks the advantages of the use of these composed multilayers to overcome the usual problems inherent to these protective layers.

The target audience of this book is a multidisciplinary collective composed by scientific and academic communities connected somehow with material science, thin-film technologies, nano and microtechnology and nanoscience. This includes researchers all around the world and students and scholars from all Universities and Colleges where the fields of materials, their properties and the technological applications of them are subjected to study. It will help them to definitively understand some basic concepts of the functional coatings and find the new advances occurred in the field. Also, the book offers recommendations to researchers on how to address their research in the future in order to revolutionize the current state of the art and open new possibilities and applications that cannot be imagined nowadays. The book covers the lack of specific recoplation of data about this subject and collects all the newest information around this topic increasing the visibility of the research performed in this area. The fact of having a complete book dedicated to functional coatings writen by experts in the area is the best way of diseminating the knowledge around this topic and encouraging other researchers and students to join this fascinating and constantly evolving subject.

Finally, we would like to thank all the authors, reviewers and advisory board members for participating in the development of this book despite their busy agendas. Special thanks to Professor Mato Knez for employing his valuable time to write the foreword and to Rachel Ginder and Katherine Shearer for assisting us during the preparation of the book.

Ador & San Sebastián, November 2015

Maica Morant
CIC nanoGUNE Consolider, Spain

Ana Zuzuarregui
CIC nanoGUNE Consolider, Spain

Chapter 1
Functional Coatings: A Rapidly and Continuously Developing Field

Ana Zuzuarregui
CIC nanoGUNE Consolider, Spain

Maria Carmen Morant-Miñana
CIC NanoGUNE Consolider, Spain

ABSTRACT

Traditionally a coating is defined as a material layer applied onto a surface for protection or decoration. Usually it protects the surface from a variety of environmental problems such as corrosion of solder joints, moisture and mildew, fuels and process solvents, service temperatures and dust, dirt and physical damage from handling. Since 2010, the rapid advancement of the micro and nanotechnology has collided with the coatings field resulting in new coatings with novel properties and functions that differ from the traditional ones. Functional coatings consists of organic, inorganic or hybrid materials and can be prepared using a huge amount of techniques depending on the properties of the materials, the substrate and the final application of the coating. Therefore it is possible to find coatings with optical properties, thermal capabilities, structural and mechanical features, physico-chemical properties, magnetic and electric characteristics and biological purposes that cannot be found in the starting materials.

INTRODUCTION

Mankind has been using coatings as protective and decorative layers since the beginning of times. Apart from the paints and decorative drawings that Neanderthal and Cro-Magnon men used in the Prehistoric era, coatings have been present in the human society for centuries and even millenniums (Makhlouf & Tiginyanu, 2011). Ancient Egyptian used mixed organic - inorganic coatings based on bees wax, arabic gum and egg whites to develop coatings implemented with asphalt and balsam exudations to waterproof protection of boats, wooden ships and bronze sculptures from year 3000 BC to around 1000 BC (Ghoniem, 2014; Gooch, 2002). Besides this, other Asian cultures utilized varnishes and lacquers to cover artwork, furnishings and buildings during the second century BC in China, Japan and Korea (Stillman, 1924). Since then coatings have been used in different cultures and societies around the world to protect and preserve tools and gadgets with any kind of application.

DOI: 10.4018/978-1-5225-0066-7.ch001

The coatings technology , as it is known at present, was firstly described in 1773 by Watin and the first patent regarding varnish was released in 1763 (Watin, 1773). However, the first coatings factory (to produce varnish) was established in England in 1790 (Seymour & Mark, 1990). After that pioneer business, other factories started their activity operating in Europe and in the United States. During the twentieth century the coatings industry and research increased continuously, leading thousands of systems that varied from single layers to complex structures applied onto a surface for protection or decoration.

Nowadays, with the development of micro and nanotechnology and their extension both in the academic and in the industrial field of coatings, these coatings have experienced a new revolution, becoming an essential element in the protective layers utilized in commercial devices as well as in the most selected research areas. In this context, the term of functional coatings has appeared. Functional coatings are systems that besides the classical properties of the coatings (protection and decoration) have an additional functionality (Wulf, Wehling, & Reis, 2002). A functional coating can be classified in several categories depending on the nature of the coating, the methodology or techniques used to fabricate it and its functional characteristics. Moreover, whatever application or functionality they have, functional coatings are expected to possess various properties: durability and reproducibility, cost effectiveness and easy application, adequate surface morphology and somehow environmental sustainability (Mathiazhagan & Joseph, 2011). The historical evolution of the coatings is summarized in Figure 1.

The increase of interest on functional coatings and their applications can be clearly observed by the number of papers published on this field on the last two decades. The evolution of the amount of papers released on micro- and nanoscale functional coatings is even more impressive. Figure 2 shows the number of papers found on the Web of Science regarding functional coatings (black bars) in general and the ones focused on micro and nanoscale coatings (grey bars). The increase of the articles published on the topic of micro- and nanoscale functional coatings follows almost a linear trend, reaching a maximum of 467 papers on 2013. The number of papers on functional coatings in general is around five times larger; however, this topic seemed to reach the top a few years ago and now the release of research related to this field remains somehow constant. The progression of the number of papers published suggests a large and exciting future of research and development on micro- and nanoscale functional coatings.

The field of micro- and nanoscale functional coatings has had a broad expansion all over the research community during the last 20 years. This fact can be easily seen in Figure 3 where the citations of the articles related to functional coatings (black bar) and the micro- and nanoscale functional coatings (grey

Figure 1. Historical evolution of coatings from the beginning of times since nowadays

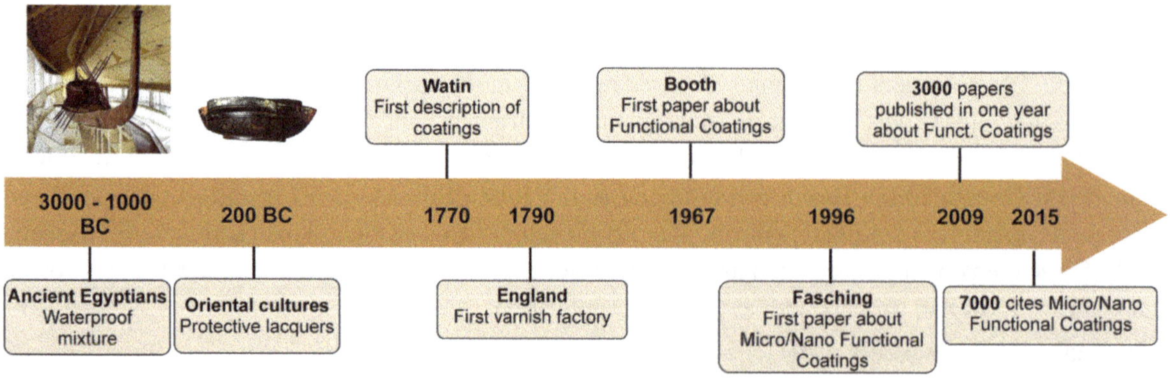

Figure 2. Number of articles per year published on micro- and nanoscale functional coatings (grey bars) and functional coatings divided by five (black bars). Source: Web of Science accessed on May 5, 2015

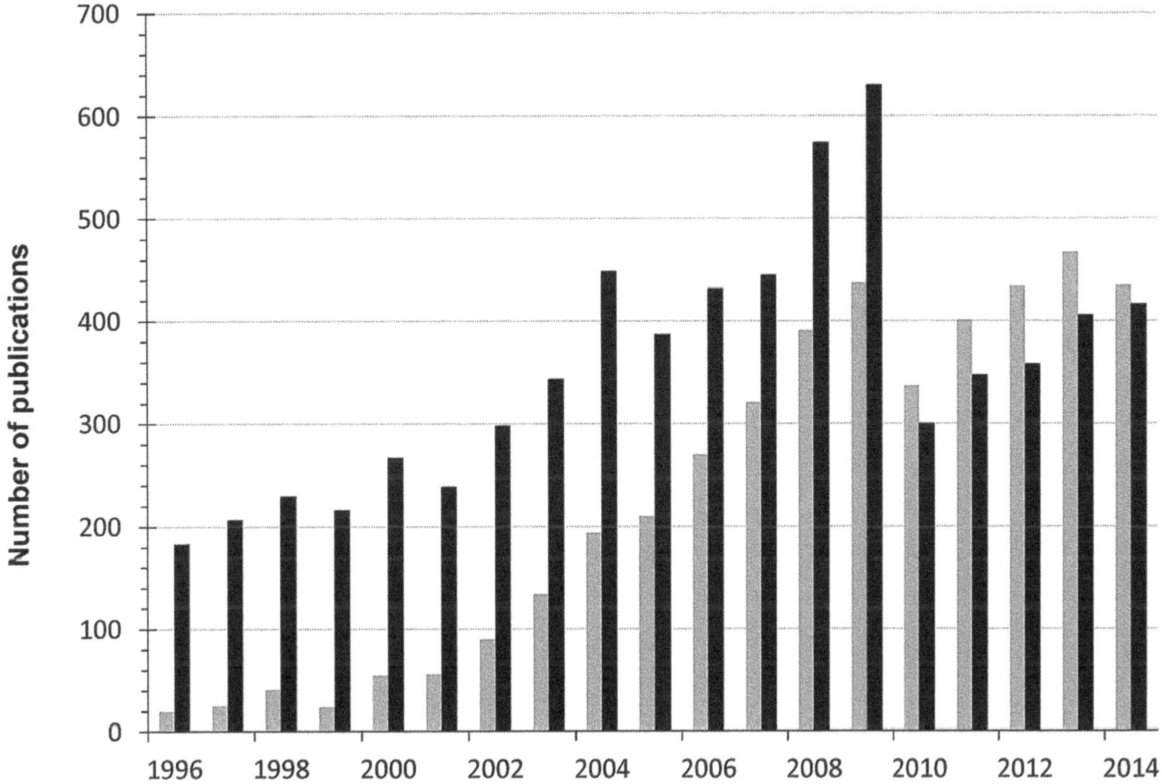

bar) are depicted. Although both numbers of citations have the same trend (taking into account that the ones regarding functional coatings totally are seven times higher), in the particular case of micro- and nanoscale layers, the slope of the curve augurs a much rapid increase of the relevancy of such publications. The papers cited only 15 years ago were slightly more than 100 and now this number has grown to more than 7000.

However, the impact of micro- and nanoscale coatings does not have only grown on research field or the academic community. Several market reports confirm the increase of this kind of coatings not only for protection but also for many more industrial applications (Future Markets, 2015; Global Industry Trends, 2014; Research and Markets, 2015). Research and Markets affirms in its report that the size of the nanocoatings market will grow in the future almost a 25% every year reaching the value of $14 billion by 2019. In this report it is said that both North America and the Asia-Pacific region are the ones with larger share of this continuously growing market. The report made by Global Industry Trends released on October 2014, was not as optimistic as the one from Research and Markets but stated that the market of nanocoatings demanded 255 kilotons in 2012 (doubled this 2015) with benefits of almost 1500 millions of USD and they anticipate an increase of these benefits to 8,17 billion dollars in 2020. Besides this, the report conducted by Future Markets confirms the growth of the nanocoatings market during the last decade with $2-2.5 billion per year at the moment and predicts an increase of $9,7 billions per year by 2025.

Figure 3. Number of citations per year of the articles published on micro- and nanoscale functional coatings (grey bars) and functional coatings divided by seven (black bars). Source: Web of Science accessed on May 5, 2015

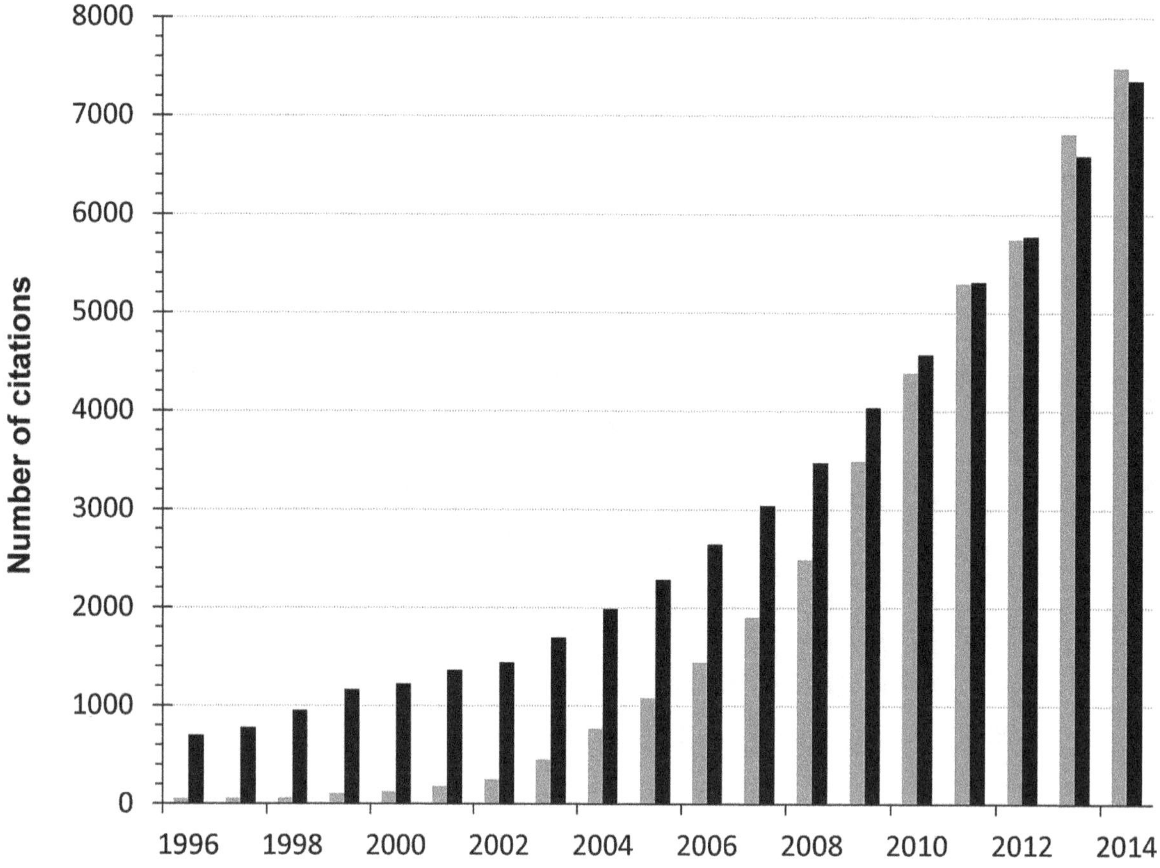

Hence our proposal in this chapter is to give a general overview of the topic of the micro and nanofunctional coatings. Greater emphasis is placed highlighting the huge amount of applications, materials and techniques that are related to the field of functional coatings. The chapter is divided into three sections. At first, general composition and nature of functional coatings is described. Traditionally inorganic materials and polymeric materials have been employed. Nowadays the combination of materials of different nature called hybrid materials is becoming more important in the field. The different methods that can be used to prepare the coatings are summarized in the second section. Finally, examples of different applications published by different research groups around the world are summarized in the third section.

TYPE, NATURE AND COMPOSITION OF FUNCTIONAL MICRO AND NANOCOATINGS

The structure of the functional coatings varies depending on the nature of the substrate, the kind of material and the function that it is going to develop. In general, functional coatings present a sandwich

Figure 4. General structure of functional coatings (a) and different configurations found in the literature: (b) structured coatings, (c) crosslinked coatings, (d) mixed coatings, (e) multilayered coatings, (f) coatings with nanoparticles, (g) hybrid nanocomposites, (h) coatings with self-assembled monolayers and (i) single-layer coatings

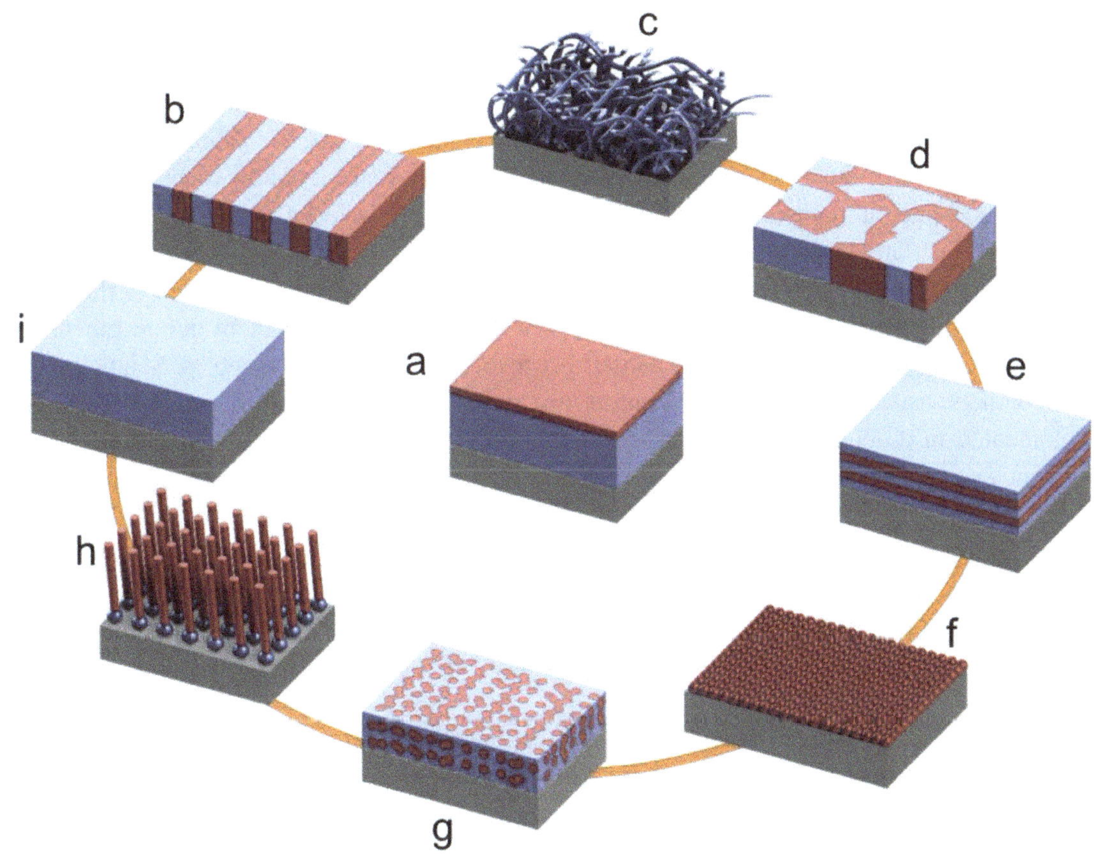

configuration with three common elements: (i) the adhesion film; (ii) the functional layer and (iii) the protection layer (Figure 4a).

The adhesion film is usually deposited immediately prior to the functional coating in order to assure the cleanliness and enhance the interaction between the surface layer and the substrate. In this way, it is possible to decrease the differences of the materials that form the functional coating. The adhesion film is often applied just after the cleaning and preparation of the substrate and it usually has a thickness of few nanometers. The roughness of this first layer has to be closely analyzed as it can have a huge impact on the shape and structure of the functional coating.

The next element in the system is the layer that will provide the functionality to the system. The functional layer it is applied immediately after the adhesion film to avoid the degradation of the former film. The functional layer can be composed by one single layer (Figure 4i) or by several layers (multilayer, Figure 4e), which enables a wide range of different configurations summarized in Figure 4.

Among the compounds and substances that could be applied as functional layers, micro and nanoscale inorganic coatings include thin layers that consist of inorganic elements such as metals, semiconductors or ceramics (Lackner & Waldhauser, 2010a). Metallic coatings have been used as protectors of other noble metals or sacrifice anodes for decades (Carter, 1977). However, metallic thin films have many more functionalities than only anticorrosion protection. For instance, copper coatings applied on Silicon nanowires have been used to study the volume expansion and morphological changes during lithiation and delithiation in Li-ion batteries (McDowell, Woo Lee, Wang, & Cui, 2012). Another approach is the use of $Ru/ZrO_2/Ru$ or Pt as an electrochemical gas sensor at high temperatures (Fesik, Grebnev, Zarazhevskii, & Mal'chikov, 2014). Regarding the field of semiconductors, ZnO appears to be the most used material to fabricate functional coatings (Spanhel, 2006). The majority of the functional layers that include ZnO show photoluminescence or photocatalytic activity that can be utilized in solar technologies and optoelectronics (Grasset et al., 2005).

Ceramic coatings could stand as an individual group of functional compounds due to the amount of research works published so far and their impact in the market, both of them due to the singular properties related to this kind of layers (Vilardell Navarro, 2014; Zeuner, Pagano, & Schnick, 2011). Ceramic coatings, usually composed of titanium, tin, aluminum and chromium oxides and carbides, are implemented in different fields of application as decorative coatings, protective layers on harsh environments, communications and electronics, catalysis, bone implants and energy technologies (De Keukeleere et al., 2015; Dosta, Torrell, Cano, & Guilemany, 2012; Levashov, Pogozhev, & Kurbatkina, 2011; Syed, Hutt, Shah, & Edge, 2015). TiO_2 stands as one of the most used ceramic coatings due to its interesting properties as well as the multiple techniques available for its fabrication. Functional titania coatings have been fabricated for years in order to used them in gas sensors, as biocompatible coatings and for devices that make use of its photocatalytic activity (Kaya, Kaya, Su, Thomas, & Boccaccini, 2005; Mathur, Shen, & Altmayer, 2007).

On the other hand, the field of organic functional coatings is not so broad in terms of composition. Organic functional coatings, that in almost all the cases are polymeric coatings, can be a simple (a single component or substance) or a mixture of different compounds to acquire the specifications needed (Mathiazhagan & Joseph, 2011). The main component of organic functional coatings is the resin used to bind the volatile elements to the substrate, which is responsible for the more basic physico-chemical properties of the coatings. This polymeric resin can be from a quite wide variety of types including acrylic polyurethane, polyvinyl butyral, epoxy, vinyl or phenolic ones (Hu, Zhang, Bu, Lin, & Song, 2012). Regarding the final coating itself, numerous functional polymer layers have been implemented in the last years taking advantage of the novel synthesis techniques discovered and developed (Yang, Neoh, Kang, Teo, & Rittschof, 2014). The nature of the coatings vary from hydrophilic, hydrophobic and amphiphilic polymers according to the properties of the coating and the intended functionality and application (Imbesi et al., 2012; Lejars, Margaillan, & Bressy, 2012; Schilp et al., 2009).

Finally, during the last years a new kind of functional coatings has become relevant, the so-called hybrid functional coatings or composites. Hybrid coatings are composed of a mixture of two elements whose properties differ completely from the properties of the starting compounds. In the field of hybrid inorganic coatings, several advances have been made in the fabrication of magnesium ceramic alloys filled with oxide nanoparticles. The resulting biocompatible coatings have been implemented in orthopedic implants endowing them with corrosion resistance (Madhankumar et al., 2014). However, most of the examples described in the literature introduces nanometric inorganic compounds into polymer matrices for improving mechanical, thermal, electrical and gas barrier properties of the polymer compared to individual bulk materials (Balazs, Emrick, & Russell, 2006). For instance, different types of inorganic materials such as silicon (Peng et al., 2014), semiconductors (Siyanbola et al., 2015), or carbon compounds (Gaddam, Kantheti, Narayan, & Raju, 2015) have been introduced into a polyurethane matrix. Among the range of applications of these hybrid coatings, corrosion protection and anti-scratch coatings stand out as the most studied ones (M. Han, Mang, & Seo, 2010; Seo & Han, 2011).

Finally, the top layer is settled in the stack. This top layer has a protective function and isolates the functional coating from the humidity and the oxygen present in the ambient. For all this, the upper layer must be free of cracks and pinholes that could appear in the deposition process to avoid the presence of channels that could allow the diffusion of molecules that could damage the functional coating. This layer, normally known as encapsulation, can be preset or not in the system and if it is, it can provide an extra functionality to the coating.

FUNCTIONAL COATINGS FABRICATION TECHNIQUES

The ways of synthesizing or fabricating functional micro and nanoscale coatings have also changed and improved with the different technological advances. There are a wide variety of methods to obtain functional coatings but all of them can be grouped into three categories regarding the aggregation state of the precursors or reactants: vapor phase methods, liquid methods and solid methods (J. Singh & Wolfe, 2005).

Vapor Phase Fabrication Methods

Among the numerous techniques that employ reactants or precursors on vapor phase, Physical Vapor Deposition (PVD), also known as sputtering, is one of the most used ones not only in the fabrication of functional coatings but in many other applications such us microelectronics (Campbell, 2001). PVD is used to deposit both conductor and insulating materials by means of an inert or reactive gas plasma in vacuum. In brief, a solid material is evaporated by bombardment with ions in high vacuum conditions (Figure 5a). Alternatively, a reactive gas can be introduced from a compound with the metal vapor and deposited as a highly adherent coating on a substrate. Functional coatings implemented by PVD have been mostly used to increase the lifetime of tools and components (lubrication and thermal barrier properties) but there are several papers that claim the use of these kind of protective layers in implants and smart packaging applications (Benelmekki, Torrell, Xuriguera, Vaz, & Teixeira, 2012; Czechowski et al., 2005; Lackner & Waldhauser, 2010b; Rabinovich et al., 2010; Ulrich et al., 2005).

Chemical Vapor Deposition (CVD) is based on the deposition of the targeted material by means of a chemical reaction that takes place in the vapor phase. The reaction of the two precursors is accomplished by thermal activation under vacuum conditions. The temperature of the reaction can be decreased using some plasma. After the reaction occurs the product condenses and is deposited on top of the surface of the substrate (Figure 5b). The way of controlling of the thickness and the deposition rate is mainly done with the process time and the temperature of the chamber. CVD is mostly used to deposit oxides and nitrides that can be used as passivation layers. In the field of functional coatings, the applications of these CVD layers vary from antimicrobial polymers and water repellent surfaces to implants and industrial tools protection (Dobrzański & Mikuła, 2005; Lackner & Waldhauser, 2010b; Martin et al., 2007; Y. Wu et al., 2004).

Figure 5. Examples of vapor phase fabrication methods: a) Physical vapor deposition or sputtering (PVD), b) Chemical vapor deposition (CVD), c) Atomic layer deposition (ALD)

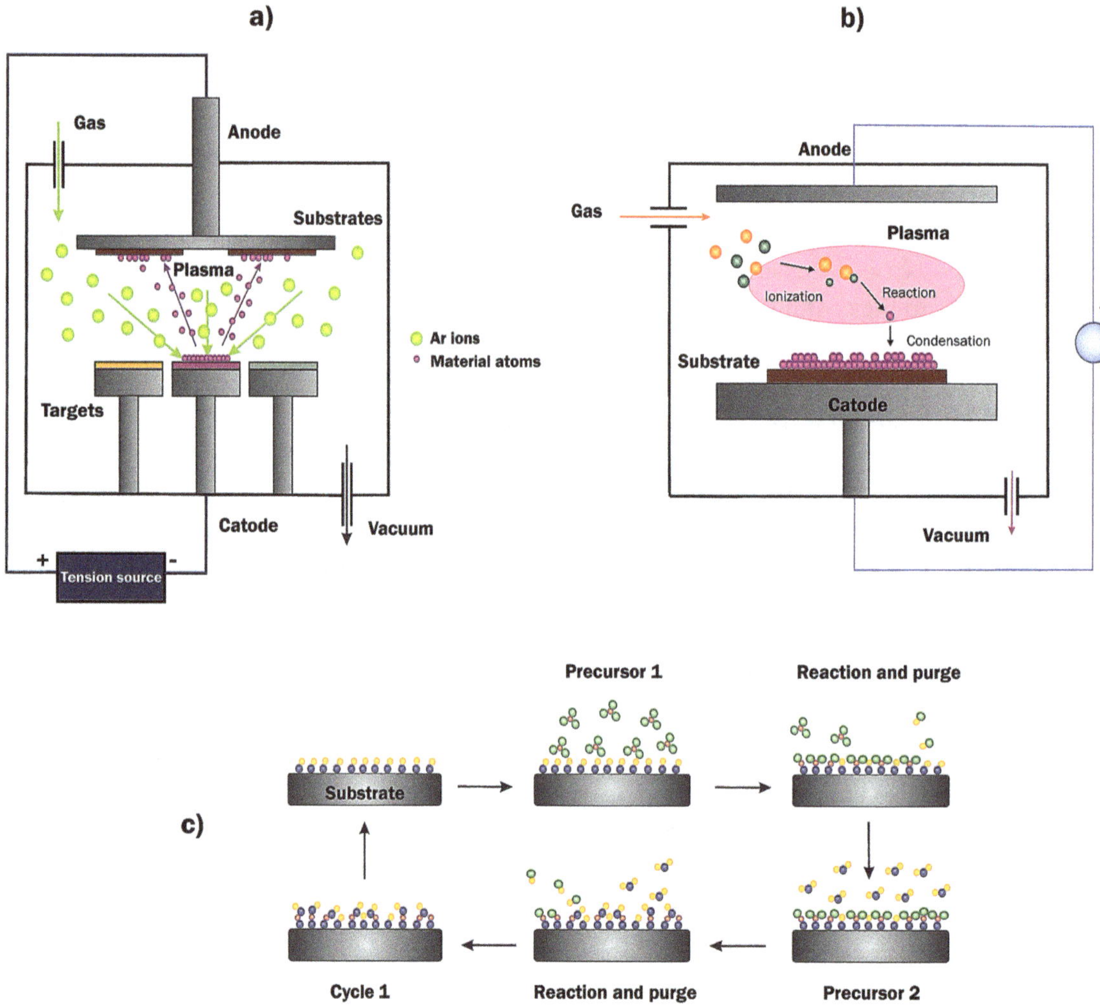

Atomic layer deposition (ALD) is a particular variant derived from CVD that shows significant differences (Figure 5c). The most important one is that the ALD precursors interact separately with the surface of the substrate whereas CVD reactants react first with each other and subsequently with the substrate (Niinistö & Niinistö, 2012). The chemical reaction is separated into two half-reactions in which the first precursor is chemisorbed on the surface of the substrate and then the excess is purged from the chamber; immediately after the second precursor enters the chamber and reacts with the absorbed species creating a monolayer of the targeted material (Knez, 2012). Repeating this cycle the thickness of the resulting coating can be controlled in an extremely more accurate way than in the CVD processes. ALD is applied in a wide variety of areas including nanotechnology due to its ability to controllably coat 3D structures in the nanoscale (Knez, 2008). In the particular case of micro and nanoscale functional coatings, the ALD techniques has been used to develop layers with hydrophobic, superhydrophillic and photocatalytic properties and to implement conductive materials and ultracapacitors (Herrmann, Fabreguette, Finch, Geiss, & George, 2005; Jur, Sweet, Oldham, & Parsons, 2011; Lee, Kim, Son, Choi, & Kim, 2009; Woo Kim, Kim, Won Park, Kim, & Hyung Shim, 2014).

Besides these methods, which have been used in the microelectronics and microsystems field for years, novel methods are starting to stand out as efficient and improved techniques to fabricate functional coatings in the micro and nanoscale. For instance, Ashurts and his coworkers have fabricated functional coatings using vapor phase hydrolysis and condensation (Ashurst, Carraro, & Maboudian, 2003). With their organic solvent free techniques they could fabricate anti-stiction layers for MEMS application and settled the basis for the development of new vapor phase approaches on functional coating implementation. Other research groups described the use of vapor phase polymerization, consisting on an oxidative synthesis, to fabricate both conductive and semiconducting polymers for several industrial applications (Bhattacharyya, Howden, Borrelli, & Gleason, 2012). Finally, Glavan and her collaborators described the functionalization of paper in the nanoscale to confer the cellulose omniphobic properties using vapor phase silanization (Glavan et al., 2014).

Liquid Phase Fabrication Methods

There are various methods available to fabricate functional coatings in the nano and microscale, but dip-coating has been the most used one along the last years due to its simplicity and easy way to operate (Grosso, 2011). In this method, the substrate or device that is going to be coated is dipped several times into the solution and then withdrawn at a constant speed, obtaining coating with a thickness that ranges from 10 to 100 nm (Figure 6b). Although it presents some drawbacks, as any other deposition technique, dip-coating is a waste-free, low energy consumption and low cost technique easy to scale-up. Among the wide variety of functional coatings that can be prepared using this liquid phase method, during the past few years several papers have been published explaining the fabrication of functional polymeric, silicon based and metallic oxide layers for antifogging and antireflection films, anticorrosion systems and biomimetic superhydrophobic surfaces (Kirtay, 2014; X. Lu et al., 2011; Xue, Ji, Zhang, Ma, & Jia, 2015). A more recently developed technique is spin-coating. In this particular approach centrifugal and viscous forces tend towards a uniform film. For that purpose a suspension of the coating is needed. The main advantage of spin coating is that the properties of the dispersion (polarity and microstructures) can be easily transferred to the coating at room temperature (Morant-Miñana & Korthals, 2011).

Layer by layer deposition (LbL) could be considered a variation of dip-coating technique but it has enough singular parameters and characteristics that make this method a single case of study. This deposition process was developed in the early nineties by Decher et al. when they tried to prepare multilayer thin films (Decher, Hong, & Schmitt, 1992). The basis of this method consists in the dipping of a positively charged substrate on a solution containing an anionic component, most times a polyelectrolyte, which interacts with the surface of the target reversing the charge. After rinsing the negatively charged system, it is immersed in a cation-containing bath. The sequential repeating of these steps allows the formation of the functional multilayer (W. Chen & McCarthy, 1997). Some of the disadvantages of this method can be solved using the spray-assisted layer by layer coating, a slightly different technique that uses the benefits of applying the coating in the shape of a spray (Zhao, Yan, Fang, Wang, & Wang, 2015). As it can be inferred from the polyelectrolytes used in this deposition process, almost the majority of the functional coatings developed by means of these methods are polymeric coatings for several applications: antifouling surfaces, impermeable films, surfaces with changeable wetting properties and flameproof textiles (Huang & Zacharia, 2015; Michel, Toniazzo, Ruch, & Ball, 2012; Yang et al., 2012; Zhao et al.,

Figure 6. Examples of liquid and solid phase fabrication methods: a) Sol-gel, b) Dip coating, c) Spray coating

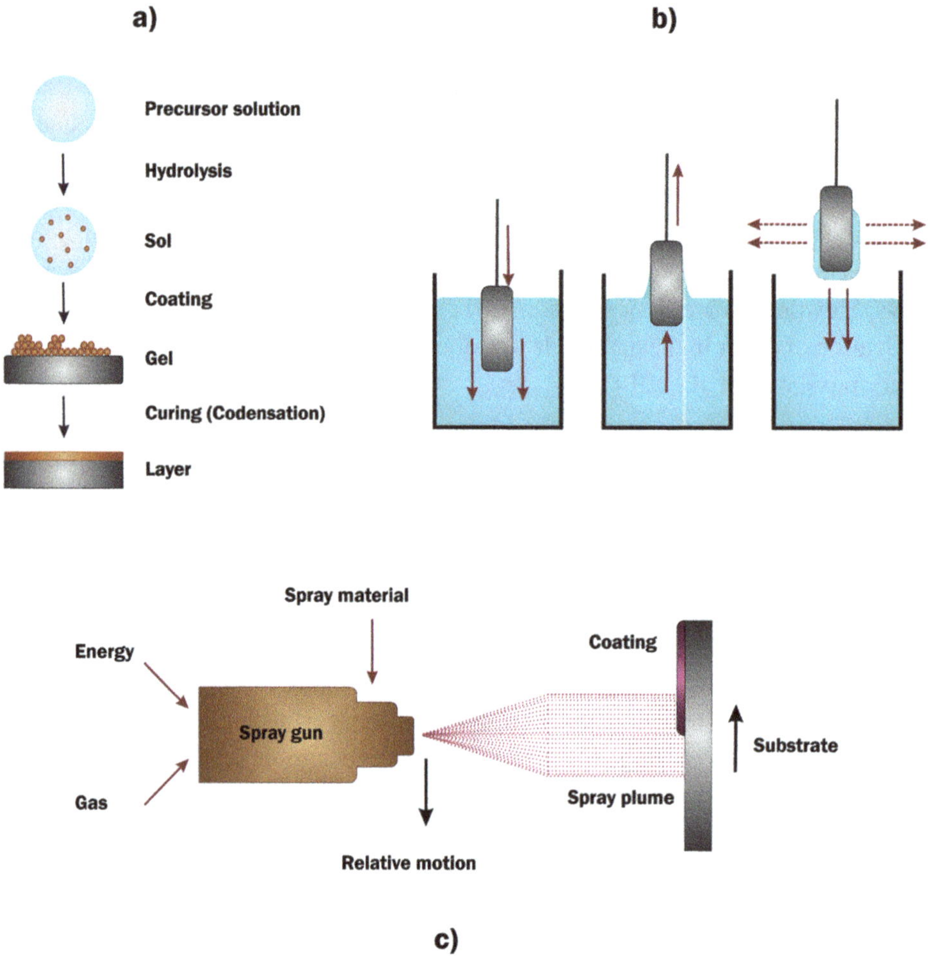

2015). However, this process has also been applied for the fabrication of catalytic functional coatings based on metallic oxides (W. Liu, Devaraju, Yin, & Sato, 2010).

Another essential liquid phase technique for the production of functional coatings is the sol-gel method. The first report of the use of a sol-gel based material was published in 1845 by Ebelmen (Pierre, 1998). The sol-gel method basically consists of preparing a homogenous solution called sol that is converted into a gel by polycondensation (Figure 6a). After that the gel is shaped to the final solid material. (Attia, Wang, Wu, Shen, & Ma, 2002). If the sol-gel based coatings are composed by inorganic compounds as metal oxides and silica based layers a final sintering step at 500 ºC can be necessary for implementing them for several applications as varied as biocompatible films, biocorrosion protective coatings, antibacterial and fungicidal elements, smart thermal barriers and camouflaging textiles (Brzeziński et al., 2012; Das et al., 2015; Pin et al., 2010; Rojaee, Fathi, & Raeissi, 2014; Sharifnabi, Fathi, Eftekhari Yekta, & Hossainalipour, 2014). Besides this metal oxides, several studies have been released showing the fabrication of hybrid coatings by sol-gel methods for multifunctional films (superhydrophobic and anticorrosion or antireflective and superhydrophobic) (Mo, Zheng, Wang, & Mo, 2015; Zhang et al.,

2012). In the most recent works, sol-gel techniques have been also used to perform self-cleaning hydrophobic coatings with polymeric components (Kumar et al., 2015).

Electrodeposition or electroplating is another liquid phase method for functional coatings fabrication that has to be taken into account. This technique, which has been used for more than 2000 years in decorative elements, is based on the employment of an electrical current to reduce cations of a desired material from a solution and form a thin film onto a conductive surface (Schwarzacher, 2006). This wet process can be used to coat arbitrary three-dimensional substrates without high energetically needs and it allows the fabrication of nanomaterials with extremely exciting properties and functionalities (Bartlett, 2004). Among all the different coatings that can be implemented by electrodeposition processes, the one that stands out is the fabrication of metal oxides (ZnO, TiO_2) for both anticorrosion and superhydrophobic biomimetic films (Hao, Sirong, & Xiangxiang, 2015; Vijayalakshmi, Chellappa, Anjaneyulu, Manivasagam, & Sethu, 2015). Besides these usual coatings, it is easy to figure out that metallic layer and alloys are one of the easier compounds to produce using electroplating. In this particular field, several papers have been published claiming the implementation of anticorrosive functional coatings of different metals and alloys (Pavithra & Hegde, 2012; Vijayalakshmi et al., 2015). However, these are not the only metal based materials that can be implemented using electrodeposition techniques; some researchers have succeeded in fabricating graphene and polymeric functional coatings for antimicrobial and cytotoxic applications as well as smart materials with self-healing and anticorrosion properties (Grigoriev, Schenderlein, Akcakayiran, & Shchukin, 2014; Santos et al., 2012).

The printing technology allows the dispersion of a wide variety of colloids, solvents, UV fluids and other liquids onto a huge number of substrates. Based on this definition, it is possible to find at least three techniques that are the most employed ones in the worldwide labs. The most employed one is the inkjet technology. This system uses droplets to create an image without direct contact between the device and the substrate. Depending on the configuration of the equipment we found continuous or drop-on demand inkjet methods. The most critical component is the ink, which is the final responsible of the quality and the drop ejection characteristics (Le, 1998). Various applications of inkjet printing methods are related to thin-film transistors, light-emitting diodes, solar cells, memory and magnetic films, contact and conductive structures, sensors and detectors, patterning and biological and pharmaceutical applications. (M. Singh, Haverinen, Dhagat, & Jabbour, 2010). Another approach based mainly on liquid dielectric and conductive inks and pastes is screen-printing technology. It is a low cost thick film process used for the fabrication of electrodes on electronic circuits and sensors. The process consists of squeezing an ink or paste through a patterned screen onto an inert substrate (Jubete et al., 2009). Many of these fabrication methods can be adapted to roll-to-roll processes, a technology able to handle the fabrication of flexible substrates (polymeric or paper). With this effective and cheap system a high number of devices can be patterned on large flexible substrates to create thin film transistors or resistors among other components.

Solid Phase Fabrication Methods

The functional coatings fabrication methods that employ only solid precursors are, usually, the most unknown ones in the academic or research areas but they are well known and optimized techniques utilized in industry and commercial fields.

Laser cladding was one of the most promising solid phase functional coatings deposition methods in the nineties. This technique is based on the following procedure: a wire or powdered feedstock of the coating material is melted and consolidated by means of the energy of a laser source to coat the substrate surface (Vilar, 1998). In those years several papers were published reporting the implementation of metal matrix composites of Sn and TiC with wear resistant and hardness enhanced functional coatings (X. Wu, 1999; Yakovlev, Bertrand, & Smurov, 2003). Recently, another research group has shown the fabrication of Ti-Al composites for functional graded coatings with this method (Shishkovsky & Smurov, 2012).

Besides this method, some Japanese researchers have developed a novel technique for functional coating fabrication that they call mechanical coating technique. They claim that this technique can be used to coat ceramic substrates with metallic functional coatings by friction, impact and adhesion processes (Yoshida, Lu, Nakayama, & Hirohashi, 2009). This group has reported the implementation of metallic titanium and titania based composites as functional coatings for photocatalytic applications (Yun Lu, Guan, Hao, & Yoshida, 2015).

The design of self-cleaning surfaces, surfaces with improved corrosion resistance or enhanced cell adhesion are examples of coatings that can be fabricated using femtosecond laser micro-nanomachining. The ultrafast laser treatment can produce a rich variety of surface structures. The main advantages of the technique are that the resulting coatings present mechanical and thermal stability as well as the absence of toxic residues on the surface that make them effective for environment-friendly applications (Rodríguez et al., 2015).

In spite of these methods that do not show a wide influence in the field of functional nano and micro scale coatings, there is another process that has indeed, spraying deposition (Figure 6c). Sprayed functional coatings are mostly fabricated employing the thermal spray coatings method, where a heated material (feedstock or precursor) is sprayed into the substrate surface. The coatings component is heated by electrical (plasma) or chemical energy (flame). Due to the particular characteristics of the system, most of the coatings implemented with this procedure are metallic, metal oxides and other inorganic materials. As it has been mentioned before, thermal spraying might not be as popular in the academic area as in the industry but several studies have been recently published reporting the use of this technique. Different research groups along the world claim the excellent thermal barrier properties of their functional metallic and oxide coatings fabricated by thermal, plasma or air-plasma spraying (X. Chen et al., 2015; Gupta & Satapathy, 2015; Lamuta, Di Girolamo, & Pagnotta, 2015; Weeks, Subramanian, Vaidya, & Mumm, 2015). However, there are other functionalities that can be achieved with these coatings such as antibacterial behavior, conductive and protective properties and superhydrophobicity (X. Chen et al., 2015; Gao et al., 2015; S. J. Han, Chen, & Sampath, 2014). Apart from this thermal spraying, a slightly modified process was developed in the middle eighties, the cold spray materials deposition process (Champagne, 2007). This method basically uses the energy of a pressured gas or a laser to accelerate the solid particles instead of heating them. Taking advantage of the benefits of this technique, some catalytic and corrosion and wear protection functional coatings of metals and metal oxides have been developed (Olakanmi & Doyoyo, 2014; F. Wang, Qi, Wang, & Cui, 2014).

APPLICATIONS OF FUNCTIONAL COATINGS

In general, functional coatings can be classified regarding the functional properties that are inherent to the components of the layer. In this way, it is possible to find functional coatings with optical properties, thermal capabilities, structural and mechanical features, physico-chemical properties, magnetic and electric characteristics and for biological purposes. Some examples of the coatings described below are summarized in the Figure 7. When functional coatings respond to light generated stimuli, they are described as coatings with optical properties. Among this group of coatings, there are some systems, named as photochromic coatings, that can change some of their features in a reversible way from a stable shape to a less stable one that presents an absorption spectrum that differs from the former one (Pardo, Zayat, & Levy, 2011). Based on this principle self-erasable and rewritable materials controlled by the dynamic non-equilibrium aggregation of photorresponsive metals (Au or Ag) nanoparticles embedded in a flexible organogel film have been prepared by Klajn and his coworkers (Klajn, Wesson, Bishop, & Grzybowski, 2009). After irradiation with UV light the trans-azobenzene group located in the shell of the nanoparticles isomerize to cis-azobenzene and the nanoparticles aggregates into metastable supra-spherical assemblies that gradually self-erase. Many of these kind of functional coatings are composed of hybrid inorganic-organic materials fabricated trough simple methods, which allow the implementation of reproducible and sustainable functional coatings with photoluminescent features. For instance, highly adherent, durable and partially transparent photoluminiscent coatings were prepared dispersing photoluminiscent pigments ($SrAl_2O_4$:Eu^{2+}, Dy^{3+}) in a silicone epoxy hybrid binder hardened with a bifunctional silane which possesses a reactive primary amino and hydrolysable ethoxysilyl groups (Barletta, Rubino, Tagliaferri, & Vesco, 2014). Optical functional coatings also comprise highly reflective surfaces fabricated on polymeric substrates to concentrate as much light as possible in the receptors of solar plants. Under this approach, atomic layer deposited TiO_2 on FEP (Fluoroethylenepropylene) provides the adhesion needed for the following reflecting layer (silver) while enhances the reflectivity of the whole device (Zuzuarregui et al., 2015). At the same time there is a high demand for highly efficient, durable and cost effective antireflective coatings. Based on several structures found in the nature two strategies can be used to fabricate antireflection coatings. The first one is based on the decrease of the refracting index by introducing porosity into the film and the second one is based on gradient-index lenses that impart a smooth transition of refractive index form a higher value to the lower value of air (Raut, Ganesh, Nair, & Ramakrishna, 2011). Moreover, wavelength selective reflectors can be prepared by preparing multilayers alternating ceramic materials (ZrO_2) and colloidal sols (SiO_2) on soda-lime glass discs using a spinning system for deposition (Kundu, Biswas, & Ganguli, 1989).

Figure 7. Different applications of functional coatings: (a) structural, (b) biological, (c) thermal, (d) physico-chemical, (e) optical and (f) electrical and magnetic

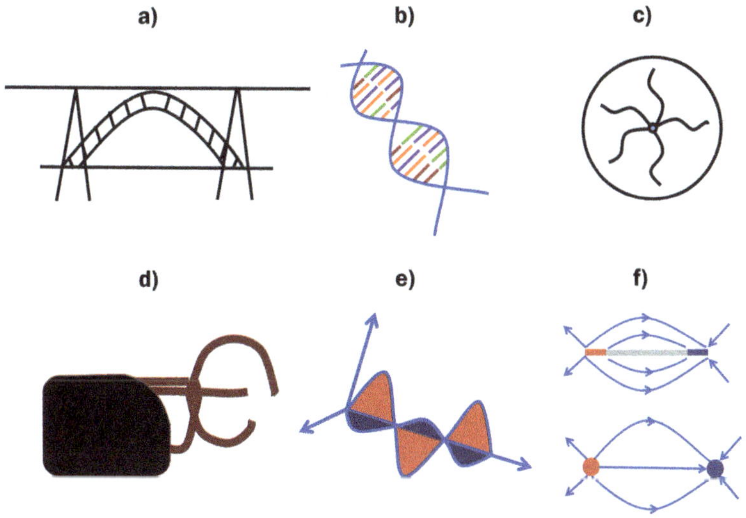

Resistance to high temperatures is a characteristic that can be also achieved using functional coatings. Industrial environments, such as turbines operating at high gas temperatures, usually request tools and systems that are able to work and maintain their performance under extreme conditions without losing their properties. Nowadays, the lifetime of these tools can be increased using thermal barrier coatings. One of the most used materials is YSZ in its metastable tetragonal-prime structure. YSZ has an excellent shock resistance, low thermal conductivity and a high thermal expansion coefficient (Clarke & Phillpot, 2005). Recently, in spite of this well-known group of coatings, there are much more thermal functionalities that can be accomplished through functional coatings. For instance, there are some hybrid coatings made of NBR (Acrylonitrile-Butadiene Rubber) or SBR (Styrene butadiene rubber) with dispersed halloysite and phthalocyanine nanotubes that show an improvement on the thermic properties and increase the heat capacity of the system making it suitable for the fabrication of fire retardant elements (Rybiński, Pająk, Janowska, & Jóźwiak, 2015).

Regarding the structural and mechanical properties that can be attributed to functional coatings, titanium alloys are the most used ones in many industrial applications. Some of these coatings are implemented in multilayers of TiN/Ti on Ti6Al4V alloys that provide an improvement in the hardness of the substrate. This enhancement of the mechanical properties for the PIRAC-N (powder immersion reaction assisted coating-nitriding) and PVD multilayers is even better than the one obtained with coatings fabricated by means of sputtering or PVD. Functional coatings are also used to enlarge the lifetime of titanium components such as joints and turbines in terms of abrasive wear and erosion (Avelar-Batista Wilson, Wu, Gotman, Housden, & Gutmanas, 2015). Sometimes, the aim such functional coatings can be to provide temporary protection. In this context several peelable coatings can be applied. Vinyl chloride and vinyl acetate copolymers are employed for several companies for these purposes. In other applications, the disintegration of the coating after the operation is of great interest. Using a wax encapsulation system that can be triggered by heat, Park et al. are able to sequester and release encapsulated acid microdroplets on demand obtaining a simple approach for device destruction (Park et al., 2015).

Another key point on the development of functional coatings is the fabrication of protective layers that are able to change the physico-chemical properties of the substrate as the wettability. Wettability is defined as the ability of a liquid to wet a substrate. It is very difficult to have a normal wetting of the surface and indeed a good adhesion (Rodríguez et al., 2015). This inherent property of several materials such as Teflon or the compounds known as anti-graffiti, can be modified by means of functional coatings which can tune the hydrophobicity or hydrophilicity characteristics of the surface (Liu et al., 2015). Self-cleaning surfaces are based on the combination of micro/ nanomorphologies which confere the superhydrophobicity. By combining TiO_2 nanoparticles coated with perfluorooctyltriehoxysilane it is possible to create a superhydrophobic surface that maintains its robust resistance to oil contamination. This makes it suitable for being implanted in harsh and oily environments (Y. Lu et al., 2015).

Functional coatings have also been widely used to prevent corrosion in harsh environments such as industrial plants, marine platforms and shipbuilding. The corrosion process is a spontaneous reaction where metals in contact with acidic or salty ambient produce ionic species as oxides and hydroxides. Apart from the use of noble metals as sacrifice anodes, there are plenty of functional coatings developed to protect from corrosion. Some of the most recent research projects on this topic are focused on the fabrication of functional coatings with polymers as polyethylene and polyethylene terephthalate. In order to enhance the barrier properties of the polymeric surface, several approaches have been described. In some of these studies, some inorganic particles or nanocontainers are inserted in the polymeric matrix acting as a healing element when the corrosion reaches them (Abu-Thabit & Makhlouf, 2015). In some other cases a ceramic coating is deposited on top of the polymer in order to protect it from the corrosive environment. Bare natural based polymers are used also as corrosion protective layers, the organic component is usually amphiphilic and reorders the water molecules present in the surface preventing the organization of the molecules into ice crystals and avoiding their formation on top of the targeted material (Balakrishnan, Sathiyanarayanan, & Mayavan, 2015). Finally, an alternative approach to prevent corrosion is to use protective coatings that can respond to damage derived from the presence of microcracks. These coatings can recover functionality automatically and retard corrosion of the substrates. For this particular application double-layered polyuria microcapsules containing hexamethylene diisocyanate (HDI) have been synthesized via emulsion polymerization incorporated in epoxy coatings showing their self-healing functionality of the released HDI in scratched areas during the accelerated corrosion process (Sun, An, Wu, & Yang, 2015).

Sometimes increasing the roughness of the coating has alternative functionalities. For instance, the use of inhomogeneous coatings is an alternative to avoid fingerprint contamination of optical coatings. This takes place due to the fact that the rough coatings imitate the properties of a fingerprint. On the other hand, the use of superhydrophilic and super wetting surfaces has a clear application on the field of antifogging surfaces (Drelich, Chibowski, Meng, & Terpilowski, 2011). The water spreads on the surface to form a thin-film instead of droplets. These surfaces can be prepared by adding on the surface densely packed molecular structures such as alkanothiols ($HS-(CH_2)_n-OH$, $HS-(CH_2)_n-COOH$) on gold, synthetic polymers (polyethylene glycol) or biomacromolecules (albumin or heparin).

In the field of electrical properties, functional coatings are also demanded to change the insulating or conductive character of the substrate while protecting it from external aggressions. Polymeric layers as polyvinyl chloride have been widely applied as antistatic functional coatings due to their properties. The ability of this kind of plastics to accumulate electric charges in the surface of the coatings makes them a useful tool for this kind of application (H. Wang et al., 2015). In some special projects and for specific purposes, the conductivity of this in principle insulating coatings has to be enhanced. In these particular circumstances silver nanowires or graphene layers are embedded in the polymeric layer obtaining an increase of the electric constant while the transition temperature of the system (T_g) is also improved (C. Chen, Jing, Pi, Zhu, & Shen, 2015).

Other fields of study that implies the use of functional coatings are, for instance, organic electronics and electrically tunable injection barriers. In the concrete area of non-volatile memories, big advantages have been described recently. The combination of the properties of ferroelectric materials (bistability and non-volatility) with the ones that belong to semiconductors (rectification and conductivity) has been possible with the mixture of different polymers. The association of the characteristics of the electrically insulating ferroelectric poly(vinylidenefluorideco-trifluorethylene) (P(VDF–TrFE)) with the semiconducting regio-random poly(3-hexylthiophene) (P3HT), has led to a system that could be applicable to numerous organic electronic devices as diodes or switches (Asadi, de Leeuw, de Boer, & Blom, 2008). Other functionalities that can be accomplished in the field of electronics are the ones implying protection against radar. Different dielectric coatings or paints can be used in order to make metal pieces, even airplanes, invisibles to radar (Katsenelenbaum, 2014).

For some years now, functional coatings have been considered not only as a tool to improve the biocompatibility of several substrates but also as a way to add additional functionalities such as drug delivery or the enhancement of the cell adhesion to the material that has been coated (Zelikin, 2010). Furthermore, the modification of surfaces by means of different biological materials (tissue, cell receptors, aptamers, nucleic acids, proteins) has prompted the boom of biosensors for almost any kind of application. For instance, modifying gold surfaces with nucleic acids has allowed the implementation of electrochemical biosensors that are able to detect *Campylobacter jejuni* from real samples (Morant-Miñana & Elizalde, 2015). Similarly, recent studies have shown that zinc oxide (ZnO) has antimicrobial properties against both Gram-positive and Gram-negative bacteria. In this way, homogenous functional coatings consisting of a hexagonal structure of wurzite are able to deactivate 90% of the present microorganisms in the first two treatment hours. This behavior responds to the ability of the ZnO to produce hydrogen peroxide (H_2O_2), which produces the damage of the cell membrane while, inhibits the formation of biofilms (Okyay et al., 2015).

CONCLUSION

Functional coatings can be easily found in our daily life. They are made of organic, inorganic or composite materials combined in different configurations. Several efforts have been done to develop different fabrication techniques that allow the deposition of coatings on substrates of different nature. These techniques can be used to produce easy to apply films of different thickness (from micro to nanometers), long lifetimes, highly resistance to chemical and atmospheric factors and with the desired properties. The advances of the micro and nanotechnology have endowed the coatings field with new properties that led to an important advance not only in this specific area but also in the fabrication process and applications. These applications have changed from the traditional ones (protection and decoration) to a huge variety of implementations that form part of many fields of knowledge and technology. However, currently, the development of new materials and the use of the existing ones in new areas is one of the main challenges for the scientists and engineers worldwide. This area has an enormous potential and the creation of multidisciplinary work and research teams, the public and private investment of the companies and governments and the ability to share and transfer the quality knowledge can result in new systems with surprising properties that cannot be imagined today.

REFERENCES

Abu-Thabit, N., & Makhlouf, A. S. (2015). Recent approaches for designing nanomaterials-based coatings for corrosion protection. Handbook of Nanoelectrochemistry, 1, 1–20. Switzerland, Springer International Publishing

Asadi, K., de Leeuw, D. M., de Boer, B., & Blom, P. W. M. (2008). Organic non-volatile memories from ferroelectric phase-separated blends. *Nature Materials*, *7*(7), 547–550. doi:10.1038/nmat2207 PMID:18552851

Ashurst, W. R., Carraro, C., & Maboudian, R. (2003). Vapor phase anti-stiction coatings for MEMS. *IEEE Transactions on Device and Materials Reliability*, *3*(4), 173–178. doi:10.1109/TDMR.2003.821540

Attia, S., Wang, J., Wu, G., Shen, J., & Ma, J. (2002). Review on Sol—Gel Derived Coatings: Process, Techniques and Optical Applications. *Journal of Materials Science and Technology*, *18*(3), 211–218.

Avelar-Batista Wilson, J. C., Wu, S., Gotman, I., Housden, J., & Gutmanas, E. Y. (2015). Duplex coatings with enhanced adhesion to Ti alloy substrate prepared by powder immersion nitriding and TiN/Ti multilayer deposition. *Materials Letters*, *157*, 45–49. doi:10.1016/j.matlet.2015.05.054

Balakrishnan, T., Sathiyanarayanan, S., & Mayavan, S. (2015). Advanced Anticorrosion Coating Materials Derived from Sunflower Oil with Bifunctional Properties. *ACS Applied Materials & Interfaces*, *150826103253006*. 7(35), 19781–19788 PMID:26292971

Balazs, A. C., Emrick, T., & Russell, T. P. (2006). Nanoparticle Polymer Composites: Where Two Small Worlds Meet. *Science*, *314*(5802), 1107–1110. doi:10.1126/science.1130557 PMID:17110567

Barletta, M., Rubino, G., Tagliaferri, V., & Vesco, S. (2014). Design and manufacture of photoluminescent coatings on stainless steel substrates. *Colloids and Surfaces. A, Physicochemical and Engineering Aspects*, *455*(1), 147–155. doi:10.1016/j.colsurfa.2014.04.044

Bartlett, P. N. (2004). *Electrodeposition of Nanostructured Films Using Self-Organizing Templates.* In Electrochemical Society Interface (pp. 28–33).

Benelmekki, M., Torrell, M., Xuriguera, E., Vaz, F., & Teixeira, V. (2012). Structure and Properties of Silver Clusters Implanted in PET by PVD Sputtering for Active Packaging Applications. *Journal of Nano Research*, *18-19*, 105–116. doi:10.4028/www.scientific.net/JNanoR.18-19.105

Bhattacharyya, D., Howden, R. M., Borrelli, D. C., & Gleason, K. K. (2012). Vapor phase oxidative synthesis of conjugated polymers and applications. *Journal of Polymer Science. Part B, Polymer Physics*, *50*(19), 1329–1351. doi:10.1002/polb.23138

Brzeziński, S., Malinowska, G., Kowalczyk, D., Kaleta, A., Borak, B., Jasiorski, M., & Tracz, A. et al. (2012). Antibacterial and fungicidal coating of textile-polymeric materials filled with bioactive nano- and submicro-particles. *Fibres and Textiles in Eastern Europe*, *90*(1), 70–77.

Campbell, S. A. (2001). *The science and engineering of microelectronic fabrication*. Oxford, Oxford University Press.

Carter, V. E. (1977). Metallic Coatings for Corrosion Control. United Kingdom, Elsevier Ltd.

Champagne, V. K. (2007). The Cold Spray Materials Deposition Process: Fundamentals and applications. United kingdom, Woodhead Publishing Books

Chen, C., Jing, M., Pi, Z., Zhu, S., & Shen, X. (2015). Preparation and Properties of Double-Sided AgNWs/PVC/AgNWs Flexible Transparent Conductive Film by Dip-Coating Process. *Nanoscale Research Letters*, *10*(1), 315-321. doi:10.1186/s11671-015-1022-0 PMID:26245859

Chen, W., & McCarthy, T. J. (1997). Layer-by-Layer Deposition: A Tool for Polymer Surface Modification. *Macromolecules*, *30*(1), 78–86. doi:10.1021/ma961096d

Chen, X., Gong, Y., Suo, X., Huang, J., Liu, Y., & Li, H. (2015). Construction of mechanically durable superhydrophobic surfaces by thermal spray deposition and further surface modification. *Applied Surface Science*. 356(30), 639-644

Clarke, D. R., & Phillpot, S. R. (2005). Thermal barrier coating materials. *Materials Today*, *8*(6), 22–29. doi:10.1016/S1369-7021(05)70934-2

Czechowski, K., Królicka, B., Szlosek, P., Smuk, B., Wszołek, J., Kurleto, A, & Kasina, J. (2005). Effect of nitride nano-scale multilayer coatings on functional properties of composite ceramic cutting inserts. *Bulletin of the Polish Academy of Sciences, Technical Sciences, 53*(4), 425-431.

Das, S., Nayak, G. C., Sahu, S. K., Routray, P. C., Roy, K., & Baskey, H. (2015). Titania Coated Magnetite and Ni-Ferrite nano-composite based RADAR Absorbing Materials for Camouflaging Application. *Polymer-Plastics Technology and Engineering*, 54(14), 1483-1493

De Keukeleere, K., Pollefeyt, G., Feys, J., De Roo, J., Rijckaert, H., Lommens, P., & Van Driessche, I. (2015). Chemical solution deposition of functional ceramic coatings using ink-jet printing. *Pure and Applied Chemistry*, *87*(3), 231–238. doi:10.1515/pac-2014-0938

Decher, G., Hong, J. D., & Schmitt, J. (1992). Buildup of ultrathin multilayer films by a self-assembly process: III. Consecutively alternating adsorption of anionic and cationic polyelectrolytes on charged surfaces. *Thin Solid Films*, 210-211(Part 2), 831-835 doi:10.1016/0040-6090(92)90417-A

Dobrzański, L., & Mikuła, J. (2005). The structure and functional properties of PVD and CVD coated Al 2O3 + ZrO2 oxide tool ceramics. *Journal of Materials Processing Technology*, *167*(2-3), 438–446. doi:10.1016/j.jmatprotec.2005.05.034

Dosta, S., Torrell, M., Cano, I. G., & Guilemany, J. M. (2012). Functional colored ceramic coatings obtained by thermal spray for decorative applications. *Journal of the European Ceramic Society*, *32*(14), 3685–3692. doi:10.1016/j.jeurceramsoc.2012.04.026

Drelich, J., Chibowski, E., Meng, D. D., & Terpilowski, K. (2011). Hydrophilic and superhydrophilic surfaces and materials. *Soft Matter*, *7*(21), 9804. 9804-9828 doi:10.1039/c1sm05849e

Fesik, E. V., Grebnev, V. V., Zarazhevskii, V. I., & Mal'chikov, G. D. (2014). Ruthenium coatings on zirconium dioxide ceramic: Physicochemical and functional properties. *Russian Journal of Applied Chemistry*, *87*(5), 591–595. doi:10.1134/S1070427214050097

Gao, J., Zhao, C., Zhou, J., Li, C., Shao, Y., Shi, C., & Zhu, Y. (2015). Plasma sprayed rutile titania-nanosilver antibacterial coatings. *Applied Surface Science*, *355*, 593–601.

Ghoniem, M. A. (2014). A bronze Osiris statuette from the Egyptian museum in Cairo: Microstructural characterization and conservation. *Mediterranean Archaeology and Archaeometry*, *14*(1), 37–49.

Glavan, A. C., Martinez, R. V., Subramaniam, A. B., Yoon, H. J., Nunes, R. M. D., Lange, H., & Whitesides, G. M. et al. (2014). Omniphobic "rF paper" produced by silanization of paper with fluoroalkyltrichlorosilanes. *Advanced Functional Materials*, *24*(1), 60–70. doi:10.1002/adfm.201300780

Global Industry Trends. (2014). Nanocoatings Market (Anti-fingerprint, Anti-microbial, Anti-fouling & Easy-to-clean, Self-cleaning and Others) for Medical & Healthcare, Food Manufacturing, Packaging, Marine, Water Treatment, Electronics, Building & Construction, Automotive, Energy and Other End-user Industries: Global Industry Analysis, Size, Share, Growth, Trends and Forecast. And Segment Fore.

Gooch, J. W. (2002). Lead-Based Paint handbook. New York, Kluwer Academic Publishers.

Grasset, F., Starukh, G., Spanhel, L., Ababou-Girard, S., Su, D. S., & Klein, A. (2005). From ZnO colloids to nanocrystalline colored ZnxTi yOw-zNz spinel films. *Advanced Materials*, *17*(3), 294–297. doi:10.1002/adma.200400031

Grigoriev, D., Schenderlein, M., Akcakayiran, D., & Shchukin, D. (2014). Protective organic coatings with anticorrosive and other feedback active features: Micro- and nanocontainers based approach. *Corrosion*, *70*(5), 446–463. doi:10.5006/0976

Grosso, D. (2011). How to exploit the full potential of the dip-coating process to better control film formation. *Journal of Materials Chemistry*, *21*(43), 17033-17038 doi:10.1039/c1jm12837j

Gupta, G., & Satapathy, A. (2015). Processing and Characterization of Plasma Spray Coatings of Glass Microspheres Premixed with Al_2O_3 Particles. *Particulate Science and Technology*, *33*(2), 145–149. do i:10.1080/02726351.2014.947658

Han, M., Mang, J., & Seo, J. (2010). Syntheses of novel sol-gel precursor containing anti-corrosive functional group and their uses in organic-inorganic hybrid coatings. *Polymer Korea*, *34*(5), 405–409.

Han, S. J., Chen, Y., & Sampath, S. (2014). Role of process conditions on the microstructure, stoichiometry and functional performance of atmospheric plasma sprayed La(Sr)MnO3 coatings. *Journal of Power Sources*, *259*, 245–254. doi:10.1016/j.jpowsour.2014.02.073

Hao, L., Sirong, Y., & Xiangxiang, H. (2015). Preparation of a biomimetic superhydrophobic ZnO coating on an X90 pipeline steel surface. *New Journal of Chemistry*, *39*(6), 4860–4868. doi:10.1039/C5NJ00576K

Herrmann, C. F., Fabreguette, F. H., Finch, D. S., Geiss, R., & George, S. M. (2005). Multilayer and functional coatings on carbon nanotubes using atomic layer deposition. *Applied Physics Letters*, *87*(12), 1–3. doi:10.1063/1.2053358

Hu, R.-G., Zhang, S., Bu, J.-F., Lin, C.-J., & Song, G.-L. (2012). Recent progress in corrosion protection of magnesium alloys by organic coatings. *Progress in Organic Coatings*, *73*(2-3), 129–141. doi:10.1016/j.porgcoat.2011.10.011

Huang, X., & Zacharia, N. S. (2015). Functional polyelectrolyte multilayer assemblies for surfaces with controlled wetting behavior. *Journal of Applied Polymer Science*. *132*(45), 42767-42778

Imbesi, P. M., Gohad, N. V., Eller, M. J., Orihuela, B., Rittschof, D., Schweikert, E. A., & Wooley, K. L. et al. (2012). Noradrenaline-functionalized hyperbranched fuoropolymer - poly(ethylene glycol) cross-linked networks as dual-mode, anti-biofouling coatings. *ACS Nano*, *6*(2), 1503–1512. doi:10.1021/nn204431m PMID:22276525

Jubete, E., Loaiza, O. A., Ochoteco, E., Pomposo, J. A., Grande, H., & Rodríguez, J. (2009). Nanotechnology: A Tool for Improved Performance on Electrochemical Screen-Printed (Bio)Sensors. *Journal of Sensors*, 842575, 1–13. doi:10.1155/2009/842575

Jur, J. S., Sweet, W. J. III, Oldham, C. J., & Parsons, G. N. (2011). Atomic layer deposition of conductive coatings on cotton, paper, and synthetic fibers: Conductivity analysis and functional chemical sensing using “all-fiber” capacitors. *Advanced Functional Materials*, *21*(11), 1993–2002. doi:10.1002/adfm.201001756

Katsenelenbaum, B. Z. (2014). Antiradar protection of a metal body with a dielectric coating. *Journal of Communications Technology and Electronics*, *59*(11), 1095–1100. doi:10.1134/S1064226914110084

Kaya, C., Kaya, F., Su, B., Thomas, B., & Boccaccini, R. (2005). Structural and functional thick ceramic coatings by electrophoretic deposition. *Surface and Coatings Technology*, *191*(2-3), 303–310. doi:10.1016/j.surfcoat.2004.03.042

Kirtay, S. (2014). Characterization of SiO2-TiO2 Hybrid Corrosion Protective Coatings on Mild Steel. *Journal of Materials Engineering and Performance*, *23*(12), 4309–4315.

Klajn, R., Wesson, P. J., Bishop, K. J. M., & Grzybowski, B. A. (2009). Writing Self-Erasing Images using Metastable Nanoparticle "Inks.". *Angewandte Chemie International Edition*, *48*(38), 7035–7039. doi:10.1002/anie.200901119 PMID:19533698

Knez, M. (2008). ALD- A versatile tool for nanostructuring. *Material Matters*, *3*(2), 28–33.

Knez, M. (2012). Diffusion phenomena in atomic layer deposition. *Semiconductor Science and Technology*, *27*(7), 074001. doi:10.1088/0268-1242/27/7/074001

Kumar, D., Wu, X., Fu, Q., Ho, J. W. C., Kanhere, P. D., Li, L., & Chen, Z. (2015). Hydrophobic sol–gel coatings based on polydimethylsiloxane for self-cleaning applications. *Materials & Design*, *86*, 855–862. doi:10.1016/j.matdes.2015.07.174

Kundu, D., Biswas, P., & Ganguli, D. (1989). Sol-gel preparation of wavelength-selective reflecting coatings in the system ZrO2-SiO2. *Journal of Non-Crystalline Solids*, *110*(1), 13–16. doi:10.1016/0022-3093(89)90176-2

Lackner, J. M., & Waldhauser, W. (2010). Inorganic PVD and CVD Coatings in Medicine — A Review of Protein and Cell Adhesion on Coated Surfaces. *Journal of Adhesion Science and Technology*, *24*(5), 925–961. doi:10.1163/016942409X12598231568023

Lamuta, C., Di Girolamo, G., & Pagnotta, L. (2015). Microstructural, mechanical and tribological properties of nanostructured YSZ coatings produced with different APS process parameters. *Ceramics International*, *41*(7), 8904–8914. doi:10.1016/j.ceramint.2015.03.148

Le, H. P. (1998). Progress and trends in ink-jet printing technology. *The Journal of Imaging Science and Technology*, *42*(1), 49–62.

Lee, C. S., Kim, J., Son, J. Y., Choi, W., & Kim, H. (2009). Photocatalytic functional coatings of TiO2 thin films on polymer substrate by plasma enhanced atomic layer deposition. *Applied Catalysis B: Environmental*, *91*(3-4), 628–633. doi:10.1016/j.apcatb.2009.06.037

Lejars, M., Margaillan, A., & Bressy, C. (2012). Fouling release coatings: A nontoxic alternative to biocidal antifouling coatings. *Chemical Reviews*, *112*(8), 4347–4390. doi:10.1021/cr200350v PMID:22578131

Levashov, E. A., Pogozhev, Y. S., & Kurbatkina, V. V. (2011). Advanced Ceramic Target Materials Produced by Self-Propagating High-Temperature Synthesis for Deposition of Functional Nanostructured Coatings - Part 1 : Four Elements and Less Systems. *Advances in Ceramics - Synthesis and Characterization- Processing and Specific Applications.*

Liu, W., X., Devaraju, M. K., Yin, S., & Sato, T. (2010). Calcium-doped ceria/titanate tabular functional nanocomposite by layer-by-layer coating method. *Journal of Solid State Chemistry*, *183*(7) 1545-1549

Liu, Y., Wang, X., Fei, B., Hu, H., Lai, C., & Xin, J. H. (2015). Bioinspired, Stimuli-Responsive, Multifunctional Superhydrophobic Surface with Directional Wetting, Adhesion, and Transport of Water. *Advanced Functional Materials*. *25*(31), 5047-5056

Lu, X., Wang, Z., Yang, X., Xu, X., Zhang, L., Zhao, N., & Xu, J. (2011). Antifogging and antireflective silica film and its application on solar modules. *Surface and Coatings Technology*, *206*(6), 1490–1494. doi:10.1016/j.surfcoat.2011.09.031

Lu, Y., Guan, S., Hao, L., & Yoshida, H. (2015). Review on the Photocatalyst Coatings of TiO2: Fabrication by Mechanical Coating Technique and Its Application. *Coatings*, *5*(3), 425–464. doi:10.3390/coatings5030425

Lu, Y., Sathasivan, S., Song, J., Crick, C. R., Carmalt, C. J., & Parklin, I. P. (2015). Robust self-cleaning surfaces that function when exposed to either air or oil. *Science*, *347*(6226), 1132–1135. doi:10.1126/science.aaa0946 PMID:25745169

Madhankumar, A., Thangavel, E., Ramakrishna, S., Obot, I.B., Jung, H.C., Shin, K S., Kim, D.-E. (2014). Multi-functional ceramic hybrid coatings on biodegradable AZ31 Mg implants: electrochemical, tribological and quantum chemical aspects for orthopaedic applications. *RSC Advances, 4*(46), 24272.

Makhlouf, A. S. H., & Tiginyanu, I. (2011). *Nanocoatings and ultra-thin films*. Woodhead Publishing Limited doi:10.1533/9780857094902

Future Markets. (2015). *The Global Nanocoatings Market*.

Martin, T. P., Lau, K. K. S., Chan, K., Mao, Y., Gupta, M., Shannan O'Shaughnessy, W., & Gleason, K. K. (2007). Initiated chemical vapor deposition (iCVD) of polymeric nanocoatings. *Surface and Coatings Technology, 201*(22-23 SPEC. ISS.), 9400–9405.

Mathiazhagan, A., & Joseph, R. (2011). Nanotechnology-A New Prospective in Organic Coating. *International Journal of Chemical Engineering and Applications*, *2*(4), 228–237.

Mathur, S., Shen, H., & Altmayer, J. (2007). Nanostructured Functional Ceramic Coatings By Molecule-Based Chemical Vapor Deposition. *Review of Advanced Materials Science*, *15*, 16–23.

McDowell, M. T., Woo Lee, S., Wang, C., & Cui, Y. (2012). The effect of metallic coatings and crystallinity on the volume expansion of silicon during electrochemical lithiation/delithiation. *Nano Energy*, *1*(3), 401–410. doi:10.1016/j.nanoen.2012.03.004

Michel, M., Toniazzo, V., Ruch, D., & Ball, V. (2012). Deposition Mechanisms in Layer-by-Layer or Step-by-Step Deposition Methods: From Elastic and Impermeable Films to Soft Membranes with Ion Exchange Properties. *ISRN Materials Science*, *2012*, 1–13. doi:10.5402/2012/701695

Mo, C., Zheng, Y., Wang, F., & Mo, Q. (2015). A Simple Process for Fabricating Organic / TiO 2 Super-Hydrophobic and Anti-Corrosion Coating, *Int. J. Electrochem. Sci.* 10, 7380–7391.

Morant-Miñana, M. C., & Elizalde, J. (2015). Microscale electrodes integrated on COP for real sample Campylobacter spp. detection. Biosensors & Bioelectronics, 70, 491–497. doi:10.1016/j.bios.2015.03.063 PubMed doi:10.1016/j.bios.2015.03.063 PMID:25889259

Morant-Miñana, M. C., & Korthals, B. (2011). Ultrathin films of variable polarity and crystallinity obtained from 1,2-polybutadiene nanoparticle dispersions. Langmuir, 27(12), 7516–7523. doi:10.1021/la200334m PMID:21591698

Niinistö, J., & Niinistö, L. (2012). Introduction: Basic Features and Historical Development of Atomic Layer Deposition. Atomic Layer Deposition of Nanostructured Materials. Wiley-VCH;

Okyay, T. O., Bala, R. K., Nguyen, H. N., Atalay, R., Bayam, Y., & Rodrigues, D. F. (2015). Antibacterial properties and mechanisms of toxicity of sonochemically grown ZnO nanorods. RSC Advances, 5(4), 2568–2575. doi:10.1039/C4RA12539H

Olakanmi, E. O., & Doyoyo, M. (2014). Laser-assisted cold-sprayed corrosion- and wear-resistant coatings: A review. Journal of Thermal Spray Technology, 23(5), 765–785. doi:10.1007/s11666-014-0098-x

Pardo, R., Zayat, M., & Levy, D. (2011). Photochromic organic-inorganic hybrid materials. Chemical Society Reviews, 40(2), 672–687. doi:10.1039/c0cs00065e PMID:21229130

Park, C. W., Kang, S.-K., Hernandez, H. L., Kaitz, J. A., Wie, D. S., Shin, J., . . . White, S. R. (2015). Thermally Triggered Degradation of Transient Electronic Devices. Advanced Materials, 27(25), 3783–3788. doi:10.1002/adma.201501180 PMID:25991389

Pavithra, G. P., & Hegde, C. (2012). Magnetic property and corrosion resistance of electrodeposited nanocrystalline iron-nickel alloys. Applied Surface Science, 258(18), 6884–6890. doi:10.1016/j.apsusc.2012.03.124

Pierre, A. C. (1998). Introduction sol-gel processing. Kluwer Academic Publishers.

Pin, L., Ansart, F., Fenech, J., Lours, P., Bonino, J. P., Sniezewski, J., & Le Maoult, Y. (2010). Advances in the Field of New Smart Thermal Barrier Coatings. Advances in Science and Technology, 66, 136–141. doi:10.4028/www.scientific.net/AST.66.136

Rabinovich, G. S., Yamamoto, K., Aguirre, M. H., Cahill, D. G., Veldhuis, S. C., Biksa, A., . . . Shuster, L. S. (2010). Multi-functional nano-multilayered AlTiN/Cu PVD coating for machining of Inconel 718 superalloy. Surface and Coatings Technology, 204(15), 2465–2471. doi:10.1016/j.surfcoat.2010.01.024 doi:10.1016/j.surfcoat.2010.01.024

Raut, H. K., Ganesh, V. A., Nair, A. S., & Ramakrishna, S. (2011). Anti-reflective coatings: A critical, in-depth review. Energy & Environmental Science, 4(10), 3779-3804. doi:10.1039/c1ee01297e

Nanocoatings Market by Type, by Application, and by Region - Trends & Forecasts to 2019. (2015). Research and Markets.

Rodríguez, A., Morant-Miñana, M. C., Dias-ponte, A., Martínez-calderón, M., Gómez-aranzadi, M., & Olaizola, S. M. (2015). Femtosecond laser-induced periodic surface nanostructuring of sputtered platinum thin films. Applied Surface Science, 351, 135–139.

Rojaee, R., Fathi, M., & Raeissi, K. (2014). Comparing Nanostructured Hydroxyapatite Coating on AZ91 Alloy Samples via Sol-gel and Electrophoretic Deposition for Biomedical Applications. IEEE Transactions on Nanobioscience, 13(4), 409–414.

Rybiński, P., Pająk, A., Janowska, G., & Jóźwiak, M. (2015). Effect of hybrid filler (HNTs-phthalocyanine) on the thermal properties and flammability of diene rubber. *Journal of Applied Polymer Science, 132*(40).

Santos, C. M., Mangadlao, J., Ahmed, F., Leon, A., Advincula, R. C., & Rodrigues, D. F. (2012). Graphene nanocomposite for biomedical applications: Fabrication, antimicrobial and cytotoxic investigations. *Nanotechnology, 23*(39), 395101-395110. doi:10.1088/0957-4484/23/39/395101 PMID:22962260

Schilp, S., Rosenhahn, A., Pettitt, M. E., Bowen, J., Callow, M. E., Callow, J., & Grunze, M. (2009). Physicochemical properties of (ethylene glycol)-containing self-assembled monolayers relevant for protein and algal cell resistance. *Langmuir, 25*(17), 10077–10082. doi:10.1021/la901038g PMID:19469528

Schwarzacher, W. (2006). Electrodeposition: A Technology for the Future. *The Electrochemical Society Interface, 15*(1), 32–35.

Seo, J. Y., & Han, M. (2011). Multi-functional hybrid coatings containing silica nanoparticles and anticorrosive acrylate monomer for scratch and corrosion resistance. *Nanotechnology, 22*(2), 25601-25609 doi:10.1088/0957-4484/22/2/025601 PMID:21135477

Seymour, R. B., & Mark, H. F. (1990). Organic coatings: their origin and development. *Proceedings of the International Symposium on the History of Organic Coatings*. Florida, Elsevier.

Sharifnabi, a., Fathi, M. H., Eftekhari Yekta, B., & Hossainalipour, M. (2014). The structural and biocorrosion barrier performance of Mg-substituted fluorapatite coating on 316L stainless steel human body implant. *Applied Surface Science, 288*, 331–340.

Shishkovsky, I., & Smurov, I. (2012). Titanium base functional graded coating via 3D laser cladding. *Materials Letters, 73*, 32–35. doi:10.1016/j.matlet.2011.12.099

Singh, J., & Wolfe, D. E. (2005). Nano and macro-structured component fabrication by electron beam-physical vapor deposition (EB-PVD). *Journal of Materials Science, 40*, 1–26.

Singh, M., Haverinen, H. M., Dhagat, P., & Jabbour, G. E. (2010). Inkjet Printing-Process and Its Applications. *Advanced Materials, 22*(6), 673–685. doi:10.1002/adma.200901141 PMID:20217769

Spanhel, L. (2006). Colloidal ZnO nanostructures and functional coatings: A survey. *Journal of Sol-Gel Science and Technology, 39*(1), 7–24. doi:10.1007/s10971-006-7302-5

Stillman, J. M. (1924). *The Story of Alchemy and Early Chemistry*. Dover Publications, Inc.

Sun, D., An, J., Wu, G., & Yang, J. (2015). Double-layered reactive microcapsules with excellent thermal and non-polar solvent resistance for self-healing coatings. *J. Mater. Chem. A, 3*(8), 4435–4444. doi:10.1039/C4TA05339G

Syed, M., Hutt, N. J., Shah, N., & Edge, J. (2015). Hydroxyapatite ceramic-coated femoral components in young patients followed up for 17 to 25 years: An update of a previous report. *The Bone & Joint Journal, 97-B*(6), 749–754. doi:10.1302/0301-620X.97B6.35278 PMID:26033053

Ulrich, S., Holleck, H., Leiste, H., Niederberger, L., Nold, E., Sell, K., … Sattel, S. (2005). Nano-scale, multi-functional coatings in the material system B-C-N-H. *Surface and Coatings Technology, 200*(1-4), 7–13.

Vijayalakshmi, U., Chellappa, M., Anjaneyulu, U., Manivasagam, G., & Sethu, S. (2016). Influence of Coating Parameter and Sintering Atmosphere on the Corrosion Resistance Behaviour of Electrophoretically Deposited Composite Coatings. *Materials and Manufacturing Processes*. 31(1), 95-106.

Vilar, R. (1999). Laser cladding. *Journal of Laser Applications*, *11*(2), 64. doi:10.2351/1.521888

Vilardell Navarro, M. (2014). *Inkjet printing : A flexible manufacturing of functional ceramic coatings by Chemical Solution Deposition*. Universitat Autonoma de Barcelona.

Wang, F., Qi, B., Wang, G., & Cui, W. (2014). Catalyst coating deposition behavior by cold spray for fuel reforming. *International Journal of Hydrogen Energy*, *39*(25), 13852–13858. doi:10.1016/j.ijhydene.2014.01.133

Wang, H., Xie, G., Fang, M., Ying, Z., Tong, Y., & Zeng, Y. (2015). Electrical and mechanical properties of antistatic PVC films containing multi-layer graphene. *Composites. Part B, Engineering*, *79*, 444–450.

Watin, M. (1773). *L'Art du peintre, doreur et vernisseur*.

Weeks, M. D., Subramanian, R., Vaidya, A., & Mumm, D. R. (2015). Surface & Coatings Technology Defining optimal morphology of the bond coat – thermal barrier coating interface of air-plasma sprayed thermal barrier coating systems. *Surface and Coatings Technology*, *273*, 50–59. doi:10.1016/j.surfcoat.2015.02.012

Woo Kim, J., Kim, B., Won Park, S., Kim, W., & Hyung Shim, J. (2014). Atomic layer deposition of ruthenium on plasma-treated vertically aligned carbon nanotubes for high-performance ultracapacitors. *Nanotechnology*, *25*(43), 435404. 435404-435412 doi:10.1088/0957-4484/25/43/435404 PMID:25299427

Wu, X. (1999). In situ formation by laser cladding of a TiC composite coating with a gradient distribution. *Surface and Coatings Technology*, *115*(2-3), 111–115. doi:10.1016/S0257-8972(99)00045-6

Wu, Y., Bekke, M., Inoue, Y., Sugimura, H., Kitaguchi, H., Liu, C., & Takai, O. (2004). Mechanical durability of ultra-water-repellent thin film by microwave plasma-enhanced CVD. *Thin Solid Films*, *457*(1), 122–127. doi:10.1016/j.tsf.2003.12.007

Wulf, M., Wehling, A., & Reis, O. (2002). Coatings with Self-Cleaning Properties. *Macromolecular Symposia*, *187*(1), 459–467. doi:10.1002/1521-3900(200209)187:1<459::AID-MASY459>3.0.CO;2-Q

Xue, C.-H., Ji, X.-Q., Zhang, J., Ma, J.-Z., & Jia, S.-T. (2015). Biomimetic superhydrophobic surfaces by combining mussel-inspired adhesion with lotus-inspired coating. *Nanotechnology*, *26*(33), 335602. 335602-335611 doi:10.1088/0957-4484/26/33/335602 PMID:26222622

Yakovlev, A., Bertrand, P., & Smurov, I. (2003). Laser cladding of wear resistant metal matrix composite coatings. *Thin Solid Films*, 453-454, 133-138

Yang, W. J., Neoh, K. G., Kang, E. T., Teo, S. L. M., & Rittschof, D. (2014). Polymer brush coatings for combating marine biofouling. *Progress in Polymer Science*, *39*(5), 1017–1042.

Yang, W. J., Pranantyo, D., Neoh, K. G., Kang, E. T., Teo, S. L. M., & Rittschof, D. (2012). Layer-by-layer click deposition of functional polymer coatings for combating marine biofouling. *Biomacromolecules*, *13*(9), 2769–2780. doi:10.1021/bm300757e PMID:22924814

Yoshida, H., Lu, Y., Nakayama, H., & Hirohashi, M. (2009). Fabrication of TiO2 film by mechanical coating technique and its photocatalytic activity. *Journal of Alloys and Compounds*, *475*(1-2), 383–386. doi:10.1016/j.jallcom.2008.07.059

Zelikin, A. N. (2010). Drug releasing polymer thin films: New era of surface-mediated drug delivery. *ACS Nano*, *4*(5), 2494–2509. doi:10.1021/nn100634r PMID:20423067

Zeuner, M., Pagano, S., & Schnick, W. (2011). Nitridosilicates and oxonitridosilicates: From ceramic materials to structural and functional diversity. *Angewandte Chemie International Edition*, *50*(34), 7754–7775. doi:10.1002/anie.201005755 PMID:21774043

Zhang, D., Yu, W., Hao, D., Li, L., Liu, H., & Lu, Z. (2012). Functional nanostructured surfaces in hybrid sol–gel glass in large area for antireflective and super-hydrophobic purposes. *Journal of Materials Chemistry*, *22*(33), 17328. 17328-17331 doi:10.1039/c2jm33557c

Zhao, L., Yan, H., Fang, Z., Wang, J., & Wang, H. (2015). On the flameproof treatment of ramie fabrics using a spray-assisted layer-by-layer technique. *Polymer Degradation & Stability*, *121*, 11–17. doi:10.1016/j.polymdegradstab.2015.08.007

Zuzuarregui, A., Coto, B., Rodríguez, J., Gregorczyk, K. E., Ruiz de Gopegui, U., Barriga, J., & Knez, M. (2015). Highly reflective polymeric substrates functionalized utilizing atomic layer deposition. *Applied Physics Letters*, *107*(6), 061602. 61602-61607. doi:10.1063/1.4928375

Chapter 2
Single Source Precursors for Semiconducting Metal Oxide-Based Films

Carolina Egler Lucas
University of Heidelberg, Germany

Carlos Lizandara Pueyo
ICIQ, Spain

ABSTRACT

Materials possessing interesting properties for current and future everyday products always have a valuable place in the scientific research. Accordingly, semiconductor materials are the foundation of modern electronics. In most of the applications, metal oxide based semiconductors are at present generally useful in the form of thin films. Metal–alkoxo complexes constitute a very important group of precursors for oxide materials. The most common chemical approaches are well described in the literature and the most relevant pathways will be summarized later in this chapter.

INTRODUCTION

Materials possessing interesting properties for current and future everyday products always have a valuable place in the scientific research. Accordingly, semiconductor materials are the foundation of modern electronics. There exists a large variety of semiconductor materials and they are classified according to their nature (inorganic or organic). In the case of inorganic semiconductors, two subclasses can be differentiated: elemental and compound semiconductors. Elemental inorganic semiconductors of group IV such as Silicon (Si) and Diamond are highly valuable due to their high stability and can be classified as excellent thermal conductor mediums.(Sze & Ng, 2006)

Among compound semiconductors, some interesting materials can be found based on combinations of elements such as Titanium dioxide, which is commonly used as pigment and photocatalyst and is classified in group II-VI, commonly named as group of the oxides.(Chen & Mao, 2007; Fujishima & Zhang, 2006) Another promising compound semiconductor of the same group is Zinc oxide (ZnO), which has

DOI: 10.4018/978-1-5225-0066-7.ch002

a wide variety of applications like UV-LEDs,(Look et al., 2002; Meyer et al., 2004) varistor,(Clarke, 1999) UV-photodetector material,(Monroy, Omnes, & Calle, 2003) or UV-blocker in sunscreens.(Carlos Lizandara-Pueyo et al., 2011) Bismuth oxide (Bi_2O_3) is also a semiconducting material possessing interesting properties as gas sensors(Devi, Manorama, & Rao, 1999) or as an effective UV light photocatalyst.(Heidong, Q. Wei, W. Xiaohong, D. Xianbo, & C. Long, 2007) It is also an important component of several ferroelectric(B. H. Park et al., 1999) multiferroic(Singh, Yang, C.G. Takoudis, & 2009) and superconducting oxide materials. (Michel et al., 1987) In most of the above-mentioned applications, metal oxide based semiconductors are generally used in the form of films. The most common chemistry-based methods to fabricate films are chemical vapor deposition (CVD) or sol-gel techniques. Both techniques usually employ metal-organic compounds of high purity in order to form films with the desired composition and high quality without high input of external energy. (Malik, Afzaal, & O'Brien, 2010)

One approach for the chemical vapor deposition of many metal oxides is to use two or more separate precursors that react individually on the growing film. This procedure makes it difficult to control film stoichiometry and usually high deposition temperatures are required. Also in some cases, toxic or pyrophoric materials are used in the production of the films. An alternative to these materials is to use Single-Source Precursors.(Malik et al., 2010)

The prime example of Single-Source Precursor for oxides was found more than 160 years ago. At that time it was observed that the hydrolysis of tetraethyl orthosilicate (TEOS) under acidic conditions yielded a glass-like material, Silicon Dioxide.(Ebelmen, 1846) In the case of metal alkoxide precursors, the advantage to metal oxide materials is the same as in the case of TEOS and SiO_2, the precursors already have metal–oxygen bonds established. Thanks to that, their thermal decomposition can be performed at relatively low temperatures and maintains the metal–oxygen core. The metal oxides derived from alkoxide Single-Source Precursors are highly pure and have specific properties like high hardness, chemical and mechanical resistance, and high temperature stability.

Alkoxides $M(OR)_x$ (M = metal cation of valence x; R = alkyl or aryl group) are formed by the replacement of the hydroxylic hydrogen of an alcohol (ROH) by a metal cation. It is also necessary to note that understanding the nature of alkoxide precursors (showing no analogy with, for example, the carbonyl compounds) and a good knowledge of the Inorganic Chemistry in general are very important in applying the new design and synthetic approaches.(L. John & Sobota, 2010)

The different coordination possibilities of the alkoxo ligands are represented in Figure 1. These possibilities are due to three unpaired electrons of the oxygen atom, which can be coordinated to the metal center in a terminal or bridging way. (Szafert, John, & Sobota, 2008)

Metal–alkoxo complexes constitute a very important group of precursors for oxide materials. There are a number of well-known applicable routes for the preparation of oxide materials using inorganic sources. The most common chemical approaches are well described in the literature and the most relevant pathways will be summarized later in this chapter. (Szafert et al., 2008)

SINGLE SOURCE PRECURSORS

The Single Source Precursors implies that all the elements (M-E) required in the film are incorporated into one compound (e.g. M = Ga and E = O for Ga_2O_3) bonded at the core of the molecule, with various other ligands attached to each of the elements, that behave as a precursor. The reaction pathway involves

Figure 1. Coordination modes of an alkoxo ligand. (Bradley, Mehrotra, & Gaur, 1978)

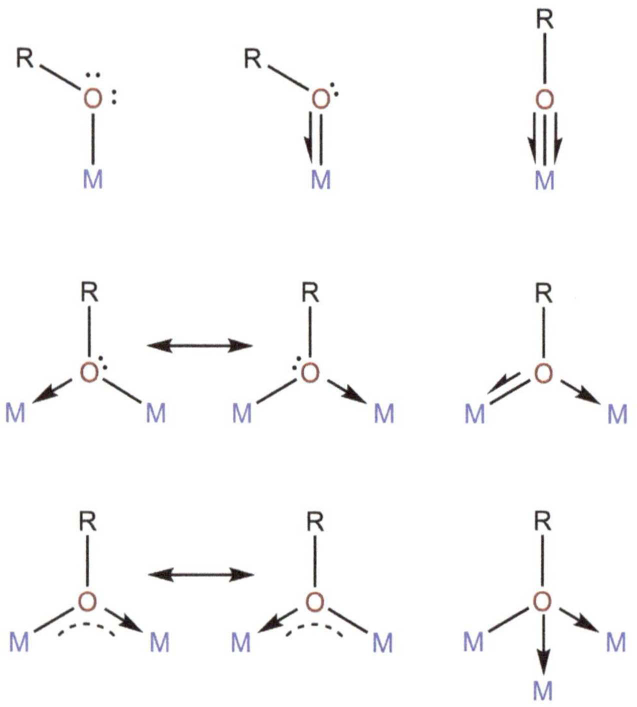

adsorption of the precursor without breaking the M-E bond but with loss of the ancillary ligands, allowing a film of the desired material.

Apart of metal oxides in which this chapter is focused on, it is important to mention that investigation on single source precursors yielding crystalline metal sulfides has attracted extensive investigation due to their exclusive optical, magnetic and electrical properties and the developments in photovoltaic solar cells, biological labeling and medical diagnostics.(Shen et al., 2011) For example ZnS has a wide direct band gap of 3.65 eV (bulk), CdS has a direct band gap of 2.42 eV at 300 K and they have potential uses for optoelectronic device applications such as optical coatings, solid-state solar cell windows, electro-optic modulators, photoconductors, field-effect transistors, sensors and photochemical catalysis.(Shukla, Gaur, & Rai, 2015) In 2005 some new examples of precursors for ZnS films like $Zn(S_2CNEtBu)_2$ or $Zn(S_2CNepp)_2$ were proposed. For example $Zn(S_2CNEtBu)_2$ was obtained from the reaction of sodium hydroxide in methanol with N-ethylbutylamine, carbon disulfide and Zinc chloride. Both precursors were used for CVD experiments in a temperature range from 300 °C to 500 °C and ZnS films were obtained. (Kwon, Seo, Yoon, Lee, & Shem, 2005)

The chemical composition can be controlled by changing the oxidation state of the metal ion. For example using $Fe(DDTC)_3$, Fe_3S_4 nanoparticles were obtained, when using $Fe(DDTC)_2(phen)$, Fe_7S_8 hexagonal nanoplates were obtained and when the concentration of the $Fe(DDTC)_2(phen)$ was increased $Fe_{1.2}S$ nanoribbons were obtained. In the case of copper, depending on the stoichiometry of the reactants Cu_2S QDs or Cu_7S_4 nanoparticles can be obtained. The influence of the R group in the thiocarboxylate ion on the solid-state structures can be observed when in $[\{(Ph_3P)_2Cu\}_2(SC\{O\}Me)_2]$ the obtained material shows an eight-membered Cu_4S_4 ring but under similar conditions a distorted cubane like neutral cluster is obtained from $[(Ph_3P)_3Cu_4(SC\{O\}Ph)_4]$. Some single precursor routes were developed to synthesize films of β-In_2S_3, $CuInS_2$, and $AgIn_5S_8$ as well as highly monodispersed colloidal nanocrystals of ZnSe, CdSe, CdS (water soluble), Ag_2Se_2 and $Cu_{2-x}Se$.

Taking as an example ZnO for the synthesis of single source precursor, a special precursor system like the alkyl- alkoxy- zinc heterocubanes has been recently presented. The strong impact of the alkoxide group was ascertained by the reaction mechanism of H_2O with $[MeZnOR]_4$. In this work it is shown that the alkoxide group becomes cleaved prior to the Zn-CH_3 group. The occurrence of a relatively stable intermediate in which represents one edge of the Zn_4O_4 core is postulated in this work.(C. Lizandara-Pueyo, Berg, Toni, Goes, & Polarz, 2008) The use of this precursor system has allowed the formation of ZnO nanoparticles with different size, shape and crystal structure by adjusting the alkoxide group of the $[MeZnOR]_4$ heterocubane core. (Carlos Lizandara-Pueyo et al., 2012; Pueyo et al., 2010)

The Single-Source Precursor approach has a great number of advantages. First, the stoichiometry of the precursor can be retained in the film. Second, the high deposition temperatures are generally lowered. Third, the precursor delivery system can be simplified. Fourth, better homogeneity is possible since the desired elements are effectively pre-mixed at the molecular scale. Thus, Single-Source Precursors provides a simple and clean route for the synthesis of materials, eliminating mixtures of precursors which can often be toxic and/or expensive, as well as involving different kinetics which can result in the formation of non-stoichiometric and, therefore are ideal for the synthesis of films.(Bloor, Carmalt, & Pugh, 2011)

There are some disadvantages in the use of Single-Source Precursors, for example, such precursors often have low vapor pressures, contamination of the films can result from the decomposition of the ligand and the stoichiometry of the precursor is not always retained in the film. However, advances in variants of deposition techniques, such as Chemical Vapor Deposition, have gone some way to overcome many of these disadvantages (Marchand & Carmalt, 2013). Metal alkoxide precursors need specific conditions for their precursor handling. However, there are several advantages of these systems like an easy fine tuning of reactivity, volatility, solubility and composition at the molecular scale, which result in mild-coating formation conditions with all their additional advantages such as cost efficiency or substrate compatibility.

The use of molecular Single Source Precursors for gas phase synthesis or sol-gel routes can be classified depending on different factors, such as the number of metallic centers that the starting materials have. Therefore the single source precursors can be classified into two different cases, Mono Metallic and Hetero Metallic Single Source Precursor Systems:

Mono Metallic Single Source Precursor Systems like $Ti(OR)_4$ alkoxide or the $[R'ZnOR]_4$ alkyl-alkoxide ZnO heterocubane are the most elucidated systems and it has been shown that the early hydrolysis products or thermolysis in the gas phase are labile, while the late ones often can be further reacted into gels, aerogels or particles with most of their structural features intact. Moreover, these processes have been also studied for building functional films for catalytic or gas sensing applications. The reactivity and structure of these precursors is usually given by the alkyl-, aryl or alkoxide-rests bonded to the metals. (Dilger, Lizandara-Pueyo, Krumm, & Polarz, 2012)

Hetero Metallic Single Source Precursors, which have two or more metals homogeneously dispersed in the final semiconducting material, are frequently used for the preparation of semiconducting materials. The complexity degree of these precursor systems increase with the number of metals added and the formed structures can be hardly predicted. The formation of intermediates controlled either by hydrolysis-condensation reactions in the case of the sol-gel approaches or by the first steps of the thermolysis becomes more important in order to fine tune the final materials than in the case of the Mono Metallic Precursors (Westin, 1997). There are different strategies for the preparation of Hetero Metallic Single Source Precursors. The most used route for the formation of these precursors is via exchange of a metal center. This synthesis requires the mixture of Homo Metallic Single Source Precursors in solution or by melting the precursors (Equation 1).

$$R'_xM_{y-n}M'_n(OR)_z \quad (1)$$

$$M'_{x-n}R'_nM_y(OR)_z \quad (2)$$

$$R'_xM_y(OR_nM'_{w-n})_z \quad (3)$$

Alternatively, more synthetic approaches have been also reported in the literature such as using the substitution of an organometallic group with a metal containing fragment (Equation 2) or by exchange of an alkoxy group by a metal containing fragment (Equation 3).(Dreher, Krumm, Lizandara-Pueyo, & Polarz, 2010)

METAL ALKOXIDES

General Methods of Synthesis

There are many methods of synthesis of metal alkoxides or aryloxides and the choice depends on the ionization energy of the metal, whose alkoxide is formed. The less electronegative alkali metals react spontaneously with alcohols, but for instance magnesium, aluminium etc. need some so-called activation agent. For other metals more complex reactions need to be applied.

One example as mentioned above is the reaction of a metal with an alcohol, but is limited to less electronegative alkali metals and based on hydroxyl hydrogen substitution by metal cation followed by intense heat and H_2 evolution (Equation 4):

$$M + (1+x)ROH \rightarrow 1/y\{MOR{\cdot}xROH\}_y + 1/2H_2\uparrow \quad (4)$$

Well soluble alkoxides can be obtained in direct reaction using sterically ramified alcohols (where R = ^{t}Bu, CF_3, aryl etc.). The use of alcohols with “bulky” R groups prevents oligomerization, and the resulting alkoxides are well soluble in common aliphatic hydrocarbons.

Another synthetic strategy employed for metals with high ionization energy like Y, Ti, Zr, Nb, Ta, Mo, W, Cu, Ge and Sn is the electrochemical method, producing extremely pure products. Due to the oxidation of metal at the anode, the formed electron and alcohol create hydrogen radical H· and alkoxide anion. Molecular hydrogen is exuding at the cathode.(L. John & Sobota, 2010; Kovsman et al., 1994; Lehmkhul & Eisenbach, 1975; Szilard, 1906)

The reaction between alcohol and metal halide leads to the substitution of halide anion into RO- group forming appropriate metal alkoxide (Equation 5).

$$MCl_z + (x + y)ROH \rightarrow MCl_{z-x}(OR)_x(ROH)_y + {}_xHCl\uparrow \quad (5)$$

Different compounds can be obtained depending on the solvent, molar ratio of reagents, and temperature. The reaction of $TiCl_4$ with iPrOH in CH_2Cl_2, where $[Ti(\mu\text{- }Cl)_2Cl_2(O^iPr)_4]$, $[TiCl_3(O^iPr)(HO^iPr)_2]$, and $[Ti_2Cl_4(\mu\text{-}O^iPr)_2(O^iPr)_2(HO^iPr)_2]$ are formed is a classic example.(Wu, Ho, Lin, & Gau, 1996)

Also ligands exchange and metathesis reactions are used to obtain homo- and heteroleptic alkoxide complexes in order to obtain metal alkoxide with high tendency to build oligomer structures. (Equation 6 and 7).(Bradley, Chakravarti, Chatterjee, Wardlaw, & Whitley, 1958; Mehrotra, 1953, 1954)

$$M(OR)_x + yR'OH \rightarrow M(OR)_{x-y}(OR')_y + yROH \quad (6)$$

$$mM'Cl_x + nM(OR)_y \rightarrow mM'(OR)_x + nMCl_y \quad (7)$$

Dialkyl amides $M(NR_2)_x$ (R = Me, Et, $SiMe_3$) with metals having high affinity for oxygen atoms, react with alcohols according to the Equation 8 forming highly pure alkoxide products. (Boyle, Hernandez-Sanchez, Baros, Brewer, & Rodriguez, 2007; M. H. Chisholm, Huffman, Kirkpatric, Leonelli, & Folting, 1981; Matchett, Chiang, & Buhro, 1990)_

$$M(NR_2)_x + xR'OH \rightarrow M(OR')_x + xR_2NH\uparrow \quad (8)$$

For the preparation of mono- and mixed-metal alkoxides reactions like shown in Equations. 9-11 are extremely useful and used (S. Basharat et al., 2007; Basharat, Carmalt, Barnett, Tocher, & Davies, 2007; Jerzykiewicz, Utko, & Sobota, 2006; Neumüller, 2003; Polarz, Pueyo, & Krumm, 2010; Sobota, Utko, Sztajnowska, Ejfler, & Jerzykiewicz, 2000) but the organometallics are air/moisture sensitive.

$$MR_2 + R'OH \rightarrow 1/x[MOR']_x + RH\uparrow \quad (9)$$

$$MR_3 + R'OH \rightarrow 1/x[R2MOR']_x + RH\uparrow \quad (10)$$

$$R_2MX + M'OR' \rightarrow 1/x[R2MOR']_x + M'X \quad (11)$$

Direct reaction of organometallic complexes with alkoxides that bear free alcohols at the metal site is a modification of this reaction (Equation 12).

$$M(OR)_x(ROH)_y + yM'R'_z \rightarrow M'R'_{z-1}M(OR)_{x+y} \quad (12)$$

Here an organometallic is driven the hydrogen abstraction of the hydroxyl group from the metal coordination sphere, leading to the subsequent evolution of simple hydrocarbons like methane or ethane and results in a linkage of RO ligands, obtaining highly pure products.(Å. u. John et al., 2008)

Precursors and Geometry

Although in the literature there are a large number of examples of metal alkoxides, it would be very difficult to formulate a precise rule that could fully explain the process of a specific geometry formation and predict the final geometry of forming alkoxide complex. In Table 1 some important alkoxide molecular precursors are summarized. (Bloor et al., 2011; Carmalt & King, 2006; L. John & Sobota, 2010; V. Kessler, 2004; Mehring, 2007)

The most common core type in the chemistry of metal alkoxides is the M_4X_{16}, a fragment of hexagonal close packing of metal cations and oxygen atoms.(V. Kessler, 2004) It is the best candidate for precursors with 1:1 metal ratios, for example, those of perovskite materials. This core type is observed usually for the derivatives of primary alcohols, and can be achieved by using chelating ligands such as *β*-diketones (acetylacetone, Hacac, for smaller cations or tetramethylheptanedione, Hthd, for the bigger

Table 1. Summary of some important alkoxide molecular precursors

Complex	Structural Motive
$[Me_4Zn_4(\mu_3\text{-}O^tBu)_4]$	$Zn_4(\mu_3\text{-}O)_4$core
TDMEDA-$Zn(eacac)_2$	ZnO
$[W_2(OCMe_2CMe_2O)_3]$	$O_3W{\equiv}WO_3$core
$[Ga_2(\mu\text{-}O^tBu)_2{}^tBu_4]$	$Ga_2(\mu\text{-}O)_2$core
$[{}^tBu_2Ga(O^tBu)]_2$	Ga_2O_2 ring, planar
$[{}^tBu_2Ga(O^nC_5H_{11})]_2$	Ga_2O_2 ring, planar
$[Ga(Me(OEt)_2]_n$	GaO_4R core trigonal bipyramidal
$[Ga(OEt)_3]$	GaO_5 core trigonal bipyramidal
$[{}^tBu_2In(OEt)]_2$	In_2O_2 ring, planar
$[{}^tBu_2In(OMe)]_2$	In_2O_2 ring, planar
$[In(\mu\text{-}O^tBu)\ (O^tBu)_2]_2$	In_2O_2 ring
$[In(OCMe_2(CF_3))_3]_2$	In_2O_2 ring slightly folded
$[In(OCH_2CH_2NMe_2)_3]_2$	In_2O_2 ring
$[In(oz)_3]$	$In_2O_3N_3$ core, octahedral
$[Bi(OEt)_3]$	-
$[Bi(OC_6H_3\text{-}2,6\text{-}iPr_2)_3]$	-
$[Bi(OCMe_2CH_2OMe)$	Bi_2O_3
$Ba_2Ti_2(thd)_4(OEt)_8(EtOH)_2$	M_4X_{16} hex, X = OR, *β*-dik/2
$Sr_2Ti_2(thd)_4(Oi\ Pr)_8$	M_4X_{16} hex, X = OR, *β*-dik/2
$[Ti(O^iPr)_2(acac)_2]$	TiO_2
$[Ti(OiPr)_2(deacam)_2]$	TiO2

ones) or aminoalcohols. Some results of the use of MSDC (molecular structure design concept) are summarized in Table 1.(V. Kessler, 2004)

In Table 1, some examples of Single-Source Precursor having Zn as metal center among others are shown. ZnO compound is well known to have a great variety of applications (Polarz et al., 2010) and therefore the ZnO films have been very studied and obtained using different techniques. One of the most used methods for ZnO films is the metal organic chemical vapor deposition process (MOCVD). It is known that this method provides high quality films with less impurities and under mild conditions in comparison for example with atomic-layer deposition (ALD) or spray pyrolysis (SP) among others. Most of the ZnO films obtained by MOCVD use precursors like liquid dimethyl-Zn, diethyl-Zn or solid acetate. In 2009 a new Single-Source Precursor for ZnO was presented. The complex can be used at lower temperatures ~ 450 °C. The Zn adduct TMEDA-Zn(eacac)$_2$ (eacac = ethyl acetoacetate, TMEDA = N,N,N',N'-tetramethyethylendiamide) was obtained from the reaction of $ZnCl_2$ and TMEDA. The TMEDA-$ZnCl_2$ adduct was isolated and then reacted with sodium ethylacetoate. The TMEDA-Zn(eacac)$_2$ precursor was obtained as colorless microcrystals. The analysis of the structure revealed a distorted octahedral geometry around the unique six-coordinate Zn^{2+} cation with four oxygen atoms from two asymmetric bidentate ligands, eacac, and two nitrogen atoms from the TMEDA ligand. The precursor was tested through the MOCVD method at relatively mild conditions ~ 450 °C, ZnO films were successfully grown on Si (100) substrates.(J. P. Park, Kim, Park, Ok, & Shim, 2009)

As previously mentioned in this chapter, the use of a special molecular precursor system like the alkyl- alkoxy zinc heterocubanes enables the synthesis of ZnO nanoparticles with new properties, which can be formulated into functional films. This fact is due to the special geometry and reactivity of these precursors. It has been demonstrated that the reactivity of the precursors could be tuned in such a way that the entire process becomes slow enough to use standard analytical tools for the in situ investigation of ZnO nucleation and growth. As a result, several interesting findings were reported.(C. Lizandara-Pueyo et al., 2008) The strong impact of the alkoxide group was ascertained by the reactions mechanism of water with $[MeZnOR]_4$. Unexpectedly, the alkoxide group becomes cleaved prior to the alkyl group. The occurrence of a relatively stable intermediate in which OH represents one edge of the "Zn_4O_4" core was postulated in this publication. Moreover, it was found that the increasing amount of intermediate species led to the formation of the first Zn-O-Zn bonds and this, in turn, led to the rapid formation of ZnO particles. Accordingly it was demonstrated that the size and shape of the nanocrystals could be tuned by adjusting parameters such as temperature, concentration of reactants, solvent and, as previously mentioned in this chapter, precursor architecture. The possibility to control particle nucleation and growth of these precursors allowed the assembly of small nanoparticles into ZnO aerogel-like materials,(Krumm, Pueyo, & Polarz, 2010) which were able to be processed as films (Krumm, Pawlitzek, Weickert, Schmidt-Mende, & Polarz, 2012). Moreover, it has been reported that the conversion of the alkyl- alkoxy- zinc heterocubanes precursor under kinetically controlled conditions resulted in a metastable polymorph of ZnO with a structure resembling α-boron nitride. It was shown that this is a metastable phase that transforms directly into the most stable ZnO structure, the Wurzite Structure. Interestingly, the obtained ZnO with a structure resembling α-boron nitride is closer to the Wurzite Structure than to the heterocubane structure, suggesting that the formation of this structure can be performed using other precursor systems. This raises the question: is the only function of the heterocubane precursor "Zn_4O_4" to enable the rapid formation at low temperatures or more? Accordingly, after the first hydrolysis and condensation step, an octameric compound "Zn_8O_8" is proposed. (C. Lizandara-Pueyo et al., 2008; Pueyo et al., 2010) Since the latter compound is a reactive intermediate, this can lead to the formation of further intermediate species closer to the ZnO with a structure resembling α-boron nitride. Thus, this is a fitting example for the Ostwald's step rule for the synthesis of metastable polymorphs using single source precursors with defined geometries. Moreover, using these alkyl- alkoxy zinc heterocubanes precursors systems, nanoparticles with different shapes and properties were synthesized in organic solvents. Accordingly, the formation of nanoplates, nanobells, nanorods and complex structures has been obtained via colloidal stabilization of the ZnO particles during growth. The formation of nanoparticles preceded by the formation of the crystallization intermediate (metastable polymorph of ZnO with a structure resembling α-boron nitride) regardless of the additive used, suggesting that the ZnO formation in organic solvents is displaced in the kinetic pathway of the crystallization. Crystallization of ZnO with macromolecular additives revealed that the mechanism of formation of nanostructures follows the non-classical crystallization route when the ZnO nanoparticles are grown under high selective particle stabilization conditions (stabilizing the (002) crystal lattice). The ZnO nanostructures formed were very likely polycrystalline particles formed by iso-oriented particles.(Carlos Lizandara-Pueyo et al., 2012) The formation of these nanomaterials was possible thanks to the use of single source precursors and organic solvents as crystallization medium.

Besides these findings, a new and versatile method for the preparation of colloidal ZnO nanoparticles with anisotropic shape and doped nanoparticles using these precursors was presented.(Carlos Lizandara-Pueyo et al., 2014; Carlos Lizandara-Pueyo et al., 2011) This method combines the advantages of a water-in-oil emulsion with alkyl- alkoxy- heterocubanes as precursors. Thus, the polar groups of the additives such as polyglyceryl head-group (P3P) are located at the droplet interface, where the hydrolysis and condensation processes happens, and this resulted in the formation of a novel ZnO nanostructure, the trigonal prismatic shape. Moreover, the presented synthetic approach allows adjustment of the aspect ratio of the trigonal nanoprisms via external parameters. Interestingly, the absorption edge of the novel structure is significantly red-shifted in comparison to bulk ZnO, which could be explained due to the quantum-confined Stark effect. Additionally, it was observed that sudden changes in the reaction kinetics resulted in the formation of well-defined nanorods. In addition to these more practical aspects, some insights into the formation of anisotropic crystalline materials could be gathered. It was recognized that the interaction with the additives starts as soon as a crystalline phase forms. Interestingly, a rather stable amorphous ZnO species without any special morphology were detected. The amorphous nature of the crystallization intermediates proved advantageous for the introduction of Li into the ZnO nanorod structure. Thus it was shown that the addition of Li sources in the emulsion mixture resulted in the homogeneous introduction of Li in the anisotropic structure. The mild crystallization conditions and control over the growth of the amorphous particles allowed the synthesis of these materials. Interestingly, the absorption edge of the doped ZnO nanorods was significantly red-shifted in comparison to bulk ZnO. Deviations observed in the optical properties were attributed as a result of the specific morphology and the incorporation of Li into the ZnO structure. Therefore this versatile method for the preparation of colloidal ZnO nanoparticles established a starting point for the synthesis of new anisotropic materials in colloidal form. These colloids have been used for the formation of UV-protective films as well as gas-sensing materials, after film deposition and organic removal.(Carlos Lizandara-Pueyo et al., 2014)

Gallium oxide (Ga_2O_3) and indium oxide (In_2O_3) films are of interest due to their range of applications, which are related to their particular properties. Gallium and indium alkoxides are often used as molecular precursors to films of gallium and indium oxide.(Marchand & Carmalt, 2013) Gallium oxide films have the monoclinic β–form and has a good chemical and thermal stability and interesting electrical properties. At room temperature it is an electrical insulator but it becomes semiconducting above 500 °C. For the synthesis of gallium (III) and indium (III) alkoxides various methods have been used, one of the most common for gallium is the reaction of trivalent organometallics with alcohols at elevated temperatures. In general the structures of these compounds are monomeric or dimeric via alkoxy bridges, obtained from the sp^2 hybridation of two M atoms bonded by two alkoxide ligands and forming a M_2O_2 ring.

Compounds $[^tBu_2Ga(O^tBu)]_2$ and $[^tBu_2Ga(O^nC_5H_{11})]_2$ were obtained via treatment of R_3Ga with excess of R-OH. The lack of further reaction was attributed to the strong electron donating alkoxides in complexes of the group 13 that generally reduces the reactivity of the M-R bond towards protonolysis. (Carmalt & King, 2006) Other gallium tris(alkoxides) like $[Ga(OR)_3]$ being R: Me, Et, *i*Pr, *n*Bu, and CH_2CH_2OMe were synthesized electrochemically using a metallic gallium anode and the corresponding alcohol. The cathode was a platinized platinum electrode and Bu_4NBr, Et_3BzNCl or LiCl were used as an electroconductive additive in the alcohol. The X-ray obtained for this compound showed a 1-D polymer with zig-zag chains of gallium centers in a distorted 5-coordinated trigonal bipyramidal geometry, this was bound to four bridging and a single terminal OEt group.(Bloor et al., 2011)

Indium complexes were prepared using other synthetic procedures. For example $[^tBu_2In(OEt)]_2$ was obtained via salt elimination from the reaction of $[^tBu_2InCl]_2$ with 2 equiv. of Li(tmp) in diethyl ether. It was assumed that lithium ethoxide was produced from the reaction of Li(tmp) with diethyl ether. Complex $[^tBu_2In(OMe)]_2$ was obtained from the reaction of $[^tBu_2In\{P(SiMe_3)_2\}]_2$ with methanol, the crystal structure revealed that the complex is isomorphs to its gallium analogue. The In-O bond distances of $[^tBu_2In(OEt)]_2$ and $[^tBu_2In(OMe)]_2$, 2.165(5) and 2.14(5) Å for $[^tBu_2In(OEt)]_2$ and 2.153(2) Å for $[^tBu_2In(OMe)]_2$, and O-In-O angles 75.2(2)° for $[^tBu_2In(OEt)]_2$ and 75.2(1)° for $[^tBu_2In(OMe)]_2$ are comparable.

A series of other indium tris(alkoxide) compounds $[In(OR)_3]_n$ were prepared by alcoholysis of $[In\{N(SiMe_3)_2\}_3]$ like for example donor-functionalized indium alkoxide complex $[In(OCH_2CH_2NMe_2)_3]_2$. X-ray analysis showed that two $[InO_6]$ octahedra share a common edge therefore adopting a dimeric structure in the solid state. The In-O bond distances (2.056(4)–2.189(4) Å) are longer than for alkoxides where indium is only four coordinate. The In-N bond distances (2.324(6)–2.477(5) Å) are in the range observed for In N coordination bonds.(Carmalt & King, 2006) Also complex $[In(OCMe_2(CF_3))_3]_2$ was obtained in the same synthetic pathway. After the reaction of $[In\{N(tBu)SiMe_3\}_3]$ with the alcohol $(CF_3)Me_2COH$, the CF_3 complex was isolated. Crystal structure of $[In(OCMe_2(CF_3))_3]_2$ showed how in the dimeric complex, the indium center is four-coordinate and the In_2O_2 ring is slightly folded. Complex $[In(oz)_3]$ was obtained via the reaction of $In(NO_3)_3{\cdot}5H_2O$ with the corresponding ligand in methanol/water and in the presence of an aqueous base. The X-ray analysis determined that the indium center adopts distorted octahedral coordination geometry. (Carmalt & King, 2006)

Homoleptic metal alkoxides are among the most important starting compounds for the synthesis of bismuth oxide containing materials prepared by the CVD or the sol–gel process.

In Table 1 some examples of these compounds are given. The methods for the synthesis of these and other bismuth alkoxides include metathesis, alcohol–amine exchange, alcohol–alcohol exchange, benzene elimination starting from $BiPh_3$ and alcohols, electrochemical oxidation and alcoholysis of freshly prepared bismuth hydroxide.(Mehring, 2007)

For example bismuth alkoxides like $[Bi(OR)_3]$ (R= Me, Et, iPr) have been prepared by Mehrotra and Rai in 1966 using the metathesis route. They are not good precursors for materials synthesis due their low solubility, poor volatility and due to the pronounced tendency of the Bi(III) center to expand the coordination sphere and form polymeric aggregates. This limits their use for CVD applications. Only the application of $[Bi(OEt)_3]$ obtained by the metathesis route for the synthesis of Bi-containing high T_c superconductors has been reported.(Yamane, Kurosawa, & Hirai, 1988)

Complex [Bi(O^{t}Bu)$_3$] was synthesized thought metathesis and was designed to avoid the oligomerization, it was obtained in good yields. The complex is partially soluble in organic solvents and the complex was found to be volatile in comparison with [Bi(OEt)$_3$]. The complex can be handled without decomposition under inert atmosphere, it has interesting properties for the synthesis of superconducting materials and is well suited for the phase-selective deposition of thin bismuth oxide films.(Mehring, 2007)

In 2001 the volatile bismuth alkoxide tris(1-methoxy-2-methyl-2-propanolato)bismuth(III) Bi(OCMe$_2$CH$_2$OMe)$_3$ which can be synthesized through metathesis or alcohol-amine exchange was proposed as an excellent precursor for the deposition of bismuth oxide and related materials. It was synthesized from the reaction of 1-methoxy-2-methyl-2-propanol (HOCMe$_2$CH$_2$OMe) and Bi[N(SiMe$_3$)$_2$]$_3$ in diethyl ether solution. The complex was found to be monomeric and contains three chelating [OCMe-$_2$CH$_2$OMe] ligands forming a disorder octahedron of oxygen atoms around the Bi atom. The complex found to be a good precursor for MOCVD of bismuth oxide in a range of 250 °C to 550 °C showing high purity. Analysis of the film showed basically pure Bi$_2$O$_3$.(Williams et al., 2001)

In the case of Wolfram, the complex [W$_2$(OCMe$_2$CMe$_2$O)$_3$] was obtained by the reaction of W$_2$(OtBu) and 3 equation of pinacol. It crystallizes from THF as 1:1 solvent complex. The molecule has virtual D$_3$ symmetry because of the tris-bridging pinacolate ligands which form M$_2$O$_2$C$_2$ six membered rings in the skew-boat conformation.(Malchom H. Chisholm, Folting, Hampden-Smith, & Smith, 1987)

Titanium oxide films have interesting applications in metal oxide semiconductor field effect transistors (MOSFETS) and is a component of a number of ferroelectric oxides used for example in infrared detectors, or dye-sensitized solar cells among others.(Anthoni C. Jones, 2002) The classical titanium precursors have been extensively used, but they have shown several inconvenient. Titanium tetrachloride [TiCl$_4$] it is a known precursor for the formation of TiO$_2$ films. The problem of this precursor is its high toxicity that requires special equipment and also the chloride contamination in CVD films leading to side-reactions during the process.(Goshtagore, 1970) Other Titanium oxide precursors such as titanium isopropoxide, [Ti(O^{i}Pr)$_4$] are unstable in air moisture and have to be handle in special conditions. The high reactivity of this complex is due to the unsaturated four-coordinate Titanium center, a possible solution for that has been found with the preparation of precursors having chelating β-diketonate such [Ti(O^{i}Pr)$_2$(acac)$_2$] which has been used for the MOCVD of TiO$_2$. The reaction conditions were optimized varying the acetylacetonate vapor pressure, the deposition temperature and the individual flow rate of the gases.(Balong & Schieber, 1972) Other less unstable TiO$_2$ precursors were found to be [Ti(O^{i}Pr)$_3$(dmae)] and [Ti(O^{i}Pr)$_2$(dmae)$_2$]. They were used to deposit TiO$_2$ at 300-450 °C by liquid injection MOCVD. Complex [Ti(O^{i}Pr)$_3$(dmae)] is a mixture of monomer and dimmers at room temperature and at elevated temperatures the monomer predominates, whereas [Ti(O^{i}Pr)$_2$(dmae)$_2$] is a monomeric complex with pseudo-octahedral coordination at the metal center. The TiO$_2$ films have a similar purity compared to films grown by MOCVD and using [Ti(OR)$_4$] as precursor.(Jones et al., 1998)

Recently A. Devi et al have shown a new titanium precursor, [Ti(O^{i}Pr)$_2$(deacam)$_2$] (deacam = *N,N*-diethylacetoacetamide). The precursor was obtained from the addition of a diluted solution of deacam in hexane, to a solution of [Ti(O^{i}Pr)$_4$]. The mixture was refluxed for 12 h at 68 °C. The precursor was obtained as white crystals, the complex is monomeric, with an octahedral geometry. The two β-ketoamides ligands and the two alkoxy ligands are arranged *cis* to each other, and the two amide moieties arranged *trans* with respect to the alkoxy ligands. The [Ti(OiPr)$_2$(deacam)$_2$] single-source precursor was used to fabricate TiO$_2$ films in a temperature range of 500-800 °C, with and without O$_2$ introduction in the reaction environment. Well-adherent TiO$_2$ films were growth dependent on the conditions, with a homogeneous formation of TiO_2 with a uniform nitrogen distribution, confirming [$Ti(OiPr)_2(deacam)_2$] as a new Single-Source Precursor for TiO_2 films.(Kim et al., 2015)

$Ba_2Ti_2(thd)_4(OEt)_8(EtOH)_2$ was obtained in a high yield reacting alkoxides and β-diketonates and adding a β-diketone to a mixture of alkoxides. It crystallized on addition of 1 equivalent of tetramethylheptanedione to ethanol solutions of barium and titanium ethoxides in a 1:1 molar ratio. (G. V. Kessler, Hubert-Pfalzgraf, Daniele, & Gleizes, 1994) Complex [$Sr_2Ti_2(thd)_4(OiPr)_8$] was obtained from the mixture of $Ti(OiPr)_4$ and $Sr(thd)_2$ dissolved in hexane. After 4 h at room temperature the reaction medium was concentrated and the complex was isolated. X-ray diffraction was obtained, all metals bear chelating β-diketone ligands and no adducts were formed in the presence of the alcohol, the isopropoxide ligands were found to be of three types, two are triply bridging, four are doubly-bridging connecting to strontium and titanium and finally two are terminal on the titanium atoms.

CONCLUSION

It can be concluded that Metal alkoxides are valuable molecular precursors for the preparation of metal oxide based films. The advantage of such molecular sources is that they allow tuning the properties of the precursors and storing information about the final materials at the molecular scale. The modification of the organic part and the inorganic core of the precursors allow influencing important key properties of materials like composition, crystal structure, particle size and shape. Furthermore, the molecular features enable the application of these precursors in material synthesis at mild conditions in order to ensure metastable characteristics. Film synthetic approaches involving well defined colloids, liquid state and gas-phase can be applied.

REFERENCES

Balog, M., & Schieber, M., Patai, S., & Michman, M. (1972). Thin films of metal oxides on silicon by chemical vapor deposition with organometallic compounds. *Journal of Crystal Growth*, *17*, 298-301.

Basharat, S., Betchley, W., Carmalt, C. J., Barnett, S., Tocher, D. A., & Davies, H. O. (2007). Synthesis of Group 13 Sesquialkoxides and Their Application as Precursors to Crystalline Oxide Films. *Organometallics*, *26*(2), 403–407. doi:10.1021/om0608657

Basharat, S., Carmalt, C. J., Barnett, S. A., Tocher, D. A., & Davies, H. O. (2007). Aerosol Assisted Chemical Vapor Deposition of In2O3 Films from Me3In and Donor Functionalized Alcohols. *Inorganic Chemistry*, *46*(22), 9473–9480. doi:10.1021/ic701372b PMID:17914814

Bloor, L. G., Carmalt, C. J., & Pugh, D. (2011). Single-source precursors to gallium and indium oxide thin films. *Coordination Chemistry Reviews*, *255*(11-12), 1293–1318. doi:10.1016/j.ccr.2010.12.018

Boyle, T. J., Hernandez-Sanchez, B. A., Baros, C. M., Brewer, L. N., & Rodriguez, M. A. (2007). Influencing the Morphology and Phase of Calcium Ceramic Nanoparticles Based on the Calcium Alkoxide Precursors' Characteristics. *Chemistry of Materials*, *19*(8), 2016–2026. doi:10.1021/cm0700091

Bradley, D. C., Charkravarti, B. N., Chatterjee, A. K., Wardlaw, W., & Whitley, A. (1958). Niobium and tantalum mixed alkoxides. *Journal of the Chemical Society*, 99. doi:10.1039/jr9580000099

Bradley, D. C., Mehrotra, R. C., & Gaur, D. P. (1978). Metal Alkoxides. *Zeitschrift für Chemie*, *20*(10), 396–396.

Carmalt, C. J., & King, S. J. (2006). Gallium(III) and indium(III) alkoxides and aryloxides. *Coordination Chemistry Reviews*, *250*(5-6), 682–709. doi:10.1016/j.ccr.2005.05.028

Chen, X., & Mao, S. S. (2007). Titanium Dioxide Nanomaterials: Synthesis, Properties, Modifications, and Applications. *Chemical Reviews*, *107*(7), 289. doi:10.1021/cr0500535 PMID:17590053

Chisholm, M. H., Folting, K., Hampden-Smith, M., & Smith, C. A. (1987). Preparation and characterization of hexacyclohexoxido-and trispinacolatoditungsten. *Polyhedron*, *6*(9), 1747.

Chisholm, M. H., Huffman, J. C., Kirkpatric, J., Leonelli, J., & Folting, K. (1981). Facile cleavage of the molybdenum.tplbond.molybdenum bond in reactions between Mo2(OBut)6 and aryl azides and molecular oxygen. *Journal of the American Chemical Society*, *103*(5), 1305–1306. doi:10.1021/ja00395a092

Clarke, D. R. (1999). Varistor Ceramics. *Journal of the American Ceramic Society*, *82*(3), 485–502. doi:10.1111/j.1151-2916.1999.tb01793.x

Devi, G. S., Manorama, S. V., & Rao, V. J. (1999). SnO2:Bi2O3 based sensor: Laser-Raman, temperature programmed desorption and X-ray photoelectron spectroscopic studies. *Sensors and Actuators. B, Chemical*, 1999, 98. doi:10.1016/S0925-4005(99)00164-1

Dilger, S., Lizandara-Pueyo, C., Krumm, M., & Polarz, S. (2012). Hierarchical zinc oxide materials with multiple porosity prepared by ultrafast temperature gradient chemical gas-phase synthesis. *Advanced Materials*, *24*(4), 543–548. doi:10.1002/adma.201103557 PMID:22213088

Dreher, A. M., Krumm, M., Lizandara-Pueyo, C., & Polarz, S. (2010). Chromium containing zinc oxide materials from organobimetallic precursors. *Dalton Transactions (Cambridge, England)*, *39*(9), 2232. doi:10.1039/b916253b PMID:20162196

Ebelmen. (1846). *Annal. Chem. Pharm., 57*, 334.

Fujishima, A., & Zhang, X. T. (2006). Titanium dioxide photocatalysis: Present situation and future approaches. *Comptes Rendus. Chimie*, *9*(5-6), 750–760. doi:10.1016/j.crci.2005.02.055

Goshtagore, R. N. (1970). Mechanism of heterogeneous deposition of thin film rutile. *Journal of the Electrochemical Society*, *117(4)*, 529. doi:10.1149/1.2407561

Heidong, H., Wei, Q., Xiaohong, W., Xianbo, D., & Long, C. (2007). The photocatalytic properties of bismuth oxide films prepared through the sol-gel method. *Thin Solid Films*, *515*(13), 5362–5365. doi:10.1016/j.tsf.2007.01.031

Jerzykiewicz, L. B., Utko, J., & Sobota, P. (2006). Structural Characterization of Manganese(II) Compounds with Methylalumonaxe (MAO) and AlMe2 Units. *Organometallics*, *25*(20), 4924–4926. doi:10.1021/om060207m

John, Å., Utko, J., Szafert, S., Jerzykiewicz, L. B., Kępiński, L., & Sobota, P. (2008). Synthesis and Characterization of Mixed-Metal Aryloxo-Organometallic Precursors for Oxides Ceramic Materials. *Chemistry of Materials*, *20*(13), 4231–4239. doi:10.1021/cm702262n

John, L., & Sobota, P. (2010). Alkoxide Molecular Precursors for Nanomaterials: A One Step Strategy for Oxide Ceramics. Intech Open Access.

Jones, A. C. (2002). Molecular design of improved precursors for the MOCVD of electroceramic oxides. *Journal of Materials Chemistry*, *12*(9), 2576–2590. doi:10.1039/b202675a

Jones, A. C., Leedham, T. J., Wright, P. J., Crosbie, M. J., Fleeting, K. A., Otway, D. J., & Pemble, M. E. et al. (1998). Synthesis and characterisation of two novel titanium isopropoxides stabilised with a chelating alkoxide: Their use in the liquid injection MOCVD of titanium dioxide thin films. *Journal of Materials Chemistry*, *8*(8), 1773–1777. doi:10.1039/a802933d

Kessler, G. V., Hubert-Pfalzgraf, G. L., Daniele, S., & Gleizes, A. (1994). Single-source precursors for BaTiO3: Synthesis and Characterization of b-diketonato alkoxides. *Chemistry of Materials*, *6*, 2336–2342. doi:10.1021/cm00048a020

Kessler, V. (2004). Geometrical Molecular Structure Design Concept in Approach to Homo- and Heterometallic Precursors of Advanced Materials in Sol-Gel Technology. *Journal of Sol-Gel Science and Technology*, *32*(1-3), 11–17. doi:10.1007/s10971-004-5757-9

Kim, S. J., Dang, V.-S., Xu, K., Barreca, D., Maccato, C., Carraro, G., et al. (2015). MOCVD of TiO2 thin films from a modified titanium alkoxide precursor. *Physica status solidi (a)*, *212*(7), 1563-1570.

Kovsman, E. P., Andruseva, S. I., Solovjeva, L. I., Fedyaev, V. I., Adamova, M. N., & Rogova, T. V. (1994). Electrochemical Synthesis of Metal Alkoxides. Prospects of commercial alkoxides production. *Journal of Sol-Gel Science and Technology*, *2*(1-3), 61–66. doi:10.1007/BF00486214

Krumm, M., Pawlitzek, F., Weickert, J., Schmidt-Mende, L., & Polarz, S. (2012). Temperature-Stable and Optically Transparent Thin-Film Zinc Oxide Aerogel Electrodes As Model Systems for 3D Interpenetrating Organic-Inorganic Heterojunction Solar Cells. *ACS Applied Materials & Interfaces*, *4*(12), 6522–6529. doi:10.1021/am302458n PMID:23194020

Krumm, M., Pueyo, C. L., & Polarz, S. (2010). Monolithic Zinc Oxide Aerogels from Organometallic Sol-Gel Precursors. *Chemistry of Materials*, *22*(18), 5129–5136. doi:10.1021/cm1006907

Kwon, W., Seo, K. W., Yoon, S. H., Lee, S. S., & Shem, I.-W. (2005). Preparation of Silver Nanoparticles in Cellulose Acetate Polymer and the Reaction Chemistry of Sylver Complexes in the Polymer. *Bulletin of the Korean Chemical Society*, *26*(10).

Lehmkhul, H., & Eisenbach, W. (1975). Elektrochemische Synthesen organischer Metallverbindungen, VII. Elektrosynthese von Alkoholaten und Acetylacetonaten des Eisens, Kobalts und Nickels. *Annual Chemistry*, *4*, 672–691. doi:10.1002/jlac.197519750410

Lizandara-Pueyo, C., van denBerg, M.W.E., DeToni, A., Goes, T., & Polarz, S. (2008). Nucleation and growth of ZnO in Organic Solvents- an in Situ Study. *Journal of the American Chemical Society*, *130*(49), 16601–16610. doi:10.1021/ja804071h PMID:19554730

Lizandara-Pueyo, C., Dilger, S., Wagner, M. R., Gerigk, M., Hoffmann, A., & Polarz, S. (2014). Li-doped ZnO nanorods with single-crystal quality - non-classical crystallization and self-assembly into mesoporous materials. *CrystEngComm*, *16*(8), 1525–1531. doi:10.1039/C3CE41670D

Lizandara-Pueyo, C., Morant-Miñana, M. C., Wessig, M., Krumm, M., Mecking, S., & Polarz, S. (2012). Biomomietic crystallization of anisotropic zinc oxide nanoparticles in the homogeneous phase: Shape control by surface additives applied under thermodynamic or kinetic control. *RSC Advances*, *2*(12), 5298. doi:10.1039/c2ra20343j

Lizandara-Pueyo, C., Siroky, S., Wagner, M. R., Hoffmann, A., Reparaz, J. S., Lehmann, M., & Polarz, S. (2011). Shape Anisotropy Influencing Functional Properties: Trigonal Prismatic ZnO Nanoparticles as an Example. *Advanced Functional Materials*, *21*(2), 295–304. doi:10.1002/adfm.201000997

Look, D. C., Reynolds, D. C., Litton, C. W., Jones, R. L., Eason, D. B., & Cantwell, G. (2002). Characterization of homoepitaxial p-type ZnO grown by mononuclear beam epitaxy. *Applied Physics Letters*, *81*(10), 1830. doi:10.1063/1.1504875

Malik, M. A., Afzaal, M., & O'Brien, P. (2010). Precursor Chemistry for Main Group Elements in Semiconducting Materials. *Chemical Reviews*, *110*(7), 4417–4446. doi:10.1021/cr900406f PMID:20481563

Marchand, P., & Carmalt, C. J. (2013). Molecular precursor approach to metal oxide and pnictide thin films. *Coordination Chemistry Reviews*, *257*(23-24), 3202–3221. doi:10.1016/j.ccr.2013.01.030

Matchett, M. A., Chiang, M. Y., & Buhro, W. E. (1990). Soluble and volatile alkoxides of bismuth. The first structurally characterized bismuth trialkoxide: [Bi(.mu.-.eta.1-OCH2CH2OMe)2(.eta.1-OCH2OMe)].infin. *Inorganic Chemistry*, *29*(3), 358-360.

Mehring, M. (2007). From molecules to bismuth oxide-based materials: Potential homo- and heterometallic precursors and model compounds. *Coordination Chemistry Reviews*, *251*(7-8), 974–1006. doi:10.1016/j.ccr.2006.06.005

Mehrotra, R. C. (1953). Aluminium alkoxides. *Journal of the Indian Chemical Society*, *30*, 585–591.

Mehrotra, R. C. (1954). *Journal of the Indian Chemical Society*, *31*, 904–910.

Meyer, B. K., Alves, H., Hofmann, D. M., Kriegseis, W., Forster, D., & Bertram, F. et al.. (2004). Idntification of bound exciton complexes in ZnO. *Physica Status Solidi. B, Basic Research*, *241*, 231.

Michel, C., Hervieu, M., Borel, M. M., Grandin, A., Deslandes, F., Provost, J., & Raveau, B. (1987). Superconductivity in the Bi-Sr-Cu-O system. *Zeitschrift für Physik B, Condensed Matter*, *68*(4), 421–423. doi:10.1007/BF01471071

Monroy, E., Omnes, F., & Calle, F. (2003). Wide-bandgap semiconductor ultraviolet photodetectors. *Semiconductor Science and Technology*, *18*(4), R33–R51. doi:10.1088/0268-1242/18/4/201

Neumüller, B. (2003). Organometallic sesquialkoxides of aluminium, gallium and indium. *Chemical Society Reviews*, *32*, 50–55. doi:10.1039/b205568f PMID:12596545

Park, B. H., Kang, B. S., Bu, S. D., Noh, T. W., Lee, J., & Jo, W. (1999). Lanthanum-substituted bismuth titanate for use in non-volatile memories. *Nature*, *401*(6754), 682–684. doi:10.1038/44352

Park, J. P., Kim, S., Park, J.-Y., Ok, K. M., & Shim, I.-W. (2009). Preparation of ZnO thin films using Zn/O-containing single source precursor through MOCVD method. *Bulletin of the Korean Chemical Society*, *30*(1), 114–118. doi:10.5012/bkcs.2009.30.1.114

Polarz, S., Pueyo, C. L., & Krumm, M. (2010). The molecular path to inorganic materials Zinc oxide and beyond. *Inorganica Chimica Acta*, *363*(15), 4148–4157. doi:10.1016/j.ica.2010.06.053

Pueyo, C. L., Siroky, S., Landsmann, S., Berg, M. W. E. d., Wagner, M. R., & Reparaz, J. S. et al.. (2010). Monolithic zinc Oxide Aerogels from Organometallic Sol-Gel Precursors. *Chemistry of Materials*, *22*(14), 4263.

Shen, S., Zhang, Y., Peng, L., Xu, B., Du, Y., Deng, M., & Wang, Q. et al. (2011). Generalized synthesis of metal sulfide nanocrystals from single-source precursors: Size, shape and chemical composition control and their properties. *CrystEngComm*, *13*(14), 4572–4579. doi:10.1039/c0ce00982b

Shukla, S., Gaur, P., & Rai, N. (2015). Complexes of tetraethylthiuram disulphide with group 12 metals: Single-source precursor in metal sulphide nanoparticles synthesis. *Applied Nanoscience*, *5*(5), 583–593.

Singh, M. K., Yang, Y., & Takoudis, C. G. (2009). Synthesis of multifunctional multiferroic materials from organometallics. *Coordination Chemistry Reviews*, *253*(23-24), 2920–2934. doi:10.1016/j.ccr.2009.09.003

Sobota, P., Utko, J., Sztajnowska, K., Ejfler, J., & Jerzykiewicz, L. B. (2000). Polynuclear magnesium and magnesium-titanium species. *Inorganic Chemistry*, *39*(2), 235–239. doi:10.1021/ic9905421 PMID:11272530

Szafert, S., John, L., & Sobota, P. (2008). Well-defined polynuclear heterobimetallic complexes as precursors for ceramic and magnetic materials. *Dalton Transactions (Cambridge, England)*, (46): 6509–6520. doi:10.1039/b806167j PMID:19030610

Sze, S., & Ng, K. K. (2006). *Physics of semiconductor devices*. John Wiley & sons. Inc. doi:10.1002/0470068329

Szilard, B. (1906). *Electrochemistry*, *12*, 373–378.

Westin, G. (1997). A Study of the Hydrolysis-Condensation Pathways of a Hetero Bimetallic Alkoxide, Ni2Sb4(OEt)16. *Journal of Sol-Gel Science and Technology*, *8*(23-28).

Williams, P. A., Jones, A. C., Crosbie, M. J., Wright, P. J., Bickley, J. F., Steiner, A., & Critchlow, G. W. et al. (2001). Crystal structure of Bi(OCMe2CH2OMe)3 and its use in the MOCVD of Bi2O3. *Chem. Vap. Deposition*, *7*(5), 205. doi:10.1002/1521-3862(200109)7:5<205::AID-CVDE205>3.0.CO;2-9

Wu, Y.-T., Ho, Y.-C., Lin, C.-C., & Gau, H.-M. (1996). Stepwise Reactions of TiCl4 and Ti(OiPr)Cl3 with 2-Propanol. Variable-Temperature NMR Studies and Crystal Structures of [TiCl2(OiPr)(HOiPr)(Î1/4-Cl)]2 and [TiCl2(OiPr)(HOiPr)(Î1/4-OiPr)]2. *Inorganic Chemistry*, *35*(20), 5948–5952. doi:10.1021/ic960147k

Yamane, H., Kurosawa, H., & Hirai, T. (1988). Preparation of Bi-Sr-Ca-Cu-O films by chemical vapor deposition with metal chelate and alkoxide. *Chemistry Letters*, 9, 1515–1516. doi:10.1246/cl.1988.1515

Chapter 3
Synergism at the Nanoscale:
Photoactive Semiconductor Nanoparticles and their Organic Ligands

Raquel Eugenia Galian
University of Valencia, Spain

Julia Pérez-Prieto
University of Valencia, Spain

ABSTRACT

Photoactive nanoparticles are smart systems that exhibit unique optical properties. In general, their intrinsic properties are size dependent. The degree and type of response to size are both related to their composition. Nanoparticles usually require to be capped with organic ligands in order to be dispersible in an aqueous or organic media, thus leading to nanoparticle colloidal dispersions and enhancing the processability of the material. The organic ligand also plays a key role in their preparation. In addition, the high surface-to-volume ratio of the nanoparticles combined with the affinity of the ligands for the nanoparticle surface can be used to place a large number of functional molecules at their periphery. The purpose of this chapter is to understand the synergism between nanoparticles and organic ligands with regard to their preparation, performance, and applicability.

INTRODUCTION

The rapid development of novel technologies and new instruments over the last years has opened up the opportunity to investigate material properties at the nanoscale level. Studies on the preparation, photophysical properties, and applicability of photoactive nanomaterials have increased exponentially since their novel properties, compared with those of bulk material, were discovered.

Photoactive nanoparticles are light-responsive nanomaterials which, depending on the nature of the material, can exhibit a broad-absorption spectrum (in the UV-visible or near infrared range) as well as narrow Stokes or anti-Stokes emission. In addition, their optical properties can depend not only on their size but also on their aggregation state. Most of nanoparticles (NPs) are capped with organic molecules,

DOI: 10.4018/978-1-5225-0066-7.ch003

usually termed surfactants, capping agents, or ligands, which provide them solubility, processability, and functionality, as well as prevent their aggregation. The unique properties of these nanosystems can be advantageously used for their application in molecular recognition, bio-imaging, drug-delivery, photodynamic therapy, and in the manufacture of optical devices, among others.(Bechet et al., 2008; Bera, Qian, Tseng, & Holloway, 2010; Huang, Barua, Sharma, Dey, & Rege, 2011; Prabhakaran, Kim, Lee, & Prasad, 2012; Rhee, Chung, & Diau, 2013; Wolfbeis, 2015)

There are different types of photoactive NPs which intrinsic properties mainly depend on their composition, such as: i) semiconductor NPs comprising elements from the II-VI (CdSe, CdTe, CdS, ZnSe), III-V (InP, InAs), and IV-VI (PbSe) groups; ii) organometal halide perovskite NPs, which contain an anionic metal-halogen semiconducting framework and charge-compensating organic cations; iii) metal NPs, such as gold, silver, and copper NPs; iv) upconversion NPs, an unusual type of NPs consisting of a matrix doped with rare-earth ions (e.g., $NaYF_4$ (Yb^{3+}, Er^{3+}); and v) carbon-dots. Here we will focus on semiconductor NPs, specifically CdSe-based NPs (CdSe core and CdSe/ZnS core-shell NPs) and organolead halide perovskite nanomaterials. While the former are all-inorganic systems and their preparation and optical properties have extensively studied, the latter are hybrid organic-inorganic systems currently under study. Both of them will be used to explain the role of the organic ligand in their preparation and optical properties.

Semiconductor NPs based on CdSe can be decorated with organic ligands, bounded to their surface by different anchoring groups (e.g., thiolates, amine, phosphine oxides, and carboxylates), and these ligands may also have a functional group such as a photoactive chromophore, chiral compound, macrocycle, etc.(Galian & Guardia, 2009) The ligand can enhance the NP emissive efficiency, be used to detect organic analytes/drugs and oxygen, and be useful in the preparation of fluorescent organogels, among other applications. (Agudelo-Morales, Galian, & Pérez-Prieto, 2012; Aguilera-Sigalat, Casas-Solvas et al., 2012; Delgado-Pérez, Bouchet, de la Guardia, Galian, & Pérez-Prieto, 2013; Gonzalez-Carrero, Agudelo-Morales, Guardia, Galian, & Perez-Prieto, 2015; González-Carrero, de la Guardia, Galian, & Pérez-Prieto, 2014; Wadhavane et al., 2012) In addition, the organic ligand has proved to be crucial in the reversible phase transfer of CdSe NPs capped with long alkyl chain amines by making use of the interdigitation between the alkyl chain of the ligand and those of a surfactant.(Francés-Soriano, Pocoví-Martínez, González-Béjar, & Pérez-Prieto, 2014)

Regarding to organolead halide perovskites, they are organic-inorganic hybrids with the general formula $A_mPb_nX_q$, where A is an ammonium cation and X is halide (Cl, Br, I). Those with the $CH_3NH_3PbX_3$ formula were firstly prepared as nanoparticulate material by using the pores of mesoporous films. {Kojima, 2009 #93} It has recently been demonstrated that long alkyl chain ammonium halide salts can confine the growth of the $CH_3NH_3PbX_3$ perovskite into the nanoscale, eventually providing nanoparticles dispersible in organic solvents, i.e., long alkyl chain ammonium halide salts enable the preparation of colloidal nanoparticles via a non-template method (Schmidt et al., 2014). The colloids have been used to prepare thin films on quartz substrates. Lately, this type of nanoparticles has been prepared with a size close to the Bohr radius of the material, and their emission color has been controlled via their composition as well as via their size.(Huang, Susha, Kershaw, Hung, & Rogach, 2015b; Zhang et al., 2015) Lead halide hybrid perovskites with other stoichiometry are prepared by using ammonium salts with a longer carbon-chain; these ammonium salts can act as both the organic component of the perovskite structure and the capping ligand, thus allowing their preparation as colloidal nanoparticles. Currently, there is a great interest in organolead halide $CH_3NH_3PbX_3$ perovskite nanoparticles for their application in photovoltaic devices and as luminescent materials for light-emitting devices. Hybrid perovskites with other stoichiometry are also promising as luminescent materials.

The main goal of this chapter is to give an overview of the mutual benefit between the NP and its organic capping. It has to be taken into account that the organic ligand plays a key role in different stages of a NP, such as in its preparation, dispersibility in a medium, optical performance, and application in different areas. (Pérez-Prieto, 2013)The NP can act as a carrier of functional organic molecules, making it possible to increase the local concentration of a functional molecule hugely. In addition, the NP and its organic ligand can cooperate to encapsulate functional molecules close to the NP surface. Examples of functional photoactive nanoparticles will be offered throughout this chapter to demonstrate the synergistic effect between the NP and its organic ligand to provide a smart nanosystem.

BACKGROUND

The properties of many inorganic materials depend on their dimensionality (D) which can vary from three to zero. Bulk 3D materials are those with all their dimensions longer than 100 nanometers (x, y, z > 100 nm), whereas bulk low-dimension (2D and 1D) materials are those with one or two dimensions, respectively, shorter than 100 nm (Figure 1). The genuine nanomaterials are those with all three dimensions in the nanoscale (González-Carrero, Galian, & Pérez-Prieto, 2015; Mitzi, 2004; Mitzi, Chondroudis, & Kagan, 2001; Mitzi, Feild, Harrison, & Guloy, 1994) (Billing & Lemmerer, 2009; Scaife, Weller, & Fisher, 1974)

In bulk materials, the dimensions of the semiconductor crystal are much larger than the exciton Bohr radius (average physical separation between the electron and hole), thus allowing the exciton (electron-hole pair) motion to extend to its natural limit. The motion of the charge carriers in bulk low-dimensional semiconductor materials is confined in one dimension (2D materials) or in two dimensions (1D materials), whereas it is confined in the three dimensions for nanoparticles (0D materials). If the size of the nanoparticle is close to the Bohr radius, the confinement causes the transition from continuous to discrete energy levels and the electronic excitation shifts to higher energy as the nanoparticle size is reduced, i.e., the NP optical absorption and emission become size-dependent. (Alivisatos, 1996; Bera

Figure 1. Schematic representation of the confinement of materials: no confinement (three dimensional, 3D, material; confinement in one dimension, 2D material; confinement in two dimensions, 1D material; confinement in the three dimensions, 0D material)
Adapted from González-Carrero et al., 2015

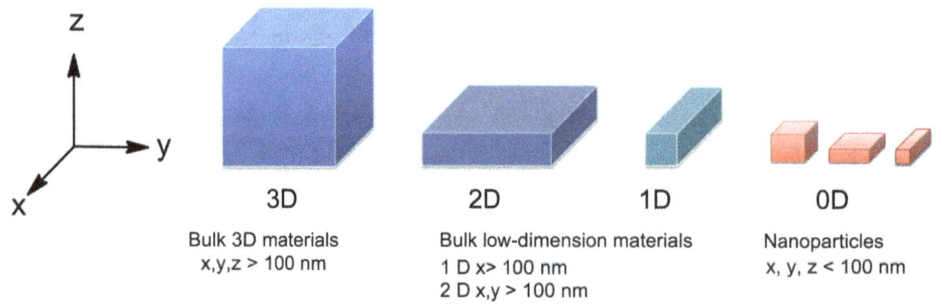

et al., 2010) A striking example of the quantum confinement effect is that of crystalline silicon, which is optically inactive, but it can be converted into an optically active by reducing the particle size to the nanoscale. (Huang, Lindgren, & Chelikowsky, 2005)

The presence of an organic ligand is crucial in the preparation of NPs, since it controls the nanoparticle growth. The organic ligand is eventually attached to the NP surface and stabilizes its dispersion in a medium, preventing agglomeration of the NPs, and passivating the NP surface. The term NP colloidal dispersion is applied to nanomaterials dispersed in a continuous phase (water or organic solvents). The dispersibility of a nanoparticle offers good processability to the corresponding material, thus enhancing its applicability. A proper surface functionalization is a requirement for any possible application of the NP and its interaction with the environment. The capping ligand can be exchanged in order to modify the nanosystem solubility and/or introduce additional functionality.

The high surface/volume ratio allows the decoration of the nanoparticle periphery with a high local concentration of a functional molecule (e.g., a chromophore, a fluorophore, etc.). Furthermore, the interaction between the functional organic capping and the NP can result in mutual changes of their intrinsic properties. (Wolfbeis, 2015) For example, the thermal responsiveness of a polymer used as an organic capping can change due to its interaction with the NP surface, and the emission of a NP, such as CdSe, can be increased or decreased and either blue- or red-shifted as a result of the interaction of its surface with its organic capping. In addition, the combined action of the organic capping and the inorganic core can exert an encapsulating effect, i.e. the organic capping can establish specific interactions with nearby molecules and this would enable the molecules to approach and interact with the NP surface or be involved in photo-induced energy/transfer processes. (Figure 2) (Agudelo-Morales et al., 2012; Aguilera-Sigalat, Casas-Solvas et al., 2012; Delgado-Pérez et al., 2013; Gonzalez-Carrero, Agudelo-Morales et al., 2015; Tomasulo, Yildiz, Kaanumalle, & Raymo, 2006) Therefore, the NP core and the ligand can work together, thus providing the overall system with new properties and capacities. It is important to emphasize that the organic ligand does not only play a key role at the different stages of the NP building and it is relevant for its optical properties, but it can be crucial for its application going from sensing and catalysis to energy conversion.

Figure 2. Schematic representation of the combined action of the nanoparticle and the organic ligand

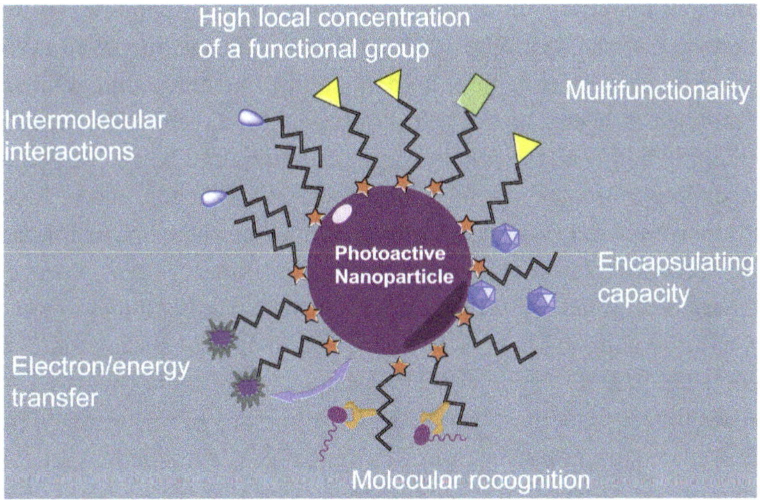

MAIN FOCUS OF THE CHAPTER

Issues, Controversies, Problems

Semiconductor Nanoparticles (CdSe and CdSe/ZnS)

Semiconductor CdSe quantum dots (QDs) are crystalline materials with a diameter between 1 nm and 10 nm. In this nanometer scale the physical properties of the material change in comparison to those of the bulk material, and their optical properties become size-dependent due to the quantum confinement effect. (Alivisatos, 1996; Bawendi, Steigerwald, & Brus, 1990) The QDs can be classified according to the periodic table group that their constituent elements belong. The most widely used group is II-VI semiconductors, which consist of cations from group II and anions from group VI. Although In and Ga are more environmentally friendly than Cd, Pb, and Hg, the stronger covalent bonding of their precursors introduces some difficulties in their synthesis and the most used cation has been Cd. (Rogach, 2008)

There are several methodologies to obtain QDs but they can be organized in two main approaches: top-down and bottom-up approach.

Briefly, the *top-down approach* is based on the semiconductor material thinning to form QDs. The most common techniques are electron beam lithography, reactive ion etching, and wet chemistry etching that produce 0D materials of ca. 30 nm. Another alternative is the focus ion or laser beam to get nanomaterials, however some limitations were found such as the presence of additional impurities added during the process, structural imperfections, and the control of the nanoparticle size. (Valizadeh et al., 2012)

The *bottom-up approach* can also be divided into two main categories: i) wet-chemical methods that involve microemulsion, sol-gel, hot solution decomposition and electrochemistry; and ii) vapor-phase methods, such as self-assembly of nanostructures using molecular beam epitaxy liquid metal ion source and aggregation of gaseous monomers.

In this chapter, we will focus on the synthesis of colloidal QDs that can be obtained by bottom-up process and in particular by the hot-solution decomposition process. This involves high temperature to ensure the pyrolysis of the precursors followed by the conventional precipitation method (Bera et al., 2010).

The growth of the QD involves two steps: nucleation, consisting of the spontaneous formation of the nanocrystal from the assembly of dispersed atoms in solution, and the growth process itself. (Rogach, 2008) The organic ligand is relevant in the whole process of the QD preparation, since i) it participates in the formation of active precursors, ii) it limits the QD growth, iii) it is important in the shape control of the QD, mainly because it is needed to achieve high monomer concentrations, iv) it acts as a passivating agent to reduce the number of defects on the QD surface, and v) it eventually permits the dispersibility of the QD in a medium. In addition, the pristine organic ligands can be exchanged by other ligands to improve the QD dispersibility in a specific medium, emission efficiency, and/or functionality.

The surface to volume ratio increases hugely from the bulk material to the nanomaterial, and therefore the surface properties become relevant due to the large proportion of atoms at the crystal surface that are incompletely bonded within the crystal lattice. The surface passivation is focused on the reduction or total elimination of those dangling bounds to remove the localized surface states from the band gap of the material, keeping the HOMO (highest energy occupied molecular orbital)-LUMO (lowest energy unoccupied molecular orbital) properties. Theoretical calculations have suggested that "hydrogenlike" pseudoatoms can be used for surface passivation of II-VI and III-V semiconductor nanoparticles. (Huang et al., 2005) Electronic passivation of QDs with a larger band gap material was demonstrated for example in CdS/HgS/CdS heterostructures. The smaller band gap of HgS compared to CdS allows the confinement of the electronic excitations and a similar behavior was suggested for colloidal nanoparticles (Alivisatos, 1996).

The attachment of specific ligands on the surface of the nanoparticle strongly affects the fluorescence properties of highly crystalline semiconductor colloidal nanoparticles, since they can passivate the dangling bonds, which act as charge carrier trap sites as was mentioned above. Consequently, great effort has been made over the last years to improve the passivation of the QD surface with organic ligands in order to reduce the number of defects on their surface. The key role of organic ligands to passivate Cd during the synthesis will be described in the following section. The passivation of Se dangling bonds with organic ligands seems to be less effective, but fortunately the growth of an inorganic shell, such as ZnS on the CdSe core surface, passivates both types of dangling bonds, giving rise to colloidal core-shell QDs (CSQDs). Though this inorganic shell can be beneficial for the QDs producing excellent emission properties and high photostability, it can be detrimental for the QD performance depending on its application. It has to be taken into account that these shells could reduce the electronic communication between the QD surface and the environment for applications that require charge transfer process. (Hammer, Emrick, & Barnes, 2007)

There are two different types of core-shell QDs depending on the band gap of the shell added to the core QDs (van Embden, Jasieniak, Gómez, Mulvaney, & Giersig, 2007).

- **Type I CSQDs:** In which the electron and the hole are confined within a potential well created by the shell. This configuration allows the charge recombination to occur at the band edge and exhibits a high fluorescence response. The shell growth induces a small red shift on the absorption and emission spectra, due to the partial distribution of the exciton into the shell material. The most common examples are CdSe/ZnS and CdSe/CdS CSQ Ds.
- **Type II CSQDs:** In which either the electron or the hole is confined to the core by the shell. In this case, the unconfined charge carrier can be scavenged or the charge separation can result in a significant increase in excited state lifetimes. Therefore, these materials can be beneficial for photovoltaic devices. The shell growth can also provoke a red shift of the emission maximum of the QDs and their fluorescence can be tuned working with different shell thickness. Some examples of this type of core-shell QDs are CdTe/CdSe and CdSe/ZnTe, which have near infrared emission.

Organic ligands not only control the QD growth of the nanoparticle but also provide colloidal stability, by preventing their aggregation, and good processability for their further application.

Usually, the organic ligand is formed by three components: i) an anchoring group with good to high affinity for the metal surface; ii) a linker, in general a carbon chain, and iii) a functional group in the other end to provide QDs with solubility in organic or aqueous media or to make their further functionalization possible.

The most common anchoring groups to bind to the QD surface are thiolate, amine, carboxylic and phosphine groups, among others. The binding strength of the organic ligands to the QD surface depends on the nature of the anchoring group, but in general is not strong enough to avoid binding-unbinding reversible processes. Therefore, multiple washing steps after the QD synthesis produce almost naked nanoparticles and, as a consequence, the QD precipitation. (Sperling & Parak, 2010) In the case of thiolate, a covalent bond Cd-S is produced, but depending on the structure of the thiol it can reduce the stability or photostability of the nanoparticles. Figure 3A shows the most common organic ligands used during the synthesis of QDs dispersible in organic media: hexadecylamine (HDA), trioctylphosphine oxide (TOPO), trioctylphosphine (TOP) and oleic acid (OLA). By contrast, the capping of QDs with polar or charged ligands, such as mercaptopropionic acid (MPA) and mercaptosuccinic acid (MSA), produces water-soluble colloidal QDs stabilized by electrostatic repulsion. It is noteworthy that long chain mercapto-alcohols can be used as organic ligands to produce organic-soluble QDs, thus demonstrating the contribution of the carbon chain into the final solubility of the nanosystem (Figure 3B). (Gill, Zayats, & Willner, 2008)

Herein, the synergy between the organic capping and the optically active CdSe nanopartice, specifically core CdSe and core-shell CdSe/ZnS QDs, will be addressed. These nanoparticles can be obtained with

Figure 3. Schematic representation of the role of the organic ligand in the QD solubility. A) lipophilic QDs and B) lipophilic versus water soluble QDs, using mercapto-containing compounds with a different carbon-chain length

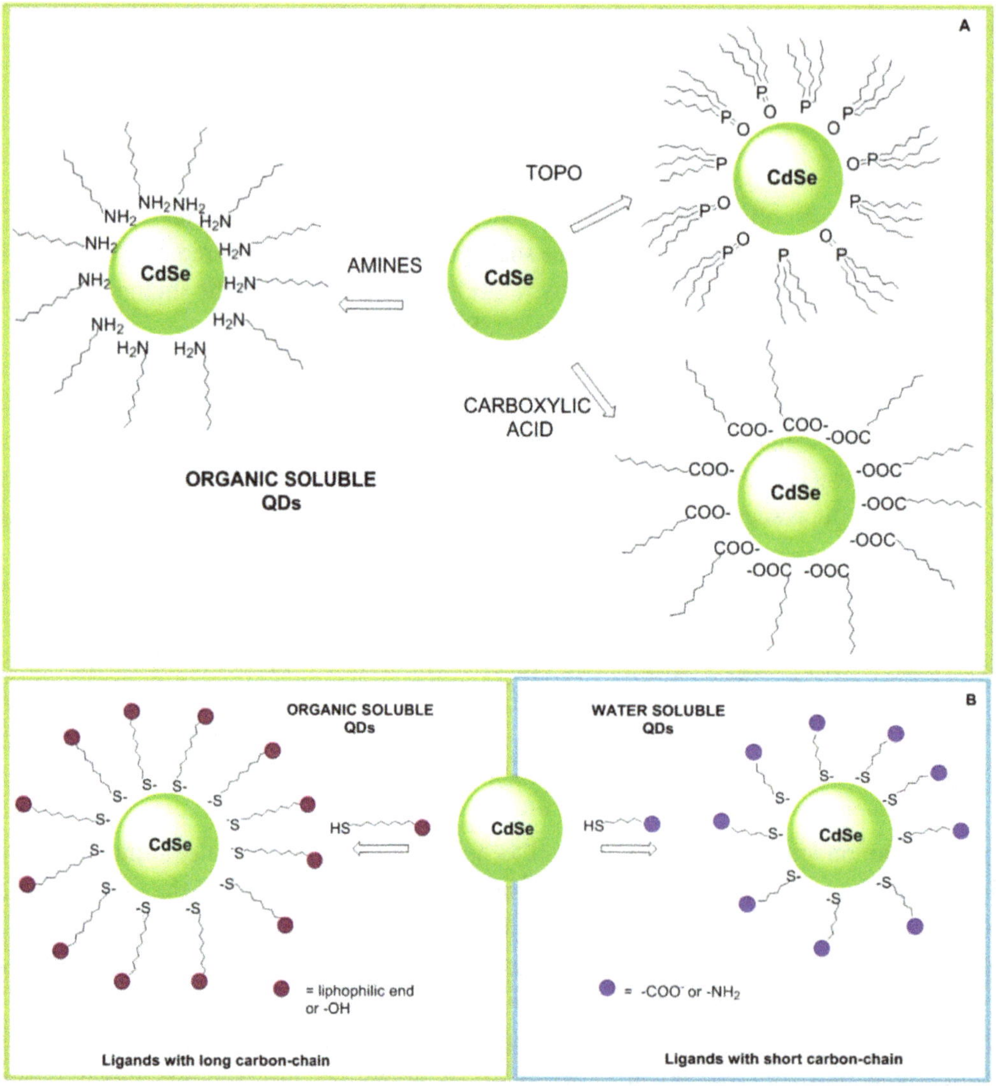

emission in the 480-650 nm range; in particular the core-shell QDs have a good fluorescence quantum yield (which is the percentage of emitted photons related to the absorbed photons).

There are a vast number of papers reporting on the synthesis of CdSe-based QDs and on their application ranging from sensing and bioimaging to the development of solar cells but the ligand-surface interaction was not analyzed. The surface functionalization is crucial for those applications and a clear understanding of the QD surface-ligand interaction would help to design new materials based on these QDs combined with other materials. The role of the organic ligand in the performance of CdSe-based QDs will be further discussed in the following sections.

Ligands in the Synthesis of Lipophilic QDs

The nanoparticles can be capped with a considerable number of ligands, which have an anchoring atom at one end with affinity for the NP surface and the other end provides the NP periphery with the hydrophobicity or hydrophilicity needed to give rise to stable lipophilic and aqueous NP colloidal solutions, respectively. The ligand can play a key role during the QD synthesis as well on the eventual solubility and stability of the colloidal QD.

In 1993 Murray et al. reported that CdE nanoparticles (E = S, Se Te) could be produced with the injection of a cadmium organometallic reagent, such as $Cd(Me)_{2,}$ and an E precursor into a hot coordinating solvent, such as trioctylphosphine oxide. (Murray, Norris, & Bawendi, 1993) The main advantage of TOPO is its high boiling point that allows the reaction to reach 350ºC, which is the temperature required for the decomposition of the precursors. Interesting, TOPO acts as both solvent and capping ligand in the preparation of CdSe core QDs and it also controls the slow growth and annealing of the nanocrystals. The interaction between TOPO and the CdSe surface occurs through the lone electron pair of the phosphine oxide, i.e., via a dative bond. (Mattoussi, Cumming, Murray, Bawendi, & Ober, 1996) Theoretical (ab initio) calculations confirmed the preferred binding of TOPO to the QD surface via the Cd sites. (Puzder et al., 2004) TOPO provides better stability to the QD than other ligands, such as thiols and dithiocarbamates. (Radhakrishnan, Lo, Knobler, Garcia-Garibay, & Monbouquette, 2011) Trioctylphosphine is also used in the synthesis of QDs to form the Se reactant, trioctylphosphine selenide (TOPSe), and it also coordinates to the QD surface, therefore, TOP has a dual role.

This strategy was later improved by adding hexadecylamine (HDA) to the TOPO/TOP mixture. (Talapin, Rogach, Kornowski, Haase, & Weller, 2001) The three-component mixture (HDA/TOPO/TOP) allowed a better control over the particle size distribution during the QD growth, thus obtaining highly monodisperse CdSe nanocrystals.

Peng et al. reported that $Cd(Me)_2$ can be replaced by CdO for the synthesis of CdSe QDs. (Peng & Peng, 2001a)This approach also uses TOPO/TOP, but introduces hexylphosphonic acid (HPA) or tetradecylphosphonic acid (TDPA) to coordinate cadmium and form cadmium-HPA or Cd-TDPA complexes. It was demonstrated that the QD shape is controlled by the monomer concentration. The addition of a high concentration of HPA or TDPA enabled a high monomer concentration and quantum rods instead of spherical quantum dots were produced. (Peng & Peng, 2001b)

Although TOPO ligand is the most used organic ligand for the synthesis of colloidal QDs, the reproducibility on the quality and, consequently, the optical properties of the QDs is rather low. Several investigations have been carried out to understand the key role of impurities in the commercial TOPO ligands, such as significant concentration of other phosphorous compounds. Surprisingly, some impurities can have beneficial effects on the growth of CdSe nanoparticles and can directly affect their shape.

(Wang, Tang, & Buhro, 2008) Thus, changes in the growth kinetics and morphology of hyperbranched CdSe or CdTe QDs have been observed for different batches and TOPO supplied by different suppliers. (Kanaras, Sönnichsen, Liu, & Alivisatos, 2005)

Following the Peng et al. methodology and using a recrystallized TOPO, QDs with irregular, elongated and spherical shapes were prepared. (Peng & Peng, 2001a) In the presence of di-*n*-octylphosphine oxide a reduction in the QD growth produced a near-isotropic nanocrystal shape. In addition, di-*n*-octylphosphine oxide and mono-*n*-octylphosphinic acid gave rise to uniform quantum rod morphologies. Finally, the presence of di-*n*-octylphosphinic acid assisted the growth of quantum wires morphologies. (Wang et al., 2008)

Regarding to CdSe/ZnS core-shell QDs, they can be prepared using the same phosphine ligands (TOPO/TOP) by following a one-flask, two-step method or a two-step procedure (Dabbousi et al., 1997; Hines & Guyot-Sionnest, 1996). In the first approach, the CdSe QDs were synthesized via pyrolysis of organometallic precursors, $Cd(Me)_2$ and TOPSe, in the coordinating solvent TOPO at ~350 °C, growing the QDs at ~300 °C. For the shelling process, dimethylzinc $Zn(Me)_2$ or diethylzinc $Zn(Et)_2$ and hexamethyldisilathiane $(TMS)_2S$ were used as the Zn and S precursors. The resulting CdSe/ZnS exhibited a fluorescent quantum yield of 50%. In the second approach, the CdSe QDs were isolated before the shelling process by a selective precipitation in methanol and, then, they were re-dispersed in hexane. In this case, the photoluminescence quantum yield was increased from 5-10% for CdSe core to 30-50% for the QDs over-coated with ZnS.

In principle, TOPO could interact with the Cd^{2+} or Zn^{2+} at the surface of CdSe and CdSe/ZnS QDs, respectively, via a dative, neutral interaction. To determine the nature of the interaction, the cleavage of the TOPO ligands from the QD surface was performed by using reactive silicon-chalcogen and silicon-chlorine bonds. Alivisatos et al. demonstrated that the nanocrystal surface of CdSe and CdSe/ZnS QDs can be terminated by the anionic binding of alkylphosphonate ligands to the cation layer. Thus, core CdSe have a surface layer of Cd- phosphonates and this interaction is favored in metal-rich surfaces (Owen, Park, Trudeau, & Alivisatos, 2008).

Long chain amines are better coordinating organic ligands than phosphines due to the stronger interaction between the amino lone electron pair with the Cd, thus improving the luminescence response of the QD. Different techniques, such as FT-IR spectroscopy and ^{1}H-NMR, have been used to determine the local environment of amine ligands of different length (C8 to C18) at the QD surface. Although broadening of the proton signals limits the use of NMR, the FT-IR allowed the study of the amine interaction with the QD surface. The splitting of the N-H scissoring amine anchoring group indicated the binding of the amine to either Cd or Se sites, confirming its suitability as QD ligand (Cooper, Franco, Gul, Corrado, & Zhang, 2011).

Organic ligands with carboxylate as the anchoring group have been proposed as an alternative phosphine-free synthetic route. For example, oleic acid was used to dissolve CdO instead of TDPA, and octadecene was used as solvent. This synthetic route leads to nanoparticles with a zinc-blende structure instead of the wurtzite structure obtained by other methods.(Deng, Cao, Tang, & Zou, 2005) The combination of carboxylic acids ligands such as oleic acid and stearic acid, used to form the Cd-carboxylate precursor, with hexadecylamine in the presence of octadecene gives rise to extremely small, bright, blue-luminescent CdSe/ZnS core-shell QDs (Čapek et al., 2009).

Nanoparticles can be also capped with organic ligands by using the deprotonated mercapto group as the anchoring group. The eventual solubility of the nanosystem depends on the length and lipohilicity of the carbon chain as well as on the functional group at the other end. Thus, CdSe/ZnS QDs capped with 11-mercapto-undecanol are soluble in toluene, whereas 2-mercaptoethanol produces water-soluble QDs (Rogach, Kornowski, Gao, Eychmüller, & Weller, 1999).

Ligand Exchange

In the *ligand exchange strategy*, the original organic capping of a QD (usually soluble in organic solvents) can be replaced by a ligand with a better anchoring group and a hydrophilic and/or a functional group in the other extreme of the alkyl chain(s), thus providing the QD with new properties, solubility, and/or functionality. Alternatively, the native ligands of the QD, such as TOPO, TOP, TDPA, and HDA, can be removed from the QD surface by refluxing in pyridine for a long period of time. Pyridine binds to the QD surface weakly and therefore facilitates the subsequent ligand exchange. (Ji, Copenhaver, Sichmeller, & Peng, 2008; Li et al., 2006; Liu et al., 2008)

The new ligand can provide the nanosystem periphery with functional groups, such as amino, carboxylic, mercapto, azide, alkene, alkyne, and thiols. Thus, the QD can then be conjugated with functional material, such as biomolecules, via amide bonding (carboxylic and amino groups) or click chemistry (alkyne and azide moieties). (Blanco-Canosa et al., 2014; Sperling & Parak, 2010) Biomolecules, such as DNA, protein, enzymes and so on, are typically bounded to the QD ligand via amide bound, offering good stability compared to other strategies such as direct binding via thiolate or electrostatic adsorption to the QD surface.

The ligand exchange strategy presents some limitations related to the inefficient loading of the new ligands on the QDs surface and poor long-term stability in solution due to their oxidation. (McDowell, Wright, & Hammer, 2010)

Ligands for Aqueous-Soluble QDs

The most common organic ligands for water-soluble QDs are short carbon chain mercaptocarboxylic compounds and mercapto alcohols. They can be directly introduced during the synthesis or via ligand exchange. Rogach et al. (Rogach et al., 1999) described the synthesis and characterization of some aqueous-soluble thiol-stabilized CdSe nanocrystals by using cadmium acetate and sodium selenosulphate and mercapto alcohols (1-thioglycerol and 2-mercaptoethanol) or mercaptocarboxylic acids (thioglycolic acid and thiolactic acid) as organic ligands. Similarly, Chen *et al.* prepared high-quality CdSe QDs capped with thiocarboxylic acids (Chen, Hutchison, Dobson, & Wakefield, 2009). Interestingly, the photoluminescence quantum efficiency of the as-synthesized water-soluble nanoparticles improved up to 22% after sunlight or normal laboratory illumination.

The use of *N*-acetyl-*L*-cysteine as organic capping for the hydrothermal synthesis of CdSe QDs eventually produces CdSe/CdS core-shell structures. Therefore, the cysteine derivative acts not only as the capping agent, but it also releases S^{2-} to the medium to form CdS onto the CdSe surface, thus increasing the particle diameter and improving the passivation of surface defects. Nanocrystals with about a 27% fluorescence quantum yield and good stability can be obtained by using the proper amount of the stabilizer related to the Cd precursor (specifically, a 3:1 ratio) (Wei et al., 2015).

It is noteworthy that the QDs agglomerate by decreasing their colloidal stability and a red-shifted and diminished luminescence is observed. Glutathione has been successfully used as capping agent for CdSe and CdTe QDs as a good stabilizing agent producing good luminescence response of 25% and 35%, respectively. (Bäumle, Stamou, Segura, Hovius, & Vogel, 2004; Zheng, Gao, & Ying, 2007)

Bidentate ligands, such as dihydrolipoic acid, polyethylene glycol ligands with two thiol-anchoring groups, and water-soluble dendrons with carboxylate anchoring groups have been used to prepare water-soluble QDs with high colloidal stability and quantum yields. (Liu et al., 2007; Medintz et al., 2003) In addition, thiolate-multigrafting has been used to prepare photoluminescent nanohybrids based on CdSe QDs capped with pH-responsive and thermoresponsive polymers (Liras, Peinado, Cañamero, Quijada-Garrido, & García, 2014).

Another strategy to obtain water soluble-QDs is by the *phase transfer* strategy. The QD is covered with a new organic layer by the interaction of its lipophilic chains with that of the original ligand. One advantage of this method is that the native layer is preserved, thus preventing the inorganic core oxidation, but it results in the increase of the QD size. This strategy has also been used for transferring lipophilic QDs reversibly between an organic solvent and a water solution, by using a switchable surfactant, specifically *N'*-oleyl-*N,N*-dimethylacetamidine, and an external stimulus (carbon dioxide/N_2 bubbling). (Francés-Soriano et al., 2014) In addition, amphiphilic polyacrylic polymers with hanging alkyl chains have been used to solubilize tri-*n*-octylphosphine oxide/hexadecylamine-capped CdSe/ZnS QDs in water by making use of the interdigitation between the alkyl chains of the QD ligand and those of the polymer.(Anderson & Chan, 2008)

The Role of the Capping Ligand in the QDs Optical Properties

The optical properties of the QDs can be modulated by changing their size, shape, and composition. However, a more recent method to tune their optical properties, such as absorption and fluorescence spectra and their emissive efficiency, is via surface chemistry. Again, the high ratio surface/volume of the QDs has a crucial effect on their optical properties due to the presence of surface defects. The energy of those surface defects usually lies in between the band gap energy of the QDs and they can act as trapping agents of electron and holes, i.e., as reducing and oxidizing agents, respectively (Bera et al., 2010).

Surface defects quench the radiative recombination and reduce the luminescence quantum yield. Therefore, surface passivation has been used to confine the charge carriers inside the core, thus improving the QD optical properties. The excited state dynamics and the manipulation of the surface chemistry have been carefully analyzed in a recent review, showing that the nature of the anchoring group of the organic capping is crucial in the QD properties (Hines & Kamat, 2014).

Phosphine

The fluorescence quantum yield of high quality colloidal nanoparticles based on CdSe QDs capped with phosphines was around 20%.(Peng & Peng, 2001a). A luminescence quantum yield as high as 85% was obtained by controlling the ratio between the Cd and Se precursors.(Qu & Peng, 2002) Whereas cation-rich QDs (Cd^{2+}-rich facets) show great affinity to basic ligands and exhibit good luminescent properties, anion-rich QDs (Se^{2-}-rich facets) do not bind to the majority of the basic ligands and the nanoparticle is sensitive to hole-trap states located on the selenium sp^3 orbitals and, as a consequence, these QDs are low emissive (Jasieniak & Mulvaney, 2007).

Interaction between the TOPO alkyl chains of TOPO-capped CdSe QDs and the moieties of a pseudopeptidic macrocycle has led to highly fluorescent organogel-QDs hybrids. The role of the capping ligand in the interaction with the macrocycle was demonstrated by NMR and FTIR techniques. Interestingly, the presence of QDs reduced the critical concentration required to form the organogel and offered luminescent properties to the new material. The CdSe QD fluorescence quantum yields were enhancement up to 268% and 528% for TOPO capped CdSe QDs (TC-QD, λ_{em}= 533 nm, FWHM = 22 nm and Φ_F=3%) and octadecylamine capped CdSe QDs (AC-QD, λ_{em}= 527 nm, FWHM = 24 nm Φ_F=4%), respectively.

In addition a moderate increase in the fluorescence average lifetime was observed for both nanoparticles in the presence of the organogelator, going from 48.7 ns to 76.1 ns for TC-QD and from 81.2 ns to 125.3 ns for AC-QD. Therefore, this is a clear example of symbiosis at the nanoscale, where the cooperative effects between the core QD and the organogelator lead to a highly transparent, stable, and luminescent hybrid material. Effective passivation of the QD surface, as was the case for core-shell QDs, reduces the effect of the organogelator on the optical properties of the systems (Wadhavane et al., 2012).

Amines

Long alkyl chain primary amines have proved beneficial for the increase of the QD emission from TOPO-capped QDs with luminescent quantum yield of 2% to 15-20%. This has been attributed to the closer packing of the amine ligand on the QD surface, thus covering almost 100% of the surface compared to the 30% coverage in the case of TOPO due to steric effects. It has also been postulated that the presence of the amine can cause lattice contraction during the growth process (Bullen & Mulvaney, 2006; Qu & Peng, 2002; Zhang, Wang, Xiao, Qu, & Peng, 2002).

The role of the number of substituents bounded to the nitrogen atoms (amine primary, secondary, and tertiary amines) has also been studied. (Bullen & Mulvaney, 2006) Comparatively, primary long-alkyl chain amines lead to more emissive QDs and the luminescence clearly follows this trend: primary > secondary > tertiary amines. In addition, butylamine-capped QDs are more luminescent than those with ligands which possess the same carbon chain length but with a higher degree of substitution, i.e., bulky substituents produce lower luminescence response. Ligand exchange in TOPO-capped CdSe QDs (TC-QD, λ_{em}= 490-586 nm, FWHM = 25-30 nm Φ_F=5%) with tertiary aliphatic amines (triethylamine) and bicyclic amines produced quenching of the QD luminescence. The approach of the amine to the QD surface will favor the electron transfer process from the amine to the hole of the nanoparticle (Galian & Scaiano, 2009).

The beneficial effect of the ligand exchange of TOPO-capped CdSe QDs with amines on the QD emission was demonstrated by comparing the effect of 2,2,6,6-tetramethyl-1-piperidinyloxyl radical (TEMPO) with that of 4-amino-TEMPO. The quenching caused by TEMPO was attributed to electron transfer or paramagnetic-assisted spin relaxation. (Laferriere, Galian, Maurel, & Scaiano, 2006) The higher emission of 4-amino TEMPO was ascribed to the high association constant of the quencher to the QD surface through the amine group. The ligand exchange occurred via two steps: a rapid binding of the amine ligand to the surface sites was followed by a slower displacement of TOPO (Maurel, Laferrière, Billone, Godin, & Scaiano, 2006).

Thiols

The response of the QD fluorescence to thiolate anchoring groups depends on the QD composition. Thus, mercapto-propionic acid quenches the emission of CdSe but not that of CdTe QDs. This has been attributed to the whole transfer from the CdSe QD to the thiol (electron transfer from the thiol to the hole originated after the QD excitation), due to the higher energy of the thiol HOMO related to the semiconductor valence band. By contrast, the CdTe valence band is located above the thiol HOMO, therefore, the trapping of the hole by the thiol compound is not expected (Wuister, de Mello Donegá, & Meijerink, 2004).

There are some controversies about the effect of thiol ligands on the luminescent properties of core-shell QDs. It has been shown that deprotonated thiols (thiolates) can strongly bind to the QD surface but it caused a drastic fluorescence quenching. This was ascribed to the exothermic reaction between the thiolate and Zn and S, thus producing a significant alteration of the shell (Pong, Trout, & Lee, 2008). Consistently, we have demonstrated that the ligand exchange of pristine amine-capped CdSe/ZnS core-shell QDs with thiol ligands can preserve the QD luminescence if performed under mild conditions (chloroform reflux for 48 h under nitrogen atmosphere). The initial luminescent quantum yield of commercial amine-capped CSQDs was Φ_F=50% (λ_{em}= 531 nm, FWHM = 45 nm) increased up to 75% upon ligand exchange with 11-mercaptoundecyl 2-(3-benzoylphenyl)propanoate preserving the maximum of emission In addition, the luminescence slightly decreased from Φ_F=63% to 54% and 45% when the 11-mercaptoundecanoic acid and mercaptopropionic acid, respectively, was used as the capping ligands (Aguilera-Sigalat, Rocton, Galian, & Pérez-Prieto, 2011; Aguilera-Sigalat, Rocton, Sanchez-Royo, Galian, & Perez-Prieto, 2012).

In addition, it has been reported that the exchange of TOPO and TOP ligands of CdSe/ZnS QDs (FWHM=33.7, Φ_F=6-13.5% depending on the number of ZnS monolayers) with thiols (1,4-benzenedimethanethiol, 1,16-hexadecanedithiol, 1,11-undecanedithiol, biphenyl-4,4′-dithiol, 11-mercapto-1-undecanol and 1,8-octanedithiol) enhances the QD photoluminescence and increases the QD storage stability. In the case of 1,4-benzenedimethanethiol the photoluminescence increase more than 5 times the original intensity. It was proposed than rather surface passivation with the thiol ligands, the enhanced response could be due to the cross-linked assembly to form dimer and trimer clusters, i.e., the QD ligands could act as molecular linkers between the QDs (Zhu et al., 2014).

Finally, the low photochemical stability of water-soluble CdSe QDs capped with thiols has been suggested to involve: i) the photocatalytic oxidation of the thiol in the QD surface to produce thiyl radicals, which eventually lead to disulfides, followed by adsorption of water molecules on the QD surface, which increases the QD luminescence; ii) the oxidation of the QD in the presence of oxygen and formation of selenium oxide (SeO_2), which acts as a protective shell reducing the defect on the QDs surface; and iii) the QD photocorrosion, where SeO_3^{2-} and Cd^{2+} are desorbed from the QD introducing new defects on its surface, thus resulting in the QD precipitation (Aldana, Wang, & Peng, 2001; Carrillo-Carrion, Cardenas, Simonet, & Valcarcel, 2009).

Bidentate Ligands

Multidentate ligands with anchoring groups such as amines and thiols have strong influence in the QD polydispersity, fluorescent quantum yield, and photostability. In fact, a higher number of binding sites usually provides a better colloidal stability, though it can also quench the QD fluorescence. (Zhang & Clapp, 2011) Common multichelating ligands possess thiolate groups. An interesting example of emission enhancement that combines a multichelating ligand and the additional functionality of the ligand is that of dihydrolipoic acid-capped CdSe/ZnS core-shell QDs conjugated with proteins via a non-covalent interaction. For example, the maltose binding protein was modified by genetic fusion with polyhistidine and it was bounded to the negatively charged QDs via electrostatic interactions, thus resulting in considerably emissive QDs (Clapp, Goldman, & Mattoussi, 2006).

In addition, it has been shown that multidentate polymers can provide a compact organic coating. Poly(acryl acid) was functionalized with mercapto groups and the multichelating polymer was strongly anchored to QD via ligand exchange from the original hydrophobic QD. The resulting water-soluble, thin film polymer-capped QDs possessed higher photoluminescence quantum yields than the initial hydrophobic QDs and high stability (Liu, Guo, Li, & Zhong, 2010).

Multichelating ligands can also be involved in energy/electron transfer reactions as shown in naphthalimide-capped CdSe/ZnS QDs. The naphtalimide was linked to a thiolate group through an ethylene chain. It was demonstrated that the mercapto group and naphtalimide units acted as bidentate ligands, thus producing a drastic quenching of the QD fluorescence. The thiolate decreased the radiative recombination of the delocalized charge, whereas the carbonyl moiety of the naphtalimide increased the non-radiative recombination of the localized charge (Aguilera-Sigalat et al., 2013).

Non-Innocent Ligands

Ligands that couple to metal centers creating new electronic states are well studied in coordination chemistry and this concept can also occur in QDs capped with organic ligands. Thus, organic ligand such as phenyldithiocarbamates can couple electronically to the delocalized states of QDs (CdSe, CdS and PbS), thus reducing the confinement energy of the exciton carriers and the optical band gap of the QD. Adsorption of π-donating aromatic dithiocarbamates onto the QDs surface produced a bathochromic shift from 542 nm to 567 nm in the absorption spectra of CdSe QDs (Frederick, Amin, & Weiss, 2013).

The Role of Capping Ligand in the QDs Functionality

Surface chemistry is not only of interest to solubilize and passivate the QD but also to create hybrid materials and to introduce new functionalities in the nanosystem. The nanosystem can be of interest in sensing, photocatalysis, and in biomedical applications (such as fluorescent probes for tracking and imaging in both in vivo and in vitro) (Bilan, Fleury, Nabiev, & Sukhanova, 2015).

Different methodologies have been followed for adding functionality to the QDs: ligand exchange, capping the QD with an additional organic layer, and direct derivatization of the QD ligand are alternative strategies for this aim (Hammer et al., 2007). The most used methodology to add new functionality to CdSe-based QDs is by ligand exchange. Thus, QDs have been decorated with photoactive functional groups, such as pyrene, ketoprofen, and naphtalimide chromophores (Agudelo-Morales et al., 2012; Aguilera-Sigalat, Casas-Solvas et al., 2012; Aguilera-Sigalat et al., 2011; Aguilera-Sigalat, Rocton et al., 2012) by ligand exchange followed by amide bond formation. Figure 4 shows examples of systems used in chiral recognition and sensing of non-chiral analytes by means of either a supramolecular system based on a host-guest interaction or the formation of a pyrene excimer in the QD periphery. (Agudelo-Morales et al., 2012; Aguilera-Sigalat, Casas-Solvas et al., 2012; Delgado-Pérez et al., 2013) On the other hand, an example in which the QD ligand (TOPO) decreases the critical concentration of gelator needed to form stable and thermoreversible organogels, eventually leading to highly luminescent QDs, is included. (Schmidt et al., 2014)

Figure 4. Examples of synergism between quantum dots and their organic ligands

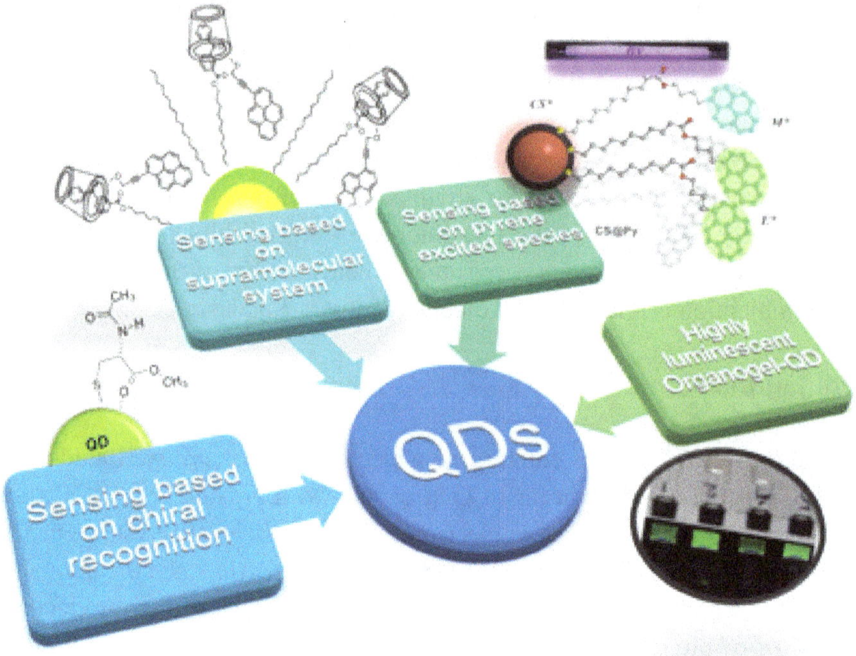

Sensing

Quantum dots have been used for sensing application in the detection of different organic and biochemical compounds by making use of interactions between the organic ligand and the analyte or the analyte and the nanoparticle. (Galian & Guardia, 2009) Most of the functional ligands for sensing application are introduced by the ligand exchange methodology.

The ligand can be: i) a photoactive chromophore that interacts with the analyte, ii) a chiral molecule for recognition of chiral molecules, or iii) it can participate in the building of a supramolecular system.

Photoactive ligands, such as pyrene thiol derivatives, were covalently bounded to the CdSe QD surface via the thiolate group. Interesting a high pyrene local concentration at the QD periphery was obtained due to the high surface/volume ratio of the spherical nanoparticle, thus leading to the efficient formation of the pyrene excimer at quite low concentration of this chromophore. The sensing properties of the pyrene-capped QDs were studied for different nitroaromatic compounds [*m*-nitroaniline and *p*-nitroaniline] and nitrobenzenes [nitrobenzene, *p*-nitrotoluene, 2,4-dinitrotoluene, and 2,6-dinitrotoluene]. The QD-pyrene hybrid presents two-luminescent independent channels, the pyrene monomer and the excimer, both deactivated by nitroaromatic compounds. The encapsulation capacity of the nanohybrid also played a key role in their sensitivity to the different analytes. (Agudelo-Morales et al., 2012)

Moreover, pyrene-capped CdSe/ZnS core-shell QDs were used as ratiometric sensors of oxygen in non-aqueous media and for the monitoring of oxygen diffusion through the solvent.(González-Carrero et al., 2014) This nanohybrid was also used as triple sensor for the rapid quantification of trinitrotluene (TNT) in binary and ternary mixture of nitroaromatic compounds, by using the three independent emissive channels, i.e., the pyrene monomer, the pyrene excimer and the nanoparticle. (Gonzalez-Carrero, Agudelo-Morales et al., 2015)

The application of QDs in chiral recognition has been shown by using chiral ligands. Thus, *N*-acetyl-*l*-cysteine was bounded to the nanoparticle surface of CdSe/ZnS QDs. This ligand was anchored to the QD surface via the thiolate (strong bound) and the carbonyl group of the ester (weak bound). Circular dichroism changes (a new broad redshift band) indicated that attachment of the chiral ligand to the QDs surface alter the ligand conformational equilibrium. Chiral recognition of anti-inflammatories compounds, specifically aryl propionic acid such as ketoprofen, naproxen, flurbiprofen and ibuprofen were assayed by following the fluorescence quenching. A remarkable positive allosteric effect was observed for the *R*-enantiomer of ketoprofen and flurbiprofen (Delgado-Pérez et al., 2013).

Interestingly, supramolecular chemistry sensing can be established on the nanoparticle surface. For example, water-soluble CdSe QDs have been prepared with a *β*-cyclodextrin as a "green" ligand, by ligand exchange of the native QD ligand (oleic acid) in alkaline media. The cyclodextrin was covalently bound to the QD surface as confirmed by FTIR spectroscopy, and consequently it was not removed by successive washing cycles. The ligand exchange caused a partial quenching of the QD fluorescence (Φ_F decreased from ca. 24% to 8.3%). The presence of a broad and symmetrical band with a tail at longer wavelengths was indicative of the presence of surface trap states. The fluorescence of the supramolecular system was quenched by anions derived from polyprotic acids, with a quenching efficiency decreasing in the order H_2PO^{4-}> HPO_4^{2-}>>PO_4^{3-}. The formation of an inclusion complex between the cyclodextrin and PO_4^{3-} was observed by ^{31}P-NMR, thus explaining the insignificant effect of the PO_4^{3-} on the QD fluorescence. Cations, such as Ag^+, Hg^+, and Co^{2+} also acted as quenchers of the QD fluorescence, with a linear response between the change of the fluorescence intensity and the analyte concentration (Shang, Hu, Wang, & Jin, 2011)

Another interesting supramolecular system is that built by using ketoprofen-capped CdSe/ZnS QDs and a *β*-cyclodextrin with an appended QD quencher (specifically, pyrene). The supramolecular system was used for the sensing of different analytes (phenol, indole, adamantane, and benzophenone). Addition of the pyrene-modified cyclodextrin to the QD induced an immediate fluorescence quenching due to the pyrene unit pointing toward the QD surface. Selective recovery of the QD fluorescence was observed after addition of analytes with good affinity for the cyclodextrin cavity. After the molecular recognition, all the components of the assay were recovered individually, which is beneficial from the economic and environmental point of view (Aguilera-Sigalat, Casas-Solvas et al., 2012).

The Role of the Ligand in the Formation of Hybrid Materials

Organic ligands can act as molecular linkers between QDs and metal oxides (TiO_2) to build a hybrid system with enhanced charge separation. The most common ligand used for this purpose has been 3-mercapto-propionic acid, but it creates charge carrier trapping states on the CdSe QD surface and interfere in the electron injection from the QD to TiO_2. Passivation of CdSe QDs with β-alanine reduced the formation of trap states and, consequently, enhanced the electron transfer process. The QDs surface was partially and fully functionalized with this ligand and the results were compared with the unfunctionalized surface with the aim to understand the role of the amine group in the electron transfer process. The ability of amine group (Lewis base) to fill surface trap sites decelerated the electron transfer rate from k_{eT} = 2.4 10^9 s^{-1} to k_{eT} = 0.64 10^9 s^{-1} when going from partial to the total surface passivation and was considerably slower than the directly bounded CdSe-TiO_2 system (k_{eT} = 150 10^9 s^{-1}) (Hines & Kamat, 2013).

Therefore, a right election of the organic ligands (bifunctional ligands as linker) is crucial to obtain CdSe-TiO_2 hybrid with efficient electron transfer between the components (Hines & Kamat, 2014). The dynamic and efficiency of the interfacial electron transfer process probed to be dependent of the QD size, the interfacial energy, and the capping agent used as linker between both semiconductors (Watson, 2010).

An example of the role of the anchoring group to introduce functionality to the QD is that of amine-capped CdSe/ZnS core shell QDs. The amino group made it possible the approach of Au salts to the QD surface. Depending on the size of the shell, the formation of a ternary or quaternary hydrids was produced at room temperature in the darkness. The mechanism proposed for the formation of CdSe/ZnS/Au hybrids involves the reduction of Au^{3+} to Au^{+}, assisted by the amine groups. The Au^{+} has a good affinity for the S^{-} in the shell and can progressively interact with the core to be reduced to Au^{0}. The remaining cation interact with the shell to form the quaternary CdSe/ZnS/Au_2S/Au hybrid (Galian et al., 2015).

Photosensitization and Photocatalysis

One of the main advantages of a photochemical process is the use of light as an innocuous reagent able to generate reactive species without using chemical reagents (Galian & Perez-Prieto, 2010). A photocatalyst has been classified as homogeneous or heterogeneous depending if its acts in solution or solid state. A homogeneous photocatalyst is soluble in organic or water media and after light absorption either transfers photon energy to a substrate in the solution (a non-absorbing molecule) or can react with a substrate. A heterogeneous photocatalyst is in the solid state and, therefore, is easily isolated after the photocatalytic reaction.

Nature provides excellent examples of photocatalytic processes as photosynthesis, synthesis of adenosine triphosphate (ATP) by retinal photochemistry, and photoactivation of enzymes. These biological processes have inspired the development of artificial photocatalysts as semiconductor nanoparticles, organic chromophores, supramolecular systems, and photoactive polymer molecules.

In the case of semiconductor nanoparticles, they combine the advantages of their dispersibility in solutions (as homogeneous photocatalysts) and facility to be separated from the reaction media after the reaction concludes (as heterogeneous photocatalysts). After the absorption of light, the QDs can act as an oxidizing and a reducing agent, with a size-dependent catalytic activity.

It has been recently reviewed the effect of the capping agents on the catalytic application of colloidal nanoparticles. Controversial results have been reported during the last years. In some cases the capping ligand can act as a physical barrier to restrict the access of the substrate or reactant to the catalytic nanoparticle or they can also be used to promote nanocatalysis. However, most of the examples are related to metal nanoparticles, where the ligands produce a decrease in the activity but cause a great improvement in the selectivity. Use of capping ligands still needs quantitative explorations in order to introduce new functionality to the colloidal nanoparticles. (Niu & Li, 2014)

The role of the ligand in the coupling of two semiconductors has also been studied in order to facility the formation of nanohybrids. Coupling of two semiconductors can enhance the charge separation yield, improving the photocatalytic activity of the system. Thus, the degradation of methyl viologen by CdSe was improved in the CdSe-TiO_2 hybrid, that it was consistent with a synergistic effect between both semiconductors. (Harris & Kamat, 2009)The understanding of these charge transfer events is important for the design of new photocatalyst or environmental remediation.

Colloidal core/shell semiconductor nanocrystals such as CdSe/CdS have been studied for the photocatalytic hydrogen (H_2) production from water and the role of ligands has been investigated using different numbers of functional thiol groups, such as poly(acrylic acid), 3-mercaptopropionic acid and 2,3-dimercaptosuccinic acid. The results indicate that the photocatalytic activity increases with the number of thiol groups in the ligand, while the photoluminescence intensity and average fluorescence lifetime decrease. Surface coating is propose to mediate the surface trap-related charge separation efficiency and, as consequence, it is the responsible of the ligand-dependent performance in the photocatalytic process (Wang, Zhang, He, Xu, & Jin, 2015).

In addition, conjugation of CdSe/ZnS core-shell QDs with biomolecules able to act as electron donors resulted in photosensitized nanoparticles, which can oxidize or reduce molecules like water and molecular oxygen, leading to the formation of reactive oxygen species (ROS, Figure 5). These photosensitized conjugates may be used as agents for photodynamic therapy (PDT) of superficial cancers. In this example, the organic ligand (OLA) from the native QDs was exchanged with mercaptopropionic acid, thus allowing its conjugation to dopamine via the (1-ethyl-3-(3-dimethylaminopropyl)-carbodiimide hydrochloride)-mediated reaction. The functional ligand acted as the photosensitizer of the nanoparticle to form ROS species, such as singlet oxygen. The hydrophilic nature of the dopamine-capped nanohybrids made them ideal for uptake into inflamed tissues, often a barrier for their successful application (Cooper, Dimitrijevic, & Nadeau, 2010).

Regarding to organic synthesis, the potential of CdSe QDs as photocatalysts has been demonstrated in the reduction of several aryl azides, which led to a selective and efficient formation of anilines (Warrier, Lo, Monbouquette, & Garcia-Garibay, 2004). The substrate should be at the nanoparticle surface within the lifetime of the charge-separated excited state. In this case, the organic ligand was an aminothiol bounded to the QDs by the thiolate anchoring group whereas the amine allowed its solubility in aqueous media.

Solar Cell Application

Colloidal QDs are interesting candidates for solar cell applications due to their solution processability and optical properties that allow their use as light harvesters in different type of photovoltaics devices (Emin, Singh, Han, Satoh, & Islam, 2011). Organic ligands play a key role in the charge transport for the construction of solar cell and photovoltaic devices. Combination of QDs with polymer has been used to

Figure 5. Cooperation between the QD and the functional ligand to lead to singlet oxygen

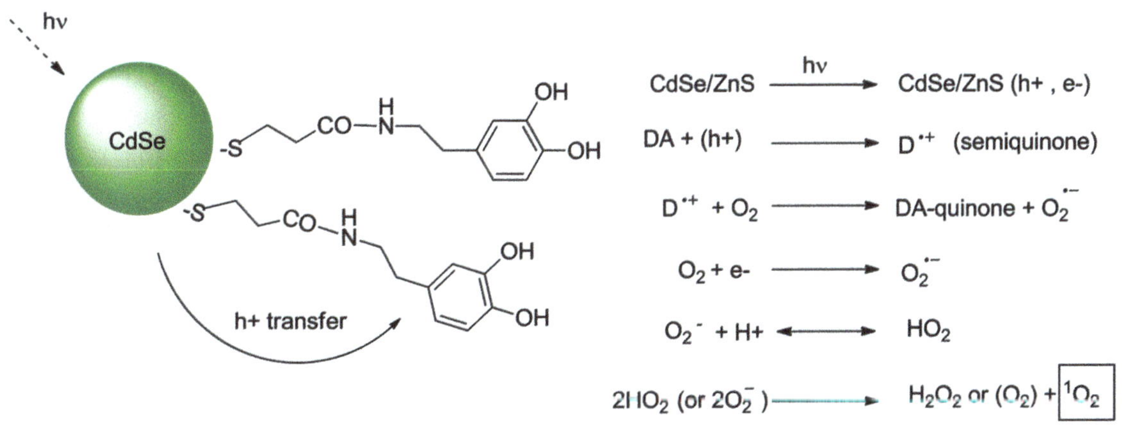

improve their performance and processability, but the use of trioctylphosphine oxide ligands should be avoided since the polymer hinders the hole transport.

Attachment of CdS QDs to polymers, such as poly (3-hexylthiophene), through solvent-assisted grafting and ligand exchange led to power conversion efficiency of 4.1%. In this case the original ligand (*n*-butylamine) of the CdS was exchange by ethanedithiol (EDT) before its treatment with the polymer. The CdSe/polymer film showed small agglomerates that reduce the distance between the nanoparticles, thus increasing the charge recombination rate constant.

The photovoltaic performance of a solar cell based on PbS nanoparticles was enhanced after treatment with acetic acid, which was explained as due to ligand exchange of oleic acid by the shorter acetic acid ligand, thus reducing the distance between the QD and the polymer and favoring the charge transfer in the device. (Hines & Kamat, 2014)

The effect of the length of the thiol chain in the electron transfer process was shown in QD-sensitized solar cells (QDSSC) based on ZnO material. Thus, CdSe QDs were anchored to ZnO nanorods through two thiol ligands, specifically, thioglycolic acid (TGA) and mercaptopropionic acid (MPA). The electron transfer process was faster in TGA-capped QDs due to the shorter chain of the ligand (Sun et al., 2010).

Recently, Förster resonance energy transfer from the QDs to adjacent organic molecules has been applied to extend the photoresponse of QD solar cells into the near infrared. The QDs were linked to a mesoscopic TiO_2 film by means of a bifunctional linker that absorbed red or near infrared light (a squaraine dye functionalized with carboxylic and thiol groups). The active ligand facilitated the two-steps energy transfer from de CdSe QDs to the TiO_2, thus enhancing the power conversion over the dye-TiO_2 (Choi, Santra, & Kamat, 2012).

Toxicity of QDs

One of the main concerns for the bio-application of CdSe-based QDs is their toxicity related to the Cd^{2+} release. In addition other limitations associated with tissue penetration, bio-distribution profiles, metabolism and excretion have reduced their use in clinical imaging. (Huang et al., 2011) Other factor that can also limit their bio-application is the clearance of QDs from cells, which has been associated to the cysteine interaction with the nanoparticle surface. (Cao, Cao, Liu, Wang, & Zhu, 2009)

It is well known that synthetic methods and surface modifications of QDs can drastically affect their photophysical properties and its interactions with the cellular membrane. Different types of cells produce different response towards the QD-induced toxicity. Different parameters has been associated with the cytotoxicity of QDs such a cell number, cell growth, apoptosis, cellular morphology or metabolic activity change. These differences have complicated the real comparison to give a consistent conclusion on their toxicity. Cytotoxicity associated to release of Cd^{2+} was studied in CdTe core and CdTe/CdS and CdTe/CdS/ZnS core-shell structures. They observed that bio-distribution of QDs and nanoscale effects could also affect the cytotoxicity in cell medium. As will be expected, different surface properties of aqueous and liphophilic QDs may lead to different cytotoxicity (Chen et al., 2012).

The toxicity of CSQDs was studied in different media, amoeba, plants, and animals (Valizadeh et al., 2012). In the amoeba, the QDs did not show any effect on the growth and cellular mobility of Dictyostelium discoideum cells. The results obtained in plants demonstrated that oxidative stress occurred taking into account the reduced glutathione levels compared to the oxidized glutathione. Endothelial toxicity of mercaptosuccinic acid-capped CSQDs was reported as consequence of activation of the mitochondrial death pathway and final endothelial apoptosis.

Different alternatives have been used to reduce the cell toxicity of CdSe-based QDs: i) to coat the QD with an inert shell, ii) to dope the QD with other metal, and iii) to avoid the use of cadmium. For example, coating with DNA has proved to be useful to reduce the CdS toxicity. (Rosenthal, Chang, Kovtun, McBride, & Tomlinson, 2011) Doping with Mn has been applied to the preparation of Mn-doped QDs, such as Mn-ZnS or Mn-ZnSe QDs, thus leading to highly luminescent QDs which combined with the absence of Cd^{2+} results in less toxic QDs (Maity et al., 2013) Finally, Cd-free QDs, such as InAs/InP/ZnSe and $CuInS_2$/ZnS with NIR fluorescence are less toxic and they are relevant for in vivo applications (Gao et al., 2010).

Organolead Halide Perovskite Nanoparticles

In general, the term "perovskite" is used to describe any material with the same structure as $CaTiO_3$. Organometal halide perovskites are considered as hybrid materials because they combine inorganic and organic properties in the same material, thus exhibiting the plastic mechanical properties of the organic component and the optical and charge mobility of the inorganic material.

Pure organolead halide perovskites have the general formula of $APbX_3$, where A is a small ammonium cation (methyammonium, MA, or formamidinium, FA) that resides at the eight corners of a cubic unit, while Pb is located at the center and is surrounded by six halide (X) anions forming an octahedral cluster $[PbX_6]^{4-}$. (Mathews, 2014) A relatively small size is necessary for the organic cation (MA or FA) to fit into the corner-sharing lead halide octahedral 3D framework (Figure 6). If the cation is too large, the dimensionality of the inorganic framework changes to 2D, 1D, or 0D (Figure 6). In the 0D hybrid perovskite structure, the PbX_6 octahedra are isolated, while in the 1D the PbX_6 octahedra are chain-connected, and in the 2D the PbX_6 the octahedra are connected at the corners in layered sheets. The stoichiometry of the perovskite changes with the dimension of the inorganic framework. Thus, the $APbX_3$, A_2PbX_4, A_6PbX_5, and A_4PbX_6 stoichiometry possesses a 3D, 2D, 1D, and 0D framework, respectively. Consequently, ammonium-lead halide perovskites with different optical and electronic properties can be prepared by changing the molar ratio between the components.

In addition, as mentioned above, the optical and electronic properties of semiconductors differ with the material dimensionality (3D, 2D, 1D and 0D, Figure 1). (González-Carrero et al., 2015)

Ammonium lead halide perovskites with a 3D framework are very promising in the nanoscale dimension, (0D) where they have demonstrated they produce a better optical and electronic response than the

Figure 6. Framework classification of ammonium lead halide perovskites. Red balls indicate halide atoms, green balls are methyl ammonium salts, and two connected balls are ammonium salts with a carbon chain longer than that of methyl ammonium salt
Adapted from González-Carrero et al., 2015

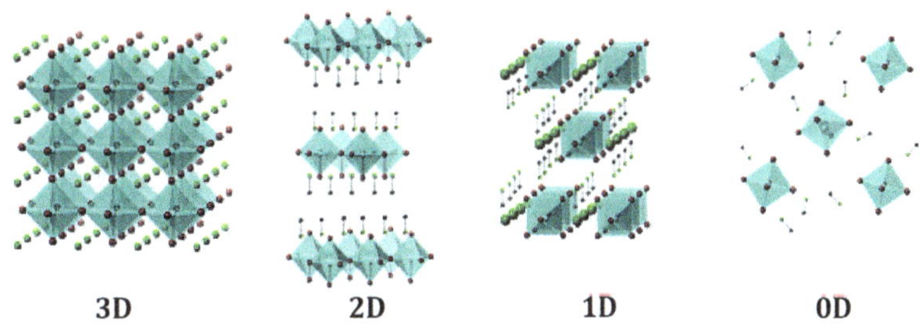

bulk 3D material. Here we mainly focus on perovskite nanoparticles (i.e., 0D materials) and specifically on those with the $APbX_3$ stoichiometry, though organolead halide perovskites with other stoichiometries are also mentioned.

Role of the Organic Ligand in the Preparation of the Perovskite Nanomaterial

Hybrid perovskites have been firstly prepared as nanoparticulate material by using a *template strategy.* Park N. G. et al. (Im, Lee, Lee, Park, & Park, 2011) demonstrated the formation of 2–3 nm sized $MAPbI_3$ perovskite QDs on nanocrystalline TiO_2 surface by spin-coating an equimolar mixture of MAI and PbI_2 in γ-butyrolactone solution. In addition, Miyasaka T. et al. showed the formation of considerably emissive $MAPbBr_3$ nanoparticulated material using porous Al_2O_3 as a template. (Kojima, Ikegami, Teshima, & Miyasaka, 2012) Thus, a solution of the $MAPbBr_3$ perovskite precursors (MABr and $PbBr_2$) was poured on the alumina film and the $MAPbBr_3$ perovskite was formed in the alumina pores by mild heating. The nanomaterial showed enhanced luminescence compared with that of the bulk material.

It has recently been demonstrated the preparation of colloidal $MAPbBr_3$ perovskite nanoparticles by a *non-template method* that uses the capacity of organic ligands to control the growth of the nanoparticle. The synthesis of the perovskite nanoparticles was carried out by first adding the long alkyl chain ammonium salt (octylammonium bromide, OCABr, or octadecylammonium bromide, ODABr) to a warm solution of oleic acid (OLA) in 1-octadecene (ODE), and then adding methylammonium bromide (MABr) and $PbBr_2$. Fast precipitation with acetone gave rise to a yellow precipitate of the perovskite nanoparticle, which was dispersible in toluene as well as in other organic solvents. (Schmidt et al., 2014)

Regarding other organo-lead halide stoichiometries, Boissière et al. (P. Audebert, G. Clavier, V. Alain-Rizzo, E. Deleporte, S. Zhang, J.-S. Lauret, G. Lanty, C. Boissière, Chemistry of Materials 2009, 21, 210-214) reported the preparation of spherical nanoparticles with the formula A_2PbX_4 (X = Br, I; A= phenylethylammonium, cyclohexylmethylammonium, and 1-adamantylmethylammonium). Perovskite solutions in DMF were prepared by mixing stoichiometric amounts of the ammonium salt with the corresponding lead halide. *Fast nebulization and subsequent lyophilization* of perovskite DMF solutions led to polydispersed samples with nanoparticles of an average size of 60 nm. Therefore, the optical features of these nanoparticles were similar to those of the bulk 2D material.

Blue-luminescent and photostable $(ODA)_2PbBr_4$ nanoparticles have been efficiently prepared by using ODABr, $PbBr_2$, and ODE (ODABr/$PbBr_2$ and ODE/$PbBr_2$ molar ratios of 2:1 and 62.6:1.0, respectively). The addition of acetone induced the precipitation of the perovskites as a white solid, which was dispersible in toluene and was reverted into its precursors by using DMF. The use of other ammonium salts with a shorter chain, such as hexyl- and octylammonium bromide, was less efficient in the preparation of A_2PbBr_4 colloidal nanomaterial.

In addition, Couris et al. (I. Papagiannouli, E. Maratou, I. Koutselas, S. Couris, The Journal of Physical Chemistry C 2014, 118, 2766-2775) reported the preparation of A_4PbX_6 perovskites, which possess a 0D framework, by diluting bulk 4-fluorophenethylammonium(FpA)–lead iodide perovskites with a 2D framework. The dilution breaks the 2D material up into smaller units, which consist of FpA-capped PbI_6 octahedra. The UV-vis absorption spectrum of these FpA-capped PbI_6 nanocrystals exhibits a peak at 365 nm.

Furthermore, Papavassiliou et al. reported on the preparation of toluene-dispersible $A(A')_2Pb_2X_7$ nanocrystalline/microcrystalline materials based on $Pb(Br_xCl_{1-x})_3$, $Pb(Br_xI_{1-x})_3$, and $Pb(Cl_xI_{1-x})_3$ (x=0–1) units. (Papavassiliou et al., 2012)These materials were prepared by injection of solutions of $Pb_m(Br_xCl_{1-x})_{3m+1}$,

$Pb_m(Br_xI_{1-x})_{3m+1}$, and $Pb_m(Cl_xI_{1-x})_{3m+1}$ (m > 2) bulk compounds with a quasi-2D framework, or their precursors, into neat toluene.

Finally, Papavassiliou et al. reported the preparation of (MA)(1-napthylmethylammonium)$_2Pb_2Cl_7$ nanocrystalline/microcrystalline (100-200 nm) materials via injection of successive DMF aliquots of the bulk material with a quasi-2D framework (prepared from an equimolar mixture of $CH_3NH_3PbCl_3$ and (1-napthylmethyl ammonium)$_2PbCl_4$) into toluene-PMMA or neat toluene.

Role of Capping Agent in the Optical Properties of the Nanomaterial

Regarding the emissive properties of $MAPbBr_3$ perovskite nanoparticles, the optical features of the material can be controlled by adjusting the molar ratio between the components (ammonium salts and $PbBr_2$). The first synthesis of colloidal nanoparticles was performed in the presence of ODE and OLA and either ODABr or OCABr as the capping ligand. Six-nm sized nanoparticles with a narrow photoluminescence peak and emission quantum yield of ca. 20% was obtained by using an OCABr/MABr and total ammonium salt/$PbBr_2$ molar ratios of 0.6:0.4 and 1:1, respectively. Homogeneous thin films of these nanoparticles, prepared by spin-coating on a quartz substrate, exhibited a similar quantum yield (Figure 7). (Schmidt et al., 2014) Under similar conditions but using ODABr instead of OCABr led to considerably less emissive material.

More emissive nanoparticles were further obtained by testing different molar ratios of all the components, which either formed part of the framework (MABr and $PbBr_2$) or acted as the organic capping (OCABr) or solvent (ODE and OLA). These studies led to the efficient preparation (chemical yield of 88%) of highly luminescent nanoparticles (83% quantum yield) in the absence of OLA. These nanoparticles exhibited a high luminescence lifetime (ca. 600 ns) and photostability (<18% emission loss after 8.5 h irradiation in air atmosphere). (Gonzalez-Carrero, Galian et al., 2015)

Comparison between the optical properties of the perovskites, such as $MAPbBr_3$, and CdSe QDs reveals important differences between them (Figure 8). Thus, $MAPbBr_3$ nanoparticles can be prepared with an extraordinary emission quantum yield without adding an inorganic shell, they have a long average photoluminescence lifetime (hundreds of nanoseconds) and they are considerably more photostable. Between the analogies, both of them exhibit small Stokes shifts and narrow emissions spectra(Figure 8).

Figure 7. Highly fluorescent hybrid lead halide perovskite nanoparticles with a 3D (left) and a 2D (right) framework

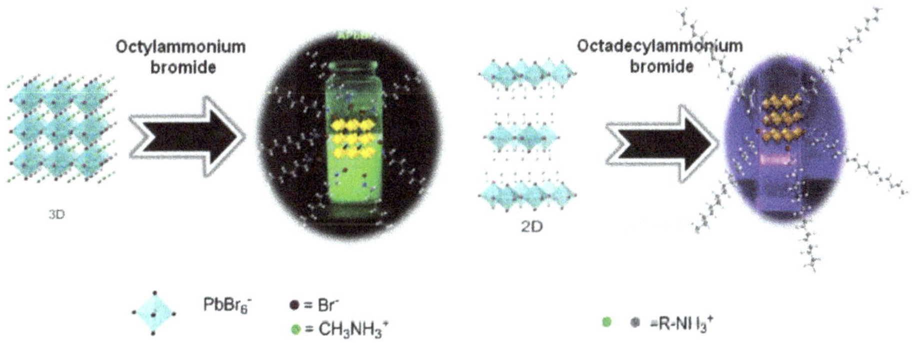

Figure 8. Absorbance and fluorescence emission spectra of A) CdSe/ZnS QDs capped with octadecylamine (λ_{ex} = 400 nm) and B) $MAPbBr_3$ perovskite nanoparticle capped with octylammonium bromide (λ_{ex} = 350 nm). C) Comparison between the typical kinetic trace of a CdSe/ZnS QD (■) and that of OCA-capped $MAPbBr_3$ perovskite nanoparticle (♦) recorded at 560 nm and 520 nm, respectively

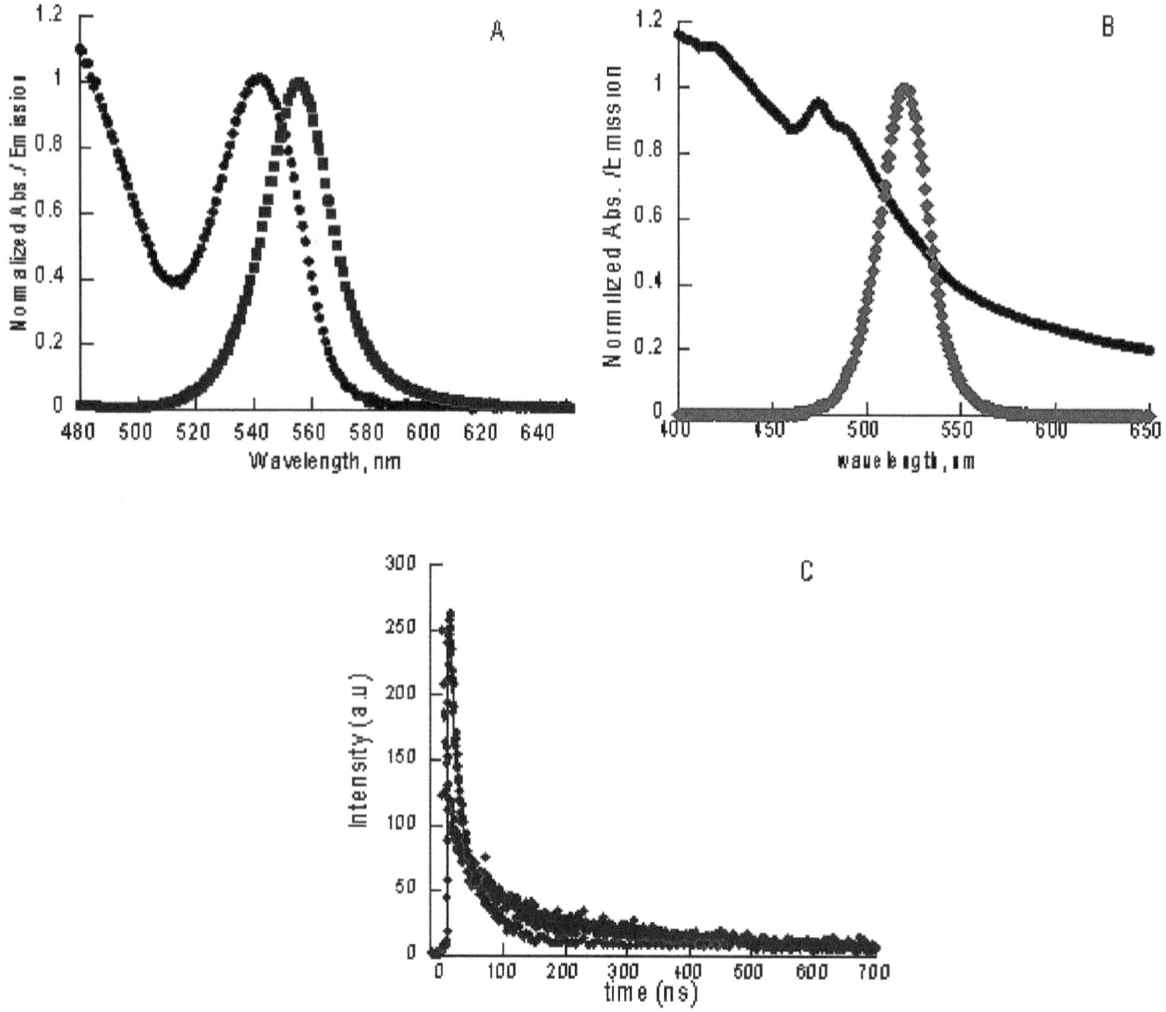

The exciton Bohr radius of the $MAPbBr_3$ material (Tanaka et al., 2003) has been predicted to be 2 nm and, consequently, the nanoparticles with a diameter of less than 4 nm would exhibit enhanced quantum confinement effects, i.e., they would behave as quantum dots (QDs). Zhang et al. (Zhang et al. 2015) described the preparation of 3.3 nm-sized $MAPbBr_3$ perovskite nanoparticles with high luminescence (70% and biexponential lifetime with a short lifetime (τ) of 6.6 ns and a long one of 18 ns) by using the *ligand-assisted re-precipitation technique*. This technique consists of the dropwise addition into toluene of a DMF solution of the capping ligands (octylamine, OA and OLA) and the perovskite precursors (MABr and $PbBr_2$). In addition, Rogach et al. (Huang, Susha, Kershaw, Hung, & Rogach, 2015) have recently reported the preparation of QDs with a remarkable quantum yield (93%) and average decay lifetimes (τ_{av}) of around 18 ns by slight modifications to this strategy. In particular, the DMF solution was added to toluene at 60^{0}C. They stated that the synthetic yield of $MAPbBr_3$ nanoparticles by the *ligand-assisted re-precipitation method* is limited due to the formation of bulk material by-products along with the desired nanoparticles.

Interestingly, OA-capped $MAPbBr_3$ organometal halide perovskite nanoplatelets of varying thicknesses have been synthesized by controlling the ratio of the organic cations, specifically OA and MABr, up to 4:1.(Sichert et al., 2015) Photoluminescence spectra showed the progressive blue-shift of the peak from 519 nm for the 0% OA sample to 502 nm for the 60% OA sample, and to 469 nm for the 80% OA sample. The most emissive samples (30% quantum yield relative to coumarin 6 dye) were those with a 70% OA.

Regarding A_2PbX_4 nanoparticles, toluene dispersions of $(ODA)_2PbBr_4$ have been prepared with a considerable luminescence (quantum yield over 20%), radiative rate constant of up to 80×10^6 s^{-1}, and high photostability under UV light (Figure 7).(Gonzalez-Carrero, Espallargas, Galian, & Perez-Prieto, 2015)

Finally, the use of functional ammoniums can modify the emissive properties of the perovskite. Thus, excitation of the inorganic component (at ca. 350 nm) of MA(1-napthylmethylammonium)$_2Pb_2Cl_7$ nanocrystalline/microcrystalline material produced a broad phosphorescence band (at ca. 500, 536, and 577 nm), which was ascribed to the organic chromophore.(Papavassiliou, Pagona, Mousdis, & Karousis, 2013) Comparatively, excitation of (1-napthylmethyl ammonium)$_2PbCl_4$ and $MAPbCl_3$ under the same conditions led to emission at ca. 341 nm and 395 nm, respectively, whereas suspensions containing equimolar amounts of both compounds again showed the phosphorescence emission of the organic chromophore. The enhanced emission in comparison to that of the $PbCl_4$ material and the fact that it occurred at room temperature was attributed to enhanced energy transfer from the excitonic level of the $PbCl_3$ moieties to the triplet excited state of the chromophore.

Sinergism in the Nanohybrid Functionality

To date, there are very few examples of the application of perovskite colloidal nanomaterials. One of them is related to their applicability in sensing. Specifically, the emission of OCA-capped $MAPbBr_3$ perovskite nanoparticles has proved to be highly selective and sensitive to the presence of 2,4,6-trinitrophenol (TNP, picric acid), both in the solution and vapor state. Comparatively, these nanoparticles proved to be less sensitive to other nitroaromatic explosives, such as 2,4-dinitrotoluene (DNT) and 2,4,6-trinitrotoluene (TNT), and even to phenolic compounds, namely 4-nitrophenol and phenol. The selective response to TNP has been attributed to the interaction between the TNP hydroxyl group and adjacent nitro groups with the nanoparticle surface. These interactions would facilitate the approach of the analyte to the nanoparticle surface and as a consequence, the electron transfer from the nanoparticle to the TNP. (Muthu, Nagamma, & Nair, 2014)

Organo-metal halide perovskites have emerged as very promising materials in the field of photovoltaics due to their remarkable conducting material and excellent electrical and optical properties. Kojima et al. showed in 2009 that self-assembled $APbX_3$ perovskites on mesoporous TiO_2 films sensitize TiO_2 for visible-light conversion in photoelectrochemical cells; the photoconversion efficiency (PCE) was 3.81%.(Kojima, Teshima, Shirai, & Miyasaka, 2009) Their efficiency in solar cell devices was later enhanced to 9.7% by using organic-hole transport materials instead of liquid-electrolytes. Great efforts have been made to increase the efficiency and stability of the material in solution-processed solar cells. (Im, Jang, Pellet, Grätzel, & Park, 2014; Kim et al., 2012; Malinkiewicz, Roldán-Carmona et al., 2014; Malinkiewicz, Yella et al., 2014; Niu, Guo, & Wang, 2015; Roldan-Carmona et al., 2014).

Regarding colloidal perovskites, OA-capped $MAPbX_3$ perovskite nanoparticles have been used to build a light-emitting electrochemical cell via an up-scalable spray-coating technique.(Aygüler et al., 2015) However, the presence of organic capping on the colloidal nanoparticles can strongly limit their performance in applications where charge carrier transport plays a key role.

FUTURE RESEARCH DIRECTIONS

We have mentioned above the crucial role of organic ligands in the preparation and colloidal stability of nanoparticles, and also as a means for providing the nanosystem with new functionalities. Examples of the synergic effect between the nanoparticle and the organic capping and its relevance in the nanosystem performance have also been included.

Most of the organic ligands used during the synthesis of nanocrystals are electrical insulators and this introduces an important limitation for charge transportation application in, for example, photovoltaics, thermoelectrics, and catalysis.

Long-chain insulating organic ligands have to be removed for the integration of the nanoparticles into solids with tunable electronic properties, as required for solar cells and light-emitting diodes, among others. Consequently, alkylamines and alkylthioles with a short chain and aromatic thiols have been used for the effective passivation of QDs while reducing their interparticle spacing. Thus, the organic ligand of OLA-capped PbSe QDs was first exchanged with octylamine and networks of PbSe QDs were formed after benzenedithiol treatment due to the strong affinity of the thiol group for the Pb atoms.(Koleilat et al., 2008) The diffusion of electrons and holes over hundreds of nanometers through the PbSe colloidal QD solid was demonstrated.

Lately, there is an increasing number of publications reporting on the beneficial effect of inorganic capping ligands for obtaining colloidal nanoparticles of relevance in photovoltaics. Sargeant et al. reported the first inorganic ligand-passivated colloidal QDs capped with an inorganic ligand to minimize the interparticle distance, and so promote carrier transport while maintaining their luminescence performance. The inorganic ligand-capped QDs were prepared from oleate-capped QDs by means of ligand exchange. The colloidal QDs were assayed in photovoltaic applications. The solar cell power conversion was 6%, which is promising for improving the charge transportation on film substrates.(Tang et al., 2011)

In addition colloidal halide-capped QDs have been prepared by a straightforward method. Thus, hexane or heptane dispersions of amine-capped CdSe QDs (or PbS QDs) were mixed with a methanol solution of the corresponding methylammonium halide. This led to the transfer of the halide-capped CdSe QDs to the polar solvent. The anion was attached to the QDs by electrostatic interaction between the halogen ion and the Cd or Pb of the QDs. The effect of the different kind of halogens in the re-dispersion in polar solvent proved to be proportional to the coordination strength. Coordination between Cd^{2+}or Pb^{2+} (soft acids) and Cl^- (hard acid) was much weaker than with the other halogen ions (soft bases). The CdSe capped with iodide ligands were adsorbed to a positive TiO_2 to obtain an assembled QD film used for photovoltaic devices, which showed better photocurrent efficiency compared with that of the CdSe capped with TOPO ligands.(Niu et al., 2013)

Films of Cl-capped $FAPbI_3$ perovskites have been synthesized via a two-step solution process. First, a PbI_2 solution was spin-coated on an ITO substrate and then a mixture of FA and MA ammonium salts was spin-coated and annealed at 343.15 K, thus providing a good film quality with enhanced interfacial opto-electric property due to the chloride incorporation. (10.1021/acs.nanolett.5b02126)

Also, compact halide (NH_4I, NH_4Br, NH_4Cl) and pseudohalide (N_3^- or CN^-) ligands were used to shorten the interparticle distance of QDs, thus improving their assembly and, consequently, enhancing the carrier mobility. These QDs have promising applications in electronic and thermoelectric devices. (Zhang, Jang, Liu, & Talapin, 2014)

Moreover, Cadavid et al. reported about a simple, general, and scalable procedure to displace the organic ligand (TOPO) with a matched metal salt solution. Thus, the addition of formamide solution of $AgNO_3$ to a chloroform solution of TOPO-capped Ag_2Te nanoparticles eventually led to the transfer of Ag_2Te to the formamide phase. The Ag-capped Ag_2Te nanoparticles showed thermoelectrical properties. (Cadavid et al., 2013)

Interestingly, halometallate-capped QDs have been obtained by ligand exchange with lead halide perovskites ($MAPbX_3$, X= I- or Br -) and other MX_n metal halide complexes (M = Pb, Cd, In, Zn, Fe, Bi and Sb). Ligands such as $MACdBr_3$ and $(NH_4)_2ZnCl_4$ highly retained the luminescence quantum yield of the original CdSe/CdS core-shell QDs (50-65% vs. 72-75%). However, the use of chalcogenidometallate ligands ($K_4Sn_2S_6$ and K_3AsS_4) reduced the fluorescent response to 10%. Halometallates can extract some cations from the nanocrystal due to the good solubility of the corresponding metal halide and, therefore, isocation ligands were used in the case of CdSe to avoid cation exchange and keep the Cd-rich stoichiometry of CdSe nanocrystals.(Dirin et al., 2014)

CONCLUSION

The main focus of this chapter was to understand the role of the organic ligands in the different phases of the nanomaterial from its synthesis to its application. The understanding of the interaction between ligand and nanoparticle surfaces could help in the future development of new smart systems with improved performance or new applications. Multidisciplinary collaboration between different areas of science is suggested in order to obtain multifunctional systems that combine the individual properties of the materials in the same nanosystem.

Alternatively to colloidal organic capped-nanoparticles, there are increasing examples of dispersible inorganic-capped nanoparticles. Each type of nanoparticles, organic- or inorganic-capped photoactive nanoparticles, has its own field of application. While organic capping is advantageous to provide the nanosystem with interesting capacities (such as encapsulation, molecular recognition, and functionalization for further conjugation with (bio)molecules, among others), most of them are electrical insulators and this is a limitation for charge transportation application, such as in photovoltaics, thermoelectrics, and catalysis. However, in charge transportation, inorganic-capping is advantageous since it makes it possible to shorten the interparticle spacing and facilitate such process. Consequently, the choice of one or the other as the nanoparticle capping depends on the target application.

REFERENCES

Agudelo-Morales, C. E., Galian, R. E., & Pérez-Prieto, J. (2012). Pyrene-Functionalized Nanoparticles: Two Independent Sensors, the Excimer and the Monomer. *Analytical Chemistry*, *84*(18), 8083–8087. doi:10.1021/ac302276j PMID:22924743

Aguilera-Sigalat, J., Casas-Solvas, J. M., Morant-Minana, M. C., Vargas-Berenguel, A., Galian, R. E., & Perez-Prieto, J. (2012). Quantum dot/cyclodextrin supramolecular systems based on efficient molecular recognition and their use for sensing. *Chemical Communications*, *48*(20), 2573–2575. doi:10.1039/C1CC15312A PMID:22080219

Aguilera-Sigalat, J., Pais, V. F., Doménech-Carbó, A., Pischel, U., Galian, R. E., & Pérez-Prieto, J. (2013). Unconventional Fluorescence Quenching in Naphthalimide-Capped CdSe/ZnS Nanoparticles. *The Journal of Physical Chemistry C*, *117*(14), 7365–7375. doi:10.1021/jp3128252

Aguilera-Sigalat, J., Rocton, S., Galian, R. E., & Pérez-Prieto, J. (2011). Fluorescence enhancement of amine-capped CdSe/ZnS quantum dots by thiol addition. *Canadian Journal of Chemistry*, *89*(3), 359–363. doi:10.1139/V10-160

Aguilera-Sigalat, J., Rocton, S., Sanchez-Royo, J. F., Galian, R. E., & Perez-Prieto, J. (2012). Highly fluorescent and photostable organic- and water-soluble CdSe/ZnS core-shell quantum dots capped with thiols. *RSC Advances*, *2*(4), 1632–1638. doi:10.1039/C1RA01005K

Aldana, J., Wang, Y. A., & Peng, X. (2001). Photochemical Instability of CdSe Nanocrystals Coated by Hydrophilic Thiols. *Journal of the American Chemical Society*, *123*(36), 8844–8850. doi:10.1021/ja016424q PMID:11535092

Alivisatos, A. P. (1996). Semiconductor Clusters, Nanocrystals, and Quantum Dots. *Science*, *271*(5251), 933–937. doi:10.1126/science.271.5251.933

Anderson, R. E., & Chan, W. C. W. (2008). Systematic Investigation of Preparing Biocompatible, Single, and Small ZnS-Capped CdSe Quantum Dots with Amphiphilic Polymers. *ACS Nano*, *2*(7), 1341–1352. doi:10.1021/nn700450g PMID:19206301

Aygüler, M. F., Weber, M. D., Puscher, B. M. D., Medina, D. D., Docampo, P., & Costa, R. D. (2015). Light-Emitting Electrochemical Cells Based on Hybrid Lead Halide Perovskite Nanoparticles. *The Journal of Physical Chemistry C*, *119*(21), 12047–12054. doi:10.1021/acs.jpcc.5b02959

Bäumle, M., Stamou, D., Segura, J.-M., Hovius, R., & Vogel, H. (2004). Highly Fluorescent Streptavidin-Coated CdSe Nanoparticles: Preparation in Water, Characterization, and Micropatterning. *Langmuir*, *20*(10), 3828–3831. doi:10.1021/la0363940 PMID:15969364

Bawendi, M. G., Steigerwald, M. L., & Brus, L. E. (1990). The Quantum Mechanics of Larger Semiconductor Clusters ("Quantum Dots"). *Annual Review of Physical Chemistry*, *41*(1), 477–496. doi:10.1146/annurev.pc.41.100190.002401

Bechet, D., Couleaud, P., Frochot, C., Viriot, M.-L., Guillemin, F., & Barberi-Heyob, M. (2008). Nanoparticles as vehicles for delivery of photodynamic therapy agents. *Trends in Biotechnology*, *26*(11), 612–621. doi:10.1016/j.tibtech.2008.07.007 PMID:18804298

Bera, D., Qian, L., Tseng, T.-K., & Holloway, P. H. (2010). Quantum Dots and Their Multimodal Applications: A Review. *Materials (Basel)*, *3*(4), 2260–2345. doi:10.3390/ma3042260

Bilan, R., Fleury, F., Nabiev, I., & Sukhanova, A. (2015). Quantum Dot Surface Chemistry and Functionalization for Cell Targeting and Imaging. *Bioconjugate Chemistry*, *26*(4), 609–624. doi:10.1021/acs.bioconjchem.5b00069 PMID:25710410

Billing, D. G., & Lemmerer, A. (2009). Inorganic-organic hybrid materials incorporating primary cyclic ammonium cations: The lead bromide and chloride series. *CrystEngComm*, *11*(8), 1549–1562. doi:10.1039/b819455f

Blanco-Canosa, J. B., Wu, M., Susumu, K., Petryayeva, E., Jennings, T. L., Dawson, P. E., & Medintz, I. L. et al. (2014). Recent progress in the bioconjugation of quantum dots. *Coordination Chemistry Reviews*, *263–264*, 101–137. doi:10.1016/j.ccr.2013.08.030

Bullen, C., & Mulvaney, P. (2006). The Effects of Chemisorption on the Luminescence of CdSe Quantum Dots. *Langmuir*, *22*(7), 3007–3013. doi:10.1021/la051898e PMID:16548550

Cadavid, D., Ibanez, M., Shavel, A., Dura, O. J., Lopez de la Torre, M. A., & Cabot, A. (2013). Organic ligand displacement by metal salts to enhance nanoparticle functionality: Thermoelectric properties of Ag2Te. *Journal of Materials Chemistry A*, *1*(15), 4864–4870. doi:10.1039/c3ta01455j

Cao, M., Cao, C., Liu, M., Wang, P., & Zhu, C. (2009). Selective fluorometry of cytochrome c using glutathione-capped CdTe quantum dots in weakly basic medium. *Mikrochimica Acta*, *165*(3-4), 341–346. doi:10.1007/s00604-009-0140-8

Čapek, R. K., Lambert, K., Dorfs, D., Smet, P. F., Poelman, D., Eychmüller, A., & Hens, Z. (2009). Synthesis of Extremely Small CdSe and Bright Blue Luminescent CdSe/ZnS Nanoparticles by a Prefocused Hot-Injection Approach. *Chemistry of Materials*, *21*(8), 1743–1749. doi:10.1021/cm900248b

Carrillo-Carrion, C., Cardenas, S., Simonet, B. M., & Valcarcel, M. (2009). Quantum dots luminescence enhancement due to illumination with UV/Vis light. *Chemical Communications*, (35): 5214–5226. doi:10.1039/b904381k PMID:19707627

Chen, N., He, Y., Su, Y., Li, X., Huang, Q., Wang, H., & Fan, C. et al. (2012). The cytotoxicity of cadmium-based quantum dots. *Biomaterials*, *33*(5), 1238–1244. PMID:22078811

Chen, X., Hutchison, J., Dobson, P., & Wakefield, G. (2009). Highly luminescent monodisperse CdSe nanoparticles synthesized in aqueous solution. *Journal of Materials Science*, *44*(1), 285–292. doi:10.1007/s10853-008-3055-6

Choi, H., Santra, P. K., & Kamat, P. V. (2012). Synchronized Energy and Electron Transfer Processes in Covalently Linked CdSe–Squaraine Dye–TiO2 Light Harvesting Assembly. *ACS Nano*, *6*(6), 5718–5726. doi:10.1021/nn301982e PMID:22658983

Clapp, A. R., Goldman, E. R., & Mattoussi, H. (2006). Capping of CdSe-ZnS quantum dots with DHLA and subsequent conjugation with proteins. *Nature Protocols*, *1*(3), 1258–1266. doi:10.1038/nprot.2006.184 PMID:17406409

Cooper, D. R., Dimitrijevic, N. M., & Nadeau, J. L. (2010). Photosensitization of CdSe/ZnS QDs and reliability of assays for reactive oxygen species production. *Nanoscale*, *2*(1), 114–121. doi:10.1039/B9NR00130A PMID:20648372

Cooper, J. K., Franco, A. M., Gul, S., Corrado, C., & Zhang, J. Z. (2011). Characterization of Primary Amine Capped CdSe, ZnSe, and ZnS Quantum Dots by FT-IR: Determination of Surface Bonding Interaction and Identification of Selective Desorption. *Langmuir*, *27*(13), 8486–8493. doi:10.1021/la201273x PMID:21631120

Dabbousi, B. O., Rodriguez-Viejo, J., Mikulec, F. V., Heine, J. R., Mattoussi, H., Ober, R., & Bawendi, M. G. et al. (1997). (CdSe)ZnS Core−Shell Quantum Dots: Synthesis and Characterization of a Size Series of Highly Luminescent Nanocrystallites. *The Journal of Physical Chemistry B*, *101*(46), 9463–9475. doi:10.1021/jp971091y

Delgado-Pérez, T., Bouchet, L. M., de la Guardia, M., Galian, R. E., & Pérez-Prieto, J. (2013). Sensing Chiral Drugs by Using CdSe/ZnS Nanoparticles Capped with N-Acetyl-L-Cysteine Methyl Ester. *Chemistry (Weinheim an der Bergstrasse, Germany)*, *19*(33), 11068–11076. doi:10.1002/chem.201300875 PMID:23813622

Deng, Z., Cao, L., Tang, F., & Zou, B. (2005). A New Route to Zinc-Blende CdSe Nanocrystals: Mechanism and Synthesis. *The Journal of Physical Chemistry B*, *109*(35), 16671–16675. doi:10.1021/jp052484x PMID:16853121

Dirin, D. N., Dreyfuss, S., Bodnarchuk, M. I., Nedelcu, G., Papagiorgis, P., Itskos, G., & Kovalenko, M. V. (2014). Lead Halide Perovskites and Other Metal Halide Complexes As Inorganic Capping Ligands for Colloidal Nanocrystals. *Journal of the American Chemical Society*, *136*(18), 6550–6553. doi:10.1021/ja5006288 PMID:24746226

Emin, S., Singh, S. P., Han, L., Satoh, N., & Islam, A. (2011). Colloidal quantum dot solar cells. *Solar Energy*, *85*(6), 1264–1282. doi:10.1016/j.solener.2011.02.005

Francés-Soriano, L., Pocoví-Martínez, S., González-Béjar, M., & Pérez-Prieto, J. (2014). Reversible phase transfer of quantum dots by gas bubbling. *Green Materials*, *2*(2), 62–68. doi:10.1680/gmat.13.00018

Frederick, M. T., Amin, V. A., & Weiss, E. A. (2013). Optical Properties of Strongly Coupled Quantum Dot–Ligand Systems. *The Journal of Physical Chemistry Letters*, *4*(4), 634–640. doi:10.1021/jz301905n PMID:26281879

Galian, R., Diaz, P., Ribera, A., Rincón-Bertolín, A., Agouram, S., & Pérez-Prieto, J. (2015). Controlled building of CdSe@ZnS/Au and CdSe@ZnS/Au2S/Au nanohybrids. *Nano Research*, *8*(7), 2271–2287. doi:10.1007/s12274-015-0738-6

Galian, R. E., & Guardia, M. (2009). The use of quantum dots in organic chemistry. *TrAC Trends in Analytical Chemistry*, *28*(3), 279–291. doi:10.1016/j.trac.2008.12.001

Galian, R. E., & Perez-Prieto, J. (2010). Catalytic processes activated by light. *Energy & Environmental Science*, *3*(10), 1488–1498. doi:10.1039/c0ee00003e

Galian, R. E., & Scaiano, J. C. (2009). Fluorescence quenching of CdSe quantum dots by tertiary amines and their surface binding effect. *Photochemical & Photobiological Sciences*, *8*(1), 70–74. doi:10.1039/B807580H PMID:19247532

Gao, J., Chen, K., Xie, R., Xie, J., Lee, S., Cheng, Z., Peng, X., & Chen, X. (2010). Ultrasmall Near-infrared Non-cadmium Quantum Dots for in vivo Tumor Imaging. *Small*, 6(2), 256-261. Weinheim an der Bergstrasse, Germany.

Gill, R., Zayats, M., & Willner, I. (2008). Semiconductor Quantum Dots for Bioanalysis. *Angewandte Chemie International Edition*, *47*(40), 7602–7625. doi:10.1002/anie.200800169 PMID:18810756

Gonzalez-Carrero, S., Agudelo-Morales, C., Guardia, M., Galian, R. E., & Perez-Prieto, J. (2015). Three independent channel nanohybrids as fluorescent probes. *RSC Advances*, *5*(109), 90065–90070. doi:10.1039/C5RA18028G

González-Carrero, S., de la Guardia, M., Galian, R. E., & Pérez-Prieto, J. (2014). Pyrene-Capped CdSe@ ZnS Nanoparticles as Sensitive Flexible Oxygen Sensors in Non-Aqueous Media. *ChemistryOpen*, *3*(5), 199–205. doi:10.1002/open.201402021 PMID:25478315

Gonzalez-Carrero, S., Espallargas, G. M., Galian, R. E., & Perez-Prieto, J. (2015). Blue-luminescent organic lead bromide perovskites: Highly dispersible and photostable materials. *Journal of Materials Chemistry A*, *3*(26), 14039–14045. doi:10.1039/C5TA01765C

Gonzalez-Carrero, S., Galian, R. E., & Perez-Prieto, J. (2015). Maximizing the emissive properties of CH3NH3PbBr3 perovskite nanoparticles. *Journal of Materials Chemistry A*, *3*(17), 9187–9193. doi:10.1039/C4TA05878J

González-Carrero, S., Galian, R. E., & Pérez-Prieto, J. (2015). Organometal Halide Perovskites: Bulk Low-Dimension Materials and Nanoparticles. *Particle & Particle Systems Characterization*, *32*(7), 709–720. doi:10.1002/ppsc.201400214

Hammer, N. I., Emrick, T., & Barnes, M. D. (2007). Quantum dots coordinated with conjugated organic ligands: New nanomaterials with novel photophysics. *Nanoscale Research Letters*, *2*(6), 282–290. doi:10.1007/s11671-007-9062-8

Harris, C., & Kamat, P. V. (2009). Photocatalysis with CdSe Nanoparticles in Confined Media: Mapping Charge Transfer Events in the Subpicosecond to Second Timescales. *ACS Nano*, *3*(3), 682–690. doi:10.1021/nn800848y PMID:19226135

Hines, D. A., & Kamat, P. V. (2013). Quantum Dot Surface Chemistry: Ligand Effects and Electron Transfer Reactions. *The Journal of Physical Chemistry C*, *117*(27), 14418–14426. doi:10.1021/jp404031s

Hines, D. A., & Kamat, P. V. (2014). Recent Advances in Quantum Dot Surface Chemistry. *ACS Applied Materials & Interfaces*, *6*(5), 3041–3057. doi:10.1021/am405196u PMID:24506801

Hines, M. A., & Guyot-Sionnest, P. (1996). Synthesis and Characterization of Strongly Luminescing ZnS-Capped CdSe Nanocrystals. *Journal of Physical Chemistry*, *100*(2), 468–471. doi:10.1021/jp9530562

Huang, H., Susha, A. S., Kershaw, S. V., Hung, T. F., & Rogach, A. L. (2015). Control of Emission Color of High Quantum Yield CH3NH3PbBr3 Perovskite Quantum Dots by Precipitation Temperature. *Advanced Science*.

Huang, H.-C., Barua, S., Sharma, G., Dey, S. K., & Rege, K. (2011). Inorganic nanoparticles for cancer imaging and therapy. *Journal of Controlled Release*, *155*(3), 344–357. doi:10.1016/j.jconrel.2011.06.004 PMID:21723891

Huang, X., Lindgren, E., & Chelikowsky, J. R. (2005). Surface passivation method for semiconductor nanostructures. *Physical Review B: Condensed Matter and Materials Physics*, *71*(16), 165328. doi:10.1103/PhysRevB.71.165328

Im, J.-H., Jang, I.-H., Pellet, N., Grätzel, M., & Park, N.-G. (2014). Growth of CH3NH3PbI3 cuboids with controlled size for high-efficiency perovskite solar cells. *Nat Nano*, *9*(11), 927–932. doi:10.1038/nnano.2014.181 PMID:25173829

Im, J.-H., Lee, C.-R., Lee, J.-W., Park, S.-W., & Park, N.-G. (2011). 6.5% efficient perovskite quantum-dot-sensitized solar cell. *Nanoscale*, *3*(10), 4088–4093. doi:10.1039/c1nr10867k PMID:21897986

Jasieniak, J., & Mulvaney, P. (2007). From Cd-Rich to Se-Rich – the Manipulation of CdSe Nanocrystal Surface Stoichiometry. *Journal of the American Chemical Society*, *129*(10), 2841–2848. doi:10.1021/ja066205a PMID:17309253

Ji, X., Copenhaver, D., Sichmeller, C., & Peng, X. (2008). Ligand Bonding and Dynamics on Colloidal Nanocrystals at Room Temperature: The Case of Alkylamines on CdSe Nanocrystals. *Journal of the American Chemical Society*, *130*(17), 5726–5735. doi:10.1021/ja710909f PMID:18396878

Kanaras, A. G., Sönnichsen, C., Liu, H., & Alivisatos, A. P. (2005). Controlled Synthesis of Hyperbranched Inorganic Nanocrystals with Rich Three-Dimensional Structures. *Nano Letters*, *5*(11), 2164–2167. doi:10.1021/nl0518728 PMID:16277446

Kim, H.-S., Lee, C.-R., Im, J.-H., Lee, K.-B., Moehl, T., Marchioro, A., & Park, N.-G. et al. (2012). Lead Iodide Perovskite Sensitized All-Solid-State Submicron Thin Film Mesoscopic Solar Cell with Efficiency Exceeding 9%. *Scientific Reports*, *2*, 591. doi:10.1038/srep00591 PMID:22912919

Kojima, A., Ikegami, M., Teshima, K., & Miyasaka, T. (2012). Highly Luminescent Lead Bromide Perovskite Nanoparticles Synthesized with Porous Alumina Media. *Chemistry Letters*, *41*(4), 397–399. doi:10.1246/cl.2012.397

Kojima, A., Teshima, K., Shirai, Y., & Miyasaka, T. (2009). Organometal Halide Perovskites as Visible-Light Sensitizers for Photovoltaic Cells. *Journal of the American Chemical Society*, *131*(17), 6050–6051. doi:10.1021/ja809598r PMID:19366264

Koleilat, G. I., Levina, L., Shukla, H., Myrskog, S. H., Hinds, S., Pattantyus-Abraham, A. G., & Sargent, E. H. (2008). Efficient, Stable Infrared Photovoltaics Based on Solution-Cast Colloidal Quantum Dots. *ACS Nano*, *2*(5), 833–840. doi:10.1021/nn800093v PMID:19206479

Laferriere, M., Galian, R. E., Maurel, V., & Scaiano, J. C. (2006). Non-linear effects in the quenching of fluorescent quantum dots by nitroxyl free radicals. *Chemical Communications*, (3): 257–259. doi:10.1039/B511515A PMID:16391726

Li, Q., Sun, B., Kinloch, I. A., Zhi, D., Sirringhaus, H., & Windle, A. H. (2006). Enhanced Self-Assembly of Pyridine-Capped CdSe Nanocrystals on Individual Single-Walled Carbon Nanotubes. *Chemistry of Materials*, *18*(1), 164–168. doi:10.1021/cm0512575

Liras, M., Peinado, E., Cañamero, P., Quijada-Garrido, I., & García, O. (2014). Smart photoluminescent nanohybrids based on CdSe quantum dots capped with multidentate thiolated pH-responsive and thermoresponsive polymers for nanosensing. *Journal of Polymer Science. Part A, Polymer Chemistry*, *52*(21), 3087–3095. doi:10.1002/pola.27358

Liu, I. S., Lo, H.-H., Chien, C.-T., Lin, Y.-Y., Chen, C.-W., Chen, Y.-F., & Liou, S.-C. et al. (2008). Enhancing photoluminescence quenching and photoelectric properties of CdSe quantum dots with hole accepting ligands. *Journal of Materials Chemistry*, *18*(6), 675–682. doi:10.1039/b715253a

Liu, L., Guo, X., Li, Y., & Zhong, X. (2010). Bifunctional Multidentate Ligand Modified Highly Stable Water-Soluble Quantum Dots. *Inorganic Chemistry*, *49*(8), 3768–3775. doi:10.1021/ic902469d PMID:20329710

Liu, Y., Brandon, R., Cate, M., Peng, X., Stony, R., & Johnson, M. (2007). Detection of Pathogens Using Luminescent CdSe/ZnS Dendron Nanocrystals and a Porous Membrane Immunofilter. *Analytical Chemistry*, *79*(22), 8796–8802. doi:10.1021/ac0709605 PMID:17939743

Maity, A. R., Palmal, S., Basiruddin, S. K., Karan, N. S., Sarkar, S., Pradhan, N., & Jana, N. R. (2013). Doped semiconductor nanocrystal based fluorescent cellular imaging probes. *Nanoscale*, *5*(12), 5506–5513. doi:10.1039/c3nr00549f PMID:23674276

Malinkiewicz, O., Roldán-Carmona, C., Soriano, A., Bandiello, E., Camacho, L., Nazeeruddin, M. K., & Bolink, H. J. (2014). Metal-Oxide-Free Methylammonium Lead Iodide Perovskite-Based Solar Cells: the Influence of Organic Charge Transport Layers. Advanced Energy Materials, 4(15).

Malinkiewicz, O., Yella, A., Lee, Y. H., Espallargas, G. M., Graetzel, M., Nazeeruddin, M. K., & Bolink, H. J. (2014). Perovskite solar cells employing organic charge-transport layers. *Nature Photonics*, *8*(2), 128–132. doi:10.1038/nphoton.2013.341

Mathews, T. C. S. N. (2014). Advancements in perovskite solar cells: Photophysics behind the photovoltaics. *Energy & Environmental Science*, *7*(8), 2518–2534. doi:10.1039/C4EE00673A

Mattoussi, H., Cumming, A. W., Murray, C. B., Bawendi, M. G., & Ober, R. (1996). Characterization of CdSe nanocrystallite dispersions by small angle x-ray scattering. *The Journal of Chemical Physics*, *105*(22), 9890–9896. doi:10.1063/1.472854

Maurel, V., Laferrière, M., Billone, P., Godin, R., & Scaiano, J. C. (2006). Free Radical Sensor Based on CdSe Quantum Dots with Added 4-Amino-2,2,6,6-Tetramethylpiperidine Oxide Functionality. *The Journal of Physical Chemistry B*, *110*(33), 16353–16358. doi:10.1021/jp061115d PMID:16913763

McDowell, M., Wright, A. E., & Hammer, N. I. (2010). Semiconductor Nanocrystals Hybridized with Functional Ligands: New Composite Materials with Tunable Properties. *Materials (Basel)*, *3*(1), 614–637. doi:10.3390/ma3010614

Medintz, I. L., Clapp, A. R., Mattoussi, H., Goldman, E. R., Fisher, B., & Mauro, J. M. (2003). Self-assembled nanoscale biosensors based on quantum dot FRET donors. *Nature Materials*, *2*(9), 630–638. doi:10.1038/nmat961 PMID:12942071

Mitzi, D. B. (2004). Solution-processed inorganic semiconductors. *Journal of Materials Chemistry*, *14*(15), 2355–2365. doi:10.1039/b403482a

Mitzi, D. B., Chondroudis, K., & Kagan, C. R. (2001). Organic-inorganic electronics. *IBM Journal of Research and Development*, *45*(1), 29–45. doi:10.1147/rd.451.0029

Mitzi, D. B., Feild, C. A., Harrison, W. T. A., & Guloy, A. M. (1994). Conducting tin halides with a layered organic-based perovskite structure. *Nature*, *369*(6480), 467–469. doi:10.1038/369467a0

Murray, C. B., Norris, D. J., & Bawendi, M. G. (1993). Synthesis and characterization of nearly monodisperse CdE (E = sulfur, selenium, tellurium) semiconductor nanocrystallites. *Journal of the American Chemical Society*, *115*(19), 8706–8715. doi:10.1021/ja00072a025

Muthu, C., Nagamma, S. R., & Nair, V. C. (2014). Luminescent hybrid perovskite nanoparticles as a new platform for selective detection of 2,4,6-trinitrophenol. *RSC Advances*, *4*(99), 55908–55911. doi:10.1039/C4RA07884E

Niu, G., Guo, X., & Wang, L. (2015). Review of recent progress in chemical stability of perovskite solar cells. *Journal of Materials Chemistry A*, *3*(17), 8970–8980. doi:10.1039/C4TA04994B

Niu, G., Wang, L., Gao, R., Li, W., Guo, X., Dong, H., & Qiu, Y. (2013). Inorganic halogen ligands in quantum dots: I-, Br-, Cl- and film fabrication through electrophoretic deposition. *Physical Chemistry Chemical Physics*, *15*(45), 19595–19600. doi:10.1039/c3cp52678j PMID:23958930

Niu, Z., & Li, Y. (2014). Removal and Utilization of Capping Agents in Nanocatalysis. *Chemistry of Materials*, *26*(1), 72–83. doi:10.1021/cm4022479

Owen, J. S., Park, J., Trudeau, P.-E., & Alivisatos, A. P. (2008). Reaction Chemistry and Ligand Exchange at Cadmium–Selenide Nanocrystal Surfaces. *Journal of the American Chemical Society*, *130*(37), 12279–12281. doi:10.1021/ja804414f PMID:18722426

Papavassiliou, G. C., Pagona, G., Karousis, N., Mousdis, G. A., Koutselas, I., & Vassilakopoulou, A. (2012). Nanocrystalline/microcrystalline materials based on lead-halide units. *Journal of Materials Chemistry*, *22*(17), 8271–8280. doi:10.1039/c2jm15783g

Papavassiliou, G. C., Pagona, G., Mousdis, G. A., & Karousis, N. (2013). Enhanced phosphorescence from nanocrystalline/microcrystalline materials based on (CH3NH3)(1-naphthylmethyl ammonium)2Pb2Cl7 and similar compounds. *Chemical Physics Letters*, *570*, 80–84. doi:10.1016/j.cplett.2013.03.041

Peng, Z. A., & Peng, X. (2001a). Formation of High-Quality CdTe, CdSe, and CdS Nanocrystals Using CdO as Precursor. *Journal of the American Chemical Society*, *123*(1), 183–184. doi:10.1021/ja003633m PMID:11273619

Peng, Z. A., & Peng, X. (2001b). Mechanisms of the Shape Evolution of CdSe Nanocrystals. *Journal of the American Chemical Society*, *123*(7), 1389–1395. doi:10.1021/ja0027766

Pérez-Prieto, J. (2013). Symbiosis Between Photoactive Nanoparticles and their Organic Ligands. *Photochemistry and Photobiology*, *89*(6), 1291–1298. doi:10.1111/php.12109 PMID:23742191

Pong, B.-K., Trout, B. L., & Lee, J.-Y. (2008). Modified Ligand-Exchange for Efficient Solubilization of CdSe/ZnS Quantum Dots in Water: A Procedure Guided by Computational Studies. *Langmuir*, *24*(10), 5270–5276. doi:10.1021/la703431j PMID:18412382

Prabhakaran, P., Kim, W. J., Lee, K.-S., & Prasad, P. N. (2012). Quantum dots (QDs) for photonic applications. *Optical Materials Express*, *2*(5), 578–593. doi:10.1364/OME.2.000578

Puzder, A., Williamson, A. J., Zaitseva, N., Galli, G., Manna, L., & Alivisatos, A. P. (2004). The Effect of Organic Ligand Binding on the Growth of CdSe Nanoparticles Probed by Ab Initio Calculations. *Nano Letters*, *4*(12), 2361–2365. doi:10.1021/nl0485861

Qu, L., & Peng, X. (2002). Control of Photoluminescence Properties of CdSe Nanocrystals in Growth. *Journal of the American Chemical Society*, *124*(9), 2049–2055. doi:10.1021/ja017002j PMID:11866620

Radhakrishnan, C., Lo, M. K. F., Knobler, C. M., Garcia-Garibay, M. A., & Monbouquette, H. G. (2011). Capping-Ligand Effect on the Stability of CdSe Quantum Dot Langmuir Monolayers. *Langmuir*, *27*(6), 2099–2103. doi:10.1021/la104244x PMID:21329375

Rhee, J. H., Chung, C.-C., & Diau, E. W.-G. (2013). A perspective of mesoscopic solar cells based on metal chalcogenide quantum dots and organometal-halide perovskites. *NPG Asia Mater*, *5*(10), e68. doi:10.1038/am.2013.53

Rogach, A. (Ed.). (2008). *Semiconductor Nanocrystal Quantum Dots* (1st ed.). Vienna: Springer-Verlag Wien. doi:10.1007/978-3-211-75237-1

Rogach, A. L., Kornowski, A., Gao, M., Eychmüller, A., & Weller, H. (1999). Synthesis and Characterization of a Size Series of Extremely Small Thiol-Stabilized CdSe Nanocrystals. *The Journal of Physical Chemistry B*, *103*(16), 3065–3069. doi:10.1021/jp984833b

Roldan-Carmona, C., Malinkiewicz, O., Soriano, A., Minguez Espallargas, G., Garcia, A., Reinecke, P., & Bolink, H. J. et al. (2014). Flexible high efficiency perovskite solar cells. *Energy & Environmental Science*, *7*(3), 994–997. doi:10.1039/c3ee43619e

Rosenthal, S. J., Chang, J. C., Kovtun, O., McBride, J. R., & Tomlinson, I. D. (2011). Biocompatible Quantum Dots for Biological Applications. *Chemistry & Biology*, *18*(1), 10–24. doi:10.1016/j.chembiol.2010.11.013 PMID:21276935

Scaife, D. E., Weller, P. F., & Fisher, W. G. (1974). Crystal preparation and properties of cesium tin(II) trihalides. *Journal of Solid State Chemistry*, *9*(3), 308–314. doi:10.1016/0022-4596(74)90088-7

Schmidt, L. C., Pertegás, A., González-Carrero, S., Malinkiewicz, O., Agouram, S., Mínguez Espallargas, G., & Pérez-Prieto, J. et al. (2014). Nontemplate Synthesis of CH3NH3PbBr3 Perovskite Nanoparticles. *Journal of the American Chemical Society*, *136*(3), 850–853. doi:10.1021/ja4109209 PMID:24387158

Shang, Z. B., Hu, S., Wang, Y., & Jin, W. J. (2011). Interaction of β-cyclodextrin-capped CdSe quantum dots with inorganic anions and cations. *Luminescence*, *26*(6), 585–591. doi:10.1002/bio.1274 PMID:21268229

Sichert, J. A., Tong, Y., Mutz, N., Vollmer, M., Fischer, S., Milowska, K. Z., & Feldmann, J. et al. (2015). Quantum Size Effect in Organometal Halide Perovskite Nanoplatelets. *Nano Letters*, *15*(10), 6521–6527. doi:10.1021/acs.nanolett.5b02985 PMID:26327242

Sperling, R.A., & Parak, W.J. (2010). Surface modification, functionalization and bioconjugation of colloidal inorganic nanoparticles. *Philosophical Transactions of the Royal Society of London A: Mathematical, Physical and Engineering Sciences*, 368(1915), 1333-1383.

Sun, X.W., Chen, J., Song, J.L., Zhao, D.W., Deng, W.Q., & Lei, W. (2010). Ligand capping effect for dye solar cells with a CdSe quantum dot sensitized ZnO nanorod photoanode. *Optics Express*, *18*(2), 1296–1301. doi:10.1364/OE.18.001296 PMID:20173955

Talapin, D. V., Rogach, A. L., Kornowski, A., Haase, M., & Weller, H. (2001). Highly Luminescent Monodisperse CdSe and CdSe/ZnS Nanocrystals Synthesized in a Hexadecylamine–Trioctylphosphine Oxide–Trioctylphospine Mixture. *Nano Letters*, *1*(4), 207–211. doi:10.1021/nl0155126

Tanaka, K., Takahashi, T., Ban, T., Kondo, T., Uchida, K., & Miura, N. (2003). Comparative study on the excitons in lead-halide-based perovskite-type crystals CH3NH3PbBr3 CH3NH3PbI3. *Solid State Communications*, *127*(9–10), 619–623. doi:10.1016/S0038-1098(03)00566-0

Tang, J., Kemp, K. W., Hoogland, S., Jeong, K. S., Liu, H., Levina, L., & Sargent, E. H. et al. (2011). Colloidal-quantum-dot photovoltaics using atomic-ligand passivation. *Nature Materials*, *10*(10), 765–771. doi:10.1038/nmat3118 PMID:21927006

Tomasulo, M., Yildiz, I., Kaanumalle, S. L., & Raymo, F. M. (2006). pH-Sensitive Ligand for Luminescent Quantum Dots. *Langmuir*, *22*(24), 10284–10290. doi:10.1021/la0618014 PMID:17107034

Valizadeh, A., Mikaeili, H., Samiei, M., Farkhani, S., Zarghami, N., kouhi, M., & Davaran, S. et al. (2012). Quantum dots: Synthesis, bioapplications, and toxicity. *Nanoscale Research Letters*, *7*(1), 1–14. doi:10.1186/1556-276X-7-480 PMID:22929008

van Embden, J., Jasieniak, J., Gómez, D. E., Mulvaney, P., & Giersig, M. (2007). Review of the Synthetic Chemistry Involved in the Production of Core/Shell Semiconductor Nanocrystals. *Australian Journal of Chemistry*, *60*(7), 457–471. doi:10.1071/CH07046

Wadhavane, P. D., Galian, R. E., Izquierdo, M. A., Aguilera-Sigalat, J., Galindo, F., Schmidt, L., & Luis, S. V. et al. (2012). Photoluminescence Enhancement of CdSe Quantum Dots: A Case of Organogel–Nanoparticle Symbiosis. *Journal of the American Chemical Society*, *134*(50), 20554–20563. doi:10.1021/ja310508r PMID:23214451

Wang, F., Tang, R., & Buhro, W. E. (2008). The Trouble with TOPO; Identification of Adventitious Impurities Beneficial to the Growth of Cadmium Selenide Quantum Dots, Rods, and Wires. *Nano Letters*, *8*(10), 3521–3524. doi:10.1021/nl801692g PMID:18754691

Wang, P., Zhang, J., He, H., Xu, X., & Jin, Y. (2015). The important role of surface ligand on CdSe/CdS core/shell nanocrystals in affecting the efficiency of H2 photogeneration from water. *Nanoscale*, *7*(13), 5767–5775. PMID:25757912

Warrier, M., Lo, M. K. F., Monbouquette, H., & Garcia-Garibay, M. A. (2004). Photocatalytic reduction of aromatic azides to amines using CdS and CdSe nanoparticles. *Photochemical & Photobiological Sciences*, *3*(9), 859–863. doi:10.1039/b404268a PMID:15346187

Watson, D. F. (2010). Linker-Assisted Assembly and Interfacial Electron-Transfer Reactivity of Quantum Dot–Substrate Architectures. *The Journal of Physical Chemistry Letters*, *1*(15), 2299–2309. doi:10.1021/jz100571u

Wei, C., Li, J., Gao, F., Guo, S., Zhou, Y., & Zhao, D. (2015). One-Step Synthesis of High-Quality Water-Soluble CdSe Quantum Dots Capped by N-Acetyl-L-cysteine via Hydrothermal Method and Their Characterization. *Journal of Spectroscopy*, *2015*, 7. doi:10.1155/2015/369145

Wolfbeis, O. S. (2015). An overview of nanoparticles commonly used in fluorescent bioimaging. *Chemical Society Reviews*, *44*(14), 4743–4768. doi:10.1039/C4CS00392F PMID:25620543

Wuister, S. F., de Mello Donegá, C., & Meijerink, A. (2004). Influence of Thiol Capping on the Exciton Luminescence and Decay Kinetics of CdTe and CdSe Quantum Dots. *The Journal of Physical Chemistry B*, *108*(45), 17393–17397. doi:10.1021/jp047078c

Zhang, F., Zhong, H., Chen, C., Wu, X.-g, Hu, X., Huang, H., & Dong, Y. et al. (2015). Brightly Luminescent and Color-Tunable Colloidal CH3NH3PbX3 (X = Br, I, Cl) Quantum Dots: Potential Alternatives for Display Technology. *ACS Nano*, *9*(4), 4533–4542. doi:10.1021/acsnano.5b01154 PMID:25824283

Zhang, H., Jang, J., Liu, W., & Talapin, D. V. (2014). Colloidal Nanocrystals with Inorganic Halide, Pseudohalide, and Halometallate Ligands. *ACS Nano*, *8*(7), 7359–7369. doi:10.1021/nn502470v PMID:24988140

Zhang, J.-Y., Wang, X.-Y., Xiao, M., Qu, L., & Peng, X. (2002). Lattice contraction in free-standing CdSe nanocrystals. *Applied Physics Letters*, *81*(11), 2076–2078. doi:10.1063/1.1507613

Zhang, Y., & Clapp, A. (2011). Overview of Stabilizing Ligands for Biocompatible Quantum Dot Nanocrystals. *Sensors (Basel, Switzerland)*, *11*(12), 11036–11055. doi:10.3390/s111211036 PMID:22247651

Zheng, Y., Gao, S., & Ying, J. Y. (2007). Synthesis and Cell-Imaging Applications of Glutathione-Capped CdTe Quantum Dots. *Advanced Materials*, *19*(3), 376–380. doi:10.1002/adma.200600342

Zhu, H., Hu, M. Z., Shao, L., Yu, K., Dabestani, R., Zaman, M. B., & Liao, S. (2014). Synthesis and Optical Properties of Thiol Functionalized CdSe/ZnS (Core/Shell) Quantum Dots by Ligand Exchange. *Journal of Nanomaterials*, *2014*, 14.

Chapter 4
Functional Polymeric Coatings: Synthesis, Properties, and Applications

Peter Zarras
NAWCWD, USA

Paul A Goodman
NAWCWD, USA

John D Stenger-Smith
NAWCWD, USA

ABSTRACT

Functional polymeric coatings comprise a diverse array of organic/inorganic functional groups, properties, and applications. This chapter attempts to provide a comprehensive review on the science and engineering of functional polymer coatings for both the novice and experienced chemist/chemical engineer. Functional polymer coatings are coatings that are designed through careful selection of organic/inorganic functional groups and controlled synthesis on the substrate's surface. The design leads to changes in the surface properties (e.g., adhesion, wettability, corrosion resistance, wear resistance, anti-fouling, non-stick, and anti-scratch) of a particular substrate (metal, glass, wood, or plastic). Functional polymeric coatings represent the "next generation" of coatings. These coatings dynamically respond to their environment, as opposed to a static coating that is used as a simple barrier, or for decorative purposes. This review is divided into three sections: (1) functionalization of polymers, (2) polymer properties and surface analytical techniques, and (3) applications.

INTRODUCTION

Coatings have been around since the dawn of civilization, and they have helped shape man's development and history for millennia. Primitive man used natural materials to paint or mark areas that he inhabited. The ancient Egyptians, Greeks, and Romans used various natural materials to coat, decorate, or impregnate various fabrics, tombs, and vessels. Since World War II, advances in the variety and availability of chemicals and improved methods of applying such materials has led to the development of

DOI: 10.4018/978-1-5225-0066-7.ch004

functional polymer coatings. In today's modern world, coatings are ubiquitous. In our daily lives, many objects, devices, and structures are coated with films or paints. Objects, structures, and devices such as pots, pans, tools, houses, bridges, automobiles, aircraft, televisions, mobile telephones, and computers are "coated materials."

These organic-polymeric functional coatings have been developed over the past several decades not only to impart decorative and/or protective properties, but also to add functionality to the coating. This functionalization can be applied to numerous surfaces with profound effects on a surface's morphology, properties, and eventual application. For example, by imparting functionality to a surface, researchers can selectively tune the properties of the film to meet a specific application or to allow for the investigation and understanding of physiological, biological, or biomimetic phenomena.

In contrast to functional coatings, static coatings, which include decorative and barrier coatings, do not possess the properties necessary to "sense and respond" to their environment, as shown in Figure 1. The coatings found in the aerospace industry are normally multi-layered coatings consisting of a pretreatment, a primer, and a topcoat layer. In the automotive industry, such coatings can be four to six different layers. Each layer provides a specific function to the overall performance of the coating. Adhesion between layers, barrier properties to corrosion and environmental stimuli, and decorative properties are typically observed for static, barrier, or decorative coatings.

A fundamental shift has occurred over the past few decades, and current coatings research focuses on tailoring a coating's ability to react to external stimuli. The term functional coatings now describes coating systems that possess not only the properties of a static coating, barrier and decorative, but the unique property of functionality. Functionality imparts specific properties to a coating that normally depends on the specific end-use or application desired. A comprehensive review of functional polymer coatings is provided by Wu and Baghdachi (2015). Several functional, or smart coatings, have been commercialized such as the anti-fogging polyester PET coating known under the trade name Visgard® (Munikata, Yamauchi, Hirono, T., Kodama, S.-I., & Matsuo, M, 2001; FSI Coating Technologies, 2015).

Brady (2007); Baghdachi, Perez, and Shah (2010); Kantheti, Narayan, and Raju (2015); Anastasiadis (2013); Cohen Stuart, et al. (2010), and Koberstein (2004) have published excellent reviews of "smart coatings." These reviews have provided guidelines on the important properties necessary to impart functionalization on a surface. These properties include (1) surface structure, (2) surface segregation, and (3) surface reorganization. In order to meet current coating industry requirements, researchers are focused on controlling the chemical surface composition on a molecular scale while controlling the morphology on the micrometer and nanometer scales.

Most reviews have focused on either the functionalization methods necessary to impart specific surface properties on a substrate and/or the applications associated with a specific surface treatment. This chapter will examine polymer functionalization of surfaces, synthetic methods, and techniques used to characterize polymer surfaces and potential applications. This comprehensive review will allow the

Figure 1. Schematic diagram of static coating

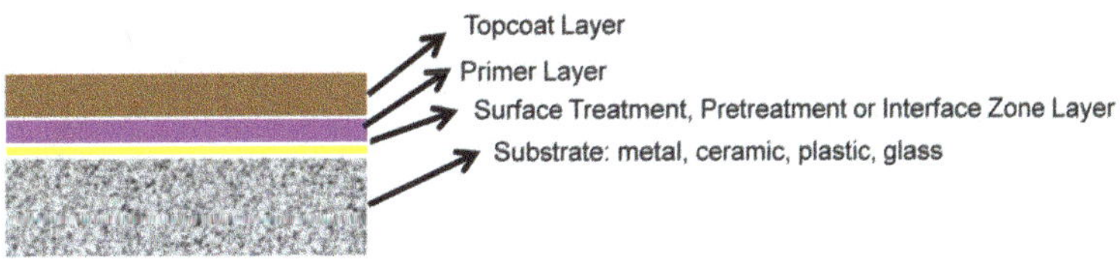

reader to have a deeper understanding of this subject matter. In addition, it is hoped that the reader will be informed as to the complexity of surface functionalization, the enormous economic benefits that these coatings offer, and the commercial products derived from them that benefit the consumer.

BACKGROUND

The term "functional polymer coating" is defined as a polymer coating containing specifically selected functional groups on the surface of a substrate that can result in a particular function, or desired property, for the modified surface. Functional coatings can perform a specific function by either physical, mechanical, thermal, or chemical properties, or a combination thereof. As an example, chemically active functional coatings exhibit unique properties either at the coating-substrate interface, in the bulk of the coating, or at the air-coating interface. At the coating-substrate interface, one could use this type of functionality for anti-corrosion or anti-fouling applications. Potential applications for the bulk of the coating could be for fire-resistant coatings. At the air-coating interface, antibacterial or self-cleaning applications could be envisioned for this latter system. Important features for functional polymer coatings include (1) durability, (2) tailored surface morphology for desired application(s), (3) reproducibility, (4) easy application, and (5) cost effectiveness. These coatings should also be environmentally green and from sustainable materials. Several types of functional coatings are shown in Figure 2 with their potential applications. The inclusion of added functionality into a coating's properties can alter and improve the application and performance of a coating. The diversity of functionality incorporated into a coating can impart new features and change the application required for a particular coating. Examples of functional coatings include anti-static, anti-corrosion, anti-fouling, anti-bacterial, self-cleaning, anti-scratch, and anti-abrasive. The diversity of functional groups that can be added to a coating's properties is quite broad, and this review article will provide the reader with a comprehensive understanding of the topic.

In order to fully appreciate the diversity of functionality associated with polymer coatings, one must begin with several fundamental concepts. Koberstein (2004) suggests that three properties are necessary for a functional polymer coating: (1) surface structure, (2) surface segregation, and (3) surface

Figure 2. Several types of functional coatings/potential applications

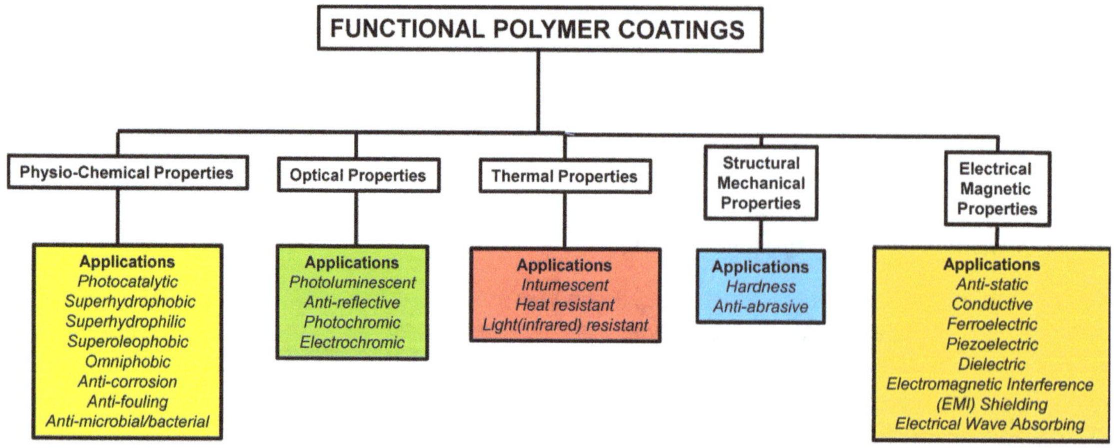

reorganization. These surface properties are described in further detail below. Briefly, surface structure concerns the types and amount of functional groups present on a surface; surface segregation relates to the structure of the coating on the substrate before the application of an external stimulus; and surface reorganization describes the change in surface structure upon application of the external stimulus.

Surface structure, the choice of the functional groups in the polymer, influences the surface segregation of a coating because of the inherent differences in surface energies between the polymer backbone and the functional groups that are attached to it. For example, tuning the difference in surface energies between the functional group and backbone can result in a non-stick coating that adheres strongly to the substrate, but not to substances that are applied to the coating's surface.

In the example given above, a non-stick coating's structure generally does not change when a substance is applied. In other examples, a structural reorganization upon stimulus application is desired. For instance, functional groups that may be present on the coating's surface after deposition may reorient themselves when exposed to different conditions such as a change in pH or temperature (O'Rourke-Muisener, Koberstein, & Kumar, 2003 and Koberstein et al., 1997). These reorganization phenomena can produce reversible functional polymer coatings that specifically react and respond to different environmental stimuli depending on the desired application.

The design of functional polymer surfaces on substrates can be accomplished by a variety of methods. The target polymers are often inspired by nature, and biomimetics is a common theme in this area of research. Guo, Liu, and Su (2011) have reviewed various methods available to produce functionalized polymer surfaces. One of the most famous nature inspired functional polymer surfaces is the "lotus leaf" with its self- cleaning properties. The lotus leaf is a water-repellent surface covered by papillae that are approximately 5 to 10 microns high and 10 to 15 microns wide with a nanometer scale covering of epicuticular waxes, which result in its superhydrophobic properties (Li, Zhang, Xu, & Wong, 2006). Additional functional polymer surfaces such as superoleophobicity and superhydrophilicity are also based on nature-inspired materials such as rice leaves, rose petals, cicada wings, butterfly wings, snail shells, and fish scales (Nishimoto & Bhushan, 2013).

The synthetic strategies that have been investigated to produce functional polymer surfaces include (1) wet chemical reactions, (2) hydrothermal reactions, (3) electrochemical syntheses, (4) electropolymerization, (5) self-assembly, (6). layer-by-layer deposition, (7) plasma etching, (8) chemical vapor deposition, (9) sol-gel reactions, (10) polymerization/template, and (11) electrospray. In each of these methods, the organic reactions leading to stable surface functionalization must maintain these key elements: (1) stable linkages on the surface, (2) minimization of unwanted crosslinking reactions between reactive functional groups, (3) high reaction yield, (4) no side or degradation products, and (5) mild reaction conditions that minimize side product formation or inadvertent substrate degradation.

SURFACE FUNCTIONALIZATION WITH POLYMERS

In this section, the authors will review synthetic approaches with the purpose of describing in detail the various methods used to produce functional polymer surfaces on various substrates, including specific examples. Functionalization of a surface is necessary to impart properties that can be tailored to a specific application.

The wet chemical reaction method is a straightforward and facile process to make rough surfaces. This method is used to control several features of the substrate, including size, shape, and microscale and nanoscale dimensionality (Liu, Szunerits, Pisarek, Xu, & Boukherroub, 2009). Metal surfaces such as copper, zinc, steel, and aluminum have been modified via chemical etching using acidic (Wang, Feng, & Jiang, 2006) or basic (Qu, Zhang, Song, Chen, Zhang, & Junyan, 2007) solutions to generate superhydrophobic surfaces directly onto the metal substrates. As an example, nanoporous aluminum oxide surfaces containing a hexagonal arrangement of monodisperse nanopores have been fabricated via both mild and hard anodization processes (Lee, Ji, Gosele, & Nielsch, 2006). These processes involve exposure of aluminum alloys to various acid baths such as sulfuric, oxalic, and phosphoric acids under varying potentials. Numerous metals can undergo these procedures producing microstructured and nanostructured substrates. Chemical modification with acidic solutions followed by reaction with poly(dimethylsiloxane) vinyl terminated materials generated stable superhydrophobic surfaces on copper substrates (Guo, Fang, Wang, & Liu, 2007). Hu, Graaf, & Moeller (2013) developed new ultraviolet (UV)-crosslinkable di-block copolymer coatings for microelectrode array sensors that were protected with pinacol groups. The protection scheme that was used for this di-block copolymer allowed for an effective and stable surface coating, and these arrays could be site-selectively deprotected, transesterfied, and functionalized via Cu(II)-catalyzed reactions, enabling binding to proteins at low concentrations.

Hydrothermal reactions are used to produce functionalized materials with different patterns and morphologies via a "bottom-up" approach. The hydrothermal approach involves controlled heating under pressure. A recent report by Cai et al. (2013) used the hydrothermal method and surface functionalization of hydrothermally produced iron oxide nanoparticles coated with branched polyethyleneimine for biomedical applications. The specific application for this system was magnetic resonance imaging therapy. Both the wet chemical reaction and hydrothermal methods are simple, easy to use processes that are scalable and cost-effective. They can be applied to various materials, including metals or plastics of varying shapes and sizes.

The electrochemical deposition method is an old technique that has recently been used to produce structures at the microscale and nanoscale level on various surfaces. A recent report by Benea (2015) described the electro-co-deposited dispersed phases of ZrO_2 and CeO_2 and ultra-high molecular weight polyethylene during cobalt or nickel electroplating. These coatings produced innovative functional surfaces with corrosion and tribocorrosion inhibitive properties. Further work by Krukiewicz et al. (2014) investigated functionalized multi-walled carbon nanotubes coated with polypyrrole via electrochemical-co-deposition techniques. Their studies showed that the surface morphology of the multi-walled carbon nanotubes influenced the orientation of the growing polypyrrole chains.

Electropolymerization of conductive polymers has been investigated for several decades. This technique is the most common method for polymerizing aniline, pyrrole and thiophene monomers, and their derivatives. A review of the electropolymerization of conductive polymers is described by Audebert and Miomandre (2007). These monomers have been electropolymerized on a variety of ferrous and nonferrous metal surfaces with controlled thickness and morphology. Several conductive polymers such as polypyrrole (Ashrafi, Golozar, & Mallakpour, 2003), poly(N-methylpyrrole) and polyaniline (Branzoi V, Golgovici, & Branzoi F., 2008) have been electrochemically polymerized on either steel or aluminum substrates. However, polyaniline is the only commercialized conductive polymer used as a corrosion-inhibiting coating. A review on the electropolymerization methods used to form conductive polymer films and their corrosion-inhibiting properties is described by Zarras and Stenger-Smith (2014). Electropolymerization is a facile, cost effective method that has been employed for nearly three decades to coat various metal/alloy surfaces with conductive polymers. This technique is widely used by academia and industry to coat various substrates and can be employed as a non-line-of-sight method for coating complex geometries.

The self-assembly technique often referred to as self-assembled monolayers (SAMs) have been investigated over several decades and represent an extremely flexible and versatile method for tailoring surface properties on a variety of substrates, including glass, metals/alloys, metal oxides, nanoparticles, and semiconductors (Love, Estroff, Kriebel, Nuzzo & Whitesides, 2005). SAMs are comprised of organic molecules that form organized assemblies. These assemblies form as a result of selective adsorption of specific groups that consist of a functional head group. The head group has a specific affinity to a certain substrate and the main chain, or tail, which is organized by Van der Waals interaction. Typically, a terminal group preset on the surface of the tail end can undergo further reaction or additional surface chemistry (Zhou & Huck, 2006). However, SAMs have several limitations, namely, it is impossible to obtain large area monolayer surfaces that are free from defects. Since these SAMs layers are nanometers in thickness, they are very fragile and not mechanically robust. This represents a severe limitation and disadvantage to using SAMs to induce polymer functionalization on various surfaces. While highly effective at producing functional polymer surfaces, this technique does not offer any commercialization prospects at this time.

Polymer brushes are a versatile and more robust approach to developing SAMs; these brushes represent an alternative methodology to producing functional polymer surfaces onto various substrates. This technique is more versatile and cost effective and provides numerous advantages over SAMs technique. Various functional polymer brushes include homopolymers, block copolymers, mixed, and surface-initiated polymer brushes as shown schematically in Figure 3 (Luzinov, Minko, & Tsukruk, 2008). These polymer brushes that are deposited on surfaces can be grown from initiator-terminated SAMs in which monolayers of macromolecules can be achieved with variable packing densities (Maidenberg, Zhang, Luo, Akhavein, & Koberstein, 2013) (Zengin, & Caykara, 2012).

Figure 3. Schematic representation of polymer brushes

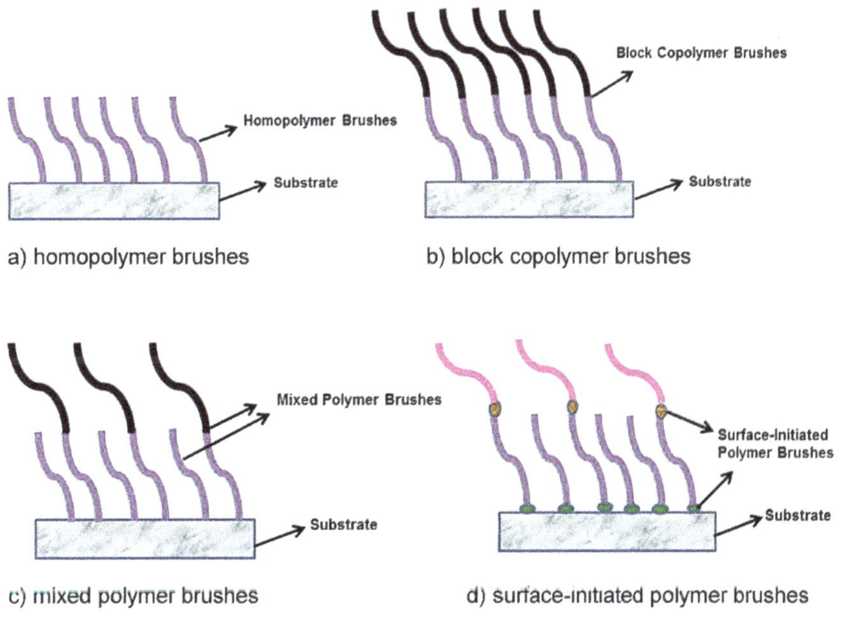

An additional method of film formation is the layer-by-layer (LbL) approaches such as Langmuir-Blodgett (LB) and alternating polyelectrolyte deposition (APD) techniques that can be used to fabricate multilayer thin films with varying surface chemistries.

Langmuir-Blodgett films are composed of one or more monolayers of an organic compound deposited on the surface of a liquid. Typically, one uses amphiphilic molecules such as a surfactant that is deposited onto a water surface. The surfactant molecules contain both a hydrophobic tail and hydrophilic head that arrange themselves at the water-air interface. This arrangement is due to surface energy features of the molecules interacting with the water-air interface. The tails are hydrophobic, and their exposure to air is favored over water; whereas, the hydrophilic heads are favored to the water over air exposure (Figure 4). Monolayers are formed with precise control of the thickness of the resulting films (Langmuir & Blodgett, 1935).

Yang, Dou, Fang, & Jiang (2013) prepared superhydrophobic coatings with hierarchical microscale and nanostructures. These coatings were inspired by the binary periodic structures found on the superhydrophobic compound eyes of mosquitoes and moths. LB films consisted of surface modification with fluorosilanes and a "bottom-up" approach of alternating assembles of silica sphere-produced films with water contact angles greater than 150 degrees. Superhydrophobic-superoleophobic polythiophene LB-electropolymerization processed films were prepared by Pernites, Ponnapati, and Advincula (2011). These films showed potential-induced reversible wetting from >150 to <60 degrees. This superhydrophobic-superoleophobic reversibility was accomplished over several cycles. Electrochromic switching from orange to dark green over several cycles was also observed with these polythiophene films. Al-Taher, Al-Telegdi, and Kalman (2007) prepared LB films of hydroxamic amphiphiles that were used as coating barriers for the corrosion-inhibition of copper surfaces. These multi-molecular LB films were tested in both acidic and basic media and showed effective corrosion inhibition with increasing numbers of monolayers (Telegdi, Rigo, & Kalman, 2005). Igari, Oyamada, Nanao, and Mori (2000) studied the microtribological effects of partially fluorinated carboxylic acid LB films under low loads and found that the frictional coefficients were inversely related to the elastic modulus.

The APD technique can produce thin films via alternating LbL. The process involves the deposition of alternating polyelectrolytes (polyanion followed by polycation or vice versa) with a rinse step between each alternating polyelectrolyte layer (Figure 5) (Shchukin & Moehwald, 2012). The multi-layer buildup

Figure 4. Schematic representation of LB film

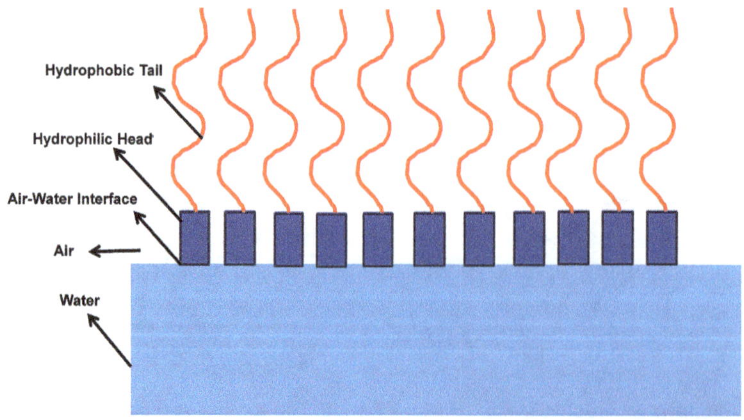

is accomplished via multiple attractive forces working in tandem. This allows for alternative systems to be employed, such as hydrogen bonding (Lee & Hong, 2005) and nanoparticles (Cassagneau, 2004). The LbL technique of APD offers several unique advantages, including inexpensive process, facile methodology, high degree of control over thickness of fabricated films, and numerous coating methods such as spin-, spray-, dip-, and flow-based techniques (Rubner, 2003). Control of the thickness of the bilayers is a unique feature of this method. There is linear growth of these films with successive bilayers, and each bilayer can be as thin as 1 nanometer.

Peniredd et al. (2015) developed polymer films by LbL fabrication of alternating polyelectrolytes. The polyelectrolytes were comprised of histidine grafted poly(allyamine) as the polycation to bind to metal ions (Cu^{+2}) and methyl ester containing polyanions for crosslinking. The method of fabrication was via both nonimprinted LbL and imprinted LbL systems. Crosslinking was accomplished at mild temperatures for improved stability in sea water, and these coatings showed anti-fouling activity to settlement of the *barnacle larvae*. Biocidal films based on APD of poly(2,2,6,6-tetramethyl-4-piperidyl methacrylate-co-acrylic acid potassium salt) and poly(2,2,6,6-tetramethyl-4-piperidylmethacrylate-co-trimethyl-2-methacryloxyethylammonium chloride) showed effective treatment against *Staphylococcus aureus* and *Escherichia coli* within 15 minutes of contact time (Cerkez, Kocer, Worley, Broughton, & Huang, 2011). Zhou, Li, Yu and Li (2007) developed highly reflective optical coatings via an LbL process employing polyelectrolyte/colloidal TiO_2 and polyelectrolyte/colloidal SiO_2 alternating composite films.

Plasma etching is a type of plasma processing that can be used to fabricate engineered materials for polymers, metals, or elastomers (Satyaprassad, Jain, & Nema, 2007). Plasma etching is based on the formation of plasma that is directed in pulses at a sample. Plasma etching can change the surface properties (e.g., wettability) of a substrate from hydrophilic to hydrophobic or vice versa (Zia, Wang, & Lee, 2015). Liu, Niu, and Sung (2014) modified moth eye-like surfaces using argon-oxygen plasma treatment on a UV-nano-imprinted poly(ethylene terephthalate) substrate. Coating this substrate with perfluorodecyltrichlorosilane solution demonstrated superhydophobic properties. Kim, Shin, Shin, Kim, and Cho (2011) demonstrated control on the degree of superhydrophilicity and superhydrophobicity on silicon nanowires via electroless etching and subsequent fluorine carbon coating.

Chemical vapor deposition (CVD) is a chemical process in which a substrate is exposed to one or more volatile precursors that either react or decompose on the substrate to produce the desired product or effect. Three methods were employed: (1) atmospheric pressure CVD (APCVD), (2) low-pressure CVD

Figure 5. Schematic representation of APD process

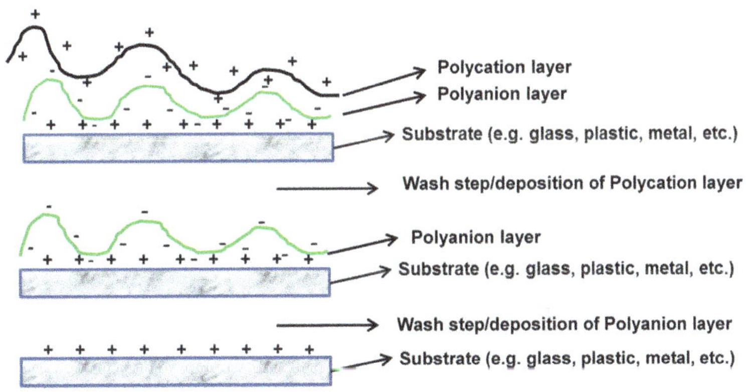

(LPCVD), and (3) ultrahigh vacuum CVD (UHVCVD) (Nuernberg, Stolle, Wahl, & Raic, 2003). Rezaei, Manoucheri, Moradian, and Pourabbas (2014) developed tetraethylorthosilicate (TEOS) and vinyltrimethoxysilane (VTMS) surface modified superhydrophobic silica nanoparticles via a facile, one-step CVD process. Bao et al. (2014) coated ZnO and other metal oxide nanoparticles with polydimethylsiloxane (PDMS) via CVD on various substrates to produce a superhydrophobic surface.

Sol-gel coatings are fabricated via two distinct methods (1) hydrolytic process in aqueous media, and (2) nonhydrolytic process in organic media (Livage, 2000). The sol-gel process allows control over the surface properties on various substrates (e.g. metals, glass, silicon wafer, polymers, and textiles) through adjustments of the reactive species, organic functionalization, temperature, time, and pH (Li, Duan, & Cai, 2007). The sol-gel technique is a highly versatile, facile, and inexpensive method that allows one to control and manipulate composition and microstructure of a coating at the molecular level under ambient conditions. The sol-gel technique provides a method to coat various surfaces with functionalized polymer hybrids. This system has been commercialized by several industrial paint companies and is currently used in the commercial aerospace industry as a pretreatment coating for improved adhesion between metal-pretreatment and primer-pretreatment layers. Vasiljevic et al. (2014) developed a two-component inorganic-organic hybrid sol-gel coating that exhibited multifunctional properties including simultaneous superhydrophobicity/oleophobicity and passive antibacterial activity. Hybrid nanostructured sol-gel coatings have also been investigated to inhibit biocorrosion on stainless steels. These functional coatings consisted of an inner compact layer of titanium oxide and an outer dense layer of poly(vinyl-N-hexylpyridinium) brushes (Yuan, Pehkonen, Ting, Neoh, & Kang, 2009).

The polymerization reaction is a versatile and useful method for polymerizing monomers and macromers on the surface of a substrate. This method was addressed earlier in this section; however more specific examples are provided. Grafting, electropolymerization, Michael addition polymerization, and click chemistry are widely used to functionalize various surfaces. Surface grafting was successfully accomplished on fibrous silicate through immobilization of reactive end functional groups of silane coupling agents. Chain growth via radical polymerization of N-vinylpyrrolidone (NVP) and vinyltrimethoxysilane (DB171) was conducted, and Fourier transform infrared spectroscopy (FTIR), X-ray photoelectron spectroscopy (XPS), and scanning electron microscopy (SEM) were used to quantify the degree of graft loading (Yuan, Li, Guan, Xiao, & Zhang, 2009). Wang, Zheng, Hong, Liu, and Pan (2006) successfully synthesized novel hyperbranched poly(amido amines) containing tertiary amines on the backbone and acryl or secondary amines as the surface groups via Michael addition polymerizations. Polymerizations via a template method are also quite useful and efficient routes to modify surface chemistries. A template-directed sol-gel process was used by Wang et al. (2009) to fabricate a wide range of metal oxide/polymer nanowires. These polymer modified nanostructures showed conductive, magnetic, and photocatalytic properties.

Finally, electrospray, sometimes referred to as electrospray deposition (ESD), is a versatile method for forming thin films on various substrates. There are several advantages to using this fabrication process such as (1) multiple solvents or polymer systems can be used, and (2) polymer thin films with controlled nano- and micro-scaled structures are deposited (Tanioka, 2010). In the electrospray process, liquid flow is injected into a nozzle as an electric field is applied between the nozzle tip and the ground plate. Microscale monodisperse fine droplets are generated due to repulsive forces (Wilhelm, Madler, & Pratsinis, 2003). Hu, Liu, Li, Yang, Fan, and Cui (2013) developed a superhydrophobic biodegradable coating using a one-step electrospray of poly(L-Lactide) for potential anti-adhesion of bacteria on medical devices. Butt et al. (2014) fabricated devices consisting of electrosprayed liquid droplets deposited onto superhydrophobic and/or superoleophobic surfaces.

Each of the above mentioned techniques has been employed over the past several decades to functionalize various surfaces for specific needs and applications. The next section will focus on synthesis methods that are currently the state-of-the-art in functionalization of surfaces using polymers. These methods have broad acceptance and are being used by academia and industry; these synthesis methods have also been commercialized. By employing the proper functionalization, various properties can be tailored for those surfaces, resulting in unique applications.

SYNTHESIS AND PROPERTIES OF FUNCTIONAL POLYMER COATINGS

In this chapter, the authors have reviewed the various methods employed to form functional polymer coatings on a variety of substrates. The synthetic methods used to obtain a functional polymer and several unique properties of these materials will be discussed in this section.

Functionalization of polymers deposited on surfaces plays a critical role in numerous engineering applications such as adhesion, lubrication, and wettability control. In order to develop these innovative polymer coatings, synthetic strategies that can maximize surface coverage and functionality must be investigated and used. These polymers must not only change the surface properties that they reside on but also respond and react to external stimuli present in the environment (Zhang, Zeng, & Liu, 2012). Two methods of polymerizing and controlling the molecular weight of polymer systems directly on a surface substrate will be described. Both of these surface deposition methods are considered state-of-the-art techniques and have been investigated by both academia and industry for commercialization.

The first method that will be discussed in this section is atom transfer radical polymerization (ATRP), which is a state-of-the-art method. The ATRP is an example of a reversible-deactivation radical polymerization in which carbon-carbon bonds are formed through a transition metal catalyst producing polymers with narrow molecular weights. Typically, ATRP requires monomers with substituents that can stabilize the propagating radical (e.g., styrenes, methacrylates), initiators such as alkyl halides, transition metal complex as a catalyst (e.g., Cu, Fe, Ru, Ni, Os), and solvent. The ATRP method offers several advantages, including (1) ease of preparation, (2) commercially available compounds, (3) inexpensive catalysts, (4) pyridine based ligands, and (5) initiators. Kato, Kamigaito, Sawamoto, and Higashimura (1995) and Wang and Matyjaszewski (1995) independently discovered this living radical polymerization method (Figure 6).

Figure 6. Synthetic example of ATRP reaction

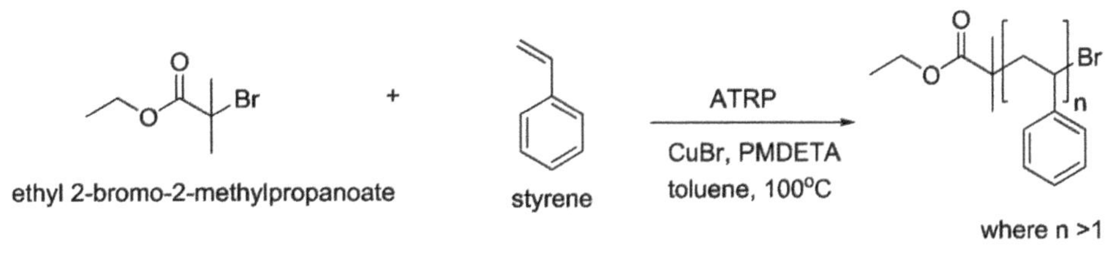

PMDETA – N,N,N,'N,N pentamethyldiethylenetriamine

Developing functional polymer coatings requires surface modification. In this regard, tethering polymer brushes to a surface can be accomplished via 'grafting to' or 'grafting from' a substrate. The 'grafting from' approach enables the formation of dense, stable polymer brush coatings via a surface initiated controlled radical polymerization (Hansson et al. 2013). Incorporating the surface-initiated (SI) approach with ATRP technique provides a method to introduce specific functionality with well-defined, controlled architecture and narrowly dispersed polymer brushes on numerous substrates (Xu, Neoh, & Kang, 2009). Both ATRP and click chemistries such as azide-alkyne, thiol-ene and activated ester-amine reactions are useful methods for the convenient, and efficient, reduction/elimination of side reactions, reactions proceeding under mild conditions, rapid kinetics, and innovative macromolecular engineering that can be achieved for preparing polymer brushes as functionalized active surface coatings. Several representative commercial monomers used in SI-ATRP polymerizations such as a nonionic hydrophilic monomer, 2-hydroxyethylmethacrylate (HEMA); nonionic hydrophobic monomer, 2,3,4,5,6-pentafluorostyrene (PFS); anionic monomer, 4-styrenesulfonic acid sodium salt (NaSS); and zwitterionic monomer, N,N'-dimethyl-(methylmethacrylolyl ethyl) ammonium propanesulfonate (DMAPS) are shown in Figure 7.

Another method of reversible-deactivation radical polymerization that is used to grow graft polymers on substrate surfaces is the reversible addition-fragmentation chain transfer (RAFT) polymerization technique. The RAFT polymerization uses a chain transfer reagent; specifically, thiocarbonylthio compounds such as dithioesters, thiocarbamates, and xanthates. These specific chain transfer compounds for RAFT polymerizations mediate the radical polymerization process through a reversible chain-transfer process allowing control over molecular weight and producing polymers with narrow polydisperity indexes (Keddie, Moad, Rizzardo, & Thang, 2012, and Chiefari, et al, 1998). In addition, complex architectures such as block co-polymers; comb-like, star, brush polymers; dendrimers; and crosslinked networks are formed using this technique. The RAFT polymerizations are very tolerant of the functionality of the monomer solvent system, including aqueous processes and wide temperature range. The initiators most often used are thermal initiators such as 2,2'-azobis(2-methylpropionitrile) (AIBN) and/or 4,4-azobis(4-cyanovaleric acid) (ACVA) (Figure 8).

The SI-RAFT methods, just as in SI-ATRP, have been used to form polymer brushes and other complex architectures on various substrates. Magnetic polyacrylonitrile nanofibers were prepared by Niu

Figure 7. Several representative SI-ATRP monomers

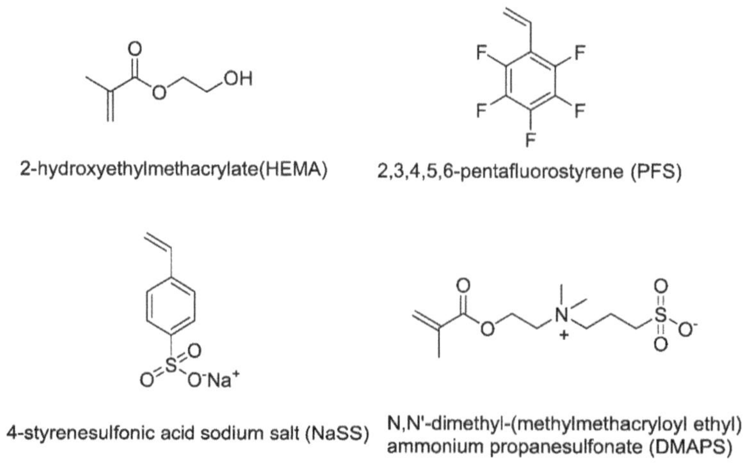

Figure 8. Example of RAFT polymerization

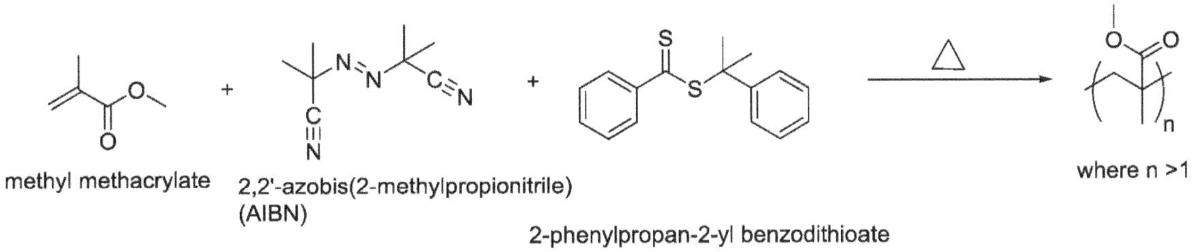

et al. (2013) using a combination of SI-RAFT polymerization and coaxial electrospinning. The RAFT polymerization of the monomer acrylonitrile was carried out on the surface of immobilized Fe_3O_4 nanoparticles, and these aligned fibers were also fabricated using an electrospinning method. Yuan, Huang, Li, Li, and Shen (2013) successfully grafted zwitterionic polysulfobetaine brushes onto cellulose membranes via a SI-RAFT polymerization process; the results displayed anti-bacterial properties. Superhydrophobic cotton fabric was synthesized using a facile SI-RAFT method of amphiphilic triblock azide copolymers containing poly(ethylene glycol) (PEG) and poly(2,2,3,4,4-hexafluorobutyl acrylate) blocks. The hydrophilic cotton fabric was changed to a superhydrophobic surface with the inclusion of the fluorinated polymer blocks and had a contact angle (CA) of 155 degrees with high stability and chemical durability (Li et al., 2010).

This chapter has discussed in detail the methods to change the surface properties of various substrates via chemical functionality, various architectures, and surface morphologies. These chemical changes on a substrate's surface can have profound effects on their physical properties and can impart such unique features as (1) superhydrophobicity, (2) superhydrophilicity, (3) superoleophobicity, and (4) omniphobicity. The functionalization methods discussed above can be tailored to alter the properties of a polymer coating. By precise control of the surface morphology and chemical or functional groups present on the surface, a wide variety of unique physical properties can be achieved. These properties can have a profound influence on the direct application of a coating's use. Several important physical properties that are found in coatings have been exploited by both academia and industry to develop coatings that can and do impact our daily lives.

Hydrophobicity refers to surfaces or materials that can repel water, causing droplets to form. Hydrophobic (water-fearing) materials or surfaces are therefore defined by the geometry of a water droplet on a flat surface; this geometry is the angle between the droplets edge and the surface underneath it. This angle is called the CA; and for a hydrophobic material or surface, the CA is >90 degrees. The CA of a standing water droplet is a measure of its static hydrophobicity; whereas, CA hysteresis is a dynamic measure in which surface heterogeneity is measured from an advancing and receding water droplet on a surface (Johnson & Dettre, 1964). The phenomenon of hydrophobicity is a result of surface chemistry; specifically, these surfaces are coated with low surface energy organic/polymer functional groups. The effect of hydrophobicity can be amplified through nanopatterns. Most naturally occurring surfaces that are superhydrophobic contain a hierarchical structure, which means that there is more than one length of scale (Liu, Wang & Jiang, 2013; Bhushan & Jung, 2010). These surfaces normally have a finer textured nanopattern feature embedded in a coarser length scale micropattern. This patterning combined with hydrophobic chemical functionality leads to the amplified version of hydrophobicity commonly referred to as "superhydrophobicity." This terminology, therefore, defines any surface that is highly hydrophobic and difficult or nearly impossible to wet. These types of surfaces have a CA of water droplet >150 degrees, and the roll-off angle or CA hysteresis <10 degrees. This phenomenon is also referred to as the "lotus effect," named after the superhydrophobic leaves of the lotus plant which was discussed earlier in this review (J. Li et al., 2006 and Cai & Xiao, 2013).

The opposite effect of hydrophobicity is hydrophilicity, in which a material or surface has a special affinity or attraction for water. Hydrophilic (water-loving) materials or surfaces allow water droplets to spread across the surface, thereby maximizing contact. The water droplet can spread across the surface wetting a large surface area and results in the CA <90 degrees (Koch & Barthlott, 2009). As described for superhydrophobic surfaces, hydrophilic materials and surfaces can be amplified via nanopatterning to produce materials and surfaces that are termed "superhydrophilic" where the CA is ~0 degrees (Drelich & Chibowski, 2010). Superhydrophilic coatings have several unique commercial applications such as defogging glass and enabling oil spots to be removed from surfaces with water (Si, Zhang, Zhao, Chen, & Xu, 2011).

Oleophobicity is defined as the physical property (fat-fearing) that a molecule or surface exhibits when exposed to an oil, and a superoleophobic surface is any surface that exhibits a CA >150 degrees and low CA hysteresis <5 degrees for low surface tension liquids such as oils and alcohols (Kota, & Tuteja, 2012). Oleophobic surfaces are rare in nature (Helbig, Nickerl, Neinhuis, & Werner, 2011) with the only naturally occurring organism being the spring tail, an anthropod that lives in soil, in decaying materials, and on plants, who has evolved unique anti-adhesive skin patterns. Most surfaces that repel lower surface tension liquids such as oils and alcohols also tend to repel higher surface tension liquids such as water. Therefore, a superoleophobic surface is also superhydrophobic. A recent study was conducted regarding a composite surface that is both superoleophobic and superhydrophilic (Yang et al., 2012a). However, no known superoleophobic surfaces can be found in nature, but chemists, chemical engineers, and materials scientists are investigating and developing new materials and coatings exhibiting these unique properties for various applications (Jiang, Guo, & Liu, 2015). Surfaces and coatings that display both superhydrophobic and superoleophobic properties are termed superomniphobic (Campos, Haddad, & Mabry, 2009).

In all of the above cases, external stimuli in the form of water droplets, oil droplets, and other matter can be attached or removed from a functionalized polymer surface. By design through chemical and morphological surface properties, specific surface effects can be imparted to a coating. These coatings depend on the physical properties of the surface that can respond to the external stimuli and either effectively remove the material or wet it. The next section will present methods to effectively determine the surface properties of a coating.

SURFACE ANALYTICAL TECHNIQUES FOR INVESTIGATING FUNCTIONAL POLYMER COATINGS

In the previous sections, the authors described in detail the synthesis leading to functional polymer surfaces and their unique properties. In order to evaluate these properties, various surface characterization techniques are used that allow for the visualization and chemical/physical determination of these properties. Several spectroscopic techniques are available to characterize functionalized polymer surfaces. These techniques include (1) attenuated total reflectance-Fourier transform infrared spectroscopy (ATR-FTIR), (2) XPS, and (3) secondary ion mass spectrometry (SIMS) (Elsner & Buchmeiser, 2006). Each of these techniques will be described in detail, and all of these techniques represent the current state-of-the-art in surface analytical techniques. These techniques are employed by various investigators to determine the chemical functionality of surfaces, morphology, topography, as well as the elements present on the surface of a functional polymer coating.

The ATR-FTIR technique is a simple, easy technique that allows coatings to be measured directly without any sample preparation. The ATR-FTIR analysis can provide information about inter- and intramolecular interactions, crystallinity, conformation, in addition to the orientation of molecules and depth profiles for various functional polymer coatings. The XPS technique is used to determine the elemental composition of a polymer surface. This spectroscopic technique is based on the emission of electrons from irradiation of the polymer surface by a beam of monochromatic X-rays. The kinetic energy of the emitted photoelectrons is unique for each of the different elements and is sensitive to the chemical state of the atoms. The XPS technique is very versatile and offers high sensitivity with sampling depths of 30Å for polymeric coatings. The SIMS technique is also an effective method for providing information on the molecular surface composition of a coating. This technique can be used to identify all elements including hydrogen as well as atomic and molecular ions and extremely high mass fragments at very low concentrations. Depth profiling of surfaces between 1 nanometer and 1 micron in thickness can be observed, and this technique allows for the determination of covalent bonding of molecules to surfaces, wettability, adhesiveness, and biological activity.

There are several microscopic techniques available for evaluating a polymer surface coating, such as scanning electron microscopy (SEM), scanning tunneling microscopy (STM), and atomic force microscopy (ATM) (Kimura, 2006; Jagtap & Ambre, 2005). The SEM method provides valuable information on surface morphology by employing a primary beam of electrons that interact with a substrate's surface in a vacuum environment. This interaction results in the emission of different types of electrons and electromagnetic waves, and the secondary electrons ejected from the coating surface are collected and displayed to provide a high resolution micrograph. The SEM technique can be combined with other analysis techniques, such as energy dispersive X-ray spectroscopy (EDS), to determine elemental distribution. In addition, IR and Raman spectroscopy can also be used to monitor surface modifications. The STM method can provide a 3-dimensional image of polymer surface topography. These 3-dimensional images are obtained by monitoring the tunneling current flowing between an extremely sharp conductive probe and the polymer surface. This method requires no sample preparation, but the technique is somewhat limited due to the non-conductive nature of most polymer systems. This difficulty can be overcome by using a very thin coating of conductive metal to coat the polymer surface, but resolution is somewhat compromised. The AFM method also can provide 3-dimemsional images of functional polymer coatings, via scanning tip attached to the end of a cantilever that moves across the polymer surface. By monitoring the minute forces of interaction between the functionalized polymer surface and probe, one can obtain high resolution images of the polymer surface topography. A unique advantage in using AFM is that polymer coating topographies and surface roughness can be determined without any polymer surface treatment and can be acquired in air, vacuum, or liquid environments. The AFM polymer coating samples require no preparation prior to imaging. In other microscopic imaging techniques, sample preparation does require a metal coating for imaging; this method can result in reduction of image resolution or polymer surface damage. Several modes in AFM technique are specifically tailored to the type of polymer surface under study; these modes include contact, non-contact, tapping, and phase imaging (Sato, Morita, & Ozaki, 2005).

The ability to obtain surface features not related to surface spectroscopic techniques, such as wetting or CA measurements, will be described in further detail. Thermodynamic parameters such as the interfacial free energy and enthalpy of adsorption or displacement can be determined via CA experiments (Shizuma et al., 2014). The CA is the angle normally measured through the liquid, where a liquid/vapor interface meets a solid surface. The CA quantifies the wettability of a solid surface by a liquid according to Young's equation, $0 = \gamma_{SG} - \gamma_{SL} - \gamma_{LG} \cos\Theta_c$, where, γ_{SG} is the solid-vapor interfacial energy parameter, γ_{SL} is the solid-liquid interfacial energy, γ_{LG} is the liquid-vapor interfacial energy (surface tension), and Θc is the equilibrium contact angle. Any liquid, solid, or vapor system at any given temperature and pressure has a unique static CA. However, a CA hysteresis is always observed in these systems, which include the advancing (maximal) and receding (minimal) CA. The term wetting used in this section describes the ability of a liquid to maintain contact with a solid surface. This degree of wetting is determined by both adhesive and cohesive forces. As a droplet spreads out (increases) over a flat solid surface, the CA decreases. Thus, CA and wettability are inversely related (Table 1) (Tadmor, 2011). There are several methods available to measure CA, including the static sessile drop, dynamic sessile drop, dynamic Wilhelmy, single-fiber Wilhelmy, and Washburn's equation capillary rise method (Barberis, 2013; Grundke et al., 2014).

APPLICATIONS OF FUNCTIONAL POLYMER COATINGS

Functionalized polymer coatings can produce a variety of surfaces with unique chemical and physical properties. These properties can advance the state of science in fundamental research and practical applications (Lai, Chen, & Lin, 2011). In this chapter, the authors focus on several types of functional polymer coatings and the applications specific to each coating. The application of functional polymer coatings to producing useful commercial products is quite diverse and is dependent on the functional groups present on the surface. While the definitions of hydrophobicity, oleophobicity, and hydrophilicity were discussed in earlier sections, the focus of this section will be to demonstrate how functional polymer surfaces can be tailored to specific applications.

As an example, functional polymer brushes in which living polymerization reactions are incorporated onto various surfaces provides a unique method to develop "smart surfaces" that can adapt to changes in an environment. In such polymer brush systems, surfaces can be switched from one morphological form to another, and monolayers with interesting topographies can be achieved. Kobayashi et al. (2012) developed high density polymer brushes with nonionic and ionic functional groups on silicon wafers

Table 1. CA and Degree of Wettability

CA	Degree of Wettability
$\Theta = 0^{\circ}$	Perfect wetting
$0 < \Theta < 90^{\circ}$	High wettability
$90^{\circ} \leq \Theta \leq 180^{\circ}$	Low wettability
$\Theta = 180^{\circ}$	Perfect non-wetting

using SI-ATRP methods. Polyelectrolyte polymer brushes repelled both air bubbles and hexadecane in water. These superhydrophilic polyelectrolyte brushes also had low adhesive forces between the brush and oil; this result has important applications in both anti-fouling and self-cleaning. Yang et al. (2012b) investigated functional polymer brushes incorporating the monomers HEMA, META, and NaSS DMAPS. The SI-ATRP method was executed on glass slides, producing dense and uniform polymer brushes containing either nonionic hydrophilic, nonionic hydrophobic, cationic, anionic, or zwitterionic groups. The best performing system in reducing marine biofouling was a combination of nonionic hydrophilic and zwitterionic polymer brushes. In addition to marine biofouling prevention, researches have grafted polymer brushes to various surfaces to prevent protein fouling in biomedical environments. Zuo, Yeh, Rossi, Brooks, and Kizhakkedathu (2010) synthesized polymer brushes of varying graft densities from SI-RAFT polymerization of poly(N-[2,3-dihydroxypropyl)acrylamide (PDHPA) containing latent aldehyde groups. The high graft density PDHPA brushes successfully prevented the nonspecific protein adsorption. Poly(sulfobetaine methacrylate) (polySBMA) brushes were grown via SI-ATRP on glass slides and showed anti-fouling activity against marine algae. Settled spores such as the green marine algae, Ulva, and sporelings showed low adhesive strength on polySBMA coatings (Zhang et al., 2009).

LbL deposition using click chemistry produced multifunctional polymer marine anti-fouling coatings as reported by Pranantyo, Neoh, Kang, Teo, and Rittschof (2012). Poly(ethylene glycol)methyl ether methacrylate containing azido functional groups and alkynyl functionalized 2-(methacryloyloxy) ethyl trimethyl ammonium chloride were assembled LbL via click chemistry. The multifunctional LbL polymer coatings were resistant to bacterial adhesion, and this coating was a bactericide to marine Gram-negative *Pseudomonas sp.* NCIMB 2021. Adhesion of barnacle cyprids was reduced by 80% using these multifunctional LbL coatings.

Bacterial/microbial contamination on the surfaces of medical devices, wound dressings, and food packages is of increasing concern due to the cost, health, and safety issues involved in removing the infected part and replacing it. Antibacterial/microbial coatings based on polymer/nanosilver composites were recently reviewed by Guo, Yuan, Lu, and Li (2013). Nanosilver has been incorporated into polymer matrix either by surface attachment via covalent bonding (Shi, Neoh, & Kang, 2004) and LbL self-assembly methods (Li, Lee, Sheng, Cohen, & Rubner, 2006). Nanosilver composite coatings have shown bactericidal properties for *Escherichia coli* and *Staphylococcus epidermidis* in laboratory studies.

Superhydrophobic fluoropolymer corrosion-inhibiting coatings have been developed by S. Yuan et al. (2011) for copper substrates via a surface grafting polymerization of 2,2,3,4,4,4-hexafluorobutyl acrylate (HFBA). The resulting polymer was superhydrophobic with a CA >150 degrees and exhibited corrosion resistance of copper substrates in corrosive solutions. N-[2-(8-heptadecenyl)-4,5-dihydro-1H-imidazole-1-ethyl]-2-bromoisobutyramide, a new heterocyclic initiator, was synthesized by Lu, Li, Lu, & Xu (2010). This initiator was immobilized on the surface of ferrous substrates, and methyl methacrylate was grafted onto the ferrous substrate via SI-ATRP. High quality polymer brush films were obtained with good control of molecular weight, $M_w/M_n < 1.32$, and the grafting density was 0.48 chains/nm^2. These polymer brush coatings demonstrated corrosion resistance on ferrous metals. Superhydrophobic coatings based on diblock copolymers of poly(heptadecafluorodecylacrylate-co-acrylic acid) as the first block with poly(acrylonitrile) as the second block showed good adhesion onto AA2024-T3 as well as excellent corrosion resistance when exposed to acetic acid spray for 200 hours (Grignard et al., 2011). Biocorrosion inhibition on stainless steel was investigated by Yuan et al. (2009). Their work focused on innovative, inorganic-organic hybrid coatings. These coatings were based on poly(vinyl-N-hexylpyridinium) deposited onto a nanostructured titanium oxide layer. The multilayer coatings were prepared using both LbL sol-gel deposition process for the titanium oxide, followed by SI-ATRP of 4-vinylpyridine, and then by the addition of hexyl bromide to produce ammonium salts on the surface of the substrate. The resulting multilayer coating showed excellent antimicrobial efficiency and inhibited corrosion as determined via Tafel and electrochemical impedance spectroscopy measurements. A superhydrophobic self-healing coating was prepared by spraying LbL polyelectrolyte complexes of poly(allyamine hydrochloride)-sulfonated poly(ether ether ketone), poly(acrylic acid). Several self-healing agents were added to this coating; specifically, perfluorooctanesulfonic acid lithium salt and 1H, 1H, 2H, 2H-perfluorooctyltriethyoxysilane (Li, Chen, Wu, & Sun, 2014). These self-healing agents incorporated into the LbL coating could automatically repair the damaged superhydrophobic coating at room temperature with slight humidity without any stimulus necessary.

The "self-cleaning" phenomenon that is found in nature, such as the lotus leaf, has been duplicated by numerous researchers in industrial, governmental, and academic laboratories. Superhydrophobic self-cleaning Si substrate surfaces containing poly(dimethylsiloxane) functionalized microstructures and nanostructures were developed by Zhang et al. (2013). A superhydrophobic self-cleaning polyprolyene surface having a CA of 157 $\pm$1.7 degrees and a sliding angle of 1.6 degrees was prepared by Z. Yuan et al. (2011) using a facile fabrication method. Elimination of 99% of the contaminant particles on the superhydrophobic polypropylene surface by water droplets was achieved by these researchers. A low-temperature superhydrophobic self-cleaning coating was prepared by coating silver-polydimethylsiloxane composite with 1H, 1H, 2H, 2H-perfluorodecanethiol. The coated composite showed self-cleaning ability in removing activated carbon powder on the surface at -10°C (Wang, Shen, Li, Li, & Yan, 2014). A superhydrophobic hybrid nanocomposite that demonstrated anti-icing properties was achieved by Zhan, Yan, Zhang, and Chen, 2014. This nanocomposite was formed by grafting fluorinated polymer chains onto silica nanoparticles via SI-ATRP, and this nanocomposite exhibited a static CA of 170.3 degrees and a small CA hysteresis <3 degrees as well as depressed the crystallization point of water for effective anti-icing applications. A hydrogen-bonded-assisted LbL assembly of poly(vinyl alcohol) and poly(acrylic acid) functionalized with poly(ethylene glycol methyl ether) segments enhanced the antifogging and frost-resistant behavior of this coating (Lee, Alcaraz, Rubner & Cohen, 2013). Recent work on multi-functional, self-healing hydrogel coatings were based on dynamic disulfide exchange reactions as reported by Yang et al (2015). These coatings provided self-healing ability and were prepared via surface-initiated thiol-ene photopolymerization. In addition to the self-healing properties, these coatings exhibited both reduced protein and bacterial absorption.

FUTURE RESEARCH DIRECTIONS

The authors have reviewed the methods, properties, and application of functional polymer coatings. A recent review on future trends in functional polymer coatings was described by Uhlmann et al., 2007. Their investigation, sponsored by the German Ministry of Education and Science (BMBF), proposed that innovative coatings are needed that are not merely decorative or barrier. These new functionalities include (1) easy-to-clean and self-cleaning properties; (2) scratch and mar-resistance; (3) adjustment to tribological properties; (4) recycling of coated materials; (5) switchable color, electrochromy, electro-luminescence; (6) integrated product identity; (7) antenna function; (8) heatable layers; (9) ultraviolet/infrared (UV/IR) absorption properties; (10) biological activity such as anti-microbial, anti-bacterial, and anti-fouling; and (11) photovoltaic activity. Many of these key features are described in this paper.

The theme for the *Research Perspectives on Functional Micro- and Nanoscale Coatings Handbook* is represented by the author's chapter contribution. Serious consideration to the focus of microscale and nanoscale coatings, preparation, properties, and applications has been thoroughly described and presented with real world applications.

CONCLUSION

The research conducted on functional polymer coatings is both exhaustive and compelling for fundamental science, which is geared to understanding unique biomimetic phenomena. From these laboratory studies, transition to real world applications becomes a reality. The work focusing on these fundamental scientific problems will no doubt yield innovative and commercial products that will have a profound effect on everyday living. By documenting the ongoing research in functional polymer coatings, the authors hope to educate and inform the novice and experienced scientist/engineer about the variety and robustness of this field.

ACKNOWLEDGMENT

This review article was supported by the Office of Naval Research, Discovery and Innovation Program, Code 333; Dr. Airan J. Perez is gratefully acknowledged.

REFERENCES

Al-Taher, F., Telegdi, J., & Kalman, E. (2007). Effect of divalent cations in Langmuir-Blodgett films on the protection of copper against corrosion. *Materials Science Forum*, 537-538, 9-14.

Al-Telegdi, J., Rigo, T., & Kalman, E. (2005). Molecular layers of hydroxamic acids in copper corrosion inhibition. *Journal of Electroanalytical Chemistry*, *582*(1-2), 191–201. doi:10.1016/j.jelechem.2005.03.045

Anastasiadis, S. H. (2013). Development of functional polymer surfaces with controlled wettability. *Langmuir*, *29*(30), 9277–9290. doi:10.1021/la400533u PMID:23789943

Ashrafi, A., Golozar, M. A., & Mallakpour, S. (2003). Electropolymerization of pyrrole and N-methylpyrrole on mild steel. *Iranian Polymer Journal*, *12*(6), 485–490.

Audebert, P., & Miomandre, F. (2007). Electrochemistry of conducting polymers. In T. A. Skotheim and J. R. Reynolds (Eds.), Handbook of Conducting Polymers 3rd Edition Conjugated Polymers Theory, Synthesis, Properties and Characterization (pp. 18–1-18–40). Boca Raton, Florida: CRC Press.

Baghdachi, J., Perez, H., & Shah, A. (2010). Design and development of self-healing polymers and coatings. In J. Baghdachi and T. Provder (Eds.), Smart Coatings III, *ACS Symposium Series* 1050 (pp. 3-20). Washington, DC: American Chemical Society. doi:10.1021/bk-2010-1050.ch001

Bao, X.-M., Cui, J.-F., Sun, X.-X., Liang, W.-D., Zhu, Z.-Q., An, J., & Li, A. et al. (2014). Facile preparation of superhydrophobic surfaces based on metal oxide nanoparticles. *Applied Surface Science, 303*, 473–480. doi:10.1016/j.apsusc.2014.03.029

Barberis, F. (2013). Contact angle measurements over macro, micro and nano dimensional systems. *Progress in Colloid and Interface Science*, 3, 1-35.

Benea, L. (2015). Nanocomposite layers obtained by electro-co-deposition: corrosion and tribocorrosion properties. *Solid State Phenomena,* 227, 1-5.

Bhushan, B., & Jung, Y. C. (2011). Natural and biomimetic artificial surfaces for superhydrophobicity, self-cleaning, low adhesion, and drag reduction. *Progress in Materials Science*, *56*(1), 1–108. doi:10.1016/j.pmatsci.2010.04.003

Brady, R. F., Jr. (2007). Twenty-first century materials: coatings that interact with their environment. In T. Provder, & J. Baghdachi (Eds.), *Smart Coatings ACS Symposium Series 957* (pp. 3-11). Washington, DC: American Chemical Society. doi:10.1021/bk-2007-0957.ch001

Branzoi, V., Golgovici, F., & Branzoi, F. (2008). Electronically conductive polymers type polypyrrole and polyqaniline obtained by electropolymerization onto aluminum substrate. *Revue Roumaine de Chemie*, *53*(7), 553–561.

Butt, H.-J., Vollmer, D., Deng, X., Klapper, M., Praven, M., Schuster, T., & Papadopoulos, P. (2014). *Process and device for particle synthesis on a supehydrophobic or superoleophobic surface*. European Patent Number 2,677,332.

Cai, H., An, X., Cui, J., Li, J., Wen, S., Li, K., & Shi, X. et al. (2013). Facile hydrothermal synthesis and surface functionalization of poyethyleneimine-coated iron oxide nanoparticles for biomedical applications. *ACS Applied Materials & Interfaces*, *5*(5), 1722–1731. doi:10.1021/am302883m PMID:23388099

Cai, X.-S., & Xiao, X.-Y. (2013). Progress in superhydrophobic surface coatings. *Xiandai Huagong*, *33*(1), 22–25.

Campos, R., Haddad, T. S., & Mabry, J. (2009). Toward mechanically-robust omniphobic composite coatings using fluoroalkyl-modified silica. *ACS Polymer Preprints*, *50*(2), 717–718.

Cassagneau, T. P. (2004). Layer-by-layer self-assembly of metal nanoparticles on planar substrates. Fabrication and properties. *Colloids and Colloid Assemblies*, 398-436.

Cerkez, I., Kocer, H. B., Worley, S. D., Broughton, R. M., & Huang, T. S. (2011). N-halamine biocidal coatings via a layer-by-layer assembly technique. *Langmuir*, *27*(7), 4091–4097. doi:10.1021/la104923x PMID:21348478

Chiefari, J., Chong, Y. K., Ercole, F., Kristina, J., Jeffrey, J., Le, T. P., & Thang, S. H. et al. (1998). Living free-radical polymerization by reversible addition-fragmentation chain transfer: The RAFT process. *Macromolecules*, *31*(16), 5559–5562. doi:10.1021/ma9804951

Drelich, J., & Chibowski, E. (2010). Superhydrophobic and superwetting surfaces: Definition and mechanism of control. *Langmuir*, *26*(24), 18621–18623. doi:10.1021/la1039893 PMID:21090661

Elsner, C., & Buchmeiser, M. R. (2006). Surface analysis on polymers. *Nachrichten aus der Chemie, 54*(2), 141–144.

Grignard, B., Vaillant, A., Connick, J., & de, . (2011). Electrospinning of functional perfluorinated block copolymer as a powerful route for imparting superhydrophobbicity and corrosion resistance to aluminum substrates. *Langmuir, 27*(1), 335–342. doi:10.1021/la102808w PMID:21141949

Grundke, K., Poschel, K., Synytska, A., Frenzel, R., Drechsler, A., Nitschke, M., & Welzrl, P. B. et al. (2014). Experimental studies of contact angle hysteresis phenomena on polymer surfaces-toward the understanding and control of wettability for different applications. *Advances in Colloid and Interface Science*. PMID:25488284

Guo, L., Yuan, W., Lu, Z., & Li, C. M. (2013). Polymer/nanosilver composite coatings for antibacterial applications. *Colloids and Surfaces. A, Physicochemical and Engineering Aspects*, *439*, 69–83. doi:10.1016/j.colsurfa.2012.12.029

Guo, Z., Liu, W., & Su, B.-L. (2011). Superhydrophobic surfaces: From natural to biomimetic to functional. *Journal of Colloid and Interfacial Science*, *353*(2), 335–355. doi:10.1016/j.jcis.2010.08.047 PMID:20846662

Guo, Z. G., Fang, J., Wang, L., & Liu, W. (2007). Fabrication of superhydrophobic copper by wet chemical reaction. *Thin Solid Films*, *515*(18), 7190–7194. doi:10.1016/j.tsf.2007.02.100

Hansson, S., Trouillet, V., Tischer, T., Goldmann, A. S., Carlmark, A., Barner-Kowollik, C., & Malmstroem, E. (2013). Grafting efficiency of synthetic polymers onto biomaterials: A comparative study of grating-from versus grating-to. *Biomacromolecules*, *14*(1), 64–74. doi:10.1021/bm3013132 PMID:23043441

Helbig, R., Nickerl, J., Neinhuis, C., & Werner, C. (2011). Smart skin patterns protect springtails. *PLoS ONE*, *6*(9), e25105. doi:10.1371/journal.pone.0025105 PMID:21980383

Hu, C., Liu, S., Li, B., Yang, H., Fan, C., & Cui, W. (2013). Micro-/nanometer rough structure of a superhydrophobic biodegradable coating by electrospraying for initial anti-bioadhesion. *Advanced Healthcare Materials*, *2*(10), 1314–1321. doi:10.1002/adhm.201300021 PMID:23554405

Hu, L., Graaf, M. D., & Moeller, K. D. (2013). The use of UV-cross-linkable di-block copolymers as functional reaction surfaces for microelectrode arrays. *Journal of the Electrochemical Society*, *160*(7), G3020–G3029. doi:10.1149/2.004307jes

Igari, T., Oyamada, N., Nanao, H., & Mori, S. (2000). Relationship between the structure of organic thin films and their microtribological property. *Toraiborojisuto*, *45*(5), 414–420.

Jagtap, R. N., & Ambre, A. H. (2005). Atomic Force microscopy (AFM): Basics and its important applications for polymer characterization: An overview. *Journal of Polymer Materials*, *22*(1), 1–26.

Jiang, T., Guo, Z., & Liu, W. (2015). Biomimetic superoleophobic surfaces: Focusing on their fabrication and applications. *Journal of Materials Chemistry A. Materials for Energy and Sustainability*, *3*(5), 1811–1827.

Johnson, R. E. Jr, & Dettre, R. H. (1964). Contact angle hysteresis. III. Study of an idealized heterogeneous surface. *Journal of Physical Chemistry*, *68*(7), 1744–1750. doi:10.1021/j100789a012

Kantheti, S., Narayan, R., & Raju, K. V. S. N. (2015). The impact of 1,2,3-triazoles in the design of functional coatings. *RSC Advances*, *5*(5), 3687–3708. doi:10.1039/C4RA12739K

Kato, M., Kamigaito, M., Sawamoto, M., & Higashimura, T. (1995). Polymerization of methyl methacrylate with the carbon tetrachloride/dichlorotris-(triphenylphosphine) ruthenium (II)/methylaluminum bis(2,6-di-tert-butylphenoxide) initiating system: Possibility of living radical polymerization. *Macromolecules*, *28*(5), 1721–1723. doi:10.1021/ma00109a056

Keddie, D. J., Moad, G., Rizzardo, E., & Thang, S. H. (2012). RAFT agent design and synthesis. *Macromolecules*, 45(13), 5321-5342.

Kim, B. S., Shin, S., Shin, S. J., Kim, K. M., & Cho, H. H. (2011). Control of superhydrophilicity/superhydrophobicity using silicon nanowires via electroless etching method and fluorine carbon coating. *Langmuir*, *27*(16), 10148–10156. doi:10.1021/la200940j PMID:21728376

Kimura, S.-I. (2006). Recent progress of surface analysis of polymers. 1. Introduction. *Nippon Gomu Kyokaishi*, *79*(6), 250–256. doi:10.2324/gomu.79.250

Kobayashi, M., Terayama, Y., Yamaguchi, H., Tereda, M., Murakami, D., Ishihara, K., & Takahara, A. (2012). Wettability and Antifouling Behavior on the Surfaces of Superhydrophilic Polymer Brushes. *Langmuir*, *28*(18), 7212–7222. doi:10.1021/la301033h PMID:22500465

Koberstein, J. T. (2004). Molecular design of functional polymer surfaces. *Journal of Polymer Science. Part B, Polymer Physics*, *42*(16), 2942–2956. doi:10.1002/polb.20157

Koberstein, J. T., Hu, W., Duch, D., Bhatia, R., Lenk, T., Smith, D. A., & Gallot, Y. et al. (1997). Molecular design concepts for creating smart polymer surfaces. *Proceedings of the 20th Annual Meeting of the Adhesion Society* (pp. 283-285).

Koch, K., & Barthlott, W. (2009). Superhydrophobic and superhydrophilic plant surfaces: an inspiration for biomimetic materials. *Philosophical transactions. Series A, Mathematical, physical, and engineering sciences*, 367(1893), 1487-1509.

Kota, A. K., & Tuteja, A. (2012). Superoleophobic surfaces. In *Advances in fluorine-containing polymers, ACS Symposium Series* (*Vol. 1106*, pp. 171-185), American Chemical Society, Washington, DC.

Krukiewicz, K., Herman, A. P., Turczyn, R., Szymanska, K., Koziol, K. K. K., Boncel, S., & Zak, J. K. (2014). A role of nanotube dangling pyrrole and oxygen functions in the electrochemical synthesis of polypyrrole/MWCNTs hybrid materials. *Applied Surface Science*, *317*, 794–802. doi:10.1016/j.apsusc.2014.08.165

Lai, Y.-K., Chen, Z., & Lin, C.-J. (2011). Recent progress on the superhydrophobic surfaces with special adhesion: From natural to biomimetic to functional. *J. of Nanoengineering and Nanomanufacturing*, *1*(1), 1–17. doi:10.1166/jnan.2011.1007

Langmuir, I., & Blodgett, K. B. (1935). A new method for investigating unimolecular films. *Kolloid-Zeitschrift*, *73*, 258–263.

Lee, H., Alcaraz, M. L., Rubner, M. F., & Cohen, R. E. (2013). Zwitter-wettability and antifogging coatings with frost-resistant capabilities. *ACS Nano*, *7*(3), 2172–2185. doi:10.1021/nn3057966 PMID:23360374

Lee, J. Y., & Hong, S.-Y. (2005). Fabrication of an alternating multilayer film of poly(ethylene-alt-maleic anhydride) and poly(4-vinyl pyridine) by layer-by-layer self-assembly method. *Polymer(Korea)*, *29*(4), 392–398.

Lee, W., Ji, R., Gosele, U., & Nielsch, K. (2006). Fast fabrication of long-range ordered porous alumina membranes by hard anodization. *Nature Materials*, *5*(9), 741–747. doi:10.1038/nmat1717 PMID:16921361

Li, G., Zheng, H., Wang, Y., Wang, H., Dong, Q., & Bai, R. (2010). A facile strategy for the fabrication of highly stable superhydrophobic cotton fabric using amphiphilic fluorinated triblock azide copolymers. *Polymer*, *51*(9), 1940–1946. doi:10.1016/j.polymer.2010.03.002

Li, J., Zhang, Z., Xu, J., & Wong, C. P. (2006). Self-cleaning materials-lotus effect surfaces. In S. Arza (Ed.), Kirk-Othmer Encyclopedia of Chemical Technology (5th ed., Vol. 22, pp. 108-127). doi:10.1002/0471238961.selfwong.a01

Li, Y., Chen, S., Wu, M., & Sun, J. (2014). All spraying processes for the fabrication of robust, self-healing, superhydrophobic coatings. *Advanced Materials*, *26*, 3344–3348. PMID:24687272

Li, Y., Duan, G., & Cai, W. P. (2007). Controllable superhydrophobic and lipophobic properties of ordered pore indium oxide array films. *Journal of Colloid and Interface Science*, *314*(2), 615–620. doi:10.1016/j.jcis.2007.05.061 PMID:17583718

Li, Z., Lee, D., Sheng, X., Cohen, R. E., & Rubner, M. F. (2006). Two-level antibacterial coating with both release-killing and contact-killing capabilities. *Langmuir*, *22*(24), 9820–9823. doi:10.1021/la0622166 PMID:17106967

Liu, C.-H., Niu, P.-L., & Sung, C.-K. (2014). Integrating antireflection and superhydrophobicity of moth-eye-like surface morphology on large–area flexible substrate. *Journal of Physics. D, Applied Physics*, *47*(1), 015401/1–015401, 5. doi:10.1088/0022-3727/47/1/015401

Liu, H., Szunerits, S., Pisarek, M., Xu, W., & Boukherroub, R. (2009). Preparation of superhydrophobic coatings on zinc, silicon, and steel by a solution-immersion technique. *ACS Applied Materials & Interfaces*, *1*(9), 2086–2091. doi:10.1021/am900488w PMID:20355837

Liu, M., Wang, S., & Jiang, L. (2013). Bioinspired multiscale surfaces with special wettability. *MRS Bulletin*, *38*(5), 375–382. doi:10.1557/mrs.2013.100

Livage, J. (2000). The sol-gel process. *Verre (Versailles)*, *6*(5), 12–16.

Love, J. C., Estroff, L. A., Kriebel, J. K., Nuzzo, R. G., & Whitesides, G. M. (2005). Self-assembled monolayers of thiolates on metals as a form of nanotechnology. *Chemical Reviews*, *105*(4), 1103–1169. doi:10.1021/cr0300789 PMID:15826011

Lu, G., Li, Y.-M., Lu, C.-H., & Xu, Z.-Z. (2010). Corrosion protection of iron surface modified by poly(methyl methacrylate) using surface-initiated atom transfer radical polymerization (SI-ATRP). *Colloid & Polymer Science*, *288*(14-15), 1445–1455. doi:10.1007/s00396-010-2283-x

Luzinov, I., Minko, S., & Tsukruk, V. V. (2008). Responsive brush layers: From tailored gradients to reversibly assembled nanoparticles. *Soft Matter*, *4*(4), 714–725. doi:10.1039/b718999k

Maidenberg, Y., Zhang, S., Luo, K., Akhavein, N., & Koberstein, J. T. (2013). Mixed silane monolayers for controlling the surface area density of click-reactive alkyne groups: A method to assess preferential surface adsorption on flat substrates and a method to verify compositional homogeneity on nanoparticle. *Langmuir*, *29*(38), 11959–11965. doi:10.1021/la402517m PMID:23985021

Munikata, S., Yamaauchi, M., Hirono, T., Kodama, S.-I., & Matsuo, M. (2001). Towering developments. *Asia Pacific coatings Journal*, 14(6), pp. 8, 10, 14.

Anti-Fog PET Film Products. (n. d.). FSI Coating Technologies. Retrieved from http://www.fsicti.com/index.php?pid=19

Nishimoto, S., & Bhushan, B. (2013). Bioinspired self-cleaning surfaces with superhydrophobicity, superoleophobicity, and superhydrophilicity. *RSC Advances*, *3*(3), 671–690. doi:10.1039/C2RA21260A

Niu, S., Zhang, L., Wang, N., Zhu, J., Zhang, W., Cheng, Z., & Zhu, X. (2013). Fabrication of magnetic nanofibers via a surface-initiated RAFT polymerization and coaxial electrospinning. *Reactive & Functional Polymers*, *73*(11), 1447–1454. doi:10.1016/j.reactfunctpolym.2013.07.011

Nuernberg, A., Stolle, R., Wahl, G., & Raic, K. T. (2003). Plasma activated CVD of Co-C-O-layers. deposition rate phenomena. *Advances in Science and Technology*, 32, 91-98.

O'Rourke-Muisener, P. A. V., Koberstein, J. T., & Kumar, S. (2003). Optimal chain architectures for the molecular design of functional polymer surfaces. *Macromolecules*, *36*(3), 771–781. doi:10.1021/ma011969q

Peniredd, S. R., Janczewski, D., Go, D. P., Zhu, X., Guo, S., Teo, S. L. M., & Vancso, G. J. et al. (2015). Imprinting of metal receptors into multilayers polyelectrolyte films: Fabrication and applications in marine antifouling. *Chemical Sciences*, *6*(1), 372–383.

Pernites, R. B., Ponnapati, R. R., & Advincula, R. C. (2011). Superhydrophobic-superoleophobic polythiophene films with tunable wetting and electrochromism. *Advanced Materials*, *23*(28), 3207–3213. doi:10.1002/adma.201100469 PMID:21638343

Pranantyo, D., Neoh, K.-G., Kang, E.-T., Teo, S. L.-M., & Rittschof, D. (2012). Layer-by-Layer Click Deposition of Functional Polymer Coatings for Combating Marine Biofouling. *Biomacromolecules*, *13*(9), 2769–2780. doi:10.1021/bm300757e PMID:22924814

Qu, M., Zhang, B., Song, S., Chen, L., Zhang, J., & Junyan, C. X. (2007). Fabrication of superhydrophobic surfaces on engineering materials via a solution-immersion process. *Advanced Functional Materials*, *217*(4), 593–596. doi:10.1002/adfm.200600472

Rezaei, S., Manoucheri, I., Moradian, R., & Pourabbas, B. (2014). One-step chemical vapor deposition and modification of silica nanoparticles at the lowest possible temperature and superhydrophobic surface fabrication. *Chemical Engineering Journal (Amsterdam, Netherlands)*, *252*, 11–16.

Rubner, M. F. (2003). PH-controlled fabrication of polyelectrolyte multilayers: assembly and applications. In G. Decher & J.B. Schlenoff (Eds.), Multilayer Thin Films (pp. 133-154). Weinheim, Germany: Wiley-VCH Verlag GmbH & Co. KGaA.

Satayaprassad, A., Jian, V., & Nema, S. K. (2007). Deposition of superhydrophobic nanostructured Teflon-like coating using expanding plasma arc. *Applied Surface Science, 253*(12), 5462–5466. doi:10.1016/j.apsusc.2006.12.085

Sato, H., Morita, S., & Ozaki, Y. (2005). Polymer Surface explored by infrared spectroscopy and atomic force microscopy. *Nippon Setchaku Gakkaishi, 41*(5), 183–195.

Shchukin, D. G., & Moehwald, H. (2012). Multilayer polyelectrolyte assembly in feedback active coatings and films. In G. Decher & J. B. Schlenoff (Eds.), *Multilayer Thin Films, 2* (pp. 1039–1052). Weinheim, Germany: Wiley-VCH Verlag GmbH & Co. doi:10.1002/9783527646746.ch43

Shi, Z., Neoh, K. G., & Kang, E. T. (2004). Surface-grafted viologen for precipitation of silver nanoparticles and their combined bactericidal activities. *Langmuir, 20*(16), 6847–6852. doi:10.1021/la049132m PMID:15274594

Shizuma, M., Shimomura, O., Moriwaki, K., Yamauchi, A., Sato, H., Kawano, S., & Ono, D. (2014). Evaluation of surface free energy by means of contact angle measurements. *Kagaku to Kogyo, 88*(5), 193–199.

Si, F., Zhang, L., Zhao, N., Chen, L., & Xu, J. (2011). Preparation method and application of superhydrophilic surfaces. *Huaxue Jinzhan, 23*(9), 1831–1840.

Stuart, M.A., C., , Huck, W.T.S., Genzer, J., Müller, M., Ober, C., Stamm, M., & Sukhorukov, G.B. (2010). Emerging applications of stimuli-responsive polymer materials. *Nature Materials, 9*(2), 101–113. doi:10.1038/nmat2614 PMID:20094081

Tadmor, R. (2011). Approaches in wetting phenomena. *Soft Matter, 7*(5), 1577–1580. doi:10.1039/C0SM00775G

Tanioka, A. (2010). Electrospray in the future. *Maku, 35*(3), 100–104.

Uhlmann, P., Frenzel, R., Voit, B., Mock, U., Szyszka, B., Schmidt, B., & Roths, K. et al. (2007). Research agenda surface technology: Future demands for research in the field of coating materials. *Progress in Organic Coatings, 58*(2-3), 122–126. doi:10.1016/j.porgcoat.2006.08.020

Vasiljevic, J., Tomsic, B., Jerman, I., Orel, B., Aska, G., Kovac, J., & Simoncic, B. (2014). Multifunctional superhydrophobic/oleophobic and flame-retardant cellulose fibres with improved ice-releasing properties and passive antibacterial activity prepared via the sol-gel method. *Journal of Sol-Gel Science and Technology, 70*(3), 385–399. doi:10.1007/s10971-014-3294-8

Wang, D., Zheng, Z., Hong, C., Liu, Y., & Pan, C. (2006). Michael addition polymerizations of difunctional amines (AA') and triacrylamides (B3). *Journal of Polymer Science. Part A, Polymer Chemistry, 44*(21), 6226–6242. doi:10.1002/pola.21716

Wang, F., Shen, T., Li, C., Li, W., & Yan, G. (2014). Low temperature self-cleaning properties of superhydrophobic surfaces. *Applied Surface Science, 317*, 1107–1112. doi:10.1016/j.apsusc.2014.08.200

Wang, H., Patil, A. J., Liu, K., Petrov, S., Mann, S., Winnik, M. A., & Manners, I. (2009). Fabrication of continuous and segmented polymer/metal oxide nanowires using cylindrical micelles and block comicelles as templates. *Advanced Materials, 21*(18), 1805–1808. doi:10.1002/adma.200803015

Wang, J.-H., & Matyjaszewski, K. (1995). Controlled/"living" radical polymerization atom transfer radical polymerization in the presence of transition-metal complexes. *Journal of the American Chemical Society*, *117*(20), 5614–5615. doi:10.1021/ja00125a035

Wang, S., Feng, L., & Jiang, L. (2006). One-step solution-immersion process for the fabrication of stable bionic superhydrophobic surfaces. *Advanced Materials*, *18*(6), 767–700. doi:10.1002/adma.200501794

Wilhelm, O., Madler, L., & Pratsinis, S. E. (2003). Electrospray evaporation and deposition. *Journal of Aerosol Science*, *34*(7), 815–836. doi:10.1016/S0021-8502(03)00034-X

Wu, L., & Baghdachi, J. (2015). *Functional Polymer Coatings: Principles, Methods, and Applications*. Hoboken, New Jersey: John Wiley & Sons, Inc. doi:10.1002/9781118883051

Xu, F., Neoh, K. G., & Kang, E. T. (2009). Bioactive surfaces and biomaterials via atom transfer radical polymerization. *Progress in Polymer Science*, *34*(68), 719–761. doi:10.1016/j.progpolymsci.2009.04.005

Yang, H., Dou, X., Fang, Y., & Jiang, P. (2013). Self-assembled biomimetic superhydrophobic hierarchical arrays. *Journal of Colloid and Interface Science*, *405*, 51–57. doi:10.1016/j.jcis.2013.05.040 PMID:23786830

Yang, J., Zhang, Z., Xu, X., Zhu, X., Men, X., & Zhou, X. (2012a). Superhydrophilic-superoleophilic coatings. *Journal of Materials Chemistry*, *22*(7), 2834–2837. doi:10.1039/c2jm15987b

Yang, J. Y., Neoh, K.-G., Kang, E.-T., Lee, S. S. C., Teo, S. L.-M., & Rittschof, D. (2012b). Functional polymer brushes via surface-initiated atom transfer radical graft polymerization for compacting marine biofouling. *Biofouling: The Journal of Bioadhesion and Biofilm Research*, *28*(9), 895–912. doi:10.1080/08927014.2012.719895 PMID:22963034

Yang, W. J., Tao, X., Zhao, T-T., Weng, L., Kamg, E-T., & Wang, L. (2015). Antifouling.

Yuan, J., Huang, X., Li, P., Li, L., & Shen, J. (2013). Surface-initiated RAFT polymerization of sulfobetaine from cellulose membranes to improve hemocompatibility and antibiofouling property. *Polymer Chemistry*, *4*(19), 5074–5085. doi:10.1039/c3py00565h

Yuan, S. J., Pehkonen, S. O., Liang, B., Ting, Y. P., Neoh, K. G., & Kang, E. T. (2011). Superhydrophobic fluoropolymer-modified copper surface via surface graft polymerisation for corrosion protection. *Corrosion Science*, *53*(9), 2738–2747. doi:10.1016/j.corsci.2011.05.008

Yuan, S. J., Pehkonen, S. O., Ting, Y. P., Neoh, K. G., & Kang, E. T. (2009). Inorganic-organic hybrid coatings on stainless steels by layer-by-layer deposition and surface- initiated atom-transfer-radical polymerization for combating biocorrosion. *ACS Applied Materials & Interfaces*, *1*(3), 640–652. doi:10.1021/am800182d PMID:20355986

Yuan, X., Li, C., Guan, G., Xiao, Y., & Zhang, D. (2009). Surface grafting modification of fibrous silicates with polyvinylpyrrolidone and its application in nanocomposites. *Journal of Applied Polymer Science*, *111*(1), 566–575. doi:10.1002/app.29109

Yuan, Z., Wu, J., Hao, X., Wang, X., Xu, X., & Zhang, H. (2011). Preparation and characterization of superhydrophobic polypropylene surface with excellent self-cleaning property. *Advanced Materials Research*, 295-297, 1613-1616.

Zarras, P., & Stenger-Smith, J. D. (2014). Electro-active polymer (EAP) coatings for corrosion protection of metals. In H. M. Abdel Salam (Ed.), *Handbook of Smart Coatings for Materials Protection* (pp. 328–369). Kidlington, United Kingdom: Woodhead Publishing Limited. doi:10.1533/9780857096883.2.328

Zengin, A., & Caykara, T. (2012). RAFT-mediated synthesis of poly[oligoethylene glycol) methyl ether acrylate] brushes for biological functions. *Journal of Polymer Science. Part A, Polymer Chemistry*, *50*(21), 4443–4450. doi:10.1002/pola.26250

Zhan, X., Yan, Y., Zhang, Q., & Chen, F. (2014). A novel superhydrophobic hybrid nanocomposite material prepared by surface-initiated AGET ATRP and its anti-icing properties. *Journal of Materials Chemistry A. Materials for Energy and Sustainability*, *2*(24), 9390–9399.

Zhang, L., Zeng, H., & Liu, Q. (2012). Probing molecular and surface interactions of comb-type polymer polystyrene-graft-poly(ethylene oxide) (PS-*g*-PEO) with an SFA. *The Journal of Physical Chemistry C*, *116*(33), 17554–17562. doi:10.1021/jp303561p

Zhang, X.-S., Zhu, F.-Y., Han, M.-D., Sun, X.-M., Peng, X.-H., & Zhang, H.-X. (2013). Self-cleaning poly(dimethylsiloxane) film with functional micro/nano hierarchical structures. *Langmuir*, *29*(34), 10769–10775. doi:10.1021/la4023745 PMID:23906343

Zhang, Z., Finlay, J. A., Wang, L., Gao, Y., Callow, J. A., Callow, M. E., & Jiang, S. (2009). Polysulfobetaine-grafted surfaces as environmentally benign ultralow fouling marine coatings. *Langmuir*, *25*(23), 13516–13521. doi:10.1021/la901957k PMID:19689148

Zhou, F., & Huck, W. T. S. (2006). Surface grafted polymer brushes as ideal building blocks for "smart" surfaces. *Physical Chemistry Chemical Physics*, *8*(33), 3815–3823. doi:10.1039/b606415a PMID:19817040

Zhou, L., Li, X., Yu, H., & Li, W. (2007). Layer-by-layer assembled highly reflective optical coatings. *Proceedings of SPIE-The International Society for Optical Engineering* (Vol. 6722, pp.1-4).

Zia, A. W., Wang, Y-Q., & Lee, S. (2015). Effects of physical and chemical plasma etching on surface wettability of carbon fiber-reinforced polymer composites for bone plate applications. *Advances in Polymer Technology*, 34(1), 1-4.

Zou, Y., Yeh, P.-Y. J., Rossi, N. A. A., Brooks, D. E., & Kizhakkeedathu, J. N. (2010). Nonbiofouling polymer brushes with latent aldehyde functionality as a template for protein micropatterning. *Biomacromolecules*, *11*(1), 284–293. doi:10.1021/bm901159d PMID:20000794

KEY TERMS AND DEFINITIONS

Atom Transfer Radical Polymerization (ATRP): Reversible-deactivation radical polymerization technique in which carbon-carbon bonds are formed through a transition metal catalyst producing polymers with highly controlled molecular weight and narrow molecular weights.

Biomimetic: Mimicking the naturally occurring structures and materials of living organisms via modern synthetic methods and process engineering techniques to develop nature-inspired structural materials.

Contact Angle: The angle that is measured through the liquid where the liquid/vapor interface meets a solid surface and this value quantifies the wettability of a solid surface by a liquid via the Young equation.

Functional Polymer Coating: A coating that can dynamically respond to their environment via surface modification through synthetic or morphological changes.

Reversible Addition-Fragmentation Chain Transfer (RAFT) Polymerization: A reversible-deactivation radical polymerization that is used to grow graft polymers on substrate surfaces using a chain transfer reagent, specifically, thiocarbonylthio compounds such as dithioesters, thiocarbamates and xanthates which enables precise control over the molecular weight and narrow polydisperity indexes.

Superhydrophobic: Surfaces that are highly hydrophobic and are extremely difficult to wet with contact angles > 150 degrees and contact hysteresis < 10 degrees.

Superhydrophilic: Surface is the opposite of a superhydrophobic surface in which water droplets do not form any contact angle on a surface (~ 0 degrees) producing near perfect wettability.

Superoleophobic: Any surface that exhibits a CA >150 degrees and low CA hysteresis < 5 degrees for low surface tension liquids such as oils and alcohols and can also exhibit superhydrophobic properties.

Chapter 5
Non-Hydrolyzed Resins for Organic-Inorganic Hybrid Coatings:
Functional Coating Films by Moisture Curing

Stefan Holberg
Danish Technological Institute, Denmark

ABSTRACT

This chapter focuses on resins based on non-hydrolyzed, monomeric and polymeric alkoxysilanes. As alternative to classical sol-gel processing, the resins are applied to a surface without a preceding hydrolysis step. Only after application, hydrolysis and condensation of the alkoxysilyl groups occur by means of atmospheric moisture to result cross-linked organic-inorganic hybrid coatings. While the use of non-hydrolyzed silanes is well established, for example by applying polyethyl silicate as binder for zinc-rich anti-corrosive primers, this chapter describes the chemical structures of various novel organic-inorganic hybrid precursors that have significantly extended the area of application to adhesives and scratch-resistant, repellent, or anti-fouling coatings. At present, individual resins are produced and applied at industrial scale in the fields of protective coatings and automotive topcoats.

INTRODUCTION

As an alternative to classical sol-gel processing, organic-inorganic hybrid coatings can be prepared from alkoxysilyl-functional resins that hydrolyze and condense first after application to a surface, generally by means of atmospheric moisture. A complete coating composition usually comprises various components such as resins, pigments, fillers, solvents and diverse additives. The present chapter focuses specifically on the chemistry of the resins as the main component. A compilation of various chemical buildings blocks is provided and possible applications are highlighted to make knowledge available, but especially to inspire other researchers in the field of functional coatings.

DOI: 10.4018/978-1-5225-0066-7.ch005

The present materials represent a sub-group of the field of organic-inorganic hybrid materials. The term "non-hydrolyzed" indicates that monomeric or polymeric alkoxysilanes are applied to a surface without a hydrolysis step prior to application, which distinguishes them from other silane-derived materials. Only after application, atmospheric moisture diffuses into the coating to provide the water necessary for hydrolysis and condensation. Subsequently, a cross-linked network forms as shown in Figure 1.

In a sol-gel process, for comparison, silanes, metal alkoxides or salts are in a first step hydrolyzed and partly condensed by addition of water to form a colloidal solution, a sol. To prepare coatings, the sol is in a second step applied to a surface and forms a gel and possibly subsequently a densified film.

Coatings prepared from non-hydrolyzed silanes have been in focus of both academic research and industrial development. Following early, pioneering work, moisture curing of silanes has seen a significant increase in research and development activities in the past decade and is quite active at present. The number of US-patents filed for epoxy-siloxane hybrid coatings per year as shown in Figure 2 (Witucki, 2013) is a good example for the described development. The results of these activities are novel coatings and, after intense optimization work, commercial products produced in industrial scale, such as, for example, scratch-resistant automotive top coats that do not compromise flexibility or protective coatings. Applications as adhesion primer or repellent coating are as well considered.

The cured films differ from solely organic coatings by including a cross-linked siloxane network. Various investigations report individual inorganic siloxane domains in an organic matrix (Groenewolt et al., 2008; Metal finishing, 2012; Rekondo, Fernándes-Berridi, & Irusta, 2006; Sangermano, Vitale & Dietliker, 2014), but two interpenetrating networks or organic domains in an inorganic network are possible as well. Such material can also be prepared by sol-gel processing. However, the application of non-hydrolyzed resins gives the researcher novel opportunities. On the one hand, there are no restrictions to apply large polymeric backbones that are immiscible with water or that would gel too fast in a sol-gel

Figure 1. Curing by hydrolysis and condensation in an applied coating film

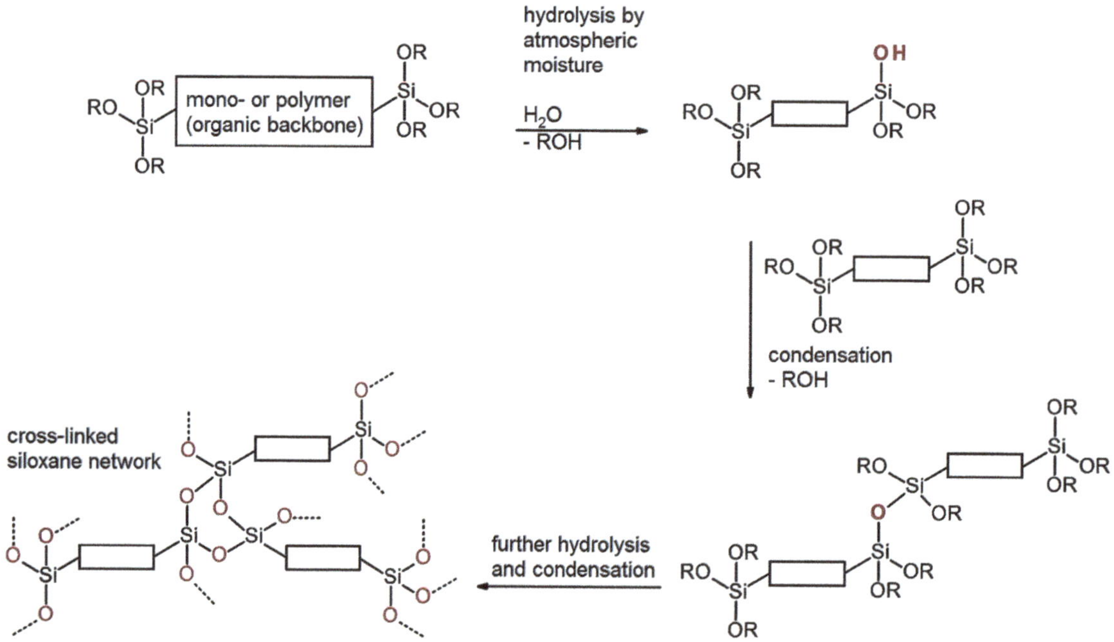

Figure 2. Patent activity in the field of epoxy-siloxane hybrid coatings. Data according to Witucki (2013)

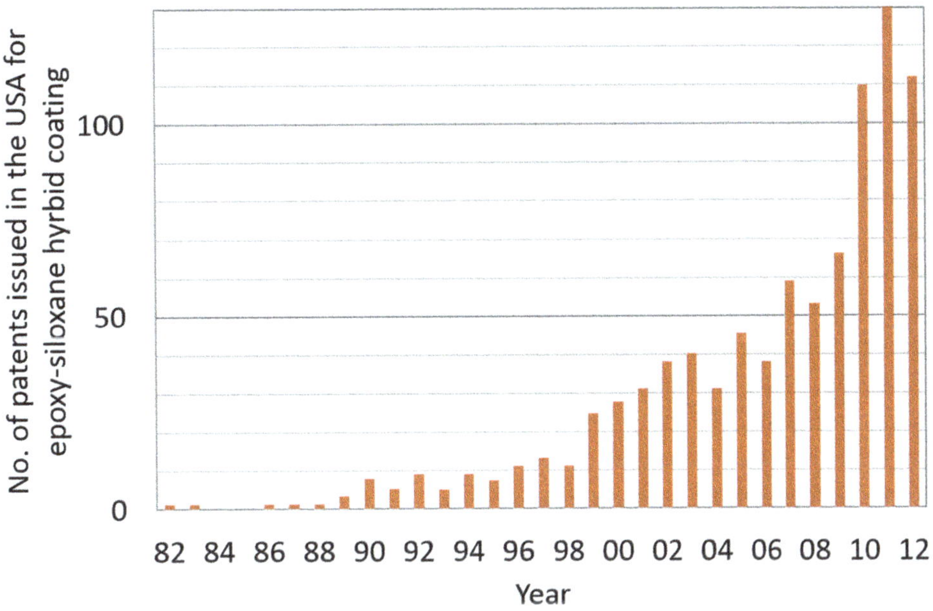

process. This expands the applicability of hybrid coatings to fields that were previously dominated by organic coatings due to a required flexibility that was difficult to match for sol-gel systems. On the other hand, solvent-free coatings can be formulated when applying smaller silane precursors with low viscosity.

This chapter outlines the general concepts and provides examples of suitable resins. In general, the present coatings do not provide one single, outstanding, unique property. This fact should not make the work any less relevant. Technical coatings have to meet complex property and cost requirements. It is by an overall improvement of a balanced property profile, that some of the coatings presented herein could outperform competing organic or sol-gel based coatings.

BACKGROUND

Technical Terms

Both "hybrid" and "sol-gel" are key words with varied definitions. The following orientation shall help the understanding, but not define who is right. Judeinstein & Sanchez (1996) and Schottner (2001) divide organic-inorganic hybrid materials in "class I" material where organic and inorganic phase are connected by weak bonds, such as van der Waals, ionic, or hydrogen bonds, and "class II" materials, where the phases are connected by a stronger covalent or partly covalent bond. The elements representing the inorganic phase may be chosen from nearly the complete periodical table. However, the inorganic part of the majority of coatings discussed in this chapter is solely based on silicon and oxygen. Most of the materials described herein represent class II hybrid where Si-C bonds of organosilanes connect organic and inorganic phase. The term "phase" is also often quite liberally interpreted. Some scientists apply a broad definition of organic-inorganic hybrid materials that includes, amongst other materials, all polysiloxanes with organic substituents, that means all materials comprising inorganic Si-O chains or networks and organic groups connected to the silicon atoms through Si-C bonds, thereby including all silicones.

Besides the silicone synthesis, an important process for the wet-chemical synthesis of organic-inorganic hybrid material is the sol-gel process (Judeinstein & Sanchez, 1996; Schmidt, Jonschker, Goedicke & Mennig, 2000; Schottner, 2001; Schubert, Huesing & Lorenz, 1995). The process includes the stages of a colloidal solution, a sol, which becomes a gel. A liberal definition according to Pierre (1998) includes any colloidal route comprising at least one of the stages of a sol or a gel. The coatings described in this chapter do not necessarily go through the steps of a colloidal solution or a gel. It is possible, that they stay homogeneous during curing, similar to organic thermoset resins. Therefore, the moisture curing process of coatings prepared from non-hydrolyzed resins is herein not regarded as sol-gel process. However, both routes share precursor material and may lead to similar coatings. Therefore, many scientists, for example Jonschker (2014, pp. 118-122), understandably include moisture-cured coatings in the group of sol-gel coatings. There can also be an overlap between both routes in case a sol-gel system is hydrolyzed with a lower amount of water than is necessary for complete condensation. In that case, atmospheric moisture can enable further cross-linking during curing.

To describe the resins discussed in chapter, various technical terms are common and used in parallel by the scientific community. Monomers with two or more alkoxysilyl groups are, for example, referred to as dipodal or polypodal silanes, bifunctional or polyfunctional silanes, or bis- or polysilanes. Larger polymers bearing alkoxysilyl groups are, for example, referred to as silane-terminated polymers (STP) or silanized polymers. To avoid long, impractical terms, organofunctional alkoxysilanes are usually solely named according to their organic function. For example, "aminosilane" identifies any silane bearing an aminoalkyl group and alkoxy groups, even though the amino group is not directly connected to the silicon atom.

Early Attempts for Moisture-Curing of Silanes

Cross-linking of resins comprising silicon-bound, hydrolysable groups by atmospheric moisture has been applied for decades, both to cure ethyl silicate binders and to cure silicone rubber. Ethyl silicate binders are primarily used to prepare inorganic zinc-rich anti-corrosive primers (Parashar, Srivastava & Kumar, 2001; Schweitzer, 2006, pp. 118-120). Hackford, Shaw and Smith (1946) were amongst the first to propose a binder based on tetraethyl orthosilicate (TEOS) that was hydrolyzed with a substoichiometric amount of water. Moisture curing is also one out of various options to cure silicone rubber. Silanol-terminated polydimethylsiloxane (PDMS) is blended with a large stoichiometric excess of a silane with hydrolysable groups. The silane acts as cross-linker when in contact with moisture (Gelest 2013, pp. 19-20; Li et al., 2006). Even though moisture-curing of non-hydrolyzed silanes has significantly profited from knowledge generated in the field of silicones, silicone rubber and silicone resins represent well-established technical fields that are not included in this chapter unless applied in combination with organic polymeric backbones.

First attempts to prepare organic, polymeric backbones bearing alkoxysilyl functions and cross-link these polymers by moisture, as shown in Figure 1, date back several decades as well. Polyacrylates with pendent alkoxysilyl groups were prepared by Plueddemann (1969) by a free-radical polymerization of blends of organic acrylates and alkoxysilanes bearing vinyl or methacryloxypropyl groups. Cross-linking resulted acrylate rubbers. Chang, Hartman & Scriven (1977) synthesized in a similar manner polyacrylates bearing methoxysilyl groups and proposed their application as hard coating films; curing at room temperature was catalyzed by dibutyltin dilaurate (DBTL). Rizk, Hsieh & Mazzeo (1985) prepared weathering stable coatings from a resin obtained by reacting a hydroxyl-functional resin with (3-isocyanatopropyl)trimethoxysilane, while Rizk & Hsieh (1986) prepared a sealer from a resin that was obtained by addition of amino- or mercaptosilanes to polyisocyanates.

THE CURING REACTION OF ALKOXYSILYL GROUPS

Basic Mechanisms and Reactivity

Hydrolysis and condensation have been the object of intense research in the past, especially due to their importance for the sol-gel process. The relevant mechanisms and determining factors have been reviewed in detail for sol-gel processing, for example, by Brinker (1988), Brinker & Scherer (1990) or Jonschker (2014, pp. 94-111), and apply largely as well for moisture curing. Therefore, this chapter solely summarize a few, selected, relevant relations.

Hydrolysis of silicon-bound alkoxy groups (≡Si-OR) occurs by nucleophilic attack of either water or hydroxid ions at the silicon atom. Silanol groups (≡Si-OH) are formed. Hydrolysis is slowest at neutral pH as the mechanism is either acid or base catalyzed. Condensation occurs according to a similar acid or base catalyzed mechanism as hydrolysis. Instead of water or a hydroxid ion, a silanol or a silanolate group attacks another silicon atom. Silanol groups condense either with alkoxysilyl group, setting free the respective alcohol, or with other silanol groups, setting free water. Figure 3 shows a reaction scheme for acid and base catalyzed hydrolysis and condensation.

The reactivity of silicon-bond hydrolysable groups depends on the nature of the hydrolysable group itself, on the further substituents of the silicon atom and, if applied, on amount and kind of a catalyst. The rate of condensation depends, in addition to the factors relevant for hydrolysis, on the probability that two reactive groups meet each other. Thus, the condensation rate depends both on the density of reactive groups and on the resin viscosity. The latter obviously increases significantly during curing. Therefore, one should expect that in general even cured films do not provide 100% condensation and thus comprise remaining silanol groups.

Under completely dry conditions, no hydrolysis occurs. Even though the availability of moisture obviously has an influence on the hydrolysis rate, it is in praxis under typical conditions and film thicknesses not necessarily considered as the main rate-determining factor. For example, moisture cured silicone rubber used to prepare molds for casting can be applied even at thicknesses above 1 mm. Despite

Figure 3. Reaction scheme of hydrolysis and condensation of alkoxysilanes

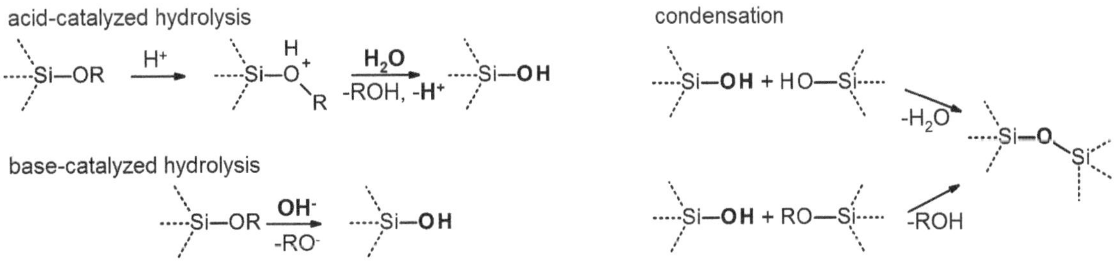

the necessary diffusion of water, fast reactions are possible. De Paz, Chemtob, Croutxé-Barghorn, Le Nouen & Rigolet (2012) followed the hydrolyses of a 0.9 µm thick poly(dimethyl silicate) film at room temperature and 30% relative humidity. Starting at t=0, a photo-latent acid was subsequently activated by UV-radiation. Already after 11 s, half of the methoxy groups were hydrolyzed.

Hydrolysable Groups

The by far most common hydrolysable groups are methoxy and ethoxy. Ethanol has the advantage of being a less toxic release product than methanol. Methoxy groups are, on the other hand, more reactive. Brand et al. (1999) investigated the hydrolysis rates of various commercial trialkoxysilanes in water/acetone mixtures at pH 4 and found that methoxysilanes react between 2.6 to 66 times faster than the respective ethoxysilanes. Larger alkoxy groups, especially when being part of an organic polymer, hydrolyze magnitudes slower than ethoxy groups. Such Si-O-R groups, with R being polymeric, can actually be part of weathering stable coatings without significant hydrolysis. This is the case for some siloxane-epoxy hybrid resins described further below or for silicone-modified polyol resins. Polyols are organic, hydroxyl-functional resins.

For special coatings, silazanes comprising silicon-bound aminogroups are applied, usually as rings and chains of the general formula $[SiR_2\text{-NH-}]_n$. The amino groups are more reactive than alkoxygroups and set free ammonia during hydrolysis. Further hydrolysable groups are common for the cross-linking of silicone by atmospheric moisture, such as, for example, acetoxy, setting free acetic acid, methylethenoxy, setting free acetone, 1-methyl-1-propanonoximino, setting free butanonoxime, and alkylamino setting free alkylamines (Gelest, 2013, p. 18). In industrial synthesis, the precursors for most alkoxysilanes are silanes bearing chloride and hydride groups. Chlorides react vigorously with water setting free corrosive hydrogen chloride and are generally not consider for use in moisture curing coatings. Hydrid groups (Si-H) may be present in coating resins, for example in combination with amino groups. Besides the possible addition to carbon-carbon or carbon-oxygen double bonds, hydrid groups react with hydroxyl-functional molecules including water, alcohols or silanol groups under evolution of hydrogen. The reactivity depends on the applied catalysts (Gelest, 2013, pp. 13-14).

Depending on the number of hydrolysable groups, different condensation products are obtained. Silicates ($Si(OR)_4$) lead to inorganic silica (SiO_2), while alkoxysilanes lead to siloxanes. The condensations products of trialkoxysilanes are due to their sum formula $[SiRO_{1.5}]_n$ also called silsesquioxanes. Sesqui is the latin expression for one-and-one half times. Silsequioxanes can form various structures such as, for example, cages, ladders or random amorphous structures. A random structure is illustrated by Figure 4. Dialkoxysilanes lead to rings or chains; monoalkoxysilanes lead to disiloxanes.

Figure 4. Condensation of a trialkoxysilane to form a random silsesquioxane structure

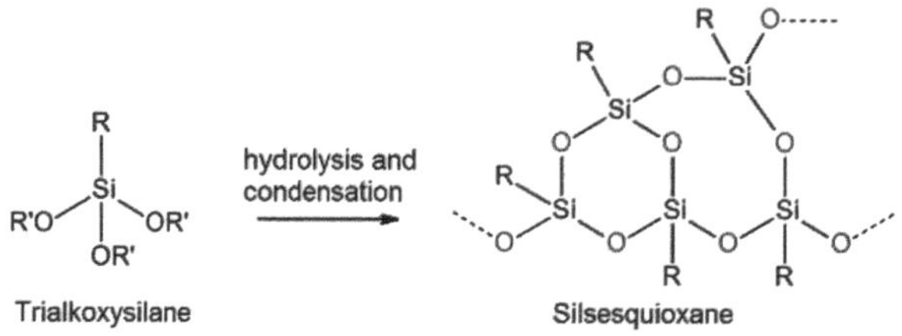

Substituent Effects

Substituents with electron drawing and electron pulling effects on the silicon atom influence the reactivity. However, the effects on reactivity depend on the underlying mechanism and are without this knowledge not readily predictable. Sterically demanding substituents decrease the reactivity. (Brinker & Scherer, 1990, pp. 119-123 and pp. 142-143). Substituents comprising Lewis bases, such as, for example, an amino group, can both act as catalyst by lowering the pH, and can directly interact with an Si-atom to increase the reactivity (Beari et al., 2001). The catalytic effect of basic groups depends on the basicity of the basic group, but also on the position. Many commercial organosilanes comprise an oxygen or nitrogen atom in γ-position to the Si-atom, such as, for example, (3-aminopropyl)trimethoxysilane (H_2N-$(CH_2)_3$-$Si(OCH_3)_3$). Oxygen or nitrogen atoms in α-position to the Si-atom, with only one methylene group between Si- and O- or N-atom, as for example in (phenylamino)methyltrimethoxysilane (Ph-NH-CH_2-$Si(OCH_3)_3$, provide a significantly stronger catalytic effect; the free electron pair of the N- or O-atom interacts stronger with the silicon atom (Wacker Chemie, 2012a). Examples for the effects of hydrolysable groups (methoxy or ethoxy) and the position of oxygen or nitrogen atoms in the organic side chain on hydrolysis rate and curing behavior are provided by Figures 5 and 6.

Catalysts and Water Scavengers

Coatings comprising alkoxysilanes bearing an alkylamino function have been cured without any additional catalyst, even at room temperature. For other alkoxysilanes, the rates of hydrolysis and condensation without catalyst are in most cases considered too low for their application in technical coatings. Thus, frequently catalysts are added.

The catalysts can be divided into acids, bases, nucleophilic catalysts and metalorganic complexes, each with different effects on hydrolysis and condensation. Brinker & Scherer (1990) compared acidic and basic catalysis when investigating the sol-gel process of TEOS. While hydrolyses is faster than condensation under acidic conditions, this is reversed under basic conditions. Even though the basicity of the fluoride ion (F^-) is low, it effectively catalyzes both hydrolysis and condensation as nucleophilic catalyst (Brinker & Scherer, 1990; Ellsworth & Novak, 1991).

Figure 5. Rate constant k in h^{-1} for the hydrolysis of diluted solutions of the respective silanes in a water/acetone mixture adjusted with HCl_{aq} to pH 4 according to Brand et al. (1999) and Wacker Chemie (2012a), showing the influence of the hydrolysable group and the position of the O-atom in the non-hydrolysable substituent

Me = methyl, Et = ethyl

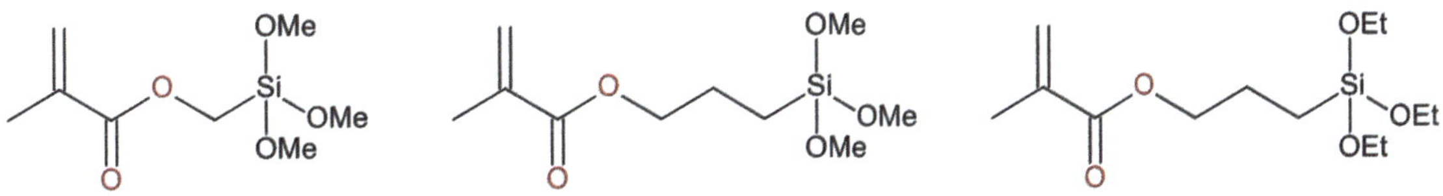

~21 (Wacker Chemie 2012a) ~6.9 (Brand et al. 1999) ~1.5 (Wacker Chemie 2012a) ~0.44 (Brand et al. 1999)

Figure 6. Skin formation time of complete adhesives formulations including catalysts and further ingredients, when exposed to atmospheric moisture. The formulations are based on linear polymer resins with terminal alkoxysilyl groups with different reactivity depending on the hydrolysable groups and the proximity of the N-atom in the polymer chain relative to the Si-atom (Wacker Chemie, 2012a)

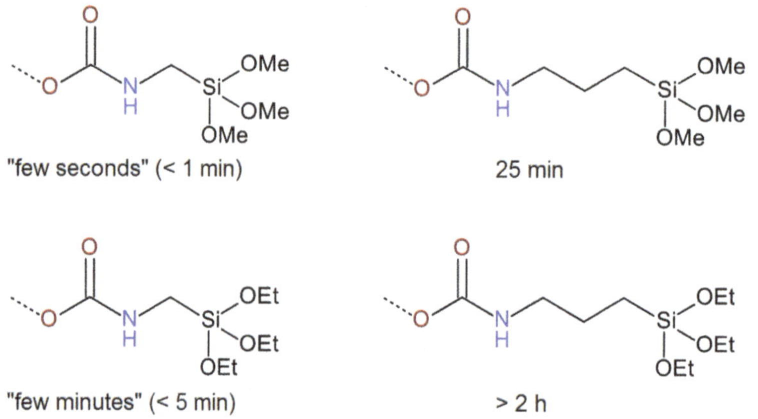

Lee, Ma, Hsu & Wu (2005) cured precursors comprising aminosilanes bearing either one, two, or three ethoxy groups without catalyst or by adding 1% of either the Lewis acid borontrifluoride monoethylamine complex, the base tetrabutylammonium hydroxide or metalorganic DBTL. Curing was investigated by ^{29}Si-NMR. Borontrifluoride was the most effective catalyst for the condensation of di- and trialkoxysilyl groups. Under the present conditions, a degree of condensation, based on the total number of hydrolysable groups, of about 91% was achieved, while the not-catalyzed resin reached a degree of condensation of about 76%. Rekondo et al. (2006) compared p-toluenesulfonic acid with metalorganic DBTL. The acid strongly catalyzes the hydrolysis reaction. Different from the mechanism shown in Figure 1, many silanol groups (Si-OH) are formed before significant condensation occurs. DBTL, on the other hand, has a minor effect on the hydrolysis, but catalysis the condensation; silanol groups are instantly consumed as indicated by Figure 1.

The multiplicity of catalysts that have been applied cannot be completely summarized herein; solely individual examples are provided. Acid catalysts include strong Brønsted acids such as p-toluenesulfonic acid or trifluoroacetic acid. Optionally, thermally activated acids can be used, such as, for example, blocked phosphoric acids esters (Groenewolt et al., 2008) or blocked sulfonic acids (Speier, Plogmann, Lomoelder & Hoppe, 2003). Examples for Lewis acids are boron trifluoride complexes or metal alkoxides and β-dicarbonyl complexes, such as aluminum acetylacetonate (Laryea, Thurn, Sepeur & Schlick, 2008a), aluminum sec-butoxide/ethylacetoacetate (Poppe, Klein, Brosseit, Wermelt & Möllers, 2006), cerium(III) acetylacetonate (Holberg, Bischoff, Poulsen & Tabrizian, 2013), tetraisopropyl titanate (Plueddemann, 1969), diisopropoxy-bis(ethylacetoacetato)titanate (Bischoff & Holberg, 2012), or zirconium acetylacetonate (Laryea, Thurn & Sepeur, 2008b).

Ammonium hydroxides, amines or amidines (Sangermano at al., 2014) have been applied as basic catalysts. The respective paragraph on UV-curing provides references for photo-latent acids and bases. DBTL has frequently been applied as metalorganic catalyst (Plueddemann, 1969; Chang et al., 1977). Gelest (2013, pp. 60-61) considers further tin-based metalorganic catalysts and further titanium complexes. However, tin-based catalysts are controversial due to their negative impact on health and environment.

Impurities of water from solvents, fillers or further coating ingredients can, under unfavorable conditions, lead to gelation of a coating resin before it has been applied, especially when effective catalysts are present. To avoid this, most adhesives and some coatings apply water scavengers such as triethylorthoformate (Dowbenko, Stewart, Hartman, Gorman & Thomas, 1985) or monomeric silanes with a higher reactivity than the alkoxysilyl groups of the resin, such as, for example, vinyltrimethoxysilane or silanes with an O- or N-atom in α-position of the silicon atom (Wacker Chemie, 2012a).

APPLICATION EXAMPLES

The present examples of organic-inorganic hybrid coatings prepared from non-hydrolyzed resins are arranged according to their respective organic backbone.

Epoxy-Amine and Acrylate-Amine Systems

Organic, two-component, amine-cured epoxy coatings are widely used, especially as protective coatings (Schweitzer, 2006, pp. 109-110). There are various options to modify these coatings with silanes. One way is to add an amino- or epoxy-functional silane to the respective amine or epoxy component. Van Ooij, Suryanarayanan & Metroke (2007) discus both the addition of non-hydrolyzed and of hydrolyzed silanes to a bisphenol A-based epoxy coating to improve the corrosion protection of galvanized steel. The integration of silanes in the coating resin is an alternative to applying a silane-primer prior to applying the organic epoxy coating and does not only improve binding at the metal-coating interface, but also the hydrophobicity of the coating. Brusciotti et al. (2013) modified a bisphenol A-based epoxy coating by addition of about 7% non-hydrolyzed aminosilanes, based on solids. The modified coatings protected magnesium alloy AZ31 such that no relevant degradation occurred during immersion in 0.05 M NaCl for one month.

A second option is to prepare, in a first step, an alkoxysilyl-terminated resin by reacting epoxides with aminosilanes. In a second step, the precursor, either neat or in solution, is applied to a surface and cured by atmospheric moisture to cross-link the alkoxy-silane functions. As a side reaction, alkoxysilyl groups might condense with alcohol groups that are part of the epoxide or that are formed by the epoxide-amine reaction. This may lead to gelation and explain, why certain precursors are dissolved in an excess of silanes or in alcohols to repress this condensation. Laryea et al. (2008a) prepared a precursor by reacting 1 mol (3-Glycidyloxypropyl)triethoxysilane with 1 mol (3-Aminopropyl)triethoxysilane. By applying the precursor solution to aluminum, followed by curing at 120 °C, a scratch and corrosion resistant coating was prepared.

Monkiewicz, Edelmann, Edinger & Borup (2006) reacted a bisphenol-A-based epoxide comprising 1 mol epoxy groups, dissolved in ethanol and tetraethyl orthosilicate, with either 0.95 or 1.9 mol (3-aminopropyl)triethoxysilane. The storage stability of the latter solutions is according to Degussa (n. d.) greater than one year. A coating film of 3.3 μm dry film thickness on aluminum, prepared from the solution comprising 0.95 mol aminosilane, was cured at 22 °C. After 7 days, the film provided a pencil hardness of 4H according to ISO 15184. Adhesion after 2 days immersion in water was still perfect as tested by a cross-cut and tape test according to DIN 53151. Lee et al. (2005) investigated the curing behavior of similar precursors.

Chattopadhyay, Muehlberg & Webster (2008), Chattopadhyay & Webster (2009a) and Chattopadhyay, Zakula & Webster (2009b) prepared a precursor from an epoxy-functional urethane and various aminosilanes. In a preceding step, the epoxy-functional urethane was prepared by reacting a polyisocyanate with glycidol. Optionally, TEOS was added to the precursors. Moisture-cured coatings of the resins on aluminum provided good adhesion.

An alternative to epoxy-amine precursors are acrylate-amine precursors. Amines react in a Michael-analogous addition with unsaturated acrylic acid esters, as shown in Figure 7. Degussa (n.d.) describes a precursor prepared by reacting a tetra-functional aliphatic urethane acrylate comprising 1 mol acrylate groups, dissolved in ethanol and tetraethyl orthosilicate, with 1 mol (3-aminopropyl)triethoxysilane for 4 h at 20-30 °C. Applied films with 3 µm dry film thickness cure within 12 h at room temperature. Monkiewicz et al. (2006) describe similar acrylate-amine precursors and the optional addition of perfluoroalkylsilanes to theses precursors to prepare anti-graffiti coatings.

The most significant commercial success for hybrid epoxy-amine and acrylate-amine systems was the introduction of alkoxyfunctional siloxane resins. These siloxane resins, also called silicone resins, are oligomers or polymers formed by reaction of chlorosilanes with water and alcohol, primarily comprising both linear [SiR_2-O-] and branched [SiR-$O_{3/2}$-] units with terminal, hydrolysable methoxy or ethoxy groups. R is typically methyl or phenyl.

Foscante, Gysegem, Martinich & Law (1981) prepared films from blends of an epoxy resin, an alkoxyfunctional siloxane resin and [3-(2-aminoethylamino)propyl]trimethoxysilane. During curing, the amino groups of the silane react with the epoxy groups. Under availability of atmospheric moisture, the alkoxy groups of the aminosilane co-condense with alkoxygroups of the siloxane resin. The aminosilane thus connects the siloxane network with the epoxy-amine network. Hinzmann, Klotzbach & Herrwerth (2012), Howard, Hallack, Schlemann & Bartek (2008), Witucki (2006), and Wood, Schiemann & Hallack (2005) outline further development, especially focusing on the application of aliphatic epoxides based on hydrogenated bisphenol-A and on the formulation of pigmented, room-temperature-curing two-component coatings. The resin components are blends of an epoxy resin, an alkoxy-functional siloxane resin, pigments and additives; the hardeners are aminosilane. An example for such resin is shown in Figure 8 in the first row. In an analogous attempt to the siloxane-epoxy-systems, Kelly, Birkert & Andrews (1998) cured blends of alkoxyfunctional siloxane resins and various organic acrylates with (3-aminopropyl)triethoxysilane.

Different variations of these resins are possible. For example, the resin blend may already comprise an epoxy-siloxane hybrid polymer connected by a chemical bond. According to Hinzmann et al. (2012), such chemical bond is formed by a reaction of hydroxyl groups, which are present in epoxy resins oligomers, and alkoxygroups of the siloxane resin. The reaction is illustrated in the second row of Figure 8. An option to adjust the rate of the amine-epoxide reaction during curing is according to Foscante et al. (1981) the addition of ketone solvents. The ketones condense with primary amines to form ketimines and water. The primary amine function is blocked; the water is consumed by hydrolysis and condensation of

Figure 7. Addition of primary amines to acrylates (left) and reversible formation of ketimines (right)

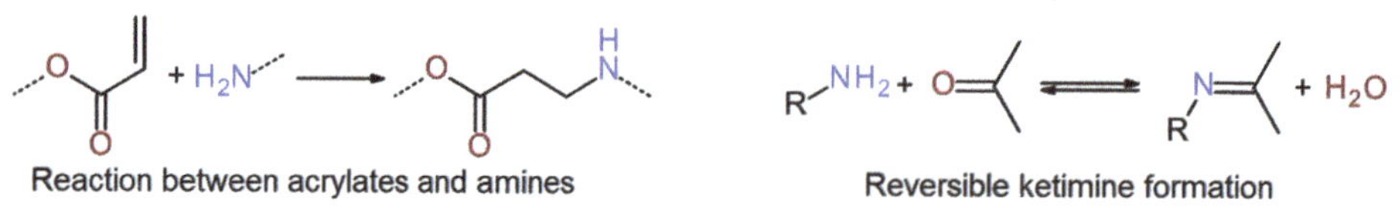

Figure 8. Possible resin and hardener combinations for two-component epoxy-siloxane hybrid coatings

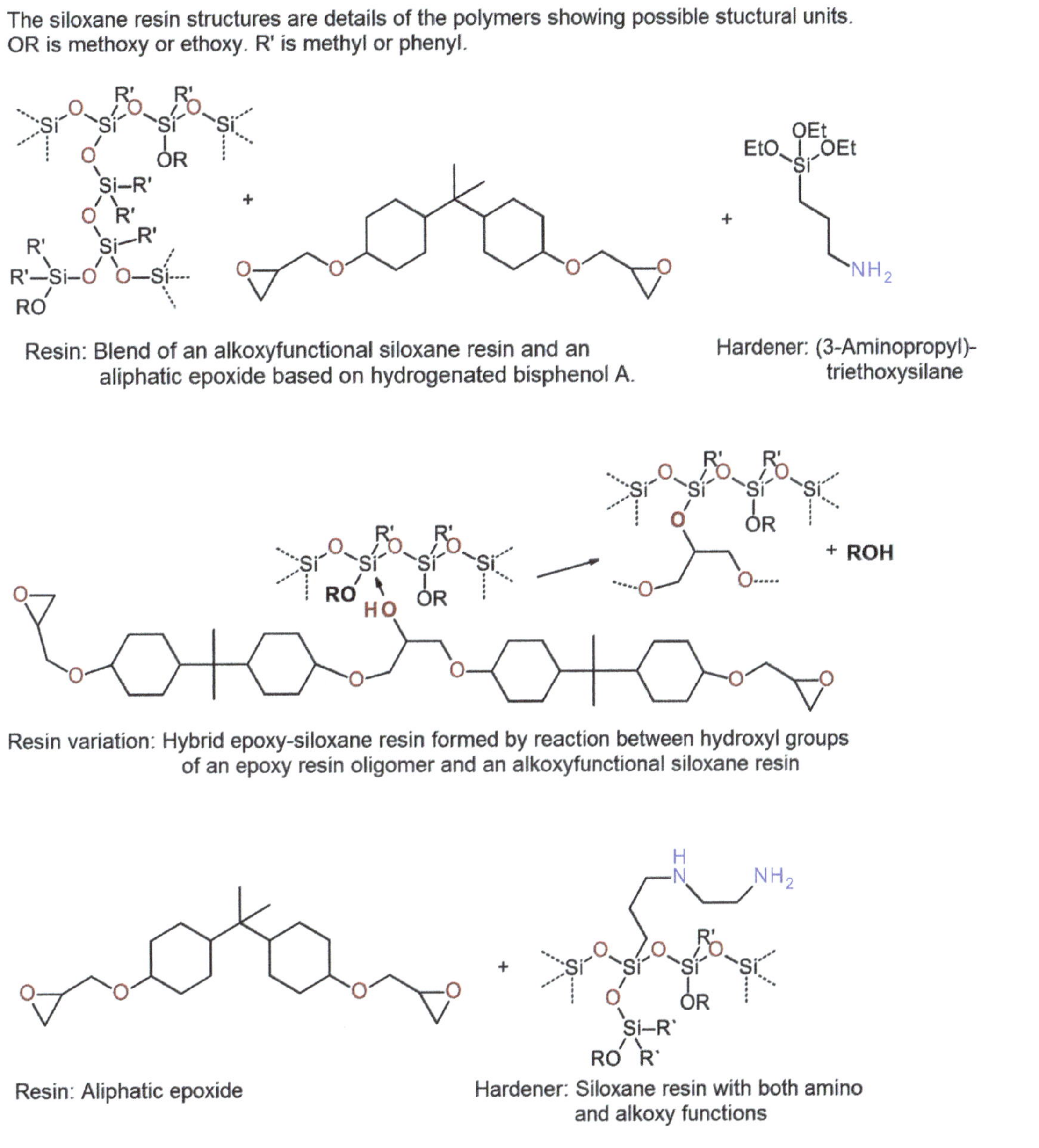

alkoxygroups. The reaction is reversible. Atmospheric moisture will gradually hydrolyze the ketimines to make the amine function available, see Figure 7 (right). Laubender & Greene (2006) and Wacker Chemie (2012b) report a further alternative by using solely an aliphatic epoxide as resin component, but a siloxane resin as hardener that is both amino- and alkoxysilyl-functional, see bottom row in Figure 8.

Both acrylate-siloxane and epoxy-siloxane hybrid coatings gained commercial success due to their specific property profile. Their main field of application are surfaces requiring both corrosion protection and good optical appearance, such as construction (e.g. buildings, bridges and industrial facilities), marine use or railroads. In these applications, typically one layer of an organic polyurethane top coat providing weathering stability is combined with an organic epoxy base coat providing chemical and corrosion resistance. Solely one layer of a high-solids hybrid coating can replace both organic top and base coat. (Wood et al. 2005). Even though the raw materials for the hybrid coatings are more costly, the overall application costs can be reduced by replacing one work step and reducing the total film thickness.

Laubender & Greene (2006), Wacker Chemie (2012b) and Wood et al. (2005) provided results of tests comparing epoxy-siloxane hybrid coatings with both organic epoxy coatings and organic polyurethane coatings in long-term QUV-B weathering. The organic epoxy coatings, as expected, fail after 200-500 h test duration. The epoxy-siloxane coatings perform at start likewise polyurethane coatings, but outperform polyurethane coatings in terms of gloss retention and color retention, depending on the respective test systems, after 1000 to 4000 h. According to Wood et al. (2005), epoxy-siloxane hybrid coatings provide also better resistance to acids and bases than polyurethane coatings, show no loss in gloss after removal of graffiti and allow the formulation high solids coatings with a low amount of volatile organic compounds (VOC). A sample formulation comprises solely 5% solvent. VOC differs from solid content in that it includes both solvents and cleavage products that evaporate during curing, such as, for example, alcohols when curing alkoxysilanes. The commercial resins are provided, for example, as 'silicone epoxy hybrid' or 'aminofunctional, methyl phenyl silicone resin', complete coating systems as 'polysiloxane topcoat', 'polysiloxane enamel' or 'acrylic polysiloxane'.

Isocyanate-Based Systems

Isocyanate-based alkoxysilane precursors are prepared by reaction between silanes and a di- or polyfunctional monomer or polymer. Either the silane or the organic mono- or polymer bears isocyanate groups (-NCO) that react with either hydroxyl groups (-OH), thiol groups (-SH) or amino groups ($-NH_2$/-NRH) of the counterpart to result the respective urethanes, thiourethanes or ureas. In commercial application, these precursors compete with organic coatings, especially aliphatic, thermoset polyurethane coatings prepared from polyol resins and isocyanate hardeners. The polyol resins are usually hydroxyl-functional polyester or polyacrylate resins. The organic polyurethane coatings are known for weathering stability and scratch resistance (Schweitzer, 2006, pp. 113-115) and outperform organic and hybrid epoxy coatings with respect to scratch resistance.

Coatings prepared from isocyanate-derived alkoxysilanes maintain weathering stability and a certain flexibility of the aliphatic polyurethane coatings, while even further improving scratch resistance and improving chemical resistance by a hybrid polymer that, after curing, partly comprises siloxane structures. As the hybrid coatings are more costly than their solely organic counterparts are, the hybrid coatings are mainly used in applications that are demanding with respect to optical appearance, for example as automotive top coat or as scratch-proof surfaces. Further motivation to apply isocyanate-based hybrid coatings is the opportunity to achieve solvent-free coatings and/or isocyanate-free coating resins that cause a lower health risk for the applicator due to complete conversion of the isocyanate groups during resin synthesis. Accordingly, the respective resins are marketed, for example, as 'isocyanate free silicon cross-linking urethanes'. Novel options to incorporate an additional functionality, such as, for example, a non-stick effect have also contributed to the interest in isocyanate based hybrid coatings.

Today, isocyanate-derived silane precursors are commercially applied in different fields, for example, as part of top coats with improved scratch resistance in automotive mass production (Metal Finishing, 2012). Besides the application in coatings, silane terminated polyurethanes and polyethers are applied as a main binder-component for moisture curing structural adhesives and sealants (Oertli & Jocker 2013; Schindler

Figure 9. Example of a silane-terminated polypropylene glycol according to Oertli & Jocker (2013) and Schindler & Schwiebacher (2009)

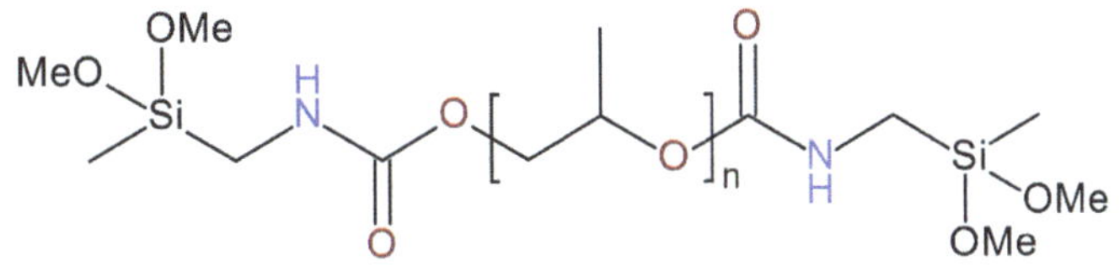

& Schwiebacher, 2009; Stanjek, Bachmeier & Zander, 2014; Wacker Chemie, 2012a). An example for such binder is polypropylene glycol with a molecular weight of 10.000 to 30.000 g/mol, end-capped by reaction with an isocyanatosilane, as shown in Figure 9. The sealants and adhesives further comprise plasticizers, inorganic fillers, catalysts and monomeric alkoxysilanes that act as adhesion promoters.

The reaction of amines with isocyanates occurs almost instantly under generation of heat and requires thus no catalyst. The reaction of hydroxyl or thiol groups is significantly slower and thus often catalyzed, for example by organotin catalysts and/or tertiary amine bases. The hydroxyl group can undergo a side reaction by reacting with a silicon-bound alkoxygroups in a transestrification reaction, setting free the respective alcohol. Suitable catalysts selectively promote either of the two the reaction (Unkelhäußer, Görlitzer, Hallack & Lomölder, 2014). For example, Stanjek, Jung & Schauer (2004) were able to react (isocyanatomethyl)triethoxysilane, which comprises especially reactive alkoxy groups due to the isocyanato group in α-position to the Si-atom, with a hydroxy-functional resin without any detectable products of a side reaction between hydroxyl groups of the resin and alkoxygroups of the silane.

Rekondo et al. (2006) compared urea with urethane groups. Poly(propylene glycol) was either reacted with an isocyanatosilanes to result terminal urethane-alkoxysilyl groups, or in a first step reacted with a di-functional isocyanate, and in a second step with an aminosilane to result terminal urea-alkoxysilyl groups. The curing of films by moisture was investigated by FTIR-spectroscopy. For both systems, the hydrogen-bridging interactions increase during curing and lead to self-association of urea and urethane groups in the polymer network. However, the urea groups lead to stronger hydrogen bridging than urethane groups. This affects the physical properties of the film. For example, urea groups lead to a higher glass transition temperature.

Groenewolt et al. (2008), Groenewolt & Müller (2012a) and Speier et al. (2003) describe two-component polyurethane coatings with improved scratch resistance and chemical resistance, where solely a part of the isocyanato groups of the hardener component was reacted with an aminosilane. Polyol resins mixed with these hardeners cure according to a dual-cure mechanism. Isocyanate groups react with the hydroxyl-functional polyol, while the cross-linking of alkoxysilyl groups is induced by moisture. According to Groenewolt et al. (2008), a good combination of scratch resistance and weathering stability, for example for automotive applications, is obtained by applying a hardener component based on an isocyanurate-trimer of hexamethylene diisocyanate, where of all initial isocyanate groups ~70% remained unreacted, ~9% had reacted with [3-(N-butylamino)proyl]trimethoxysilane and ~21% had reacted with bis[3-(trimethoxysilyl)propy)lamine. Figure 10 illustrates one out of various possible structures of the hardener.

Groenewolt, Klein, Hilge, Austrup & Huebner (2012b) found for similar coatings as described by Groenewolt et al. (2008), that it is especially easy to remove scratches by polishing, while Groenewolt, Hesener, Stubbe, Niemeier & Westhoff (2009) propose to react isocyanate resins with mercaptosilanes

Figure 10. Possible structure of a modified isocyanate hardener for two-component polyurethane coatings according to Groenewolt et a. (2008)

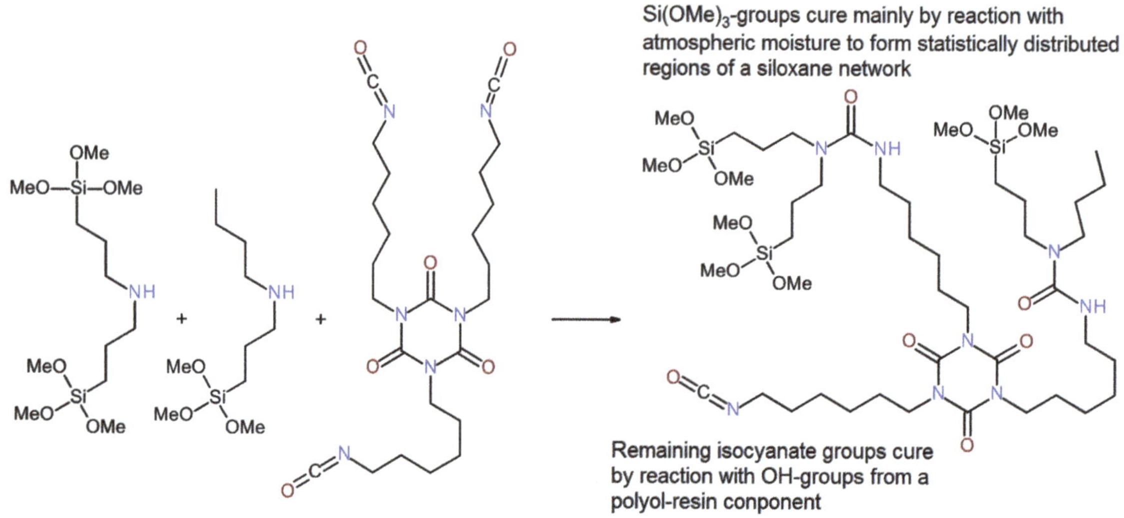

instead of aminosilanes. The synthesized thiourethanes provide a higher refractive index than ureas and lead to coating films with increased gloss.

When applying solely an alkoxysilyl-functional isocyanate resins without addition of a polyol resin, both the alkoxysilyl function and the isocyanate function both crosslink by moisture to form a coating film. The isocyanates cure to form urea groups via the intermediate formation of amines. Solvent-free resins have been prepared by reaction of a stoichiometric excess of the difunctional isocyanate isophorone diisocyanate with various precursor resins comprising both hydroxyl and alkoxysilyl functions. Allauddin, Narayan & Raju (2013) and Gireesh et al. (2010) prepared precursors by reaction of an epoxysilane with the acid groups of polyester resins comprising both terminal acid and hydroxyl groups. Gireesh et al. (2010) prepared the polyester resin from castor oil and succinic anhydride. Castor oil is a bio-based raw material, a triglyceride with hydroxyl-functions in the fatty acid chains. Shaik, Narayan & Raju (2014) applied blends of castor oil and a diol prepared by reaction of 1 mol aminosilane with 2 mol epoxysilane as precursor. Besides improving the mechanical properties, the addition of the silane-based diol to the castor oil also significantly improves solvent resistance and reduces bacterial growth.

Arun, Kantheti, Gadaam, Narayan & Raju (2014) describe a similar route to alkoxysilyl/isocyanate resins incorporating TiO_2. Dry TiO_2 nanoparticles were modified with a melamine alkoxysilane, an organic diol resin was added and the blend was reacted with an excess of a difunctional isocyanate. Compared to the pure organic polyurethane, the addition of the silane-modified TiO_2 improves the corrosion resistance of mild steel and incorporates hydrophilicity, leading to reduced bacteria growth.

The complete conversion of iscyanate groups during resin synthesis leads to resins that cure solely via the alkoxysilyl function. Fenn & Davies (1993) and Poppe et al. (2006) propose blends of two resins, where the first resin is the reaction product of an aminosilane and an organic isocyanate, and the second resin is a polyacrylate/polymethacrylate polymer. By incorporating silanes with double bonds in the polymerization of the polyacrylate/polymethacrylate, these resins comprise alkoxysilyl groups as well. Poppe et al. (2006) obtained flexible transparent films with excellent scratch resistance.

As an alternative to achieve scratch resistant top coats, but with a low amount of VOC, Gregorovich, Hazan & Wilczek (2001), Laryea et al. (2008a), Lomoelder, Görlitzer, Hallack & Unkelhäußer (2013), and Sepeur, Laryea, Goedicke & Groß (2008, pp. 139-143) prepared moisture-curing resins by reaction of isocyanatosilanes with di-or polyfunctional alcohols or polyol resins. Also with focus on low-VOC, Laryea et al. (2008b) applied the reaction product of 1,6-hexanediol with (3-isocyanatopropyl)triethoxysilane as powder coating. Choice of specific polyol resins allows the introduction of further functionalities. Laryea et al (2008b) and Unkelhäußer et al. (2014) applied fluorinated polyols, such as, for example, a perfluoropolyether, to prepare hydrophobic and oleophobic coating films with water contact angle of up to 113°. Laryea et al. (2008b) prepared also hydrophilic coatings with an anti-fog effect by applying poly(tetrahydrofuran) as polyol.

Hydrophobic, repellent coatings were in the focus of further work. Bischoff & Holberg (2012) report a method to chemically cross-link silanol-terminated polydimethylsiloxane (PDMS) to aminosilanes prior to the reaction of aminosilanes with an isocyanate resin. On a cured film, the PDMS provides a hydrophobic and oleophobic surface and avoids crude-oil-derived fouling, for example on heat exchangers. Huang et al. (2012a, 2012b) formulated repellent coatings from various moisture curing resins prepared by reacting an isocyanatosilane with either a hydroxy-functional polybutadiene or a polyester polyol. To achieve repellency, either a perfluoroalkylsilane was added to the obtained alkoxysilyl-functional resin, or some of the alkoxy groups of the obtained resin were condensed with silanol-terminated PDMS. 300 μm thick films cured at room temperature provided water contact angles of up to 113° and were recommended against marine fouling. In a pseudo-barnacle test, where a dolly glued to the coating surface simulates a barnacle, adhesion was between 0.2 and 0.5 MPa.

Reaction products of isocyanates and amines are also reported to provide wet-adhesion to aluminium (Plassmann, 2001) or to be suitable as primer for polycabonate (Hofacker, Mechtel, Hovestadt & Kobusch, 2002). Holberg & Bischoff (2014) applied an aminosilane-isocyante adduct both as moisture cured coating and, comprising additional methyltriethoxysilane, as hydrolyzed sol-gel coating. While both coatings provide wet-adhesion to aluminum and stainless steel, corrosion protection of aluminum, as tested by salt-spray test, was significantly better for the sol-gel system. However, according to Holberg, Bischoff, Poulsen & Tabrizian (2013), filiform corrosion under the moisture cured coating can be inhibited by replacing a titanium complex applied as curing-catalyst by cerium(III) acetylacetonate.

UV-Cured Systems

Three main types of initiators are applied to UV-cure organic coatings: radical, cationic (acidic) and basic initiators. The free-radical polymerization of carbon-carbon double bonds is the route that is most frequently applied by the industry. This mechanism is also suitable to cross-link unsaturated side chains of silanes, such as, for example, vinyl or or methacryloxy groups. However, radical initiators do not catalyze the curing of alkoxysilyl groups by moisture. Besides the polymerization of carbon-carbon double bonds, these double bonds can also react in a radical addition with thiols to form sulfides. Due to its selectivity and mild conditions, the thiol-ene reaction is part of the so-called click chemistry (Hoyle, Lowe & Bowman, 2010).

Cationic photoinitiators, commonly aromatic iodonium or sulfonium salts, split of a phenyl cation when exposed to UV-radiation. Via subsequent reactions, a proton, and thus a very strong Brønsted acid is formed, such as, for example, hexafluorophosphoric acid (Dow Chemical Company, 2001). The acid catalyzes the polymerization of organic epoxides, especially cycloaliphatic epoxides, or vinylethers to form polyether chains. The acid also effectively catalyzes the curing of silicon-bound alkoxy groups by atmospheric moisture (Chemtob, De Paz-Simon, Crouxté-Barghorn & Rigolet, 2014a). Thus, both epoxy and alkoxysilyl function can be cured simultaneously. A variety of suitable resins can be combined.

Organic and inorganic network can be chemically connected via epoxysilanes, but this is not mandatory and depends on the chosen precursors. The simplest resins are sole epoxysilanes; an example is shown in Figure 11 (Chemtob, Versace, Belon, Croutxé-Barghorn & Rigolet 2008). Further options are blends of epoxy silanes with organic epoxides and/or further silanes or silicates, or blends of organic epoxides with silanes or silicates without an epoxy group (Amerio, Sangermano, Malucelli, Priola & Voit, 2005; Belon et al., 2010a; Croutxe-Barghorn, Chemtob, Moreau & Bouder, 2013a). Combinations of organic acrylates and silanes bearing an acrylate or methacrylate group can as well be cured simultaneously when combining a radical initiator to cross link the (meth)acryl functions and a cationic initiator to catalyze cross-linking of the alkoxysily-functions as shown in Figure 12 (Belon et al., 2010b).

Basic photoinitiators are initially neutral or very weak bases. When exposed to UV radiation, they decompose. Subsequent reactions result strong amine or amidin bases (Sangermano et al., 2014). In organic coatings, these catalysts are applied, for example, to catalyze Michael-additions (Noomen, 1997) or to react organic thiols with isocyanates or epoxides. Basic photoiniators are, as well, suitable to catalyze the moisture-induced hydrolysis and condensation of alkoxysilanes (Bachon 2005; Yang, Semonick, Moore, Boardman & Craton 2012). According to Chemtob et al. (2014a), photobases are, when compared to cationic catalysts, less effective. However, in a UV-induced formation of silica films by moisture, photobases led to more homogeneous films than cationic initiators (Chemtob et al., 2014b). Like cationic initiators, latent photobases can as well be combined with radical initiators (Chemtob, Courtecuisse, Croutxé-Barghorn & Rigolet, 2011).

The general advantage of UV-curing is curing on demand without using energy for heating. An additional advantage is often the absence of organic solvents due to the application of organic monomers and silanes with low viscosity. Thus, UV/moisture-cured inorganic-organic hybrid coatings have been investigated for various applications. Solvent-free, cationically cured coatings comprising anti-corrosive pigments were applied from different resins as environmental friendly protection for aluminum. Up to 3000 h salt spray test did not lead to corrosion. For this purpose, Croutxé-Barghorn, Chemtob, Ni, De-

Figure 11. Schematic of synchronous cationic curing of both epoxides and alkoxysilyl groups of a silane Chemtob et al. (2008)

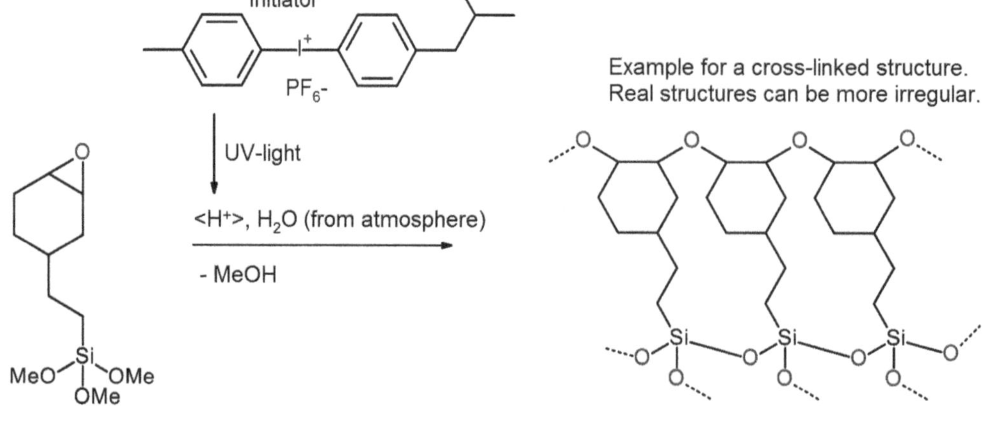

Figure 12. Schematic of synchronous curing of acrylate groups by a radical mechanism and alkoxysilyl group by acid catalysis according to Belon (2010b)

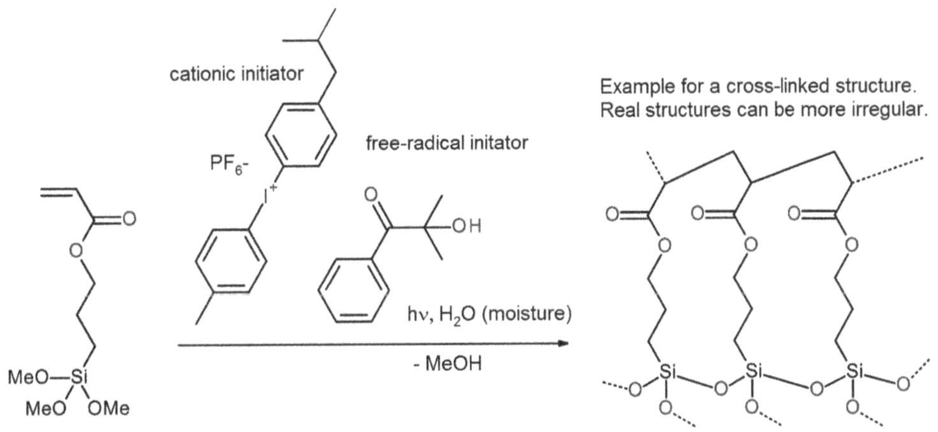

Paz-Simon & Belon (2013c) applied epoxysilanes and organic epoxides, and propose the application as primer in the aviation industry. Croutxe-Barghorn et al. (2013a) applied blends of poly(dimethylsilicate), alkoxysilanes and aliphatic epoxides, while Croutxe-Barghorn, Chemtob, Ni, Moreau & Bouder (2013b) applied blends of an acryloxy- or methacryloxy silane, an alkylsilane and an organic acrylate resin and added an additional radical initiator.

By combining radical and cationic initiators, an organic acrylate and octyltrimethoxysilane, Croutxé-Barghorn et al. (2012) prepared repellent coatings with a water contact angle of 100.5°. The polymer formed by condensation of octyltrimethoxysilane merges towards the coating surface due to its low surface tension. Çakmakçı, Mülazim, & Kahraman (2013) combined the cross-linking of alkoxysilyl-functions with the radical-based thiol-ene reaction by curing blends of an organic triacrylate, (3-mercaptoproyl) trimethoysilane and a radical initiator. An optionally added perfluoroalkylsilane leads to strong water- and oil repellency with a water contact angle of 113°. Belon et al. (2011) prepared scratch resistant coatings by combining an organic acrylate resin with (3-methacryloxypropyl)trimethoxysilane and both a radical and a cationic initator. Chemtob et al. (2012) combined trimethoxy silanes bearing longer C12 to C16 alkyl chains with an epoxy silane and a cationic photocatalyst. In this combination, the long-chain alkylsilanes lead to the formation of macro pores in the obtained coating films.

Sangermano, Amerio, Priola, Di Gianni & Voit (2006) prepared films with in-situ formed nano-TiO_2 domains from a blend of a radical initiator, acrylate-functional PDMS, (2-hydroxyethyl)acrylate and tetraisopropyl titanate. The reactive titanate does not require a catalyst for moisture curing. The formed TiO_2 shows strong interactions with the polyacrylate network via hydroxyl groups. Chung et al. (2011) prepared 150 nm thick transparent barrier films to protect encapsulated organic solar cells or organic LEDs from moisture and oxygen. A blend of an urethane acrylate, (3-methacryloxypropyl)trimethoxysilane, 30-100 nm sized SiO_2-nanoparticles and solely a radical initiator, was in a first step by UV-curing pre-polymerized to a resin, then applied to the device and in a second step cured by further UV radiation. No catalyst to cross-link the alkoxysilyl functions was mentioned.

Sangermano et al. (2014) cured blends of a polythiol resin, an epoxy resin and TEOS (tetraethyl orthosilicate) with a latent photobase leading to transparent films comprising silicate clusters in the size of 100 nm. By adding a mercaptosilane connecting inorganic and organic phase, smaller silicate/siloxane clusters in the size of 3-5 nm were formed. Instead of TEOS, zinc acetate can be applied as a precursor for nano zinc oxide. Zinc oxide is known for anti-bacterial properties and UV absorbance.

Further Organic Backbones

Organic chemistry provides a large variety of building blocks. One can speculate that future coatings will take advantage of that and provide further organic backbones for inorganic-organic hybrid coatings. This paragraph provides examples that are not covered by the other headlines. Silanized polyacrylates were among the first organic polymers to be equipped with alkoxysilyl groups. The work of Chang et al. (1977) and Plueddemann (1969), the combination of silanized polyacrylates with urea-polymers described by Fenn & Davis (1993) and Poppe et al. (2006) and the in-situ, UV-initiated synthesis of polyacryl chains during curing have already been mentioned.

Dowbenko et al. (1985) describe both pigmented base coats and top coats prepared by moisture-curing of polyacrylate polymers with alkoxysilyl groups using DBTL as catalyst. In a radical polymerization of a blend of methylmetacrylate, butylmethacrylate and sytrene, (3-methacryloxypropyl)trimethoxysilane was applied as co-mononer and (3-mercaptopropyl)trimethoxy silane was applied as chain-transfer reagent to form polymers bearing alkoxysilyl groups with peak molecular weights of about 6800 or 17400 g/mol.

Cumurcu, Yazgan & Erciyes (2011) introduced a natural raw material. Sunflower oil was transesterified with glycerol to obtain partial glycerides that were subsequently reacted with maleic anhydride. In a second step, the obtained unsaturated ester oligomers were co-polymerized in a radical polymerization with styrene and vinyltrimethoxysilane. Thereby, mainly the double bond derived from maleic anhydride reacts and not the unsaturated fatty acid chains derived from sunflower oil. Figure 13 (left) shows a possible structural unit of the resin. After application, the resin cures in a dual-cure reaction. Atmospheric moisture induces the cross-linking of alkoxysilyl groups, while atmospheric oxygen induces oxidative coupling of the polyunsatured fatty acid chains derived from sunflower oil as shown in Figure 13 (right). The latter is the common crosslinking mechanism of alkyd resins that are widely applied in paints. Incorporation of the silane improved alkali resistance and thermal stability.

Hansen, Hays, Iyer, Frank & Kumar (2013) prepared a polyamide resin by reacting oxalic acid diethlyester subsequently with aminoalkyl-terminated PDMS and with (3-aminopropyl)triethoxysilane, optionally in combination with ethylene diamine. The obtained polyamides comprise PDMS blocks

Figure 13. Possible structural unit of resin according to Cumurcu et al. (2011) representing all educts (left) and oxidative curing mechanism (right)

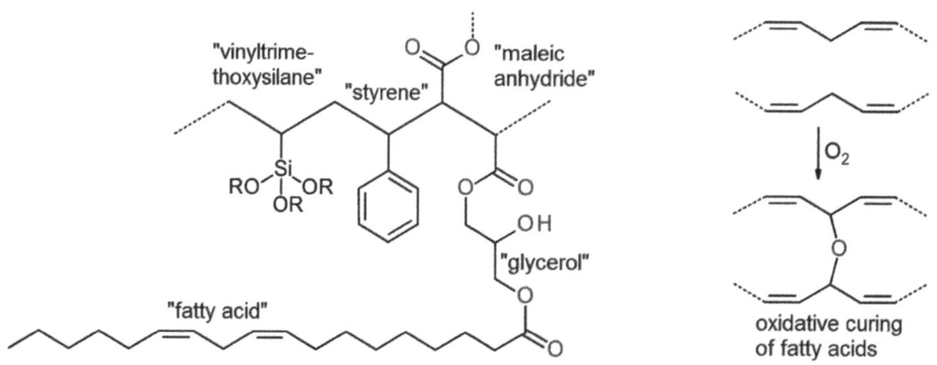

and terminal alkoxysilyl groups. They were cured by moisture on scissor blades to provide repellency to adhesive tape.

Silazane Systems

While silazanes mainly serve as precursors for inorganic, ceramic coatings prepared by pyrolysis, this paragraph outlines applications at lower curing temperatures to result organic-inorganic hybrid coatings. For this purpose, blends of cyclosilazanes and polysilazanes are applied. Two typical compositions, in this paragraph labelled S1 and S2, are shown in Figure 14.

Vu, Osterod, Brand & Ryan (2008) describe applications of polysilazane-based coatings as graffiti protection that is not wetted by paint, as weathering stable clear coat or as corrosion protection when directly applied to aluminum surfaces. Due to high reactivity, coatings cure within one hour at ambient temperature by reaction of atmospheric moisture with both amino and hydrogen function under evolution of ammonia and hydrogen to from an organosiloxane network. Even though ammonia is toxic and its evolution in general is not desired, it is inorganic and thus not regarded as volatile organic compound (VOC). According to Intersolia Sweden (2012), solutions of S1 are commercially applied as graffiti protection.

S1 has also been a starting point for modifications. Nguyen (2011) functionalized S1 by reacting Si-H groups in a hydrosilylation reaction with allyl groups of allyl-terminated polyethylene glycol (PPG). PPG, also called polyethylene oxide (PEO), is a hydrophilic polymer. Coatings prepared from the functionalized silazane, especially those where PPG has a molecular weight of ~350 g/mol, provide antibacterial properties and are interesting for marine applications requiring an anti-fouling effect. In a different approach, Marceaux, Bressy, Perrin, Martin & Margaillan (2014) modified S1 with PDMS. While PDMS without reactive groups can be extracted with hexane from the cured coating, hydroxyalkyl-functional PDMS is better connected with the coating and increases the hydrophobicity of the already hydrophobic coatings prepared from S1. The water contact angle rises from 95-100° (no PDMS) to 100-110° (with

Figure 14. Structures of two common silizane resins

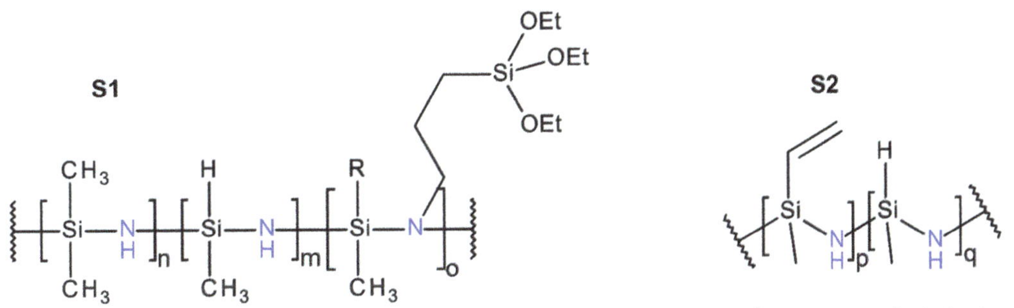

Cyclosilazanes, dimethyl, methyl hydrogen, polymers with dimethyl, methyl hydrogen silazanes, reaction products with 3-(triethoxysilyl)-1-propanamine

CAS-No.: 475645-84-2

typical composition (Marceaux et al. 2014):
n=0.29, m=0.59, o=0.12, R is either H or CH_3

Cyclosilazanes, methyl vinyl, methyl hydrogen, polymers with methyl vinyl, methyl hydrogen silazanes

CAS No.: 503590-70-3

PDMS). The PDMS modified coatings provide also good anti-bacterial properties in seawater and are thus interesting as marine anti-fouling coatings. Fish (2014) prepared mold-release coatings from blends of PDMS without reactive groups with either S1 or mixtures of S1 and S2. The release coatings are suitable for the casting of polymer composites. Even though PDMS is solely physically bonded, it is in this application not transferred to the molded object.

Vu et al. (2008) and Kion Defense Technologies (n.d.) describe the combination of polysilazanes, especially S2, with organic epoxy, phenolic or isocyanate resins. Such resins can be cured at 100-150° to form corrosion protective coatings. The isocyanate reaction results in urea-groups within the silazane chain. For these coatings, the evolution of hydrogen or ammonia is an undesired side reaction.

No Volatile Byproducts: Alkoxy Groups as Monomers

An interesting concept for coatings that are not solely solvent-free, but do also not generate any volatile methanol or ethanol during curing, is the application of silanes bearing special alkoxygroups that do not evaporate, but become part of the cured coating. A further advantage of this concept is that it can reduce tension in a cured film as no shrinkage occurs. The concept has not yet been applied to prepare technical coatings, but is described herein due to its fundamental relevance.

Ellsworth & Novak (1991), Novak & Davies (1991), and Ellsworth & Novak (1993) synthesized silanes with alkoxy-groups bearing double bonds, such as, for example, silicic acid esters of (2-hydroxyethyl) acrylate. The silanes were hydrolyzed in a conventional sol-gel process applying NaF as catalyst. The cleaved alcohols were polymerized by free-radical polymerization. In an alternative route, alkoxides bearing a cylcloalkene structure were polymerized by Ru^{2+} or Ru^{3+}-catalyzed ring-opening metathesis polymerization (ROMP). Both attempts resulted in interpenetrating organic-inorganic composites. Uygun et al. (2011) synthesized as well a silicic acid ester of (2-hydroxyethyl) acrylate and prepared 1.4 mm thick, practically transparent films. The films were either prepared by sol-gel processing of the silane followed by subsequent UV-induced curing of the acrylate side chain, or in reverse order by UV-induced cross-linking of the acrylate groups of the non-hydrolyzed silane followed by subsequent curing of the alkoxysilyl groups by atmospheric moisture. Figure 15 outlines the relevant curing reactions.

Figure 15. Curing of silanes with polymerizable alkoxygroups according to Ellsworth & Novak (1991) (top) and Uygun et al. (2011) (bottom)

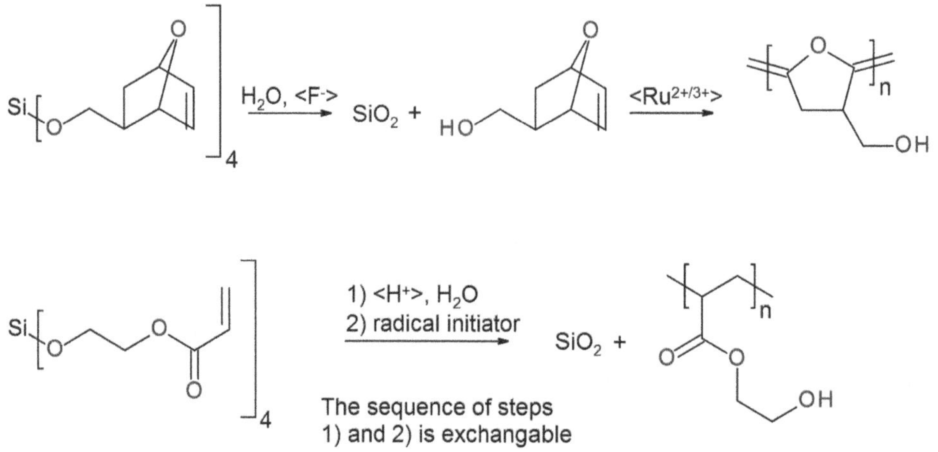

Grund, Kempe, Baumann, Seifert & Spange (2007), Kempe et al. (2014) and Löschner et al. (2014), Spange & Grund (2009a), Spange et al. (2009b), and Seifert & Spange (2013) synthesized silanes prepared by transesterification of common methoxy- or ethoxysilanes with furfuryl alcohol or 2-hydroxybenzyl alcohol. With an acid catalyst, such as, for example, trifluoroacetic acid, both a siloxane/silicate network and an organic polymer, such as a phenolic resin or poly(furfuryl alcohol) are simultaneously formed. No moisture is required for curing; the formation of poly(furfuryl alcohol) even sets free water. The reaction is called twin polymerization as one monomer forms two different polymers that are not necessarily chemically connected, see Figure 16. Transparent nano-composites can be formed that proof, that the domains of each polymer are smaller than the wavelength of visible light.

Arnebold, Schorsch, Beckmann & Hartwig (2013) synthesized 2,2-dimethyl-4H-1,3,2-benzodioxasilin, that was also applied by Seifert & Spange (2013), in an alternative approach from 2-hydroxybenzyl alcohol and dichorodimethylsilane. The compound was, together with the cycloaliphatic epoxide 3,4-epoxycyclohexylmethyl-3',4'-epoxycyclohexane carboxylate, polymerized in a concomitant, acid catalyzed polymerization. Spectroscopic investigations showed that three polymers are formed. Besides the formed PDMS, the two organic polymers, a phenolic novolak resin and an epoxy-derived polyether, are basically not chemically connected. The otherwise brittle epoxy-polymer is effectively toughened by addition of the silane.

FUTURE RESEARCH DIRECTIONS

Besides continuation and optimization of present work, the attempt to work on problems that have not been adequately covered or not been solved will drive future research. The author describes below such problems und subjects according to his own expectations.

- **Fundamental Knowledge:** Selected non-hydrolyzed resins have gained industrial importance. These resins are a promising origin for further investigations leading to fundamental knowledge

Figure 16. Reactions involved in twin polymerizations according to Seifert & Spange (2013)

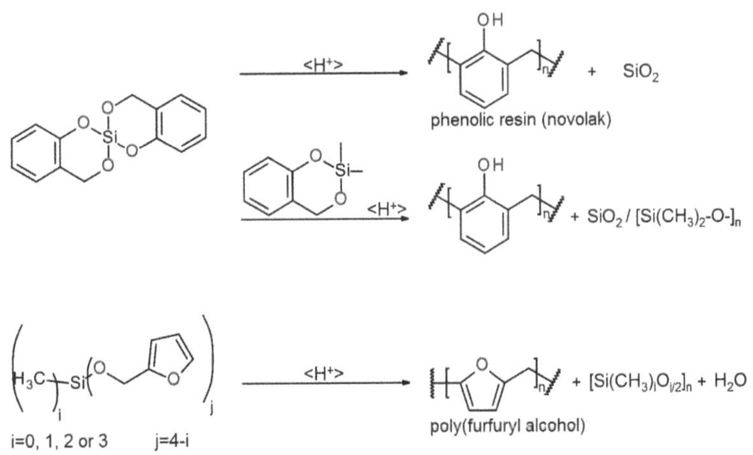

of the underlying reactions and properties. Such investigations can provide a benefit for both scientific research and further industrial development at the same time.

- **Novel Functions Beyond Classical Industrial Coating Technology:** Coatings based on nanotechnology and sol-gel processing received attention as key-enabling technology in various technical fields such as, for example mechanics, biology, catalysis, surface-physics, electronics or optics, providing a vast variety of functionalities (Sakka 2005; Sepeur et al. 2008; Stenzel & Rehfeld 2011). For some specific functionalities, organic-inorganic hybrid coatings based on non-hydrolyzed resins should facilitate the implementation. In comparison with both organic coatings and sol-gel processing, the non-hydrolyzed application method is on the one hand a simple and elegant way to incorporate water-immiscible organic components such as, for examples, specific polymers or dyes, homogeneously in a dense, organic-inorganic network. On the other hand, inorganic, nanoscale domains can be formed in-situ in an organic matrix as an alternative to the addition of nanoparticles. Many present investigations dealing with non-hydrolyzed resins have covered common subjects of industrial coating technology, such as mechanical and chemical protection. An example is the successful exploitation of inorganic nano-domains for scratch resistant automotive coatings. However, novel functional coatings are expected from interdisciplinary collaborations involving further technical fields. Present work on sol-gel based coatings can serve as supply for ideas. Non-hydrolyzed resins extend the toolbox available to implement these ideas.
- **VOC-Free Coatings:** The majority of the present work is carried out with methoxy- or ethoxy silanes available in industrial scale. Concepts to apply silanes with alternative alkoxy groups that are split off during curing, but do not evaporate and instead become part of the coating matrix, have not yet achieved wide attention. Such concepts would make a step forward and provide desired properties of organic-inorganic hybrid coatings not only as low-VOC, but as zero-VOC technology. When assuming that the final properties can compete with established solutions, such coatings would be a reputable achievement and a strong marketing argument with respect to growing environmental and work safety concern.

CONCLUSION

The application of resins comprising alkoxysilanes has significantly extended the field of organic-inorganic hybrid coatings. Research in the past decade led to both success in industrial applications such as, for example, scratch-resistant, automotive coatings, protective coatings or anti-graffiti coatings, and to the availability of a wide variety of novel chemical building blocks for functional coatings. The coatings have so far gained only limited recognition for entirely new effects, but have provided outstanding overall performance in selected fields in competition with organic coatings and coatings prepared by a classical sol-gel approach. The present lively research activities suggest further innovations in the future.

REFERENCES

Allauddin, S., Narayan, R., & Raju, R. V. S. N. (2013). Synthesis and properties of alkoxysilane castor oil and their polyurethane/urea-silica hybrid coating films. *ACS Sustainable Chemistry & Engineering*, *1*(8), 910–918. doi:10.1021/sc3001756

Amerio, E., Sangermano, M., Malucelli, G., Priola, A., & Voit, B. (2005). Preparation and characterization of hybrid nanocomposite coatings by photopolymerization and sol-gel process. *Polymer, 46*(25), 11241–11246. doi:10.1016/j.polymer.2005.09.062

Arnebold, A., Schorsch, O., Beckmann, J., & Hartwig, A. (2013). Concomitant cationic polymerization of a hybrid monomer and an epoxy resin. *Reactive & Functional Polymers, 73*(12), 1625–1631. doi:10.1016/j.reactfunctpolym.2013.09.004

Arun, M., Kantheti, S., Gaddam, R. R., Narayan, R., & Raju, K. V. S. N. (2014). Surface modification of TiO_2 nanoparticles with 1,3,5-triazine based silane coupling agent and its cumulative effect on the properties of polyurethane composite coating. *Journal of Polymer Research, 21*(12), 600. doi:10.1007/s10965-014-0600-7

Bachon, T. (2005). Compositions based on silane-terminated polymers and hardened by radiation and humidity, production and use thereof. *Patent WO 2005/100482 A1*.

BASF iGloss: A breakthrough clearcoat for automotive applications. (2012). *Metalfinishing.com*. Retrieved from http://www.metalfinishing.com/view/25005/basf-igloss-a-breakthrough-clearcoat-for-automotive-applications/

Beari, F., Brand, M., Jenkner, P., Lehnert, R., Metternich, H. J., Monkiewicz, J., & Siesler, H. W. (2001). Organofunctional alkoxysilanes in dilute aqueous solution: New accounts on the dynamic structural mutability. *Journal of Organometallic Chemistry, 625*(2), 208–216. doi:10.1016/S0022-328X(01)00650-7

Belon, C., Chemtob, A., Croutxé-Barghorn, C., Rigolet, S., Le Houérou, V., & Gauthier, C. (2011). A simple method for the reinforcement of UV-cured coatings via sol-gel photopolymerization. *Macromolecular Materials and Engineering, 296*(6), 506–516. doi:10.1002/mame.201000350

Belon, C., Chemtob, A., Croutxé-Barghorn, C., Rigolet, S., Le Houréou, V., & Gauthier, C. (2010b). Combination of radical and cationic photoprocess for the single-step synthesis of organic-inorganic hybrid films. *Journal of Polymer Science, 48*, 4150–4158.

Belon, C., Chemtob, A., Croutxé-Barghorn, C., Rigolet, S., Schmitt, M., Bistac, S., & Gauthier, C. et al. (2010a). Nanocomposite coatings via simultaneous organic-inorganic photo-induced polymerization: Synthesis, structural investigation and mechanical characterization. *Polymer International, 59*, 1175–1186.

Bischoff, C., & Holberg, S. (2012). Repellent coating composition and coating, method for making and uses thereof. *Patent WO 2012/083970 A1*.

Brand, M., Frings, A., Jenker, P., Lehnert, R., Metternich, H. J., Monkiewicz, J., & Schramm, J. (1999). NMR-spectroscopic investigations on the hydrolysis of functional trialkoxysilanes. *Zeitschrift für Naturforschung, 54b*, 155–164.

Brinker, C. J. (1988). Hydrolysis and condensation of silicates: Effects on structure. *Journal of Non-Crystalline Solids, 100*(1-3), 31–50. doi:10.1016/0022-3093(88)90005-1

Brinker, C. J., & Scherer, G. W. (1990). *Sol-gel science*. San Diego, CA: Academic Press.

Brusciotti, F., Snihirova, D. V., Xue, H., Montemor, F., Lamaka, S. V., & Ferreira, M. G. S. (2013). Hybrid epoxy-silane coatings for improved corrosion protection of Mg alloy. *Corrosion Science*, *67*, 82–90. doi:10.1016/j.corsci.2012.10.013

Çakmakçı, E., Mülazim, Y., & Kahraman, M. V. (2013). UV-curable fluorine-containing hybrid coatings via thiol-ene "click" reaction and an in situ sol–gel method. *Polymer Bulletin*, *70*(3), 1037–1048. doi:10.1007/s00289-012-0871-2

Chang, W., Hartman, M. E., & Scriven, R. L. (1977). Ambient temperature, moisture-curable acrylic-silane coating compositions having improved potlife. *Patent US 4043953*.

Chattopadhyay, D. K., Muehlberg, A. J., & Webster, D. C. (2008). Organic–inorganic hybrid coatings prepared from glycidyl carbamate resins and amino-functional silanes. *Progress in Organic Coatings*, *63*(4), 405–415. doi:10.1016/j.porgcoat.2008.06.009

Chattopadhyay, D. K., Zakula, A. D., & Webster, D. C. (2009b). Organic-inorganic hybrid coatings prepared from glycidyl carbamate resin, 3-aminopropyl trimethoxy silane and tetraethoxyorthosilicate. *Progress in Organic Coatings*, *64*(2-3), 128–137. doi:10.1016/j.porgcoat.2008.09.008

Chattopadhyay, & D. K., Webster, D. C. (2009a). Hybrid coatings from novel silane-modified glycidyl carbamate resins and amine crosslinkers. *Progress in Organic Coatings, 66,* 73-85.

Chemtob, A., Courtecuisse, F., Croutxé-Barghorn, C., & Rigolet, S. (2011). Simultaneous sol–gel and anionic photopolymerization of 3-(glycidyloxypropyl)trimethoxysilane via photobase catalysis. *New Journal of Chemistry*, *35*(9), 1803–1808. doi:10.1039/c1nj20320g

Chemtob, A., De Paz-Simon, H., Croutxé-Barghorn, C., & Rigolet, S. (2014a). UV-activated silicone oligomer cross-linking through photoacid and photobase organocatalysts. *Journal of Applied Polymer Science*, *131*(3), 39875. doi:10.1002/app.39875

Chemtob, A., De Paz-Simon, H., Dietlin, C., Croutxé-Barghorn, C., Salmi, H., Allonas, X., & Rigolet, S. et al. (2014b). A highly reactive photobase catalytic system for sol-gel polymerization. *Thin Solid Films*, *550*, 177–183. doi:10.1016/j.tsf.2013.10.146

Chemtob, A., Ni, L., Dietlin, C., Croutxé-Barghorn, C., Kitzmann, P., Brogly, M., & Vidal, L. (2012). Spontaneous photoinduced formation of hybrid polymer films with functionalized macroporous patterns. *Surface and Coatings Technology*, *209*, 64–72. doi:10.1016/j.surfcoat.2012.08.030

Chemtob, A., Versace, D.-L., Belon, C., Croutxé-Barghorn, C., & Rigolet, S. (2008). Concomitant organic-inorganic UV-curing catalyzed by photoacids. *Macromolecules*, *41*(20), 7390–7398. doi:10.1021/ma801017k

Chung, M.-H., Lin, J.-S., Hsieh, T.-E., Chen, N.-P., Juang, F.-S., Chen, C.-M., & Liu, L.-C. (2011). Preparation of organic/inorganic hybrid nanocomposite by ultraviolet irradiation and their packaging applications for organic optoelectronic devices. *Applied Surface Science*, *257*(21), 9142–9151. doi:10.1016/j.apsusc.2011.05.117

Croutxe-Barghorn, C., Chemtob, A., Moreau, N., & Bouder, T. (2013a). Radiation curable composition, and method for preparing a hybrid sol-gel layer on a surface of a substrate using said composition. *Patent WO 2013/171582 A1*.

Croutxé-Barghorn, C., Chemtob, A., Ni, L., De-Paz-Simon, H., & Belon, C. (2013c, June). Nanocomposite coatings by UV driven sol-gel process and photopolymerization. *Paper presented at Coating Science International*, Noordwijk, The Netherlands.

Croutxe-Barghorn, C., Chemtob, A., Ni, L., Moreau, N., & Bouder, T. (2013b). Radiation radically and cationically curable composition, and method for preparing a hybrid sol-gel layer on a surface of a substrate using said composition. *Patent WO 2013/171579 A1*.

Croutxé-Barghorn, C., De Brito, M., Allonas, X., Belon, C., Chemtob, A., Ni, L., & Dietlin, C. et al. (2012). Organic and hybrid interpenetrated polymer networks for advanced materials. *Journal of Photopolymer Science and Technology*, *25*, 131–135. doi:10.2494/photopolymer.25.131

Cumurcu, A., Yazgan, P., & Eriyes, T. (2011). Synthesis and properties of alkoxysilane-functionalized styrenated oil based polyester. *Progress in Organic Coatings*, *71*(2), 147–152. doi:10.1016/j.porgcoat.2011.01.007

Cyracure cycloaliphatic epoxides - cationic UV cure [brochure]. (2001Dow Chemical Company. Midland, MI: Dow Chemical Company.

De Paz, H., Chemtob, A., Croutxé-Barghorn, C., Le Nouen, D., & Rigolet, S. (2012). Insights into photoinduced sol–gel polymerization: An in situ infrared spectroscopy study. *The Journal of Physical Chemistry B*, *116*(17), 5260–5268. doi:10.1021/jp212386e PMID:22486534

Dowbenko, R., Stewart, R. S., Hartman, M. E., Gorman, B., & Thomas, S. J. (1985). Colour plus clear coating method utilizing addition interpolymers containing alkoxy silane and/or alcoxy silane groups. *Patent US 4499150*.

Ellsworth, M. W., & Novak, B. M. (1991). Mutually interpenetrating inorganic-organic networks. New routes into nonshrinking sol-gel composite materials. *Journal of the American Chemical Society*, *113*(7), 2756–2758. doi:10.1021/ja00007a062

Ellsworth, M. W., & Novak, B. M. (1993). "Inverse" organic-inorganic composite materials. 3. High glass content "nonshrinking" sol-gel composites via poly(silicic acid esters). *Chemistry of Materials*, *5*(6), 839–844. doi:10.1021/cm00030a020

Fenn, D. R., & Davies, S. P. (1993). Silane functional oligomer. *Patent EP 0571073 A2*.

Fischer, W., & Yuva, N. (2011). Self-healing UV-coatings. *Paper presented at RadTech Europe*, Basel, Switzerland. Retrieved from www.bayercoatings.com

Fish, C. (2014). High performance silicon based coating compositions. *Patent US 2014/0011009 A1*.

Foscante, R. E., Gysegem, A. P., Martinich, P. J., & Law, G. H. (1981). Interpenetrating polymer network comprising epoxy polymer and polysiloxane. *Patent US 4250074*.

Reactive silicones [brochure]. (2013). *Gelest, Inc.* Retrieved from http://www.gelest.com/goods/PDF/brochures/reactivesilicones.pdf

Gireesh, K. B., Jena, K. K., Allauddin, S., Radhika, K. R., Narayan, R., & Raju, K. V. S. N. (2010). Structure and thermo-mechanical properties study of polyurethane-urea/glycidoxypropyltrimethoxysilane hybrid coatings. *Progress in Organic Coatings*, *68*(3), 165–172. doi:10.1016/j.porgcoat.2010.02.008

Gregorovich, B. V., Hazan, I., & Wilczek, L. (2001). Coatings that contain silicone oligomers. *Patent EP 1123951 A2*.

Groenewolt, M., Hesener, S., Stubbe, W., Niemeier, M., & Westhoff, E. (2009). Verwendung teilsilanisierter Verbindungen auf Polyisocyanatbasis als Vernetzungsmittel in Beschichtungszusammensetzungen und Beschichtungszusammensetzung enthaltend die Verbindungen. *Patent DE 102008030304 A1*. [Use of partially silanized polyisocyanate-based compounds as crosslinking-agents in coating compositions, and coating compositions comprising the compounds]. *Patent US 8658752 B2*.

Groenewolt, M., Klein, G., Hilge, O., Austrup, B., & Huebner, K. (2012b). Coating agent compositions, coatings made therefrom and exhibiting high scratch resistance and good polishability, and use thereof. *Patent WO 2012/168079 A1*.

Groenewolt, M., & Müller, J. (2012a). Coating composition having a high solids content and good levelling and also multilayer surface coatings produced therefrom and their use. *Patent US 20140065313 A1*.

Groenewolt, M., Poppe, A., Klein, G., Niemeier, M., Westhoff, E., Stübbe, W., & Hesener, S. (2008). Coating agents having high scratch resistance and weathering stability. *Patent US 8569438 B2*.

Grund, S., Kempe, P., Baumann, G., Seifert, A., & Spange, S. (2007). Nanocomposites prepared by twin polymerization of a single-source monomer. *Angewandte Chemie International Edition*, *46*(4), 628–632. doi:10.1002/anie.200504327 PMID:17152104

Hackford, J. E., Shaw, C., & Smith, W. E. (1946). Improvements in or relating to liquids comprising silicyl esters. *Patent GB 575752*.

Hansen, R. G., Hays, D. S., Iyer, S., Frank, J. W., & Kumar, R. C. (2013). Moisture-curable siloxanes and siloxane polymers. *Patent US 2013/0011673 A1*.

Hinzmann, D., Klotzbach, T., & Herrwerth, S. (2012). Ein starkes Netzwerk gegen Rost. *Farbe + Lack*, *118*(9), 22–26.

Hofacker, S., Mechtel, M., Hovestadt, W., & Kobusch, C. (2002). Two-component polyurethane binders as primers. *Patent US 2002/0142169 A1*.

Holberg, S., & Bischoff, C. (2014). Application of a repellent urea-siloxane hybrid coating in the oil industry. *Progress in Organic Coatings*, *77*(10), 1591–1595. doi:10.1016/j.porgcoat.2013.10.019

Holberg, S., Bischoff, C., Poulsen, T., & Tabrizian, N. (2013, September). Versatile urea-siloxane hybrid coatings: Corrosion protection of aluminium and reduction of crude oil fouling. *Poster presented at Eurocorr 2013*, Estoril, Portugal.

Howard, C., Hallack, M., Schlemann, U., & Bartek, A. (2008). Novel high-solids systems based on silicone epoxy resins. *Paint & Coating Industry*, *2008*(October), 42–48.

Hoyle, C., Lowe, A. B., & Bowman, C. N. (2010). Thiol-click chemistry: A multifaceted toolbox for small molecule and polymer synthesis. *Chemical Society Reviews*, *39*(4), 1355–1387. doi:10.1039/b901979k PMID:20309491

Huang, Y., Chen, H., Roper, J., III, Tucker, C., Vandezande, G., Li, Y., & Zhang, Y. (2012a). Moisture curable anti-fouling coating compositions. *Patent US 2012/0269976 A1.*

Huang, Y., Chen, H., Roper, J., III, Tucker, C., Vandezande, G., Li, Y., & Zhang, Y. (2012b). Two-component moisture curable coating compositions. *Patent US 2012/0269977 A1.*

Innovative sol-gel coatings with sivento silanes [brochure]. (n. d.). Degussa. Retrieved from www.dynasylan.com

Safety Data Sheet AGS 3550 (Trion Tensid AB). (2012). *Intersolia Sweden.* Retrieved from http://cloud.intersolia.com/pdf/610/%7B693D0CF5-5ED3-42DC-B2DE-2668D571D32E%7D/100/AGS_3550_EN

Jonschker, G. (2014). *Sol-Gel-Technology in Praxis.* Hanover, Germany: Vincentz Network.

Judeinstein, P., & Sanchez, C. (1996). Hybrid organic-inorganic materials: A land of multidisciplinarity. *Journal of Materials Chemistry*, *6*(4), 511–525. doi:10.1039/jm9960600511

Kelly, S. A. M., Birkert, C. R., & Andrews, A. F. (1998). Coating compositions. *Patent EP 0941290 B1.*

Kempe, P., Löschner, T., Auer, A. A., Seifert, A., Cox, G., & Spange, S. (2014). Thermally induced twin polymerization of 4H-1,3,2-benzodioxasilines. *Chemistry (Weinheim an der Bergstrasse, Germany)*, *20*(26), 8040–8053. doi:10.1002/chem.201400038 PMID:24849004

TB3: KiON Polysilazanes - Reactivity with Isocyanates. (n. d.). Kion Defense Technologies. Retrieved in 2014 from www.kiondefense.com/bulletins/TB3.pdf

Laryea, N., Thurn, C., & Sepeur, S. (2008b). Verfahren zur Herstellung eines Beschichtungsmaterials. *Patent DE 102007020404* A1. [Method for the production of a coating material. *Patent US 2010/0092686 A1].*

Laryea, N., Thurn, C., Sepeur, S., & Schlick, G. (2008a). Silanbeschichtungsmaterial und Verfahren zur Herstellung eines Silanbeschichtungsmaterials. *Patent DE 102006044310 A1.* [Silane coating material and a process to produce silane coating. *Patent US 2009/0326146 A1].*

Laubender, T., & Greene, J. (2006). Adding value to coatings by using unique organofunctional silicone hybrids. *Polimery*, *51*(2), 138–142.

Lee, T.-Z., Ma, C.-C. M., Hsu, C.-W., & Wu, H.-L. (2005). Effect of molecular structures and mobility on the thermal and dynamical mechanical properties of thermally cured epoxy-bridged polyorganosiloxanes. *Polymer*, *46*(19), 8286–8296. doi:10.1016/j.polymer.2005.06.069

Li, Z., Han, W., Kozodaev, D., Brokken-Zijp, J. C. M., de With, G., & Thüne, P. C. (2006). Surface properties of poly(dimethylsiloxane)-based inorganic/organic hybrid materials. *Polymer*, *47*(4), 1150–1158. doi:10.1016/j.polymer.2005.12.057

Lomoelder, R., Görlitzer, H., Hallack, M., & Unkelhäußer, T. (2013). Beschichtungsmittel mit hoher Kratzbeständigkeit. *Patent EP 2676982 A1.*

Löschner, T., Mehner, A., Grund, S., Seifert, A., Pohlers, A., Lange, A. ... Spange, S. (2012). A modular approach for the synthesis of nanostructured hybrid materials with tailored properties: The simultaneous twin polymerization. Angewandte Chememie International Edition, 51, 3258–3261.

Marceaux, S., Bressy, C., Perrin, F.-X., Martin, C., & Margaillan, A. (2014). Development of polyorganosilazane-silicone marine coatings. *Progress in Organic Coatings*, *77*(11), 1919–1928. doi:10.1016/j.porgcoat.2014.06.020

Monkiewicz, J., Edelmann, R., Edinger, A., & Borup, B. (2006). Air-drying silane coating compositions. *Patent US 7026398 B2.*

Nguyen, T. D. H. (2011). *Revêtements polysilazane a activités antibactériennes*. (PhD thesis). Université de Toulon, La Garde, France. Retrieved from tel.archives-ouvertes.fr/tel-00769942/document

Noomen, A. (1997). Applications of Michael addition chemistry in coatings technology. *Progress in Organic Coatings*, *32*(1-4), 137–142. doi:10.1016/S0300-9440(97)00070-2

Novak, B. M., & Davies, C. (1991). "Inverse" organic-inorganic composite materials. 2. Free-radical routes into nonshrinking sol-gel composition. *Macromolecules*, *24*(19), 5481–5483. doi:10.1021/ma00019a044

Oertli, M., & Jucker, B. (2013). Moisture-curing compositons comprising at least two polymers having silane groups. *Patent US 8372514 B2.*

Parashar, G., Srivastava, D., & Kumar, P. (2001). Ethyl silicate binders for high performance coatings. *Progress in Organic Coatings*, *42*(1-2), 1–14. doi:10.1016/S0300-9440(01)00128-X

Pierre, A. C. (1998). *Introduction to Sol-Gel Processing*. Norwell, MA: Kluwer Academic Publishers. doi:10.1007/978-1-4615-5659-6

Plassmann, W. (2001). Coating compositions having improved adhesion to aluminum substrates. *Patent WO 01/98393 A1.*

Plueddemann, E. P. (1969). Room temperature curable acrylate rubbers. *Patent US 3453230.*

Poppe, A., Klein, G., Brosseit, A., Wermelt, K., & Möllers, M. (2006). Hochkratzfeste und hochelastische Beschichtungsmittel auf Basis von Alkoxysilan-funktionellen Komponenten. *Patent DE 102004050746 A1.* [Highly scratch-resistant and highly elastic coating agents based on alkoxysilane functional components. *Patent US 7772320 B2].*

Rekondo, A., Fernándes-Berridi, M. J., & Irusta, L. (2006). Synthesis of silanized polyether urethane hybrid systems. Study of the curing process through hydrogen bonding interactions. *European Polymer Journal*, *42*(9), 2069–2080. doi:10.1016/j.eurpolymj.2006.03.024

Rizk, S. D., & Hsieh, H. W. S. (1986). Moisture curable polyurethane polymers. *Patent US 46235012.*

Rizk, S. D., Hsieh, H. W. S., & Mazzeo, M. P. (1985). Acrylic resin having pendant silane groups thereon, and methods of making and using the same. *Patent US 4491650.*

Sakka, S. (Ed.). (2005). *Applications of sol-gel technology*. In Handbook of sol-gel science and technology: processing, characterization, and application (Vol. 3). Norwell, MA: Kluwer Academic Publishers.

Sangermano, M., Amerio, E., Priola, A., Di Gianni, A., & Voit, B. (2006). Preparation and characterization of acrylic resin/titania hybrid nanocomposite coatings by photopolymerization and sol-gel process. *Journal of Applied Polymer Science*, *102*(5), 4659–5664. doi:10.1002/app.24707

Sangermano, M., Vitale, A., & Dietliker, K. (2014). Photolatent amines producing a strong base as photocatalyst for the in-situ preparation of organic-inorganic hybrid coatings. *Polymer*, *55*(7), 1628–1635. doi:10.1016/j.polymer.2014.02.045

Schindler, W., & Schwiebacher, E. (2009). Alkoxysilane cross-linked polymers having improved elastic recovery properties. *Patent US 20090131591 A1*.

Schmidt, H., Jonschker, G., Goedicke, S., & Mennig, M. (2000). The sol-gel process as a basic technology for nanoparticle-dispersed inorganic-organic composites. *Journal of Sol-Gel Science and Technology*, *19*(1/3), 39–51. doi:10.1023/A:1008706003996

Schottner, G. (2001). Hybrid sol-gel-derived polymers: Applications of multifunctional materials. *Chemistry of Materials*, *13*(10), 3422–3435. doi:10.1021/cm011060m

Schubert, U., Huesing, N., & Lorenz, A. (1995). Hybrid inorganic-organic materials by sol-gel processing of organofunctional metal alkoxides. *Chemistry of Materials*, *7*(11), 2010–2027. doi:10.1021/cm00059a007

Schweitzer, P. A. (Ed.). (2006). *Paint and coatings - application and corrosion resistance*. Boca Raton, FL: CRC Press.

Seifert, A., & Spange, S. (2013). Aus eins mach zwei: Die Zwillingspolymerisation. *Nachrichten aus der Chemie*, *61*(10), 1028–1032. doi:10.1002/nadc.201390319

Sepeur, S., Laryea, N., Goedicke, S., & Groß, F. (2008). *Nanotechnology - Technical Basics and Applications*. Hannover, Germany: Vincentz Network.

Shaik, A., Narayan, R., & Raju, K. V. S. N. (2014). Synthesis and properties of siloxane-crosslinked polyurethane-urea/silica hybrid films from castor oil. *Journal of Coatings Technology and Research*, *11*(3), 397–407. doi:10.1007/s11998-013-9548-5

Spange, S., & Grund, S. (2009a). Nanostructured organic-inorganic composites by twin polymerisation of hybrid monomers. *Advanced Materials*, *21*(20), 2111–2116. doi:10.1002/adma.200802797

Spange, S., Kempe, P., & Seifert, A. Auer, A. A., Ecorchard, P., Lang, H., … Kaskel, S. (2009b). Nanocomposites with structure domains of 0.5 to 3 nm by polymerization of silicon spiro compounds. Angewandte Chemie International Edition, 48, 8254–8258.

Speier, P., Plogmann, F., Lomoelder, R., & Hoppe, D. (2003). Nonaqueous thermosetting two-component coating composition. *Patent US 2003/0027921 A1*.

Stanjek, V., Bachmeier, B.-J., & Zander, L. (2014). Cross-linkable materials based on organyl oxysilane-terminated polymers. *Patent US 8907027 B2*.

Stanjek, V., Jung, S., & Schauer, F. (2004). Silanvernetzbare Beschichtungsformulierungen. *Patent DE 10237270 A1*.

Stenzel, V., & Rehfeld, N. (2011). *Functional Coatings*. Hannover, Germany: Vincentz Network.

TB4. KiON Polysilazanes - Reactivity with Phenolic Resins. (n. d.). Kion Defense Technologies. Retrieved from www.kiondefense.com/bulletins/TB4.pdf

TB5. KiON Polysilazanes - Reactivity with Epoxy Resins. (n. d.). Kion Defense Technologies. Retrieved from www.kiondefense.com/bulletins/TB5.pdf

Unkelhäußer, T., Görlitzer, H., Hallack, M., & Lomölder, R. (2014). The best of two worlds - Silane-urethane hybrid crosslinkers create scratch-resistant clearcoats. *European Coatings Journal, 2014*(07/08), 21-25.

Uygun, M., Cook, W. D., Moorhoff, C., Chen, F., Vallo, C., Yagci, Y., & Sangermano, M. (2011). Photopolymerization kinetics and dynamic mechanical properties of silanes hydrolyzed without evolution of byproducts. Tetrakis (methacryloyloxy)ethoxy]silane–diethylene glycol dimethacrylate. *Macromolecules, 44*(7), 1792–1800. doi:10.1021/ma102903q

Van Ooij, W. J., Suryanarayanan, K., & Metroke, T. L. (2007). Integral epoxy resin-silane primer system for hot-dip galvanized steel. In K. L. Mittal (Ed.), *Silanes and other coupling agents* (Vol. 4, pp. 275–286). Boca Raton, FL: CRC Press. doi:10.1163/ej.9789067644525.i-410.122

Vu, C., Osterod, F., Brand, S., & Ryan, K. (2008). Silicon and nitrogen. *European Coatings Journal, 2008*(02).

For Powerful Connections [brochure]. (2012a). Wacker Chemie. Retrieved from http://www.wacker.com/cms/media/publications/downloads/6085_EN.pdf

Silres HP 2000 [brochure]. (2012b). Wacker Chemie. Retrieved from http://www.wacker.com/cms/media/publications/downloads/6870_EN.pdf

Witucki, G. L. (2006). Silicone organic hybrid coatings for the industrial maintenance market. Retrieved from http://www.dowcorning.com/content/publishedlit/26-1469-01.pdf

Witucki, G. L. (2013). *Practical applications of polysiloxane coatings* [PowerPoint slides]. Retrieved from http://www.paintsquare.com/education/branding_images/gwwebinar.ppt

Wood, N., Schiemann, U., & Hallack, M. (2005). New Resin Hybrid. *Paint & Coating Industry, 2005*(April), 70–74.

Yang, Y., Semonick, M. A., Moore, G. G. I., Boardman, L. D., & Craton, M. A. (2012), Curable-on-demand polysiloxane coating composition. *Patent WO 2012/003153 A1*.

KEY TERMS AND DEFINITIONS

Alkoxysilane: Silicon compound bearing at least one silicon-bond alkoxy group and optionally further silicon-bond groups. When directly referring to silanes bearing four alkoxy groups and polymers thereof, these are often referred to as silicates (silicic acid esters).

Condensation: In chemistry, this refers to the formation of chemical bonds that releases a third compound.

Hydrolysis: Cleavage of chemical bonds by addition of water.

Polysiloxane: Polymer consisting of silicon-oxygen chains or a silicon-oxygen network. The silicon atoms bear at least one substituents that is either organic or hydrogen.

Silazane: Organosilicon compound with a Si-NH-Si linkage.

Silicate: A compound containing an anionic silicon compound. Also applied for silicic acid esters.

Silicone: Any polysiloxane with organic substituents. Also applied when specifically referring to poly(dimethylsiloxane) (PDMS).

Siloxane: Organosilicon compound with a Si–O–Si linkage.

VOC: Volatile organic compound.

Chapter 6
Research Perspectives on Functional Micro and Nano Scale Coatings:
New Advances in Nanocomposite Coatings for Severe Applications

Jaime Andrés Pérez Taborda
Universidad Tecnológica de Pereira, Colombia & Instituto de Microelectrónica de Madrid-CSIC, Colombia

Elvis O. López
Centro Brasileiro de Pesquisas Físicas, Brazil

ABSTRACT

Research topics related to the production of nanocomposites are the most important directions of development of new semiconductor engineering, ensuring high nanocomposites obtaining useful properties in the scope of biophysical characteristics, biomedical and piezoelectric applications. We present two case studies as Hydroxyapatite are in medical applications and aluminum nitride as acoustic wave sensor. Hydroxyapatite, is the main inorganic structure of the tooth enamel and bone and is a biomaterial that is commonly used in biomedical applications that involve bone substitution, drug delivery and bone regeneration because of its excellent biocompatibility, high bioactivity and good osseoconductivity. Since the past decade. Aluminum nitride (AlN), an electrical insulating ceramic with a wide band gap of 6.3 eV, is a potentially useful dielectric material very important in fields such as optoelectronic and micro electronics.

INTRODUCTION

In recent years, there has been a breakthrough in the research of new materials with increasingly complex structures with properties of ionic and electronic conduction (Morkoc, 2013). Such materials have applications in important areas of technology as components of many optoelectronic devices - among the new materials studied highlights the metal nitrides - and hydroxyapatite nanostructured films fabricated by RF magnetron sputtering for biomedical applications (Zhang, 2013).

DOI: 10.4018/978-1-5225-0066-7.ch006

In general, semiconductor nitrides are the group of inorganic compounds consisting of nitrogen and other more electropositive element, except for halogens and hydrogen. Nitrides of elements such as boron, titanium, zirconium, aluminum, and silicon are hard, stable at high temperatures, and are resistant to chemical agents. At a much smaller scale, coatings are used in numerous electronic products (both consumer and industrial electronics) and biomedical products.

Currently, our society is increasingly immersed in nanotechnology. They are shown, two approaches to nanocoatings applications that impact our daily lives. In section 1, a new method of magnetic confinement of plasmas RF hydroxyapatite has shown to produce a high crystalline quality for biomedical applications. In section 2, aluminum nitride nanostructures grown via pulsed laser deposition - PLD that has been used as substrates for the manufacture of acoustic wave devices - SAW are highly demanded by the telecommunications industry.

1. HYDROXYAPATITE NANOSTRUCTURED FILMS FABRICATED BY RF MAGNETRON SPUTTERING FOR BIOMEDICAL APPLICATIONS

Hydroxyapatite - HA ($Ca_{10}(PO_4)_6(OH)_2$),(Morgan, Wilson, Elliott, Dowker, & Anderson, 2000) is the main inorganic structure of the bone and enamel teeth and also is the main bioceramic that is commonly used in biomedical applications that involve drug delivery, (Hollister, 2005; Vallet-Regí & Ruiz-Hernández, 2011) bone repair and bone regeneration due its excellent properties of osteoinduction, osteoconduction and osseointegration (Astala & Stott, 2008; Paul & Sharma, 2006). Since past decades, the HA has been used as a bioactive coatings to improve the osteoconductivity of metallic implants for the orthopaedic applications. Metals as titanium and alloys have not the bioactive capacity, different deposition techniques to produce calcium phosphate (CaP) coatings has been developed to improve the osseointegration, osteoconduction and osteoinduction of the metallic implants (Dos Santos et al., 2010; Macaskie et al., 2005; Socol et al., 2004; Ueda, Narushima, Goto, Taira, & Katsube, 2007; Wang et al., 2006; Wen, Xu, Hu, & Hodgson, 2007).

Chemical and physical deposition processes are used to synthesize HA as coatings over a metallic surface, these deposition techniques are: biomimetics, (Uchida, Oyane, Kim, Kokubo, & Ito, 2004) sol-gel routes, (Melde & Stein, 2002; Padilla, Roman, Carenas, & Vallet-Regı, 2005; K. Zhang, Francis, Yan, & Stein, 2005) electrochemical deposition,(Liu, Savino, & Yates, 2011; Narayanan, Seshadri, Kwon, & Kim, 2008; Wang, Eliaz, & Hobbs, 2011) and others,(Li, Hong, Yu, & Qi, 2010; Li & Qi, 2008) as well as plasma spray, (Dong, Khor, Quek, White, & Cheang, 2003; Ellies, Nelson, & Featherstone, 1992; Khor, Dong, Quek, & Cheang, 2000) Pulsed Laser Deposition (PLD), (Carradò, 2011; Joanni et al., 2004) Ion and Electron Beam Deposition (IBD, EBD),(Jeong, Choe, & Eun, 2011; Rabiei et al., 2006) Electron Cyclotron Resonance deposition (ECR)(Akazawa & Ueno, 2013, 2014) and Radio Frequency Magnetron Sputtering (RF-MS) (Akazawa & Ueno, 2014; Boyd, Meenan, & Leyland, 2006). The challenge is produce a thin, dense, adherent, stoichiometric and crystalline HA coating over the implant surface, thus the HA coatings will be stable in a biological environments,(Bauer, Schmuki, von der Mark, & Park, 2013) and not to degrade along the time when inside of the human body. Besides, all deposition techniques has been adapted to produce thin coatings with thickness less than of 500 nm to improve the adhesion to the substrate and to preserve the original topography of the metallic implants (Akazawa & Ueno, 2013). Nevertheless, it is difficult produce a thin HA coating without the formation of undesirables amorphous calcium phosphate phases (ACP) (Ferro, Barinov, Rau, Teghil, & Latini, 2005; Hahn et al., 2011; Surmenev, 2012; Yonggang, Wolke, Yubao, & Jansen, 2007; Zhang, 2013). Usually, CaP coatings are commonly treated at temperatures up to 900° C to transform the amorphous phases (ACP) in a HA structure (Akazawa & Ueno, 2013; Surmenev, 2012; S Zhang et al., 2007). In addition, these thermal treatment may produce undesirables phases mainly as tricalcium phosphate (α,β –TCP) and calcium oxide (CaO) and induces the crack and release of the coatings (Jeong et al., 2011; Surmenev, 2012; S Zhang et al., 2007). The tricalcium phosphate ($Ca_3(PO_4)_2$) is a CaP material with a dissolution rate greater than the hydroxyapatite and the CaO is a material that produce high toxicity levels when in contact with proteins and cell of the body.

On the other hand, to optimize the properties of osteoinduction, osteoconduction and osseointegration of the hydroxyapatite materials (in bulk) is necessary functionalize the chemical-structure with adequate quantities of ions substitutions as for example: Zn^{2+}, Sr^{2+}, Mn^{2+}, Mg^{2+}, Fe^{3+}, $(CO_3)^{2-}$, $(SiO_4)^{4-}$ and F^- by Ca^{2+}, $(PO_4)^{3-}$ and $(OH)^-$ (Miao et al., 2008; Pereiro et al., 2012; Surmeneva et al., 2013). In processes of coatings fabrication by chemical deposition is possible functionalize the HA structure with many other ions substitution, controlling mainly the stoichiometry and crystallinity but not the density, roughness, adhesion and thickness (Kim & Choe, 2011; Surmeneva et al., 2013). The physical deposition using plasma processes, produce a functional HA structure is not very common due to the complication of tune the plasma properties and control the deposition parameters to form this complex structure ($Ca_{10-m}M_m(PO_4)_{6-n}X_n(OH)_{2-q}Z_q$). In addition, several studies have shown that the Radio Frequency Magnetron Sputtering technique (RF-MS) have potential for applications in biomedical implant, that due to the use of a stable plasma along the time it could produce a nanostructured coatings with low surface roughness of high density and show good adhesion to the substrate (Nelea, Morosanu, Iliescu, & Mihailescu, 2003; Snyders et al., 2008; Snyders et al., 2007). Nevertheless, inadequate sputtering processes may produce low temperatures, energies and diffusion in the plasma that may inhibit the growth of a nanocoatings with a correct stoichiometry and high crystalline order (Kim & Choe, 2011; Nelea et al., 2003; Shi, Chen, Yu, & Zhang, 2008; Snyders et al., 2007; Van Der Wal et al., 2006).

In other works is possible observe that the thermodynamic properties of the plasma is the main responsible of coating formation (Han, 2009; Shaginyan, Han, Shaginyan, & Musil, 2006). About this, is very important understand know control and tuning the thermodynamic properties of the plasma for each plasma configuration or geometry. In this section is discussed the production of a high crystallinity and stoichiometric nanocoatings of hydroxyapatite and functionalized hydroxyapatite with fluorine ions obtained from the tuning of a highly magnetically confined plasma by orthopedic implant applications (Hong et al., 2007; López et al., 2013; López et al., 2015; Mello et al., 2007).

1.1. Deposition Techniques of CaP Coatings by Biomedical Applications

1.1.1. Plasma Spray

Plasma spray is the first commercial deposition technique to fabricate coating of CaP on metallic implants due the low cost and practicality. This technique consists in produce a plasma jet from a discharge at high frequency together with a gas injection (Khor et al., 2000). The gases commonly used in this process to produce the plasma jet are: the acetylene, butane and argon, which these gases limits the plasma temperature. After ignition of the plasma jet is aggregate the CaP powder. The plasma produces the ionization of the CaP particles that later are ejected to the substrate to form the coating. With this deposition technique is possible obtained CaP coatings with high hardness, low fracture strength, low coefficient of thermal expansion, high porosity, thick, low adhesion between coating and substrate (Khor et al., 2000). To produce HA coatings for this technique is commonly observed additional CaP phases as: TCP and CaO (Dong et al., 2003; Ellies et al., 1992). In addition, the crystallinity, density, thickness and roughness of the coatings are difficult to control producing not adequate mechanical properties[55] that in implant applications are not efficient.

1.1.2. Sol-Gel Deposition

This chemical deposition technique is extensively used for deposit CaP coatings over metallic implants mainly due to control of the HA stoichiometry and low cost (Guipont et al., 2010; Padilla et al., 2005; K. Zhang et al., 2005). The technique consist in polymerize an organometallic in an alcohol or water solution (Sol) together with an acid or base solvent (Gel) up to produce the bioceramic (Belabbes, de Carvalho, Schleife, & Bechstedt, 2011; Schwertfeger et al., 1994). With this technique is possible produce coating, fibers and powder (Schwertfeger et al., 1994). In coatings fabrication the process can form particles size of HA equal and less than the production by plasma spray (S Zhang et al., 2007). Commonly, deposition of HA and HA substituted with other ions may produce thick, porous and cracked coatings with weak adhesion and to form amorphous phases at room temperature (Chung et al., 2006; Jallot et al., 2005; Kim & Choe, 2011; Layrolle, Ito, & Tateishi, 1998; Tredwin, Young, Neel, Georgiou, & Knowles, 2014; S Zhang et al., 2007).

1.1.3. Electrodeposition

Several works has been performed using this technique to produce HA coatings. The mainly characteristics of these CaP coatings are low temperature processing, thick coatings ($\Lambda \approx 50$ μm), cracked, non-dense, high crystallinity, easy to coat complex parts, control of the composition and microstructure and long time to depositions. The technique involve the deposit over a metallic substrate by electrolysis controlling: the composition into the aqueous environment, the temperature, length between electrodes, voltage and current applied on the metallic substrate.(Dos Santos et al., 2010) From this technique is possible produce microcrystals of HA (5 μm of size) oriented along the *c* axis of the HA structure (Dos Santos et al., 2010; Zhang, 2013).

1.1.4. Pulsed Laser Deposition

The technique involve in ionize the particles of a target from the use of a pulsed laser of high energy density. The most used laser to deposited CaP films is the KrF ($\lambda = 248$ nm), (Carradò, 2011; Socol et al., 2004) (Iliescu, Nelea, Werckmann, & Mihailescu, 2004; Pelletier, Carradò, Faerber, & Mihailescu, 2011; Rau et al., 2010) but others are also used as: Nd:YaG ($\lambda = 355$ nm) and ArF ($\lambda = 193$ nm)(Jedynski, Hoffman, Mroz, & Szymanski, 2008; Mróz, Jedyński, Prokopiuk, Ślósarczyk, & Paszkiewicz, 2009; Saju, Reshmi, Jayadas, James, & Jayaraj, 2009).

When the laser beam of high energy strike the target surface the firsts micrometric layers may absorbed this energy producing a high atomic and molecular vibration, dissociating the atomic bond and consequently producing the ionization and ablations of atoms, radicals, clusters and agglomerates of particles that will be deposited over a substrate. The PLD technique may reproduce HA coatings with the same target stoichiometry, as well as, can reduce the thickness up to 50 nm. Commonly, these films produce a combination of crystalline and amorphous phases with high roughness and porous but with a better adhesion between substrate and coating (K. Zhang et al., 2005; Sam Zhang, 2013; S Zhang et al., 2007). In addition, is common perform a threat thermal *in-situ* on the substrate (TT ~ 350° C - 600° C --> 350 °C - 600 °C) to induce a formation of only crystalline HA phase (Surmenev, 2012; Surmenev et al., 2011).

1.1.5. Radio Frequency Magnetron Sputtering (RFMS)

This is the principal physical deposition technique to produce CaP nanocoatings (Akazawa & Ueno, 2013, 2014; Boyd, Duffy, McCann, Cairns, & Meenan, 2007; Feddes, Wolke, Jansen, & Vredenberg, 2003; Padilla et al., 2005; Snyders et al., 2007; Surmenev, 2012; Surmenev et al., 2011; Van Der Wal et al., 2006; Xu, Long, Sim, Diong, & Ostrikov, 2005; Yamaguchi, Tanaka, & Ide-Ektessabi, 2006). This technique consist in ionize an inert gas, commonly argon (Ar), producing a plasma which is confined by a magnetic field generated by the magnetrons. The Ar ions of the plasma are accelerated in direction to the cathode\target striking the target surface and consequently producing the pulverization of the target atoms and finally the deposition of these atoms over the substrate surface.

Most of coatings produced at room temperature by RFMS are composed by a lower crystalline order (some only of amorphous phase) and non-stoichiometric (Ca/P $\neq$ 1.67). Threat thermal at T > 500° C --> 500 °C (*in-situ* and *ex-situ*) are necessary to induce the crystallinity of the coatings and transform the ACP to HA coating (Boyd et al., 2007; Boyd et al., 2006; Ozeki, Aoki, & Masuzawa, 2010; Snyders et al., 2008; Snyders et al., 2007; Van Der Wal et al., 2006; Van der Wal, Wolke, Jansen, & Vredenberg, 2005). Nevertheless, in some cases the heat processes endanger the coating stability and the adhesion between coating and substrate.

On the other hand, all works using this technique has tried adjusting the deposition parameters for to get the fabrication of unique phase of crystalline and stoichiometric HA coatings. In this direction were produced and characterized CaP films using bias at high voltage and heaters at high temperatures on the substrate holder, which make the deposition system more expensive and difficult to operate. Despite the large number of publications on this theme, few studies has been carried out over the mechanisms of CaP films formation. Most of the research is limited to the knowledge of the relationship between the structural characteristics of the films with the deposition parameters such as: power density, voltage, gas pressure, and target - substrate distance, deposition time (rate), bias and temperature applied to the substrate (Surmenev, 2012; Surmenev et al., 2011; Surmeneva et al., 2013; Sam Zhang, 2013). In the case of physical vapor deposition techniques (PVD), have not still detailed studies involving the influence of the thermodynamic properties of the plasma with the films properties deposited on different substrates (Surmenev, 2012; Surmenev et al., 2011; Surmeneva et al., 2013).

1.2. Right Angle Magnetron Sputtering System (RAMS)

This sputtering system was adapted from the conventional magnetron sputtering geometry. This system exhibited two magnetrons positioned face-to-face (*in-axis*) with a substrate holder kept in a floating electric potential positioned in right angle to the magnetrons (*off-axis*). This modification was called right angle magnetron sputtering (RAMS) (López et al., 2013; Mello et al., 2007; López et al., 2015). The two magnetrons of the RAMS system work at 13.56 MHz of radio frequency (RF). The magnets into the magnetrons are composed for NbFeB (grade N42) and produce a high intensity magnetic field

($\mathbf{B} \approx 0.15$ T) which confined the plasma inside the central region of HA targets, as observed in Figure. 01 (López et al., 2013).

The high magnetic confinement of the plasma increases the ionization rate, the diffusion processes and the plasma temperature of the argon (Ar), oxygen (O_2) (5 mTorr and 1 mTorr, respectively) and the atoms that conform the targets (Ca, P and H). The substrates are positioned at distance of $\mathbf{Z} = 29$ mm from the center of the plasma bulk (see Figure. 01) which in this position there have a plasma trapped between magnetic mirror. The substrate holder was maintained electrically insulate with the RAMS chamber, producing a potential of 2 V relative to the ground, approximately. This floating potential and the deposition of ions (Ca^+, PO^-, O^-, H*) induces a great mobility of these species (ions and electrons) over the substrate surface, increasing the ions diffusion in the film growing as well as producing crystalline thin films conformed by HA nanodomains (in lower deposition time) that along the time the heat delivered by the plasma produces crystalline coatings of HA at room temperature (Figure 1).

1.3. HA Coatings Fabricated by RAMS System

Commonly, there have a routine to coatings preparation which consist in associate the structural characterization (stoichiometry, crystallinity, adhesion film to substrate, roughness and texture) with the deposition conditions as: height between substrate - target, working pressure of argon and oxygen gases, bias potential on substrate holder, *in-situ* thermal treatment on substrates and RF power delivered to magnetrons (Surmenev, 2012; Sam Zhang, 2013). Other alternative is study, understand and control the thermodynamic properties of the plasma to form the coating with adequate physical-chemical properties and to propose new models of materials formations. For this, adequate knowledge in plasma technology and plasma processing must be taken in consideration as well as diagnose the plasma from adequate

*Figure 1. (a) Scheme of the RAMS system showing the magnetic field (B) and the substrate position region (**Z**) where the **B** ≈ 0 T. (b) Picture of the interior of the sputtering chamber showing the confined plasma glow at 120 W of RF power and at Ar and O_2 gas pressures of 5 mTorr and 1 mTorr, respectively* López et al., 2013;

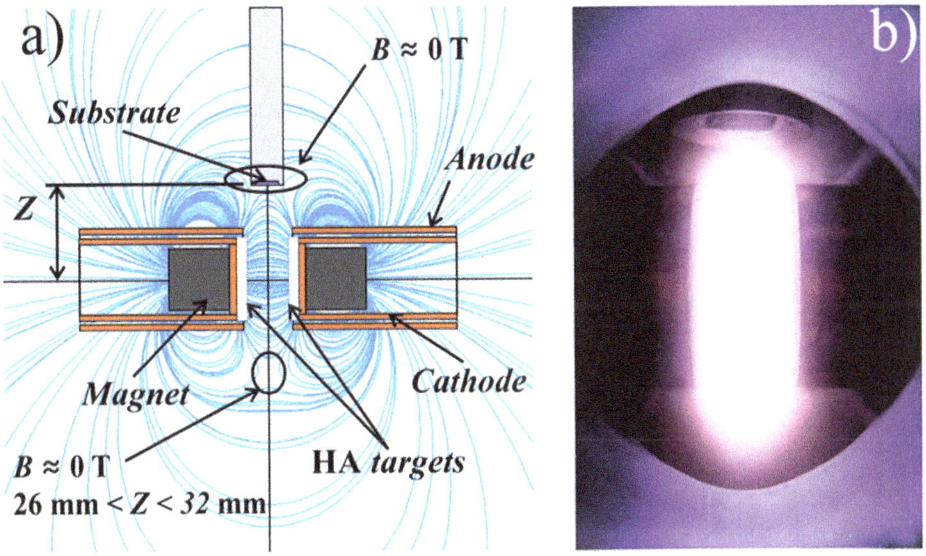

probes to measure the kinetic and dynamic of the electrons and ions in the discharge. The method most used to diagnose the plasma is from the Langmuir probe, which is possible characterize the thermodynamic properties of the electrons in the plasma discharge such as: density (n_e), temperature (T_{eff}), plasma potential (V_p), average kinetic energy ($< \varepsilon >$), diffusions (Γ) and energy delivered to film formation (E_D). (Chen & Smith, 1984; M. Lieberman & Lichtenberg, 2005; M. A. Lieberman & Lichtenberg, 1994; Shun'ko, 2009) With the RAMS system (using crystalline and stoichiometric and non-stoichiometric HA targets) was possible obtained stoichiometric (Ca/P = 1.68 ± 0.08) and crystalline (120 nm of crystallite size) nanostructured HA thin films (thickness less than 540 nm) with smooth roughness of 2 nm from the tuning of electron plasma properties in: $n_e = 2.0 \times 10^{15}$ m^{-3} and $T_{eff} = 11.8$ eV (Figure. 2 and Figure. 3) (López et al., 2013).

From these results was possible to propose a growth model of stoichiometric and crystalline HA film produce at room temperature. This growth model associates the plasma temperature at $T_{eff} = 11.8$ eV and the total energy to form one HA unit cell, since ions deposition of Ca^+, O^-, H^* and PO^- to formations of CaP aggregates, formations of CaP clusters precursors (possible Posner clusters), formation of a HA nanodomains and formation of crystalline films of HA (López et al., 2013). The energy of formation of these CaP clusters (precursors of the HA structure) were compared from a simulation made by *ab initio* in vacuum environment as proposed by other authors (Kanzaki, Treboux, Onuma, Tsutsumi, & Ito, 2001; Louis-Achille, De Windt, & Defranceschi, 2000; Treboux, Layrolle, Kanzaki, Onuma, & Ito, 2000).

The fabrication of a controlled crystallinity and stoichiometry of HA nanocoatings (at room temperature) produced without heat treatment (*in-situ* or after deposition) and bias is only possible from the knowledge and the control of the plasma properties, which will have a high efficiency in orthopedic implant applications.

*Figure 2. GIXRD (λ = 0.13775 nm) patterns at a grazing incidence of Φ = 0.5° for the films that were fabricated with thicknesses and deposition times of (a) 90 nm – 30 min, (b) 135 nm – 45 min, (c) 225 nm – 75 min and (d) 540 nm – 180 min. The HA thin films were made at a deposition rate of 3 nm/min (**Z** = 29 mm) with a plasma power of 120 W and gas pressures of 5 mTorr and 1 mTorr for argon and oxygen*
López et al., 2013

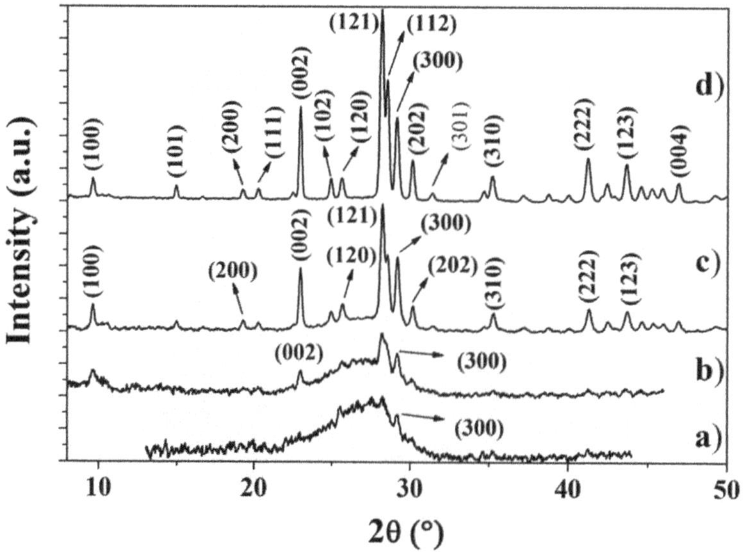

Figure 3. (a) Image obtained by scanning transmission electron microscopy - STEM (bright field imaging) - for the sample that was prepared by FIB showing the CaP thin film region with a thickness of 45 nm between the Ti and Pt films. (b) HRTEM of the CaP thin film showing many HA crystalline nanodomains. The thin film was fabricated at 120 W of RF power, Z = 29 mm, deposition time of 15 min, deposition rate of 3 nm/min and gas pressures of 5 mTorr and 1 mTorr for argon and oxygen, respectively
López et al., 2013

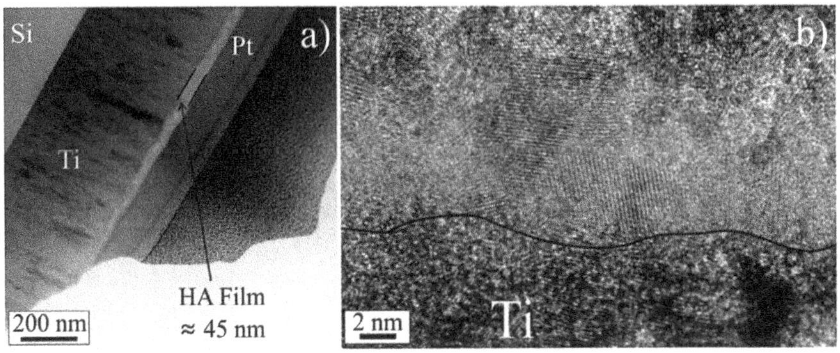

1.4. Functionalized HA Coatings by Incorporation of Fluorine Ions and Fabricated by RAMS System

HA targets with ionic substitution of OH^- by F^-, ($Ca_{10}(PO_4)_6(OH)_{2-x}F_x$, where x = 0.6) was synthesized and produced to fabricated FHA films with the finality of functionalize the film surface and produce an increase of the physic-chemical stability, increase the resistance to dissolution by acid, reduce the bacterial proliferation and increase the bioactivity, as reported by other works (Louis-Achille et al., 2000; Montazeri et al., 2013; Sam Zhang, 2013).

To produce FHA films with excellent crystallinity (in all film thickness as shown in the Figure. 04 and Figure. 05), excellent stoichiometry with adequate introduction of fluorine ions in the HA structure (Ca/P = 1.68 ± 0.04 and F/Ca = 0.12 ± 0.004), low roughness (R_{RMS} = 2 nm) and good reproducibility in different substrates surfaces, the deposition conditions was maintained at Z = 29 mm (height between substrate and magnetrons, as observed in the Figure. 01), RF power of 110 W, deposition time of 180 min and deposition rate of 3 nm/min. In this condition the plasma parameters showed an electron density and temperature of: $n_e = 1.2 \times 10^{15}$ m^{-3} and T_{eff} = 9.0 eV, respectively (López et al., 2015).

Finally, these results are evident that the right angle magnetron sputtering (RAMS) have a great potential to fabricate crystalline and stoichiometric nanocoatings ($\Lambda < 500$ nm of thickness) at room temperature of HA and FHA with adequate control of the fluorine incorporation and without the use of heater and bias on the substrate holder.

2. ACOUSTIC WAVE NANO DEVICES ALUMINUM NITRIDE OBTAINED BY PULSED LASER DEPOSITION

Semiconductors nitrides of group III-V (GaN, InN AlN) alloys have been considered for many years as ideal for manufacturing semiconductor optoelectronic devices for the performance limit blue-ultraviolet

Figure 4. TEM images of 540 nm thick FHA film produced at 110 W RF power, 180 min and Z = 29 mm. The gas pressures of Ar and O_2 were 5.0 mTorr and 1.0 mTorr, respectively. (a) Low-magnification image of the sample prepared by FIB. (b) TEM dark field image showing columnar crystals oriented perpendicularly to the substrate surface
López et al., 2015

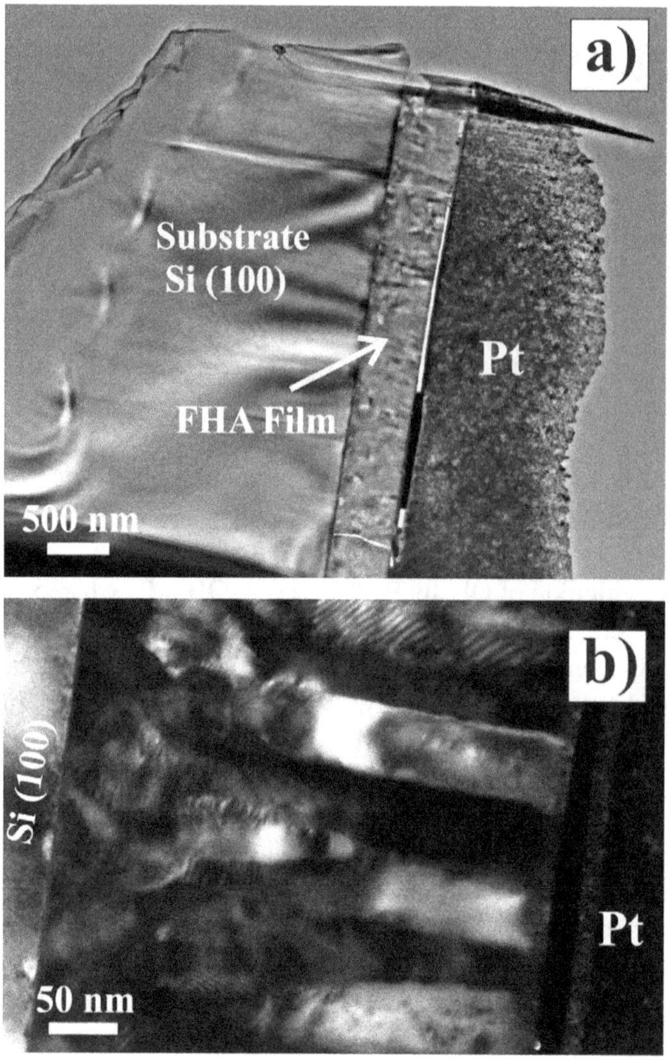

(UV). The main appeal of these materials is the fact that these have a direct energy gap covering a wide range of energies from 1.9 eV (red) InN, to 3.4 eV (UVA) of GaN, to 6.2 eV (UVC) of AIN (Strite & Morkoç, 1992). Maruska *et al.* (Maruska & Tietjen, 1969) fabricate in the decade of the 60s the first GaN crystal thin film by epitaxy techniques hydride vapor phase. The main applications of these materials relate to emitters / detectors radiation devices in the blue-ultraviolet range. Today, most optoelectronic devices on the market operate at wavelengths from infrared (IR) to green.

Figure 5. HRTEM (a) and filtered HRTEM (b) images showing the interface between an FHA film and a Si (100) substrate. The arrows in (a) and the segmented lines in (b) indicate the interface between the film and the substrate. The hkl planes are indicated in the images. The FHA film was produced at 110 W of RF power, 180 min and **Z** *= 29 mm. The gas pressures of Ar and O_2 were 5.0 mTorr and 1.0 mTorr, respectively*
López et al., 2015

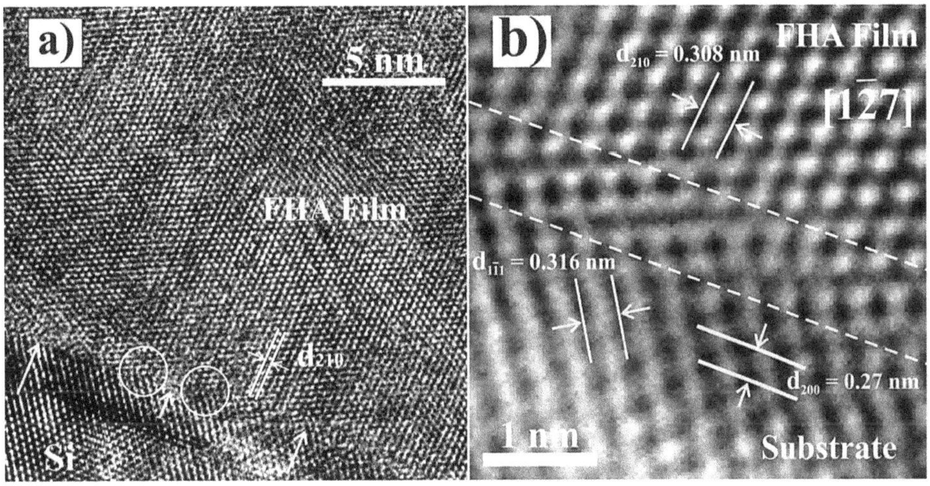

2.1. Structural Properties of Nitride Semiconductors (Group III-V)

These nitrides crystallize according to three types of structure: Wurtzite ((hexagonal close packing)), also called (*α-AlN; α-GaN and α-InN*); zinc-blende (cubic), denoted as (β-*AlN,* β-*GaN and* β-*InN*) and rock salt structure ((Sodium chloride, NaCl Structure). The first two are the most common; the Wurtzite structure is the most stable at room temperature while the zinc-blende is metastable. The Wurtzite type structure consists of two hexagonal compact packages (HCP), two types of atoms interconnected along the axis direction of hexagonal symmetry, with a spacing of 5/8*c*. The Wurtzite zinc-blende structures are similar in that each metal atom (Group - III) is coordinated with four non-metallic (Nitrogen) and vice versa. In fact, they only differ in the sequence of stacking of planes diatomic N-Metal. Wurtzite structure in sequence in the direction <0001> is *ABABAB* while in the zinc-blende type *ABCABC* stacking is in the direction <111>. When having both structures similarity relates to appear zincblende type inclusions in layers with Wurtzite structure, especially in areas in which has stacking faults (Levinshtein, Rumyantsev, & Shur, 2001; Maruska & Tietjen, 1969; Morkoc, 2013; Ruterana, Albrecht, & Neugebauer, 2006). The lattice parameter of the three binary (GaN, AlN, InN) for the different structures is shown in Table 1.

The lattice vibrations (phonons), observable in the nitrides with Wurtzite structure by Raman spectroscopy, are related in Table 2.

2.2. Aluminum Nitride Applications in Acoustic Wave Sensors

The relationship between the surface acoustic wave (SAW) designed on thin films of aluminum nitride (AlN) grown on Si_3N_4 substrates by using a Pulsed Laser Deposition and optical properties has been widely studied due to their low net mismatch and thermal stability (Cressler, 2005). Moreover the dependency

Table 1. Lattice parameters and energy gap of nitrides of group III-V at room temperature

	Lattice Parameters (Å)	EG (300K) (eV)
β-AlN	$a = 4.38$	5.11 (theoretical)
α-AlN	$a = 3.112$ $c = 4.982$	6.2
β-GaN	$a = 4.52$	3.2
α-GaN	$a = 3.189$ $c = 5.185$	3.39
β-InN	$a = 4.98$	2.2 (theoretical)
α-InN	$a = 3.548$ $c = 5.705$	1.89, 0.9

Levinshtein et al., 2001; Morkoc, 2013; Ruterana et al., 2006

Table 2. Nitrides vibrational modes in Wurtzite structure. LO: Longitudinal Optical; TO: Optical Transversal

Vibration Mode (cm^{-1})	AlN	GaN	InN
E2 – Low frequency	215	137	85
E2 – High frequency	652	592	485
A1 – *TO*	640	546	449
E1 – *LO*	883	732	587
E1 – *TO*	668	555	457
E1 – *LO*	922	741	596

Levinshtein et al., 2001; Morkoc, 2013; Ruterana et al., 2006

of both properties with a temperature of deposition. Aluminum nitride, an electrically insulating ceramic with a wide band gap of 6.3 eV, is a potentially useful dielectric material very important in fields such as optoelectronic and microelectronics. For instance, it can serve as the dielectric gate with high voltage and high power electronic devices; AlN is also being investigated as a substitute for the silicon dioxide buried layer in silicon-on-insulator (SOI) substrates due to its low thermal expansion coefficient, high breakdown dielectric strength, and high chemical and thermal stability (Clement, Vergara, Sangrador, Iborra, & Sanz-Hervás, 2004; Jiang, Shen, Ogawa, & Guo, 2003; Yamashita, Fukui, Misawa, & Yoshida, 1979). Moreover, metal nitrides based on AlN are of high interest because of their electrical, optical and acoustical properties (Clement et al., 2004). Their hardness and thermal coefficients of expansion are comparable to that of Si. AlN is used for acoustic wave devices on Si (Clement et al., 2004), optical coatings for spacecraft components, heat sinks in electronic packaging applications, as well as electroluminescent devices in the wavelength range from 215 nm to the blue end of the optical spectrum. Some optical properties near the fundamental band gap of AlN have been reported by Yamashita et al. (Yamashita et al., 1979) since 1979.

Several theoretical studies have been well documented in the past, (Jiang et al., 2003; Levinshtein et al., 2001) however, reliable experimental data has been reported just recently because both the sample quality and the spectroscopic techniques have only been substantially improved in recent years. Precise knowledge of the optical constants is particularly important in view of the use of AlN thin films in optical filters and light emitting laser diodes since they also show temperature dependence. Similar to the section 1.1.4 for the HA, Pulsed Laser Deposition (PLD) is largely applied for processing thin films and other structures. The highly no equilibrium nature of the PLD process is attractive for the synthesis of stoichiometries thin films of various metal nitrides, carbon nitrides, oxides and oxynitrides from the corresponding bulk targets. PLD appears to be a suitable method to transfer stoichiometrically complex monolayer structures from AlN ceramic structures to substrates.

Thus, from AlN stoichiometries films, many infrared (IR) measurements presented in the literature were used to estimate the film quality. The IR-spectra provided important information about the composition, homogeneity, crystallinity and the residual stresses present in the films. Moreover, the ellipsometry and reflectance measurements have been used to determine the refractive index in many AlN compounds, since the ellipsometry and reflectance are sensitive and non-destructive techniques used for studies of optical properties and microstructures of surfaces and thin films (Caicedo, Pérez, Caicedo, & Riascos, 2013; Pérez, Riascos, Caicedo, Cabrera, & Yate, 2011a; Rosenberger, Baird, McCullen, Auner, & Shreve, 2008; Zhuang et al., 2009). Additionally, single-chip front-end RF modules incorporating Surface Acoustic Wave (SAW) filters are a matter of intense research. At present, these modules are fabricated by bonding single crystal piezoelectric SAW devices onto integrated circuits. Thin films of polycrystalline Aluminum Nitride (AlN) are promising materials for the integration of SAW devices on Si_3N_4 substrates due to their good piezoelectric properties and the possibility of deposition a low temperature compatible with the manufacturing of Si integrated circuits. AlN thin films of sufficient quality for SAW applications can be a obtained by the PLD technique (Pérez et al., 2011a).

2.3 X-Ray Photoelectron Spectroscopy (XPS) Analysis

The XPS survey spectrum of AlN thin film grown at 300 °C is shown in Figure 6. The peaks at 534.4 eV, 397.3 eV, 124.4 eV, and 73.9 eV correspond to O1s, N1s, Al2s, and Al2p binding energies, respectively. Calculation of the peak areas of deconvoluted spectrum (Figure 7) without O1s contribution gives an atomic ratio of Al:N = 0.392: 0.588, which is similar to the stoichiometry of $Al_{0.40}N_{0.60}$ (Caicedo et al., 2013). The core electronic spectra carry the information about the chemical composition and bonding character of the Al–N films. The integral of N1s, Al2s, and Al2p spectra corrected by relevant sensitive factors can evaluate the concentrations of Al, and N elements in the Al–N films. The corresponding integral of the deconvoluted peaks can also be used to estimate the bond contents, which are described by the following formula:

$$C_i = \Sigma(A_i/S_i)/\Sigma(A_j/S_j) \quad (1)$$

Where -***S*** is the sensitivity factor, ***A*** is the integral of deconvoluted peaks, and ***C*** is the atomic content. The numerator is the sum of the integral of one sort of bond; the denominator is the sum of the integral

Figure 6. XPS survey spectrum of Al-N coatings deposited on Si_3N_4 at 300 °C, where different elemental signals show

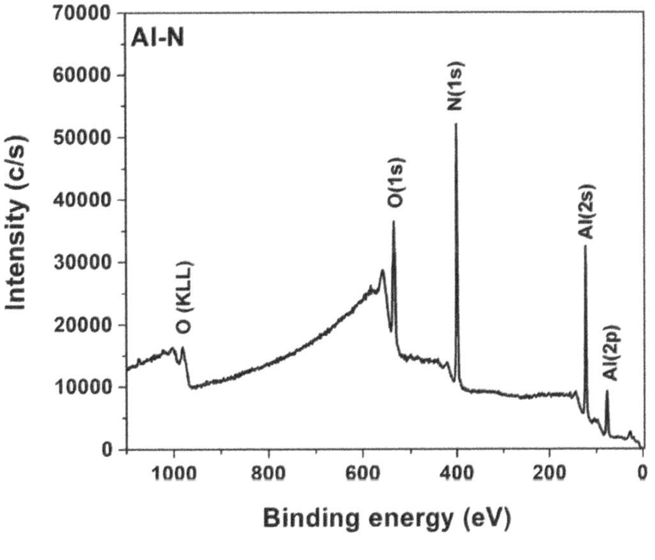

Figure 7. High-resolution XPS spectrum of: (a) Al2p and (b) N1s of Al-N film is around 300 °C

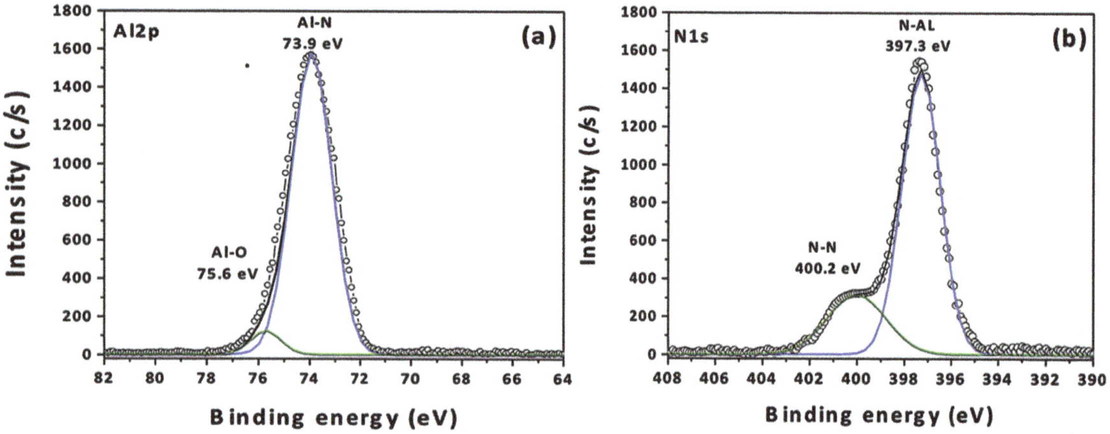

of all types of bonds decomposed from the whole peak of N1s, A2s, and Al2p in the sample. The atomic concentrations of Al, N and O elements of XPS analysis of the coating deposited at 300 °C are listed in Table 3. For all the Al-N films obtained, the O concentration is less than 3 at.%. We can speculate that the majority of the oxygen at the surface of the non-treated sample (which is smooth and almost perfect) is chemically bonded into alumina-dominated oxide- nitride material.

The high-resolution spectra of Al2p, and N1s were recorded from AlN films, as shown in Figure 7(a), the Al2p peak is composed of a shoulder separated by 1.7 eV with intense peak. The XPS spectrum of Al2p can be fitted well by two Gaussian functions. The value of the binding energy obtained for the Al2p peak was 73.9 eV and the higher value for Al2p was 75.6 eV, respectively. According to the literature(Mahmood et al., 2003; Pérez Taborda, Riascos Landázuri, Jiménez García, & Caicedo Angulo, 2010; Taborda, Caicedo, Grisales, Saldarriaga, & Riascos, 2015) for the Al2p peak, the first one (73.9 eV) and the second one (75.6 eV) can be assigned to Al–N and Al-O bonds.

Table 3. XPS analysis-derived atomic concentrations of Al, and N elements in the Al-N film synthesized by PLD with 7 mbar and substrate deposition temperatures as function of atomic percentage (at%)

Growing Conditions		Atomic Composition (at %) Al-N Films		
Substrate Temperature (°C)	Pressure (10^{-3} mbar)	Al	N	O
200	9	39.0	58.6	2.4
300	9	39.4	58.8	2.0
400	9	39.6	59.0	1.8
500	9	39.9	59.3	1.5
600	9	40.1	59.5	1.3
630	9	40.3	59.7	1.1

(Caicedo et al., 2013; Pérez, Riascos, Caicedo, Cabrera, & Yate, 2011b; PEREZ TABORDA, RIASCOS, & VERA; Taborda et al., 2015)

The appearance of the peak at 73.9 eV clearly shows that Al has reacted with N; therefore, it can be assigned to Al-N. In Figure 7(b), N1s peak is composed of spin doublets, each separated by 2.9 eV. The XPS spectrum of N1s, well fitted by two Gaussian functions, depicts the N1s spectrum with values at 397.3 eV and 400.2 eV characteristics for N-N and Al-N bonds, respectively.

The XPS results demonstrated that Al atoms bonded to N in the form of nitride, taking into account that the elemental concentration of the Al-N film was obtained by adjusting the laser incidence on Al target, and the N_2 as a working gas. In this specific case, it was found that amounts of Al (N) in the AlN film were maximum at the current establishment of process conditions and the ratio of Al and N in the film was about for 2:1. In general, formative Al (N) phase indicates that the aluminum and nitrogen activity and activation energy provided by the present deposition conditions are enough for the formation of an Al-N thin film. Although the surface temperature of the substrate during deposition of Al-N film is around 300 °C, the substrate lies in a high-density plasma region and a high ion-to-atom ratio of aluminum and nitrogen, thus behavior can induce to the formation of Al-N phase at the low temperature below 300 °C. Moreover, the correlation between temperature deposition and chemical properties due to the reduction of oxygen concentration when the increase of deposition temperature, this behavior was associated with high surface energy activity generated by high surface temperatures that induce more nitrogen reaction.

The concentrations in these depth profiles were obtained by curve fitting of high-resolution XPS spectra and measurements of the XPS peak areas. In order to see the effects of increasing the substrate temperature during the growth of thin films which would serve AlN buffer for Mo SAWs. The dependence of aluminum and nitrogen in the deposition temperature is shown in Figure 8 which represent various chemical states as described in the section 2.3 XPS in Table 3 and Figure 8.

Figure 8. Temperature dependence at the concentrations of Al and N for AlN films

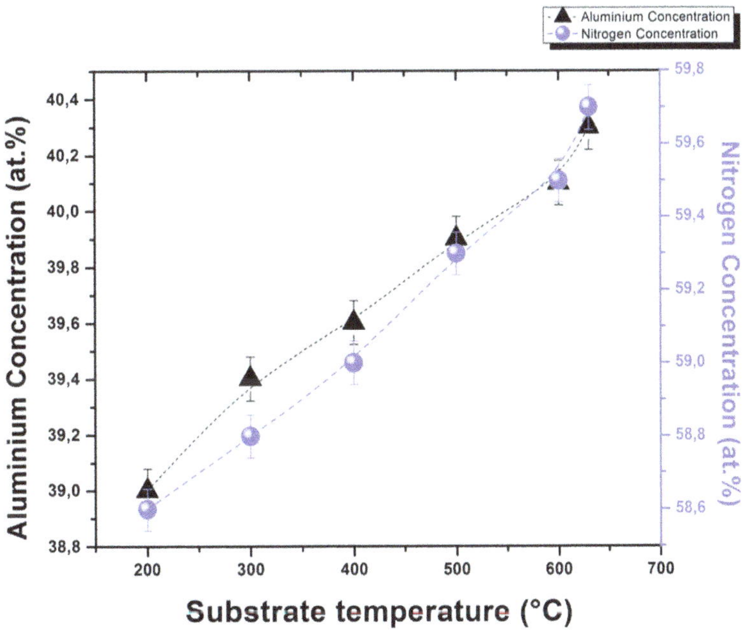

2.4 Surface Analysis of AlN Films via Atomic Force Microscope

Nevertheless, most of the energy in the propagating surface acoustic waves is contained in the region within one wavelength below the surface, so they are very sensitive to the surface state and are affected by any change, such as residual stresses and the level of roughness. Moreover, authors such as C. Zhuang *et al.(Zhuang et al., 2009)* have shown that at the same frequency and in the small roughness limits (RMS $\ll$ wavelength), the SAW velocity increases according to the decrease in the RMS roughness. Figure 9 a-c shows a 5μm × 5μm image for AlN films deposited with different deposition temperatures. The general trend in the AlN films with the decreasing temperature exhibits a decrease of the grain size and roughness. In Figure 9 obtained from AFM drops observed of which are characteristic of the PLD technique and are occasioned by the release of particulates from the Aluminum target at impact with the substrate, such generate imperfections on the surface of the films also shows micro drops on AlN surface generated by PLD energy deposition corresponding to high laser fluency. Moreover, the correlation between roughness and grain size with deposition temperature is shown in Figure 10 The quantitative values were extracted from the AFM images by means of statistical analysis Scanning Probe Image Processor (SPIP®). Each data point in the plots represents an average over AFM images. The corresponding error bar was obtained by a standard deviation of the values (Mahmood et al., 2003; PEREZ TABORDA et al.; Pérez Taborda et al., 2010; Taborda et al., 2015; Zhuang et al., 2009).

Figure 10 shows the grain size and roughness values for the AlN films, thus, in this case, a decreasing tendency in surface measurements was observed when the deposition temperature was increased. The lower grain size and roughness values obtained in this study by the Al–N films deposited on silicon substrate represented a decrease of grain size around 42% and roughly 31%, compared to that obtained from the Al–N films deposited at 200 °C. This fact is relevant since the continuity of the surface is important for optical and electrical SAW devices (Kar & Bose, 2011; Raole, Mukherjee, & John, 2005).

The variation of grain size and roughness (Figure 10) obtained in the Al–N films suggest that thermal activation generated by the increasing in the deposition temperature on surface substrates modifies the superficial morphology by increasing the energy associated with atoms on the substrate surface and/or the growing film surface which produces a relative reduction of grain size and roughness.

Figure 9. AFM images for AlN films grown onto Si_3N_4 substrate with different deposition temperatures: (a) 200 °C, (b) 400 °C and (c) 630 °C.(Caicedo et al., 2013)

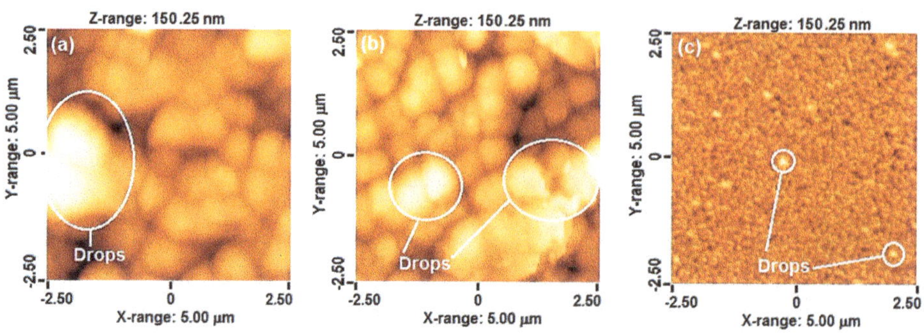

Figure 10. Surface measurements obtained via AFM and SPIP® analysis for AlN films with a total thickness of 150 nm: Correlation between grain size and roughness with deposition temperature

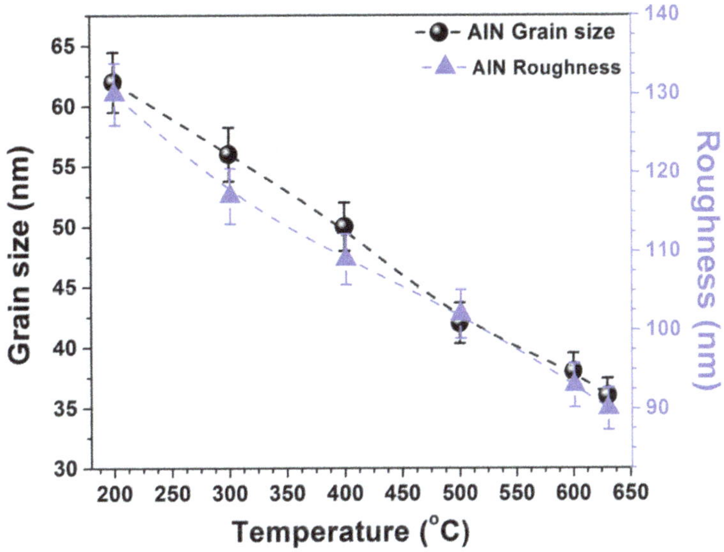

2.5 Dominant Wavelength and Color Purity Analysis

The reflectivity measure is the fractional amplitude of the reflected electromagnetic field while reflectance refers to the fraction of the incident electromagnetic power that is reflected at an interface. The reflectance is thus the square of the magnitude of the reflectivity. The reflectivity can be expressed as a complex number as determined by the Fresnel equations for a single layer, whereas the reflectance is always a positive real number.

Additionally, reflectivity is distinguished from reflectance by the fact that reflectivity is a value that applies to thick reflecting objects. When reflection occurs from thin layers of material, internal reflection effects can cause the reflectance to vary with surface thickness. Reflectivity is the limit value of reflectance as the surface becomes thick; it is the intrinsic reflectance of the surface, hence irrespective of other parameters such as the reflectance of the rear surface (Kar & Bose, 2011; Touati, Hadjoub, & Doghmane, 2006; Zayats et al., 2008). On the other hand, the dominant wavelength (of a color stimulus) is defined as: "*the wavelength of the monochromatic stimulus that, when additively mixed in suitable proportions with the specified achromatic stimulus, matches the color stimulus considered*" (Kar & Bose, 2011). The AlN films have been compared with pure aluminum because the mirror finish aluminum has the highest reflectance of any metal in the 200–500 nm and the 3000 –10000 nm (far IR) regions, while in the 500 –700 nm visible range it is slightly outdone by AlN and silver, but in the 700 –3000 nm range (near IR) it is slightly outdone by, gold, and copper materials.

The reflectance spectrum or spectral reflectance curve is the plot of the reflectivity as a function of wavelength as can be seen in Figure 11 Therefore, all the deposited films appear highly reflective with white and aluminum colored. Figure 11a. shows the optical reflectance spectra of the AlN single layers obtained at different deposition temperatures. The reflectance of the aluminum and the eye sensibility are shown for comparison. The spectra of the samples show high reflectance for long wavelengths, near to 33% for the AlN films deposited with 200 °C and close to 23% of AlN films deposited with 630 °C. These values of reflectance at these wavenumbers are in good agreement with previous reports for AlN films (Zhuang et al., 2009). A clear decrease in reflectivity for short wavelengths is seen, characteristic of a system with high free-electron density with a reflectance edge below 550 nm due to a screened plasma resonance. The white and aluminum colors of the AlN films are a result of the steep plasma reflection edge that occurs in the visible region where the reflectivity minimum is around 560 nm. The color purity refers to the degree of purity present at a color; therefore in a pure color do not exist any achromatic effect within its composition, which can vary its capacity for reflection. One of the three dimensions of color refers to the color purity that a surface can reflect. When the tone is full and is not mixed with an achromatic effect (brightness), this presents its maximum saturation or purity. On the other hand, if a particular color is combined with an achromatic effect, or with its complement (luminosity), it suffers a loss of purity or neutralization. The color purity changes with the temperature of deposition, at ranges from 0.63 to 0.66 for AlN films, away from the color purity of pure aluminum (0.90), hence confirming that the films are less white compared with the aluminum. Color coordinates of AlN single layer films are shown in Figure 11b.

Reference coordinates of pure aluminum are shown for comparison in that Figure 12. The results of color measurement indicate that all films reflect a hue slightly shifted far to the white as compared to aluminum reflectivity and with similar color purity. Likewise, dependence is observed the high dependence on reflectance percentages when the deposition temperature is varied from 200 to 630 °C. Figure 12a exhibits one constant region for wavelengths ranged from 760 to 800 nm; in those regions it is possible to appreciate the effect of deposition temperature on changes in the reflectance values. Figure 12b shows the decreasing of reflectance when the temperature is increased, which indicate that temperature

Figure 11. Dominant wavelength and color purity results (color online): (a) Optical reflectance of AlN films deposited onto Si_3N_4 (100) substrates at the different deposition temperature, also aluminum optical reflectance and eye sensibility were also plotted as the references. (b) Chromatic diagram, in the x, y coordinates, of the reflectivity for AlN films. White coordinates of an achromatic point are located at (1/3, 1/3)

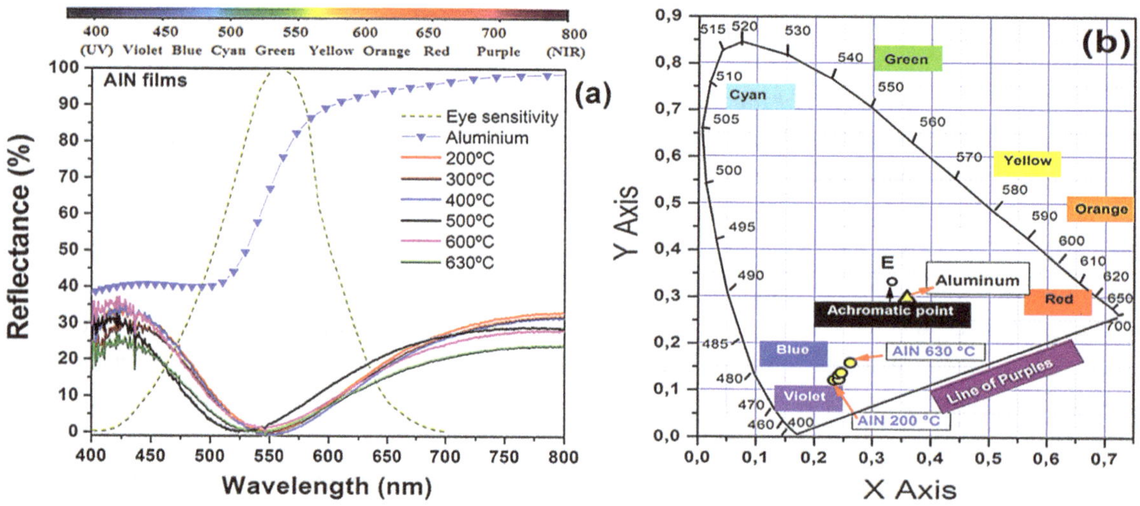

promotes the absorbance in the AlN deposited via PLD. The last behavior changes in optical properties (reflectance) can be related with changes generated by the reduction of grain size and reduction of surface roughness when the depositing temperature is increased producing an activation surface energy observed from AFM images.

From the reflectance spectra it is seen a weak but clear effect of deposition temperature on the optical properties. As the temperature is decreased, the reflectance of the films tends to be higher in the near infrared region. So, dominant wavelength and color purity have been calculated from the reflectance spectra of all films by using the *java-optics* software (Table 4). The dominant wavelength was varied for all the samples, from 562 to 570 nm, and a little close to that of pure aluminum reference (574 nm)

In. Figure 13a it can be observed the influence of deposition temperature on color purity. This graph shows the increase of purity values towards color gray purity. Figure 13b shows the differences in the dominant wavelength for all AlN films deposited with different deposition temperatures.

Figure 12. Temperature-Reflectance dependence (color online): (a) Reflectances of AlN films deposited onto Si_3N_4 (100) substrates at the different deposition temperature from 760 to 800 nm wavelength, (b) reflectance as function of deposition temperature

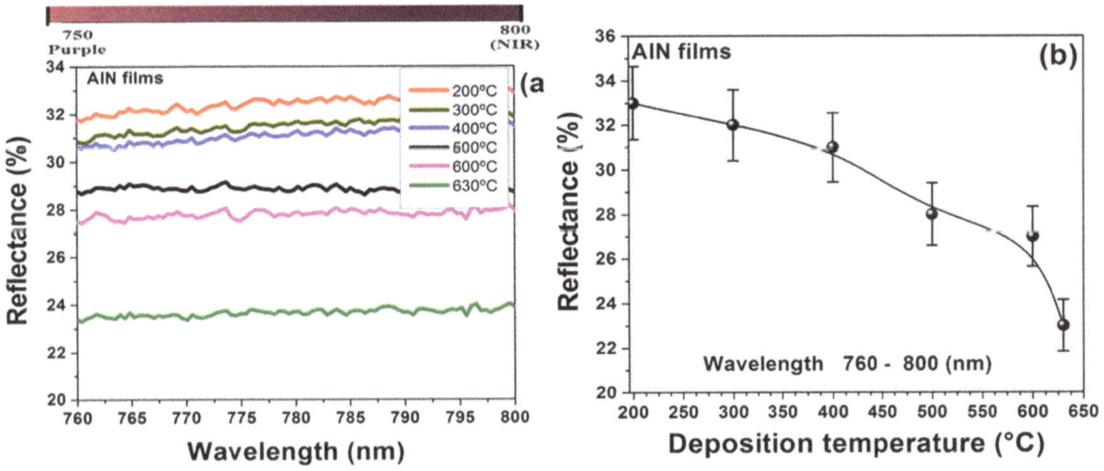

Table 4. Optical properties for AlN films deposited onto a Si_3N_4 substrate

Deposition Temperature (°C)	Dominant Wavelength (nm)	Color Purity	Axis X	Axis Y
200	562	0.63	0.234	0.121
300	567	0.63	0.242	0.123
400	569	0.63	0.240	0.132
500	569	0.64	0.246	0.137
600	569	0.65	0.262	0.158
630	570	0.66	0.302	0.270
Reference Aluminum	574	0.90	0.358	0.371

Figure 13. Temperature dependence at color results for AlN films: (a) color purity, and (b) dominant wavelength as a function of deposition temperature

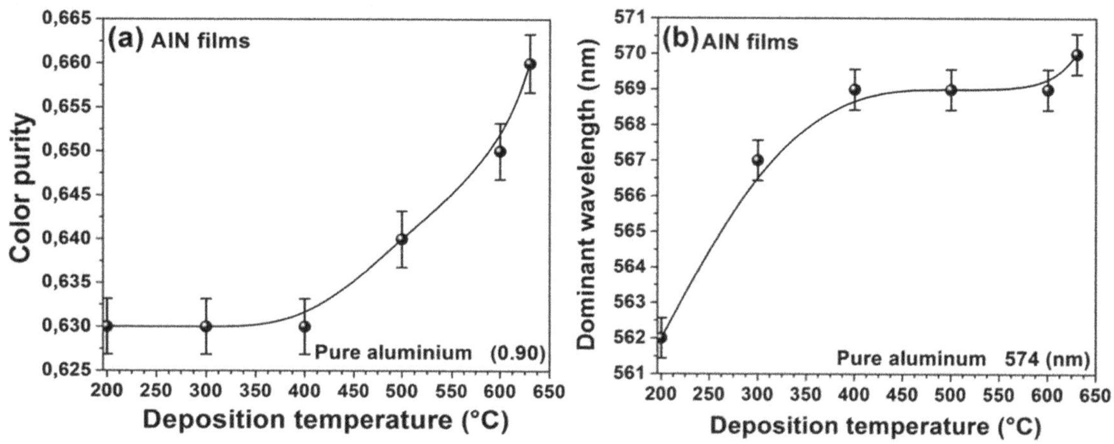

The wavelength is an important optical characteristic for different materials. When the wavelength is changed it is possible to observe that natural color is changed. This modification on optical properties is associated to changes generated by the reduction of grain size and reduction of surface roughness observed in AFM images. Moreover, the changes in chemical composition of AlN films observed in XPS spectra show that oxygen concentration is reduced when the temperature deposition is increased. Therefore, the low oxygen concentration induce changes in the energy bands generating changes at optical properties (color purity and dominant wavelength). Finally, the purity color dependence in AlN films with temperature obtained demonstrates the possibility of some purity color control.

SEM images was carried out to quantitatively study the surface morphology of our samples in relation to an increase on deposition temperature in AlN films grown onto Si_3N_4 (100). Figure 14 Shows SEM micrographs for AlN films with a random distribution of microparticles or micro drops that were analyzed on these surfaces in agreement with AFM images. Accordingly, the deposition temperature clearly affects the reduction of micro-drops; this can be possible due to great surface mobility when the temperature was varied from 200 to 630 °C. This surface mobility reduces the possibility that the micro-drops are anchored on the surface when arriving with high energy on AlN film. Otherwise, the high superficial temperature that produces surface diffusion of micro-particles or micro-drops, which can decrease the overall number of particles. The magnification image showed in Figure 15 exhibits that molybdenum lines are continuous, without any interruption or short circuit. After electrical measurements, it was possible to deduce an insulator behavior for our AlN structures, a good prerequisite for using them as surface acoustic wave (Mo/AlN/SAW) devices.

2.6 Applications: Frequency response of Mo/AlN SAW devices

Authors as Hadjoub *et al.* (Hadjoub, Touati, Doghmane, & Doghmane, 2008) been reported improvement in the Rayleigh Velocity (VR) by decreasing the thickness of the films and substrates used Si_3N_4 with a velocity of 5694 m/s value very next to that of VR (SiC) = 6810 m/s. These values are referred to substrates for fast VR (Al_2O_3) = 5680 m/s, VR (Si) = 4710 m/s, VR (SiO_2) = 3410 m / s, VR (Ti) =

Figure 14. SEM micrographs where is observed the micro-particles or micro-drops: (a) AlN films deposited with 200 °C, (b) AlN films deposited with 630 °C

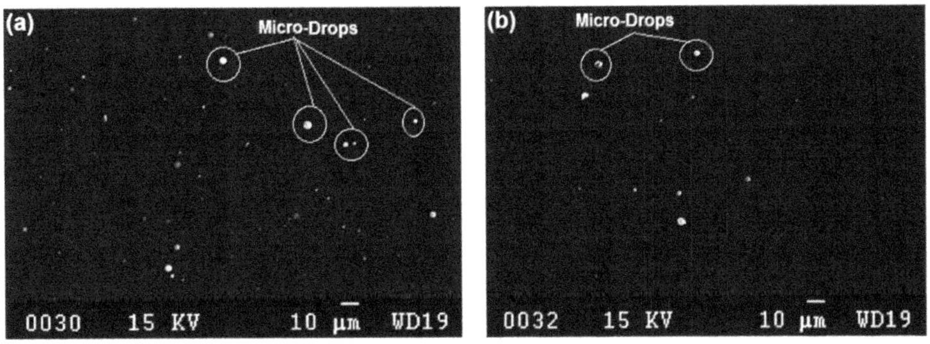

Figure 15. SEM micrographs showing the interdigital structure of Mo coating on AlN films obtained by PLD: (a) general interdigital structure, (b) magnification images where it is possible to observe the width and distribution of lines that conform the interdigital device

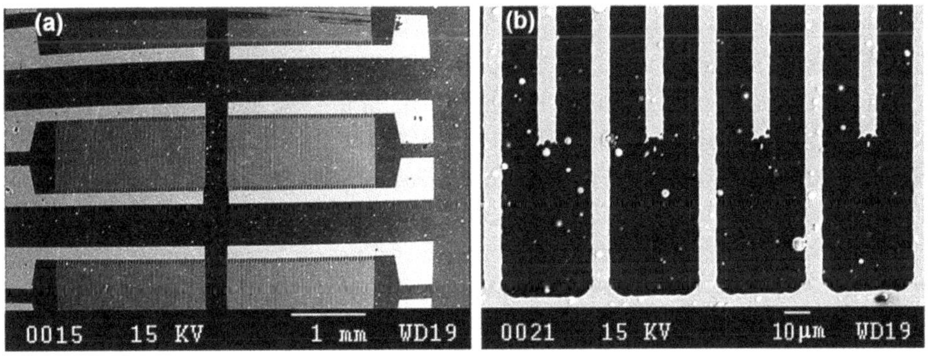

2960 m/s and VR (W) = 2670 m/s, reported in the literature (Caicedo, Zambrano, Aperador, Escobar-Alarcon, & Camps, 2011; Drafts, 2001; Hashimoto & Yamaguchi, 2001; Kao, Cheng, & Chen, 2002; Paduano, Weyburne, Jasinski, & Liliental-Weber, 2004). It clearly shows the importance of using substrates with good speed of Rayleigh and compatible with existing silicon technologies so that in our case Si_3N_4 because VR is 20% higher than that of Si.

Recently, some authors (Caicedo, Zambrano, Aperador, Escobar- Alarcon, & Camps, 2011; Drafts, 2001; Hashimoto & Yamaguchi, 2001; Kao, Cheng, & Chen, 2002; Paduano, Weyburne, Jasinski, & Liliental-Weber, 2004) have reported the relation between the h/λ and V_{SAW} by experiment or theory, where h is AlN thickness, λ is wavelength, and V_{SAW} is SAW velocity. The relationship between the thickness and the wavelength this $h/\lambda = 0.004$. The frequency response characteristics of Mo/AlN SAW devices fabricated using the different deposition conditions, such as, 200 °C, 400 °C and 630 °C, have been measured, of which results are shown in Figure 16 a-d. As can be clearly seen by comparing the wide-frequency scan spectra of those three devices (which are plotted in the insets of Figure 16c, by using the AlN buffer layers the center frequency was shifted from approximately 0.618 GHz Figure 16a to higher frequencies of 0.621 GHz (Figure 16b) and 0.625 GHz (Figure 16c). This indicates that the AlN buffer layer could be used as high phase velocity substrate in the fabrication of high-frequency SAW devices, as with Molybdenum.

Figure 16. Frequency response characteristics of fabricated Mo/AlN SAW devices with different deposition temperatures (a) 200 °C, (b) 400 °C, (c) 630 °C and the insertion loss by the use different deposition temperatures of AlN buffer for (a), (b) and (c)

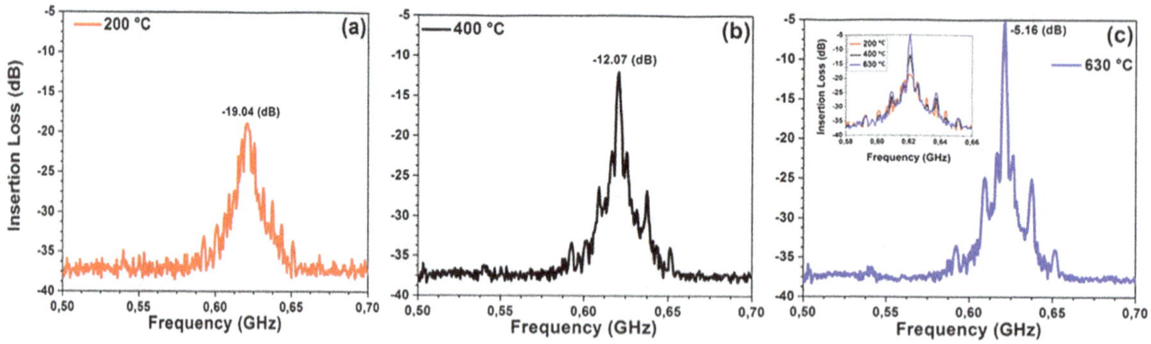

Additionally, in Figure 16d that the insertion loss was reduced by the use different deposition temperatures of AlN buffer, by using the AlN buffer layers the center frequency was shifted from approximately 0.618 GHz (Figure 16a) to higher frequencies of 0.621GHz (Figure 16b) and 0.625GHz (Figure 16c). This indicates that the AlN buffer layer could be used as high phase velocity substrate in the fabrication of high-frequency SAW devices. In addition, it appeared in Figure 16c that the insertion loss was reduced by the use of AlN buffer. To examine in more detail the dependence of insertion loss on the deposition condition of AlN-buffer, the insertion loss has been measured for all the fabricated SAW devices.

The results are plotted in Figure 15 as a function of temperature deposition of AlN films and the estimated quantitative values are also listed in Figure 16. As can be seen in Figure 17, for SAW devices, fabricated using the sputtered-AlN buffer with Mo channel the insertion loss was exactly correlated with the corresponding temperature value for all AlN films.

This can be identified by comparing the results obtained from the device with AlN films deposited on silicon nitride substrate. For the above devices, the temperature deposition was increasing systematically, as already shown in Figure 17. Moreover, the device with AlN deposited to 630 °C revealed the relatively large insertion loss in relation to frequency value (0.625 GHz). This may be attributed to the relatively low surface roughness (90 nm) AlN film used in Mo/AlN device considering that the surface roughness of AlN film may also be associated by the surface morphology and deposition temperature of AlN buffer layer, the results shown in Figure 16d. confirm that frequency response characteristics (insertion loss) can be affect by reduction of oxygen concentration, reduction of grain size, roughness, and decreasing with reflectance of AlN films together with increase of color purity and dominant wavelength thin this sense, the property of AlN buffer may strongly govern the characteristic of SAW device using the Mo/ AlN system. (Caicedo et al., 2013)

The direct relationship between increased frequency and speed in SAW devices, all this related to the morphological qualities of roughness and grain size favor in increased rate of velocity. The resonance frequency of an SAW is determined by the equation $f = V/k$, where f is the resonance frequency, V is the phase velocity of the acoustic wave and k is the acoustic wavelength, which in turn is defined by the electrode pitch in the *interdigital transducer* (IDT). The reflected signal of the Rayleigh mode in AlN films deposited to highest temperatures (630°C) is observed to be much stronger than that of the AlN films deposited to lowest temperatures (200 °C) and the phase velocity of AlN films deposited to (630

Figure 17. The measured insertion losses of fabricated Mo/AlN SAW devices, as a function of the temperature values from the deposited AlN films

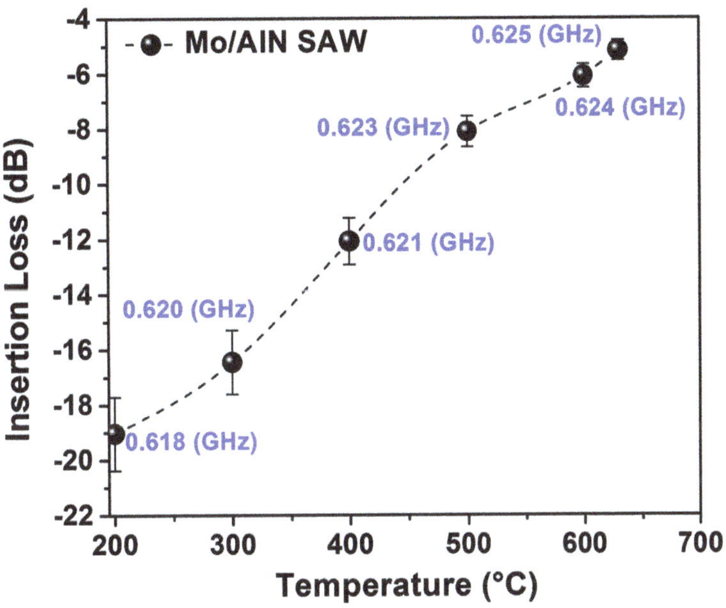

°C) is much higher. For The Mo/AlN SAW devices, the measured resonant frequencies of the Rayleigh mode waves ranges from 0.618 GHz to 0.625 GHz corresponding to phase velocities of 24720 m/s to 25000 m/s, indicating good repeatability and reproducibility (Pérez et al., 2011a).

Thus, the lowers deposition temperature of the substrate, generate a perturbation of the acoustic velocity that is related to a change in frequency of the oscillator circuit. In the Figure 17 is shown as before said to improve velocity with increased temperature of the films deposited during the growth of AlN and were used as substrates for SAW devices Mo. Figure 18 Acoustic wave velocity as a function of deposition temperature in Mo/AlN/Si_3N_4 system.

Take the last behavior is possible to observe that the quasi-static electric field associated with the SAW interacts with the charge carriers of the film. The SAW velocity is, therefore, dependent on the AlN thin film conductivity changes. These velocity changes can easily be interpreted as a shift in the resonant frequency, when the SAW delay line is incorporated in an oscillator conFigureuration. The effect of temperature is one of the easiest parameters that can be sensed because almost all physical and chemical effects depend on temperature (Kar & Bose, 2011; Morkoc, 2013).

In certain applications, the effect of temperature should be minimized in order to avoid any interference with other effects. For acoustic wave sensors, the temperature effect induces dimensional dilatations and small changes in the mechanical and electrical properties. Two kinds of interaction can thus be measured: *(i)* shift of the acoustic wave velocity due to intrinsic changes of material properties (elasticity and density) (this is the case for almost all SAW devices using metal-nitrides materials with low thickness) (Aissa et al., 2015; Caicedo et al., 2013; Clement et al., 2004; Jung, Lee, Kim, & Park, 2004; Trang et al., 2007) and *(ii)* for acoustic devices for which the resonant frequency or the velocity is dependent on the dimensional values (Lamb wave devices, thickness and shear mode), temperature can affect changes of both material properties and boundary conditions(Kar & Bose, 2011; Taborda et al., 2015).

Figure 18. Acoustic wave velocity as a function of deposition temperature in Mo/AlN/ Si3N4 system

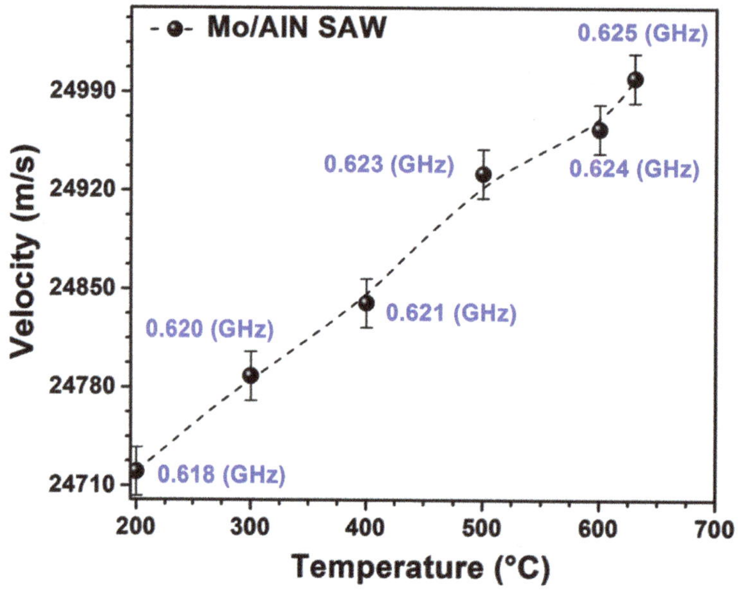

2.7 Correlation Between Optical Properties and Acoustic Wave Velocity in Mo/AlN SAW Devices

In Figure 19 shows the relationship between optical properties (reflectance), acoustic wave velocity and deposition temperature in AlN films. In Figure 19 it is clearly shown that the reduction of grain size, reduction of reflectance and the increase of the color purity when the deposition temperature decreases, is associated with an increasing of frequency response, due to the increase of acoustic wave velocity, produced by energy activation at the higher deposition temperature. From this correlation, it is possible to determine that one merit index associates to an acceptable reflectance and acoustic wave velocity at the same temperature deposition.

Is observed a direct relationship between the roughnesses of the AlN thin films and increasing the deposition temperature, the AFM results show a decrease in roughness with increasing temperature in the deposit. Figure 19b shows a clear correlation between the roughness and the Rayleigh velocity, obtaining the highest velocity for the samples grown at a higher temperature of 630 °C to deposit are the lower roughness.

Therefore, the AlN films deposited around 450 °C offer the best synergy for optical and acoustic properties with good reflectance and an acceptable acoustic wave velocity. Furthermore, the nanometric grain size and in thickness of such thin films AlN/Si_3N_4 a low surface roughness (≤120 nm), the nucleation side is characterized by a relatively small grain size (≤70 nm), which can be decreased with increased substrate temperature improving processes nucleation, hence enhancing the propagation of elastic waves as shown in Table 5.

The morphological properties of AlN are highly dependent on their polarity, since the nitrogen, finished films typically have a greater roughness than the finished aluminum, but instead its crystal quality is better. Initial reservoir conditions greatly affect the polarity of the material obtained (Levinshtein et al.,

Figure 19. Correlation between optical properties (Reflectance) and acoustic wave velocity in Mo/AlN SAW devices. In Figure 19b. There is a correlation between roughness and Rayleigh velocity as a function of temperature deposit AlN thin films

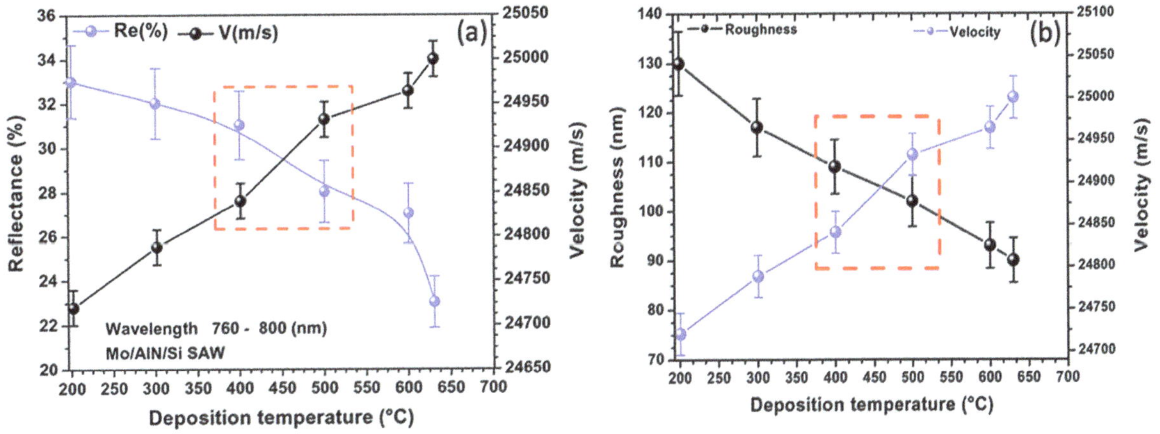

Table 5. Influence of the substrate temperature of AlN thin films on the morphological properties - roughness and the velocity of sound on the SAW

Temperature Substrate (°C)	Insertion Loss (dB)	Frequency (Ghz)	Grain Size (nm)	Roughness (nm)	SAW velocity (m/s)
200	19,04	0,618	62	130	24720
300	16,45	0,620	56	117	24788
400	12,07	0,621	50	109	24840
500	8,1	0,623	42	102	24932
600	6,09	0,624	38	93	24964
630	5,16	0,625	36	90	25000

2001). Although in principle it might be thought that all the material has the same polarity, however, is possible that their areas in which the crystals have different polarities. This may explain why the lower the roughness (due to increased substrate temperature during the deposit) on AlN films has favored an increase in the velocity of the waves in the SAW devices.

CONCLUSION

The method of plasma diagnosis by tuning the formation energy of materials formation as HA films may lead to propose and confirm experimental and theoretical models. It was demonstrated that knowledge of the plasma geometry and the corresponding energetic characteristics was a necessary pre-condition for controlling the film stoichiometry and crystallinity adequate for biomedical applications. Furthermore, the knowledge and control of plasma properties is more important than the use of heater and bias used in the conventional sputtering systems.

Likewise, investigated the role of deposition temperature in determining the optical properties of AlN films and the frequency response characteristics of Mo/AlN/Si_3N_4 SAW devices, due to AlN films were grown on Si_3N_4 (100) substrates by PLD using an Al target in N_2 as working gas. The temperature deposition was varied from 200 °C to 630 °C to analyze their effect on the optical and morphological properties of the films. X-ray photoelectron spectroscopy (XPS) confirmed the formation of the binary films. The study revealed that temperature deposition has a marked influence on the chemical and physical nature of all the AlN thin films (oxygen concentration grain size and roughness). For piezoelectric films with a smooth surface, a decrease in the RMS roughness can lead to an increase in its SAW velocity. The evidence is increased with decreasing speed of the roughness and grain size with increasing substrate temperature. It was found a decrease in the reflectance of 10%, an increasing of color purity of 4.5% and dominant wavelength around 1.5% with temperature deposition between 200 - 630 °C. Furthermore, the deposition temperature clearly affects the reduction of micro-drops; this can be possible due to great surface mobility when the temperature was varied from 200 to 630 °C. The SAW propagation velocity is strongly dependent on deposition conditions. The Rayleigh mode and can be obtained with h/λ ratio larger than 0.004. The reflection signal amplitude and phase velocity of Rayleigh wave have been increased with increasing of deposition temperature. Finally, it was possible to design an interdigital structure with Mo circuit on AlN films like an insulator, which can be used as surface acoustic wave (SAW) devices.

Recently authors like Aissa et al.(Aissa et al., 2015) have achieved a breakthrough in obtaining high quality thin films of AlN. The films were deposited by dc magnetron sputtering (dcMS) and high power impulse magnetron sputtering (HiPIMS) deposition techniques. In both cases of films deposited by dcMS and HiPIMS, the XRD results showed that the obtained films are oriented, with full width at half maximum rocking curves of around 1°. Raman spectroscopy revealed higher residual stress relaxation in the AlN epilayers grown by HiPIMS compared to AlN grown by dcMS, highlighted by a blue shift in the E2 (high) Raman mode. The SAW measurements indicated an insertion loss of AlN-SAW devices of about 53 and 35 dB for the AlN films deposited by dcMS and HiPIMS respectively. Even though the values of the insertion loss of the as-deposited AlN thin film by HiPIMS are a little bit inadequate for real applications, these frequency-response characteristics, manifestly, show the potential for high-frequency applications of AlN thin films deposited on Silicon substrates at room temperature.

REFERENCES

Aissa, K. A., Achour, A., Elmazria, O., Simon, Q., Elhosni, M., Boulet, P., & Djouadi, M. et al. (2015). AlN films deposited by dc magnetron sputtering and high power impulse magnetron sputtering for SAW applications. *Journal of Physics. D, Applied Physics*, *48*(14), 145307. doi:10.1088/0022-3727/48/14/145307

Akazawa, H., & Ueno, Y. (2013). Growth of preferentially c-axis oriented hydroxyapatite thin films on Si (100) substrate by electron-cyclotron-resonance plasma sputtering. *Applied Surface Science*, *276*, 217–222. doi:10.1016/j.apsusc.2013.03.070

Akazawa, H., & Ueno, Y. (2014). Control of composition and crystallinity in hydroxyapatite films deposited by electron cyclotron resonance plasma sputtering. *Journal of Physics and Chemistry of Solids*, *75*(1), 94–99. doi:10.1016/j.jpcs.2013.08.014

Astala, R., & Stott, M. (2008). First-principles study of hydroxyapatite surfaces and water adsorption. *Physical Review B: Condensed Matter and Materials Physics*, *78*(7), 075427. doi:10.1103/PhysRevB.78.075427

Bauer, S., Schmuki, P., von der Mark, K., & Park, J. (2013). Engineering biocompatible implant surfaces: Part I: Materials and surfaces. *Progress in Materials Science*, *58*(3), 261–326. doi:10.1016/j.pmatsci.2012.09.001

Belabbes, A., de Carvalho, L., Schleife, A., & Bechstedt, F. (2011). Cubic inclusions in hexagonal AlN, GaN, and InN: Electronic states. *Physical Review B: Condensed Matter and Materials Physics*, *84*(12), 125108. doi:10.1103/PhysRevB.84.125108

Boyd, A., Duffy, H., McCann, R., Cairns, M., & Meenan, B. (2007). The Influence of argon gas pressure on co-sputtered calcium phosphate thin films. *Nuclear Instruments & Methods in Physics Research. Section B, Beam Interactions with Materials and Atoms*, *258*(2), 421–428. doi:10.1016/j.nimb.2007.02.072

Boyd, A., Meenan, B., & Leyland, N. (2006). Surface characterisation of the evolving nature of radio frequency (RF) magnetron sputter deposited calcium phosphate thin films after exposure to physiological solution. *Surface and Coatings Technology*, *200*(20-21), 6002–6013. doi:10.1016/j.surfcoat.2005.09.032

Caicedo, J., Pérez, J., Caicedo, H., & Riascos, H. (2013). Determination of physical response in (Mo/AlN) SAW devices. *Surface Review and Letters*, *20*(2), 1350017. doi:10.1142/S0218625X13500170

Caicedo, J., Zambrano, G., Aperador, W., Escobar-Alarcon, L., & Camps, E. (2011). Mechanical and electrochemical characterization of vanadium nitride (VN) thin films. *Applied Surface Science*, *258*(1), 312–320. doi:10.1016/j.apsusc.2011.08.057

Carradò, A. (2011). Nano-crystalline pulsed laser deposition hydroxyapatite thin films on Ti substrate for biomedical application. *Journal of Coatings Technology and Research*, *8*(6), 749–755. doi:10.1007/s11998-011-9355-9

Chen, F. F., & Smith, M. D. (1984). *Plasma*. Wiley Online Library.

Chung, R. J., Hsieh, M. F., Huang, C. W., Perng, L. H., Wen, H. W., & Chin, T. S. (2006). Antimicrobial effects and human gingival biocompatibility of hydroxyapatite sol–gel coatings. *Journal of Biomedical Materials Research. Part B, Applied Biomaterials*, *76*(1), 169–178. doi:10.1002/jbm.b.30365 PMID:16080164

Clement, M., Vergara, L., Sangrador, J., Iborra, E., & Sanz-Hervás, A. (2004). SAW characteristics of AlN films sputtered on silicon substrates. *Ultrasonics*, 42(1–9), 403-407. doi:10.1016/j.ultras.2004.01.034

Cressler, J. D. (2005). *Silicon Heterostructure Handbook: Materials, Fabrication, Devices, Circuits and Applications of SiGe and Si Strained-Layer Epitaxy*. CRC press. doi:10.1201/9781420026580

Dong, Z. L., Khor, K. A., Quek, C., White, T. J., & Cheang, P. (2003). TEM and STEM analysis on heat-treated and in vitro plasma-sprayed hydroxyapatite/Ti-6Al-4V composite coatings. *Biomaterials*, *24*(1), 97–105. doi:10.1016/S0142-9612(02)00267-3 PMID:12417183

Dos Santos, E., Moldovan, M., Jacomine, L., Mateescu, M., Werckmann, J., Anselme, K., & Pelletier, H. et al. (2010). Oriented hydroxyapatite single crystals produced by the electrodeposition method. *Materials Science and Engineering B*, *169*(1), 138–144. doi:10.1016/j.mseb.2009.10.007

Drafts, B. (2001). Acoustic wave technology sensors. *IEEE Transactions on Microwave Theory and Techniques*, *49*(4), 795–802. doi:10.1109/22.915466

Ellies, L., Nelson, D., & Featherstone, J. (1992). Crystallographic changes in calcium phosphates during plasma-spraying. *Biomaterials*, *13*(5), 313–316. doi:10.1016/0142-9612(92)90055-S PMID:1318087

Feddes, B., Wolke, J., Jansen, J., & Vredenberg, A. (2003). Radio frequency magnetron sputtering deposition of calcium phosphate coatings: The effect of resputtering on the coating composition. *Journal of Applied Physics*, *93*(12), 9503–9507. doi:10.1063/1.1576894

Ferro, D., Barinov, S. M., Rau, J. V., Teghil, R., & Latini, A. (2005). Calcium phosphate and fluorinated calcium phosphate coatings on titanium deposited by Nd: YAG laser at a high fluence. *Biomaterials*, *26*(7), 805–812. doi:10.1016/j.biomaterials.2004.03.027 PMID:15350786

Guipont, V., Jeandin, M., Bansard, S., Khor, K. A., Nivard, M., Berthe, L., & Boustie, M. et al. (2010). Bond strength determination of hydroxyapatite coatings on Ti-6Al-4V substrates using the LAser Shock Adhesion Test (LASAT). *Journal of Biomedical Materials Research. Part A*, *95*(4), 1096–1104. doi:10.1002/jbm.a.32907 PMID:20878900

Hadjoub, Z., Touati, I., Doghmane, M., & Doghmane, A. (2008). Application of negative velocity dispersion curves to the distinction between layer and substrate Rayleigh waves. *Comptes Rendus Physique*, *9*(8), 903–910. doi:10.1016/j.crhy.2008.08.001

Hahn, B.-D., Lee, J.-M., Park, D.-S., Choi, J.-J., Ryu, J., Yoon, W.-H., & Kim, H.-E. et al. (2011). Enhanced bioactivity and biocompatibility of nanostructured hydroxyapatite coating by hydrothermal annealing. *Thin Solid Films*, *519*(22), 8085–8090. doi:10.1016/j.tsf.2011.07.008

Han, J. G. (2009). Recent progress in thin film processing by magnetron sputtering with plasma diagnostics. *Journal of Physics. D, Applied Physics*, *42*(4). doi:10.1088/0022-3727/42/4/043001

Hashimoto, K.-Y., & Yamaguchi, M. (2001). Excitation and propagation of shear-horizontal-type surface and bulk acoustic waves. *IEEE Transactions on* Ultrasonics, Ferroelectrics, and Frequency Control, *48*(5), 1181–1188.

Hollister, S. J. (2005). Porous scaffold design for tissue engineering. *Nature Materials*, *4*(7), 518–524. doi:10.1038/nmat1421 PMID:16003400

Hong, Z., Luan, L., Paik, S.-B., Deng, B., Ellis, D. E., Ketterson, J. B., & Rossi, A. et al. (2007). Crystalline hydroxyapatite thin films produced at room temperature—an opposing radio frequency magnetron sputtering approach. *Thin Solid Films*, *515*(17), 6773–6780. doi:10.1016/j.tsf.2007.02.089

Iliescu, M., Nelea, V., Werckmann, J., & Mihailescu, I. (2004). Transmission electron microscopy investigation of pulsed-laser deposited hydroxylapatite thin films prepared by tripod and focused ion beam techniques. *Surface and Coatings Technology*, *187*(1), 131–140. doi:10.1016/j.surfcoat.2004.01.022

Jallot, E., Nedelec, J.-M., Grimault, A. S., Chassot, E., Grandjean-Laquerriere, A., Laquerriere, P., & Laurent-Maquin, D. (2005). STEM and EDXS characterisation of physico-chemical reactions at the periphery of sol–gel derived Zn-substituted hydroxyapatites during interactions with biological fluids. *Colloids and Surfaces. B, Biointerfaces*, *42*(3), 205–210. doi:10.1016/j.colsurfb.2005.03.001 PMID:15893220

Jedynski, M., Hoffman, J., Mroz, W., & Szymanski, Z. (2008). Plasma plume induced during ArF laser ablation of hydroxyapatite. *Applied Surface Science*, *255*(5), 2230–2236. doi:10.1016/j.apsusc.2008.07.070

Jeong, Y.-H., Choe, H.-C., & Eun, S.-W. (2011). Hydroxyapatite coating on the Ti–35Nb–xZr alloy by electron beam-physical vapor deposition. *Thin Solid Films*, *519*(20), 7050–7056. doi:10.1016/j.tsf.2011.04.086

Jiang, L., Shen, W., Ogawa, H., & Guo, Q. (2003). Temperature dependence of the optical properties in hexagonal AlN. *Journal of Applied Physics*, *94*(9), 5704–5709. doi:10.1063/1.1616988

Joanni, E., Ferro, M. C., Mardare, C. C., Mardare, A. I., Fernandes, J. R. A., & Pina, S. C. A. (2004). Pulsed laser deposition of SiO2-P2O5-CaO-MgO glass coatings on titanium substrates. *Materials Research*, *7*(3), 431–436. doi:10.1590/S1516-14392004000300010

Jung, J.-P., Lee, J.-B., Kim, J.-S., & Park, J.-S. (2004). Fabrication and characterization of high frequency SAW device with IDT/ZnO/AlN/Si conFigureuration: Role of AlN buffer. *Thin Solid Films*, *447*, 605–609. doi:10.1016/j.tsf.2003.07.022

Kanzaki, N., Treboux, G., Onuma, K., Tsutsumi, S., & Ito, A. (2001). Calcium phosphate clusters. *Biomaterials*, *22*(21), 2921–2929. doi:10.1016/S0142-9612(01)00039-4 PMID:11561898

Kao, K.-S., Cheng, C.-C., & Chen, Y.-C. (2002). Synthesis and surface acoustic wave properties of AlN films deposited on LiNbO/sub 3/substrates. *Ultrasonics, Ferroelectrics, and Frequency Control. IEEE Transactions on*, *49*(3), 345–349.

Kar, J. P., & Bose, G. (2011). *Aluminum Nitride (AlN) film based acoustic devices: material synthesis and device fabrication*. INTECH Open Access Publisher.

Khor, K., Dong, Z., Quek, C., & Cheang, P. (2000). Microstructure investigation of plasma sprayed HA/Ti6Al4V composites by TEM. *Materials Science and Engineering A*, *281*(1), 221–228. doi:10.1016/S0921-5093(99)00717-0

Kim, W.-G., & Choe, H.-C. (2011). Surface characteristics of hydroxyapatite/titanium composite layer on the Ti-35Ta-xZr surface by RF and DC sputtering. *Thin Solid Films*, *519*(20), 7045–7049. doi:10.1016/j.tsf.2011.04.090

Layrolle, P., Ito, A., & Tateishi, T. (1998). Sol-Gel Synthesis of Amorphous Calcium Phosphate and Sintering into Microporous Hydroxyapatite Bioceramics. *Journal of the American Ceramic Society*, *81*(6), 1421–1428. doi:10.1111/j.1151-2916.1998.tb02499.x

Levinshtein, M. E., Rumyantsev, S. L., & Shur, M. S. (2001). *Properties of Advanced Semiconductor Materials: GaN, AlN, InN, BN, SiC, SiGe*. John Wiley & Sons.

Li, C., Hong, G., Yu, H., & Qi, L. (2010). Facile fabrication of honeycomb-patterned thin films of amorphous calcium carbonate and mosaic calcite. *Chemistry of Materials*, *22*(10), 3206–3211. doi:10.1021/cm100363a

Li, C., & Qi, L. (2008). Bioinspired fabrication of 3D ordered macroporous single crystals of calcite from a transient amorphous phase. *Angewandte Chemie International Edition*, *47*(13), 2388–2393. doi:10.1002/anie.200705403 PMID:18300213

Lieberman, M., & Lichtenberg, A. (2005). *Principles of Plasma Discharges and Plasma Processing*. John Wiley and Sons.

Lieberman, M. A., & Lichtenberg, A. J. (1994). Principles of plasma discharges and materials processing. *MRS Bulletin*, *30*, 899–901.

Liu, D., Savino, K., & Yates, M. Z. (2011). Coating of hydroxyapatite films on metal substrates by seeded hydrothermal deposition. *Surface and Coatings Technology*, *205*(16), 3975–3986. doi:10.1016/j.surfcoat.2011.02.008

López, E. O., Mello, A., Sendão, H., Costa, L. T., Rossi, A. L., Ospina, R. O., & Rossi, A. M. et al. (2013). Growth of crystalline hydroxyapatite thin films at room temperature by tuning the energy of the RF-magnetron sputtering plasma. *ACS Applied Materials & Interfaces*, *5*(19), 9435–9445. doi:10.1021/am4020007 PMID:24059686

López, E. O., Rossi, A. L., Archanjo, B. S., Ospina, R. O., Mello, A., & Rossi, A. M. (2015). Crystalline nano-coatings of fluorine-substituted hydroxyapatite produced by magnetron sputtering with high plasma confinement. *Surface and Coatings Technology*, *264*, 163–174.

Louis-Achille, V., De Windt, L., & Defranceschi, M. (2000). Electronic structure of minerals: The apatite group as a relevant example. *International Journal of Quantum Chemistry*, *77*(6), 991–1006. doi:10.1002/(SICI)1097-461X(2000)77:6<991::AID-QUA7>3.0.CO;2-Z

Macaskie, L. E., Yong, P., Paterson-Beedle, M., Thackray, A. C., Marquis, P. M., Sammons, R. L., & Hall, L. D. et al. (2005). A novel non line-of-sight method for coating hydroxyapatite onto the surfaces of support materials by biomineralization. *Journal of Biotechnology*, *118*(2), 187–200. doi:10.1016/j.jbiotec.2005.03.006 PMID:15964651

Mahmood, A., Machorro, R., Muhl, S., Heiras, J., Castillon, F., Farıas, M., & Andrade, E. (2003). Optical and surface analysis of DC-reactive sputtered AlN films. *Diamond and Related Materials*, *12*(8), 1315–1321. doi:10.1016/S0925-9635(03)00076-1

Maruska, H. P., & Tietjen, J. (1969). The preparation and properties of Vapor-Deposited single-crystalline GaN. *Applied Physics Letters*, *15*(10), 327–329. doi:10.1063/1.1652845

Melde, B. J., & Stein, A. (2002). Periodic macroporous hydroxyapatite-containing calcium phosphates. *Chemistry of Materials*, *14*(8), 3326–3331. doi:10.1021/cm020259b

Mello, A., Hong, Z., Rossi, A., Luan, L., Farina, M., Querido, W., & Webster, T. et al. (2007). Osteoblast proliferation on hydroxyapatite thin coatings produced by right angle magnetron sputtering. *Biomedical Materials (Bristol, England)*, *2*(2), 67–77. doi:10.1088/1748-6041/2/2/003 PMID:18458438

Miao, S., Cheng, K., Weng, W., Du, P., Shen, G., Han, G., & Zhang, S. et al. (2008). Fabrication and evaluation of Zn containing fluoridated hydroxyapatite layer with Zn release ability. *Acta Biomaterialia, 4*(2), 441–446. doi:10.1016/j.actbio.2007.08.013 PMID:17942380

Montazeri, L., Javadpour, J., Shokrgozar, M., Bonakdar, S., Moghaddam, M. K., & Asgary, V. (2013). The interaction of plasma proteins with nano-size fluoride-substituted apatite powders. *Ceramics International, 39*(6), 6145–6152. doi:10.1016/j.ceramint.2013.01.033

Morgan, H., Wilson, R., Elliott, J., Dowker, S., & Anderson, P. (2000). Preparation and characterisation of monoclinic hydroxyapatite and its precipitated carbonate apatite intermediate. *Biomaterials, 21*(6), 617–627. doi:10.1016/S0142-9612(99)00225-2 PMID:10701462

Morkoc, H. (2013). *Nitride semiconductors and devices* (Vol. 32). Springer Science & Business Media. doi:10.1002/9783527649006

Mróz, W., Jedyński, M., Prokopiuk, A., Ślósarczyk, A., & Paszkiewicz, Z. (2009). Characterization of calcium phosphate coatings doped with Mg, deposited by pulsed laser deposition technique using ArF excimer laser. *Micron (Oxford, England), 40*(1), 140–142. doi:10.1016/j.micron.2008.01.016 PMID:18407507

Narayanan, R., Seshadri, S., Kwon, T., & Kim, K. (2008). Calcium phosphate-based coatings on titanium and its alloys. *Journal of Biomedical Materials Research. Part B, Applied Biomaterials, 85*(1), 279–299. doi:10.1002/jbm.b.30932 PMID:17853421

Nelea, V., Morosanu, C., Iliescu, M., & Mihailescu, I. (2003). Microstructure and mechanical properties of hydroxyapatite thin films grown by RF magnetron sputtering. *Surface and Coatings Technology, 173*(2), 315–322. doi:10.1016/S0257-8972(03)00729-1

Ozeki, K., Aoki, H., & Masuzawa, T. (2010). TEM Observation of Nano-Interface between Hydroxyapatite Sputtered Film and Titanium Substrate. *Journal of the Australasian Ceramic Society, 46*(2), 59–62.

Padilla, S., Roman, J., Carenas, A., & Vallet-Regı, M. (2005). The influence of the phosphorus content on the bioactivity of sol–gel glass ceramics. *Biomaterials, 26*(5), 475–483. doi:10.1016/j.biomaterials.2004.02.054 PMID:15276355

Paduano, Q., Weyburne, D., Jasinski, J., & Liliental-Weber, Z. (2004). Effect of initial process conditions on the structural properties of AlN films. *Journal of Crystal Growth, 261*(2), 259–265. doi:10.1016/j.jcrysgro.2003.11.017

Paul, W., & Sharma, C. P. (2006). Nanoceramic matrices: Biomedical applications. *Am J Biochem Biotechnol, 2*(2), 41–48. doi:10.3844/ajbbsp.2006.41.48

Pelletier, H., Carradò, A., Faerber, J., & Mihailescu, I. N. (2011). Microstructure and mechanical characteristics of hydroxyapatite coatings on Ti/TiN/Si substrates synthesized by pulsed laser deposition. *Applied Physics. A, Materials Science & Processing, 102*(3), 629–640. doi:10.1007/s00339-010-6144-8

Pereiro, I., Rodríguez-Valencia, C., Serra, C., Solla, E., Serra, J., & González, P. (2012). Pulsed laser deposition of strontium-substituted hydroxyapatite coatings. *Applied Surface Science, 258*(23), 9192–9197. doi:10.1016/j.apsusc.2012.04.063

Pérez, J., Riascos, H., Caicedo, J., Cabrera, G., & Yate, L. (2011a). *Optical Properties Dependence with Gas Pressure in AlN Films Deposited by Pulsed Laser Ablation. Paper presented at the Journal of Physics: Conference Series*. doi:10.1088/1742-6596/274/1/012119

Pérez, J., Riascos, H., Caicedo, J., Cabrera, G., & Yate, L. (2011b). Optical Properties Dependence with Gas Pressure in AlN Films Deposited by Pulsed Laser Ablation. *Journal of Physics: Conference Series, 274*(1), 012119. doi:10.1088/1742-6596/274/1/012119

Perez Taborda, J. A., Riascos, H., & Vera, L. (n. d.). Correlation between optical, morphological and compositional properties of Aluminum Nitride thin films by Pulsed Laser Deposition.

Pérez Taborda, J. A., Riascos Landázuri, H., Jiménez García, F. N., & Caicedo Angulo, J. C. (2010). Structural and morphological analysis of aluminum nitride films of obtained by pulsed laser deposition. *Ciencia e Ingeniería Neogranadina, 20*(2), 107–115.

Rabiei, A., Thomas, B., Jin, C., Narayan, R., Cuomo, J., Yang, Y., & Ong, J. (2006). A study on functionally graded HA coatings processed using ion beam assisted deposition with in situ heat treatment. *Surface and Coatings Technology, 200*(20), 6111–6116. doi:10.1016/j.surfcoat.2005.09.027

Raole, P., Mukherjee, S., & John, P. (2005). X-ray photoelectron spectroscopic study of plasma source nitrogen ion implantation in single crystal natural diamond. *Diamond and Related Materials, 14*(3), 482–485. doi:10.1016/j.diamond.2004.11.009

Rau, J., Smirnov, V., Laureti, S., Generosi, A., Varvaro, G., Fosca, M., & Barinov, S. et al. (2010). Properties of pulsed laser deposited fluorinated hydroxyapatite films on titanium. *Materials Research Bulletin, 45*(9), 1304–1310. doi:10.1016/j.materresbull.2010.04.028

Rosenberger, L., Baird, R., McCullen, E., Auner, G., & Shreve, G. (2008). XPS analysis of aluminum nitride films deposited by plasma source molecular beam epitaxy. *Surface and Interface Analysis, 40*(9), 1254–1261. doi:10.1002/sia.2874

Ruterana, P., Albrecht, M., & Neugebauer, J. (2006). *Nitride semiconductors: handbook on materials and devices*. John Wiley & Sons.

Saju, K., Reshmi, R., Jayadas, N., James, J., & Jayaraj, M. (2009). Polycrystalline coating of hydroxyapatite on TiAl6V4 implant material grown at lower substrate temperatures by hydrothermal annealing after pulsed laser deposition. *Proceedings of the Institution of Mechanical Engineers. Part H, Journal of Engineering in Medicine, 223*(8), 1049–1057. doi:10.1243/09544119JEIM568 PMID:20092101

Schwertfeger, F., Emmerling, A., Gross, J., Schubert, U., Fricke, J., & Attia, Y. (1994). *Sol-Gel Processing and Applications* (p. 343). New York. doi:10.1007/978-1-4615-2570-7_29

Shaginyan, L. R., Han, J. G., Shaginyan, V. R., & Musil, J. (2006). Evolution of film temperature during magnetron sputtering. *Journal of Vacuum Science & Technology A, 24*(4), 1083-1090. doi:10.1116/1.2210947

Shi, J., Chen, C., Yu, H., & Zhang, S. (2008). Application of magnetron sputtering for producing bioactive ceramic coatings on implant materials. *Bulletin of Materials Science, 31*(6), 877–884. doi:10.1007/s12034-008-0140-z

Shun'ko, E. V. (2009). *Langmuir probe in theory and practice*. Universal-Publishers.

Snyders, R., Bousser, E., Music, D., Jensen, J., Hocquet, S., & Schneider, J. M. (2008). Influence of the Chemical Composition on the Phase Constitution and the Elastic Properties of RF-Sputtered Hydroxyapatite Coatings. *Plasma Processes and Polymers*, *5*(2), 168–174. doi:10.1002/ppap.200700112

Snyders, R., Music, D., Sigumonrong, D., Schelnberger, B., Jensen, J., & Schneider, J. (2007). Experimental and ab initio study of the mechanical properties of hydroxyapatite. *Applied Physics Letters*, *90*(19), 193902. doi:10.1063/1.2738386

Socol, G., Torricelli, P., Bracci, B., Iliescu, M., Miroiu, F., Bigi, A., & Mihailescu, I. et al. (2004). Biocompatible nanocrystalline octacalcium phosphate thin films obtained by pulsed laser deposition. *Biomaterials*, *25*(13), 2539–2545. doi:10.1016/j.biomaterials.2003.09.044 PMID:14751739

Strite, S., & Morkoç, H. (1992). GaN, AlN, and InN: A review. *Journal of Vacuum Science & Technology B Microelectronics and Nanometer Structures*, *10*(4), 1237–1266. doi:10.1116/1.585897

Surmenev, R. A. (2012). A review of plasma-assisted methods for calcium phosphate-based coatings fabrication. *Surface and Coatings Technology*, *206*(8), 2035–2056. doi:10.1016/j.surfcoat.2011.11.002

Surmenev, R. A., Surmeneva, M. A., Evdokimov, K. E., Pichugin, V. F., Peitsch, T., & Epple, M. (2011). The influence of the deposition parameters on the properties of an rf-magnetron-deposited nanostructured calcium phosphate coating and a possible growth mechanism. *Surface and Coatings Technology*, *205*(12), 3600–3606. doi:10.1016/j.surfcoat.2010.12.039

Surmeneva, M. A., Chaikina, M. V., Zaikovskiy, V. I., Pichugin, V. F., Buck, V., Prymak, O., & Surmenev, R. A. et al. (2013). The structure of an RF-magnetron sputter-deposited silicate-containing hydroxyapatite-based coating investigated by high-resolution techniques. *Surface and Coatings Technology*, *218*, 39–46. doi:10.1016/j.surfcoat.2012.12.023

Taborda, J. A. P., Caicedo, J., Grisales, M., Saldarriaga, W., & Riascos, H. (2015). Deposition pressure effect on chemical, morphological and optical properties of binary Al-nitrides. *Optics & Laser Technology*, *69*, 92–103. doi:10.1016/j.optlastec.2014.12.009

Touati, I., Hadjoub, Z., & Doghmane, A. (2006). Investigations of elastic properties of Al2O3 layers on sic substrate via acoustic microscopie. *Physical & chemical news*, 27, 26-30.

Trang, H., Marie-Helene, V., Jean-Sebastien, D., Philippe, R., Philippe, B., & Pierre, L.-A. (2007). Temperature-compensated structure for SAW pressure sensor in very high temperature. *Paper presented at the Frequency Control Symposium, 2007 Joint with the 21st European Frequency and Time Forum.* IEEE International. doi:10.1109/FREQ.2007.4319026

Treboux, G., Layrolle, P., Kanzaki, N., Onuma, K., & Ito, A. (2000). Existence of Posner's cluster in vacuum. *The Journal of Physical Chemistry A*, *104*(21), 5111–5114. doi:10.1021/jp994399t

Tredwin, C. J., Young, A. M., Neel, E. A. A., Georgiou, G., & Knowles, J. C. (2014). Hydroxyapatite, fluor-hydroxyapatite and fluorapatite produced via the sol–gel method: Dissolution behaviour and biological properties after crystallisation. *Journal of Materials Science. Materials in Medicine*, *25*(1), 47–53. doi:10.1007/s10856-013-5050-y PMID:24052344

Uchida, M., Oyane, A., Kim, H. M., Kokubo, T., & Ito, A. (2004). Biomimetic Coating of Laminin–Apatite Composite on Titanium Metal and Its Excellent Cell-Adhesive Properties. *Advanced Materials*, *16*(13), 1071–1074. doi:10.1002/adma.200400152

Ueda, K., Narushima, T., Goto, T., Taira, M., & Katsube, T. (2007). Fabrication of calcium phosphate films for coating on titanium substrates heated up to 773 K by RF magnetron sputtering and their evaluations. *Biomedical Materials*, *2*(3), S160–S166. doi:10.1088/1748-6041/2/3/S14 PMID:18458462

Vallet-Regí, M., & Ruiz-Hernández, E. (2011). Bioceramics: From bone regeneration to cancer nanomedicine. *Advanced Materials*, *23*(44), 5177–5218. doi:10.1002/adma.201101586 PMID:22009627

Van Der Wal, E., Oldenburg, S., Heij, T., Van Der Gon, A. D., Brongersma, H., Wolke, J., & Vredenberg, A. et al. (2006). Adsorption and desorption of Ca and PO 4 species from SBFs on RF-sputtered calcium phosphate thin films. *Applied Surface Science*, *252*(10), 3843–3854. doi:10.1016/j.apsusc.2005.06.006

Van der Wal, E., Wolke, J., Jansen, J., & Vredenberg, A. (2005). Initial reactivity of rf magnetron sputtered calcium phosphate thin films in simulated body fluids. *Applied Surface Science*, *246*(1), 183–192.

Wang, H., Eliaz, N., & Hobbs, L. W. (2011). The nanostructure of an electrochemically deposited hydroxyapatite coating. *Materials Letters*, *65*(15), 2455–2457. doi:10.1016/j.matlet.2011.05.016

Wang, H., Eliaz, N., Xiang, Z., Hsu, H.-P., Spector, M., & Hobbs, L. W. (2006). Early bone apposition in vivo on plasma-sprayed and electrochemically deposited hydroxyapatite coatings on titanium alloy. *Biomaterials*, *27*(23), 4192–4203. doi:10.1016/j.biomaterials.2006.03.034 PMID:16618502

Wen, C., Xu, W., Hu, W., & Hodgson, P. (2007). Hydroxyapatite/titania sol–gel coatings on titanium–zirconium alloy for biomedical applications. *Acta Biomaterialia*, *3*(3), 403–410. doi:10.1016/j.actbio.2006.10.004 PMID:17204459

Xu, S., Long, J., Sim, L., Diong, C. H., & Ostrikov, K. K. (2005). RF plasma sputtering deposition of hydroxyapatite bioceramics: Synthesis, performance, and biocompatibility. *Plasma Processes and Polymers*, *2*(5), 373–390. doi:10.1002/ppap.200400094

Yamaguchi, T., Tanaka, Y., & Ide-Ektessabi, A. (2006). Fabrication of hydroxyapatite thin films for biomedical applications using RF magnetron sputtering. *Nuclear Instruments & Methods in Physics Research. Section B, Beam Interactions with Materials and Atoms*, *249*(1), 723–725. doi:10.1016/j.nimb.2006.03.126

Yamashita, H., Fukui, K., Misawa, S., & Yoshida, S. (1979). Optical properties of AlN epitaxial thin films in the vacuum ultraviolet region. *Journal of Applied Physics*, *50*(2), 896–898. doi:10.1063/1.326007

Yonggang, Y., Wolke, J. G., Yubao, L., & Jansen, J. A. (2007). In vitro evaluation of different heat-treated radio frequency magnetron sputtered calcium phosphate coatings. *Clinical Oral Implants Research*, *18*(3), 345–353. doi:10.1111/j.1600-0501.2006.01332.x PMID:17298487

Zayats, N., Boĭko, V., Gentsar, P., Litvin, O., Papusha, V., & Sopinskiĭ, N. (2008). Optical studies of AlN/n-Si (100) films obtained by the method of high-frequency magnetron sputtering. *Semiconductors*, *42*(2), 195–198. doi:10.1134/S1063782608020139

Zhang, K., Francis, L. F., Yan, H., & Stein, A. (2005). Apatite Converted from 3-D Ordered Macroporous Sol–Gel Bioactive Glass (3DOM-BG) Particles. *Journal of the American Ceramic Society*, *88*(3), 587–592. doi:10.1111/j.1551-2916.2005.00143.x

Zhang, S. (2013). *Hydroxyapatite Coatings for Biomedical Applications*. CRC Press.

Zhang, S., Wang, Y., Zeng, X., Cheng, K., Qian, M., Sun, D., & Chia, W. et al. (2007). Evaluation of interfacial shear strength and residual stress of sol–gel derived fluoridated hydroxyapatite coatings on Ti6Al4V substrates. *Engineering Fracture Mechanics*, *74*(12), 1884–1893. doi:10.1016/j.engfracmech.2006.04.021

Zhuang, C., Zhao, J., Jia, F., Guan, C., Wu, Z., Bai, Y., & Jiang, X. (2009). Tuning bond contents in B–C–N films via temperature and bias voltage within RF magnetron sputtering. *Surface and Coatings Technology*, *204*(5), 713–717. doi:10.1016/j.surfcoat.2009.09.031

Chapter 7
Functionalization of Surfaces with Optical Coatings Produced by PVD Magnetron Sputtering

Walter Raniero
University of Trento, Italy

Gianantonio Della Mea
University of Trento, Italy

Matteo Campostrini
Legnaro National Laboratories, Italy

ABSTRACT

This chapter describes the functionalization of the surface of the glass through thin film inorganic coatings. Such filters called dichroic filters, following the physical principle of interference of light using the rules of optical physics. The design of the optical multilayer materials with high and low refractive index and optimizing the predetermined optical responses. The AFM quantify the real thickness and, calibrate the deposition plant, using as an input data for the simulation to evaluate the dispersion index refraction and absorption. The dichroic filters separate the incident radiation into two or more predetermined optical bands. The materials used are deposited using PVD Reactive Magnetron Sputtering, allows to increase the deposition rate, to obtain good homogeneity range of the surface deposited. The set point of deposition which allows to obtain stoichiometric oxides is analyzed with the technique RBS. These optical filters, also called dichroic are applied in the research field of the splitting photovoltaic concentration.

INTRODUCTION

An optical filter that selectively, transmits light in a particular range is known as a dichroic or interference filter. These filters exploit the interference phenomenon to form multiple reflections between the different layers of a multilayer thin film. This effect permits to increase or decrease the reflectance of an optical surface in the visible and the near infrared spectrum (Asghar, Shoaib, Placido, & Naseem,

DOI: 10.4018/978-1-5225-0066-7.ch007

Versita 2008). The dichroic filters are applied in the concentrated photovoltaic (CPV) research field in order to split the solar radiation into two or more optical bands and send them to photovoltaic cells with different energy gaps. This enables to fabricate a tandem cell system coupled with optical lenses and to increase the energy yield and the overall efficiency of the CPV module compared to conventional PV modules (A.G. Imenes, 2004).

This chapter will focus on the functionalization of glass surfaces through the application of thin film coatings. An overview of an optical multilayer coating is described outlining the design, the fabrication and the characterization of the thin film.

The design of the optical multilayer requires the evaluation of the optical characteristics of the substrate glass material, including its refractive index and the absorption coefficient. This is done applying analytical discrete functions to only slab materials. Optical multilayer coating design provides the evaluation of the optical constants of the materials used in the deposition step. Using such materials it can be simulated a stack, formed of alternating oxides layers, which allows to optimize the sequence of films with high and low refractive index optimizing the predetermined optical response (Mazuri, et al., 2013).

The PVD reactive magnetron sputtering technique is used to fabricate the optical multilayer. Single layers of TiO_2 and SiO_2 are deposited by reactive magnetron sputtering. The oxide coating is prepare from a metallic target. The poisoning of the metallic target realized by oxygen gas is controlled retroactively by a voltage system imposed on target by power supply (Lou, et al., 1999). After the fabrication the Rutherford backscattering spectrometry (RBS) and the spectrophotometric analysis (UV-Vis-NIR) are employed to analyze the chemical and physical properties of the oxide materials. Finally some typical applications of the multilayer in the field of photovoltaics are described.

PHYSICS OF THIN FILMS

In the field of geometrical optics, each optical phenomenon is connected with the interaction of electromagnetic radiation with matter. The classical description of electric and magnetic fields in matter (O.Stenzel, 2005), (Ohring, 1992) is given by Maxwell's Equations (Macleod, 1986):

$$divB = 0 \qquad B = \mu_0 \left(H + M\right) \tag{1}$$

$$\nabla \mathrm{x}\, \mathrm{E} = -\frac{\partial B}{\partial t} \tag{2}$$

$$\mathrm{divD} = 0 \tag{3}$$

$$H = \frac{\partial D}{\partial t} \qquad D = \varepsilon_0 E + P \tag{4}$$

where E and H are respectively the vectors of the electric and magnetic field strength, B, and D respectively represent the magnetic field and electric displacement, P is the polarization density, and M is the magnetization.

The polarization density P is defined as the electric dipole moment per unit volume, and is time-dependent and dependent on the field E. When calculated for an oscillating electric field in an anisotropic material, the polarization P takes the form (O.Stenzel, 2005):

$$P(t) = E(t)\varepsilon_0 \int_0^{\infty} \kappa(\xi) e^{i\omega\xi} d\xi \tag{5}$$

where κ is the homogeneity of the time from t to t' and ξ is the time difference $(t - t')$. The linear electric susceptibility χ can be defined as:

$$\chi = \int_0^{\infty} \kappa(\xi) e^{i\omega\xi} d\xi = \chi(\omega) \tag{6}$$

The polarization P is then:

$$P = \varepsilon_0 \chi E \tag{7}$$

where ε_0 is the vacuum permittivity and the electric displacement vector D assumes the relationship:

$$D = \varepsilon_0 E + P = \varepsilon_0 \left[1 + \chi(\omega)\right] E \equiv \varepsilon_0 \varepsilon(\omega) E \tag{8}$$

PLANAR INTERFACE

In this paragraph it is described a general theory of the propagation of electromagnetic waves in the given materials, focused on thin film spectra by a purely classical physical treatment.

A plate (Figure 1) irradiated with a light radiation at a certain angle of incidence displays the optical phenomena of iteration with the surface and the bulk of the material irradiated. More precisely, the optical phenomena that occur are: transmission (T), reflection (R), diffusion (D) and absorption (A). Treating the intensity as an input signal on the plate, the transmission coefficient is defined as:

$$T \equiv \frac{I_T}{I_E} \tag{9}$$

where I_T is the intensity of transmitted light and I_E is the intensity of the incoming light.

The reflection coefficient for light incoming on the plate is defined as:

$$R \equiv \frac{I_R}{I_E} \tag{10}$$

where I_R is the intensity of the reflected light.

Figure 1. Performance of optical parameters of a slab radiated

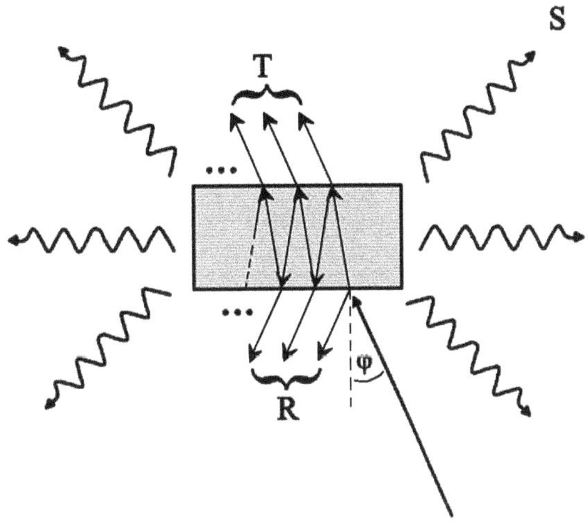

It should be stressed that the oxide material in this case does not absorb very much light; the transmitted light is equal to 100% of the incident light. In addition, in most cases such materials are subjected to light radiation diffusion due to "scattering" of light within the "bulk" or absorption of a fraction of the incoming light, depending on the structure of the bulk material. The diffusion coefficient S is defined:

$$S \equiv \frac{I_S}{I_E} \tag{11}$$

where I_S is the intensity of light scattered.

The absorption coefficient is defined as:

$$A \equiv \frac{I_A}{I_E} \tag{12}$$

where I_A is the intensity of light absorbed.

The principle of energy conservation requires that (Ohring, 1992), (Willey, 2012):

$$T + R + A + S = 1 \tag{13}$$

The sum of the diffusion coefficient D and absorption coefficient A equals the optical loss L:

$$L \equiv S + A = 1 - T - R \tag{14}$$

The values T, R, S and A, are dependent on the nature of the material, the geometry of the sample analyzed and the wavelength of the incident light (Macleod, 1986). The signals T, R, S and

A, are acquired through the use of a spectrophotometer $UV / VIS / NIR$ in which the wavelength can be varied by a monochromator in the range $\lambda = 200$ to 2500 nm. The absorbance can be measured by the spectrophotometer if the transmittance is taken into account. Indeed, the spectrophotometer measures the change in transmittance T from a reference transmittance T_{REF}. The absorbance can then be calculated as:

$$A \equiv -lg \frac{T}{T_{REF}} \tag{15}$$

It should be noted that this measurement of absorbance is not absolute, but depends on the optical properties of the chosen reference. If $T_{REF} = 1$ is assumed as a reference, at each wavelength the absorbance takes the form:

$$A \equiv -lgT = -\lg(1 - A - S - R) \tag{16}$$

The concept of absorption is also extended to the thin films but it is necessary to deposit on the substrates for reasons of mechanical strength. The absorption of thin films can be defined simply by the transmittance of the film T and the transmittance of the substrate T_{SUB} through the relationship:

$$A \equiv -lg \frac{T}{T_{SUB}} \tag{17}$$

THE FRESNEL LAW FOR PLANAR INTERFACE

The interactions of an electromagnetic wave with the surface of the film are described by the Fresnel law. In (Figure 2) an electromagnetic wave incident on a flat surface at an angle φ is shown.

The examined interface is flat and smooth, the two material n_1 and n_2 at the interface are optically homogeneous, isotropic and non-magnetic and the angle ψ (refraction angle) and the angle between the normal to the interface and the direction of the transmitted beam ψ are coplanar. The relationship between the angle of incidence at the interface φ and the refraction angle ψ is known as the Snell law:

$$\frac{\sin \varphi}{\sin \psi} = \frac{n_2}{n_1} \tag{18}$$

When thin films characterized by flat and parallel interfaces are used, the refractive index n assumes a complex shape and the complex angles of refraction appear in Snell's Law. The complex refractive index is defined as:

$$\tilde{n} = n - ik \tag{19}$$

where n indicates the phase velocity with which the wave propagates and k is the damping coefficient associated with the absorption coefficient of the system, (Ohring, 1992). As see in Figure 3, the damp-

Figure 2. Electromagnetic wave incident on an interface level

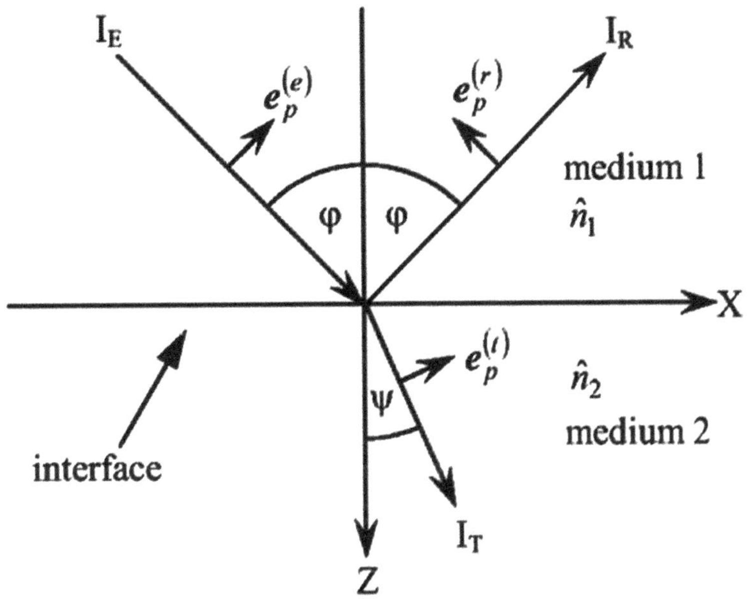

ing coefficient for the intensity of light is present only along the z axis of the xz plane which identifies the system, while the intensity along the x axis remains unchanged. Usually, the propagation direction of the electric field vector can be represented as the sum of two components: a component parallel to the incident plane (component-p) and a component normal to the incident plane (component-s) parallel to the y axis.

According to Maxwell's theory (Macleod, 1986), a light wave is composed of an electric component and a magnetic component. It is therefore possible to derive Fresnel equations for an isotropic and non-magnetic material. In the direction of the p-type polarization, the reflectance $\left(r_p\right)$ and transmittance $\left(t_p\right)$ are:

$$r_p = \frac{n_2 \cos\varphi - n_1 \cos\psi}{n_2 \cos\varphi + n_1 \cos\psi} \tag{20}$$

$$t_p = \frac{2n_1 \cos\varphi}{n_2 \cos\varphi + n_1 \cos\psi} \tag{21}$$

Instead, for the s-type polarization the reflectance $\left(r_s\right)$ and transmittance $\left(t_s\right)$ are:

$$r_s = \frac{n_1 \cos\varphi - n_2 \cos\psi}{n_1 \cos\varphi + n_2 \cos\psi} \tag{22}$$

Figure 3. Types of polarization

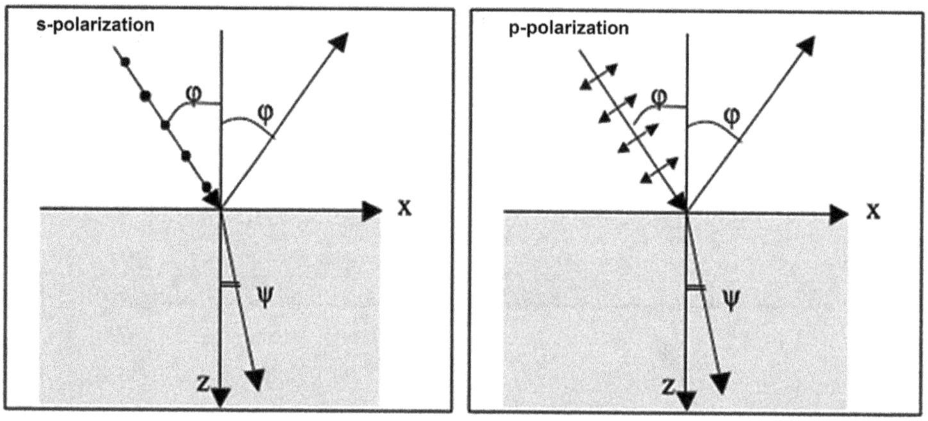

$$t_s = \frac{2n_1 \cos\varphi}{n_1 \cos\varphi + n_2 \cos\psi} \tag{23}$$

defining the reflectance as:

$$R = |r|^2 \tag{24}$$

and the total transmittance through the interface:

$$T = 1 - R = \frac{R(n_2 \cos\psi)}{R(n_1 \cos\varphi)} |t|^2 \tag{25}$$

In the simple case of normal incidence R takes the form:

$$R = \left(\frac{n_1 - n_2}{n_1 + n_2}\right)^2 \tag{26}$$

THICK SLAB

As it has been mentioned in previous sections, only the case of electromagnetic waves incident on individual interfaces was addressed. When treating the problem of thin films, it is necessary to expand the previously considered description to a wave incident on multiple parallel interfaces. In particular, to treat thin films, we will begin by examining interactions in a system of thick slabs and then reduce to a thin-film system through limiting-case assumptions.

In the case of a glass sheet composed of two parallel interfaces with high transparency, the system is considered optically limited and therefore losses are negligible. The transmittance coefficient T_{12} for a beam of light through the first interface is equal to:

$$T = 1 - R = \frac{R(n_2 \cos\psi)}{R(n_1 \cos\varphi)} |t|^2 \tag{27}$$

Light transmitted through the first interface continues on to interact with the second interface according to a transmittance coefficient T_{21}. The transmittance coefficient for the entire system is given by the product of the two individual transmittances.

$$T = T_{12} \cdot T_{21} \tag{28}$$

A fraction of incident light is reflected at each of the interfaces. The fraction reflected at the first interface is given by R_{12} and at the second interface by R_{21}. The total reflectance and transmittance are given by:

$$\begin{aligned} R &= R_{12} + T_{12}R_{21}T_{21} + T_{12}R_{21}^{3}T_{21} + \cdots \\ &= R_{12} + T_{12}R_{21}T_{21}\sum_{j=1}^{\infty}\left(R_{21}^{2}\right)^{j-1} = R_{12} + \frac{T_{12}R_{21}T_{21}}{1 - R_{21}^{2}} = \frac{2R_{12}}{1 + R_{12}} \end{aligned} \tag{29}$$

$$\begin{aligned} T &= T_{12}T_{21} + T_{12}R_{21}^{2}T_{21} + \cdots \\ &= T_{12}T_{21}\sum_{j=1}^{\infty}\left(R_{21}^{2}\right)^{j-1} = \frac{T_{12}T_{21}}{1 - R_{21}^{2}} = \frac{1 - R_{12}}{1 + R_{12}} \end{aligned} \tag{30}$$

In the case of normal incidence, T and R have the form:

$$R_{12} = R_{21} = \left(\frac{n_2 - n_1}{n_2 + n_1}\right)^2 \tag{31}$$

$$T_{12} = 1 - R_{12} = \frac{4n_1 n_2}{\left(n_2 + n_1\right)^2} = T_{21} \tag{32}$$

In the case of normal incidence on a transparent material with refractive index $\left(n_2 = n\right)$, in the air $\left(n_1 = 1\right)$, as shown in Figure 4, the coefficients are:

$$R\Big|_{\varphi=0} = \frac{\left(n-1\right)^2}{n^2+1} \tag{33}$$

$$T\Big|_{\varphi=0} = \frac{2n}{n^2+1} \tag{34}$$

If the transmittance of a transparent slab is measured, its index of refraction can then be calculated as:

$$n = T^{-1} + \sqrt{\left(T^{-2}-1\right)} \tag{35}$$

At each step of the optical path of the incident radiation in the multilayer part is reflected from the surface while a fraction through the coating. The light transmitted through a series of surfaces is affected by an intensity reduction.

The drop in intensity due to every component of penetrating light with the plate is proportional to the factor $exp^{\left(-\alpha d_s\right)}$, where d_s is the thickness of the slab geometry. For this reason, the reflectance and the transmittance take the form:

$$T_{21} \rightarrow T_{21} exp^{\left(-\alpha d_s\right)} \qquad R_{21} \rightarrow R_{21} exp^{\left(-\alpha d_s\right)} \tag{36}$$

and the Fresnel equations that depend on both the real part and the imaginary part of the refractive index. Therefore:

$$R = \frac{R_{12}\left[1 - exp^{\left(-2\alpha d_s\right)}\left(2R_{12}-1\right)\right]}{1 - R_{12}^2 exp^{\left(-2\alpha d_s\right)}} \tag{37}$$

$$T = \frac{\left(1-R_{12}\right)^2 exp^{\left(-\alpha d_s\right)}}{1 - R_{12}^2 exp^{\left(-2\alpha d_s\right)}} \tag{38}$$

These equations explain that the value of d_s is only relevant for T and R of absorbent materials. For $\alpha d_s \rightarrow \infty$ the transmittance value tends to zero instead the reflectance that reaches the value assumed by the first interface. It should be noted that the equations are no longer valid when the thickness of the sheet becomes too small. When $d = 0$ there is no slab material, the reflectance should take a null value, and the transmittance should be equal to 1. In the equations above, however, there is a nonzero signal of reflectance for a very thin plate. The equations can therefore be applied only to thick slabs and not to thin plate systems.

In the previous analysis we examined the overlapping intensity of partial waves, which imparts the phase information necessary to describe the phenomenon of interference. The sum of the intensities of the electric fields, however, is correct until the optical path difference between the partial beams is

Figure 4. Thick slab with parallel interfaces

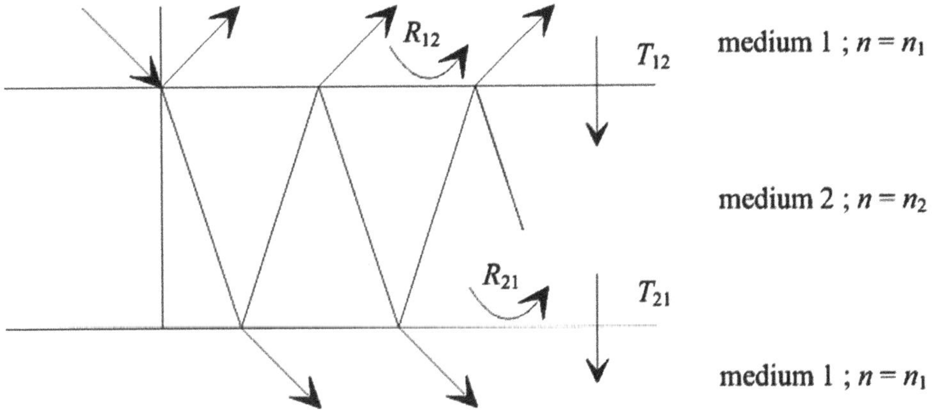

larger than the coherence length of the light. For a small bandwidth, there is a situation where the sample is too thin to be described by the equations: When the thickness is ideally zero, the sample is smaller than any of the coherence lengths. For very thin layers the use of these formulas is nonphysical, and it is therefore justified to formulate a separate policy for treating discrete cases. Analyzing the case shown in (Figure 4), multiple reflections between the first and the second interface take time t:

$$t = \frac{2nd_s}{c} > t_{coh} \tag{39}$$

where t_{coh} is the coherence time of the light and d_s is the thickness of the slab material

A finite coherence time of the light creates a spectral band over:

$$\Delta\omega = \frac{2}{t_{coh}} \qquad \Delta\omega = \frac{2\pi c}{\lambda^2}\Delta\lambda \tag{40}$$

Taking into account the reports, we obtain the condition of incoherent light:

$$d_s > \frac{\lambda^2}{2\pi n \Delta\lambda} \qquad \Delta\lambda > \frac{\lambda^2}{2\pi n d_s} \tag{41}$$

When these conditions are satisfied, it is possible to treat the case of a thick slab. If interference is visible, we use a theoretical model based on the superposition of electric fields that involves phase and not intensity of the radiation. The condition for the thin film case, therefore, is when the multiple internal reflections overlap consistently and the film thickness d is smaller than the coherence length. The thin film conditions are then:

$$d \lll \frac{\lambda^2}{2\pi n \Delta\lambda} \qquad \Delta\lambda \lll \frac{\lambda^2}{2\pi n d} \tag{42}$$

Figure 5. Reflectance and transmittance of a thin film deposited on a substrate

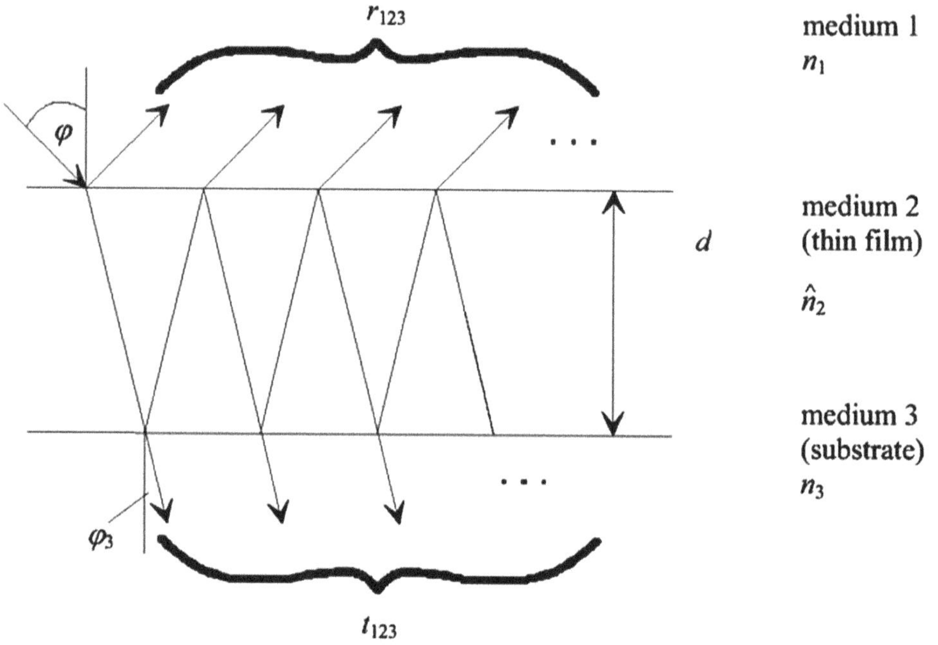

THIN FILM SPECTRA

With the appropriate physical conditions in place, it is necessary to calculate the reflectance and transmittance of thin films. The condition of Equation 42 requires that the film thickness is a few micrometers. Thin films therefore do not have mechanical strength by themselves. For this reason, the films should be deposited on a substrate to ensure appropriate mechanical strength (Figure 5).

In the calculation of the Fresnel coefficients of transmittance and reflectance of the thin films, the coefficients t_{ij} and r_{ij} are complex and carry the information about the phase. The increase in the phase difference per cycle is defined as 2δ, and the transmission and reflection coefficients are defined as:

$$\begin{aligned} r_{123} &= r_{12} + t_{12}e^{i\delta}r_{23}e^{i\delta}t_{21} + t_{12}e^{i\delta}r_{23}e^{i\delta}r_{21}e^{i\delta}r_{23}e^{i\delta}t_{21} + \cdots \\ &= r_{12} + t_{12}r_{23}t_{21}e^{2i\delta}\left(1 + r_{21}r_{23}e^{2i\delta} + \cdots\right) = r_{12} + \frac{t_{12}r_{23}t_{21}e^{2i\delta}}{1 - r_{21}r_{23}e^{2i\delta}} \end{aligned} \tag{43}$$

$$t_{123} = t_{12}e^{i\delta}t_{23}\left[1 + r_{21}r_{23}e^{2i\delta} + \left(r_{21}r_{23}e^{2i\delta}\right)^2 + \cdots\right] = \frac{t_{12}t_{23}e^{i\delta}}{1 - r_{21}r_{23}e^{2i\delta}} \tag{44}$$

The factor r_{123} relates the electric field of the reflected waves to that of the incident waves and is valid on the film surface, t_{123} gives the relationship between the electric field transmitted in the substrate and the incident one, and t_{12}, t_{23}, r_{12}, and r_{23} are the Fresnel coefficients at the interfaces between the different materials. Each time the wave enters a new layer in the film its wave function should be multiplied by the factor $e^{i\delta}$, the damping of the light into the film. Analysis of the phase shift is illustrated in Figure 6.

$$t_{12}t_{21} = 1 - r_{12}^2 \qquad r_{123} = \frac{r_{12} + r_{23}e^{2i\delta}}{1 - r_{21}r_{23}e^{2i\delta}} \tag{45}$$

In the case of $n_1 = 1$ (submersion in air), in the absence of damping, recalling that the wave number ν is the inverse of the wavelength λ, we obtain:

$$2\delta = \frac{2\pi}{\lambda}\left(n_2 a - x\right) = 2\pi\nu\left(n_2 a - x\right) \rightarrow \delta = \pi\nu\left(n_2 a - x\right)$$
$$a = \frac{2d}{\cos\psi} \qquad x = L\sin\varphi \qquad L = 2d\tan\psi \tag{46}$$

The phase shift is then:

$$\delta = \pi\nu\left(\frac{2n_2 d}{\cos\psi} - \frac{2d}{\cos\psi}\sin\psi\cos\varphi\right) \tag{47}$$

where

$$\sin\psi = \frac{\sin\varphi}{n_2} \qquad n_2\cos\psi = n_2\sqrt{1 - \sin^2\varphi} \tag{48}$$

Figure 6. Representation of the phase shift for one cycle in a layer of the film

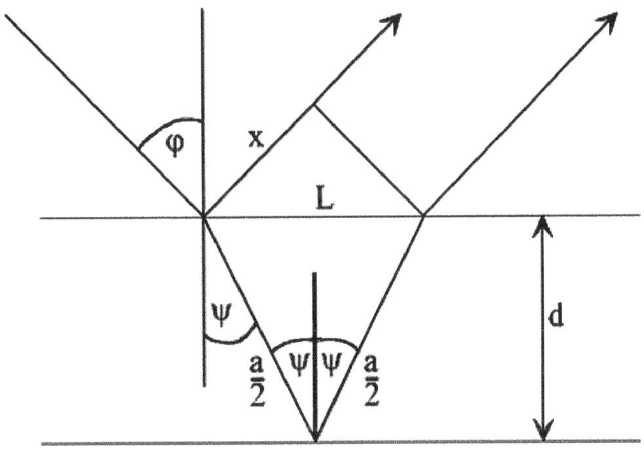

For a complex refractive index, δ becomes complex and $e^{i\delta}$ represents the damping of the light in the film. The transmittance and reflectance are:

$$T = \frac{n_3 \cos \varphi_3}{n_1 \cos \varphi}\left|t_{123}\right|^2 \qquad R = \left|r_{123}\right|^2 \tag{49}$$

a. Half-Wave Layers ($\lambda / 2$ Layers)

The following case refers to a layer of non-absorbent material with optical thickness equal to:

$$d = \frac{\lambda}{2\sqrt{n_2^2 - \sin^2 \varphi}} \tag{50}$$

In should be stressed that $4\pi \nu d\sqrt{n_2^2 - \sin^2 \varphi} = 2\pi$ and cosines in the equation (49) are equal to 1. In the case of normal incidence, the transmittance (Figure 6) takes the form:

$$T = \frac{n_3}{n_1} \cdot \frac{t_{12}^2 t_{23}^2}{\left(1 + r_{12} r_{23}\right)^2} \qquad T = \frac{4 n_1 n_3}{\left(n_1 + n_3\right)^2} \tag{51}$$

Note that, for layers of thickness $\lambda / 2$, the transmittance and reflectance do not depend on the refractive index of the film n_2. Therefore, the effect is to have the transmittance air-substrate as in the report for thick slabs and then a layer of this thickness has no effect on the transmittance and the reflectance of the system.

This effect is also valid for oblique radiation till the condition $4\pi \nu d\sqrt{n_2^2 - \sin^2 \varphi} = 2\pi$ is met. A film behaves as a half wave layer only for certain values of wavelength and when the cosines in the equation (49) are equal to 1.

$$2\delta = 4\pi \nu d\sqrt{n_2^2 - \sin^2 \varphi} = j\pi \quad , \quad j = 0, 2, 4, 6, \ldots \tag{52}$$

b. Quarter Wavelength Layers ($\lambda / 4$ Layers)

The quarter-wave layer condition occurs when the following relationship is satisfied:

$$4\pi \nu d\sqrt{n_2^2 - \sin^2 \varphi} = \pi \tag{53}$$

In this case, the optical thickness is equal $d\sqrt{n_2^2 - \sin^2\varphi} = \lambda / 4$ and cosines in the equation (49) equal -1 (O.Stenzel, 2005). In the case of normal incidence, the transmittance takes the form:

$$T = \frac{n_3}{n_1} \cdot \frac{t_{12}^2 t_{23}^2}{\left(1 - r_{12} r_{23}\right)^2} = \frac{4 n_1 n_2^2 n_3}{\left(n_1 n_3 + n_2^2\right)^2} \tag{54}$$

The equation (54) shows a hypothesis where the refractive index of the medium n_2 takes a value between the n substrate (n_3) an air (n_1) of the type:

$$n_2 = \sqrt{n_1 n_3} \tag{55}$$

And the transmittance in the equation (54) is equal to 1 and reflectance assumes zero value. It is known as the index of refraction of the film that is usually very high $\left(n_1 > n_2, n_3\right)$ and the derived relation (54) in n_2 is obtained:

$$\frac{\partial R}{\partial n_2} = -\frac{\partial T}{\partial n_2} = \frac{8 n_1 n_2 n_3}{\left(n_1 n_3 + n_2^2\right)^3}\left(n_2^2 - n_1 n_3\right) > 0 \tag{56}$$

A high refractive index of the film increases the reflectance of a quarter wave layer, and therefore layers of highly refractive material are used to optimize the total reflection. In addition, the properties described above are used to explain the phenomenon of interference and are valid for the extremes of interference when the following condition is met:

$$2\delta = 4\pi\nu d\sqrt{n_2^2 - \sin^2\varphi} = j\pi, \qquad j = 1, 3, 5, 7, \ldots \tag{57}$$

This condition describes the trend of the transmittance and reflectance of a single film without damping. In all the figures showing interference, the film behaves both as a $\lambda / 2$ layer and as $\lambda / 4$ layer: The half-wave points in the transmittance and reflectance of the system are equal to those of an uncoated substrate and are well identifiable in a $T - R$ spectrum (Figure 7).

The minima of a $T - R$ spectrum represent the quarter wave layers $(\lambda / 4)$ in which the transmittance depends on the refractive indices. In the absence of damping, the reflectance is taken as $R = 1 - T$. In the case $n_1 < n_3$ and $n_1 < n_2 < \sqrt{n_1 n_3}$ the derivative of transmittance as in equation (56) at the quarter-wave points is > 0. Therefore, the transmittance at the quarter-wave points exceeds that of the substrate and reaches the maximum value when $n_2 = \sqrt{n_1 n_3}$. The derivative becomes < 0 for $n_2 > \sqrt{n_1 n_3}$ but the transmittance should be still higher than that of the substrate until $n_2 < n_3$. If the refractive index of the film increases, the transmittance is still lower than that of the substrate and the quarter wave values correspond to the minima in the transmittance.

Figure 7. Spectrum of a TiO_2 thin film

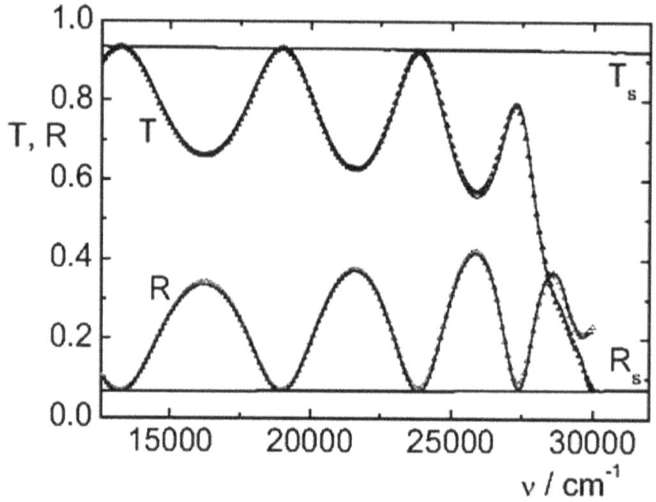

OPTICAL DESIGN

The design of the optical filters is a fundamental step to obtain multilayers of excellent quality and reproducibility. As a first step, it is important to be able to characterize thin films produced in terms of roughness and thickness, which can be done through atomic force microscopy (AFM). This analysis is of crucial importance, since the determination of the thickness through actual measurements with high sampling allows us to obtain a statistical value of real thickness with a relatively small standard deviation. This enables a high degree of control over the calibration of the deposition system.

The deposition system allows us to reproduce with high reliability layers of variable thicknesses of oxide materials in order to create a multi-layer filter with desirabled optical characteristics. Such layers of oxide materials are deposited by the PVD (physical vapor deposition) Magnetron Sputtering technique on different types of substrates.

About the oxide materials used, they must be transparent in the UV-VIS-NIR spectra. The materials most commonly used optical filter applications are TiO_2, SiO_2, Al_2O_3, Nb_2O_5, MgF_2 (Macleod, 1986) and are employed in groups of two or more different materials to realize the light interference phenomenon. In this chapter, thin films of titanium oxide and silicon oxide will be described and analyzed for the fabrication of optical films used to separate solar radiation, which are referred to as "dichroic". The materials TiO_2 and SiO_2 have good optical properties and are able to create a constructive light interference phenomenon.

The design of optical multilayers requires an iterative process of production, characterization and simulation of thin films needed to appropriately design optical multilayers with default specifications and a high level of accuracy. As a first step, it is important to fabricate samples of the same single-layer film on multiple substrates to separate the effects of the film from those of the base material. The first single layers deposited, silica or titanium oxide, are required in order to realize oxide materials with optimized stoichiometric ratios. We can measure the chemical composition of samples on substrates made of amorphous carbon through the Rutherford Back Scattering ion beam technique. That procedure, which is necessary for chemical calibration of the deposited material, is of extreme importance to achieve the poisoning tendency or the ratio of Ti/O or Si/O during the deposition step.

The single layer of oxide material is also deposited on silicon substrate with low surface roughness, so that the thickness deposited can be determined with the aid of an atomic force microscope. Such measures of statistical sampling are extremely important for calibration of the rate of the deposition system. This increases reproducibility for measurements on different layers of oxide materials deposited during multilayer production. In addition, the single layers are deposited on optical substrates such as quartz for deposits of titanium dioxide and sapphire for deposits of silicon dioxide.

These single layer depositions can be used to calculate optical properties of transmittance, reflectance and absorbance by spectrophotometric measurements UV-VIS-NIR. The average thickness measurements, with relative statistical dispersion and measures of transmittance and reflectance, can be used as an input in the design step of the single thin optical film. Design of the optical thin films using simulation software, such as Tf_Calc, allows us to simulate through the use of different dispersion functions a thin film of silica on a sapphire substrate or a thin film of titania on a quartz substrate. These simulations allow us to acquire the index of refraction $n(\lambda)$ and the absorption index $k(\lambda)$ of the deposited material. The dispersion functions used for fitting experimental curves of transmittance and reflectance of the thin film of SiO_2 are the Cauchy 1 for $n(\lambda)$ and the Exponential for $k(\lambda)$:

$$n(\lambda) = A_0 + \frac{A_1}{\lambda^2} + \frac{A_2}{\lambda^4} \tag{58}$$

$$k(\lambda) = B_1 * \exp\left(B_2\lambda^{-1}\right) \tag{59}$$

The functions of the dispersion used for fitting experimental curves of transmittance and reflectance of the thin film of TiO_2 are the Sellmeier for $n(\lambda)$ and the Exponential for $k(\lambda)$:

$$n(\lambda) = \sqrt{A_0 + \frac{A_1\lambda^2}{\lambda^2 - A_2}} \tag{60}$$

$$k(\lambda) = B_1 * \exp\left(B_2\lambda^{-1}\right) \tag{61}$$

These functions allow us to describe the dispersion of the optical response of thin films deposited (Figure 8), obtaining the optical constants as a function of wavelength (Figure 9) (Macleod, 1986).

The optical constants of the materials deposited demonstrates that new custom materials were deposited with the PVD magnetron sputtering technique. These materials can be used to simulate a multilayer composed of layers with different thicknesses and materials by alternating between high and low indices of refraction. The design of a multilayer optical device begins with the definition of an optical target, which is designed around the desired optical response in the filter. The design process involves stacking

Figure 8. Optical fitting dispersion of TiO_2 oxide material

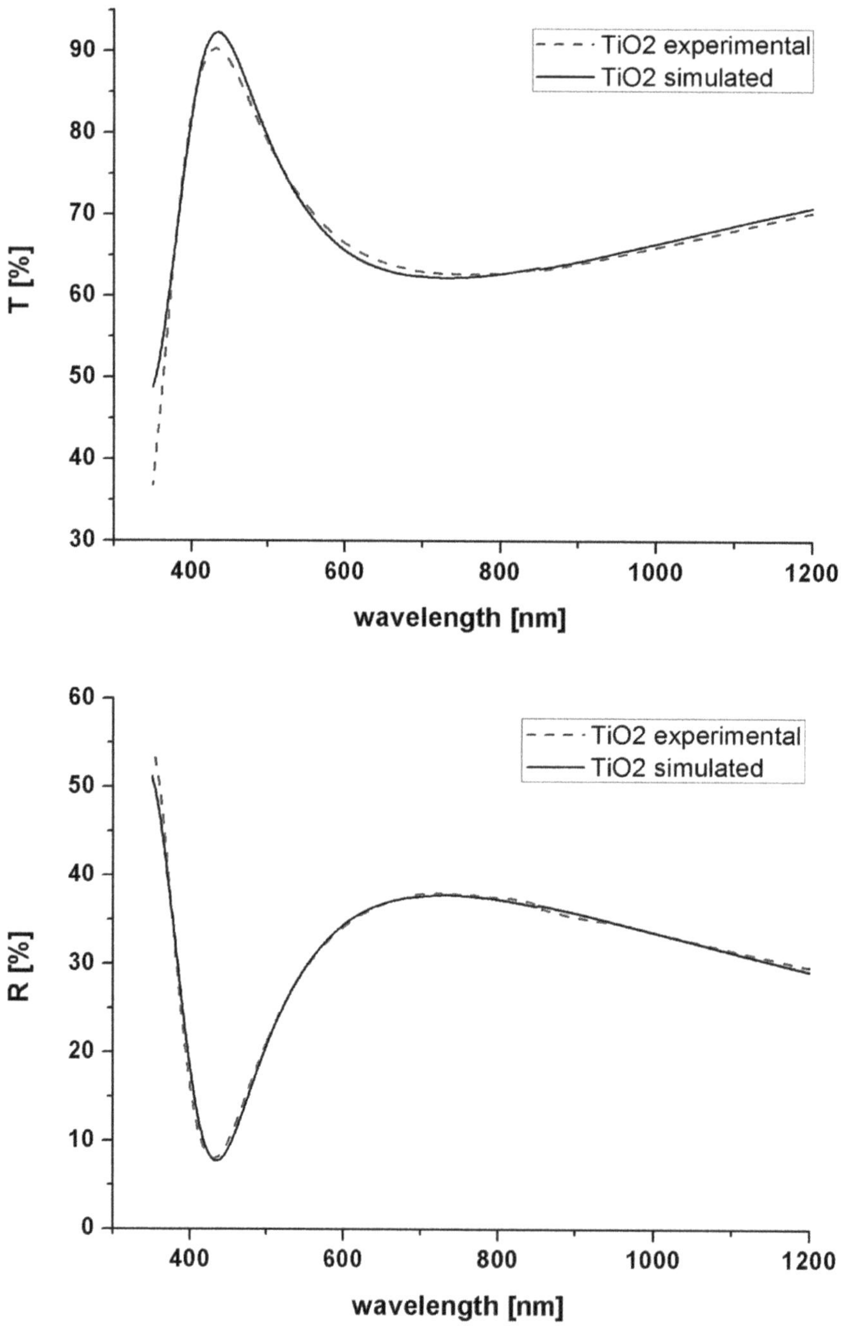

alternating layers of materials with low and high refractive indices to allow the maximum reflection, causing constructive interference when all the reflections at the interfaces of the individual layers are in phase.

The multilayer design is an iterative process which potentially leads to a convergence to a result and is usually performed in the following steps. First, two layers of TiO_2 and SiO_2 are stacked on the glass

Figure 9. Index of refraction of TiO_2 and SiO_2 oxide material

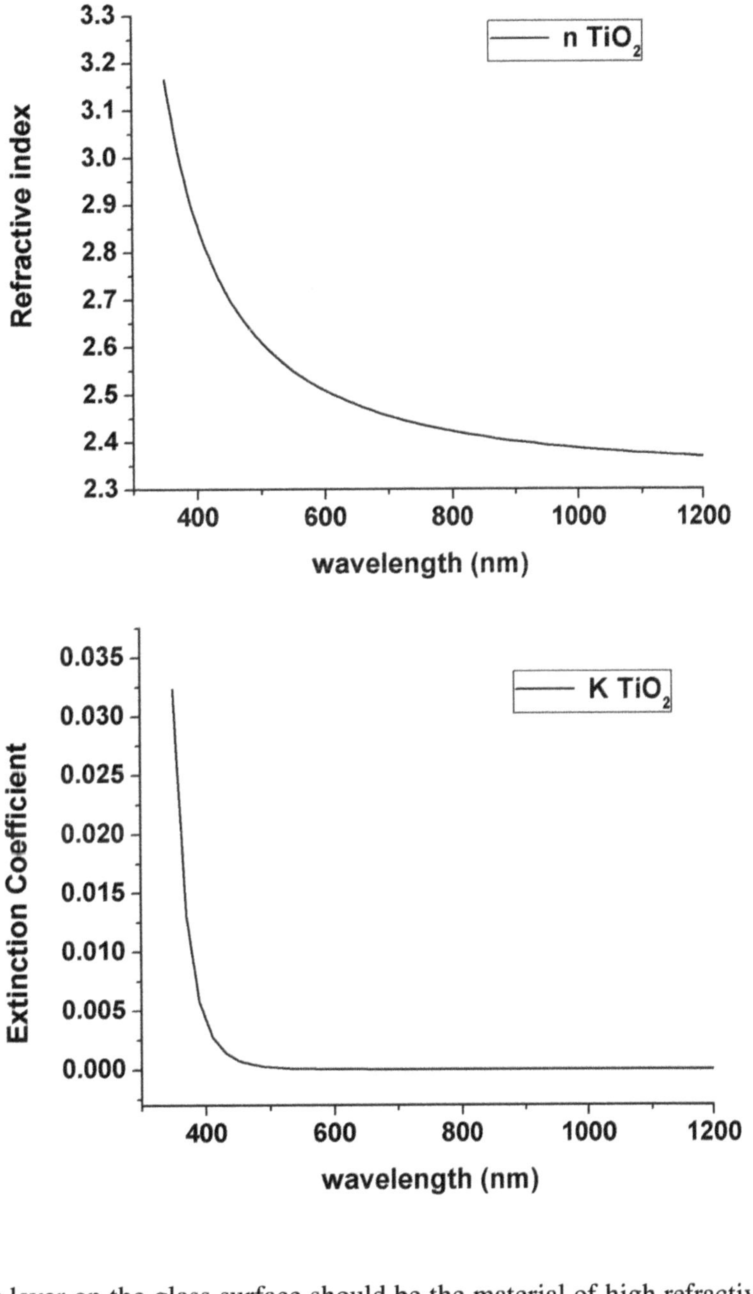

substrate (the first layer on the glass surface should be the material of high refractive index), in order to create a separate index between glass and nanometer layer, which in this case is TiO_2. In the first simulation, constraints are introduced on individual thicknesses or the total multilayers thickness, so that the processor has the ability to converge quickly. In an initial stacking sequence, an optimization algorithm is executed on the different individual layer thicknesses while keeping in mind the minimum thickness limits required to avoid a physically unachievable geometry. The optimization of a multilayer optical process requires multiple iterations around the design target, additionally changing the target in the regions of the spectrum where there are decreases in optical efficiency. The multilayer designed outputs a spectrum reflectance or transmittance as a function of wavelength. This response is related to the sequence of the materials used (Table 1), when the theoretical design is transferred to the deposition system. As a rule of thumb, it is better to fabricate a multilayer with a limited number of layers, as this will reduce the error due to the physical deposition process which compromises the reproducibility of the process.

PRODUCTION OF OPTICAL THIN FILMS

The thin films used for surface functionalization of glass are made by PVD reactive magnetron sputtering technology. This technology is based on the sputtering of a material, called the "target" in a high vacuum chamber on the order of 10^{-6} mbar. It is created through an argon plasma glow discharge in DC for materials such as titanium and pulsed DC for target materials such as heavily insulated silicon surfaces (Scholl). The glow discharge of these technologies allows sputtering of the material on the surface of the target by controlling the deposition rate as a function of the power supplied to the cathodes themselves.

The transfer of kinetic energy from an electron to a heavy particle in an elastic collision is proportional to the ratio of the electron mass to that of the heavy particle and is therefore very small. At low

Table 1. Design of multilayers dichroic show in Figure 10

Layer	Material	Thickness [nm]
1	TiO_2	79.8
2	SiO_2	89.7
3	TiO_2	79.1
4	SiO_2	162.1
5	TiO_2	74.2
6	SiO_2	147.6
7	TiO_2	69.9
8	SiO_2	157.4
9	TiO_2	89.9
10	SiO_2	89.8
11	TiO_2	67.2
12	SiO_2	95.5
13	TiO_2	43.5
14	SiO_2	97.6
15	TiO_2	70.1
16	SiO_2	89.5
17	TiO_2	37.8
18	SiO_2	77.5
19	TiO_2	105.9
20	SiO_2	113.5
21	TiO_2	42.7

Figure 10. Real optical response compared with the simulated optical response of a dichroic mirror

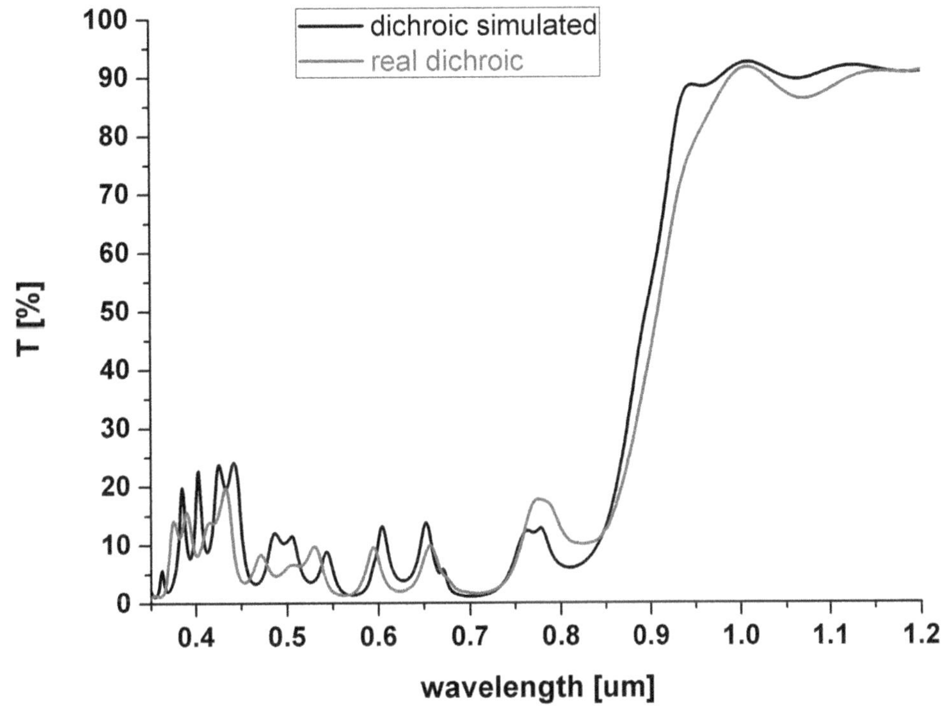

pressures, where the mean free path is greater and there are lower frequencies of collision, the electrons can accumulate enough kinetic energy to excite and ionize the heavy particles. To generate and maintain a plasma, we require a power source that produces the ionization needed by balancing the loss of ion recombination, diffusion and convection outside of the volume of the plasma. For plasma deposition, we use electrical fields produced by generators and direct current (DC) or alternating current (AC) with excitation frequencies of 100 kHz, 13.56 MHz and 2.45 GHz .

The radio frequency discharge can be triggered at pressures lower than those necessary with DC generators, due to the electron loss is reduced or high frequencies that increasing of the efficiency of ionization.

Part of the primary electron beam is rejected from the electrode with less energy and is then accelerated and remains longer in the download part of the ionizing collisions. Electrons can also borrow power from the radio frequency through the inversion of the polarity where they are reflected and increase collisions. This phenomenon is more marked as the pressure increases.

The ion bombardment is the basis of the sputtering phenomenon that allows the emission of atoms or ions from the “target”. In particular, the target mounted on the cathodes with a magnetic confinement “magnetron”, which allows us to localize a sputtering area, expanding the presence of Argon ions on the surface of the target and therefore increasing, the sputtering yield. The efficiency of the ionization process, and therefore of the sputtering, is increased by use of a magnetic field confined to the target with permanent magnets. This prevents the escape of electrons towards the walls of the room. Then, the radiofrequency discharges require lower pressures to obtain the plasma ionization with appropriate AC: typical values of pressures are comprised between 2 and 10 mTorr. To sum up, the advantage of using a magnetron cathode type is to increase the deposition rate but also to enable deposition uniformity on an area of interest (Figure 11).

One of the problems with the use of the magnetron source is the non-homogeneity of the plasma on the surface of the target. This leads to non-uniform erosion of the target and to the formation of preferential erosion traces. The trend of inhomogeneity of the film increases with the distance *d* from the target to the substrate.

The trend of the profile of the deposition is a function of the distance *d*, depends on the shape and size of the target used, on the power used and the magnetic field distribution lines in the neighborhood of the target and the working pressure.

The homogeneity of deposition is a fundamental parameter in the creation of optical filters with similar characteristics throughout the whole of interest area of the coating. Additionally, the use of magnetron technology allows us to confine very well the plasma region, thereby obtaining thin films with low thicknesses variations, of $\pm$ 3% in our specific case. (Figure 11). Moreover, the use of magnetron technology with a distance between the target and the sample near 10 cm also allows us to obtain nanometric dense films with high surface homogeneity. In particular, the functionalization of glass by thin film coating is able to change its optical properties using transparently oxides materials.

These types of materials can be deposited using PVD technology, beginning with the target bulk material oxide by using radio frequency power generators or through a reactive deposition, where there is a flushing of a reactant gas necessary to obtain compounds. The technique we used is the PVD magnetron reactive sputtering, where there is a check on the instantaneous deposition voltage (Raniero, et al., 2013). By using voltage control of the target, it is possible to regulate the reactant gas flow through mass flow controllers or piezo-valves, creating the poisoning characteristics curve of the target mate-

Figure 11. Lateral homogeneity of magnetron sputtering technique

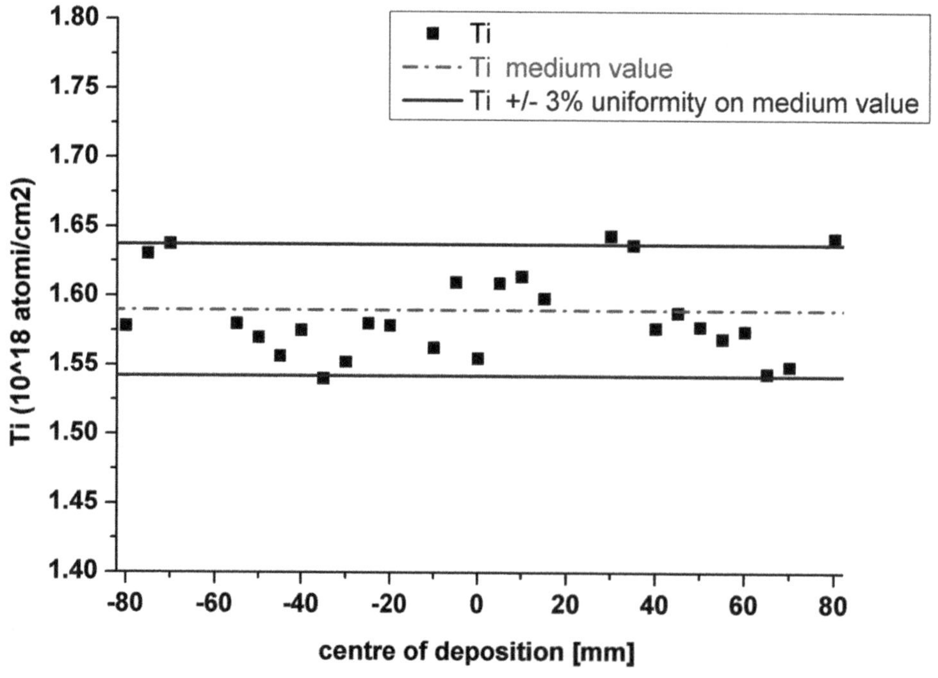

rial (Carter, Sproul, & Christie). These curves' hysteresis loop (Lou, et al., 1999) (Willey, 2012) permit us to analyze the optimum working points (Figure 12), from which it is possible to choose a set point for the realization of a thin film of stoichiometric oxide material (Raniero, Campostrini, Maggioni, & Quaranta, 2014).

The thin film coatings used to realize the dichroic filters are described in the Table 2 for the SiO_2 TiO_2 thin film, where the characteristic process PVD parameter like pressure, voltage of deposition, and the gas flow used are specified

In the Table 3 is expressed the typical poisoned characteristic, Voltage vs. gas flow curve of reactive gas flow on metal target material. In this case the poisoned of titanium material into titanium bi-oxide material is described.

In the Table 4 the poisoned curve, Voltage vs. Gas flow, of silicon material into silicon bi-oxide material is described.

The voltage of the target decreases as the flow of technician gas flux in the deposition system. The poisoning of the target is necessary to determine the hysteresis curve of the reactive sputtering, and then find the point Voltage - gas flow to achieve the stoichiometric material.

Figure 12. Reactive gas partial pressure and target voltage versus reactive gas flow for silicon oxide reactive deposition

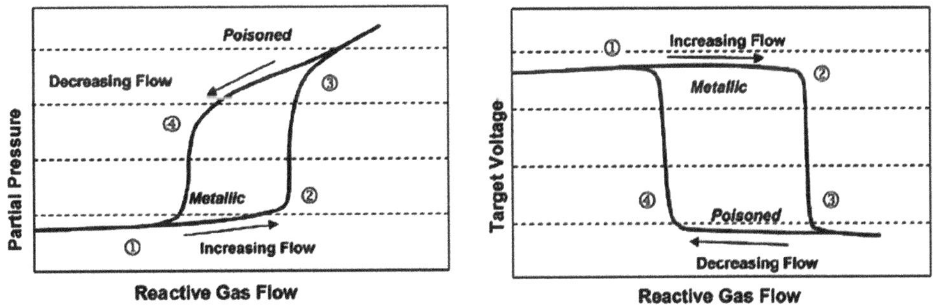

Table 2. Deposition parameter of TiO_2 and SiO_2

Parameter	TiO_2	SiO_2
Power (W)	DC - 1410	DC pulsed -1480
Voltage (V)	468 ± 0.4	370 ± 0.5
Current (A)	11 ± 0.5	4 ± 0.01
Vacuum limit (mbar)	2.8 x 10^{-6}	2.8 x 10^{-6}
O_2 flow (sccm)	11 ± 0.5	28 ± 0.5
O_2 partial pressure (mbar)	3.25 x 10^{-3}	3.3 x 10^{-3}
Ar flow (sccm)	30 ± 0.2	30 ± 0.2
Ar partial pressure (mbar)	3.18 x 10^{-3}	3.18 x 10^{-3}
Deposition rate (nm/min)	3.4 ± 0.1	11.4 ± 0.3

Table 3. Poisoning target Titanium

V [V]	Pressure [10^{-3} mbar]	SCCM O_2
380	3.216	0
444	3.28	17.8
448,8	3.88	15.9
454	3.29	14.4
459	3.294	13.1
461	3.298	12.5
463	3.304	11.7
464	3.305	11.3
464,5	3.305	11.1
465	3.305	10.9
465,5	3.306	10.7
466	3.308	10.5

Table 4. Poisoning target silicon

V [V]	Pressure [10^{-3} mbar]	sccm O_2
413	3,236	0
344,9	3,374	31,6
350,1	3,371	31
355,3	3,364	30,7
359,4	3,36	30,6
363,4	3,355	30,5
369,5	3,35	30,5

CHARACTERIZATION NUCLEAR RUTHERFORD BACK SCATTERING (RBS) OF OPTICAL MULTILAYER

The Rutherford Backscattering Spectrometry analytic technique uses the nuclear ion-matter interaction to extract precise information on the composition and thickness of a sample to be analyzed. More particularly, the RBS technique consists of measuring with a detector the number and energy of the ions coming from an accelerated beam after than these ions were scattered from the sample. The interaction between the ion beam and material, generally at energies between 400 keV to 30 MeV, and in this case 2 MeV, can be divided into two types: ion-nucleus interactions and electron-ion interactions. The ion-nucleus interaction, an elastic type of interaction, yields compositional information. The electron-ion interaction, an elastic type of interaction, instead allows one to obtain information with depth resolution.

More particularly, when an ion interacts with a nucleus, it can have elastic collisions of Rutherford type and inelelastic interactions with nuclear reactions. In this case the energy is equal to 2MeV and is insufficient to overcome the coulomb barrier, so available nuclear reactions are limited. In order to describe this technique, it is necessary to introduce some basic concepts, such as differential cross section:

$$\frac{d\sigma(\theta)}{d\text{Ù}} = \frac{number\,of\ scattered\ particles\,in\,d\ \text{Ù}}{number\,of\ incident\ particles} \tag{62}$$

The cross section $d\sigma(\theta)$ indicates the probability to detect a particle scattered at angle θ within a solid angle $d\text{Ù}$. From the equation (62) it is possible to derive the number of detected particles which in turn is proportional to the number of particles that appear in the sample.

$$Y_i = Q * \Omega * \sigma(\theta) * N_i \tag{63}$$

Where Q is the ion beam sent on the sample, θ is the solid angle subtended by the detector, $\sigma(\theta)$ is the cross section N_i is the number of atoms of the ith species

We will now briefly describe the elastic ion-nucleus interaction process that is the basis of the RBS analysis. As a first hypothesis, it is useful to consider the elastic interaction between two isolated particles (an ion and an atom on the sample surface). In that case, the ion back scattering energy can be derived directly from the conservation of kinetic energy and momentum (Mayer, 2003).

$$\begin{cases} \frac{1}{2}M_1v^2 = \frac{1}{2}M_1v_1^2 + \frac{1}{2}M_2v_2^2 \\ M_1v = M_1v_1cos(\theta) + M_2v_2cos(\phi) \\ 0 = M_1v_1sen(\theta) - M_2v_2sin(\phi) \end{cases} \tag{64}$$

achieving:

$$\frac{v_1}{v} = \frac{M_1cos(\theta) \pm \sqrt{M_2^2 + M_1^2sen^2(\theta)}}{M_1 + M_2} \tag{65}$$

$$\frac{E_1}{E_0} = \left[\frac{M_1cos(\theta) \pm \sqrt{M_2^2 + M_1^2sen^2(\theta)}}{M_1 + M_2}\right]^2 = K(M_1, M_2, \theta) \tag{66}$$

Where $K(M_1, M_2, \theta)$ is the scattering constant or kinematic factor, a function of masses of the particles involved and of the scattering angle. As a result, by knowing the geometric characteristics (scattering angle), the characteristics of the beam (E_0 and M_1), and measuring the particle scattered energy (E_1), it is possible to calculate the unknown mass M_2 of the sample that is being analyzed. Differentiating Equation (65), the maximum achievable resolution dE_1 / dM_2 is high for values of M_2 equal to M_1 with an angle of 180 °; however, the finite size of the detector prevents such a configuration generally, since for RBS analysis an angle between 160 ° and 170 ° is used, as shown in (Figure 6).

The areal density (atoms/ unit area), $(Nt)_i$, can be derived by referring to Equation (63) and to the (Figure 13c). We arrive at the equation:

$$(Nt)_i = \frac{A_i \cdot cos\theta_1}{Q \cdot \grave{U} \cdot \sigma_1(E, \theta)} \tag{67}$$

where A_i is the integral of the $i - th$ peak in the spectrum.

Figure 13. (a) Experimental Backscattering spectrometry, (b,c) Backscattering spectrum of two elements on a substrate

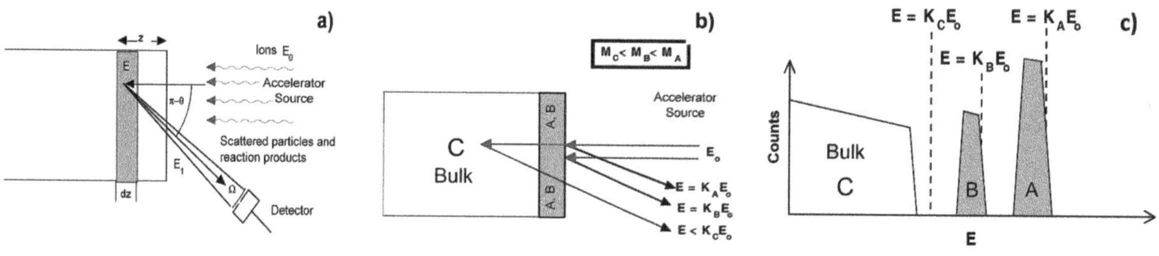

Given the small size of the bullet particle (10^{-15} m) and the relative distance between the atoms of the sample (10^{-10} m), the probability of obtaining a subsequent interaction between the particle backscattered and the sample is relatively low. The example shown in (Figure 13) contains a sample composed of a thin layer of materials A and B $(M_A > M_B)$ on a substrate C. K is a function of the mass (M_2) and the energy of the backscattered ions from the element A is greater than that of the element B. The peak reported to the substrate is not at the energy $E_C = K_C * E_0$, because the incident and back scattered ions cross the surface material slowly.

Therefore, the energy of the back scattered ions is a function of the mass (M_2) and the depth position (Z) of the element in question. Figure 14 shows a diagram of how the loss of scattered ion energy depends on the depth of the scattering event.

Considering the simple case shown in (Figure 14b), in which there is a thin layer of material A n a substrate of material B with $(M_A > M_B)$ and looking at the diagram (Figure 14a) of ions impacting on the surface $(z = 0)$ with energy E_0; it is possible to see the fraction of the beam that undergoes scattering, which will have a residual energy equal to $E = K * E_0$.

The remaining portion of ions penetrates the sample and gradually loses energy. It is practical to consider the sample as a stacking of thin layers of thickness dz. In a continuous system, the energy loss ΔE_1 at depth z is given by the integral:

Figure 14. (a) Schematic energy loss as a function of depth (z), (b) example of RBS energy loss spectra, single element layer and single compound layer

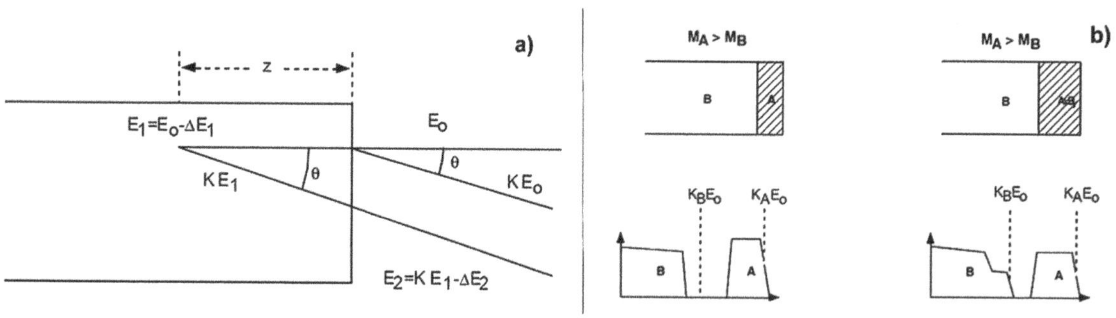

$$\Delta E_1 = \int_0^z \frac{dE}{dz} dz \tag{68}$$

where dE / dz is the "stopping power" of the material.

This loss of energy is called "energy straggling" (Figure 15) (Tesmer, Nastasi, Barbour, Maggiore, & Mayer, 1995) and is caused by several factor as: electronic energy loss straggling, due to statistical fluctuations in transfer to electron; nuclear energy loss straggling, due to statistical fluctuations in the nuclear energy loss; multiple small and large angle strangling; surface and interlayer roughness.

At the time of the ion-nucleus interactions at depth z, the incident ions will not have initial energy E_0, but an energy E_1:

$$E_1 = E_0 - \Delta E_1 = E_0 - \int_0^z \frac{dE}{dz} dz \tag{69}$$

Similarly, the ion will lose more energy in the output trajectory. The final energy ion E_2 after back-scattering will be:

$$E_2 = E_0 - \Delta E_1 - KE_1 - \Delta E_2 = E_0 - \int_0^z \frac{dE}{dz} dz - KE_1 - \int_0^{\frac{z}{\cos(\theta)}} \frac{dE}{dz} dz \tag{70}$$

Figure 15. Example of energy straggling

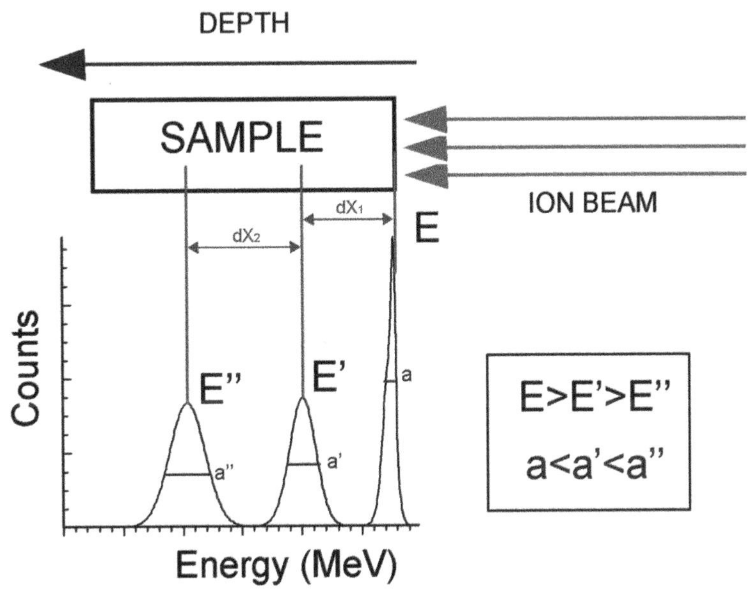

For a composite material, such as that shown in (Figure 14b), the different stopping powers of the different materials must be taken into account (Chu, Mayer, & Nicolet, 1978). The cross section of bremsstrahlung is defined as:

$$\varepsilon_A = \frac{1}{N}\left(\frac{dE}{dZ}\right) \tag{71}$$

where N is the atomic density of the material.

In the composite material A_xB_y the cross section braking is:

$$\varepsilon_{AxBy} = x\varepsilon_A + y\varepsilon_B \tag{72}$$

The energy loss in dz is:

$$\left(\frac{dE}{dz}\right)_{AxBy} = N\varepsilon_A\varepsilon_B = xN\varepsilon_A + yN\varepsilon_B \tag{73}$$

Where N in this case indicates the density of the solid, expressed in mol/cm^3.

By combining Equations (66), (68) and (73), it is possible to determine a function $E_2(z)$ that yields the energy spectrum as in terms of depth. Furthermore, the infinitesimal interval energy δE of the spectrum provides information about the material A in a thickness range δz.

$$\delta E = \left(\frac{dE_{2A}}{dz}\right)\delta z \tag{74}$$

Finally, the scattering yield of the material A in a section dz, applying Equation (62), becomes:

$$Y_i = Q\cdot\Omega\cdot\sigma_A(\theta)\cdot x\cdot N\cdot\delta z = Q\cdot\Omega\cdot\sigma_A(\theta)\cdot x\cdot N\cdot\left(\frac{\delta E}{dE_{2A}/dz}\right) \tag{75}$$

In example (Figure 14b), it is possible to observe the presence of element B as changes in the spectrum. In the spectrum an energy peak .., appears that is not seen in the spectrum of the compound. That evidences the presence of material B in this area. Moreover, both peaks of material A and B are less intense and more widened, and the presence of two elements causes an increase of the stopping power.

SILICON AND TITANIUM DIOXIDE SIGLE LAYER AND MULTILAYER RBS ANALISYS

Having in mind the theoretical concepts discussed in the previous sections, the RBS analysis of thin films of TiO_2 and SiO_2 will now be described. In Figure 16 there are two RBS spectra: one of titanium dioxide (Figure 16a) and one of Silica (Figure 15b)

The oxide films were deposited on an amorphous carbon substrate, as indicated above, and the position of the peak relative to the carbon (light element) is on the left of the peak of oxygen. This type of substrate is used to separate suitably the peaks in the spectrum acquired and thus make the measurements more accurate and simple. To obtain the stoichiometric oxide, it is necessary to use the dedicated software XRump to analyze RBS spectra. That software contains in-house functions for several characteristics (stopping power, strangling, etc.) of various materials. Through the integration of individual peaks (O and Ti or Si) the software provides the density (atoms/cm^2) of the elements in the material. This is useful because by calculating ratios of the doses, it is possible to determine the stoichiometry of the oxide deposited. The spectra shown in (Figure 16) have already been adjusted for the beam energy, energy channel, position gained, live time, and solid angle. These corrections and normalizations are calculated through analysis of standard samples.

The oxide films are made with the reactive magnetron sputtering technique to calibrate the release of oxygen gas in the system. As the gas injection system is controlled retroactively by the potential of the power supply applied to the cathode, it is necessary to identify the correct voltage and the corresponding amount of oxygen to be flushed into the deposition system to maintain an accurate stoichiometric ratio and obtain the predetermined optical properties. The Figure 17 explains the trend of the stoichiometric ratio as a function of the potential for the two oxides realized by the reactive magnetron sputtering process (Raniero, Campostrini, Maggioni, & Quaranta, 2014).

The data reported in (Figure 17) are derived from an analysis of multiple RBS measurements on samples composed of a single layer of TiO_2 and SiO_2 by varying the voltage applied to the cathode. This process allowed us to calibrate the deposition plant using 370 V for the silicon dioxide and 470 V for titanium dioxide. After the simulation and optical characterization of the dichroic filters, the RBS measurements on the optical multilayers confirm the thicknesses of the individual layers. The RBS spectrum of an oxide multilayer material is particularly complicated because the peaks of the three elements (O, Si and Ti) overlap. In order to determine the thickness of the individual layers, is important to create a simulation of the RBS spectrum through use of XRump software (Hussain, et al., 2013). In Figure 18 the spectra of the dichroic filters described in the preceding paragraphs are presented.

In Table 5 the RBS spectra simulation parameters of every one layer of dichroic mirror express in Figure 16a. The AFM thickness analysis and the areal dose analysis of single layer are used to determine the areal dose ranges of the individual layers used in a multilayer filter. This allows running a simulation of an RBS spectrum of a multilayer and comparing it with a real RBS spectrum.

The unit of measurement that is inserted in the software XRump m/cm_2 indicates the molecular dose and the composition of the material, and while the right column shows the thicknesses obtained by the simulation and used for the deposition process. The simulation was made by using the doses instead of the thicknesses directly, as the density of the material sputtered is slightly different than that of the bulk as tabulated in the software. To convert the values (from nm to m/cm_2), measurements are made using RBS and AFM in order to obtain the ratio [(m/cm_2)/nm], which can then be converted to the real thickness values used in the deposition step. A correction was applied to consider strangling of 0.5 using the

Figure 16. RBS spectrum of a single layer TiO_2 and SiO_2 deposited on carbon substrate

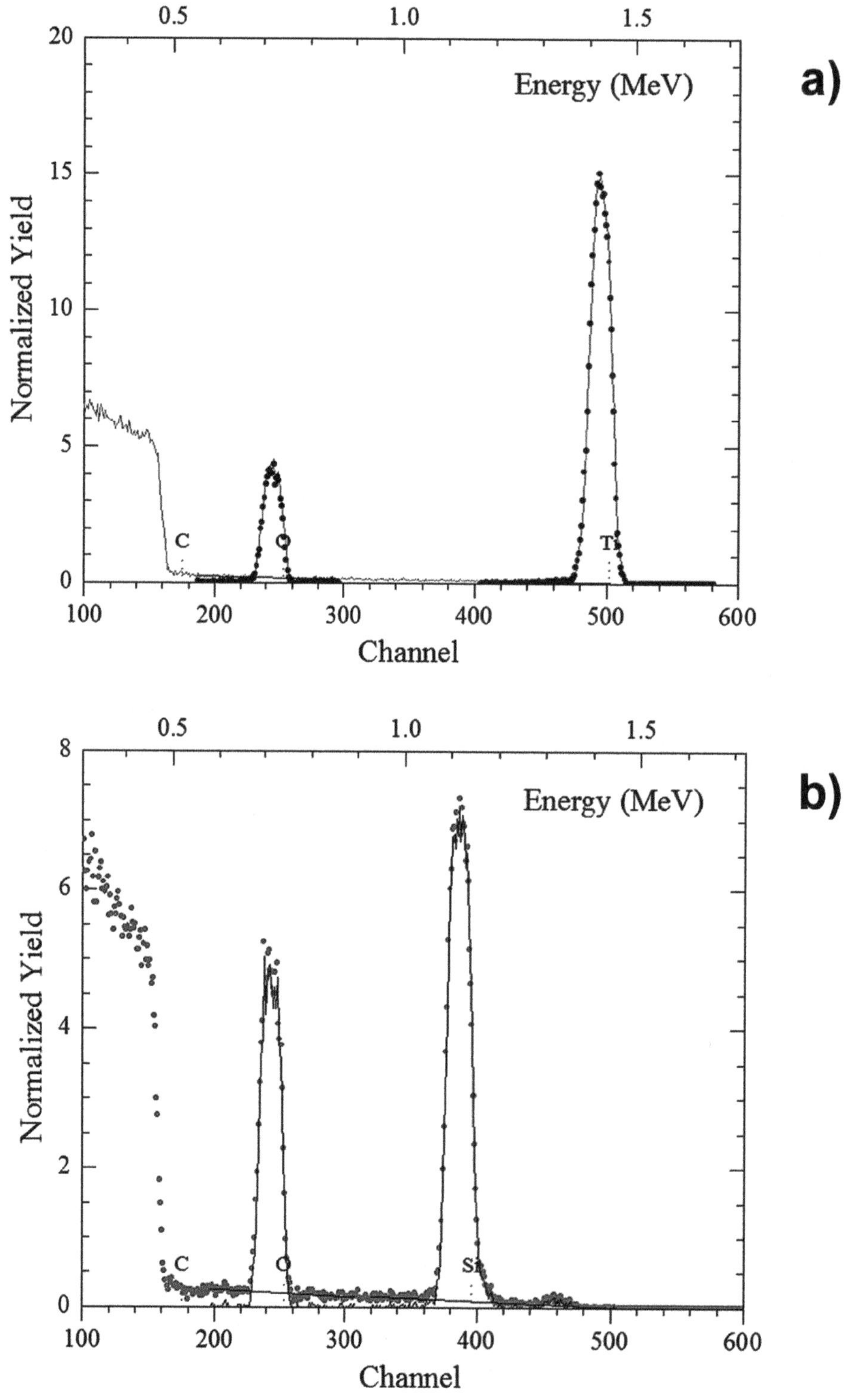

Figure 17. Stoichiometric ratio of TiO_2 and SiO_2 vs. DC-voltage of deposition

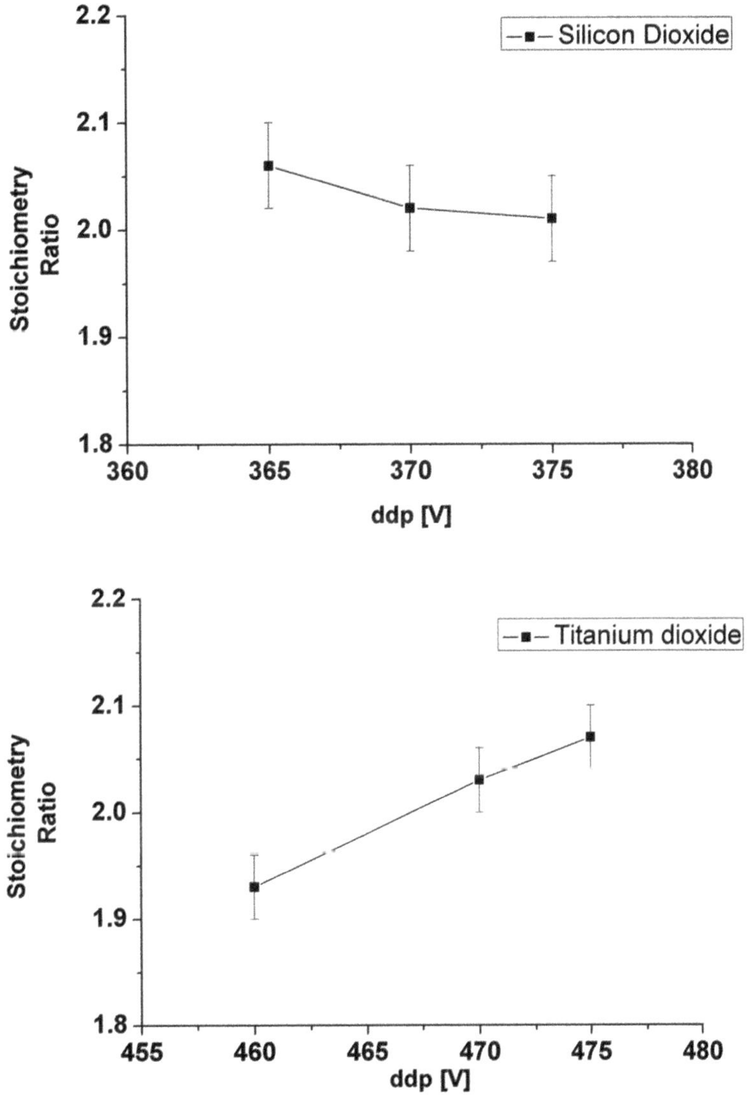

Bohr's theoretical model, which is default content in the software. The uncertainty on the measurement (error 3%) is given by the standard deviation of the AFM measurements. In that case, the relationship between dose and thickness is $TiO_2 = 2.912 \pm 3\%$ [(m/cm^2) / nm] and $SiO_2 = 2.359 \pm 3\%$ [(m/cm^2) / nm].

OPTICAL MULTILAYER APPLICATION

The multilayer optical filters can be used in varied optical applications (Ohring, 1992). Multilayer optical filters have advantages over other types of filters because there is very little absorption loss in dielectric film layers, since they rely on the effects of interference (Pulker, 1984). There are several application examples (Figure 19) of optical multilayers.

Figure 18. RBS spectrum vs. simulation, (a) dichroic 25 layers and (b) dichroic 23 layers

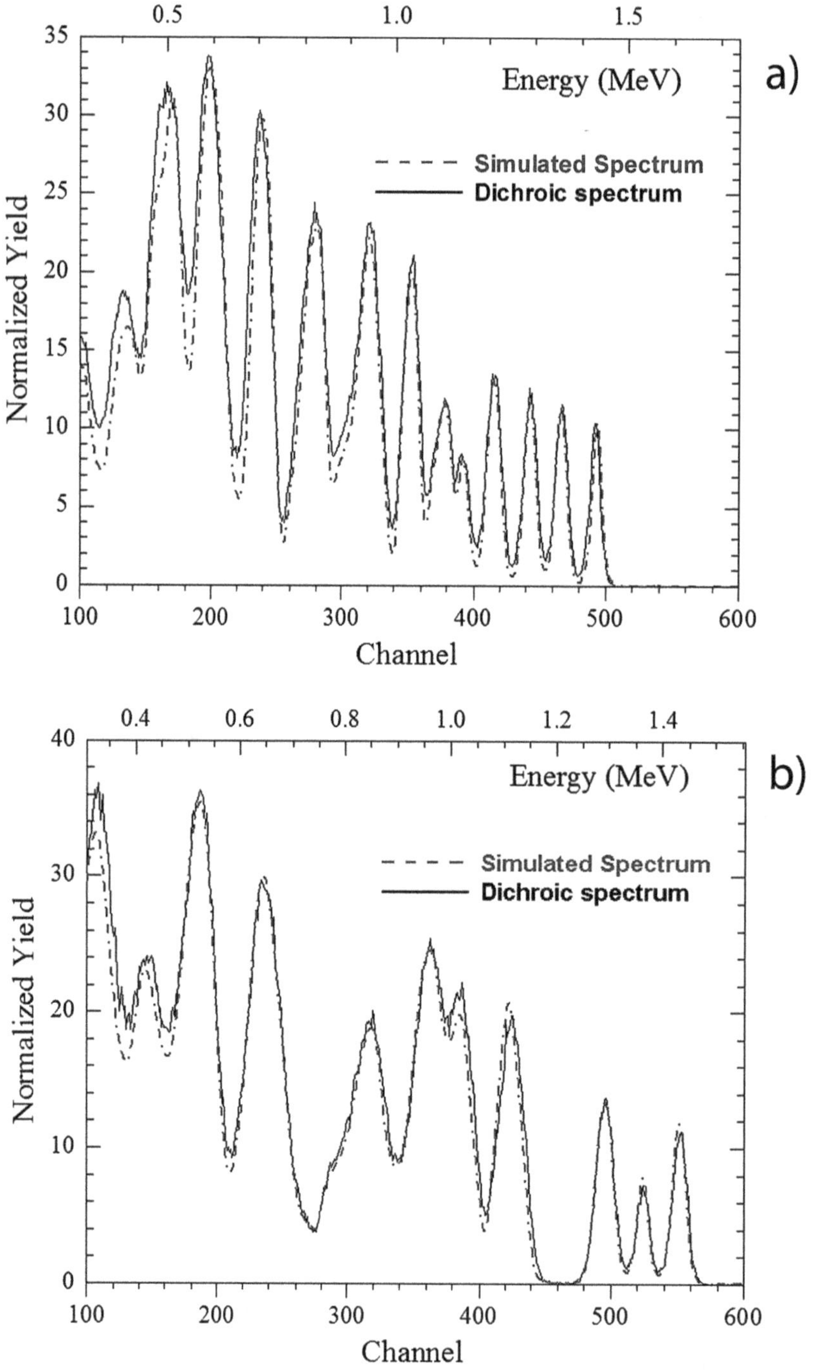

Table 5. Example with the shims inserted in the simulation in (Figure 16a).

Layer	Material	Dose (m/cm2)	Thickness (nm)
1	TiO_2	63.92	21.95
2	SiO_2	224.18	95.03
3	TiO_2	80.78	27.74
4	SiO_2	182.56	77.39
5	TiO_2	88	30.23
6	SiO_2	209.5	88.8
7	TiO_2	106	36.45
8	SiO_2	176.5	74.8
9	TiO_2	49	16.85
10	SiO_2	117.3	49.75
11	TiO_2	26.8	9.21
12	SiO_2	202	85.65
13	TiO_2	101	34.65
14	SiO_2	225.4	95.55
15	TiO_2	173.1	59.44
16	SiO_2	239.6	101.56
17	TiO_2	151.72	52.1
18	SiO_2	277.27	117.58
19	TiO_2	133.26	45.76
20	SiO_2	256.3	108.63
21	TiO_2	199.2	68.39
22	SiO_2	111.1	47.09
23	TiO_2	82.5	28.33
24	SiO_2	64.35	27.28
25	TiO_2	94.5	32.46

EDGE FILTER

The methodology of the design, characterization and production of dichroic filters permits us to realize specific optical thin film multilayers to be applied in the photovoltaic industry (Ho-Baillie & Green, 2010). Because of the interference phenomenon, the surface functionalization of the glass allows us to see the optical effect of dichroism that separates two optical bands in terms of reflectance and transmittance. These filters are used in the new generation of research in photovoltaics, especially in concentrated photovoltaic (CPV) with spectral separation of the solar radiation (A.G. Imenes, 2004). The purpose of this concentrating photovoltaic (CPV) system is its use in photovoltaic cells with different energy gaps, such as InGaP (Indium Gallium Phosphorus), GaAs (Gallium Arsenide) or Si (silicon). By analyzing the quantum efficiency EQE (Figure 19) of these cells, it is possible to control the optical behavior of an optical filter to be optimally coupled with two different cells (Figure 20).

Figure 19. Optical filter application

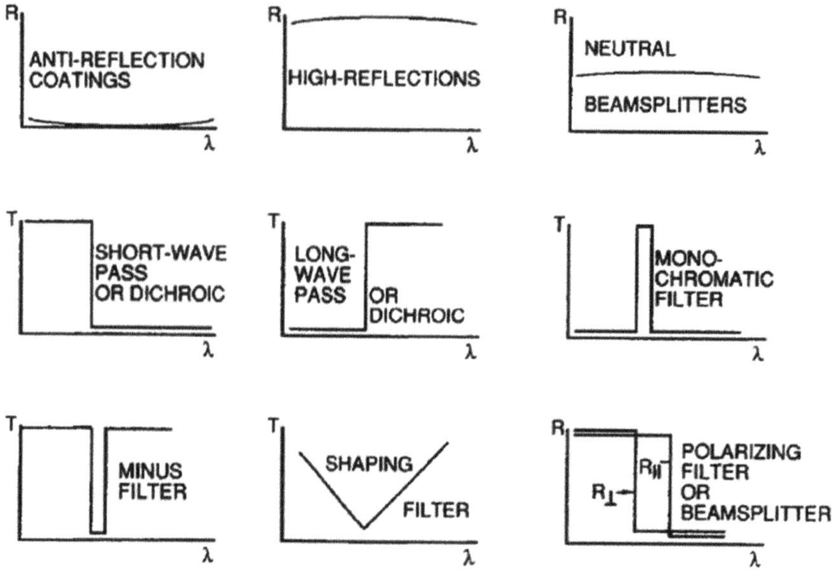

This design of a CPV system (Figure 21) using a Fresnel lens allows us to focus solar radiation on a dichroic filter that separates the radiation into two optical bands. In each optical band, we use a solar cell with a different energy gap. This coupling constitutes a greater use of the solar spectrum and, therefore, increases the energy efficiency compared to traditional photovoltaic systems.

The dichroic filter is designed to be used at 45 degrees, and in the optical system it is matched with a concentrated light and coupled with the radiation focused by the lens. A component of the solar radia-

Figure 20. EQE of different photovoltaic cells.

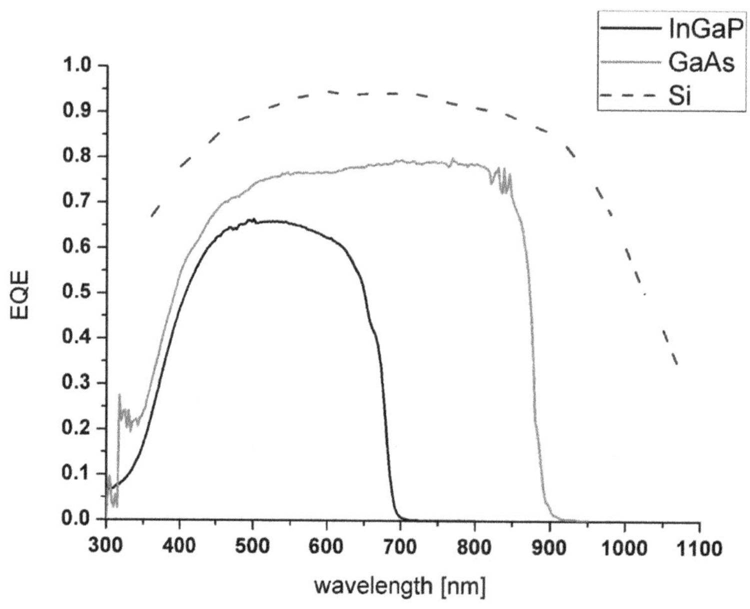

Figure 21. CPV solar spectral splitting

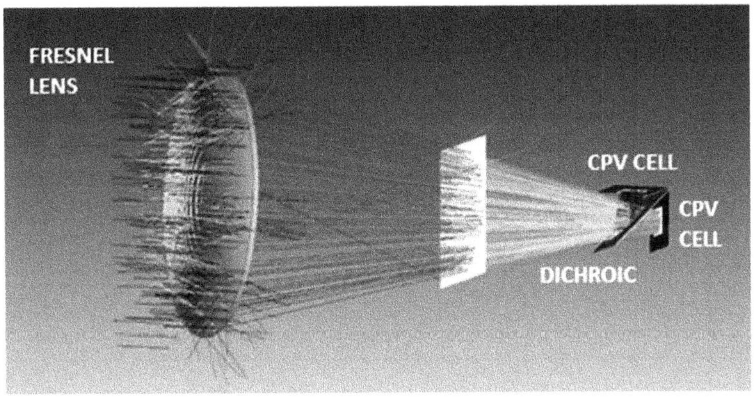

tion is reflected to a dedicated CPV cell e.g. GaAs (Gallium Arsenide) and the remaining part of the radiation is transmitted to another CPV cell e.g. Si (Silicon) cell.

ANTIREFLECTIVE COATING

The antireflection coatings shown in Figure 22 and Figure 23 are used in a wide range of industries and applications such as: Ophthalmics, optics, flexible displays, bio-engineering, energy production, decorative coatings and barrier for food and pharmaceutical products. The Coatings made from thin films can modify the electrical, mechanical or optical properties of a base material. Some common examples are scratch-resistant coatings for eyeglasses, anti-reflective coatings for lenses and windows, and transparent conductive coatings for flat-panel display (Mazuri, et al., 2013).

Figure 22. AR- Coating MgF_2 + ZrO_2 + CeF_3 on glass. (Heavend, Ring, & Smith, 1957)

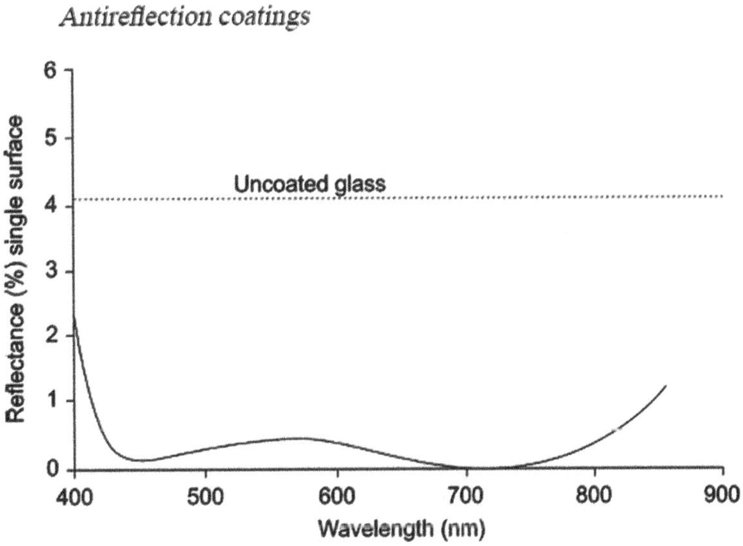

Figure 23. AR- Coating TiO_2 + SiO_2 on glass

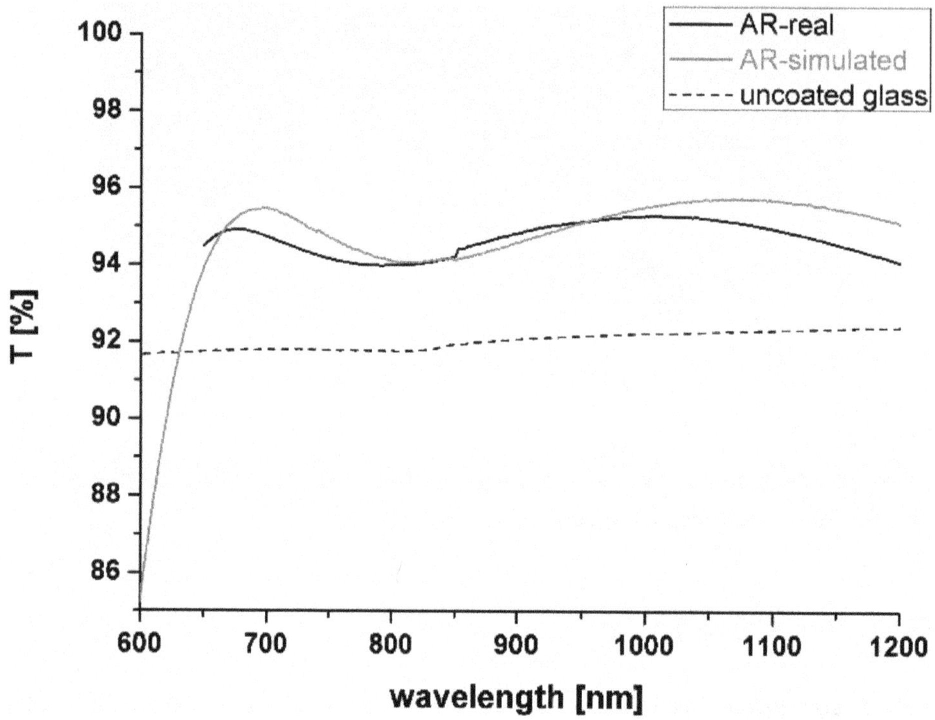

The indispensable requirements in practical applications for minimizing the reflection by coupling more light into such systems at surface and interfaces have promoted the further study of novel structure, materials and fabrication techniques of ARCs. The multilayer antireflection studied for many years, are currently used to increase the photovoltaic conversion efficiency of the Cu(In,Ga)Se2 (CIGS) solar cells. An alternative method to obtain antireflection properties stacks broadband multilayer using coatings with inhomogeneous composite layers, which results in a gradual change of the refractive index from the bottom to the upper interface of the coating (Figure 24) (Li, et al., 2014).

Figure 24. (a) ARCs- Broadband multilayer Coating on solar cell, (b) Transmittance of substrate material with and without ARCs multilayer

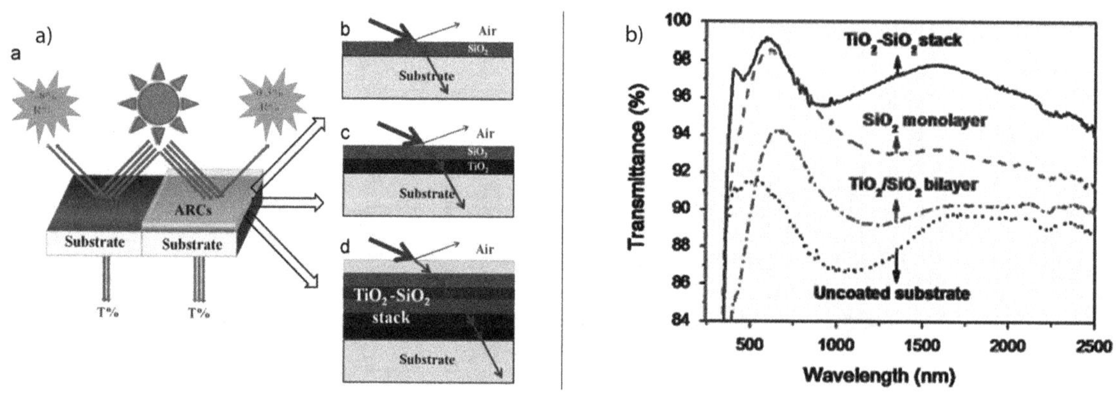

NARROW-BAND FILTER

Narrow-band is a particular optical filter that permits us to refract or transmit a precise wavelength. Figure 25 shows an example of a narrow-band filter applied in medical devices, where it is necessary to use a specific wavelength with high resolution between 414 nm and 540 nm with bandwidth equal to 30nm (Silva, et al., 2014). The technology that is employed by Image Enhanced Endoscopy (IEE) uses Narrow Band Imaging (NBI) to evaluate an early diagnosis of oropharyngeal and gastrointestinal cancer. The optimized hemoglobin absorption in these shorter wavelengths of light permit to focus the visualization of micro vessels of the superficial layer of the mucosa (415 nm) and submucosa (540 nm).

CONCLUSION

This chapter has importance to applied research in the field of thin films by providing a physical description of the theoretical thin films' optical behavior and characterizations, which is necessary to optimize optical materials that describe a practical sample of optical design and production of optical multilayers in particular dichroic filters.

Figure 25. Narrow-band optical filter about 414nm and 540nm

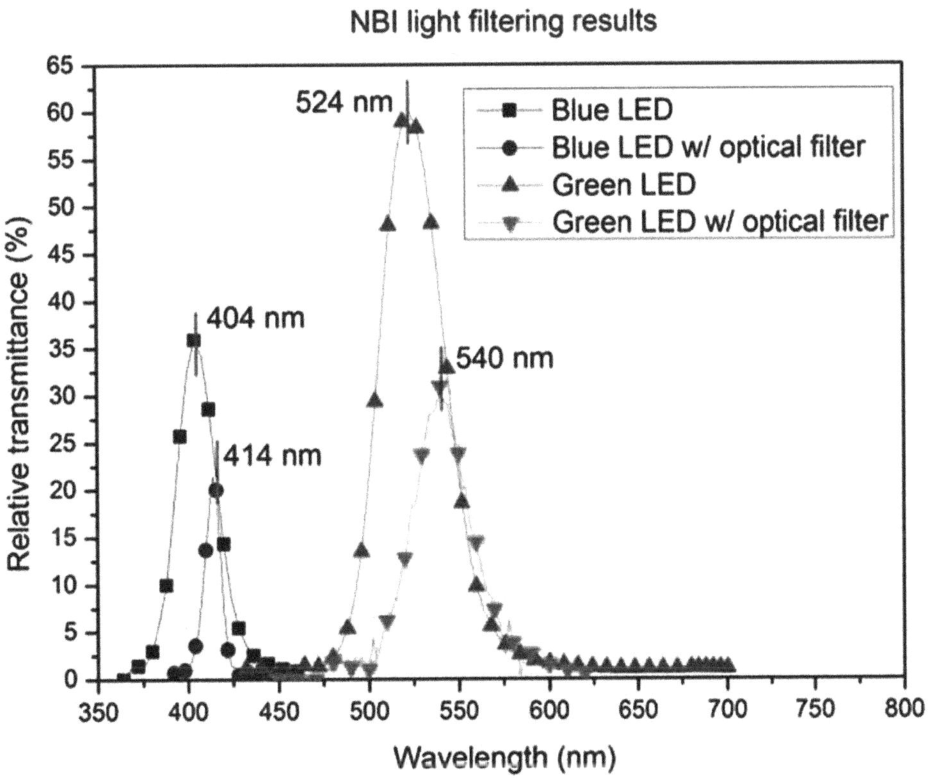

The purpose of this chapter was therefore to provide designers of thin films a simple understanding of the physics of electromagnetic waves iteration within bulk materials, and coatings of thin films aimed the functionalization of the glass surface by changing the optical properties.

To sum up, the knowledge of the analytical behavior of such materials through the use of dispersion functions is particularly important for the description of the thin films optical behavior, which offers an insight into issues related to the optical design. The analysis of the thin film and the numerical modeling allows us to learn physical characteristics of the materials used, such as the refractive index and absorption, which are of fundamental importance when trying to obtain a correspondence between optical design and experimental optical measurements.

The optical characteristics of the materials used for the realization of the coating, in particular titanium dioxide and silicon dioxide, are obtained from the elaboration of numerical experimental optical measurements such as measurements of transmittance $T(\lambda)$ and reflectance $R(\lambda)$. The realization of custom materials is a necessary condition to better describe the optical behavior obtained through the physical vapor deposition Reactive Magnetron Sputtering deposition technique.

The physical and chemical single layer characterizations, and in particular the Rutherford Back Scattering analyses, are essential. This technique allows us first to quantify the stoichiometric ratio of oxide thin film material, and second to calibrate and obtain the optimal set-point of the target poisoning point in the reactive magnetron sputtering process needed to achieve reproducible depositions.

The back scattering analyses are also needed to verify the multilayer structure of the thin films produced; comparing the thickness determined by atomic force microscopy and Rutherford backscattering compositional analysis on a single layer can realize a simulated Rutherford backscattering spectrum and compare it with the experimental spectrum. This method allows us to verify the thickness of each layer of the multilayer between the theoretical value of design and experimental measurements.

The spectrophotometric optical analyses determine the quality of the multilayers obtained when compared with the optical design by allowing us to locate the spectral optical response. These optical filters are used mainly in the light visible spectrum as antireflective coating, narrow band or edge filter.

The dichroic filters described in this chapter are applied in the field of renewable energy for applied research to a concentrated photovoltaic system with the spectral separation of the solar radiation. This technology enables one to couple the photovoltaic cells with different external quantum efficiencies, allowing one to optimize the photovoltaic conversion with respect to solar radiation.

REFERENCES

Asghar, M., Shoaib, M., Placido, F., & Naseem, S. (2008). Wide bandpass optical filters with TiO2 and Ta2O5. *Central European Journal of Physics*.

Carter, D., Sproul, W., & Christie, D. (n. d.). Effective Control for Reactive Sputtering Processes. *Advanced Energy Industries Inc.*

Chu, W., Mayer, J., & Nicolet, M. (1978). *Backscattering Spectrometry*. Academic Press.

Heavend, O., Ring, J., & Smith, S. (1957). Interference filters for the infra-red. *Spectrochimica Acta*.

Ho-Baillie, A., & Green, M. (2010). Forty three per cent composite split-spectrum concentrator solar cell efficiency. *Progress in Photovoltaics: Research and Applications*, 2010, 42–47.

Hussain, M., Mehmooda, M., Ahmada, J., Tanvir, M., Khan, A., Ali, T., & Mahmood, A. (2013). RBS depth profiling and optical characterization of multilayers of TiO2 (20 nm) and Ge (15 nm). *Materials Chemistry and Physics*, *139*(1), 17–26. doi:10.1016/j.matchemphys.2012.10.037

Imenes, A. G., & Mills, D. R. (2004). Spectral beam splitting technology for increased conversion efficiency in solar concentrating systems: A review. *Solar Energy Materials and Solar Cells*, *84*(1-4), 19–69. doi:10.1016/j.solmat.2004.01.038

Li, D., Wan, D., Zhu, X., Wang, Y., Han, Z., Han, S., & Huang, F. et al. (2014). Broadband antireflection TiO2–SiO2 stack coatings with refractive-index-grade structure and their applications to Cu(In,Ga)Se2 solar cells. *Solar Energy Materials and Solar Cells*.

Lou, L., MacDonough, G., Walde, H., Carter, D., Roche, G., & Scholl, R. (1999). *Closed loop controlled reactive dual magnetron sputtering*. Advanced Energy Insudries, Inc.

Macleod, H. (1986). *Thin film optical filters*. New York: McGraw-Hill. doi:10.1887/0750306882

Mayer, M. (2003). RBS – Rutherford Backscattering Spectrometry. *Workshop on nuclear data for science and technology: materials analysis*.

Mazuri, M., Woicieszak, D., Domaradk, J., Kaczmarek, D., Song, S., & Placido, F. (n. d.). TiO2/SiO2 multilayer as an antireflective and protective coating deposited by microwave assisted magnetron sputtering. *Versita Opto-Electronics review 21 (2)*, 233 - 238.

Ohring, M. (1992). *The material science of thin film*. Academic Press.

Pulker, H. K. (1984). *Coatings on Glass*. Elsevier Amsterdam.

Raniero, W., Campostrini, M., Maggioni, G., & Quaranta, A. (2014). Physical vapour deposition reactive magnetron sputtering for the production and application of dichroics in photovoltaic system with solar spectral splitting. *Applied Surface Science*, *308*, 170–175. doi:10.1016/j.apsusc.2014.04.130

Raniero, W., Maggioni, G., Della Mea, G., Campostrini, M., Marigo, S., & Nardo, M. (2013). Rutherford Backscattering Spectrometry (RBS) analysis of dichroic systems for optical applications. *AIP Conference Proceedings*, *1530*, 133–139. doi:10.1063/1.4812915

Scholl, R. (1997). Power System for Reactive Sputtering of Insulating Films. *Surface and Coatings Technology*, 93(1), 7-13.

Silva, M., Rodrigues, J., Oliveira, M., Fernandes, A., Pereira, S., Costa, C., . . . Correia, J. (2014). Optical Filter for Providing the Required Illumination to Enable Narrow Band Imaging. *EUROSENSORS 2014, the XXVIII edition of the conference series*. Elsevier.

Stenzel, O. (2005). *The Physics of Thin Film Optical Spectra*. Springer.

Tesmer, J., Nastasi, M., Barbour, J., Maggiore, C., & Mayer, J. (1995). *Handbook of modern ion beam material analysis*. Material Research Society.

Willey, R. R. (2012). *Practical Production of Optical Thin Films* (2nd ed.). Charlevoix, USA: Willey Optical Consultants.

Chapter 8
Analysis of Plasmonic Structures by Spectroscopic Ellipsometry

Jose Luis Pau
Autonomous University of Madrid, Spain

María Jesús Hernández
Autonomous University of Madrid, Spain

Antonio García Marín
Autonomous University of Madrid, Spain

Manuel Cervera
Autonomous University of Madrid, Spain

Juan Piqueras
Autonomous University of Madrid, Spain

ABSTRACT

This chapter focuses on the plasmonic effects that appear in the ellipsometric functions and the pseudodielectric function when metal thin films and nanoparticles are analyzed by spectroscopic ellipsometry in the visible, near infrared and ultraviolet regions of the electromagnetic spectrum. The chapter is structured in two large sections. The first section reviews the basics of total internal reflection ellipsometry (TIRE), based on the excitation of surface polaritons in metal thin films. The conditions required to excite polaritons in TIRE systems are analyzed along with the main characteristics of those electromagnetic waves. The second section of the chapter is devoted to study the optical properties of plasmonic resonances in nanostructures and the characteristics introduced in the dielectric functions. The treatment of optical anisotropies and Fano resonances in the ellipsometric models is discussed. The last section of the chapter reviews the state of the art of the technique in biosensing applications.

INTRODUCTION

Ellipsometry is a non-perturbing optical technique that enables the characterization of bulk materials, thin films and surfaces by studying the change in the state of polarization of a light beam that it is reflected on the sample under analysis or transmitted through it. The morphology and optical properties of materials and nanostructures can be investigated through the variations of the state of polarization of the reflected beam compared to the state of polarization of the incident beam. In reflection mode, ellipsometry has

DOI: 10.4018/978-1-5225-0066-7.ch008

demonstrated capabilities to sense fractions of a monolayer of organic and inorganic materials. Thus, Aspnes reported that in clean air, the surface sensitivity can make possible the determination of oxide thickness in systems like SiO_2/Si down to the 0.1 pm range (Aspnes, 2004). Furthermore, aside being an effective tool for the study of the optical properties of solid-state materials, ellipsometric analysis can be employed as a transduction method in the development of gas and biological sensors (Arwin, Poksinski, & Johansen, 2004). Metal nanoparticles (NPs) and metal thin films contribute to increase the sensitivity of ellipsometry through the excitation of localized surface plasmons and surface polaritons. However, the traditional ellipsometric fitting models face new challenges in those scenarios due to the optical anisotropies introduced by the morphology and surface distribution of the nanostructures. These anisotropies limit the use of a single dielectric function to account for the optical properties of the material, hindering the application of effective medium theories.

In metal thin films, the excitation of surface polaritons requires the use of Krestchmann configuration, in which the film is directly deposited on the coupling prism, or alternatively, on a transparent substrate that is attached to the prism by means of an index matching fluid. Under the proper incidence angle and radiation energy, internal total reflection occurs and the condition for the excitation of the interface polariton at the air-film interface can be satisfied. If the metal thickness is thin enough, part of the radiation can tunnel through the film producing the collective oscillation of surface charges. Compared to regular surface plasmon resonance (SPR) sensing techniques, total internal reflection ellipsometry presents larger sensitivity to the chemical environment due to the phase information obtained from the parallel and transverse components of the reflected beam.

In metal NPs, the resonant absorption of radiation gives rise to oscillatory processes of the electrons confined in the particle whose characteristic frequencies depend on the size and morphology of the NP. Part of that absorbed energy is reemitted as scattered light (Rayleigh scattering) and part is delivered as thermal energy to the surrounding environment. Both energy losses contribute to reduce the specular reflection and modify the amplitude and phase of the polarization components in the reflected beam. Furthermore, when those oscillatory processes occur in a particle deposited on a surface, they experience interactions with the substrate and the neighboring particles through the formation of mirror image charges. Those interactions can be interpreted through a depolarization factor that leads to obtain different polarizabilities of the nanoparticle for directions parallel and perpendicular to the substrate.

This chapter reviews the plasmonic effects on the ellipsometric functions and the pseudodielectric function obtained from spectroscopic ellipsometry in reflection mode. Reflection mode is generally more sensitive to surface changes than transmission mode, and more versatile, since it allows the measurement of samples with absorbing substrates. The chapter will include references to recent works on the subject and an overview of the technique to characterize plasmonic structures of application in nanotechnology, biochemistry, and information technology. The analysis focuses on the visible, near infrared and ultraviolet regions of the electromagnetic spectrum where metal plasmon resonances are typically found.

The chapter is structured in two large sections. The first section reviews the basics of the total internal reflection ellipsometry (TIRE), based on the excitation of surface polaritons in metal thin films. The conditions required to excite polaritons in TIRE systems are analyzed along with the main properties of those electromagnetic waves. The chapter includes the characteristic features of the ellipsometric functions found upon polariton excitation and the analysis of the perturbations introduced by the absorption in the semitransparent metal layer. The section ends up with a short discussion on the motivation for using those tools in biosensing applications and the variations introduced by the formation of a self-assembled monolayer of biological material. The second section of the chapter is devoted to study the optical properties of plasmonic resonances in nanostructures and their role on the energy dependence of the dielectric functions. The treatment of optical anisotropies in the ellipsometric models is discussed from different viewpoints. In particular, the ellipsometric response of Drude metals is compared with the response of layers with isotropic and anisotropic Lorentzian resonances associated to the deposition of metal nanostructures on top of a substrate. Later, the chapter introduces the use of the pseudodielectric function in conditions of polarization handedness inversion as a means to sense the adsorption of biomolecules, and reviews the treatment of asymmetrical resonances whose absorption departs from Lorentzian functions due to strong interaction between resonant modes in ordered metal nanostructures. The last section of the chapter aims to overview the state-of-the-art of the technique in biosensing applications.

SPR IN TOTAL INTERNAL REFLECTION ELLIPSOMETRY (TIRE)

SPR Theory

SPR is a sensing technology whose tools and protocols have been widely developed in the last 20 years to study biomolecular interactions (Homola, Yee, & Gauglitz, 1999). This technique relies on *surface plasmons*, i.e. longitudinal collective oscillations of free electrons at the interface between a negative (metal) and a positive (dielectric) permittivity material. Under light or electron excitation, these charge oscillations can propagate along the metal/dielectric interface coupled to an electromagnetic wave (*surface polariton*), whose amplitude and phase change with the adsorption of molecules to the metal interface.

The electric field of the electromagnetic wave in a surface polariton propagating along the xz-plane is described by

$$\vec{E} = \left(E_{0x}^{\pm}, 0, E_{0z}^{\pm}\right) \exp\left[i\left(k_x x \pm k_z z - \omega t\right)\right] \tag{1.1}$$

According to the coordinate system included in Figure 1, the plus superscript is valid in the dielectric region ($z > 0$) and the minus superscript, in the metal region ($z < 0$). Since the existence of the electromagnetic wave is limited to the interface, k_z must be purely imaginary. On the other hand, k_x is the momentum of the wave in the direction of propagation and it is complex, containing real and imaginary parts ($k_x = k_x' + k_x''$). It is worth remembering that the magnetic field will have an analog expression with the same frequency ω.

The excitation of the surface polariton requires that the component of the light wave vector parallel to the interface matches the real part of the polariton momentum in the direction of propagation. For light impinging the interface from the dielectric side (dielectric constant $= \varepsilon_d$) at an incidence angle ϕ_i, the parallel component of the wave vector is:

$$k_x^{light} = \frac{\omega}{c}\sqrt{\varepsilon_d}\sin\phi_i \tag{1.2}$$

The dielectric medium could be a gas or a liquid solvent like water or ethanol.

Figure 1. (a) Diagram of non-resonant external reflection on a metal surface. (b) Decay of the electric field amplitude of a polariton wave in the transverse direction. (c) Otto and Kretschmann configurations for polariton excitation

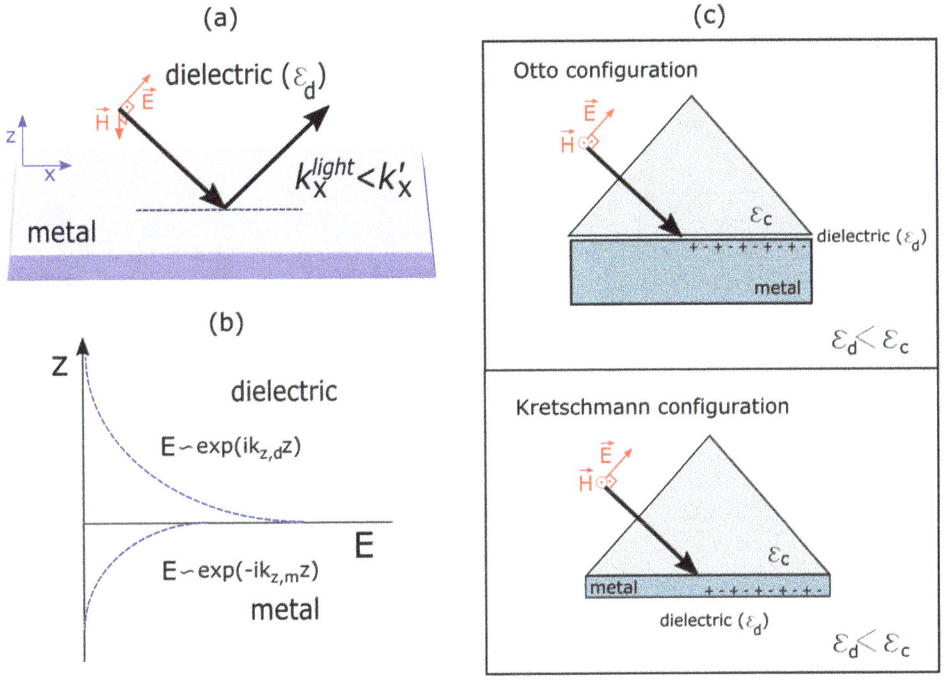

The polariton momentum k_x is given by the optical properties of the metal and the dielectric material forming the interface. As the electric field in Equation 1.1 must satisfy the Maxwell's and the continuity equations, the real part of k_x can be written as:

$$k_x' = \frac{\omega}{c}\sqrt{\frac{\varepsilon_m'\varepsilon_d}{\varepsilon_m' + \varepsilon_d}} \tag{1.3}$$

where ε_m' is the real part of the metal permittivity and ε_d is the dielectric permittivity (Raether, 2013). To get a real value of k_x' and an exponential decay of the electric field in the normal direction to the surface (bound polariton), one needs that $\varepsilon_m' < 0$ and $\left|\varepsilon_m'\right| > \varepsilon_d$. Under those conditions, the values of k_z are purely imaginary and they can be obtained for the metallic and the dielectric regions from

$$k_{z,m} = \frac{\omega}{c}\left(\frac{\varepsilon_m'^2}{\varepsilon_m' + \varepsilon_d}\right)^{1/2} \tag{1.4}$$

and

$$k_{z,d} = \frac{\omega}{c}\left(\frac{\varepsilon_d^2}{\varepsilon_m^{'} + \varepsilon_d}\right)^{1/2} \tag{1.5}$$

If $\varepsilon_m^{'} > 0$, the values of k_z are not purely imaginary and the polariton is not bound to the interface. The skin depth (Z) is defined as the value at which the field decays a e^{-1} factor in the transverse direction and it can be calculated from:

$$Z = \frac{1}{\left|k_z\right|} \tag{1.6}$$

Thus, the values of Z in the metal and in the dielectric become

$$Z_m = \frac{\lambda}{2\pi}\left|\frac{\varepsilon_m^{'} + \varepsilon_d}{\varepsilon_m^{'2}}\right|^{1/2} \tag{1.7}$$

and

$$Z_d = \frac{\lambda}{2\pi}\left|\frac{\varepsilon_m^{'} + \varepsilon_d}{\varepsilon_d^{'2}}\right|^{1/2} \tag{1.8}$$

respectively. Since $\left|\varepsilon_m^{'}\right| > \varepsilon_d$, the penetration of the wave in the metal is shallower than in the dielectric, as illustrated in Figure 1.

From Equations 1.2 and 1.3, it can be demonstrated that the coupling of light from the dielectric medium into the metal surface is not possible because k_x^{light} is always lower than $k_x^{'}$ for any ϕ_i (Raether, 2013). Therefore, the plasmon excitation must be carried out using a grating or a total internal reflection coupler. The latter is the most extensively used method in spectroscopic ellipsometry to study the plasmonic resonance in thin metal films, and relies on the deposition of a thin metal film on the interface where the evanescent wave is built upon total internal reflection in the optical coupler. The material of the optical coupler is chosen such that the travelling wave has a low phase velocity in comparison to the phase velocity in the surrounding medium. Thus, the dielectric constant of the coupler (ε_c) must be larger than ε_d to increase k_x^{light} and enable the excitation of the polariton at the interface. Transparent prisms or half-cylinders made of fused silica, BK7 or quartz are commonly used as optical couplers. The advantage of the half-cylinder in comparison to the prism is that allows to scan over a wide range of angles in absence of undesirable reflections on the inner faces.

In TIRE, two setups can be used to excite the surface plasmon: the Otto and the Kretschmann setups (Figure 1). In the Otto setup, the evanescent wave interacts with the metal surface when the metal layer and the dielectric are in optical contact. This means that the distance between both materials is shorter than a wavelength of light. As a very thin air gap can stand in-between, the evanescent wave needs to tunnel through to excite the polariton on the metal surface. In the Kretschmann setup, the metal film is evaporated directly on the optical coupler, or alternatively, onto a slide of the same material that the optical coupler and attached to it through an index-matching fluid. The evanescent wave tunnels through the metal film and, if the film is thin enough, the light can excite the surface polaritons under the appropriate incidence angle and energy. A disadvantage of the Otto configuration is the difficulty to implement in samples that can be damaged by direct contact with the prism or to study real-time surface interactions. On the other hand, the Kretschmann configuration enables the realization of in-situ real-time experiments, a property that is very useful for the analysis of adsorption kinetics during physical or chemical deposition.

The resonant condition $k_x^{light} = k_x^{'}$ causes the extinction of the reflectivity for p-polarized light and provokes the enhancement of the electric field intensity at the metal surface. In spectroscopic ellipsometry, it is possible to estimate the energy position of the resonance by plotting $k_x^{'}(c / \omega)$ against energy for the metal under study. Figure 2 shows those plots for Au and Ag in air (Figure 2a) and water (Figure 2b). The vertical line points out the energy value at which $\left|\varepsilon_m^{'}\right| = \varepsilon_d$. For larger energies, the polariton condition is not fulfilled because $k_x^{'}$ becomes imaginary. Thus, according to Equation 1.3, that line represents the maximum energy of the surface polariton at the metal-dielectric interface, remaining the excitation of the travelling wave forbidden at larger photon energies.

The resonance energy for an optical coupler of dielectric constant ε_c and an incidence angle ϕ_i is found for $k_x^{'}(c / \omega) = \sqrt{\varepsilon_c}\sin\phi_i$. Thus, the resonance energy found depends on the values of ε_c and ϕ_i. On the other hand, the incidence angle must be compatible with the total internal reflection condition in the chosen optical coupler. Besides, it is important to notice that the use of a coupler with a ε_0 equal or lower than the ε_d would not satisfy the resonant condition since $k_x^{'}(c / \omega) > \sqrt{\varepsilon_c}\sin\phi_i$. In practice, those limitations produce a narrow range of incidence angles and a small variety of coupling materials to monitor the resonance.

Figure 2. Dispersion relation of surface polaritons on Au and Ag surfaces in (a) air and (b) water. The horizontal line shows an arbitrary value of the longitudinal component of the incident wave vector. The crossing point with the dispersion relations provides an estimation of the polariton energies.
Optical data of Au and Ag were extracted from Weaver, Krafka, Lynch, and Koch (1981) and Johnson and Christy (1972).

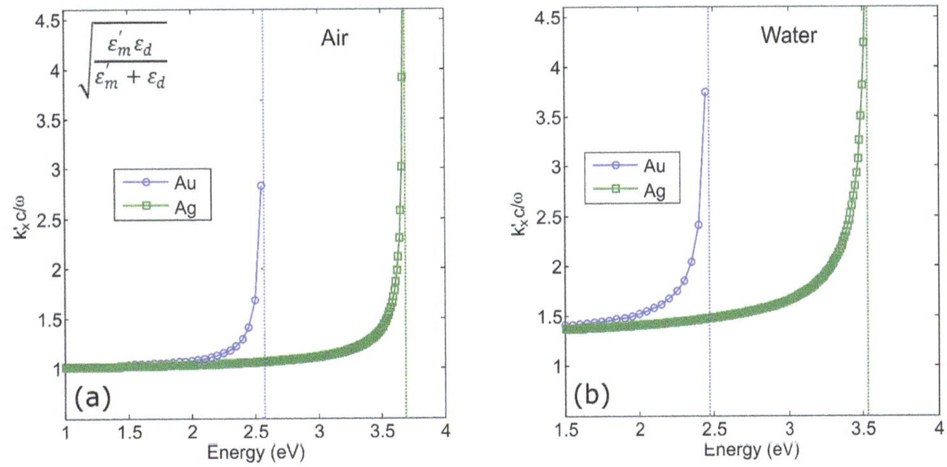

Comparing the dispersion relation in air and in water, it can be noted that the use of water will require a larger $k_x^{'}(c/\omega)$ than in air. This implies that the condition is satisfied at larger incidence angles for the same optical coupler. To prevent a large energy shift towards the mid-IR, it might be necessary to increase ε_c by using a denser material for the optical coupler.

Total Internal Reflection Ellipsometry

The electric field direction defines the state of polarization of an electromagnetic wave. When a light beam impinges a surface, the electric field vector can be decomposed in two components: a component that oscillates in the plane of incidence (*p* or *transverse magnetic* polarization) and a component that oscillates perpendicularly to that plane (*s* or *transverse electric* polarization). Spectroscopic ellipsometry analyzes the relative changes in amplitude and phase for both components as a function of the radiation energy. A broadband light source is used to produce a light beam whose polarization components are phase-modulated using a linear polarizer and an electro-optic modulator based on a piezoelectric crystal (Figure 3). The reflected beam is spectrally analyzed using a monochromator coupled to another linear polarizer called Analyzer.

Equations 1.1 and 1.3 are valid for a wave whose electric field vector oscillates in a plane that is perpendicular to the metal surface and contains the direction of propagation. Maxwell's equations predict that the existence of a surface polariton with an electric field vector oscillating on the metal surface transversely to the direction of propagation is forbidden (for details on the derivation, see Zayats, Smolyaninov, and Maradudin (2005)). For that reason, the excitation of plasmon polaritons only occurs under illumination with p-polarized light. This property makes spectroscopic ellipsometry very powerful in characterizing polaritonic structures, since the ellipsometric functions compare the amplitude and phase shifts of the reflected beams for both types of polarization. Hence, it is possible to record changes on

Figure 3. Spectroscopic ellipsometry setup for measurements in total internal reflection mode. The setup allows the determination of the ellipsometric parameters as a function of the radiation energy through the coupling with a monochromator

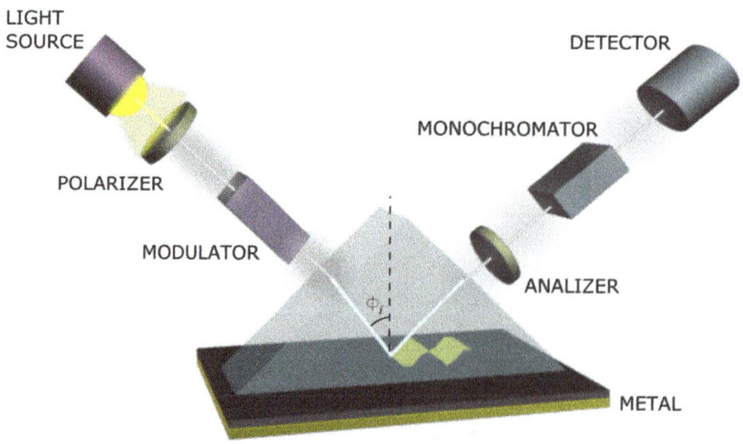

the p-polarization component while keeping the s-polarization component as a reference of the smooth variations of the reflectivity. The ellipsometric functions are defined as

$$\tan \psi = \frac{\left|R_p\right|}{\left|R_s\right|} \tag{1.9}$$

$$\Delta = \delta_p - \delta_s \tag{1.10}$$

where $\left|R_p\right|$ and $\left|R_s\right|$ are the amplitude reflection coefficients for p and s polarization. Likewise, in the Δ function, δ_p and δ_s are the phase shifts for the p- and the s-polarization components. The oscillating strain induced in the electro-optic modulator introduces a retardation between both p- and s-polarized electromagnetic waves. In this setup, the fundamental and second-harmonic signals in the detector can be synchronously measured by two separate lock-in amplifiers while the dc component is separated by a low-pass filters. The values for the three (dc, fundamental and second-harmonic) signals is used to extract ψ and Δ.

The reduction of the amplitude for p polarization at resonant conditions provokes the appearance of a characteristic V-shape in the spectrum of ψ as a function of the radiation energy. Figure 4(a) shows this behavior for a gold film of about 40 nm deposited on a glass slide and attached to a prism in the Kretschmann configuration.

Figure 4. (a) Variations of the ellipsometric functions around the resonant energy for different incidence angles. (b) Phasorial representation of the electric field components in the reflected beam

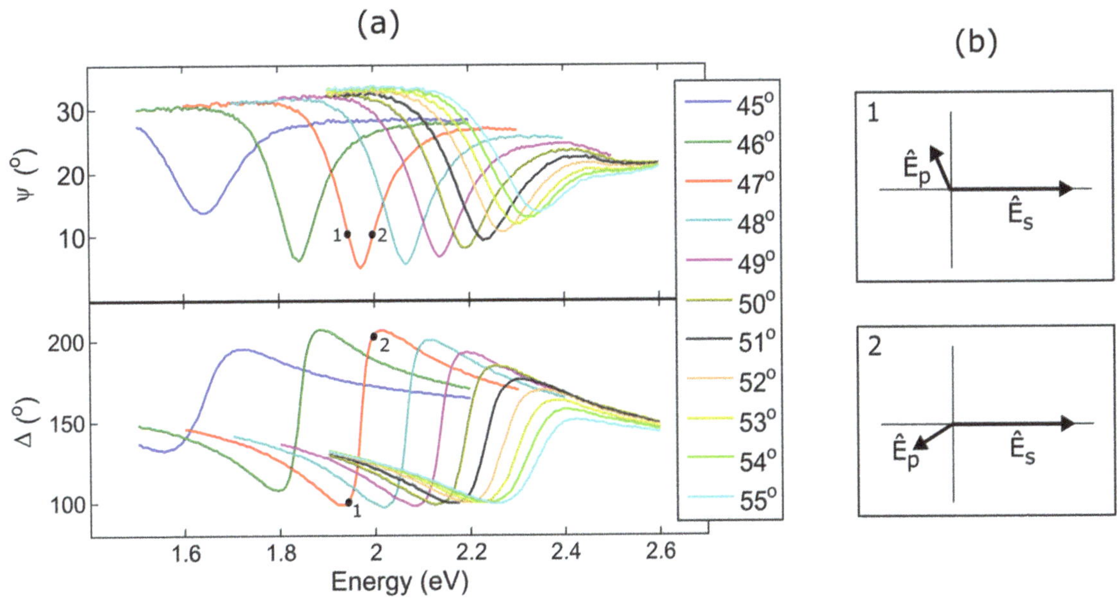

In comparison to regular SPR, ellipsometry can intrinsically provide information about the difference between the phase shifts for both types of polarizations through the Δ function. This is an important advantage of ellipsometry because the resonance produces a strong change of the phase shift difference. Figure 4(b) illustrates the phasorial representation of the p and s electric field components for the reflected beam at two energies near resonance. If the minimum of the ψ function is nullified completely by adjusting the thickness of the metal layer, the reflected beam becomes linearly polarized due to the suppression of the p component.

Taking into account the complex nature of the momentum vector ($k_x = k_x^{'} + ik_x^{''}$), the dispersion relation can be written as

$$k_x = \frac{\omega}{c}\left(\frac{\varepsilon_m \varepsilon_d}{\varepsilon_m + \varepsilon_d}\right)^{1/2} = k_x^{'} + ik_x^{''} = \frac{\omega}{c}\left(\frac{\varepsilon_m^{'} \varepsilon_d}{\varepsilon_m^{'} + \varepsilon_d}\right)^{1/2} + i\frac{\omega}{c}\left(\frac{\varepsilon_m^{'} \varepsilon_d}{\varepsilon_m^{'} + \varepsilon_d}\right)^{3/2} \frac{\varepsilon_m^{''}}{2\left(\varepsilon_m^{'}\right)^2} \tag{1.11}$$

This relation applies when the metal is assumed to be semi-infinite. The imaginary part accounts for the polariton damping as it propagates along the interface. As the intensity of the polariton decreases as $\exp(-2k_x^{''}x)$, the propagation length (the length at which the intensity decreases e^{-1}) can be estimated from $(2k_x^{''})^{-1}$. In the gold-air system, this propagation length can reach up to 10 μm at 2.0 eV.

In Kretschmann configuration, the polariton must travel through a metal semi-transparent layer before reaching the interface between the metal and the external dielectric. This can be treated as a perturbation ($k_{ST} = k_{ST}^{'} + ik_{ST}^{''}$) in the momentum vector of the polariton, which depends on the thickness and the preparation method of the metal layer (Arwin, Poksinski, & Johansen, 2008):

$$k_x = \frac{\omega}{c}\left(\frac{\varepsilon_m \varepsilon_d}{\varepsilon_m + \varepsilon_d}\right)^{1/2} + k_{ST} \tag{1.12}$$

Although the air gap between the metal and the prism is another source of perturbation in the Otto configuration, its effect is less important due to the non-absorbing properties of the pseudo-layer. Following the derivation by Raether (2013), it is possible to obtain the reflectance for a p-polarized incident beam of momentum k_x^{light}:

$$R_p = r_{0m,p} \frac{k_x^{light} - \left(k_x^{'} + k_{ST}^{'}\right) - i\left(k_x^{''} - k_{ST}^{''}\right)}{k_x^{light} - \left(k_x^{'} + k_{ST}^{'}\right) - i\left(k_x^{''} + k_{ST}^{''}\right)} \tag{1.13}$$

where $r_{0m,p}$ is the Fresnel complex-amplitude reflection coefficient for the interface between the optical coupler and the metal. The resonant condition is obtained when the wavelength and the incidence angle make that $k_x^{light} = k_x^{'} + k_{ST}^{'}$. It is worth noticing that to nullify R_p, the metal thickness and the incidence angle need to be adjusted to have $k_x^{''} = k_{ST}^{''}$. The $k_x^{''}$ term represents the internal polariton damping, and $k_{ST}^{''}$, the radiation damping as it tunnels through the semitransparent layer.

The values of the ψ and Δ functions around the resonance can be calculated from Equation 1.13, after introducing the amplitude reflection coefficient for the s-polarized component R_s. R_s is a smooth function around the resonance because of the impossibility of exciting the surface polariton. Thus, $\tan\psi$ and Δ can be written as:

$$\tan\psi = \frac{\left|r_{0m,p}\right|}{\left|R_s\right|}\sqrt{1 - \frac{4k_x^{''}k_{ST}^{''}}{Q^2 + \left(k_x^{''} + k_{ST}^{''}\right)^2}} \tag{1.14}$$

$$\Delta = \arg\frac{r_{0m,p}}{R_s} + \arctan\frac{2Qk_{ST}^{''}}{Q^2 + \left(k_x^{''} + k_{ST}^{''}\right)\left(k_x^{''} - k_{ST}^{''}\right)} \tag{1.15}$$

where Q is equal to $k_x^{light} - \left(k_x^{'} + k_{ST}^{'}\right)$. For the zero reflectance condition ($k_x^{''} = k_{ST}^{''}$), the Δ function reduces to:

$$\Delta = \arg\frac{r_{0m,p}}{R_s} + \arctan\frac{2k_{ST}^{''}}{k_x^{light} - \left(k_x^{'} + k_{ST}^{'}\right)} \tag{1.16}$$

The second term is a step function with a 180° shift around the resonance. According to Arwin et al. (2008), if a small mismatch ξ is introduced between $k_x^{''}$ and $k_{ST}^{''}$ ($\xi = k_x^{''} - k_{ST}^{''}$), the sensitivity of the Δ function to the resonant phenomenon can be estimated by calculating $d\Delta / dk_x^{light}$ at $Q = 0$. The result indicates that the derivative is approximately equal to ξ^{-1}. Since ξ can be minimized by optimizing the thickness of the metal film, this property represents a significant enhancement of the sensitivity in comparison to regular SPR methods whose sensitivity is proportional to $k_x^{'-1}$ (Yeatman, 1996), a parameter that depends on the physical properties of the metal layer and the dielectric, making its minimization difficult in a predetermined system.

Practical Use in Biosensing

Due to the narrow linewidth and the enhancement of the electric field at the metal surface found in resonant conditions, total reflection ellipsometry is a powerful technique in sensing molecules adsorbed on the surface. The field enhancement produced by the surface polariton at the metal/dielectric surface is given by the Fresnel transmission coefficient $\left|t_{cmd}^{p}\right|^2$ in the coupler/metal/dielectric system of the Kretschmann configuration. This coefficient is defined as the ratio between the magnetic field intensity on the metal surface at the dielectric side and the magnetic field intensity of the incoming light at the coupler/metal interface. Thus, its value is given by

$$\left|t_{cmd}^{p}\right|^{2}=\frac{\left|H_{dm}\right|^{2}}{\left|H_{cm}\right|^{2}}=\frac{t_{cm}^{p}t_{md}^{p}\exp(ik_{z}d_{m})}{1+r_{cm}r_{md}\exp(2ik_{z}d_{m})} \tag{1.17}$$

where t_{ik}^{p} and r_{ik}^{p} are the corresponding transmission and reflection coefficients for the binary systems and d_m is the thickness of the metal layer. The maximum of the electric field amplitude shifts from the coupler/metal interface to the metal/dielectric interface when the resonant condition is fulfilled. The maximum electric field enhancement achievable at the surface can be calculated from:

$$T_{\max}^{el}=\left(\frac{\left|E_{dm}\right|^{2}}{\left|E_{cm}\right|^{2}}\right)_{\max}=\frac{\varepsilon_{d}}{\varepsilon_{c}}\left(\frac{\left|H_{dm}\right|^{2}}{\left|H_{cm}\right|^{2}}\right)_{\max}=\frac{1}{\varepsilon_{d}}\frac{2\left|\varepsilon_{m}^{'}\right|^{2}}{\varepsilon_{m}^{''}}\frac{a}{1+\left|\varepsilon_{m}^{'}\right|} \tag{1.18}$$

with $a^{2}=\left|\varepsilon_{m}^{'}\right|\left(\varepsilon_{c}-1\right)-\varepsilon_{c}$. Table 1 shows the values for different metals at λ=600 nm in air ($\varepsilon_{d}=1$) by using a quartz optical coupler ($\varepsilon_{c}=2.13$).

Although the thickness and roughness in all these metals can be optimized to nullify the R_p component in the ellipsometric experiments, the sensitivity to the surface adsorption of molecules depends on the metal used, being more favorable the use of a metal with a large electric field enhancement and a resonance with a narrow linewidth. Ag is the metal that shows the largest electric field enhancement but its use is limited by their chemical instability in air and in aqueous environments (Manickam et al., 2012).

The most common metal used in SPR experiments is Au. Some advantages of this metal are:

- A SPR signal at convenient reflectance angles and wavelengths for BK7 or fused silica couplers, enabling the use of cost-effective visible gas lasers or solid-state light emitters.

Table 1. Maximum electric field enhancement for different metals at 600 nm and 275 nm. The last column includes the source of the optical constants for each material. Au, Al and Pt constants were taken from database of the WVASE software. Ga constants were obtained from in-house measurements

	$T_{\max}^{el}$		
	600 nm	**275 nm**	
Ag	239	--	Taft and Philipp (1961)
Au	57.7	--	WVASE database
Al	44.5	42.4	WVASE database
Cu	15.9	--	Palik (1998)
Ga	1.34	3.92	-
Pt	8.1	1.87	WVASE database

- Chemically inertness to solutions and solutes.
- Large resilience to surface oxidation.
- Wide variety of methods and functional groups available for bio-functionalization of the surface (Tiwari, Vig, Dennis, & Singh, 2011).

The regular thickness of the Au layer in commercial SPR chips is around 50 nm. An important disadvantage of Au is that the adherence to transparent dielectrics like quartz or glass requires the use of a thin film of Cr or Ti. Due to the high absorbing properties of those layers, there is a strong reduction of the electric field intensity when the light tunnels through the thin film, broadening the reflectance dip and causing the attenuation of the step in the Δ function. This perturbation can be reduced if a transparent conductive oxide like indium tin oxide or Al-doped ZnO is used as intermediate layer, improving the amplitude of the step in Δ (García-Marín et al., 2014). Thus, a shift close to the theoretical value of 180° can be achieved along with the full suppression of the p component. Although the amplitude of the step can be also affected by the roughness of the intermediate layer, a surface roughness of a few nanometers is acceptable to get a 180° step in Δ. Chang et al. (2010) also showed an improvement of the sensitivity of SPR for the detection of tumor markers through the use of a ZnO intermediate layer between the Au film and the coupling prism.

The strong singularity found in the delta function can be used for biosensing purposes. Figure 5 shows the effect of a self-assembled monolayer of glutathione polyclonal antibody on the surface of the Au layer. As the formation typically occurs in aqueous media ($\varepsilon_d = 1.77$) and the organic compounds tend to have a real dielectric constant near 2.25, the SAM formation produces an effective increase of the dielectric permittivity and a redshift in the polariton energy. This shift can be seen as the effect of the charges introduced by the organic monolayer whose dipolar moments compensate the longitudinal oscillations on the metal surface. This property can be used to monitor in real time the formation of self-assembled monolayers on the Au surface to build biosensors based on the specific recognition of analytes using antigen-antibody complexes or single-stranded DNA chains (Chang et al, 2010). As the air permittivity is significantly lower than the water permittivity, the formation of the SAM is more evident after sample drying.

ELLIPSOMETRIC NANOPLASMONICS

Localized Surface Plasmon Resonance

A good method to improve the sensitivity of ellipsometry in biosensing applications is the excitation of localized surface plasmon resonances (LSPRs) in metal nanostructures. The surface modifications of these nanoparticles produce a large change in the ellipsometric functions under the appropriate experimental conditions. To interpret the ellipsometric changes, it is necessary to deepen into the basics of plasmonics in nanoparticles.

LSPRs emerge as collective oscillatory processes of free charges in the metal nanoparticles. Their energies depend on the nanoparticle size and the material of the nanoparticle. The morphology of the nanoparticles, the interparticle coupling and the interactions with the environment induce energy differences between the resonances of the p- and s-polarizations that are revealed in the ellipsometric

Figure 5. Ellipsometric functions measured in TIRE around the resonant energy before and after antibody functionalization

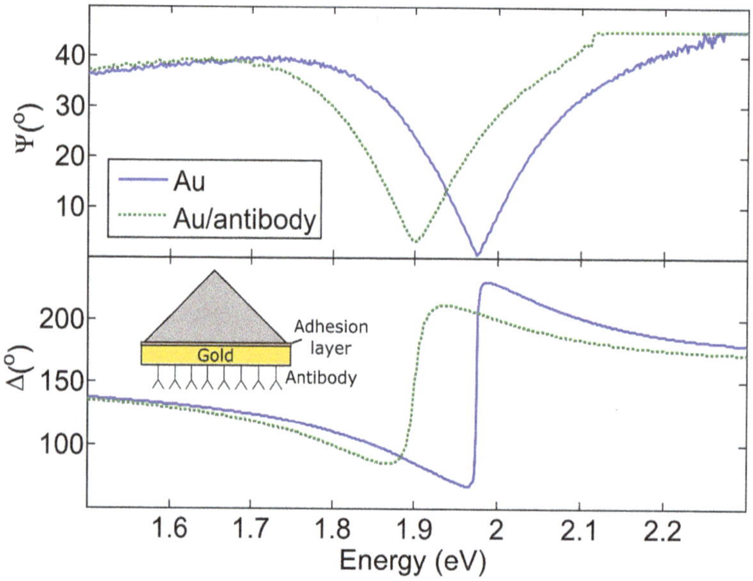

spectra. This section describes the effect of resonant charge oscillations on the dielectric function of metal nanostructures using Gaussian units. In an isotropic material, the dielectric function is a scalar and the polarization vector induced during the dielectric response of the material to an electrical stimuli is parallel to the electric field direction with a coefficient of proportionality χ_e known as the electrical susceptibility of the medium:

$$\vec{P} = \chi_e(\omega)\vec{E} \tag{2.1}$$

The relationship between χ_e and $\varepsilon(\omega)$ can be found in the relationship between the displacement vector $\vec{D}$ and the electric field $\vec{E}$:

$$\vec{D}(\omega) = \varepsilon(\omega)\vec{E}(\omega) = \vec{E}(\omega) + 4\pi\vec{P}(\omega) = (1 + 4\pi\chi_e(\omega))\vec{E}(\omega) \tag{2.2}$$

$$\varepsilon(\omega) = 1 + 4\pi\chi_e(\omega) \tag{2.3}$$

In the most general case, the motion of a particle q_i in a periodic electric field is governed by the Newton's second law of motion:

$$\vec{F}_i = q_i\vec{E}e^{-i\omega t} - b\frac{d\vec{r}_i}{dt} - K\left(\vec{r}_i - \vec{r}_{i0}\right) = m_i\frac{d^2\vec{r}_i}{dt^2} \tag{2.4}$$

where the friction and restoring forces are included. The b parameter describes friction losses and K is the spring constant of the restoring force. Assuming that $\vec{r_i} = \vec{r_{i0}} + \Delta\vec{r_i}e^{-i\omega t}$, where $\vec{r_{i0}}$ is the equilibrium position and $\Delta\vec{r_i}$ is the excursion about it, then:

$$\Delta\vec{r_i} = \frac{q_i}{K - m_i\omega^2 - i\omega b}\vec{E} \tag{2.5}$$

In a collective oscillation in which all the charges are equal to q_i and have a m_i mass, the polarization vector can be obtained from the summation of the total number of dipoles, denoted by n:

$$\vec{P} = \frac{1}{V}\sum_i q_i\Delta\vec{r_i} = \frac{nq^2}{K - m\omega^2 - i\omega b}\vec{E} \tag{2.6}$$

$$\chi_e(\omega) = \frac{nq^2}{K - m\omega^2 - i\omega b} \tag{2.7}$$

and the dielectric function becomes:

$$\varepsilon(\omega) = 1 + \frac{4\pi nq^2 / m}{\frac{K}{m} - \omega^2 - i\omega\frac{b}{m}} = 1 + \frac{\omega_p^2}{\omega_0^2 - \omega^2 - i\omega\gamma} \tag{2.8}$$

This is the most general expression of $\varepsilon(\omega)$ for a resonant process in a Lorentz oscillator. The ω_0, ω_p and γ parameters are known as the resonant frequency, the plasma frequency and the broadening parameter, respectively, and take the following expressions:

$$\omega_0 = \sqrt{\frac{K}{m}} \qquad \omega_p = \sqrt{\frac{4\pi nq^2}{m}} \qquad \gamma = \frac{b}{m} \tag{2.9}$$

It is common to add a corrective constant ε_∞ that accounts for the background electron scattering at high frequency:

$$\varepsilon(\omega) = \varepsilon_\infty + \frac{\omega_p^2}{\omega_0^2 - \omega^2 - i\omega\gamma} \tag{2.10}$$

If the restoring force is not included ($K = 0$), the expression of the dielectric function given in Equation 2.8 becomes the Drude model that explains the contribution of free electrons in a metal. The restoring force in nanoparticles is needed to take into account the effect of the ions left behind as the free charges oscillate in the nanoparticle. The value of this force and the related term ω_0 increases as the particle size becomes smaller.

Figure 6 (a) shows the real and imaginary parts of the dielectric function as a function of energy for a Lorentz oscillator with a single resonance. A maximum in the imaginary part occurs when the energy of the stimuli matches the resonant energy $\hbar\omega_0$. For small nanoparticles, the restoring force increases and the resonance shifts towards larger energies. The real part of the dielectric function takes a value of one at $\hbar\omega_0$ and the real part of the susceptibility changes its sign. Thus, in the vicinity of the resonance, the nanoparticle changes its optical behavior from a material with a positive dielectric constant (dielectric) to a material with a negative dielectric constant (metal). This behavior differs from that shown by bulk metals that, according to the Drude model, present high absorbance and a negative dielectric constant in the IR and visible ranges, only becoming transparent at high photon energies where they exhibit positive dielectric constants (Figure 6(b)).

Anisotropies in the Dielectric Function of the Nanoparticles

The real and imaginary parts of the dielectric function for a Lorentz oscillator can account for the optical behavior of the nanoparticles in the ellipsometric analysis. Thus, the dielectric function can be used to fit the ellipsometric curves along with effective medium approximations that help to take into account the discrete volume of the nanoparticles. However, the parameters of the functions can be affected by anisotropies induced by the shape of the nanoparticle, the interaction with the substrate, or the interparticle coupling, which makes necessary the use of a dielectric function tensor (Wu et al., 2007). The anisotropies can yield different resonance energies for p- and s-polarized incident light, provoking char-

Figure 6. (a) Real and imaginary parts of the dielectric function around a resonant oscillatory process. The maximum value of ε'' occurs at the resonant energy $\hbar\omega_0 = 2.48\ eV$. (b) Drude model for an oscillatory process of equal plasma frequency and damping parameter

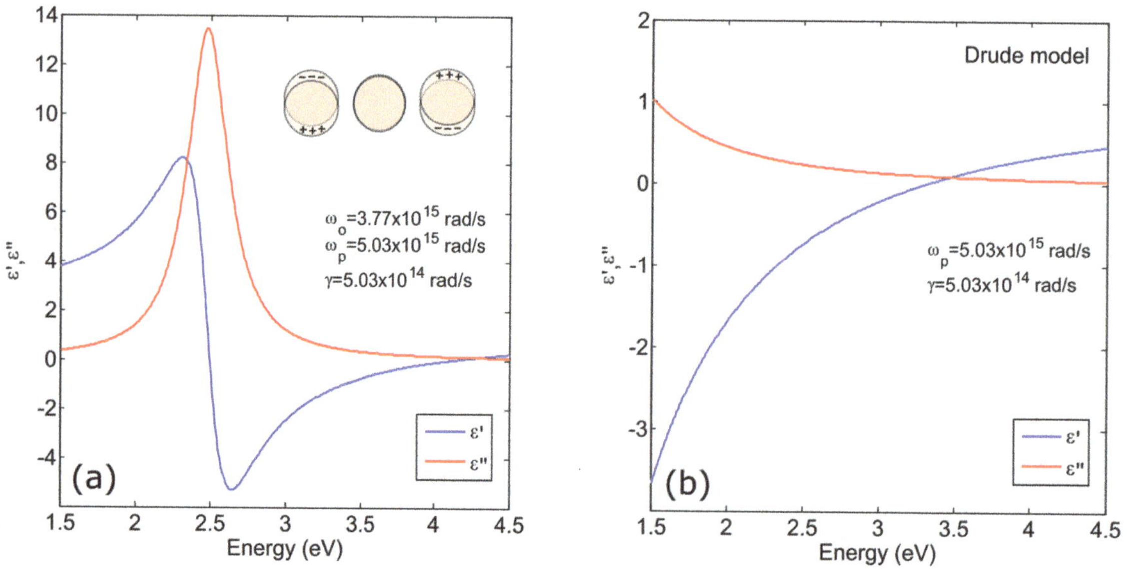

acteristic features in the ellipsometric functions (Lodewijks, Van Roy, Borghs, Lagae, & Van Dorpe, 2012). To consider these anisotropies, it is common to introduce depolarization factors that account for the polarizability of the particle in planes parallel and perpendicular to the substrate plane (Noguez, 2007). If the anisotropy is uniaxial, that approach provides two components of the dielectric tensor $\varepsilon_{\perp}$ (out-of-plane) and $\varepsilon_{\parallel}$ (in-plane) that represent the optical properties of the nanoparticle film along the directions perpendicular and parallel to the optic axis of the film. In spherical or hemispherical shaped particles, that axis is perpendicular to the substrate plane. Thus, for s-polarized light, the ordinary refractive index extracted from $\varepsilon_{\parallel}$ applies, whereas, for p-polarized light, the reflection coefficient depends on both in-plane (ordinary) and out-of-plane (extraordinary) refractive index extracted from $\varepsilon_{\parallel}$ and $\varepsilon_{\perp}$, respectively. Furthermore, if the anisotropic system is biaxial, the dielectric tensor will have three different terms along the principal axis. For the sake of simplicity, the ellipsometric analysis in this chapter will focus on uniaxial anisotropic systems. For further information on the analysis of biaxial systems, the reader can consult Azzam, Bashara, and Burns (1987). Another approach will be reviewed in Section 3.4 to deal with in-plane anisotropies.

The discrete dipole approximation (DDA) is a numerical method, based on the decomposition of a nanoparticle in a finite array of point dipoles, which enables the calculation of the absorption (Q_{abs}), scattering (Q_{sca}), and extinction efficiencies ($Q_{ext}=Q_{abs}+Q_{sca}$) for an specific metal and particle geometry (Draine & Flatau, 2013). Furthermore, it is possible to calculate the phase lag efficiency of a plane wave after propagating a distance along the nanoparticle. The polarization of the incident plane wave is an input parameter, and the particle can be arbitrarily oriented relative to the incident radiation. The technique is useful to calculate the plasmon resonance energy as a function of the light polarization (Albella et al., 2011). The extinction coefficient (imaginary part of the refractive index) can be obtained from the extinction efficiency since they are linked by an exact analytical relationship (Near, Hayden, Hunter, Thackston, & El-Sayed, 2013; Yurkin, 2014). The information that DDA provides on the phase lag and the relative extinction efficiency of p-polarized and s-polarized light beams helps to interpret the ellipsometric results and the origin of some of the features found in the ψ and Δ functions.

Once obtained $\varepsilon_{\perp}$ and $\varepsilon_{\parallel}$ constants from numerical or analytical methods, it is possible to find out the ellipsometric functions of the system adopting the model of an anisotropic film. For an isotropic ambient and substrate with refractive indexes N_0 and N_2, and a film with ordinary and extraordinary indexes $N_{1o} = \sqrt{\varepsilon_{\parallel}}$ and $N_{1e} = \sqrt{\varepsilon_{\perp}}$, the reflection coefficients of the three-phase (ambient-metal NP film-substrate) system can be obtained as

$$R_p = \frac{r_{01p} + r_{12p} e^{-i2\beta_p}}{1 + r_{01p} r_{12p} e^{-i2\beta_p}} \tag{2.11}$$

$$R_s = \frac{r_{01s} + r_{12s} e^{-i2\beta_s}}{1 + r_{01s} r_{12s} e^{-i2\beta_s}} \tag{2.12}$$

where r_{01p}, r_{12p} and r_{01s}, r_{12s} are the reflection coefficients at the ambient-film and film-substrate interfaces for p and s polarizations, respectively. It is important to notice that the phase thicknesses β_p and β_s are different for both polarizations and can be expressed as

$$\beta_p = 2\pi\left(\frac{d_1}{\lambda}\right)\left(\frac{N_{1o}}{N_{1e}}\right)\left(N_{1e}^2 - N_0^2\sin^2\phi_i\right)^{1/2} \tag{2.13}$$

$$\beta_s = 2\pi\left(\frac{d_1}{\lambda}\right)\left(N_{1o}^2 - N_0^2\sin^2\phi_i\right)^{1/2} \tag{2.14}$$

where d_1 is the effective thickness of the nanoparticle layer and λ is the free-space wavelength. The reflection coefficients can be obtained as follows:

$$r_{01p} = \frac{N_{1o}N_{1e}\cos\phi_i - N_0(N_{1e}^2 - N_0^2\sin^2\phi_i)^{1/2}}{N_{1o}N_{1e}\cos\phi_i + N_0(N_{1e}^2 - N_0^2\sin^2\phi_i)^{1/2}} \tag{2.15}$$

$$r_{12p} = \frac{-N_{1o}N_{1e}\cos\phi_s + N_2(N_{1e}^2 - N_2^2\sin^2\phi_s)^{1/2}}{N_{1o}N_{1e}\cos\phi_s + N_2(N_{1e}^2 - N_2^2\sin^2\phi_s)^{1/2}} \tag{2.16}$$

$$r_{01s} = \frac{N_0\cos\phi_i - (N_{1o}^2 - N_0^2\sin^2\phi_i)^{1/2}}{N_0\cos\phi_i + (N_{1o}^2 - N_0^2\sin^2\phi_i)^{1/2}} \tag{2.17}$$

$$r_{12s} = \frac{-N_2\cos\phi_s + (N_{1o}^2 - N_2^2\sin^2\phi_s)^{1/2}}{N_2\cos\phi_s + (N_{1o}^2 - N_2^2\sin^2\phi_s)^{1/2}} \tag{2.18}$$

The ϕ_s angle is the angle of refraction in the substrate and it is related to the incidence angle ϕ_i by Snell's law:

$$N_0\sin\phi_i = N_2\sin\phi_s \tag{2.19}$$

Effective Medium Theories

A system with nanoparticle sizes lower than the wavelength of the incident light embedded in a homogeneous dielectric matrix is optically equivalent to a homogeneous material whose effective dielectric function is dependent on the dielectric properties of the metal and the matrix. Effective medium theories provide a way to estimate the dielectric function of such systems. For a layer formed by nanoparticles, there are two main theories whose applicability depends on the morphology and the density of the nanoparticles in the matrix: the Maxwell-Garnett theory and the Yamaguchi theory (Tonova et al., 1999).

As discussed in the previous section, the film formed by the nanoparticles can be anisotropic, allowing to distinguish different dielectric functions for the directions parallel and perpendicular to the film plane if the anisotropy is uniaxial and the optical axis is perpendicular to that plane. Those dielectric functions can be analytically obtained from:

$$\varepsilon_{\parallel} - \varepsilon_m = \alpha_{\parallel} W \tag{2.20}$$

$$\frac{1}{\varepsilon_m} - \frac{1}{\varepsilon_{\perp}} = \frac{\alpha_{\perp} W}{\varepsilon_m^2} \tag{2.21}$$

where $\alpha_{\parallel}$ and $\alpha_{\perp}$ are the polarizabilities of the individual nanoparticles in the parallel and perpendicular directions, ε_m is the dielectric function of the matrix, and W is the metal filling factor.

In the Maxwell-Garnet theory, the polarizabilities depend on the dielectric function of the matrix, the dielectric function of the nanoparticle material ε_p, which can be approximated by the Drude model, and the effective depolarization factors, as follows:

$$\alpha_{\parallel,\perp} = \frac{\varepsilon_m}{F_{\parallel,\perp} + \varepsilon_m / (\varepsilon_p - \varepsilon_m)} \tag{2.22}$$

The depolarization factors comprise two terms:

$$F_{\parallel,\perp} = L_{\parallel,\perp} + Y_{\parallel,\perp} \tag{2.23}$$

The first term ($L_{\parallel,\perp}$) is a geometrical term determined by the shape of the nanoparticle, whereas the second term introduces the coupling effects between nanoparticles ($Y_{\parallel,\perp}$).

The Maxwell-Garnett theory is based on the assumption that the nanoparticles are spherical and they are three-dimensionally distributed along the layer with Lorentz-Lorenz interactions among them. In that case, the depolarization factors become:

$$F_{\parallel} = \frac{1}{3} - \frac{W}{3} \tag{2.24}$$

$$F_{\perp} = \frac{1}{3} + \frac{W}{3} \tag{2.25}$$

On the other hand, the Yamaguchi approximation is developed for a two-dimensional distribution of ellipsoidal particles on the interface between two semi-infinite media (Yamaguchi, Takahashi, & Sudoh, 1978). Assuming a static-field dipole-dipole interaction between nanoparticles and between nanoparticles and substrate, the polarizabilities can be expressed as

$$\alpha_{\parallel,\perp} = \frac{2\varepsilon_m\varepsilon_s}{\varepsilon_m + \varepsilon_s} \frac{\varepsilon_m}{F_{\parallel,\perp} + \varepsilon_m / \left(\varepsilon_p - \varepsilon_m\right)} \tag{2.26}$$

where ε_s is the dielectric function of the substrate.

The Yamaguchi model introduces the effect of image dipoles in the substrate and the neighboring nanoparticles through the anisotropic polarization factor. Thus, the depolarization factors depend on the average interparticle distance (a) and the distance between the nanoparticles and the substrate. They can be obtained from

$$F_{\parallel} = L_{\parallel} + \frac{\zeta^2}{24\eta^3} \frac{\varepsilon_m - \varepsilon_s}{\varepsilon_m + \varepsilon_s} - \frac{\pi^2}{24} \frac{2\varepsilon_m}{\varepsilon_m + \varepsilon_s} \frac{d_m}{a} \tag{2.27}$$

$$F_{\perp} = L_{\perp} + \frac{\zeta^2}{12\eta^3} \frac{\varepsilon_m - \varepsilon_s}{\varepsilon_m + \varepsilon_s} - \frac{\pi^2}{12} \frac{2\varepsilon_m}{\varepsilon_m + \varepsilon_s} \frac{d_m}{a} \tag{2.28}$$

where ζ represents the ratio between the principal axes of the ellipsoid and η is the distance between the point-dipole moment of a particle and its mirror image in the substrate, divided by the particle height.

The analytical calculation of $\varepsilon_{\parallel}$ and $\varepsilon_{\perp}$ through those models allow their use in the fitting of the ellipsometric experimental curves for ideal cases (Aspnes, 1975). However, in systems with randomly distributed nanoparticles, the determination of the geometrical parameters can be troublesome, hindering their applicability.

Anisotropic Surface Excess Function

Verre et al. (2011) proposed to study the response of ultrathin plasmonic layers (below 10 nm) using the quantity named anisotropic surface excess function (ASEF), that can be extracted directly from the measured spectroscopic ellipsometry signal. This quantity depends only on the response of the plasmonic layer alone and can be modeled using analytical methods. In systems with in-plane anisotropies, ASEF can depend on the azimuthal angle.

In multilayer structures, where multiple reflections between single layers take place, the analytical modeling of the dielectric properties of each single layer is not additive in general. But in certain systems with nanoparticles (NP) localized on the substrate surface, the pseudo-dielectric function extracted from the ellipsometric measurements can be approximated by the addition of the dielectric functions for the NP layer and the substrate. In those cases, the ASEF can be extracted from the experimental data through

$$ASEF_{\exp,j} = \frac{i\lambda\left(\varepsilon_s - 1\right)}{4\pi\varepsilon_s\sqrt{\varepsilon_s - \sin^2\phi_i}}\left(\left\langle\varepsilon_j\right\rangle - \varepsilon_s\right) \tag{2.29}$$

where $\left\langle \varepsilon_j \right\rangle$ is the pseudodielectric function for one of the in-plane principal axis $j = (x, y)$. The z direction follows the direction of the normal vector.

The experimental ASEF obtained for x and y can be fitted through the use of Lorentzian dielectric functions for the nanoparticle layer in

$$ASEF_j = d\left[\left(\Delta\varepsilon_j - \varepsilon_b\Delta\frac{1}{\varepsilon_z}\right) + \left(\frac{1}{\varepsilon_b} - \frac{\cos^2\phi_i}{\sin^2\phi_i}\right)\left(\Delta\varepsilon_j - \Delta\varepsilon_k\right)\right] \tag{2.30}$$

where d is the effective thickness of the nanoparticle layer, k is the transverse in-plane direction, and $\Delta\varepsilon_{j,k}$ and $\Delta\left(1/\varepsilon_z\right)$ represent the difference between the dielectric functions of the nanoparticle layer and the substrate in each direction. Through numerical analysis, the parameters of the Lorentzian dielectric functions for the nanoparticle layer can be extracted from the fitting of $ASEF_{\exp,j}$.

The previous ASEF functions are valid for a three phase system (substrate, plasmonic layer, and air) in which the NP layer is thin, the substrate optical response is isotropic, and the optical axes are aligned with the high symmetry crystallographic axis of the system. The formalism can be extended to multilayer structures, magnetically active materials, and anisotropic substrates using a transfer matrix (Verre et al., 2013). Formalism for this approach was formulated by Schubert (1996), who develop a complete algebra to calculate all the measurable polarization dependent parameters for arbitrarily anisotropic and homogeneous layered systems, derived from the original treatment for isotropic media. A four-vector is defined for the p and s components of the electric and magnetic fields; then, the effect of each layer can be represented by a 4×4 partial transfer matrix. By knowing the thickness and the dielectric function of each layer, the partial transfer matrix of each individual layer can be calculated and accumulated to derive the full response of the system. Thus, an expression of the complex reflection coefficient can be obtained, and compared to the experiment.

External Reflection Ellipsometry on Metal Nanostructures

Although the introduction of analytical methods for the determination of the anisotropic dielectric functions is useful, in practice, it is difficult to generalize the methods for any nanoparticle morphology and distribution. In spectroscopic ellipsometry, one possible solution to fit the ellipsometric behavior as a function of energy at a given incidence angle is to build $\varepsilon_{\|}$ and $\varepsilon_{\perp}$ functions from Lorentzian resonances, after numerical analysis using methods such as finite-difference time-domain or discrete dipole approximation. Both methods enable the determination of the absorption efficiency from the incidence angle and the polarization state, making possible the modelling of the optical response for a single nanoparticle or multiple nanoparticle arrays through Lorentzian functions that account for the in-plane and out-of-plane oscillatory processes. The in-plane dielectric function ($\varepsilon_{\|}$) can be obtained from the absorption efficiency of s-polarized light, whereas the out-of-plane function ($\varepsilon_{\perp}$) can be estimated at grazing angles using p-polarized light. Absorption efficiency is proportional to the imaginary part of the dielectric function enabling the determination of the real part of the dielectric constant through the relationship between the imaginary and the real parts of the Lorentzian function. For cases with a single resonance in each oscillatory plane, the resultant functions will have the following form:

$$\varepsilon_{\|}\left(\omega\right) = \varepsilon_{\infty} + \frac{\omega_{p,\|}^{2}}{\omega_{0,\|}^{2} - \omega^{2} - i\omega\gamma_{\|}} \tag{2.31}$$

$$\varepsilon_{\perp}\left(\omega\right) = \varepsilon_{\infty} + \frac{\omega_{p,\perp}^{2}}{\omega_{0,\perp}^{2} - \omega^{2} - i\omega\gamma_{\perp}} \tag{2.32}$$

In order to compare the effects of a Lorentzian resonance model with a simple Drude model, calculations for an ambient/film/silicon substrate system are presented using films with three different optical behaviors:

1. A dielectric function following the Drude model with $\omega_p = 5.03 \times 10^{15}$ rad/s ($\hbar\omega_p = 3.31$ eV) and $\gamma = 5.03 \times 10^{14}$ rad/s ($\hbar\gamma = 0.33$ eV).
2. An isotropic ($\varepsilon_{\|} = \varepsilon_{\perp}$) dielectric function following a Lorentzian model with $\omega_0 = 6.08 \times 10^{15}$ rad/s (= 4.0 eV), $\omega_p = 6.84 \times 10^{15}$ rad/s ($\hbar\omega_p = 4.5$ eV) and $\gamma = 1.52 \times 10^{15}$ rad/s ($\hbar\gamma = 1.0$ eV).
3. An anisotropic dielectric function with $\omega_{0,\|} = 6.08 \times 10^{15}$ rad/s ($\hbar\omega_{0,\|} = 4.0$ eV), $\omega_{p,\|} = 8.36 \times 10^{15}$ rad/s ($\hbar\omega_{p,\|} = 5.5$ eV), and $\gamma_{\|} = 1.52 \times 10^{15}$ rad/s ($\hbar\gamma_{\|} = 1.0$ eV), for the in-plane resonance, and $\omega_{0,\perp} = 3.34 \times 10^{15}$ rad/s ($\hbar\omega_{0,\perp} = 2.2$ eV), $\omega_{p,\perp} = 6.84 \times 10^{15}$ rad/s ($\hbar\omega_{p,\perp} = 4.5$ eV) and $\gamma_{\perp} = 6.84 \times 10^{14}$ rad/s ($\hbar\gamma_{\perp} = 0.45$ eV), for the out-of-plane resonance. The calculation is based on the anisotropic analysis shown through Equations 2.11-2.19.

For the sake of comparison, in all cases, the film thickness and the high-frequency dielectric constant ε_{∞} are assumed to be 60 nm and 1, respectively.

a. Drude Model Layer

In external ellipsometry, the maximum sensitivity to the deposition of overlayers on a substrate is reached under experimental conditions that produce total extinction of one of the components of the polarization. Then, the incidence angle is commonly called Brewster's angle and makes the psi function values 0 or 90°, depending on the nulled polarization component (Azzam et al., 1987). In comparison to dielectric materials, the external reflection of light on planar metal substrates is less dependent on the polarization due to their high absorbance. Only in the UV region, where the imaginary part of the dielectric constant is lower, they can exhibit a pseudo-Brewster's or polarizing angle. Figure 7 shows the ellipsometric functions as the incidence angle is varied between 30 and 70° for a material with Drude optical constants ($\omega_p = 5.03 \times 10^{15}$ rad/s, $\gamma = 5.03 \times 10^{14}$ rad/s). The differences between s and p polarizations are mainly revealed above the plasma frequency, which in this case corresponds to an energy value of $\hbar\omega_p$ =3.31 eV. In that range, the real part of the dielectric constant is positive and the material behaves as a dielectric. Hence, the monotonous behavior observed in the "metallic" range provides a poor scenario for sensing purposes in ellipsometric setups.

b. Isotropic Resonant Layer

Figure 8 shows the effects of introducing an isotropic film with a single Lorentzian resonance on a silicon substrate that can resemble the case of a metal nanostructured layer (Yamaguchi et al., 1978). The p-polarization component is almost suppressed at energies below the resonance and plasma energies, where the real part of the dielectric function is positive. The suppression of the p component in the re-

Figure 7. Calculation of the ellipsometric functions in an air-film-substrate system with a film dielectric response based on the Drude model ($\hbar\omega_p$ = 3.31 eV, $\hbar\gamma$ = 0.33 eV)

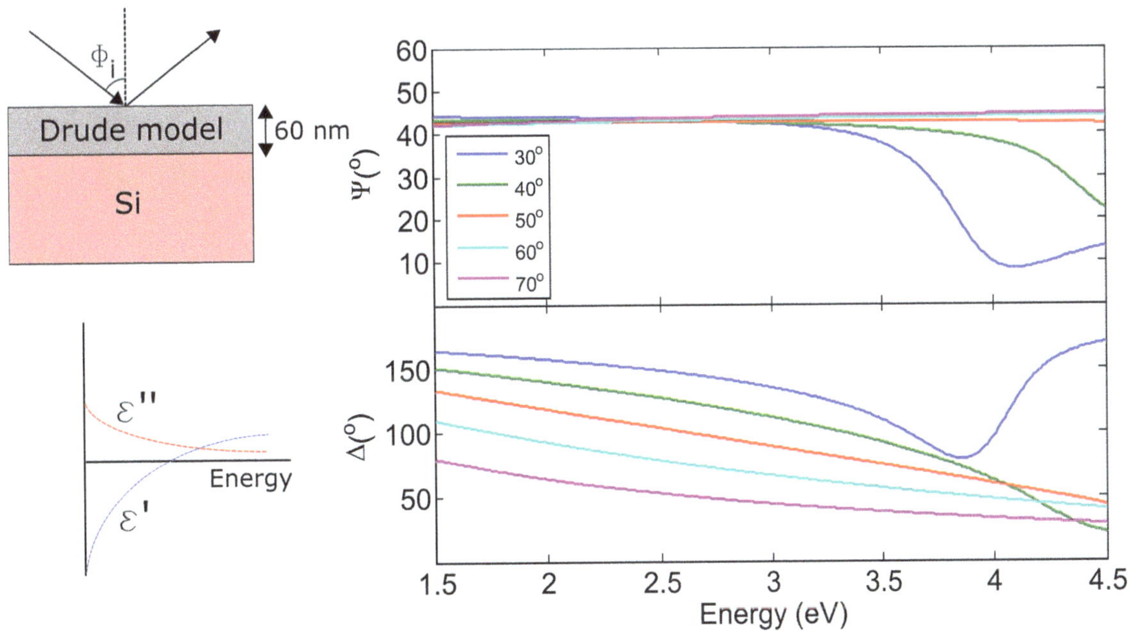

flected beam leads the ψ function to take values close to zero at energies between 3.1 and 3.5 eV. In addition, a strong angle shift is found in the Δ function at the same energy. Interestingly, the unwrapped phase shift for Δ is negative for 45, 50 and 55° but positive for the rest of angles. In addition, the maximum amplitude of the angle shift is close to ±180°. From a mathematical viewpoint, that behavior evidences the existence of a singularity in the Δ function caused by the suppression of the p component and the matching condition between the light wave vector in the plane of incidence and the real part of the momentum for the electron oscillations in the nanoparticles. This condition can be also used in bio-sensing experiments since the momentum of the electron oscillations will be greatly influenced by the surface charges on the nanoparticles.

As in TIRE mode, the full suppression of p polarization occurs at a critical thickness of the nano-structured layer that depends on the resonance frequency, the plasma frequency, the broadening, and the optical properties of the substrate. However, the variation of the minimum of ψ with the incidence angle is low in comparison to TIRE conditions. Therefore, the sensitivity to adsorbed surface species is handicapped in comparison to TIRE due to the lower value of $d\Delta / dk_x^{light}$.

c. Anisotropic Resonant Layer

This is the most common scenario in systems formed by nanoparticle layers since the non-spherical morphology or the interactions with the surrounding optical media produce differences in the different components of the dielectric function tensor. An example of this case is the coating of a substrate with Ga nanodroplets that have a hemispherical morphology causing large differences between $\varepsilon_\perp$ and $\varepsilon_\parallel$

Figure 8. Calculation of the ellipsometric functions in an air-film-substrate system with an isotropic resonant layer (RL) as intermediate film. The resonance, plasmon and broadening energies of the Lorentz oscillator are 4.0, 4.5 and 1.0 eV, respectively. In the Δ plot, the phase is unwrapped for the sake of clarity.

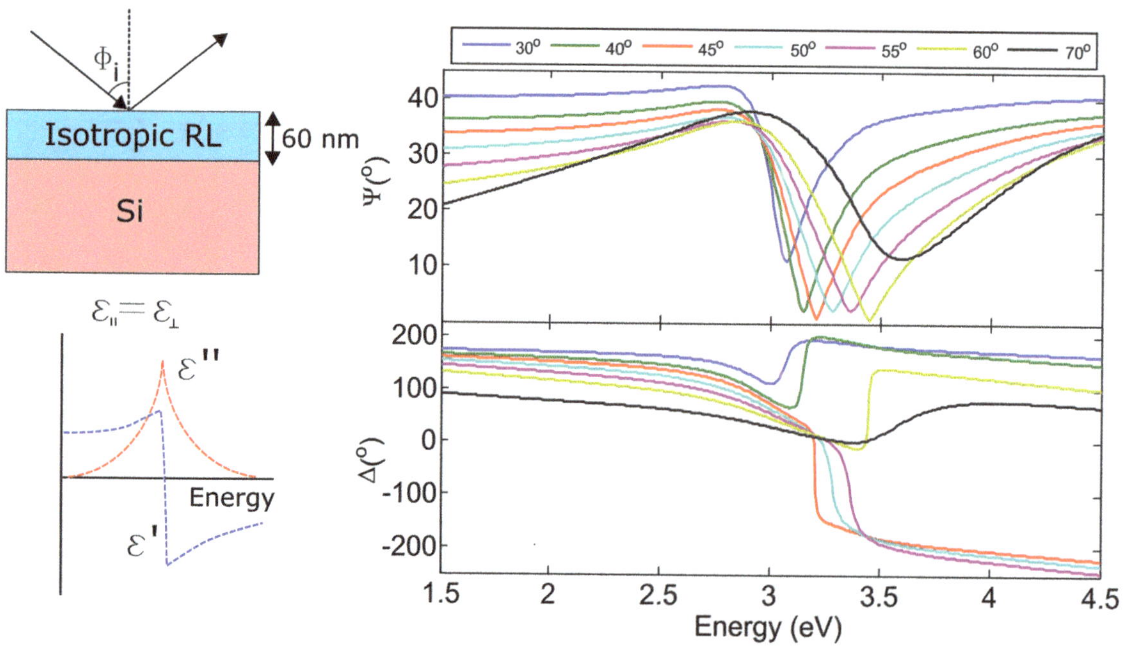

(Albella et al., 2011). To take those differences into account in a uniaxial anisotropic system, the ellipsometric functions must be calculated from Equations 2.11-2.19, introducing different parameters for the Lorentzian resonances corresponding to $\varepsilon_\perp$ and $\varepsilon_\parallel$, as sketched in Figure 9.

The energy region between both resonances provides an interesting situation: the material presents a negative real part of $\varepsilon_\perp$ but a positive real part of $\varepsilon_\parallel$. These unusual characteristics for a thin film produce remarkable effects on the ellipsometric functions. Thus, the calculations show the suppression of p polarization in the region between both resonances, evidenced by the minimization of ψ at around 2.8 eV, and a strong phase shift in Δ. In contrast to the isotropic case, the polarizing energy is hardly dependent on the incidence angle making unattractive its use in biosensing applications.

Also between both resonances but at larger energies (3.4 eV), the phase shift difference between p- and s-polarizations is 180°, while the reflection coefficients for both polarizations are similar in amplitude. Under those values of ψ and Δ, a singularity is found in the pseudodielectric function. The pseudodielectric function is defined as

$$\langle\varepsilon\rangle = \langle\varepsilon'\rangle + i\langle\varepsilon''\rangle = \sin^2\phi_i\left[1+\tan^2\phi_i\frac{\left(1-\rho\right)^2}{\left(1+\rho\right)^2}\right] \tag{2.33}$$

and represents the global optical behavior of the material system. Since $\rho = \tan\psi e^{i\Delta}$, it becomes -1 when $\tan\psi = 1$ and $\Delta = 180^\circ$, making the denominator null in Equation 2.33. From a physical view-

Figure 9. Calculation of the ellipsometric functions in an air-film-substrate system with an anisotropic resonant layer (RL) as intermediate film. For the out-of-plane Lorentz oscillator, the resonance, plasmon and broadening energies are 2.2, 4.5 and 0.45 eV, respectively. For the in-plane Lorentz oscillator, the resonance, plasmon and broadening energies are 4.0, 5.5 and 1.0 eV, respectively.

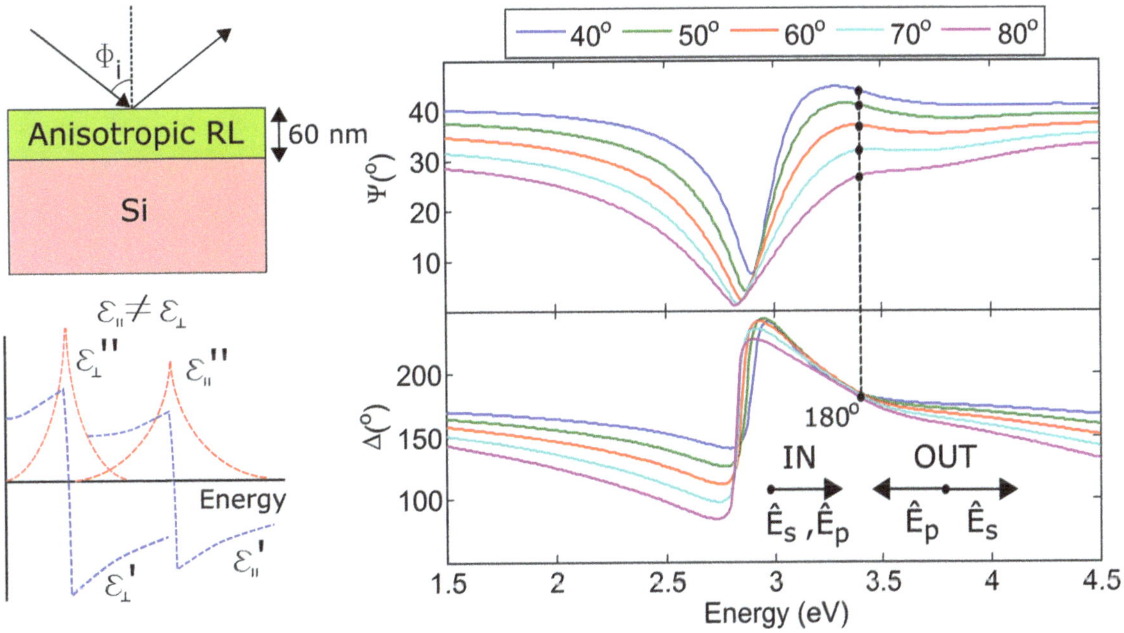

point, this condition means that, for an incident light beam with circular polarization, the reflected beam presents a polarization vector that rotates circularly in the opposite direction (reversal of the polarization handedness). In a few millielectronvolts, the pseudodielectric function changes from negative to positive as ψ and Δ varied in the energy region between the resonances of $\varepsilon_{\perp}$ and $\varepsilon_{\parallel}$ (Figure 10(a)). The amplitude of the asymmetrical function increases as the incidence angle diminishes. The inset of Figure 10(a) indicates the effect of the nanostructures on the rotation of the polarization vector in a circularly-polarized beam that impinges the surface.

The variations of the Lorentzian parameters for $\varepsilon_{\perp}$ and $\varepsilon_{\parallel}$ affect the width, amplitude and energy of the singularity associated to the reversal of the polarization handedness. For instance, the width is very dependent on the broadening factors of the resonances. The energy position strongly depends on the plasma and resonance frequencies of $\varepsilon_{\perp}$ and $\varepsilon_{\parallel}$, as illustrated in Figure 10(b) for the in-plane plasma frequency. The amplitude is especially sensitive to the separation of both resonances, increasing with the resonance splitting. Therefore, from the analysis of the pseudo-dielectric functions in that range, it is possible to extract information about the anisotropic properties of the resonant layer.

The amplitude of the $\langle \varepsilon'' \rangle$ singularity is also dependent on the dielectric properties of the substrate. As discussed above, to observe the characteristic line-shape associated to the polarization handedness, the ψ value must be close to 45° when the phase shift difference reaches 180°. Figure 11a shows that this condition is fulfilled for the resonant layer described above on silicon but not on glass where the ψ value is much lower than in silicon. Thus, the convenience of a particular substrate can be tested by analyzing the ellipsometric functions in a simple ambient-substrate system at the range of interest.

The strong variation of the amplitude with the incidence angle fosters the use of the reversal of polarization handedness in biosensing applications. Figure 11(b) includes the variation observed in a $\langle \varepsilon'' \rangle$ singularity of an array of Ga nanoparticles on a silicon substrate after functionalization with a self-assembled monolayer of glutathione polyclonal antibody. The amplitude of the singularity is considerably

Figure 10. (a) Effect of the angle of incidence on the singularity associated to the reversal of the polarization handedness. The parameters of the anisotropic resonant layer are the same as in Figure 9. The inset represents the effect on the direction of rotation of the polarization vector of a circularly-polarized beam. (b) Variation in the energy position as the plasma frequency of the in-plane resonance is increased.

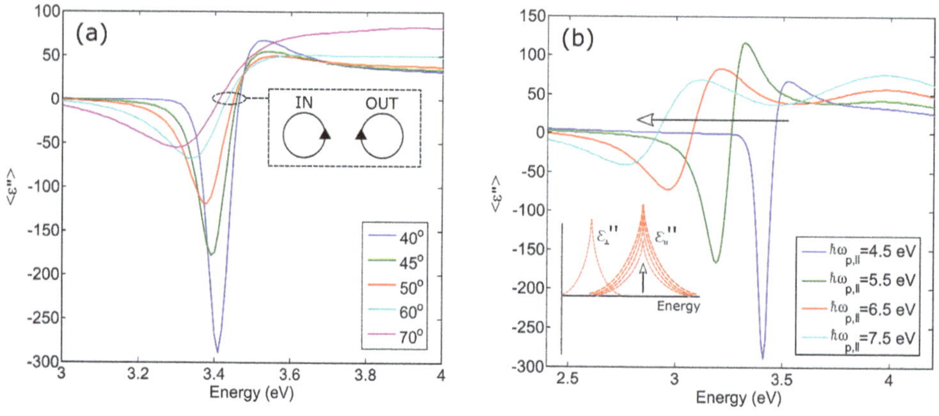

Figure 11. (a) Singularity calculated at an incidence angle of 40° for Si and glass substrates. The parameters of the anisotropic RL are the same as in Figure 9. (b) Experimental variation of the polarization handedness inversion measured on arrays of Ga nanoparticles on silicon substrates upon functionalization with an antibody monolayer. The inset figure shows a scanning electron microscopy viewgraph of the surface before biofunctionalization.

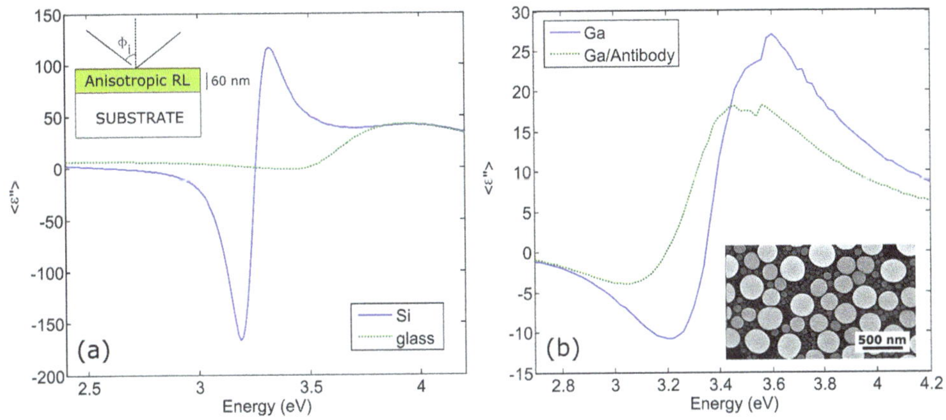

reduced corroborating the high sensitivity of the singularity to chemical changes on the surface of the nanoparticles.

Fano Resonance

LSPRs are dependent on the particle size, shape, interparticle spacing and dielectric constant of the surrounding dielectric materials (Haynes & Van Duyne, 2001; Mock, Hill, Tsai, Chilkoti, & Smith, 2012). A case of special interest is the organization in systems of close proximity nanoparticles, also called oligomers. As the nanoparticle separation decreases, the plasmons of individual nanoparticles begin to interact and hybridize with the plasmons of other particles. As an example, in a dimer, as the interaction increases, two configurations arise, antibonding, with no net dipolar moment (dark mode), and bonding with a net dipole (bright mode) (Nordlander, Oubre, Prodan, Li, & Stockman, 2004). The interaction between broad bright modes with dark ones gives rise to a new type of resonance known as Fano resonance. This comprises a characteristic asymmetric line-shape function which has been observed and described by Ugo Fano under certain conditions in scattering, transmission and reflection spectra, in contrast to the symmetric line-shape of a Lorentzian resonance (Fano, 1961). The slope in the resonance spectral region is extremely sensitive to small changes of the surface-attached species and prompts its use in sensor platforms. Previous publications have suggested that this resonance also have a great potential in chemical and biological sensing due to its high sensitivity to changes in the surrounding medium (Luk'yanchuk et al., 2010; Fan et al., 2010).

The resonance cross section (σ_t) can be described by curves that include asymmetric (σ_a) and symmetric (σ_s) contributions (Gallinet & Martin, 2011):

$$\sigma_t = \sigma_a(E) \cdot \sigma_s(E) \tag{2.34}$$

Equation 2.34 can be used to fit optical properties such as the reflectance, transmittance, and extinction in optical systems where a Fano resonance comes out. The asymmetric and symmetric contributions are expressed as:

$$\sigma_a(E) = \frac{\left(\frac{E^2 - E_a^2}{2W_aE_a} + q\right)^2 + b}{\left(\frac{E^2 - E_a^2}{2W_aE_a}\right)^2 + 1} \quad (2.35)$$

$$\sigma_s(E) = \frac{a^2}{\left(\frac{E^2 - E_s^2}{2W_sE_s}\right)^2 + 1} \quad (2.36)$$

where E_a, W_a, b and q are the resonant energy, the spectral bandwidth, the damping parameter and the asymmetry constant, for the asymmetric contribution; and E_s, W_s and a are the resonant energy, the spectral bandwidth and the resonance amplitude, for the symmetric contribution.

If $q = 1$, (Figure 12), a dip in the high energy side of σ_t appears for $E_a > E_s$, whereas, if $E_s > E_a$, the dip shows up in the low energy side. This situation is inverted when $q = -1$, that is, the dip in the low energy side is found for $E_s > E_a$, and the one in the high energy range shows up for $E_a > E_s$.

Alternatively to the use of symmetrical Lorentzian resonances, Fano resonances can be used to model the optical properties of effective layers built from ordered arrangements of nanoparticles with strong interactions among them that lead to obtain dark resonant modes. The ellipsometric functions can be extracted from the formalism depicted in Section 3.2 to compare with the experimental data. Depending on the symmetry of the nanoparticle arrays, the fitting may require developing a biaxial anisotropic model.

BIOSENSING APPLICATIONS

The use of metal films to enhance plasmonic effects in TIRE is gaining importance as an optical method in biosensing applications. TIRE presents higher surface sensitivity than regular SPR. Besides, high speed and label free analysis are intrinsic characteristics of TIRE which make it very attractive for monitoring the adsorption of biomolecules in real time. Another important advantage is that TIRE is non-invasive with the biomolecules under investigation. Thus, different biosensing applications of TIRE have been developed in the last years. These applications require the ability to immobilize ligands on a metal (typically Au) and the precise control of the ligand density.

Immunosensing experiments have been reported through the immobilization of antibody in metallic surfaces (Cho et al., 2014). Ellipsometry has been integrated with electrochemical measurements for detecting interferon-gamma (IFN-gamma), a biomarker of tuberculosis (Tsai et al., 2014). For that purpose, anti-IFN-gamma antibody is immobilized on a gold thin film to form the label-free biosensor. The

Figure 12. Overall cross-section for q=1 using the following parameters; 1) $E_a > E_s$ (black curve): E_a =3.8 eV W_a =0.05 eV, b =0.2, E_s =3.3 eV, W_s =0.2 eV and a =5; 2) $E_s > E_a$ (red curve) E_a =2.8 eV W_a =0.05 eV, b =0.2, E_s =3.3 eV, W_s =0.2 eV and a =5.

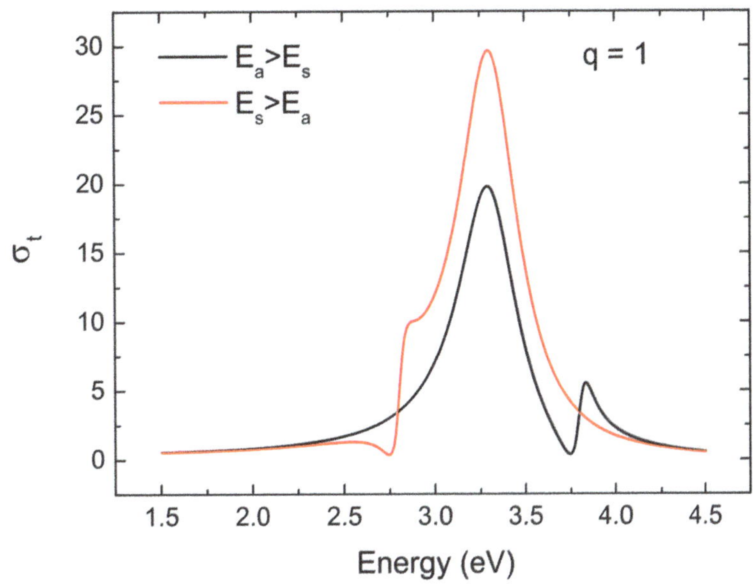

antibody is linked to the Au surface using thiolated bis-thiophene. A linear Δ shift is measured as the antibody-antigen interactions take place in the TIRE setup, allowing the ultrasensitive detection of IFN-gamma. In real-time measurements, it has been found that Δ presents higher sensitivity than ψ for monitoring the Au-protein and the protein-protein interactions at the early stages (Baleviciute, Balevicius, Makaraviciute, Ramanaviciene, & Ramanavicius, 2013). However, ψ seems more sensitive near the steady-state conditions.

To enhance the antigen antibody interaction in the experiments it is possible to use Au NPs functionalized with the antigen. Then, the antigen-Au NPs complexes bind to the antibody immobilized on the Au thin film amplifying the energy shift of the resonance in the TIRE setup by the effect of the Au NPs. This procedure has been used to fabricate glutathione (GSH) sensors in a competitive assay between GSH and a constant amount of functionalized AuNPs (García-Marín et al., 2014). A detection limit for free GSH in the nanomolar range is found.

The TIRE technique has been also introduced to monitor the hybridization of DNA chains (Keske, Atar, Üstündağ, & Çağlayan, 2014). Thus, the selective detection of Influenza A has been reported from the immobilization of a single-strand oligodeoxynucleotide (ssODN), which is complementary of the characteristic sequence of Influenza A. A linear sensor response is found in the 0.05 μM-2 μM range with high selectivity in comparison to the response to non-complementary ssODN.

The high-sensitivity of the technique can be also exploited to study weak affinity interactions among biomolecules. Imaging ellipsometry in total internal reflection mode has demonstrated to be competent to monitor the interaction between tris and lysozyme, a typical example of weak affinity interaction (Kang, Niu, & Jin, 2014). Tris is immobilized on the gold thin film to detect the presence of LZM and compared with the signal obtained from negative control solutions. From real-time analysis, the authors estimate the dissociation constant.

Reversal of polarization handedness has been recently introduced to sense glutathione on arrays of Ga nanoparticles (García-Marín et al., 2015). The detection limit outperforms those obtained in commercial detection kits. In addition, the method operates in external reflection mode in an instrumental setup simpler than the TIRE setup. Preliminary data obtained from the use of the reversal of the polarization handedness for sensing DNA hybridization show sensitivities in the sub-nanomolar range. Fano resonances are up-to-date less developed in biosensing applications but they are considered promising tools to improve the sensitivity of current optical sensor systems.

REFERENCES

Albella, P., Garcia-Cueto, B., Gonzalez, F., Moreno, F., Wu, P. C., Kim, T.-H., & Videen, G. et al. (2011). Shape matters: Plasmonic nanoparticle shape enhances interaction with dielectric substrate. *Nano Letters, 11*(9), 3531–3537. doi:10.1021/nl201783v PMID:21848270

Arwin, H., Poksinski, M., & Johansen, K. (2004). Total internal reflection ellipsometry: Principles and applications. *Applied Optics, 43*(15), 3028–3036. doi:10.1364/AO.43.003028 PMID:15176189

Arwin, H., Poksinski, M., & Johansen, K. (2008). Enhancement in ellipsometric thin film sensitivity near surface plasmon resonance conditions. *Physica Status Solidi, 205*(4), 817–820. doi:10.1002/pssa.200777899

Aspnes, D. E. (1975). Photometric ellipsometer for measuring partially polarized light. *Journal of the Optical Society of America, 65*(11), 1274–1278. doi:10.1364/JOSA.65.001274

Aspnes, D. E. (2004). Expanding horizons: New developments in ellipsometry and polarimetry. *Thin Solid Films, 455-456*, 3–13. doi:10.1016/j.tsf.2003.12.038

Azzam, R. M. A., Bashara, N. M., & Burns, D. T. (1987). *Ellipsometry and polarized light.* Amsterdam: North Holland.

Baleviciute, I., Balevicius, Z., Makaraviciute, A., Ramanaviciene, A., & Ramanavicius, A. (2013). Study of antibody/antigen binding kinetics by total internal reflection ellipsometry. *Biosensors & Bioelectronics, 39*(1), 170–176. doi:10.1016/j.bios.2012.07.017 PMID:22938839

Chang, C.-C., Chiu, N.-F., Lin, D. S., Chu-Su, Y., Liang, Y.-H., & Lin, C.-W. (2010). High-sensitivity detection of carbohydrate antigen 15-3 using a gold/zinc oxide thin film surface plasmon resonance-based biosensor. *Analytical Chemistry, 82*(4), 1207–1212. doi:10.1021/ac901797j PMID:20102177

Cho, H., Zook, J., Banner, T., Park, S.-H., Min, B.-H., Hasty, K. A., & Lindner, E. et al. (2014). Immobilization of fibrinogen antibody on self-assembled gold monolayers for immunosensor applications. *Tissue Engineering and Regenerative Medicine, 11*(1), 10–15. doi:10.1007/s13770-013-1119-4

Draine, B. T., & Flatau, P. J. (2013). User guide for the discrete dipole approximation code DDSCAT 7.3. Retrieved from http://arxiv.org/abs/1305.6497

Fan, J. A., Wu, C., Bao, K., Bao, J., Bardhan, R., Halas, N. J., & Capasso, F. et al. (2010). Self-assembled plasmonic nanoparticle clusters. *Science*, *328*(5982), 1135–1138. doi:10.1126/science.1187949 PMID:20508125

Fano, U. (1961). Effects of configuration interaction on intensities and phase shifts. *Physical Review*, *124*(6), 1866–1878. doi:10.1103/PhysRev.124.1866

Gallinet, B., & Martin, O. J. F. (2011). Influence of electromagnetic interactions on the line shape of plasmonic Fano resonances. *ACS Nano*, *5*(11), 8999–9008. doi:10.1021/nn203173r PMID:22026329

García-Marín, A., Abad, J. M., Ruiz, E., Lorenzo, E., Piqueras, J., & Pau, J. L. (2014). Glutathione immunosensing platform based on total internal reflection ellipsometry enhanced by functionalized gold nanoparticles. *Analytical Chemistry*, *86*(10), 4969–4976. doi:10.1021/ac5005212 PMID:24766219

García-Marín, A., Hernández, M. J., Ruiz, E., Abad, J. M., Lorenzo, E., Piqueras, J., & Pau, J. L. (2015). Immunosensing platform based on gallium nanoparticle arrays on silicon substrates. *Biosensors & Bioelectronics*, *74*, 1069–1075. doi:10.1016/j.bios.2015.08.002 PMID:26276543

Haynes, C. L., & Van Duyne, R. P. (2001). Nanosphere lithography: A versatile nanofabrication tool for studies of size-dependent nanoparticle optics. *The Journal of Physical Chemistry B*, *105*(24), 5599–5611. doi:10.1021/jp010657m

Homola, J., Yee, S. S., & Gauglitz, G. (1999). Surface plasmon resonance sensors [review]. *Sensors and Actuators. B, Chemical*, *54*(1-2), 3–15. doi:10.1016/S0925-4005(98)00321-9

Johnson, P. B., & Christy, R. W. (1972). Optical constants of the noble metals. *Physical Review B: Condensed Matter and Materials Physics*, *6*(12), 4370–4379. doi:10.1103/PhysRevB.6.4370

Kang, T., Niu, Y., & Jin, G. (2014). Visualization of the interaction between tris and lysozyme with a biosensor based on total internal reflection imaging ellipsometry. *Thin Solid Films*, *571*, 463–467. doi:10.1016/j.tsf.2014.02.098

Keske, A., Atar, A., Üstündağ, İ., & Çağlayan, M. O. (2014). Detection of Influenza A by Surface Plasmon Resonance Enhanced Total Internal Reflection Ellipsometry. *Journal of Computational and Theoretical Nanoscience*, *11*(4), 981–986. doi:10.1166/jctn.2014.3453

Lodewijks, K., Van Roy, W., Borghs, G., Lagae, L., & Van Dorpe, P. (2012). Boosting the figure of merit of LSPR-based refractive index sensing by phase-sensitive measurements. *Nano Letters*, *12*(3), 1655–1659. doi:10.1021/nl300044a PMID:22356465

Luk'yanchuk, B., Zheludev, N. I., Maier, S. A., Halas, N. J., Nordlander, P., Giessen, H., & Chong, C. T. (2010). The Fano resonance in plasmonic nanostructures and metamaterials. *Nature Materials*, *9*(9), 707–715. doi:10.1038/nmat2810 PMID:20733610

Manickam, G., Gandhiraman, R., Vijayaraghavan, R. K., Kerr, L., Doyle, C., Williams, D. E., & Daniels, S. (2012). Protection and functionalisation of silver as an optical sensing platform for highly sensitive SPR based analysis. *Analyst (London)*, *137*(22), 5265–5271. doi:10.1039/c2an35826c PMID:23037590

Mock, J. J., Hill, R. T., Tsai, Y. J., Chilkoti, A., & Smith, D. R. (2012). Probing dynamically tunable localized surface plasmon resonances of film-coupled nanoparticles by evanescent wave excitation. *Nano Letters, 12*(4), 1757–1764. doi:10.1021/nl204596h PMID:22429053

Near, R. D., Hayden, S. C., Hunter, R. E. Jr, Thackston, D., & El-Sayed, M. A. (2013). Rapid and efficient prediction of optical extinction coefficients for gold nanospheres and gold nanorods. *The Journal of Physical Chemistry C, 117*(45), 23950–23955. doi:10.1021/jp4082596

Noguez, C. (2007). Surface plasmons on metal nanoparticles: The influence of shape and physical environment. *Journal of Physical Chemistry, 111*(10), 3806–3819. doi:10.1021/jp066539m

Nordlander, P., Oubre, C., Prodan, E., Li, K., & Stockman, M. I. (2004). Plasmon hybridization in nanoparticle dimers. *Nano Letters, 4*(5), 899–903. doi:10.1021/nl049681c

Palik, E. D. (1998). *Handbook of optical constants of solids*. Academic press.

Raether, H. (2013). *Surface plasmons on smooth and rough surfaces and on gratings. Springer Tracts in Modern Physics* (Vol. 111). Springer Berlin Heidelberg; doi:10.1007/BFb0048317

Schubert, M. (1996). Polarization-dependent optical parameters of arbitrarily anisotropic homogeneous layered systems. *Physical Review B: Condensed Matter and Materials Physics, 53*(8), 4265–4274. doi:10.1103/PhysRevB.53.4265 PMID:9983979

Taft, E. A., & Philipp, H. R. (1961). Optical constants of silver. *Physical Review, 121*(4), 1100–1103. doi:10.1103/PhysRev.121.1100

Tiwari, P. M., Vig, K., Dennis, V. A., & Singh, S. R. (2011). Functionalized gold nanoparticles and their biomedical applications. *Nanomaterials, 1*(1), 31–63. doi:10.3390/nano1010031

Tonova, D., Patrini, M., Tognini, P., Stella, A., Cheyssac, P., & Kofman, R. (1999). Ellipsometric study of optical properties of liquid Ga Nanoparticles. *Journal of Physics Condensed Matter, 11*(10), 2211–2222. doi:10.1088/0953-8984/11/10/008

Tsai, P.-I., Lee, S.-S., Chou, S.-T., Chang, Y.-T., Lee, A. S.-Y., & Lee, C. K. (2014). Integrating SPR-ellipsometry and electrochemical measurements for performance evaluation of label-free thiophene-based biosensor. *Proc. SPIE 8933. Frontiers in Biological Detection: From Nanosensors to Systems, VI*, 89330H. doi:10.1117/12.2038955

Verre, R., Fleischer, K., Smith, C., McAlinden, N., McGilp, J. F., & Shvets, V. (2011). Probing the out-of-plane optical response of plasmonic nanostructures using spectroscopic ellipsometry. *Physical Review B: Condensed Matter and Materials Physics, 84*(8), 085440. doi:10.1103/PhysRevB.84.085440

Verre, R., Modreanu, M., Ualibek, O., Fox, D., Fleischer, K., Smith, C., & Shvets, I. V. et al. (2013). General approach to the analysis of plasmonic structures using spectroscopic ellipsometry. *Physical Review B: Condensed Matter and Materials Physics, 87*(23), 235428. doi:10.1103/PhysRevB.87.235428

Weaver, J. H., Krafka, C., Lynch, D. W., & Koch, E. E. (1981). *Optical properties of metals Pt 2*. Germany: Fachinformationszentrum Energie, Physik, Mathematik.

Wu, P. C., Kim, T.-H., Brown, A. S., Losurdo, M., Bruno, G., & Everitt, H. O. (2007). Real-time plasmon resonance tuning of liquid Ga nanoparticles by in situ spectroscopic ellipsometry. *Applied Physics Letters*, *90*(10), 103119. doi:10.1063/1.2712508

Yamaguchi, T., Takahashi, H., & Sudoh, A. (1978). Optical behavior of a metal island film. *Journal of the Optical Society of America*, *68*(8), 1039–1044. doi:10.1364/JOSA.68.001039

Yeatman, E. M. (1996). Resolution and sensitivity in surface plasmon microscopy and sensing. *Biosensors & Bioelectronics*, *11*(6-7), 635–649. doi:10.1016/0956-5663(96)83298-2

Yurkin, M. A. (2014). Comment on "Rapid and efficient prediction of optical extinction coefficients for gold nanospheres and gold nanorods". *Journal of Physical Chemistry*, *118*(37), 21738–21739. doi:10.1021/jp5054524

Zayats, A. V., Smolyaninov, I. I., & Maradudin, A. A. (2005). Nano-optics of surface plasmon polaritons. *Physics Reports*, *408*(3-4), 131–314. doi:10.1016/j.physrep.2004.11.001

KEY TERMS AND DEFINITIONS

Ellipsometry: A non-perturbing optical technique that enables the characterization of surfaces, interfaces, thin films and nanostructured materials by studying the change in the state of polarization upon reflection of polarized light.

Fano Resonance: Interaction between bright and dark oscillatory modes in ordered nanoparticle arrays.

Handedness of Polarized Light: Direction of rotation of the electric field for an entire plane that is perpendicular to the direction of propagation.

P or Transverse Magnetic (TM) Polarization: Property of an electromagnetic wave whose electric field vector is parallel to the plane of incidence.

S or Transverse Electric (TE) Polarization: Property of an electromagnetic wave whose electric field vector is perpendicular to the plane of incidence.

Surface Plasmons: Collective oscillations of free electrons at the interface between a negative (metal) and a positive (dielectric) permittivity material.

Surface Polariton: Electromagnetic wave coupled to collective charge oscillations that propagates along the metal/dielectric interface.

Chapter 9
Functional Coatings for Bone Tissue Engineering

M Tarik Arafat
University of Leeds, UK

Xu Li
*A*STAR (Agency for Science, Technology & Research), Singapore*

ABSTRACT

This chapter outlines the application of coatings for implants and/or scaffolds used for bone tissue engineering (TE). Previously orthopaedic implants and/or scaffolds were designed considering mainly their mechanical aspects. However, due to the criticality of osteointegration with the surrounding tissues after implanting, the biological aspects of implants and/or scaffolds are becoming crucial. Recent trend is to use functional coatings like anti-infective, Ca-P and biomolecules coatings onto implants and/or scaffolds to improve anti-infective, osteoconductive and osteoinductive properties, respectively; thereby improving the osteointegration of the implant and/or scaffold with the surrounding tissue. Here, the application of different types of coatings on implants and/or scaffolds for bone TE will be described. The use of coatings as a drug and gene delivery carrier will also be covered in brief. As many of the coatings are still in preclinical testing stage, challenges associated with successful clinical use will be discussed as well.

INTRODUCTION

Bone tissue engineering (TE) is a fascinating field of research with substantial research activities focusing on delivering implants and/or scaffolds that help to restore fractures, non-unions and deformities. Previously, the design of orthopaedic implants and/or scaffolds was mainly focused on the mechanical aspects. In this way, the implants and/or scaffolds provide support to the fractured location, and thus ensure less pain while using the injured limb and help to recover back to normal use. These classical implants and/or scaffolds, such as stainless steel, titanium, alumina, partially stabilized zirconia, and polyethylene scaffolds are bioinert thus having minimal interaction with the surrounding tissues when implanted. However, to accelerate bone healing process osteointegration with the surrounding is criti-

DOI: 10.4018/978-1-5225-0066-7.ch009

cal. Thus, the biological aspects of implants and/or scaffolds are becoming imperative along with its mechanical properties.

To improve the biological aspects functional coatings on the implants and/or scaffolds can be considered as the surface of the implants and/or scaffolds is the first part that interacts with the host. Coatings was introduced approximately 35 years ago as a way to modify the surface of the implants and/or scaffolds (Abukawa et al., 2006). Recently, applying functional coating onto the implants and/or scaffolds has become a common trend to improve their bioactivity and osteoconductivity properties, and thereby, improving their osteointegration with the surrounding tissue while implanted *in vivo*. To some extent, the clinical advantage of functional coatings has already been proven through early osteointegration and long term stability of the implant and/or scaffold.

This chapter will discuss the recent trends and give a future insight on the application of functional coatings for bone TE. Challenges associated to the clinical use of such coatings will also be discussed.

CLINICAL RELEVANCE OF FUNCTIONAL COATIONS

In the United States alone, delayed bone healing has been reported in approximately 600,000 fractures per year, and 100,000 progress to non-union (Bishop, Palanca, Bellino, & Lowenberg, 2012). This intensive number of patients related to delayed healing from the bone diseases and injuries created the socio-economic need for continued research on bone tissue engineering (TE), where implant and/or scaffold is an integral part of treatment. For clinical use, the implant and/or scaffold should be anti-infective, osteoconductive and osteoinductive. This performance can be effectively achieved through functional coatings on the implant and/or scaffold.

Bacterial infection on the implants and/or scaffolds surface is a major cause of the delayed healing in orthopaedic diseases. Despite strict sterilization and aseptic procedures, in United States alone, the infection rates associated with orthopaedic implants have been reported around 4.3% (Darouiche, 2004). Along with the physical pain and sufferings, bacterial infection leads to economic burden with the direct medical costs exceeding annually USD 3 billion alone (Hetrick & Schoenfisch, 2006). As the majority of the patients having implants and/or scaffolds operation are older than 65 years old (Hetrick & Schoenfisch, 2006), with the increase in life expectancy implant associated bacterial infection deserves further attention to resolve. To prevent such bacterial infection inorganic coatings such silver, zinc oxide and antibiotics incorporated coating showed great potential (Hetrick & Schoenfisch, 2006; Zhang, Myers, Wallace, Brandt, & Choong, 2014).

To accelerate bone healing at early implantation time recent developments are also focused to use functional coatings. To improve the osteoconductive and osteoinductinve properties, and thereby to improve osteointegration, implants and/or scaffolds were coated with coatings such as Ca-P and components of extracellular matrix (ECM) e.g. collagen (Surmenev, Surmeneva, & Ivanova, 2014). It is also found that composite coatings of Ca-P/ECM components offer improved biological functionality over coatings consisting of merely inorganic Ca-P or organic ECM components (Goodman, Yao, Keeney, & Yang, 2013). The biological properties of functional coatings can be improved further by adding growth factors, gene and other molecules to produce a truly osteoinductive platform (Goodman et al., 2013).

FUNCTIONAL COATINGS FOR BONE TISSUE ENGINEERING

This section presents a comprehensive study of the application of different functional coatings for bone TE application.

Anti-Bacterial Coatings

Bacterial infection is one of the most common complications associated with implants and/or scaffolds for fracture fixation, joint replacement and spine surgery. The bacteria can be induced from ambient atmosphere of operating room, clothing worn by medical professionals or patients, surgical instruments, the patient's skin or even bacteria already in the body (An & Friedman, 1996). Though with the advances of sterilization techniques the possibility of bacterial infection has been greatly reduced in hospital settings, pathogenic micro-organisms are still found at the site of approximate 90% implants (Hetrick & Schoenfisch, 2006; Nablo, Rothrock, & Schoenfisch, 2005); and may cause acute and chronic infection of the underlying bone and the adjacent soft tissues. Thus failure to eliminate bacterial infection eventually leads to patient's pain and sufferings, along with additional costs induced by the necessitate removal of the implants followed by delayed revision procedure to re-implant.

In natural environment there are two bacterial living states, these are free floating and attached states, known as planktonic and sessile states, respectively. Planktonic states allows bacterial to spread, whereas sessile state allows bacterial persist and resist adverse environment through the formation of biofilms, which are defined as biopolymer matrix-enclosed microbial populations adhering to each other and/or surface (Glinel, Thebault, Humblot, Pradier, & Jouenne, 2012). Biofilms are particularly problematic as such formed biofilms can help bacteria withstand host immune responses and can be 1000-fold more resistant than bacteria which are in floating state (Davies, 2003; Glinel et al., 2012). Prolonged uses of antibiotics at higher doses are needed to cure such infections, however, may lead to antibiotic resistance, and potentially compromise bone growth and immune system surveillance (Goodman et al., 2013). Given the difficulty associated with removing established biofilms, it is desirable to prevent bacterial in the early stages of adhesion through implant coatings.

Eventually, it is expected that the orthopaedic implant will stimulate host tissue integration and prevent bacterial adhesion. With regards to the complex interaction between host immune system and the invading bacteria during implant integration, the theory of "race to the surface" can be implied (Gristina, 1987). The theory states that the host cells and bacteria compete to adhere, replicate and colonize on the implant surface. Ideally, the race is won by the host cells by colonizing the implant surface before bacteria, thus ensures a stable interface with implant integration while defending the implant surface from bacteria through vigorous immune system. However, if bacteria colonize the implant before the arrival of the host cells, it is inevitable to avoid implant infection. Hence, it is imperative that implant surface will promote host tissue integration while preventing bacterial adhesion. However, the balance between these two requirements is often challenging as the host cells and bacterial shares almost the same adhesive mechanisms (Goodman et al., 2013). Thus, the challenge is to develop anti-infective coatings which will not compromise with surrounding tissue integration.

Inorganic Anti-Bacterial Coatings

Silver and silver ions have long been known to have antibacterial activity (Bosetti, Masse, Tobin, & Cannas, 2002; Schreurs & Rosenberg, 1982), while they are nontoxic to mammalian tissue (Bologna et al., 1999), and are well-known due to their inhibition of bacterial adhesion and sustained anti-bacterial effect over both gram-positive and gram-negative bacteria (Hasan, Crawford, & Lvanova, 2013). The antibacterial mechanism of Ag-related materials has been widely studied, and it was suggested that it inactivates bacterial replication and key metabolic processes, ultimately leading to cell death (Feng et al., 2000; G. C. Wang & Zreiqat, 2010; Zhang et al., 2014). Once the bacteria approached to the implant and/or scaffold surface, the functional groups such as imidazole, amino, carboxyl and thiol groups were electrostatically attracted to the available silver ion, and thus initiates the silver ion exchange which leads to the damage of the cell wall of the bacteria (Lim et al., 2014). The damage in the cell wall subsequently resulted in osmotic pressure difference, thus resulted in cell wall rupture and finally the death of bacteria. The process has been depicted in Figure 1 (in appendix). One important feature of using silver and silver ion is that there is no known tendency to develop anti-bacterial resistance while using it (D. Lee, Cohen, & Rubner, 2005), and thus silver releasing material possess a great potential for reducing bacterial adhesion to implant and/or scaffolds.

Metallic silver-coated fracture fixation pins have showed limited success in human studies, which could be related to the lower propensity of silver ion to be released from the implant coatings (Masse et al., 2000; Simchi, Tamjid, Pishbin, & Boccaccini, 2011). Eventually, the antibacterial performances of silver have been attributed to its oxidized form, which can be released in controlled manner from polyamide polymers (R. Kumar & Munstedt, 2005). The release of silver ion can also be controlled by crystallinity of the polymer in which it is embedded (Radhesh Kumar & Münstedt, 2005). It has also been reported that incorporation of a small amount of platinum into the coating increases the oxidation of silver through galvanic reaction, as platinum is more cathodic than silver in the galvanic series (Dowling et al., 2003; Hetrick & Schoenfisch, 2006). Long term release of silver ion has been ensured by incorporating silver in titanium oxide nano-tubes (TiO_2-NTs), with a sustained release of silver ion for over 30 days (Cheng, Li, Huo, Gao, & Xiong, 2014). The release kinetics of silver ion from TiO_2-NTs can be controlled by the tube dimension and spacing of the tubes (Peng et al., 2009). The diameter of the TiO_2-NTs and the space between nanotubes can be controlled from few nanometres to hundreds of nanometres by controlling anodization, and thus providing an efficient and direct way to control silver ion release (Cheng et al., 2014). Silver ion has also been incorporated into hydroxyapatite (HA) to bring the antibacterial properties to this osteoconductive material, and found that silver-doped HA coating exhibits antibacterial properties against common gram-positive and gram-negative bacteria without compromising the osteconductive activity (Iqbal et al., 2012; Yan, Zhang, Huang, Ding, & Pang, 2014).

Other inorganic anti-infective agents such as zinc oxide (ZnO) (Chang et al., 2012; Hu et al., 2012; Shen et al., 2014; Vargas-Reus, Memarzadeh, Huang, Ren, & Allaker, 2012), fluorine (F) (Perez-Jorge et al., 2012; Stobie, Duffy, Hinder, McHale, & McCormack, 2009; Yoshinari, Oda, Kato, & Okuda, 2001) have been studied as a coating on implants and/or scaffolds. Zn is an important constituent of bone, and it plays a vital role in enzyme activity, DNA synthesis, biomineralization and hormonal activity; while possess excellent antibacterial property (Hu et al., 2012; Tang, Chappell, Dove, Reeder, & Lee, 2009). ZnO has been incorporated into TiO_2 coatings on Ti by plasma electrolytic oxidation (PEO) (Hu et al., 2012), a popular and effective technique to prepare TiO_2 based coating on Ti (X. Liu, Chu, & Ding, 2010). Zn was homogeneously distributed over the TiO_2 coating without altering the structural details such as roughness, wettability, phase composition and chemical states of TiO_2 coatings. Enhanced anti-infective property of the ZnO incorporated TiO_2 has been reported, which is ascribed to a slow and constant release of Zn ions over a period of 15 days (Hu et al., 2012). Zn has also been incorporated on micro-rough Ti via a sol-gel process with assistance of spin-coating technique, and showed a sustained release over a period of 28 days (Shen et al., 2014). The Zn incorporated micro-rough Ti showed improved osteoblast proliferation and differentiation attributed to the micro-roughness on Ti and inhibited bacterial adhesion attributed to the antibacterial property of Zn. *In vivo* data also showed Zn incorporated micro-rough Ti implants promote new bone formation after 12 weeks of implantation in epiphysis of femur of rabbits.

Fluoride which is commonly used as an effective anticaries element in dental filed, showed antibacterial activity through the release of fluorine ions which can inhibit bacterial metabolism (Yoshinari et al., 2001). Fluorine incorporated anodized Ti implant exhibited bacterial adhesion by the fluoride containing nanostructured surfaces (Perez-Jorge et al., 2012). It should also be noted that fluoride ions are capable on inducing fluorapatite in the surrounding bone and thus improves the osteointegration property of the implants along with its anti-infective property.

Antibiotics Loaded Anti-Bacterial Coatings

Controlled local delivery of antibiotics from surface coating of an implant can be considered as the most direct approach to address implant associated bacterial infection. There are many potential advantages of local delivery of antibiotics through coating compared to the so-called systemic administration strategy where patients are treated with high concentration of dose. Some of the advantages of local delivery of antibiotics are of greater control over antibiotic dose, sustained and prolonged release of antibiotic, and less susceptibility to develop antibiotic resistance (Wu & Grainger, 2006). Such delivery system also allows for the specific selection of antibiotics toward specific pathogens associated with implant infection, and thus prevents the unnecessary exposure of the other parts of the body to antibiotics.

The effectiveness of an antibiotic loaded coating strongly depends on the controlled and sustained release profile, and can be affected by the delivery matrix. The delivery matrix can be polymeric (e.g. polyurethane (PU), poly(L-lactic acid) (PLA), poly(glycolic acid) (PGA), poly(methyl methacrylate) (PMMA)) or inorganic (hydroxyapatite) coating, and the release kinetics is governed by the interaction of antibiotic with the delivery matrix (Hetrick & Schoenfisch, 2006; G. C. Wang & Zreiqat, 2010). The release profile of the antibiotic from the polymer depends on the physico-chemical similarity between the antibiotics and polymer matrix. As an example, antibiotics and polymers having similar lipophilicity results in homogenous distribution of antibiotics into polymer matrix, which to some extent helps to attain a smooth release profile (Schierholz, Steinhauser, Rump, Berkels, & Pulverer, 1997). Ciprofloxacin loaded polyurethane (PU) matrix showed potential antibacterial activity with an antibiotic release rate greater than the minimum killing dose for about 5 days (Kwok, Wan, et al., 1999). However, the release behaviour was greatly controlled by the initial amount of antibiotic in the polymer matrix, which often may result in an initial burst, followed by a greatly reduced long term rate of release (Kwok, Wan, et al., 1999). A thin barrier of poly(n-butyl methacrylate) (PBMA) on the drug loaded polymer has been found useful in this regard, as the thin barrier limits the initial burst and helps to achieve a constant and sustained release (Kwok, Horbett, & Ratner, 1999).

Not only polymers but also inorganic coatings like Ca-P based coatings can be used as a deliver matrix of antibiotics (Grecu, Novac, Ionita, & Ungureanu, 2011; Radin, Campbell, Ducheyne, & Cuckler, 1997; Sorensen et al., 2014; Stigter, Bezemer, de Groot, & Layrolle, 2004; Stigter, de Groot, & Layrolle, 2002; B. Wang et al., 2014). Antibiotic such as tobramycin has been incorporated into carbonated hydroxyapatite coatings (CHA) applied on titanium implants through a biomimetic co-precipitation using simulated body fluid (SBF), and found that the antibiotic loaded coating is effective in inhibiting bacterial growth (Stigter et al., 2002). The study also found that the release kinetics of the antibiotic depends on the coating dissolution, loading dose and pH of the media. Different antibiotics such as cephalothin, carbenicillin, amoxicillin, cefamandol, gentamicin and vancomycin were also introduced into CHA coatings and found that with the increase in concentrations of antibiotics in the SBF a higher quantity of antibiotic was incorporated into the coatings (Stigter et al., 2004). However, it was found that antibiotics having carboxylic groups were better incorporated into the coatings, which eventually signifies the importance of chemical structure of antibiotic in the co-precipitation process. Antibiotics can also be incorporated into coatings through dipping process (Radin et al., 1997), however, it was effective to inhibit bacteria only up to 24 hours. Generally, physical adsorption of antibiotics limits the drug loading amount and shows poor release kinetics with an initial burst release. It was found that the physically adsorbed antibiotic can be susceptible to initial burst release which can be as high as 80 – 90% of the loaded antibiotic within just 1 hour (Radin et al., 1997).

Other Anti-Bacterial Coatings

Chlorhexidine is another organic antibacterial agent, and has long been used as a skin antiseptic. It has demonstrated antibacterial activity to a broad range of gram positive and gram negative bacteria, and it might be an alternative to antibiotic as bacterial resistance to it is extremely rare (Campbell et al., 2000; DeJong et al., 2001).

Nitric oxide (NO) which is a strong oxidant is another antibacterial agent which has resistance towards both gram positive and negative bacteria (Charville, Hetrick, Geer, & Schoenfisch, 2008; Nablo et al., 2005). When NO comes to contact with the cell membrane, it diffuses through the membrane and oxidize the bacterial DNA, thus damage the DNA strands of the bacteria (Hetrick & Schoenfisch, 2006). However, NO is very unstable and its half-life is only of few seconds in biological ambiance, which leads to the development of NO releasing polymers coatings (Hetrick & Schoenfisch, 2006). Sol-gel chemistry is an effective way to deposit NO-releasing coatings onto implant. Nitric oxide (NO)-releasing sol–gels coated stainless steel implant could effectively reduce the bacterial adhesion by 30 to 90% depending on the NO flux from the coatings (Nablo et al., 2005).

Beside these synthetic antibacterial agents, there are natural antibacterial peptides which are secreted by living organisms such as amphibians and fishes surrounded by pathogens. These peptides destabilize the bacterial membrane by penetrating into it to form pores, which subsequently induce cell lysis (Guyomard et al., 2008). These natural antibacterial peptides have a broad spectrum of antibacterial activity, and not likely to cause the bacterial resistance. These natural antibacterial peptides possess a great potential as antibacterial coating on implants.

Regulatory and Development Context of Ca-P Based Apatite Coatings

With the increasing trend of incorporating silver and other antimicrobials on implants and/or scaffolds, in 2007, the US Food and Drug Administration (FDA) issued a draft guidance to assist the growing industry in preparing and submitting premarket notification submissions (510(k)s) for medical devices that include antimicrobial agents ("Draft Guidance for Industry and FDA Staff - Premarket Notification

[510(k)] Submissions for Medical Devices that Include Antimicrobial Agents. (2007)., "). FDA suggests that the typical information that should be provided to get approval include concentration, action mechanism, antimicrobial activity spectrum and an evaluation of the toxicity of the antimicrobial agents, metabolites, and degradation products in the body. If the antimicrobial agent is eluting it should include the data of the release kinetics from the device, and the data should be collected using a method having an adequate level of detection. For non-eluting antimicrobial agent the testing results should demonstrate that it is permanently bound to the implant and/or scaffolds surface.

Apatite Based Coatings

Ca-P Based Apatite Coatings

Two main important constituents of natural bone extracellular matrix (ECM) are type I collagen and Ca-P based apatite (Stevens, 2008). Eventually, what defines bone as a mineralized tissue is the deposition of inorganic apatite onto collagen fibrils. The inorganic Ca-P based apatite constitutes about 50% of the bone weight, and thus forms an integral part of natural bone minerals. Initially the apatite was assumed to be hydroxyapatite (HA), however, due to the presence of significant amount of carbonate, it is better defined as carbonate hydroxyapatite (CHA). Though the clinical interest is mainly on HA or CHA, other composition to the mineral phase such as tricalcium Phosphate (TCP), bioactive glasses and their combinations were studied as well (Stevens, 2008; Surmenev et al., 2014).

Ca-P apatite coating can be coated onto implants and/or scaffolds through physical deposition techniques or wet chemical techniques (Surmenev et al., 2014). Different physical deposition techniques such as plasma spraying (De Groot, Geesink, Klein, & Serekian, 1987; Weng et al., 1993), matrix-assisted pulsed laser evaporation (Bigi et al., 2009; Miroiu et al., 2010), ion-beam-assisted deposition(Choi, Kim, & Lee, 2000), radio-frequency (RF) magnetron sputtering (van Dijk, Schaeken, Wolke, & Jansen, 1996; Yamashita et al., 1994) have been used successfully to develop HA coatings. However, the common limitation of these physical processes lies in the difficulty to incorporate organic components such as collagen during the coating process. Therefore, the recent trend is to develop wet-chemical deposition techniques which has the potential to incorporate biomolecules along with HA. Kokubo *et al.* first reported the use of simulated body fluid (SBF) as a wet-chemical deposition technique for biomimetic growth of apatite coatings on bioactive $CaO–SiO_2$ glasses by mimicking the natural biomineralization process (Kokubo et al., 1990). The proposed SBF by Kokubo *et. al.* has inorganic ion concentration similar to those of human blood plasma but does not contain any protein or cells with a pH of 7.25-7.40 at 36.5 °C (Kamitakahara, Ohtsuki, & Miyazaki, 2007). In this classical biomimetic approach the implants and/or scaffolds are immersed into 1×SBF or 1.5×SBF, with the SBF being refreshed every other day for 1–4 weeks. Therefore, though SBF was successful to develop biomimetic Ca-P apatite, the time consuming nature of the process (I. S. Park et al., 2008; Xiong, Li, Wang, Hodgson, & Wen, 2008), requirement of a constant pH and replenishment of the concentrated SBF frequently to maintain super-saturation for apatite crystal growth lead to the development of simple biomimetic process which takes only a few of hours to develop CHA on polymer/ceramic composite scaffolds through the initiation of nucleation sites (M. Tarik Arafat, Lam, Ekaputra, Wong, He, et al., 2011; M. Tarik Arafat, Lam, Ekaputra, Wong, Li, et al., 2011). Figure 2 (in appendix) demonstrates the schematic diagram of the CHA coating through biomimetic precipitation. In this process, to promote CHA mineralization on the additive manufactured poly(ε-caprolactone)/tri-calcium phosphate (PCL/TCP) scaffolds (M. T. Arafat,

Savalani, Gibson, & Asme, 2009; M. Tarik Arafat, Gibson, & Li, 2014), $CaHPO_4$ was coated onto the scaffold as a nucleation site (M. Tarik Arafat, Lam, Ekaputra, Wong, He, et al., 2011; M. Tarik Arafat, Lam, Ekaputra, Wong, Li, et al., 2011). CHA coating was subsequently coated onto the $CaHPO_4$ coated PCL/TCP through a solution based biomimetic process. This biomimetic process also showed efficacy to incorporate biological molecules like gelatin (M. Tarik Arafat, Lam, Ekaputra, Wong, Li, et al., 2011).

The quality of the Ca-P based apatite coatings largely depends on the coating deposition processes. The parameters of the coating deposition are the cause of the variations of the physical and structural properties such as crystallinity, phase composition, porosity, surface roughness and thickness of the coatings. A recent comprehensive study on Ca-P coating found the importance of surface topography of the coatings on cellular adhesion, however, unable to conclude the impact of crystallinity due to the absence of comparable studies (Surmenev et al., 2014).

The main purpose of Ca-P based apatite coatings is to stimulate and accelerate new bone formation, and thereby, promote early integration of the implant. Clinical studies found that HA coatings enhanced new bone formation on implant surface, which results in shortening the post-operative period (Habibovic, Barrere, van Blitterswijk, de Groot, & Layrolle, 2002; Surmenev et al., 2014). Recently in clinical settings, HA coatings are routinely applied to implants, and in long-term most of the clinical studies have positive results on HA-coated implants (Surmenev et al., 2014). However, there are reports showing no difference between coated and non-coated implants (J.-M. Lee & Lee, 2007; Lombardi, Berend, & Mallory, 2006). This could be due to the common failures such as delamination, fragmentation and migration associated with coatings, which are related to the thicker coatings, and is less relevant to implants and/or scaffolds with thinner coatings (M. Tarik Arafat, Lam, Ekaputra, Wong, Li, et al., 2011; Chim et al., 2006; Goodman et al., 2013).

Regulatory and Development Context of Anti-Bacterial Coatings

Along with the increased trend of using Ca-P functional coatings, FDA and International Organization for Standardization (ISO) have set the minimal requirement for the Ca-P based hydroxyapatite, and it stated that chemical purity of HA should be as high as possible ($\geq 95\%$), with a Ca/P ratio of 1.67 (ASTM-F1185–03, 2009; ISO13779–2, 2008; ISO13779–3, 2008; ISO13779–4, 2002; Surmenev et al., 2014). However, there is no agreement on the crystallinity, which can be varied from 50 to 90% (usually approximately 70%) (Surmenev et al., 2014). FDA suggest to include information on the physical characteristics of the coatings such as particle size, particle size distribution, porosity, pore volume, thickness, tolerances, surface roughness and abrasion to get the regulatory approval ("Guidance for Industry and FDA Staff - Class II Special Controls Guidance Document: Root-form Endosseous Dental Implants and Endosseous Dental Abutments. (2004)."). FDA also expects data relating to the deposition process used and the post-deposition treatment, chemical analysis of the powder before and after coating including Ca/P ratios in atomic percent and elemental analysis and x-ray diffraction pattern of the powder and the coating in terms of relative intensity versus diffraction angle ("Guidance for Industry and FDA Staff - Class II Special Controls Guidance Document: Root-form Endosseous Dental Implants and Endosseous Dental Abutments. (2004).").

Collagen and Peptides Based Coatings

Integrins are a widely expressed family of heterodimeric ($\alpha\beta$) transmembrane receptors which play a crucial role in providing anchorage forces and trigger signals regulating cell adhesion and growth (Petrie et al., 2008). As an example, $\alpha_V\beta_3$ Integrin play essential role in osteocyte-polarized mechanosensing and mechanotransduction (Thi, Suadicani, Schaffler, Weinbaum, & Spray, 2013). Eventually, during cell-implant interactions cells recognize and adhere to implant and/or scaffolds surface via integrin receptors to biomolecules present onto the implant and/or scaffolds surface. Considering the vital role of integrins in cell-implant interaction, coating implants and/or scaffolds with integrin ligands such as extracellular matrix (ECM) components and short motifs derived from ECM is a logical approach (Puleo & Nanci, 1999; Schonherr & Hausser, 2000). The most abundant component of natural bone ECM is collagen I, and due to its excellent biological features and physicochemical properties, it has been among the most widely used biomaterials for biomedical applications (Ferreira, Gentile, Chiono, & Ciardelli, 2012). In human bone it constitutes around 20 wt% of the bone elements (Suchanek & Yoshimura, 1998). Collagen coated Ti showed better osteoblast cell adhesion compared to the uncoated via integrin $\beta1$ based pathway (Geissler et al., 2000; Morra et al., 2006; Rammelt et al., 2004). Collagen coated implants also showed enhanced osteointegration in vivo (Morra et al., 2006; Morra et al., 2005).

Besides collagen chondroitin sulfate (CS), which is also an important glycosaminoglycane found in cartilage and bone, helps to mediate the binding of bone-like cells to the matrix and capture soluble molecules such as growth factors into the matrix and cell surfaces (Forster et al., 2012; Ruoslahti & Yamaguchi, 1991). *In vitro* studies revealed that collagen/CS coating on implants and/or scaffolds improves cell adhesion, proliferation and differentiation (Bierbaum et al., 2006; Rentsch, Hofmann, Breier, Rentsch, & Scharnweber, 2009). It appears that CS is promising as a further addition to biomaterials to enhance bone healing (Forster et al., 2012). *In vivo* CS modified HA-collagen cement implants showed a significantly earlier callus reaction and new bone formation compared to HA-collagen cement (Schneiders et al., 2009). Other *in vivo* study of collagen-CS coated Ti implant also found a positive effect on new bone information, however, there is still lacking of direct comparison of collagen-CS with just collagen coated samples (Stadlinger et al., 2009).

Cell adhesion and bone formation from osteoblast can also be increased by using peptides coating (Auernheimer et al., 2005; Elmengaard, Bechtold, & Soballe, 2005; Rezania & Healy, 1999; Schuler et al., 2006; Wojtowicz et al., 2010). One of the most well-known peptides is the Arg-Gly-Asp (RGD), an amino acid sequence in fibronectin which serves as a primary cell attachment cue (Petrie et al., 2008; Schaffner & Dard, 2003). Smaller synthetic peptides containing amino acid sequence RGD facilitates cell attachment similarly as its considerably larger parental molecule (Schaffner & Dard, 2003). GFOGER is another peptide which is derived from collagen and involved in osteogenesis through the binding of $\alpha2\beta1$ integrin receptor. GFOGER coated PCL scaffolds resulted in 2-fold increase in bone volume at 12 weeks in a rat model (Wojtowicz et al., 2010).

Inorganic-Organic Composite Coatings

Inorganic-organic composite coating aims to 'mimic' the composite nature of natural bone ECM. The inorganic-organic composite coatings offer improved biological functionality over coatings consisting of merely inorganic or organic components (L. de Jonge, Leeuwenburgh, Wolke, & Jansen, 2008; Surmenev et al., 2014). In this regards composite coatings made of collagen incorporated Ca-P apatite generated a great deal of interest. It should be noted here the Ca-P coating synthesizes which use extreme conditions such as high temperature or non-physiological condition are not feasible to incorporate biomolecules like collagen. However, researchers have successfully synthesized Ca-P/collagen coatings through suitable process such as spin coating (Teng et al., 2008), elecgtrophoretic deposition (Fan, Duan, & Wang, 2005; Manara et al., 2008; Miao et al., 2009), electrospray deposition (L. T. de Jonge et al., 2010), perfusion-flow technique(Antebi et al., 2013) and biomimetic processes (M. Tarik Arafat, Lam, Ekaputra, Wong, He, et al., 2011; M. Tarik Arafat, Lam, Ekaputra, Wong, Li, et al., 2011; Xia, Yu, & Wei, 2012). Among various coating approaches, the biomimetic approach acts in a more efficient and similar way to the natural system (Mann et al., 1993).

Gelatin which is a derivative of collagen has also been successfully incorporated into carbonated hydroxyapatite (CHA) coating through biomimetic co-precipitation process on additive manufactured scaffolds (M. Tarik Arafat, Lam, Ekaputra, Wong, He, et al., 2011; M. Tarik Arafat, Lam, Ekaputra, Wong, Li, et al., 2011). Figure 3 (in Appendix) shows the CHA and CHA-gelatin composite coated additive manufactured scaffolds with a uniform distribution of coating. The CHA coating showed globular apatite while CHA-gelatin coating showed plate-like apatite of slightly larger size than CHA particles. The morphological difference between the CHA coating and CHA–gelatin composite coating could be due to incorporation of gelatin in the biomimetic co-precipitation process, as known that the adsorption of protein on the mineral surface can alter crystal nucleation and growth during biomineralization (Combes & Rey, 2002). CHA-gelatin composite coated scaffolds showed the most uniform distribution of cells and tissue sheet, specifically within the interior of the scaffolds, among the non-coated, CHA coated and CHA-gelatin composite coated scaffolds groups. Thus it was proved that inorganic-organic composite coating had a significant influence on cell and tissue distribution. CHA–gelatin composite coating also showed lesser tendency of flaking from the scaffolds during the 28 days of *in vitro* study (M. Tarik Arafat, Lam, Ekaputra, Wong, Li, et al., 2011).

RECENT TRENDS

Growth Factors Delivery

Recently, apatite coatings have been used as a career to deliver growth factors, and thus integrate osteoinductive property of the growth factors with the osteoconductive property of the apatite coatings (He et al., 2012; Y. Liu, Huse, de Groot, Buser, & Hunziker, 2007; Y. Liu, Li, Hunziker, & De Groot, 2006; Saran, Zhang, & Turcotte, 2011; H. S. Yang et al., 2011). Among different growth factors bone morphogenetic protein (BMP) is the most common one, and approved by FDA for treatment of fractures (Nauth, Ristevski, Li, & Schemitsch, 2011). To evaluate the effective delivery modes and efficacy of BMP-2 in association with the implants, BMP-2 was incorporated into Ca-P coated Ti implants through either adsorption or co-precipitation process (Y. Liu et al., 2007). After studying the implants at a subcutaneous ossification site in rats for about 5 weeks, it was found that BMP-2 does not show any osteogenic property when it is directly adsorbed onto bare implant surface. Eventually, the main drawback of the direct adsorption is its inconsistent release profile due to the initial burst release. However, when BMP-2 was adsorbed onto a Ca-P coating it was osteoinductive, but not highly efficacious. On the other hand when BMP-2 was introduced in the coating process, and thus incorporated into the Ca-P coating through co-precipitation, it acts as a potent bone-inducer. Other study also supported the fact that co-precipitation is an effective strategy in securing BMP-2 in the crystal lattice structure (Y. Liu et al., 2006). Ionic interaction is another way to secure sustained release of BMP-2, where negatively charged CS can be coupled with BMP-2; and subsequently this nanocomplex of CS and BMP-2 was incorporated during the apatite coating process (Bae, Choi, Joung, Park, & Han, 2012). The immobilized BMP-2 was released in a moderate rate for 4 weeks without any initial burst (Bae et al., 2012). Heparin, a linear polysaccharide that can also be used to protect growth factors from degradation through electrostatic interaction (Sakiyama-Elbert, 2014). Heparin incorporated coating was found to be effective in immobilizing BMP-2 and extending growth factor retention time. (Hee Seok Yang et al., 2010; Hee Seok Yang et al., 2012). Ca-P coating showed to be efficient in rendering long term release of BMP-2 with retained osteogenic efficacy not only on metallic implants but also on polymeric scaffolds (H. S. Yang et al., 2011). Study on non-coated collage scaffolds showed an initial burst release of BMP-2 with around 98% release within a day; whereas, Ca-P coated scaffolds showed around 80% release over a time of 13 days, which also showed the sustained release of growth factor due to coating (H. S. Yang et al., 2011). Ca-P coating can also be used for sustained release of multiple growth factors for over 60 days with dual release profile (Suarez-Gonzalez et al., 2014). Rather than achieving sustained release of growth factors, another challenge is to increase the growth factor incorporation rate. However, the growth factor incorporation rate into Ca-P coatings can be as low as 3% to 15%, which limits to its widespread clinical use (J. Wang, Guo, Liu, Wei, & Wu, 2014). Growth factor incorporation rate can be improved through adjusting the substrate surface area to SBF volume ratio; and it was found that incorporation rate of 90% of bovine serum albumin (BSA) can be achieved with the substrate surface area to SBF ratio of 0.072 (J. Wang et al., 2014).

Gene Delivery

Gene delivery provides an alternative way to induce osteoinductivity to the implants, and it has the advantage of a longer half-life and lower expense compared to the growth factors (Partridge & Oreffo, 2004). In gene delivery approach, the transfected cells acts as a "bioreactors" by generating the proteins into their local microenvironment (J. Wang et al., 2014). Genes can be incorporated into cells through virally and non-virally (Luong, McFalls, & Kohn, 2009; J. Park et al., 2003; Partridge & Oreffo, 2004). Though viral vectors showed higher transfection efficiency compared to their non-viral counterpart, it possess the risk of mutations and immunogenic rejection (Partridge & Oreffo, 2004). Non-viral gene delivery has been achieved by co-precipitating DNA with Ca-P (Kofron & Laurencin, 2004), and the transfection efficiency of the system can be improved by incorporating plasmid DNA complexed with a cationic lipid into Ca-P apatite (Luong et al., 2009). The stability of the plasmid DNA–cationic lipid complex has been retained during the biomimetic co-precipitation process; and thus it proves the potential of Ca-P apatite coating to deliver gene therapy (Luong et al., 2009).

FUTURE CHALLENGES

The future challenges will lie on combining various favourable properties of coating into a single multifunctional coating. As an example, to integrate osteointegration and anti-infective properties in a multifunctional coating is challenging as the host cells and bacteria shares the same adhesive mechanism to attach onto the implants. In this respect, the concept of multi-functional coating has recently been explored (Bronk et al., 2014; Chen et al., 2006). However, the works related to multi-functional coatings are still in their initial stage, and extensive research is required to understand the response of cells and bacterial to different biomolecules.

Additional work is needed to determine the optimal characteristics of coating, as an example, although it is generally accepted that thinner coatings are desirable, there is no defined desired value of thickness yet. Moreover, most of coating synthesis route is not scalable. Hence, repeatable, controlled and scalable synthesis processes to form functional coating is essential to get regulatory approval for clinical application.

CONCLUSION

Along with the increase in life expectancy the number of patients who need implants and/or scaffolds to contain the bone disorder has been increasing dramatically. Old generation of orthopaedic implants were focused only on the mechanical properties of the implants and/or scaffolds. However, the biological aspects such as bacterial resistance, osteoconductivity and osteoinductivity are becoming crucial for an early recovery, and thus to reduce physical and economic burden.

Considering implant associated infection is the most common cause of complications in orthopaedic surgery, and different antibacterial coatings have been studied. Silver and silver ions, which are known for their anti-infective properties, showed good potential for antibacterial coating. Antibiotics can be incorporated into apatite coatings, and released through a controlled manner to prevent bacterial infection.

To improve the osteoconductive properties of the implants and/or scaffolds Ca-P based apatite coating can be used. Among various Ca-P based apatite coating synthesis processes, biomimetic approach allows the incorporation of ECM molecules through co-precipitation process. Such composite coatings consist of organic ECM molecules and inorganic Ca-P showed both osteoconductive and osteoinductive properties, and thereby ensure early osteointegration of the implants to the surrounding tissues. Recently, growth factors and gene has also been successfully incorporated into Ca-P based apatite coatings.

In conclusion, the functional coatings on the orthopaedic implants and/or scaffolds showed enormous potential to prevent bacterial infection and early osteointegration. To move forward, the concept of multifunctional coatings can be introduced, which will help to induce anti-infective properties of the coatings without compromising osteointegration properties. However, as many of the coatings are still in the developmental stage, extensive long term *in vivo* study is required. Researchers, surgeons and manufacturers need to work together to develop and test the coatings in clinical settings. Last but not the least sustained and scalable manufacturing route to synthesize functional coating is essential for securing the regulatory approval for clinical application.

REFERENCES

Abukawa, H., Papadaki, M., Abulikemu, M., Leaf, J., Vacanti, J. P., Kaban, L. B., & Troulis, M. J. (2006). The engineering of craniofacial tissues in the laboratory: A review of biomaterials for scaffolds and implant coatings. *Dental Clinics of North America*, *50*(2), 205–216. doi:10.1016/j.cden.2005.11.006 PMID:16530058

An, Y. H., & Friedman, R. J. (1996). Prevention of sepsis in total joint arthroplasty. *The Journal of Hospital Infection, 33*(2), 93–108. doi:10.1016/S0195-6701(96)90094-8 PMID:8808743

Antebi, B., Cheng, X., Harris, J.N., Gower, L.B., Chen, X.-D., & Ling, J. (2013). Biomimetic collagen-hydroxyapatite composite fabricated via a novel perfusion-flow mineralization technique. *Tissue Engineering Part C-Methods, 19*(7), 487-496. doi:10.1089/ten.tec.2012.0452

Arafat, M. T., Gibson, I., & Li, X. (2014). State of the art and future direction of additive manufactured scaffolds-based bone tissue engineering. *Rapid Prototyping Journal, 20*(1), 13–26. doi:10.1108/RPJ-03-2012-0023

Arafat, M. T., Lam, C. X. F., Ekaputra, A. K., Wong, S. Y., He, C., Hutmacher, D. W., & Gibson, I. et al. (2011). High performance additive manufactured scaffolds for bone tissue engineering application. *Soft Matter, 7*(18), 8013–8022. doi:10.1039/c1sm05793f

Arafat, M. T., Lam, C. X. F., Ekaputra, A. K., Wong, S. Y., Li, X., & Gibson, I. (2011). Biomimetic composite coating on rapid prototyped scaffolds for bone tissue engineering. *Acta Biomaterialia, 7*(2), 809–820. doi:10.1016/j.actbio.2010.09.010 PMID:20849985

Arafat, M.T., Savalani, M.M., Gibson, I., & Asme. (2009). *Improving the mechanical properties in tissue engineered scaffolds.*

ASTM-F1185–03. Standard specification for composition of hydroxylapatite for surgical implants. (2009).

Auernheimer, J., Zukowski, D., Dahmen, C., Kantlehner, M., Enderle, A., Goodman, S. L., & Kessler, H. (2005). Titanium implant materials with improved riocompatibility through coating with phosphonate-anchored cyclic RGD peptides. *ChemBioChem, 6*(11), 2034–2040. doi:10.1002/cbic.200500031 PMID:16206226

Bae, S. E., Choi, J., Joung, Y. K., Park, K., & Han, D. K. (2012). Controlled release of bone morphogenetic protein (BMP)-2 from nanocomplex incorporated on hydroxyapatite-formed titanium surface. *Journal of Controlled Release, 160*(3), 676–684. doi:10.1016/j.jconrel.2012.04.021 PMID:22543042

Bierbaum, S., Douglas, T., Hanke, T., Scharnweber, D., Tippelt, S., Monsees, T. K., & Worch, H. et al. (2006). Collageneous matrix coatings on titanium implants modified with decorin and chondroitin sulfate: Characterization and influence on osteoblastic cells. *Journal of Biomedical Materials Research. Part A, 77A*(3), 551–562. doi:10.1002/jbm.a.30572 PMID:16498599

Bigi, A., Boanini, E., Capuccini, C., Fini, M., Mihailescu, I. N., Ristoscu, C., & Torricelli, P. et al. (2009). Biofunctional alendronate–Hydroxyapatite thin films deposited by Matrix Assisted Pulsed Laser Evaporation. *Biomaterials, 30*(31), 6168–6177. doi:10.1016/j.biomaterials.2009.07.066 PMID:19692118

Bishop, J. A., Palanca, A. A., Bellino, M. J., & Lowenberg, D. W. (2012). Assessment of compromised fracture healing. *The Journal of the American Academy of Orthopaedic Surgeons, 20*(5), 273–282. doi:10.5435/JAAOS-20-05-273 PMID:22553099

Bologna, R. A., Tu, L. M., Polansky, M., Fraimow, H. D., Gordon, D. A., & Whitmore, K. E. (1999). Hydrogel/silver ion-coated urinary catheter reduces nosocomial urinary tract infection rates in intensive care unit patients: A multicenter study. *Urology*, *54*(6), 982–987. doi:10.1016/S0090-4295(99)00318-0 PMID:10604694

Bosetti, M., Masse, A., Tobin, E., & Cannas, M. (2002). Silver coated materials for external fixation devices: In vitro biocompatibility and genotoxicity. *Biomaterials*, *23*(3), 887–892. doi:10.1016/S0142-9612(01)00198-3 PMID:11771707

Bronk, J. K., Russell, B. H., Rivera, J. J., Pasqualini, R., Arap, W., Hook, M., & Barbu, E. M. (2014). A multifunctional streptococcal collagen-mimetic protein coating prevents bacterial adhesion and promotes osteoid formation on titanium. *Acta Biomaterialia*, *10*(7), 3354–3362. doi:10.1016/j.actbio.2014.04.005 PMID:24732634

Campbell, A. A., Song, L., Li, X. S., Nelson, B. J., Bottoni, C., Brooks, D. E., & DeJong, E. S. (2000). Development, characterization, and anti-microbial efficacy of hydroxyapatite-chlorhexidine coatings produced by surface-induced mineralization. *Journal of Biomedical Materials Research*, *53*(4), 400–407. doi:10.1002/1097-4636(2000)53:4<400::AID-JBM14>3.0.CO;2-Z PMID:10898881

Chang, Y.-Y., Lai, C.-H., Hsu, J.-T., Tang, C.-H., Liao, W.-C., & Huang, H.-L. (2012). Antibacterial properties and human gingival fibroblast cell compatibility of TiO2/Ag compound coatings and ZnO films on titanium-based material. *Clinical Oral Investigations*, *16*(1), 95–100. doi:10.1007/s00784-010-0504-9 PMID:21234622

Charville, G. W., Hetrick, E. M., Geer, C. B., & Schoenfisch, M. H. (2008). Reduced bacterial adhesion to fibrinogen-coated substrates via nitric oxide release. *Biomaterials*, *29*(30), 4039–4044. doi:10.1016/j.biomaterials.2008.07.005 PMID:18657857

Chen, W., Liu, Y., Courtney, H. S., Bettenga, M., Agrawal, C. M., Bumgardner, J. D., & Ong, J. L. (2006). In vitro anti-bacterial and biological properties of magnetron co-sputtered silver-containing hydroxyapatite coating. *Biomaterials*, *27*(32), 5512–5517. doi:10.1016/j.biomaterials.2006.07.003 PMID:16872671

Cheng, H., Li, Y., Huo, K., Gao, B., & Xiong, W. (2014). Long-lasting in vivo and in vitro antibacterial ability of nanostructured titania coating incorporated with silver nanoparticles. *Journal of Biomedical Materials Research. Part A*, *102*(10), 3488–3499. doi:10.1002/jbm.a.35019 PMID:24178451

Chim, H., Hutmacher, D. W., Chou, A. M., Oliveira, A. L., Reis, R. L., Lim, T. C., & Schantz, J. T. (2006). A comparative analysis of scaffold material modifications for load-bearing applications in bone tissue engineering. *International Journal of Oral and Maxillofacial Surgery*, *35*(10), 928–934. doi:10.1016/j.ijom.2006.03.024 PMID:16762529

Choi, J.-M., Kim, H.-E., & Lee, I.-S. (2000). Ion-beam-assisted deposition (IBAD) of hydroxyapatite coating layer on Ti-based metal substrate. *Biomaterials*, *21*(5), 469–473. doi:10.1016/S0142-9612(99)00186-6 PMID:10674811

Combes, C., & Rey, C. (2002). Adsorption of proteins and calcium phosphate materials bioactivity. *Biomaterials*, *23*(13), 2817–2823. doi:10.1016/S0142-9612(02)00073-X PMID:12059033

Darouiche, R. O. (2004). Current concepts - Treatment of infections associated with surgical implants. *The New England Journal of Medicine*, *350*(14), 1422–1429. doi:10.1056/NEJMra035415 PMID:15070792

Davies, D. (2003). Understanding biofilm resistance to antibacterial agents. *Nature Reviews. Drug Discovery*, *2*(2), 114–122. doi:10.1038/nrd1008 PMID:12563302

De Groot, K., Geesink, R., Klein, C. P. A. T., & Serekian, P. (1987). Plasma sprayed coatings of hydroxylapatite. *Journal of Biomedical Materials Research*, *21*(12), 1375–1381. doi:10.1002/jbm.820211203 PMID:3429472

de Jonge, L. T., Leeuwenburgh, S. C. G., van den Beucken, J. J. J. P., te Riet, J., Daamen, W. F., Wolke, J. G. C., & Jansen, J. A. et al. (2010). The osteogenic effect of electrosprayed nanoscale collagen/calcium phosphate coatings on titanium. *Biomaterials*, *31*(9), 2461–2469. doi:10.1016/j.biomaterials.2009.11.114 PMID:20022365

de Jonge, L. T., Leeuwenburgh, S. C. G., Wolke, J. G. C., & Jansen, J. A. (2008). Organic–inorganic surface modifications for titanium implant surfaces. *Pharmaceutical Research*, *25*(10), 2357–2369. doi:10.1007/s11095-008-9617-0 PMID:18509601

DeJong, E. S., DeBerardino, T. M., Brooks, D. E., Nelson, B. J., Campbell, A. A., Bottoni, C. R., & McManus, A. T. et al. (2001). Antimicrobial efficacy of external fixator pins coated with a lipid stabilized hydroxyapatite/chlorhexidine complex to prevent pin tract infection in a goat model. *The Journal of Trauma Injury Infection and Critical Care*, *50*(6), 1008–1013. doi:10.1097/00005373-200106000-00006 PMID:11426113

Dowling, D. P., Betts, A. J., Pope, C., McConnell, M. L., Eloy, R., & Arnaud, M. N. (2003). Anti-bacterial silver coatings exhibiting enhanced activity through the addition of platinum. *Surface and Coatings Technology*, *163–164*(0), 637–640. doi:10.1016/S0257-8972(02)00689-8

Draft Guidance for Industry and FDA Staff - Premarket Notification [510(k)] Submissions for Medical Devices that Include Antimicrobial Agents. (2007).

Elmengaard, B., Bechtold, J. E., & Soballe, K. (2005). In vivo study of the effect of RGD treatment on bone ongrowth on press-fit titanium alloy implants. *Biomaterials*, *26*(17), 3521–3526. doi:10.1016/j.biomaterials.2004.09.039 PMID:15621242

Fan, Y. W., Duan, K., & Wang, R. Z. (2005). A composite coating by electrolysis-induced collagen self-assembly and calcium phosphate mineralization. *Biomaterials*, *26*(14), 1623–1632. doi:10.1016/j.biomaterials.2004.06.019 PMID:15576136

Feng, Q. L., Wu, J., Chen, G. Q., Cui, F. Z., Kim, T. N., & Kim, J. O. (2000). A mechanistic study of the antibacterial effect of silver ions on Escherichia coli and Staphylococcus aureus. *Journal of Biomedical Materials Research*, *52*(4), 662–668. doi:10.1002/1097-4636(20001215)52:4<662::AID-JBM10>3.0.CO;2-3 PMID:11033548

Ferreira, A. M., Gentile, P., Chiono, V., & Ciardelli, G. (2012). Collagen for bone tissue regeneration. *Acta Biomaterialia*, *8*(9), 3191–3200. doi:10.1016/j.actbio.2012.06.014 PMID:22705634

Forster, Y., Rentsch, C., Schneiders, W., Bernhardt, R., Simon, J. C., Worch, H., & Rammelt, S. (2012). Surface modification of implants in long bone. *Biomatter*, *2*(3), 149–157. doi:10.4161/biom.21563 PMID:23507866

Geissler, U., Hempel, U., Wolf, C., Scharnweber, D., Worch, H., & Wenzel, K. W. (2000). Collagen type I-coating of Ti6Al4V promotes adhesion of osteoblasts. *Journal of Biomedical Materials Research*, *51*(4), 752–760. doi:10.1002/1097-4636(20000915)51:4<752::AID-JBM25>3.0.CO;2-7 PMID:10880125

Glinel, K., Thebault, P., Humblot, V., Pradier, C. M., & Jouenne, T. (2012). Antibacterial surfaces developed from bio-inspired approaches. *Acta Biomaterialia*, *8*(5), 1670–1684. doi:10.1016/j.actbio.2012.01.011 PMID:22289644

Goodman, S. B., Yao, Z. Y., Keeney, M., & Yang, F. (2013). The future of biologic coatings for orthopaedic implants. *Biomaterials*, *34*(13), 3174–3183. doi:10.1016/j.biomaterials.2013.01.074 PMID:23391496

Grecu, M., Novac, G., Ionita, D., & Ungureanu, C. (2011). Incorporation of Tobramycin Biomimetic in Hydroxyapatite Coating on CoCrMo Alloy and its Antimicrobial Acitivity. *Revista De Chimie*, *62*(3), 352–356.

Gristina, A. G. (1987). Biomaterial-centered infection-microbial adhesion versus tissue integration. *Science*, *237*(4822), 1588–1595. doi:10.1126/science.3629258 PMID:3629258

Guidance for Industry and FDA Staff - Class II Special Controls Guidance Document: Root-form Endosseous Dental Implants and Endosseous Dental Abutments. (2004).

Guyomard, A., De, E., Jouenne, T., Malandain, J. J., Muller, G., & Glinel, K. (2008). Incorporation of a hydrophobic antibacterial peptide into amphiphilic polyelectrolyte multilayers: A bioinspired approach to prepare biocidal thin coatings. *Advanced Functional Materials*, *18*(5), 758–765. doi:10.1002/adfm.200700793

Habibovic, P., Barrere, F., Blitterswijk, C. A., Groot, K., & Layrolle, P. (2002). Biomimetic hydroxyapatite coating on metal implants. *Journal of the American Ceramic Society*, *85*(3), 517–522. doi:10.1111/j.1151-2916.2002.tb00126.x

Hasan, J., Crawford, R. J., & Ivanova, E. P. (2013). Antibacterial surfaces: The quest for a new generation of biomaterials. *Trends in Biotechnology*, *31*(5), 295-304. doi:10.1016/j.tibtech.2013.01.017 PMID:23434154

He, J., Huang, T., Gan, L., Zhou, Z. K., Jiang, B., Wu, Y., & Gu, Z. W. et al. (2012). Collagen-infiltrated porous hydroxyapatite coating and its osteogenic properties: In vitro and in vivo study. *Journal of Biomedical Materials Research. Part A*, *100*(7), 1706–1715. doi:10.1002/jbm.a.34121 PMID:22447757

Hetrick, E. M., & Schoenfisch, M. H. (2006). Reducing implant-related infections: Active release strategies. *Chemical Society Reviews*, *35*(9), 780–789. doi:10.1039/b515219b PMID:16936926

Hu, H., Zhang, W., Qiao, Y., Jiang, X., Liu, X., & Ding, C. (2012). Antibacterial activity and increased bone marrow stem cell functions of Zn-incorporated TiO2 coatings on titanium. *Acta Biomaterialia*, *8*(2), 904–915. doi:10.1016/j.actbio.2011.09.031 PMID:22023752

Iqbal, N., Abdul Kadir, M. R., Nik Malek, N. A. N., Humaimi Mahmood, N., Raman Murali, M., & Kamarul, T. (2012). Rapid microwave assisted synthesis and characterization of nanosized silver-doped hydroxyapatite with antibacterial properties. *Materials Letters*, *89*, 118–122. doi:10.1016/j.matlet.2012.08.057

ISO13779–2. Implants for surgery – hydroxyapatite – Part 2: Coatings of hydroxyapatite.(2008).

ISO13779–3. Implants for surgery – hydroxyapatite – Part 3: Chemical analysis and characterization of crystallinity and phase purity. (2008).

ISO13779–4. Implants for surgery – hydroxyapatite – Part 4: Determination of coating adhesion strength. (2002).

Kamitakahara, M., Ohtsuki, C., & Miyazaki, T. (2007). Coating of bone-like apatite for development of bioactive materials for bone reconstruction. *Biomedical Materials (Bristol, England)*, *2*(4), R17–R23. doi:10.1088/1748-6041/2/4/R01 PMID:18458474

Kofron, M. D., & Laurencin, C. T. (2004). Development of a calcium phosphate co-precipitate/poly(lactideco-glycolide) DNA delivery system: Release kinetics and cellular transfection studies. *Biomaterials*, *25*(13), 2637–2643. doi:10.1016/j.biomaterials.2003.09.042 PMID:14751750

Kokubo, T., Ito, S., Huang, Z. T., Hayashi, T., Sakka, S., Kitsugi, T., & Yamamuro, T. (1990). Ca,P-rich layer formed on high-strength bioactive glass-ceramic A-W. *Journal of Biomedical Materials Research*, *24*(3), 331–343. doi:10.1002/jbm.820240306 PMID:2156869

Kumar, R., & Munstedt, H. (2005). Silver ion release from antimicrobial polyamide/silver composites. *Biomaterials*, *26*(14), 2081–2088. doi:10.1016/j.biomaterials.2004.05.030 PMID:15576182

Kumar, R., & Münstedt, H. (2005). Polyamide/silver antimicrobials: Effect of crystallinity on the silver ion release. *Polymer International*, *54*(8), 1180–1186. doi:10.1002/pi.1828

Kwok, C. S., Horbett, T. A., & Ratner, B. D. (1999). Design of infection-resistant antibiotic-releasing polymers: II. Controlled release of antibiotics through a plasma-deposited thin film barrier. *Journal of Controlled Release*, *62*(3), 301–311. doi:10.1016/S0168-3659(99)00105-4 PMID:10528068

Kwok, C. S., Wan, C., Hendricks, S., Bryers, J. D., Horbett, T. A., & Ratner, B. D. (1999). Design of infection-resistant antibiotic-releasing polymers: I. Fabrication and formulation. *Journal of Controlled Release*, *62*(3), 289–299. doi:10.1016/S0168-3659(99)00106-6 PMID:10528067

Lee, D., Cohen, R. E., & Rubner, M. F. (2005). Antibacterial properties of Ag nanoparticle loaded multilayers and formation of magnetically directed antibacterial microparticles. *Langmuir*, *21*(21), 9651–9659. doi:10.1021/la0513306 PMID:16207049

Lee, J.-M., & Lee, C.-W. (2007). Comparison of hydroxyapatite-coated non-hydroxyapatite-coated noncemented total hip arthroplasty in same patients. *The Journal of Arthroplasty*, *22*(7), 1019–1023. doi:10.1016/j.arth.2007.04.029 PMID:17920475

Lim, P. N., Chang, L., Tay, B. Y., Guneta, V., Choong, C., Ho, B., & Thian, E. S. (2014). Proposed mechanism of antibacterial action of chemically modified apatite for reduced bone infection. *ACS Applied Materials & Interfaces*, *6*(19), 17082–17092. doi:10.1021/am504716g PMID:25198775

Liu, X., Chu, P. K., & Ding, C. (2010). Surface nano-functionalization of biomaterials. *Materials Science and Engineering R Reports*, *70*(3-6), 275–302. doi:10.1016/j.mser.2010.06.013

Liu, Y., Huse, R. O., de Groot, K., Buser, D., & Hunziker, E. B. (2007). Delivery mode and efficacy of BMP-2 in association with implants. *Journal of Dental Research*, *86*(1), 84–89. doi:10.1177/154405910708600114 PMID:17189469

Liu, Y., Li, J. P., Hunziker, E. B., & De Groot, K. (2006). Incorporation of growth factors into medical devices via biomimetic coatings. *Philosophical Transactions of the Royal Society a-Mathematical Physical and Engineering Sciences, 364*(1838), 233-248. doi:10.1098/rsta.2005.1685

Lombardi, A. V. Jr, Berend, K. R., & Mallory, T. H. (2006). Hydroxyapatite-coated titanium porous plasma spray tapered stem - Experience at 15 to 18 years. *Clinical Orthopaedics and Related Research*, *453*, 81–85. doi:10.1097/01.blo.0000238872.01767.09 PMID:17312587

Luong, L. N., McFalls, K. M., & Kohn, D. H. (2009). Gene delivery via DNA incorporation within a biomimetic apatite coating. *Biomaterials*, *30*(36), 6996–7004. doi:10.1016/j.biomaterials.2009.09.001 PMID:19775750

Manara, S., Paolucci, F., Palazzo, B., Marcaccio, M., Foresti, E., Tosi, G., & Roveri, N. et al. (2008). Electrochemically-assisted deposition of biomimetic hydroxyapatite-collagen coatings on titanium plate. *Inorganica Chimica Acta*, *361*(6), 1634–1645. doi:10.1016/j.ica.2007.03.044

Mann, S., Archibald, D. D., Didymus, J. M., Douglas, T., Heywood, B. R., Meldrum, F. C., & Reeves, N. J. (1993). Crystallization at inorganic-organic interfaces biomaterials and biomimetic synthesis. *Science*, *261*(5126), 1286–1292. doi:10.1126/science.261.5126.1286 PMID:17731856

Masse, A., Bruno, A., Bosetti, M., Biasibetti, A., Cannas, M., & Gallinaro, P. (2000). Prevention of pin track infection in external fixation with silver coated pins: Clinical and microbiological results. *Journal of Biomedical Materials Research*, *53*(5), 600–604. doi:10.1002/1097-4636(200009)53:5<600::AID-JBM21>3.0.CO;2-D PMID:10984710

Miao, S., Weng, W., Li, Z., Cheng, K., Du, P., Shen, G., & Han, G. (2009). Electrolytic deposition of octacalcium phosphate/collagen composite coating on titanium alloy. *Journal of Materials Science. Materials in Medicine*, *20*(1), 131–134. doi:10.1007/s10856-008-3552-9 PMID:18704651

Miroiu, F. M., Socol, G., Visan, A., Stefan, N., Craciun, D., Craciun, V., & Ducu, C. et al. (2010). Composite biocompatible hydroxyapatite–silk fibroin coatings for medical implants obtained by Matrix Assisted Pulsed Laser Evaporation. *Materials Science and Engineering B*, *169*(1–3), 151–158. doi:10.1016/j.mseb.2009.10.004

Morra, M., Cassinelli, C., Cascardo, G., Mazzucco, L., Borzini, P., Fini, M., & Giardino, R. et al. (2006). Collagen I-coated titanium surfaces: Mesenchymal cell adhesion and in vivo evaluation in trabecular bone implants. *Journal of Biomedical Materials Research. Part A*, *78A*(3), 449–458. doi:10.1002/jbm.a.30783 PMID:16721758

Morra, M., Cassinelli, C., Meda, L., Fini, M., Giavaresi, G., & Giardino, R. (2005). Surface analysis and effects on interfacial bone microhardness of collagen-coated titanium implants: A rabbit model. *The International Journal of Oral & Maxillofacial Implants*, *20*(1), 23–30. PMID:15747670

Nablo, B. J., Rothrock, A. R., & Schoenfisch, M. H. (2005). Nitric oxide-releasing sol–gels as antibacterial coatings for orthopedic implants. *Biomaterials*, *26*(8), 917–924. doi:10.1016/j.biomaterials.2004.03.031 PMID:15353203

Nauth, A., Ristevski, B., Li, R., & Schemitsch, E. H. (2011). Growth factors and bone regeneration: How much bone can we expect? *Injury-International Journal of the Care of the Injured*, *42*(6), 574–579. doi:10.1016/j.injury.2011.03.034 PMID:21489530

Park, I. S., Choi, U. J., Yi, H. K., Park, B. K., Lee, M. H., & Bae, T. S. (2008). Biomimetic apatite formation and biocompatibility on chemically treated Ti-6Al-7Nb alloy. *Surface and Interface Analysis*, *40*(1), 37–42. doi:10.1002/sia.2666

Park, J., Ries, J., Gelse, K., Kloss, F., von der Mark, K., Wiltfang, J., & Schneider, H. et al. (2003). Bone regeneration in critical size defects by cell-mediated BMP-2 gene transfer: A comparison of adenoviral vectors and liposomes. *Gene Therapy*, *10*(13), 1089–1098. doi:10.1038/sj.gt.3301960 PMID:12808439

Partridge, K. A., & Oreffo, R. O. C. (2004). Gene delivery in bone tissue engineering: Progress and prospects using viral and nonviral strategies. *Tissue Engineering*, *10*(1-2), 295–307. doi:10.1089/107632704322791934 PMID:15009954

Peng, L., Mendelsohn, A. D., LaTempa, T. J., Yoriya, S., Grimes, C. A., & Desai, T. A. (2009). Long-term small molecule and protein elution from TiO2 nanotubes. *Nano Letters*, *9*(5), 1932–1936. doi:10.1021/nl9001052 PMID:19323554

Perez-Jorge, C., Conde, A., Arenas, M. A., Perez-Tanoira, R., Matykina, E., de Damborenea, J. J., & Esteban, J. et al. (2012). In vitro assessment of Staphylococcus epidermidis and Staphylococcus aureus adhesion on TiO2 nanotubes on Ti-6Al-4V alloy. *Journal of Biomedical Materials Research. Part A*, *100A*(7), 1696–1705. doi:10.1002/jbm.a.34118 PMID:22447745

Petrie, T. A., Raynor, J. E., Reyes, C. D., Burns, K. L., Collard, D. M., & Garcia, A. J. (2008). The effect of integrin-specific bioactive coatings on tissue healing and implant osseointegration. *Biomaterials*, *29*(19), 2849–2857. doi:10.1016/j.biomaterials.2008.03.036 PMID:18406458

Puleo, D. A., & Nanci, A. (1999). Understanding and controlling the bone-implant interface. *Biomaterials*, *20*(23-24), 2311–2321. doi:10.1016/S0142-9612(99)00160-X PMID:10614937

Radin, S., Campbell, J. T., Ducheyne, P., & Cuckler, J. M. (1997). Calcium phosphate ceramic coatings as carriers of vancomycin. *Biomaterials*, *18*(11), 777–782. doi:10.1016/S0142-9612(96)00190-1 PMID:9177855

Rammelt, S., Schulze, E., Bernhardt, R., Hanisch, U., Scharnweber, D., Worch, H., & Biewener, A. et al. (2004). Coating of titanium implants with type-I collagen. *Journal of Orthopaedic Research*, *22*(5), 1025–1034. doi:10.1016/j.orthres.2004.02.011 PMID:15304275

Rentsch, B., Hofmann, A., Breier, A., Rentsch, C., & Scharnweber, D. (2009). Embroidered and surface modified polycaprolactone-co-lactide scaffolds as bone substitute: In vitro characterization. *Annals of Biomedical Engineering*, *37*(10), 2118–2128. doi:10.1007/s10439-009-9731-0 PMID:19626441

Rezania, A., & Healy, K. E. (1999). Integrin subunits responsible for adhesion of human osteoblast-like cells to biomimetic peptide surfaces. *Journal of Orthopaedic Research*, *17*(4), 615–623. doi:10.1002/jor.1100170423 PMID:10459771

Ruoslahti, E., & Yamaguchi, Y. (1991). Proteoglycans as modulators of growth factor activities. *Cell*, *64*(5), 867–869. doi:10.1016/0092-8674(91)90308-L PMID:2001586

Sakiyama-Elbert, S. E. (2014). Incorporation of heparin into biomaterials. *Acta Biomaterialia*, *10*(4), 1581–1587. doi:10.1016/j.actbio.2013.08.045 PMID:24021232

Saran, N., Zhang, R. W., & Turcotte, R. E. (2011). Osteogenic protein-1 delivered by hydroxyapatite-coated implants improves bone ingrowth in extracortical bone bridging. *Clinical Orthopaedics and Related Research*, *469*(5), 1470–1478. doi:10.1007/s11999-010-1573-4 PMID:20878288

Schaffner, P., & Dard, M. M. (2003). Structure and function of RGD peptides involved in bone biology. *Cellular and Molecular Life Sciences*, *60*(1), 119–132. doi:10.1007/s000180300008 PMID:12613662

Schierholz, J. M., Steinhauser, H., Rump, A. F. E., Berkels, R., & Pulverer, G. (1997). Controlled release of antibiotics from biomedical polyurethanes: Morphological and structural features. *Biomaterials*, *18*(12), 839–844. doi:10.1016/S0142-9612(96)00199-8 PMID:9184747

Schneiders, W., Reinstorf, A., Biewener, A., Serra, A., Grass, R., Kinscher, M., & Rammelt, S. et al. (2009). In vivo effects of modification of hydroxyapatite/collagen composites with and without chondroitin sulphate on bone remodeling in the sheep tibia. *Journal of Orthopaedic Research*, *27*(1), 15–21. doi:10.1002/jor.20719 PMID:18634066

Schonherr, E., & Hausser, H. J. (2000). Extracellular matrix and cytokines: A functional unit. *Developmental Immunology*, *7*(2-4), 89–101. doi:10.1155/2000/31748 PMID:11097204

Schreurs, W. J. A., & Rosenberg, H. (1982). Effect of silver ions on transport and retention of phosphate by escherichia coil. *Journal of Bacteriology*, *152*(1), 7–13. PMID:6749823

Schuler, M., Owen, G. R., Hamilton, D. W., de Wild, M., Textor, M., Brunette, D. M., & Tosatti, S. G. P. (2006). Biomimetic modification of titanium dental implant model surfaces using the RGDSP-peptide sequence: A cell morphology study. *Biomaterials*, *27*(21), 4003–4015. doi:10.1016/j.biomaterials.2006.03.009 PMID:16574219

Shen, X., Hu, Y., Xu, G., Chen, W., Xu, K., Ran, Q., & Cai, K. et al. (2014). Regulation of the biological functions of osteoblasts and bone formation by Zn-incorporated coating on microrough titanium. *ACS Applied Materials & Interfaces*, *6*(18), 16426–16440. doi:10.1021/am5049338 PMID:25148131

Simchi, A., Tamjid, E., Pishbin, F., & Boccaccini, A. R. (2011). Recent progress in inorganic and composite coatings with bactericidal capability for orthopaedic applications. *Nanomedicine; Nanotechnology, Biology, and Medicine*, *7*(1), 22–39. doi:10.1016/j.nano.2010.10.005 PMID:21050895

Sorensen, J. H., Lilja, M., Astrand, M., Sorensen, T. C., Procter, P., Stromme, M., & Steckel, H. (2014). Co-precipitation of Tobramycin into Biomimetically Coated Orthopedic Fixation Pins Employing Submicron-Thin Seed Layers of Hydroxyapatite. *Current Drug Delivery*, *11*(4), 501–510. doi:10.2174/1567201811666140311154104 PMID:24611653

Stadlinger, B., Bierbaum, S., Grimmer, S., Schulz, M. C., Kuhlisch, E., Scharnweber, D., & Mai, R. et al. (2009). Increased bone formation around coated implants. *Journal of Clinical Periodontology*, *36*(8), 698–704. doi:10.1111/j.1600-051X.2009.01435.x PMID:19531092

Stevens, M. M. (2008). Biomaterials for bone tissue engineering. *Materials Today*, *11*(5), 18–25. doi:10.1016/S1369-7021(08)70086-5

Stigter, M., Bezemer, J., de Groot, K., & Layrolle, P. (2004). Incorporation of different antibiotics into carbonated hydroxyapatite coatings on titanium implants, release and antibiotic efficacy. *Journal of Controlled Release*, *99*(1), 127–137. doi:10.1016/j.jconrel.2004.06.011 PMID:15342186

Stigter, M., de Groot, K., & Layrolle, P. (2002). Incorporation of tobramycin into biomimetic hydroxyapatite coating on titanium. *Biomaterials*, *23*(20), 4143–4153. doi:10.1016/S0142-9612(02)00157-6 PMID:12182316

Stobie, N., Duffy, B., Hinder, S. J., McHale, P., & McCormack, D. E. (2009). Silver doped perfluoropolyether-urethane coatings: Antibacterial activity and surface analysis. *Colloids and Surfaces. B, Biointerfaces*, *72*(1), 62–67. doi:10.1016/j.colsurfb.2009.03.014 PMID:19427177

Suarez-Gonzalez, D., Lee, J. S., Diggs, A., Lu, Y., Nemke, B., Markel, M., & Murphy, W. L. et al. (2014). Controlled multiple growth factor delivery from bone tissue engineering scaffolds via designed affinity. *Tissue Engineering. Part A*, *20*(15-16), 2077–2087. doi:10.1089/ten.tea.2013.0358 PMID:24350567

Suchanek, W., & Yoshimura, M. (1998). Processing and properties of hydroxyapatite-based biomaterials for use as hard tissue replacement implants. *Journal of Materials Research*, *13*(01), 94–117. doi:10.1557/JMR.1998.0015

Surmenev, R. A., Surmeneva, M. A., & Ivanova, A. A. (2014). Significance of calcium phosphate coatings for the enhancement of new bone osteogenesis - A review. *Acta Biomaterialia*, *10*(2), 557–579. doi:10.1016/j.actbio.2013.10.036 PMID:24211734

Tang, Y. Z., Chappell, H. F., Dove, M. T., Reeder, R. J., & Lee, Y. J. (2009). Zinc incorporation into hydroxylapatite. *Biomaterials*, *30*(15), 2864–2872. doi:10.1016/j.biomaterials.2009.01.043 PMID:19217156

Teng, S.-H., Lee, E.-J., Park, C.-S., Choi, W.-Y., Shin, D.-S., & Kim, H.-E. (2008). Bioactive nanocomposite coatings of collagen/hydroxyapatite on titanium substrates. *Journal of Materials Science. Materials in Medicine*, *19*(6), 2453–2461. doi:10.1007/s10856-008-3370-0 PMID:18219556

Thi, M. M., Suadicani, S. O., Schaffler, M. B., Weinbaum, S., & Spray, D. C. (2013). Mechanosensory responses of osteocytes to physiological forces occur along processes and not cell body and require alphaVbeta3 integrin. *Proceedings of the National Academy of Sciences of the United States of America*, *110*(52), 21012–21017. doi:10.1073/pnas.1321210110 PMID:24324138

van Dijk, K., Schaeken, H. G., Wolke, J. G. C., & Jansen, J. A. (1996). Influence of annealing temperature on RF magnetron sputtered calcium phosphate coatings. *Biomaterials*, *17*(4), 405–410. doi:10.1016/0142-9612(96)89656-6 PMID:8938234

Vargas-Reus, M. A., Memarzadeh, K., Huang, J., Ren, G. G., & Allaker, R. P. (2012). Antimicrobial activity of nanoparticulate metal oxides against peri-implantitis pathogens. *International Journal of Antimicrobial Agents*, *40*(2), 135–139. doi:10.1016/j.ijantimicag.2012.04.012 PMID:22727529

Wang, B., Lilja, M., Ma, T., Sörensen, J., Steckel, H., Ahuja, R., & Stromme, M. (2014). Theoretical and experimental study of the incorporation of tobramycin and strontium-ions into hydroxyapatite by means of co-precipitation. *Applied Surface Science*, *314*, 376–383. doi:10.1016/j.apsusc.2014.06.193

Wang, G. C., & Zreiqat, H. (2010). Functional coatings or films for hard-tissue applications. *Materials (Basel)*, *3*(7), 3994–4050. doi:10.3390/ma3073994

Wang, J., Guo, J., Liu, J., Wei, L., & Wu, G. (2014). BMP-functionalised coatings to promote osteogenesis for orthopaedic implants. *International Journal of Molecular Sciences*, *15*(6), 10150–10168. doi:10.3390/ijms150610150 PMID:24914764

Weng, J., Liu, X., Zhang, X., Ma, Z., Ji, X., & Zyman, Z. (1993). Further studies on the plasma-sprayed amorphous phase in hydroxyapatite coatings and its deamorphization. *Biomaterials*, *14*(8), 578–582. doi:10.1016/0142-9612(93)90174-Z PMID:8399949

Wojtowicz, A. M., Shekaran, A., Oest, M. E., Dupont, K. M., Templeman, K. L., Hutmacher, D. W., & Garcia, A. J. et al. (2010). Coating of biomaterial scaffolds with the collagen-mimetic peptide GFOGER for bone defect repair. *Biomaterials*, *31*(9), 2574–2582. doi:10.1016/j.biomaterials.2009.12.008 PMID:20056517

Wu, P., & Grainger, D. W. (2006). Drug/device combinations for local drug therapies and infection pro phylaxis. *Biomaterials*, *27*(11), 2450–2467. doi:10.1016/j.biomaterials.2005.11.031 PMID:16337266

Xia, Z., Yu, X., & Wei, M. (2012). Biomimetic collagen/apatite coating formation on Ti6Al4V substrates. *Journal of Biomedical Materials Research. Part B, Applied Biomaterials*, *100*(3), 871–881. doi:10.1002/jbm.b.31970 PMID:22102365

Xiong, J., Li, Y., Wang, X., Hodgson, P., & Wen, C. (2008). Mechanical properties and bioactive surface modification via alkali-heat treatment of a porous Ti-18Nb-4Sn alloy for biomedical applications. *Acta Biomaterialia*, *4*(6), 1963–1968. doi:10.1016/j.actbio.2008.04.022 PMID:18524702

Yamashita, K., Arashi, T., Kitagaki, K., Yamada, S., Umegaki, T., & Ogawa, K. (1994). Preparation of Apatite Thin Films through rf-Sputtering from Calcium Phosphate Glasses. *Journal of the American Ceramic Society*, *77*(9), 2401–2407. doi:10.1111/j.1151-2916.1994.tb04611.x

Yan, Y., Zhang, X., Huang, Y., Ding, Q., & Pang, X. (2014). Antibacterial and bioactivity of silver substituted hydroxyapatite/TiO2 nanotube composite coatings on titanium. *Applied Surface Science*, *314*, 348–357. doi:10.1016/j.apsusc.2014.07.027

Yang, H. S., La, W.-G., Bhang, S. H., Jeon, J.-Y., Lee, J. H., & Kim, B.-S. (2010). Heparin-conjugated fibrin as an injectable system for sustained delivery of bone morphogenetic protein-2. *Tissue Engineering. Part A*, *16*(4), 1225–1233. doi:10.1089/ten.tea.2009.0390 PMID:19886733

Yang, H. S., La, W. G., Bhang, S. H., Lee, T. J., Lee, M., & Kim, B. S. (2011). Apatite-coated collagen scaffold for bone morphogenetic protein-2 Delivery. *Tissue Engineering. Part A*, *17*(17-18), 2153–2164. doi:10.1089/ten.tea.2010.0702 PMID:21529263

Yang, H. S., La, W.-G., Cho, Y.-M., Shin, W., Yeo, G.-D., & Kim, B.-S. (2012). Comparison between heparin-conjugated fibrin and collagen sponge as bone morphogenetic protein-2 carriers for bone regeneration. *Experimental & Molecular Medicine*, *44*(5), 350–355. doi:10.3858/emm.2012.44.5.039 PMID:22322342

Yoshinari, M., Oda, Y., Kato, T., & Okuda, K. (2001). Influence of surface modifications to titanium on antibacterial activity in vitro. *Biomaterials*, *22*(14), 2043–2048. doi:10.1016/S0142-9612(00)00392-6 PMID:11426884

Zhang, B. G. X., Myers, D. E., Wallace, G. G., Brandt, M., & Choong, P. F. M. (2014). Bioactive coatings for orthopaedic implants-recent trends in development of implant coatings. *International Journal of Molecular Sciences*, *15*(7), 11878–11921. doi:10.3390/ijms150711878 PMID:25000263

APPENDIX

Figure 1. Schematic diagram of the antibacterial action mechanism of silver ion

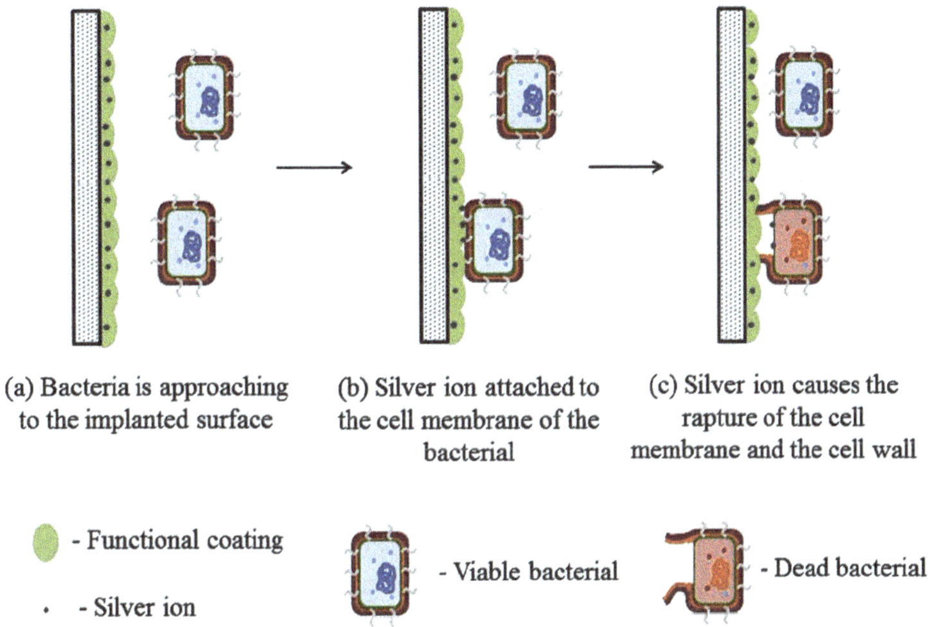

Figure 2. Schematic diagram of the biomimetic coating on polymeric/ceramic composite scaffolds

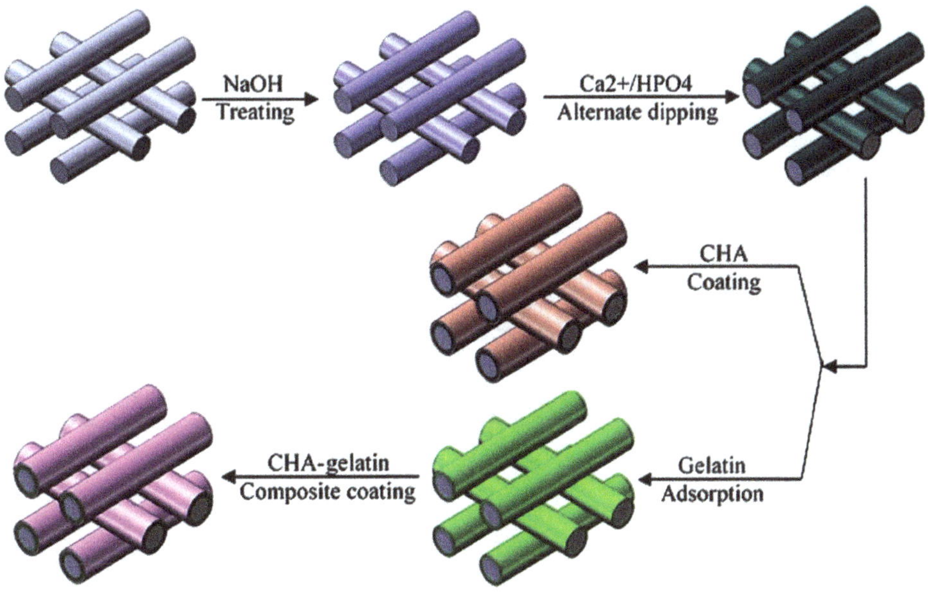

Figure 3. SEM images of (a and b) PCL/TCP, (c and d) PCL/TCP–CHA and (e and f) PCL/TCP–CHA–gelatin scaffolds. Right column images are at higher magnification of the respective left column images [adopted from M. Tarik Arafat, Lam, Ekaputra, Wong, Li, et al., 2011]

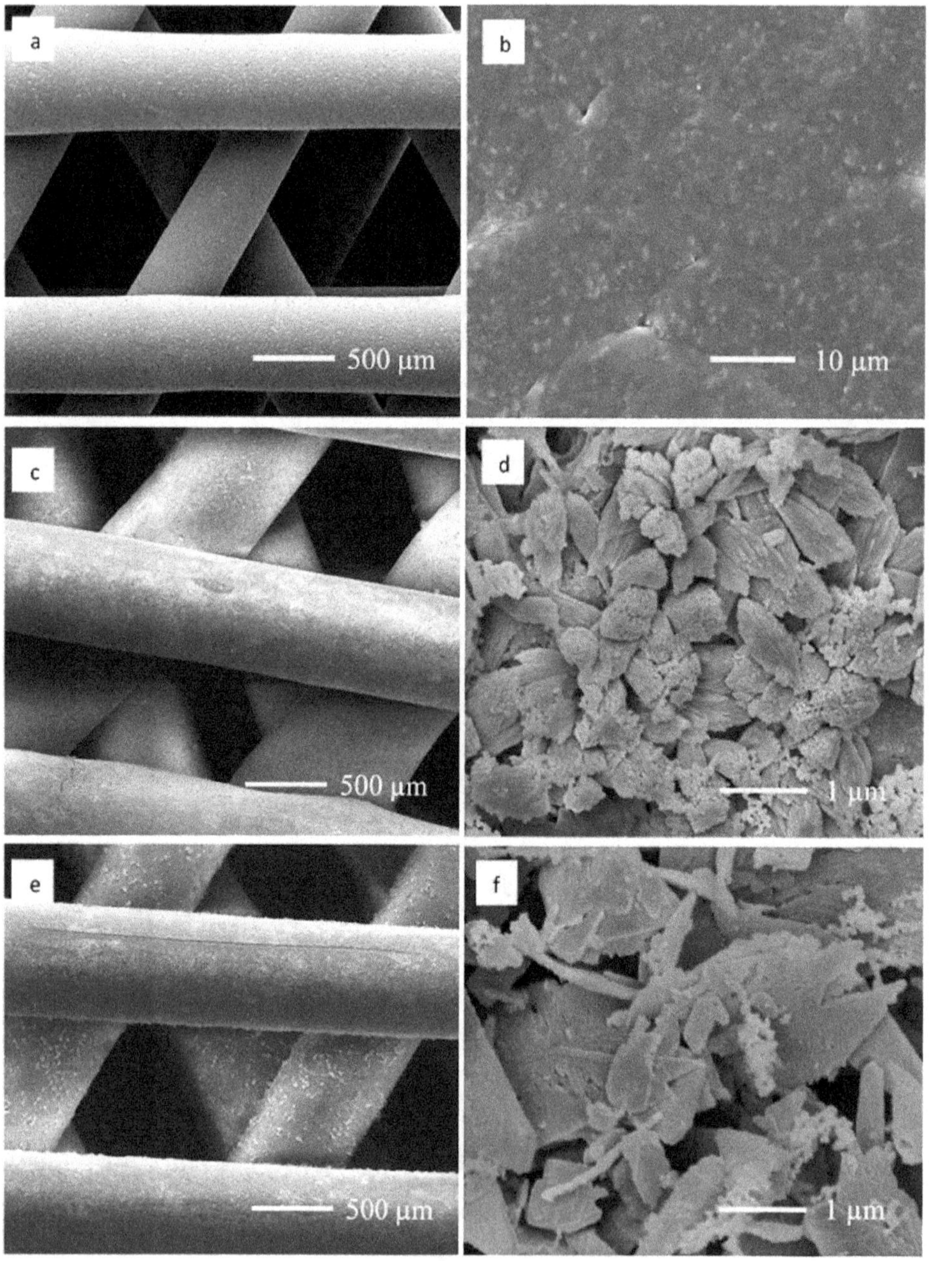

Chapter 10
Opto-Smart Systems in Microfluidics

Larisa Florea
Centre for Data Analytics, National Centre for Sensor Research, Dublin City University, Dublin 9, Ireland

Dermot Diamond
Centre for Data Analytics, National Centre for Sensor Research, Dublin City University, Dublin 9, Ireland

Fernando Benito-Lopez
Microfluidics UPV/EHU Cluster, Analytical Chemistry Department, University of the Basque Country UPV/EHU, Spain

ABSTRACT

The possibility of using photo-stimulus to control flow in microfluidics devices is very appealing as light can provide contactless stimulation, is biocompatible and can be applied in a non-invasive and highly precise manner. One of the most popular ways to achieve photo-control flow in microfluidic channels is throughout the use of photo-responsive molecules. We review here the different principles and strategies of using photo-responsive molecules to induce or control liquid motion using light, which include the use of photo-controlled polymeric actuators, photo-sensitive coatings, or photo-sensitive surfactants. We further analyse the capability of these approaches to induce flow control throughout the photo-operation of valves, photo-control of electro-osmotic flows or photo-manipulation of discrete microliter-sized droplets.

INTRODUCTION

The intrinsic features of microfluidic devices ensure two main characteristics during an analytical process: low consumption of reagents and sample as well as rapid and repeatable analysis protocols (Martinez, Phillips, Whitesides, & Carrilho, 2010). However, to date, extensive chemical and biological tasks need to be carried out outside the microfluidic device to prepare and pre-process samples prior microfluidic manipulation as conventional lab-on-a-chip devices are not yet suitable for non-ideal sample analysis (Gonzalez Crevillen, Hervas, Angel Lopez, Cristina Gonzalez, & Escarpa, 2007). These tasks include

DOI: 10.4018/978-1-5225-0066-7.ch010

sampling, pre-concentration, fluorescence labelling, filtration, mixing, sample analysis, as well as many other techniques that require manpower and are very time-consuming (Argentiere, Gigli, Mortato, Gerges, & Blasi, 2012).

Therefore, the main aim is to realise downscaling of these conventional tasks within the microfluidic devices and consequently develop innovative systems capable of preparing and/or analysis of samples "on-chip". If these functions could be integrated within the microfluidic device, then time and costs would be reduced and high throughput and high degree of automation would be realised (R. Byrne, Benito-Lopez, & Diamond, 2010). For this purpose two main ways of investigation are considered: an "evolutionary" and a "revolutionary" approach.

The "evolutionary" approach involves tremendous downscale of conventional units, and their assembly level by level to achieve the desired functionalities (Abgrall & Gue, 2007; Laser & Santiago, 2004; Psaltis, Quake, & Yang, 2006). However, microfluidic systems developed to date according to this approach, generally require complicated control systems and are by far from being trivial and widely accessible.

The "revolutionary" approach in the area of microfluidics is recently emerging and consists in the incorporation of stimuli-responsive materials into microfluidic devices, to obtain smart, functional, highly controllable components integrated in the microfluidic device (F. Benito-Lopez et al., 2010; M. Czugala, Ziolkowski, Byrne, Diamond, & Benito-Lopez, 2011; Florea, Diamond, & Benito-Lopez; Florea, Hennart, Diamond, & Benito-Lopez, 2012; Ziółkowski, Czugala, & Diamond, 2012). For instance, the use of a light stimulus in microfluidic systems is appealing because it can be applied precisely in different sections within a microdevice, in a non-invasive manner.

The possibility of controlling flow in microfluidics using opto-stimulus will offer new platforms with unprecedented flexibility and improved versatility. Some recent reviews have already been devoted to specific aspects of this field (R. Byrne et al., 2010; M. Chen et al., 2011; M. Czugala et al., 2011; Padgett & Di Leonardo, 2011). The use of holographic optical tweezers for optical manipulation, actuation and sensing in lab-on-chip systems was described by Padgett and Di Leonardo (Padgett & Di Leonardo, 2011). Other recent review by Baigl (Baigl, 2012) describes in detail strategies to photo-actuate microfluidic systems, such as the application of the chromocapillary effect, light-induced Marangoni effects, optically induced dielectrophoresis, digital optofluidics, light-induced electro-osmosis and optoelectrowetting, among others.

Here we review the possibility of controlling flow in closed microchannels, and manipulating discrete microliter-sized droplets by employing photo- and thermo-responsive materials incorporated in microfluidic units in the form of polymeric actuators, photo-responsive coatings or photo-sensitive surfactants.

OPTO-SMART SYSTEMS INTEGRATION IN MICROFLUIDIC DEVICES

The most widely used photo-sensitive molecules for photo-actuation of liquids in microfluidics are spiropyrans and azobenzenes. These are photochromic units that change their characteristics (conformation, polarity) reversibly in response to light of particular wavelengths. When incorporated in functional units, these changes in conformation/polarity of the photochromic unit induce in turn "on demand" changes in volume (Bassetti, Chatterjee, Aluru, & Beebe, 2005; Beebe et al., 2000; Kwon, Jeong, Park, Moon, & Lee, 2011; Richter, Klatt, Paschew, & Klenke, 2009), surface chemistry (Fernando Benito-Lopez et al., 2009; Garcia et al., 2007; Higuchi et al., 2004; Samanta & Locklin, 2008), surface charge, wettability (Anastasiadis, Lygeraki, Athanassiou, Farsari, & Pisignano, 2008; Dattilo, Armelao, Fois, Mistura, &

Figure 1. Reversible structural transformations of (A) spiropyran derivatives and (B) azobenzene in response to light

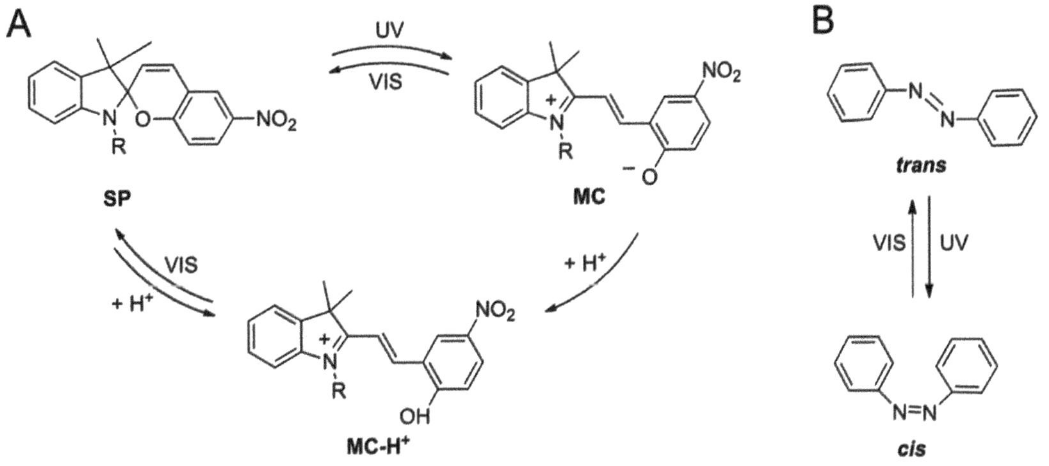

Maggini, 2007; Samanta & Locklin, 2008) or interfacial tension (Diguet et al., 2009; Diguet, Li, Queyriaux, Chen, & Baigl, 2011). Spiropyrans have been investigated extensively as active components in soft photo-actuators for microfluidic valves (F. Benito-Lopez et al., 2010; S. Sugiura et al., 2007). The light-stimulated switching of spiropyran is due to the photo-cleavage of the C_{spiro}-O bond upon irradiation with UV light (Hirshberg, 1956; Minkin, 2004). This cleavage allows a conformational rearrangement between a closed, colourless spiropyran form and an opened, colourful merocyanine (MC) form (Dürr & Bouas-Laurent, 2003; Minkin, 2004) with the colour arising from a strong absorption band in the visible spectral region (Figure 1A). In contrast, exposure of the MC isomer to visible light induces reversion to the closed spiropyran form (SP); therefore, it is possible to reversibly switch between the polar (or zwitterionic), coloured MC and colourless uncharged SP forms using light irradiation. Another interesting property of spiropyrans is its sensitivity towards pH (Doron, Katz, Tao, & Willner, 1997; Lee, Song, Lee, Ryu, & Chi, 2009), under acidic conditions, the MC isomer is converted to the protonated merocyanine ($MC-H^+$) form (Figure 1A).

Azobenzenes are another widely used class of photochromic molecules (Gaonkar, Kumar, & Neckers, 1990; Ide, Ozama, & Matsui, 2011). Under dark conditions, azobenzene preferentially exists in its *trans*, less polar configuration. Under UV light irradiation (350–370 nm) it undergoes photo-isomerisation to its *cis*, more polar form (Figure 1B). The reverse *cis* → *trans* isomerisation can be driven by visible light or occurs thermally in the dark. It is known that azobenzenes reversibly change their geometry from a planar to a non planar upon UV irradiation, generating a width decrease in the distance between the *para* carbon atoms from 9.9 Å to 5.5 Å and a corresponding increase in the dipole moment from 0.5 D to 3.1 D (Nicoletta et al., 2012). One of the most important advantages of azobenzenes is their stability over many illumination cycles, which makes them ideal components of numerous molecular devices and functional materials (Hoffmann, Resch-Genger, & Marlow, 2000; Singleton, Ramsay, Barsan, Butler, & Barrett, 2012).

Another approach to opto-actuation in microfluidics is through photo-thermal activation. The novelty of these systems relies on using light absorbance to generate localised heating (G. Chen, Svec, & Knapp, 2008; Hua, Pal, Srivannavit, Burns, & Gulari, 2008; Lo, Zhu, & Jiang, 2011).

This chapter will discuss recent trends for the incorporation of thermo- and photo-responsive moieties (in particular spiropyrans and azobenzene) into functional units in microfluidic devices capable of responding to an external opto-stimulus. The chapter will be structured into three main sections based on different microfluidic operations that can be controlled by light, *i.e.*:

- Photo-controlled polymeric actuators,
- Photo-controlled electro-osmotic flow,
- Photo-manipulation of droplets.

Specific examples will be discussed in terms of the technique used for integration of the photo- activated or photo-thermo- activated unit into the microfluidic device, the impact of the photo-stimulus and the applicability of these approaches for photo-actuation of liquids or droplets in microfluidics.

PHOTO-CONTROLLED POLYMERIC ACTUATORS

Stimuli-responsive polymers present all the advantages of polymeric materials, such as versatility, processability, low cost and amplified response to weak stimuli. Stimuli-responsive materials produce many useful effects in microfluidic systems, such as "on-demand" changes in volume (Bassetti et al., 2005; Beebe et al., 2000; Kwon et al., 2011; Richter et al., 2009), optical proprieties (Angiolini, Benelli, Giorgini, & Raymo, 2009; Ivanov, Eremeev, Wahlund, Galaev, & Mattiasson, 2002; Levy et al., 1997; Lin & Chiu, 2003; Samoladas, Bikiaris, Zorba, Paraskevopoulos, & Jannakoudakis, 2008; Stitzel, Byrne, & Diamond, 2006; Such, Evans, & Davis, 2006), permeability (Chung, Ito, & Imanishi, 1994; Park, Ito, & Imanishi, 1998) and surface chemistry (Fernando Benito-Lopez et al., 2009; Garcia et al., 2007), which in turn activate/deactivate different functions such as the capability of molecular recognition (Fernando Benito-Lopez et al., 2009; R. J. Byrne, Stitzel, & Diamond, 2006) (including capture, release and detection of analytes), autonomous control of flow rate and direction (Beebe et al., 2000; Dong & Jiang, 2007), and wettability switching (Anastasiadis et al., 2008).

The most commonly used opto-actuators in microfluidics are photo-responsive hydrogels as they have the ability to undergo volumetric changes in response to an external light stimulus (F. Benito-Lopez et al., 2010; S. Sugiura et al., 2007). A hydrogel is defined as a network of hydrophilic polymer chains that is able to swell and retain large amounts of water. In a fully-swollen hydrogel, up to 95% of its mass can consist in water that was absorbed into its three dimensional network (Alvarez-Lorenzo & Concheiro, 2008; Yoshida, Yang, Kumakura, Hagiwara, & Katakai, 1991). Hydrogels possess excellent biocompatibility and a degree of flexibility, very similar to natural tissue, due to their significant water content (He, Zuo, Xie, Huang, & Xue, 2011; Kim & Shin, 2007). Photo-responsive engineered hydrogels can be made to collapse and thereby release a percentage of their water content upon light irradiation (Figure 2).

In this context, polymeric actuators are commonly employed in microfluidic devices to provide the functionality of micro-valves. Valves are central components within microfluidic systems, being essential for directional control and movement of flows while enabling important actions, such as flow regulation, flow diversion and addition of reagents. The micro-valves developed to date can be classified in two categories, active and passive, employing mechanical, non-mechanical and external control systems (Oh & Ahn, 2006).

Figure 2. Photo-induced hydrogel shrinkage

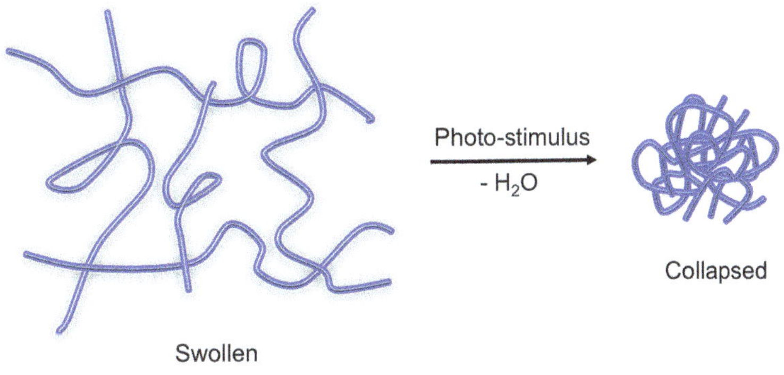

In the case of passive valves, which do not have any moving components and are easy to fabricate, hydrogel based materials have received special attention. These valves change volume in response to external stimuli to stop/open the flow (Beebe et al., 2000; G. Chen et al., 2008; Dong & Jiang, 2007; Geiger, Pisano, & Svec, 2010).

Incorporation of photo-responsive units within hydrogel-based valves allows non-invasive, external control of the valve based on the intrinsic responsiveness of the polymer to an optical stimulus. Photo-responsive polymer materials for micro-valve applications have been studied by many research groups, and many polymers and polymer gels functionalised with azobenzene (D. Kungwatchakun, 1988), leu-kochromophore (M. H. M. Irie, 1985; M.Irie, 1986), and spirobenzopyran (A. Menju, 1981; F. Benito-Lopez et al., 2010; A. M. M. Irie, K. Hayashi, 1979; S. Sugiura et al., 2007) have been examined.

This section is organised as follows: Firstly, we show how smart micro-valves can be actuated by light using a light-to-heat energy conversion mechanism. In this case, a localised temperature increase is caused by light irradiation, which in turn induces volume changes of the micro-valve (usually based on hydrogels). Then, we review how photo-responsive moieties like spiropyrans and azobenzenes are incorporated into polymers to achieve photo-precise manipulation of micro-valves.

Chen *et al.* (G. Chen et al., 2008) developed a light-actuated micro-valve via photoinitiated patterned polymerisation of *N*-isopropylacrylamide (NIPAAm) within microfluidic channels. PNIPAAm is one of the best-known temperature-sensitive hydrogels, having the ability to shrink and expand due to its lower critical solution temperature (LCST)(Harmon, Kucking, & Frank, 2003; Kuckling, Harmon, & Frank, 2002; Xu, Wei, Zhang, Cheng, & Zhuo, 2007). Below the LCST (32 °C), PNIPAAm chains are hydrated by water due to the hydrogen bondings between the amide residue and on the polymer chains and the water molecules, and therefore the gel is swollen. Above this temperature hydrogen bonds between solvent molecules and water are gradually broken and polymer-polymer interactions become thermodynamically favored compared with polymer-solvent interactions. As a consequence, the PNIPAAm chains collapse sharply from a hydrophilic coil into a hydrophobic globule that precipitates in water (Dusek, 1993; Schild, 1992) and the PNIPAAm hydrogels shrink rapidly. The novelty of the approach used in Chen's PNIPAAm micro-valve relies on heating the valve from the absorption of light provided by a quartz halogen illuminator. The pressure-tolerance of the micro-valve can be tuned by changing the mechanical strength of the polymer monolith inside the microchannel through the choice of a suitable amount of monomer and crosslinker. In addition, the response time and pressure resistance of a PNIPAAm valve can be optimised by varying the tetrahydrofuran (THF) content in the polymerisation mixture. In these conditions, opening and closing response times of the valve can be modulated in the range of 4.0 to 6.2 s. The on/off switching of the valve using the quartz halogen was repeated multiple times and show no deterioration after at least 9 irradiation cycles (G. Chen et al., 2008).

Using a different approach, Lo *et al.* (Lo et al., 2011) reported the performance of an IR light-responsive micro-valve. The valve consisted of PNIPAAm hydrogel nanocomposite incorporating glycidyl methacrylate functionalised graphene oxide (GO– GMA) sheets. The valve operated in two states – OFF (closed) - adopted when the IR source is turned off and the hydrogel completely blocks the channel, and ON (opened), adopted when the IR light is turned on, due to shrinking of the hydrogel upon absorbance of the IR light by the GO–GMA sheets. The ON state could be explained as follow: GO–GMA sheets absorbed the IR radiation converting it into heat. During this process, the temperature of the system is increased above the LCST of PNIPAAm, which triggers the hydrogel to contract, allowing for fluidic flow. When the IR light is switched off the heat dissipated to the surrounding environment, the hydrogel absorbed water and expanded its volume back to the original size, blocking the flow. The performance of the hydrogel micro-valve is shown in Figure 1. The intensity of the IR light reaching the hydrogel sample in order to induce actuation is 1.21 W cm^{-2} and causes a temperature increase of the valve from 25 °C to 40 °C after 2 min of IR irradiation, which might make this type of material not appropriate for several applications including some bioanlysis. Moreover, the reversability of the photo-induced actuation is not discussed by the authors in the present paper, suggesting that this valve might be suitable for one-use only.

Independent optical control of microfluidic valves composed of optomechanically responsive nanocomposite hydrogels (poly-*N*-isopropylacrylamide-co-acrylamide) was achieved using strongly absorbing gold nanoparticles or nanoshells embedded in the actuator polymer (Figure 4). These nanocomposite materials respond to different wavelengths of light: while gold-colloid nanocomposite hydrogel collapses in response to green light, the gold-nanoshell nanocomposite hydrogel collapses in response to near-IR light. In both cases, the valves opened completely in less than five seconds. This approach has the advantage of allowing independent control of valves formed from composites with different nanoparticles depending on the illumination wavelength. As the phase changes in these thermo-responsive polymers are reversible, these type of valves are suitable for multiple use applications (Sershen et al., 2005).

Figure 3. A micro-valve made of GO–GMA IR-responsive hydrogel. Top- views of the micro-valve are on the left. Respective images taken under a microscope are on the right. (a) GO–GMA hydrogel micro-valve before actuated by IR irradiation (OFF state) and (b) after exposure to IR light (ON state).
Reproduced by permission of The Royal Society of Chemistry

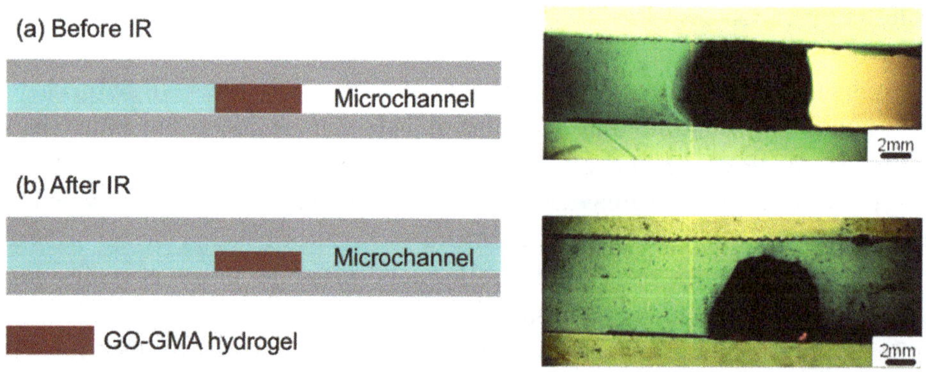

Figure 4. Two valves formed at a T-junction in a fluidics device, one made of a gold-colloid nanocomposite hydrogel and the other a gold- nanoshell nanocomposite hydrogel. The channels are 100 μm wide. a) When the entire device was illuminated with green light (532 nm, 1.6 W cm^{-2}), the gold colloid valve opened while the nano-shell valve remained closed. b) However, when the device was illuminated with near infrared light (832 nm, 2.7 W cm^{-2}), the opposite response was observed. In both cases, the valves opened within 5 s.

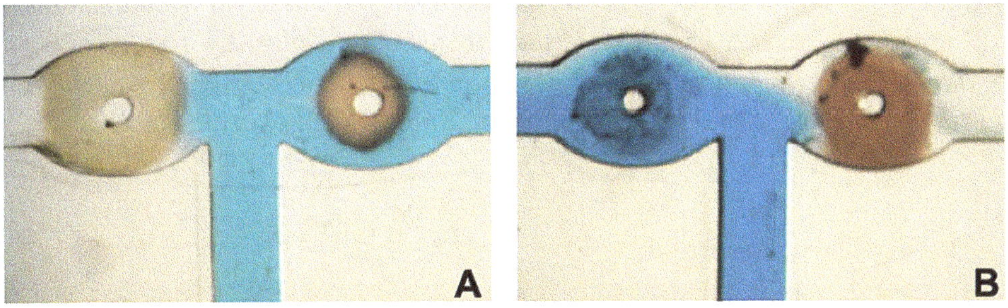

A different optically addressable micro-actuator array (microfluidic "flash memory") with latched operation was proposed by Hua *et al.* (Hua et al., 2008). The micro-actuator array consists of individual phase-change based actuators addressed by localised heating through focused light patterns provided by a modified projector or high power laser pointer. After the initial light "writing" during which the phase is temporarily changed to a liquid form, the actuated status is self-maintained by the solid phase of the actuator without power and pressure inputs. The micro-fluidic flash memory can be re-configured by a new light illumination pattern and common pressure signal providing a flexible, energy-efficient, and low cost multiplexing solution for microfluidic applications based on physical displacements. The advantage of this approach is that the phase-change wax is fully encapsulated and thus immune from contamination issues in fluidic environments. Moreover, compared to other systems previously discussed, this optically-addressed microactuator array is reusable and thermal simulation shows that the temperature rise in the microfluidic channels when the wax melts was less than 10 °C, which should not be considered an issue for most applications.

While in all the examples described so far the key of the opto-stimulation is the conversion of photo-energy to thermal energy, photo-stimulation can also be achieved by incorporation of photo-responsive moieties in microfluidic units.

In 2004, Sumaru *et al.* (Sumaru, Kameda, Kanamori, & Shinbo, 2004) presented an actuation mechanism based on the lower critical solubility temperature (LCST) of spiropyran (3 mol %) copolymers of poly(*N*-isopropylacrylamide) (pSPNIPAM). Figure 5a shows the chemical structure of pSPNIPAM and the reversible isomerisation of the photochromic unit from the protonated merocyanine (MC-H$^+$) to the closed spiropyran form (SP) under white light irradiation. In acidic conditions (0.26 mM HCl), the spiropyran unit present in the copolymer changed from SP to the MC-H$^+$ form, which presents a strong absorption band around 422 nm. When the copolymer was irradiated with light matching the absorbance of MC-H$^+$, the MC-H$^+$ is switched back to the closed hydrophobic SP form, releasing one proton in the process (Figure 3a); As result of the formation of the more hydrophobic SP isomer, dehydration of the main polymer chain occurs. A correlated change in both absorbance at 422 nm (particular to MC-H$^+$) and specific conductance was recorded when pSPNIPAAm aqueous solution (0.10 wt %) was repeatedly irradiated with white light over three cycles.

Figure 5. (A) Chemical structure of pSPNIPAAm and characteristics of its components: pNIPAAm main chain and spirobenzopyran side chain (Sumaru et al., 2004). Copyright (2004) American Chemical Society. (B) Schematic representation of a micro-fluidic platform that contains multiple pSPNIPAAm hydrogel micro-valves. These micro-valves can be independently controlled by means of local light irradiation. A - A solution containing blue dye was introduced into the main micro-channel. B - Blue light was locally irradiated to the right side pSPNIPAAm gel micro-valve. C - The right side micro-valve was opened after 18 s blue light irradiation and after 24 s from start, the center micro-valve was locally irradiated with blue light. D - The center micro-valve was opened after 30 s blue light irradiation and after 60 s from start, blue light was locally irradiated to the left side micro-valve. E - The left side micro-valve was opened after 24 s blue light irradiation and it is shown that after 144 s from start all the chambers were filled with the blue dye solution. Adapted from ref. (Shinji Sugiura et al., 2007)

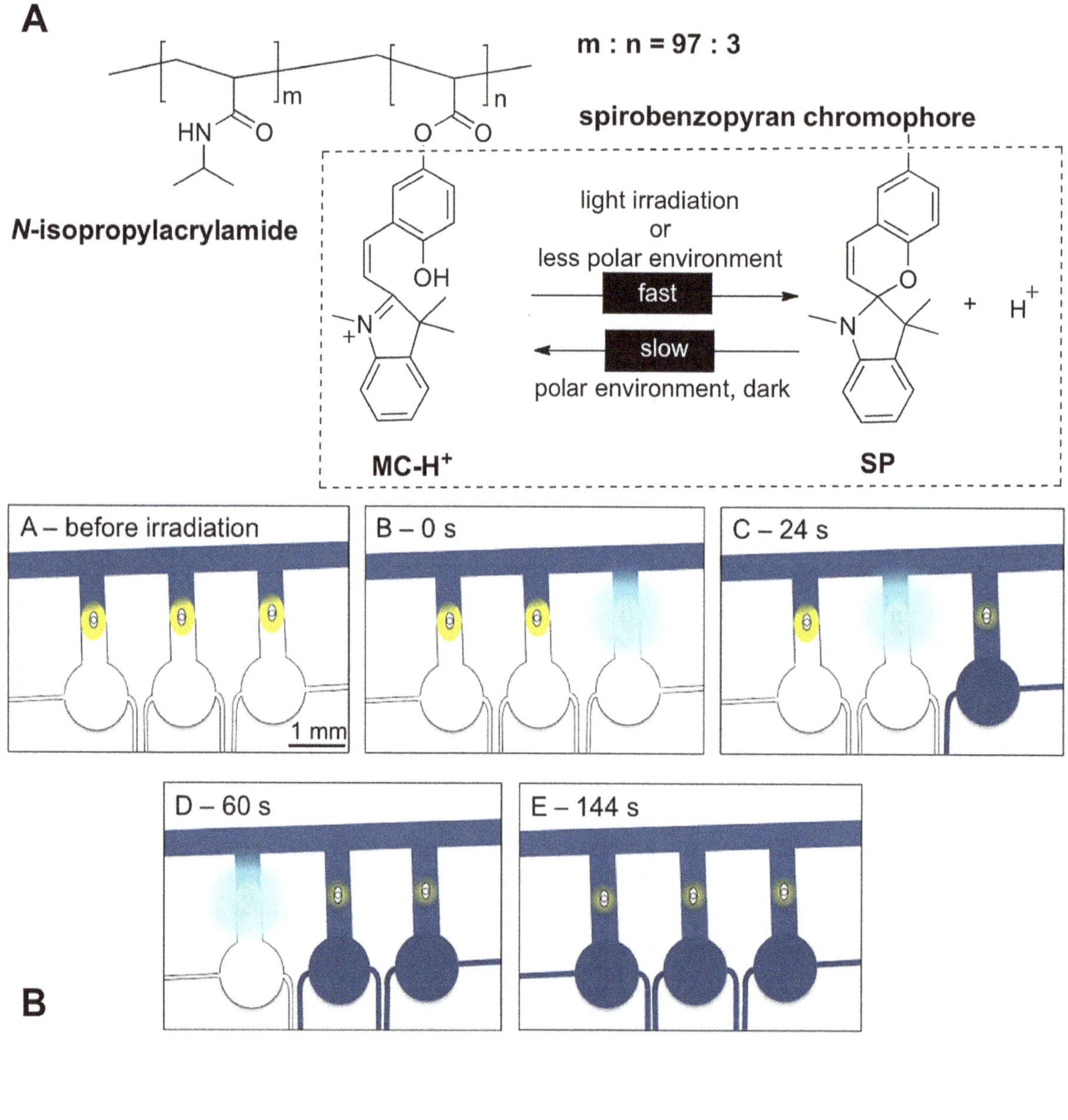

Independently controlled micro-valves made from such material, capable of photo-actuation in acidic media (0.5 mM HCl), were later demonstrated by Sugiura *et al.* (Shinji Sugiura et al., 2007). These micro-valves (Figure 5b) were fabricated through in situ polymerisation at desired positions in the micro-channels and later opened by local light irradiation. This process was found to be rather fast, with each valve opening within 18 to 30 s of light irradiation. The maximum pressure the pSPNIPAAm gel micro-valves could withstand was determined to be 30 ± 6.6 kPa. Over this pressure the gels deformed and leakage occurred.

A similar approach was presented by Benito-Lopez *et al.* (F. Benito-Lopez et al., 2010) in our laboratories, with the difference that in this case, an ionic liquid (IL) was incorporated within the pSPNIPAAm polymer matrix to produce microfluidic valves. Four different phosponium based ionic liquids incorporating different anions (dicyanoamide ($[dca]^-$), bis(trifluoro methanesulfonyl)-amide ($[NTf_2]^-$), dodecylbenzenesulfonate ($[dbsa]^-$), tosylate ($[tos]^-$)) were used and the micro-valves were produced throughout *in situ* polymerisation of the ionogels (IL + pSPNIPAAm) in micro-channels fabricated in poly(methyl methacrylate) (PMMA). After immersion of the ionogels for 2 h in 0.1 mM HCl aqueous solution, in order to allow protonation of the photochromic unit to the $MC\text{-}H^+$ form and to obtain a fully swollen ionogel, photo-induced dehydration kinetics were measured. It was found that the incorporation of different ILs causes different actuation behaviours of the ionogels. When used for the production of micro-valves, by simply varying the IL-component of the ionogel, different micro-valves can be tuned to open at different times using a single light source. Results showed that in the particular configuration used by the authors, the IL-free polymer gel micro-valve opens after 2 s of light irradiation, followed by the $[dca]^-$ based ionogel (4 s), the $[tos]^-$ based ionogel (18 s) and the $[dbsa]^-$ based ionogel (44 s). The slowest kinetics are given by the $[NTf_2]^-$ based ionogel (49 s) (Figure 6).

Recently, ionogel actuator microstructures based on pSPNIMAAm have been thoroughly characterized in our laboratories and their potential for micro-valve fabrication has been investigated. It was found that the photo-induced swelling and shrinking behaviour of the ionogels is strongly dependent on the nature of the ionic liquid. For example, when the $[NTf_2]^-$ based phosphonium ionogel was used, the ionogel reached the highest degree of swelling (up to 180% of its initial size) and showed rhe fastest shrinkage rate ($k_{sh} = 29 \pm 4 \times 10^{-2}\ s^{-1}$)(Monika Czugala et al., 2014)

An original approach for photo-control of fluid in microfluidics incorporating pSPNIMAAm hydrogels was described by Sugiura *et al.* (S. Sugiura et al., 2008; Shinji Sugiura et al., 2009). The authors demonstrated on-demand formation of micro-channels with arbitrary pathways by micro-patterned light irradiation of a 200 μm thick pSPNIMAAm hydrogel sheet prepared by *in situ* free radical polymerisation (Figure 7). The hydrogel layer was covalently attached to a glass slide while another glass slide containing multiple inlets and outlets was placed on top of the hydrogel sheet. Micro-patterned light irradiation was realised by means of computer-controlled maskless micro-pattern projection unit. As a result of micro-patterned light irradiation, the pSPNIMAAm hydrogel gel in the irradiated area between adjacent inlet/outlet ports shrank, forming a micro-channel between the two glass plates and allowing for fluid to flow. The effect was repeated for several channel configurations – straight, bent or serpentine. Furthermore, authors also demonstrated, in the same publication, independent and parallel flow control in a polydimethylsiloxane (PDMS) micro-channel network through the opening of targeted microvalves created with the same photoresponsive hydrogel sheet. In this system, a microfluidic device containing multiple photoresponsive micro-valves was constructed by stacking a PDMS micro-channel chip and a glass slide with mechanically fabricated through-holes on pSPNIPAAm hydrogel sheet attached to a glass plate. Parallel control of multiple micro-valves was also successfully demonstrated as two micro-valves

Figure 6. Performance of the ionogel microfluidic valves: (a) micro-valves closed; the applied vacuum is unable to pull the dyes through the microchannels. White light is applied for the time specified in each picture (b). 'No I.L.' valve is first to actuate followed by ionogels incorporating [dca]- (c), [tos]- (d), [dbsa]- (e), [NTf2]- (f), all valves are open. Numbers and arrows indicate when the channel is filled with the dye because of micro-valve actuation
F. Benito-Lopez et al., 2010. Reproduced by permission of The Royal Society of Chemistry

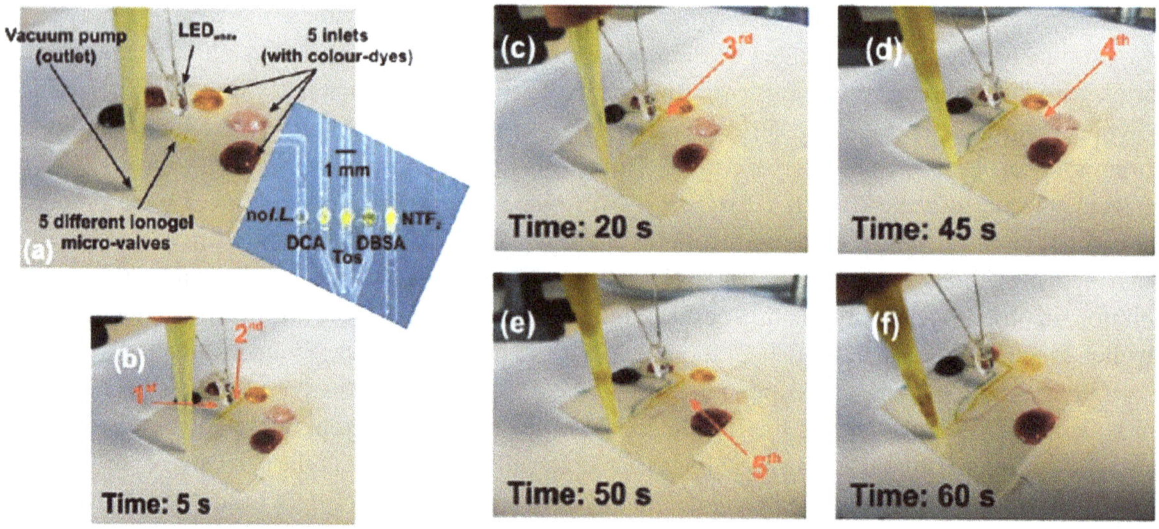

Figure 7. Schematic showing on-demand formation of microchannels with arbitrary pathways in a "universal" microfluidic system by micro-patterned light irradiation. White arrows indicate the flow direction of a fluorescently labelled latex bead suspension. A - pSPNIMAAm hydrogel sheet having one inlet and multiple outlets; B - Microchannel formation by micro-patterned light irradiation of the pSPNIPAAm hydrogel sheet; C - Latex bead suspension flowing through the microchannel after irradiation. Flow of the red coloured latex bead suspension from the inlet to the upper-left-side outlet is visible. D,E - A different sequence of micro-patterned light irradiation and microchannel formation
Adapted from S. Sugiura et al., 2008

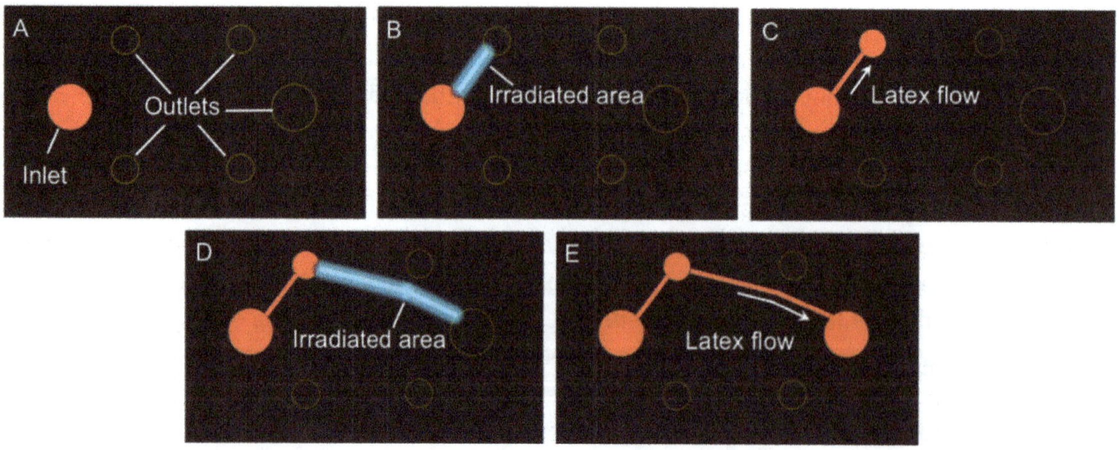

were simultaneously opened within several minutes by micro-patterned light irradiation. As this system works in "photon mode" in contrast with "heat mode", no interference between adjacent micro-valves was observed. Although, the isomerisation of the SP unit is a reversible phenomenon, the reswelling of the gel is still diffusion limited due to the requirement for re-protonation of SP for valve closing. As a consequence, the reswelling process takes more than one hour.

However, the main disadvantages of pSPNIMAAm based hydrogels and ionogels has been the long reswelling times (up to one hour) and the need for an external acidic source (typically HCl solution 10^{-3} M) in order to induced reprotonation of the spiropyran photochromic unit to induce reswelling of the gel. This has made pSPNIMAAm materials suitable mainly for single use application.

Recently, the need for an external acidic source has been eliminated by introducing acrylic acid (AA) as a comonomer in the polymeric chain (Ziółkowski, Florea, Theobald, Benito-Lopez, & Diamond, 2013). The newly synthesised p(SP-AA-NIPAAm) hydrogels work on the same principle as described above, however, the reswelling times become comparable with the shrinking times. Micro-valves based on this material are considered to be capable of multiple operations as the shrinking and reselling of the hydrogel is fully reversible. Moreover, the pH operation range has been extended to up to pH 7-8, making p(SP-AA-NIPAAm) based actuators suitable for biological applications.

In a different approach, a new micro-valve membrane actuator for microfluidic system applications was realised by using a crosslinked liquid-crystalline polymer (CLCP) incorporating azobenzene moieties (M. Chen et al., 2011). It was found that under UV light irradiation the valve is open while under white light irradiation the valve is closed. The time to open and close the valve can be as short as 8 s and 6 s, respectively.

PHOTO-CONTROL OF ELECTRO-OSMOTIC FLOW

When an electric field is applied along a micro-channel or micro-capillary having a charged wall, the motion of mobile counter-ions in the diffuse counter-ion layer near the channel wall induces a spontaneous global liquid motion. This movement is called electroosmotic flow (EOF) (Huang, Gordon, & Zare, 1988; Y. Liu, Fanguy, Bledsoe, & Henry, 2000; Rathore & Horvath, 1997). EOF constitutes an important component in certain chemical separation techniques, notably capillary electrophoresis (Huang et al., 1988; Y. Liu et al., 2000). There has been an increased interest in EOF in the recent years, in parallel with the development of microfluidic systems and EOF has been exploited in microfluidics to provide a degree of flow control. This approach offers a range of advantages compared to more established pumps based technologies, such as peristaltic pumps, as it creates a constant pulse-free flow in which the magnitude and direction of the flow are controlled externally, without the use of any mechanically moving parts.

EOF is strongly sensitive to the zeta potential of the channel wall and the key requirement to realise photo-controlled EOF is the ability to optically change the zeta potential of the micro-channel/micro-capillary wall. Moorthy *et al.* (Moorthy, Khoury, Moore, & Beebe, 2001) demonstrated this type of approach by using micro-channels (1000 μm wide, 20 μm deep) coated with titanium dioxide (TiO_2). TiO_2 is a direct semiconductor that exhibits a change in surface charge upon radiation with UV light, leading to variations in the zeta potential. In these conditions, electroosmotic flow modulation was recorded in the presence/absence of UV radiation. The magnitude of the change in electroosmotic flow rate due to UV irradiation was found to be pH dependent. Under optimal conditions (pH = 4.7), the authors report a change in EOF from 0 (no light) to -242 ± 60 μm s^{-1} (with light). Another strategy was later proposed by Oroszi *et al.* (L. Oroszi, Der, Kirei, Ormos, & Rakovics, 2006) who used micro-channels formed between PDMS and glass coated with cadmium sulfide (CdS) photoactive layer. Light illumination (provided by a mercury lamp filtered to 400 – 725 nm) resulted in a decreased electric field accompanied by a reduction of electroosmotic flow. Specifically, at a driving voltage of 100 V, photo-irradiation of the photo-conductor for periods of 30 s caused a decrease in the average flow velocity from 45 μm s^{-1} (in the dark) to 8 μm s^{-1} (under illumination). The same group later on demonstrated that introducing a pattern into the photosensitive layer enables the generation of complex flow profiles upon illumination (L. s. Oroszi, Der, Kirei, Rakovics, & Ormos, 2009).

Spiropyrans have also been used to modify the electroosmotic flow. It has recently been demonstrated in our laboratories by Scarmagnani *et al.* (Scarmagnani et al., 2010) that, under acidic conditions, electroosmotic flow (EOF) generated in a spiropyran-modified acrylate based monoliths could be modulated using light irradiation (Figure 6). In an acidic environment (HCl 1 mM) The SP-modified acrylate based monolith gained a positively charged surface due to the protonation of the SP unit to MC-H^+ (Figure 8A) and the micro-capillary adopted a yellow colour (Figure 6B). Upon irradiation with white light, MC-H^+ converted back to the non-polar SP form (Figure 6A) indicated by the change in colour of the microcapillary to colourless (Figure 6B). When the MC-H^+ form is dominant, it produces a charged surface, which enables a relatively high flow rate (up to 1.6 μL min^{-1}) to be generated under electroosmotic conditions. Upon exposure to white light, the concentration of MC-H^+ decreases due to the photo-conversion to the uncharged SP form, with up to 20% reduction of the EOF. The process is reversible, and removal of the light source results in a flow increase back to the original rate. A schematic of the microchip used in these experiments is shown in Figure 6C.

The ability to alter flow rates in microfluidic channels using light has very significant implications, as it could dramatically simplify the manner in which micro-flow systems are controlled.

PHOTO-MANIPULATION OF DROPLETS

Digital or droplet-based microfluidics involves the generation and manipulation of discrete droplets inside micro-devices (Teh, Lin, Hung, & Lee, 2008) and unlike continuous-flow systems, droplet-based microfluidics allows for independent control of each droplet. Droplet microfluidics encompasses two different approaches;

- Mechanisms that are much more similar to traditional microfluidics, in which droplets are carried by a carrier flow, and;
- More innovative mechanisms like electrowetting for droplet transport or the use of stimuli-responsive surfactants for directed droplet movement at the air-liquid interface.

Photo-manipulation of droplets for microfluidic purposes can be realised through the use of photosensitive molecules incorporated in surface (channel) coatings or surfactants.

In 1992 Chaudhury and Whitesides (Chaudhury & Whitesides, 1992) showed that a wettability gradient on a surface is able to induce droplet motion. In this context, if a wettability gradient of a microfluidic platform could be optically induced, then droplet movement could be realised. Given the fact that light intensity is easily manipulated, a gradient can be created by partially illuminating a substrate that exhibits photo-responsive wettability behaviour.

Figure 8. (A) SP acrylate switching in acidic environment between the MC-H+ form (right) and the SP form (left) when irradiated with white light. (B) Picture of the spiropyran acrylate monolith flushed with HCl 10^{-3} M switched between the MC-H+ form, in the absence of light exposure and the SP form when irradiated with white light. (C) Schematic representation of the cross section of the microchip: 1) Electrode; 2) 1.5 ml 10^{-3} M HCl reservoir; 3) PEEK tubing; 4) epoxy glue; 5) PMMA sheet 5 mm thickness; 6) layer of PSA; 7) second sheet of PMMA (0.5 mm thickness); 8) SP acrylate monolith 8 mm length; 9) white LEDs (430-760 nm)

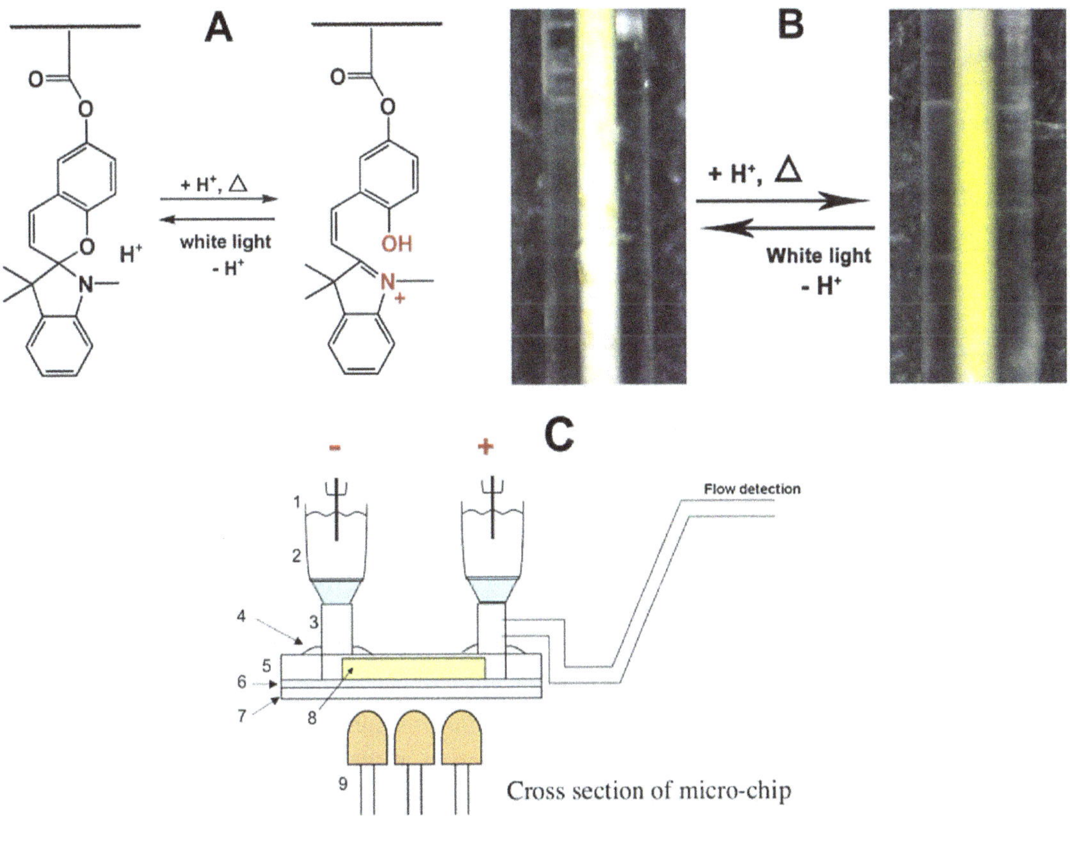

Photo-responsive surface wettability is usually achieved by functionalising with a photochromic moiety, most often azobenzenes and spiropyrans that dramatically change polarity upon photo-irradiation. Although a vast variety of photochromic coated substrates have shown to exhibit significant changes in wettability upon photo-stimulation (Anastasiadis et al., 2008; Fries, Samanta, Orski, & Locklin, 2008; Lygeraki et al., 2008; Samanta & Locklin, 2008; Wang et al., 2012), so far only a few have demonstrated droplet photo-manipulation.

Ichimura *et al.* (Ichimura, Oh, & Nakagawa, 2000) showed the movement of an oil droplet on a flat surface modified with calix[4]resorcinarene derivative having photochromic azobenzene units. Asymmetrical photo-irradiation of this surface with different UV and visible light fluxes caused a gradient in the surface free energy due to the isomerisation of the azobenzene-containing monolayer, leading to a directional droplet movement. Although the velocity of the droplet depended intrinsically on the intensity and gradient of the light, a typical speed of 35 μL min^{-1} was observed for the motion of a 2 μL olive oil droplet. Authors also demonstrated the photo-manipulation of fluid inside a surface-modified glass tube, showing the applicability of this technique for fluid manipulation in microfluidics. Although there is no theoretical limitation in movement distance for this approach, the surface-assisted method was only demonstrated for liquids that exhibited both small-hysteresis and large photo-induced changes in contact angles upon photo-isomerisation.

More recently, similar results were reported by Yang *et al.* (D. Yang et al., 2007) using a flat surface functionalised with an azobenzene monolayer, and also by Berna *et al.* (Berna et al., 2005), using a photo-sensitive rotaxane-modified substrate (flat or tilted by 12 °C). Although promising, these photo-droplet manipulations were only demonstrated for olive oil, a few organic compounds (benzonitrile, diiodomethane, 1-Methylnaphthalene, 1,1,2,2 - Tetrachloroethane), some nematic liquid crystals, but never for water. At this stage, photo-manipulation of surface wettability cannot produce fast manipulation of a large number of individual droplets, which is required for delivering real-life applications of digital microfluidics.

Another interesting means of transporting μL-sized droplets that has been explored in recent years is the use of photo-sensitive surfactants (Diguet et al., 2009; Diguet et al., 2011). These are surfactants that incorporate a photochromic moiety, most often spiropyrans (S. L. Liu, Fujihira, & Saji, 1994; H. Sakai et al., 2007; K. Sakai, Imaizumi, Oguchi, Sakai, & Abe, 2010; Takagi, Kurematsu, & Sawaki, 1995; Takumi et al., 2007) or azobenzenes (Diguet et al., 2009; Diguet et al., 2011; Eastoe, Dominguez, Wyatt, Beeby, & Heenan, 2002; L. Yang, Takisawa, Hayashita, & Shirahama, 1995; Zhang, Wang, & Lee, 2009), although other photoswitchable moeties have also been investigated (Bonini et al., 2005; Eastoe et al., 2002).

For linear surfactants functionalised with spiropyrans or azobenzenes, UV irradiation induces isomerisation of the photochromic unit to the more polar form which causes an increase in air–water interfacial tension of up to 10 mN m^{-1} (Baigl, 2012). Diguet *et al.* (Diguet et al., 2009) explored the use of a cationic photo-sensitive surfactant that incorporates an azobenzene unit for photo-manipulation of droplets. The first demonstration was performed with an oil droplet floating on a water bath containing a cationic photo-sensitive azobenzene trimethylammonium bromide surfactant AzoTAB (Figure 7A) that is initially in the trans configuration. Partial illumination of the solution containing AzoTAB with UV light, next to the droplet, resulted in a higher surface tension in the illuminated area and the droplet could be 'pushed' along by the UV light. Contrariwise, by changing the light source to blue light, the surface tension gradient formed is in the opposite direction, causing the droplet to be 'pulled' along by the blue light. This type of droplet manipulation can be performed repeatedly, with droplet velocities of up to 300 μm s^{-1}. The authors also demonstrated that greater precision of droplet manipulation can be achieved by using simultaneous two-color illumination, wherein the center part of the droplet is under blue illumination (attraction) while the droplet periphery is surrounded by a UV light (repulsion). This two-color illumination creates a 'trap' for the droplet and therefore by moving the two-color source, the droplet can be moved at will around a desired trajectory. Figure 9B shows a superposition of a set of images which follows the path of a single droplet as it is moved along a heart-shaped trajectory at a speed of about 300 μm s^{-1}. This type of precise and fast droplet manipulation is very appealing for droplet-based analytical/microfluidic applications, as now one can imagine parallel and complex manipulation of a large number of individual μL-sized droplets using two-color location control.

Figure 9. A-Molecular structure of the photo-sensitive surfactant AzoTAB. Upon illumination at 365 nm, trans-AzoTAB isomerises to cis configuration, resulting in a more polar tail
Adapted from Diguet et al., 2009. B-Photomontages of the motion of a droplet along a heart shape trajectory. Reproduced with author's permission from www.baigllab.com

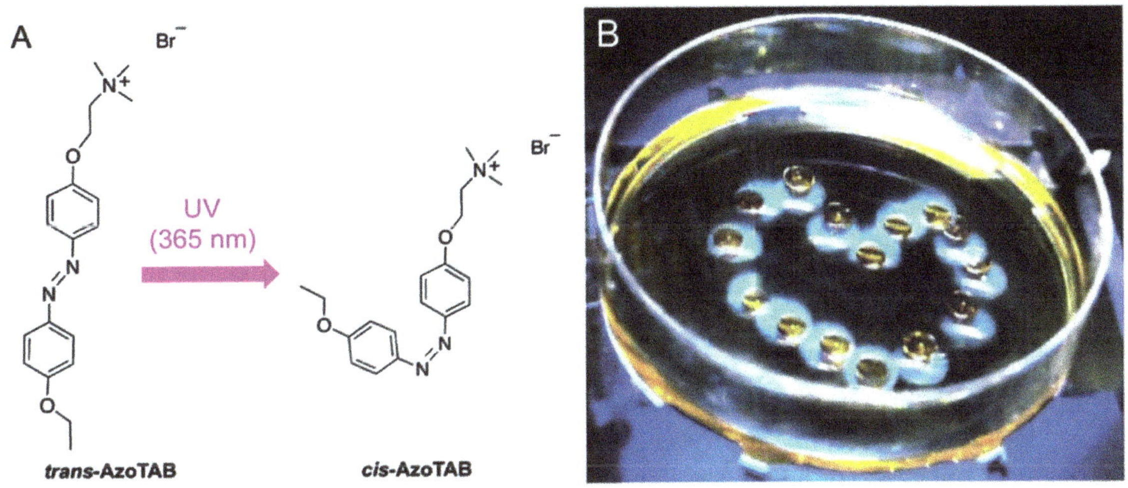

Another fascinating way of using photo-sensitive surfactants to control liquid behavior was recently demonstrated by the same group. In this case, AzoTAB was used to provide photo-induced reversible switching between a continuous two-phase laminar flow and a droplet generating regime, in microfluidic devices employing a typical water-in-oil flow focusing geometry (Diguet et al., 2011). Based on the same principle, they demonstrated that interfacial energy between flowing liquids and the microfluidic substrate could be photo-modulated. When AzoTAB was added to the aqueous phase of a two-phase laminar-flow regime (water/oil) in a microfluidic system, UV light induced fragmentation of the flow, to a regime where mono-dispersed droplets were formed. When the light stimulus was removed, the system reverted to the laminar regime. This strategy was applicable to different device morphologies and allowed many cycles of reversible and rapid switches as well as spatio-temporal control of flow regime localisation.

More recently, we have demonstrated that surfactant type molecules can be on-demand released from organic based droplets (*e.g.* droplets made of dichloromethane) through photo-manipulation of the pH of the external solution (Florea et al., 2014). This process causes a propulsion-type movement, much faster than previously reported photo-induced droplet movement with speeds up to 4000 $\mu m\ s^{-1}$. This is the first example of photodriven chemopropulsion (photo-chemopropulsion), in which photo irradiation in close proximity to the micro-vehicle (lipophilic droplet) initiates a rapid pH change in the aqueous channel fluid in which the vehicle (droplet) sits. This causes rapid solubilisation of the released lipophilic surfactant type molecule in the channel fluid and an immediate change in surface tension of the fluid surrounding the droplet resulting in fast movement (ca. 100–4000 $\mu m\ s^{-1}$) of the droplet away from the light source. Although this movement is reversible it requires a rather complex system in order to achieve photo-chemopropulsion. In this case particular chemicals are required for both the droplet and the surrounding solution making this system rather specific and not easily implemented for wide applications. Moreover, the droplet (made of dichloromethane) is susceptible to evaporation and therefore not likely to be used for long-time applications. Several approaches have been made in order to overcome these problems and chemo-propelled droplet made of ionic liquids (ILs) have been recently demonstrated by us (Francis, Fay, Florea, & Diamond, 2015). Due to the inherited no vapour pressure of the IL, these droplets are not subject to evaporation, are environmentally stable, can dissolve a vast range of chemicals and have been shown to spontaneously move in a wide range of chemical gradients in open fluidic channels, following the source of a chemoattractant (*eg*. HCl, NaCl, Na_2SO_4). Several strategies to photo-induce chemopropulsion of these IL droplets are currently under investigation in our laboratories.

CONCLUSION

The examples presented here give some indication of the potential of using an optical stimulus for photo-modulation of flow behavior in micro-fluidics by incorporating photo-responsive units into polymeric actuators, coatings, or surfactants. Although these examples can only hint at the possibilities, such as photo-modulation of flows or photo-induced droplet formation in microfluidics, the reader can gain some knowledge of the tools now available to achieve these effects. Incorporation of photo-responsive or photo-thermo-responsive molecules in microfluidics together with smart engineering of the opto-functional unit can provide switchable characteristics that have great potential for many target applications. Opto-controlled microfluidics can open new possibilities based on unprecedented flexibility, versatility, reduced costs and simplicity. Moreover, advanced features such as parallel manipulation of multiple spatially resolved photo-responsive units through the use of patterned light exposure, variable light intensities, user selectable wavelengths, and excellent temporal control, become practically realisable. This clearly opens great advantages over standard actuation methods that rely on highly complex arrays of actively moving components (pumps, valves) based on individually addressable electronic or pressure based control.

We believe that the use of opto-responsive units in microfluidics could form the basis of the new types of autonomous microfluidic-based analytical devices that are much more biomimetic in nature, and lower cost than current technologies. However, the future of opto-manipulated microfluidics relies on convincing demonstrations of their application to real scenarios in which their unique photo-controlled characteristics lead to truly disruptive platforms for studying chemical or biological processes compared to existing technologies. As for now, this field is still at an early stage and the functionality of these photo-responsive materials in real-life scenarios is yet to be demonstrated. Issues like biocompatibility, environmental stability, environmental safety and contamination are still to be addressed before these disruptive technologies can become competitive on the microfluidics market.

ACKNOWLEDGMENT

This work was partially supported by Science Foundation Ireland under INSIGHT Award SFI/12/RC/2289. F.B.-L. wants to thanks the Ramón y Cajal Programme (Ministerio de Economía y Competitividad) and the European Union's Seventh Framework Programme for research, technological development, and demonstration under Grant Agreement No. 604241.

REFERENCES

Abgrall, P., & Gue, A. M. (2007). Lab-on-chip technologies: Making a microfluidic network and coupling it into a complete microsystem - a review. *Journal of Micromechanics and Microengineering*, *17*(5), R15–R49. doi:10.1088/0960-1317/17/5/R01

Alvarez-Lorenzo, C., & Concheiro, A. (2008). Intelligent Drug Delivery Systems: Polymeric Micelles and Hydrogels. *Mini Reviews in Medicinal Chemistry*, *8*(11), 1065–1074. doi:10.2174/138955708785909952 PMID:18855723

Anastasiadis, S. H., Lygeraki, M. I., Athanassiou, A., Farsari, M., & Pisignano, D. (2008). Reversibly Photo-Responsive Polymer Surfaces for Controlled Wettability. *Journal of Adhesion Science and Technology*, *22*(15), 1853–1868. doi:10.1163/156856108X320014

Angiolini, L., Benelli, T., Giorgini, L., & Raymo, F. M. (2009). Optical and chiroptical switches based on photoinduced photon and proton transfer in copolymers containing spiropyran and azopyridine chromophores in their side chains. *Polymer*, *50*(24), 5638–5646. doi:10.1016/j.polymer.2009.10.013

Argentiere, S., Gigli, G., Mortato, M., Gerges, I., & Blasi, L. (2012). Advances in Microfluidics. In R. Kelly (Ed.), *Advances in Microfluidics*. InTech.

Baigl, D. (2012). Photo-actuation of liquids for light-driven microfluidics: State of the art and perspectives. *Lab on a Chip*, *12*(19), 3637–3653. doi:10.1039/c2lc40596b PMID:22864577

Bassetti, M. J., Chatterjee, A. N., Aluru, N. R., & Beebe, D. J. (2005). Development and modeling of electrically triggered hydrogels for microfluidic applications. *Journal of Microelectromechanical Systems*, *14*(5), 1198–1207. doi:10.1109/JMEMS.2005.845407

Beebe, D. J., Moore, J. S., Bauer, J. M., Yu, Q., Liu, R. H., Devadoss, C., & Jo, B. H. (2000). Functional hydrogel structures for autonomous flow control inside microfluidic channels. *Nature*, *404*(6778), 588–590. doi:10.1038/35007047 PMID:10766238

Benito-Lopez, F., Byrne, R., Raduta, A. M., Vrana, N. E., McGuinness, G., & Diamond, D. (2010). Ionogel-based light-actuated valves for controlling liquid flow in micro-fluidic manifolds. *Lab on a Chip*, *10*(2), 195–201. doi:10.1039/B914709H PMID:20066247

Benito-Lopez, F., Scarmagnani, S., Walsh, Z., Paull, B., Macka, M., & Diamond, D. (2009). Spiropyran modified micro-fluidic chip channels as photonically controlled self-indicating system for metal ion accumulation and release. *Sensors and Actuators. B, Chemical*, *140*(1), 295–303. doi:10.1016/j.snb.2009.03.080

Berna, J., Leigh, D. A., Lubomska, M., Mendoza, S. M., Perez, E. M., Rudolf, P., & Zerbetto, F. et al. (2005). Macroscopic transport by synthetic molecular machines. *Nature Materials*, *4*(9), 704–710. doi:10.1038/nmat1455 PMID:16127455

Bonini, M., Berti, D., Di Meglio, J. M., Almgren, M., Teixeira, J., & Baglioni, P. (2005). Surfactant aggregates hosting a photoresponsive amphiphile: Structure and photoinduced conformational changes. *Soft Matter*, *1*(6), 444–454. doi:10.1039/b511544b

Byrne, R., Benito-Lopez, F., & Diamond, D. (2010). Materials science and the sensor revolution. *Materials Today, 13*(7-8), 9–16. doi:10.1016/S1369-7021(10)70124-3

Byrne, R. J., Stitzel, S. E., & Diamond, D. (2006). Photo-regenerable surface with potential for optical sensing. *Journal of Materials Chemistry, 16*(14), 1332–1337. doi:10.1039/b516400a

Chaudhury, M. K., & Whitesides, G. M. (1992). How to make water run uphill. *Science, 256*(5063), 1539–1541. doi:10.1126/science.256.5063.1539 PMID:17836321

Chen, G., Svec, F., & Knapp, D. R. (2008). Light-actuated high pressure-resisting microvalve for on-chip flow control based on thermo-responsive nanostructured polymer. *Lab on a Chip, 8*(7), 1198–1204. doi:10.1039/b803293a PMID:18584098

Chen, M., Huang, H., Zhu, Y., Liu, Z., Xing, X., Cheng, F., & Yu, Y. (2011). Photodeformable CLCP material: Study on photo-activated microvalve applications. *Applied Physics. A, Materials Science & Processing, 102*(3), 667–672. doi:10.1007/s00339-010-6103-4

Chung, D. J., Ito, Y., & Imanishi, Y. (1994). Preparation of porous membranes grafted with poly(spiropyran-containing methacrylate) and photocontrol of permeability. *Journal of Applied Polymer Science, 51*(12), 2027–2033. doi:10.1002/app.1994.070511207

Crevillen, A., Hervas, M., Lopez, M., Gonzalez, M., & Escarpa, A. (2007). Real sample analysis on microfluidic devices. *Talanta, 74*(3), 342–357. doi:10.1016/j.talanta.2007.10.019 PMID:18371648

Czugala, M., O'Connell, C., Blin, C., Fischer, P., Fraser, K. J., Benito-Lopez, F., & Diamond, D. (2014). Swelling and shrinking behaviour of photoresponsive phosphonium-based ionogel microstructures. *Sensors and Actuators. B, Chemical, 194*, 105–113. doi:10.1016/j.snb.2013.12.072

Czugala, M., Ziolkowski, B., Byrne, R., Diamond, D., & Benito-Lopez, F. (2011). *Materials science: the key to revolutionary breakthroughs in micro-fluidic devices.*

Dattilo, D., Armelao, L., Fois, G., Mistura, G., & Maggini, M. (2007). Wetting properties of flat and porous silicon surfaces coated with a spiropyran. *Langmuir, 23*(26), 12945–12950. doi:10.1021/la7024878 PMID:18027976

Diguet, A., Guillermic, R. M., Magome, N., Saint-Jalmes, A., Chen, Y., Yoshikawa, K., & Baigl, D. (2009). Photomanipulation of a Droplet by the Chromocapillary Effect. *Angewandte Chemie International Edition, 48*(49), 9281–9284. doi:10.1002/anie.200904868 PMID:19890927

Diguet, A., Li, H., Queyriaux, N., Chen, Y., & Baigl, D. (2011). Photoreversible fragmentation of a liquid interface for micro-droplet generation by light actuation. *Lab on a Chip, 11*(16), 2666–2669. doi:10.1039/c1lc20328b PMID:21727984

Dong, L., & Jiang, H. (2007). Autonomous microfluidics with stimuli-responsive hydrogels. *Soft Matter, 3*(10), 1223–1230. doi:10.1039/b706563a

Doron, A., Katz, E., Tao, G. L., & Willner, I. (1997). Photochemically-, chemically-, and pH-controlled electrochemistry at functionalized spiropyran monolayer electrodes. *Langmuir, 13*(6), 1783–1790. doi:10.1021/la960729e

Durr, H., & Bouas-Laurent, H. (2003). *Photochromism: Molecules and Systems.* Elsevier.

Dusek, K. (1993). Responsive gels, volume transitions I and II. In D. K. (Ed.), *Adv. Polym. Sci.* (Vol. 109 & 110).

Eastoe, J., Dominguez, M. S., Wyatt, P., Beeby, A., & Heenan, R. K. (2002). Properties of a stilbene-containing gemini photosurfactant: Light-triggered changes in surface tension and aggregation. *Langmuir, 18*(21), 7837–7844. doi:10.1021/la0257384

Florea, L., Diamond, D., & Benito-Lopez, F. Polyaniline Coated Micro-capillaries for Continuous Flow Analysis of Aqueous Solutions. *Analytica Chimica Acta*. doi:10.1016/j.aca.2012.11.027 PMID:23260671

Florea, L., Hennart, A., Diamond, D., & Benito-Lopez, F. (2012). Synthesis and characterisation of spiropyran-polymer brushes in micro-capillaries: Towards an integrated optical sensor for continuous flow analysis. *Sensors and Actuators. B, Chemical, 175*, 92–99. doi:10.1016/j.snb.2011.12.055

Florea, L., Wagner, K., Wagner, P., Wallace, G. G., Benito-Lopez, F., Officer, D. L., & Diamond, D. (2014). Photo-Chemopropulsion–Light-Stimulated Movement of Microdroplets. *Advanced Materials, 26*(43), 7339–7345. doi:10.1002/adma.201403007 PMID:25236879

Francis, W., Fay, C., Florea, L., & Diamond, D. (2015). Self-propelled chemotactic ionic liquid droplets. *Chemical Communications, 51*(12), 2342–2344. doi:10.1039/C4CC09214G PMID:25562698

Fries, K., Samanta, S., Orski, S., & Locklin, J. (2008). Reversible colorimetric ion sensors based on surface initiated polymerization of photochromic polymers. *Chemical Communications*, 47, 6288–6290. doi:10.1039/b818042c PMID:19048131

Gaonkar, S. R., Kumar, G. S., & Neckers, D. C. (1990). Photochromism of azobenzene-containing polymers. 4. effect of spacer groups. *Macromolecules, 23*(24), 5146–5148. doi:10.1021/ma00226a018

Garcia, A., Marquez, M., Cai, T., Rosario, R., Hu, Z., Gust, D., & Park, C.-D. et al. (2007). Photo-, thermally, and pH-responsive microgels. *Langmuir, 23*(1), 224–229. doi:10.1021/la061632n PMID:17190508

Geiger, E. J., Pisano, A. P., & Svec, F. (2010). A Polymer-Based Microfluidic Platform Featuring On-Chip Actuated Hydrogel Valves for Disposable Applications. *Journal of Microelectromechanical Systems, 19*(4), 944–950. doi:10.1109/JMEMS.2010.2048702

Harmon, M. E., Kuckling, D., & Frank, C.W. (2003). Photo-cross-linkable PNIPAAm copolymers. 5. Mechanical properties of hydrogel layers. *Langmuir, 19*(26), 10660–10665. doi:10.1021/la030232m

He, L., Zuo, Q., Xie, S., Huang, Y., & Xue, W. (2011). Intelligent hydrogels for drug delivery system. *Recent Patents on Drug Delivery & Formulation, 5*(3), 265–274. doi:10.2174/187221111797200533 PMID:21834779

Higuchi, A., Hamamura, A., Shindo, Y., Kitamura, H., Yoon, B. O., Mori, T., & Umezawa, A. et al. (2004). Photon-modulated changes of cell attachments on poly(spiropyran-co-methyl methacrylate) membranes. *Biomacromolecules, 5*(5), 1770–1774. doi:10.1021/bm049737x PMID:15360286

Hirshberg, Y. (1956). Reversible formation and eradication of colors by irradiation at low temperatures - a photochemical memory model. *Journal of the American Chemical Society*, *78*(10), 2304–2312. doi:10.1021/ja01591a075

Hoffmann, K., Resch-Genger, U., & Marlow, F. (2000). Photoinduced switching of nanocomposites consisting of azobenzene and molecular sieves: Investigation of the switching states. *Microporous and Mesoporous Materials*, *41*(1-3), 99–106. doi:10.1016/S1387-1811(00)00277-8

Hua, Z., Pal, R., Srivannavit, O., Burns, M. A., & Gulari, E. (2008). A light writable microfluidic "flash memory": Optically addressed actuator array with latched operation for microfluidic applications. *Lab on a Chip*, *8*(3), 488–491. doi:10.1039/b712983a PMID:18305870

Huang, X. H., Gordon, M. J., & Zare, R. N. (1988). Current-monitoring method for measuring the electroosmotic flow-rate in capillary zone electrophoresis. *Analytical Chemistry*, *60*(17), 1837–1838. doi:10.1021/ac00168a040

Ichimura, K., Oh, S. K., & Nakagawa, M. (2000). Light-driven motion of liquids on a photoresponsive surface. *Science*, *288*(5471), 1624–1626. doi:10.1126/science.288.5471.1624 PMID:10834837

Ide, T., Ozama, Y., & Matsui, K. (2011). Photochemistry of azobenzene in sol-gel systems. *Journal of Non-Crystalline Solids*, *357*(1), 100–104. doi:10.1016/j.jnoncrysol.2010.09.009

Irie, M., & Hosoda, M. (1985). Photoresponsive polymers—reversible solution viscosity change of poly(N,N-dimethylacrylamide) with pendant triphenyl- methane leucohydroxide residues in methanol. *Die Makromolekulare Chemie. Rapid Communications*, *6*(8), 533–536. doi:10.1002/marc.1985.030060803

Irie, M., & Kunwatchakun, D. (1986). Photoresponsivepolymers.8.Reversiblephoto- stimulated dilation of polyacrylamide gels having triphenylmethane leuco derivatives. *Macromolecules*, *19*(10), 2476–2480. doi:10.1021/ma00164a003

Irie, M., Menju, A., & Hayashi, K. (1979). Photoresponsive polymers. reversible solution viscosity change of poly (methyl methacrylate) having spiroben- zopyran side groups. *Macromolecules*, *12*(6), 1176–1180. doi:10.1021/ma60072a032

Ivanov, A. E., Eremeev, N. L., Wahlund, P. O., Galaev, I. Y., & Mattiasson, B. (2002). Photosensitive copolymer of N-isopropylacrylamide and methacryloyl derivative of spyrobenzopyran. *Polymer, 43*(13), 3819-3823. doi:10.1016/s0032-3861(02)00191-x

Kim, B., & Shin, Y. (2007). pH-sensitive swelling and release behaviors of anionic hydrogels for intelligent drug delivery system. *Journal of Applied Polymer Science*, *105*(6), 3656–3661. doi:10.1002/app.26450

Kuckling, D., Harmon, M. E., & Frank, C. W. (2002). Photo-cross-linkable PNIPAAm copolymers. 1. Synthesis and characterization of constrained temperature-responsive hydrogel layers. *Macromolecules*, *35*(16), 6377–6383. doi:10.1021/ma0203041

Kungwatchakun, D., & Irie, M. (1988). Photoresponsive polymers—photocontrol of the phase-separation temperature of aqueous-solutions of poly N-isopropylacrylamide-co-N-(4-phenylazophenyl)acrylamide. *Die Makromolekulare Chemie. Rapid Communications*, *9*(4), 243–246. doi:10.1002/marc.1988.030090408

Kwon, G. H., Jeong, G. S., Park, J. Y., Moon, J. H., & Lee, S.-H. (2011). A low-energy-consumption electroactive valveless hydrogel micropump for long-term biomedical applications. *Lab on a Chip, 11*(17), 2910–2915. doi:10.1039/c1lc20288j PMID:21761057

Laser, D. J., & Santiago, J. G. (2004). A review of micropumps. *Journal of Micromechanics and Microengineering, 14*(6), R35-R64. doi:10.1088/0960-1317/14/6/r01

Lee, C. W., Song, Y. H., Lee, Y., Ryu, K. S., & Chi, K. W. (2009). Reversible photochromic switch ensemble and its photoimaging using H+ transfer between spiropyran and fluorescein in a polymer matrix. *Chemical Communications*, 41, 6282–6284. doi:10.1039/b913096a PMID:19826695

Levy, D., DelMonte, F., Oton, J. M., Fiksman, G., Matias, I., Datta, P., & Lopez-Amo, M. (1997). Photochromic doped sol-gel materials for fiber-optic devices. *Journal of Sol-Gel Science and Technology, 8*(1-3), 931–935. doi:10.1007/BF02436963

Lin, J. S., & Chiu, H. T. (2003). Photochromic behavior of spiropyran and fulgide in thin films of blends of PMMA and SBS. *Journal of Polymer Research-Taiwan, 10*(2), 105–110. doi:10.1023/A:1024940921678

Liu, S. L., Fujihira, M., & Saji, T. (1994). Formation of an organic thin-film by photochemical isomerization of a surfactant with a spiropyran moiety. *Journal of the Chemical Society. Chemical Communications*, 16, 1855–1856. doi:10.1039/c39940001855

Liu, Y., Fanguy, J. C., Bledsoe, J. M., & Henry, C. S. (2000). Dynamic coating using polyelectrolyte multilayers for chemical control of electroosmotic flow in capillary electrophoresis microchips. *Analytical Chemistry, 72*(24), 5939–5944. doi:10.1021/ac0009321 PMID:11140760

Lo, C.-W., Zhu, D., & Jiang, H. (2011). An infrared-light responsive graphene-oxide incorporated poly(N-isopropylacrylamide) hydrogel nanocomposite. *Soft Matter, 7*(12), 5604–5609. doi:10.1039/c1sm00011j

Lygeraki, M. I., Tsiranidou, E., Anastasiadis, S. H., Fotakis, C., Pisignano, D., Cingolani, R., & Athanassiou, A. (2008). Controlling the reversible wetting capability of smart photochromic-polymer surfaces by micro patterning. *Applied Physics. A, Materials Science & Processing, 91*(3), 397–401. doi:10.1007/s00339-008-4473-7

Martinez, A. W., Phillips, S. T., Whitesides, G. M., & Carrilho, E. (2010). Diagnostics for the Developing World: Microfluidic Paper-Based Analytical Devices. *Analytical Chemistry, 82*(1), 3–10. doi:10.1021/ac9013989 PMID:20000334

Menju, A., Hayashi, K., & Irie, M. (1981). Photoresponsive polymers. 3. Reversible solution viscosity change of poly(methacrylic acid) having spirobenzopy- ran pendant groups in methano. *Macromolecules, 14*(3), 755–758. doi:10.1021/ma50004a055

Minkin, V. I. (2004). Photo-, thermo-, solvato-, and electrochromic spiroheterocyclic compounds. *Chemical Reviews, 104*(5), 2751–2776. doi:10.1021/cr020088u PMID:15137806

Moorthy, J., Khoury, C., Moore, J. S., & Beebe, D. J. (2001). Active control of electroosmotic flow in microchannels using light. *Sensors and Actuators. B, Chemical, 75*(3), 223–229. doi:10.1016/S0925-4005(01)00557-3

Nicoletta, F. P., Cupelli, D., Formoso, P., De Filpo, G., Colella, V., & Gugliuzza, A. (2012). Light Responsive Polymer Membranes: A Review. *Membranes*, *2*(1), 134–197. doi:10.3390/membranes2010134 PMID:24957966

Oh, K. W., & Ahn, C. H. (2006). A review of microvalves. *Journal of Micromechanics and Microengineering*, *16*(5), R13–R39. doi:10.1088/0960-1317/16/5/R01

Oroszi, Laszlo, Der, Andras, Kirei, Huba, Ormos, Pal, & Rakovics, Vilmos. (2006). Control of electro-osmostic flow by light. *Applied Physics Letters*, *89*(26). doi:10.1063/1.2424648

Oroszi, L., Dér, A., Kirei, H., Rakovics, V., & Ormos, P. (2009). Manipulation of microfluidic flow pattern by optically controlled electroosmosis. *Microfluidics and Nanofluidics*, *6*(4), 565–569. doi:10.1007/s10404-008-0346-9

Padgett, M., & Di Leonardo, R. (2011). Holographic optical tweezers and their relevance to lab on chip devices. *Lab on a Chip*, *11*(7), 1196–1205. doi:10.1039/c0lc00526f PMID:21327211

Park, Y. S., Ito, Y., & Imanishi, Y. (1998). Photocontrolled gating by polymer brushes grafted on porous glass filter. *Macromolecules*, *31*(8), 2606–2610. doi:10.1021/ma9605199

Psaltis, D., Quake, S. R., & Yang, C. (2006). Developing optofluidic technology through the fusion of microfluidics and optics. *Nature*, *442*(7101), 381–386. doi:10.1038/nature05060 PMID:16871205

Rathore, A. S., & Horvath, C. (1997). Capillary electrochromatography: Theories on electroosmotic flow in porous media. *Journal of Chromatography. A*, *781*(1-2), 185–195. doi:10.1016/S0021-9673(97)00627-4 PMID:9368385

Richter, A., Klatt, S., Paschew, G., & Klenke, C. (2009). Micropumps operated by swelling and shrinking of temperature-sensitive hydrogels. *Lab on a Chip*, *9*(4), 613–618. doi:10.1039/B810256B PMID:19190798

Sakai, H., Ebana, H., Sakai, K., Tsuchiya, K., Ohkubo, T., & Abe, M. (2007). Photo-isomerization of spiropyran-modified cationic surfactants. *Journal of Colloid and Interface Science*, *316*(2), 1027–1030. doi:10.1016/j.jcis.2007.08.042 PMID:17884070

Sakai, K., Imaizumi, Y., Oguchi, T., Sakai, H., & Abe, M. (2010). Adsorption Characteristics of Spiropyran-Modified Cationic Surfactants at the Silica/Aqueous Solution Interface. *Langmuir*, *26*(12), 9283–9288. doi:10.1021/la100161a PMID:20225812

Samanta, S., & Locklin, J. (2008). Formation of photochromic spiropyran polymer brushes via surface-initiated, ring-opening metathesis polymerization: Reversible photocontrol of wetting behavior and solvent dependent morphology changes. *Langmuir*, *24*(17), 9558–9565. doi:10.1021/la8017387 PMID:18642863

Samoladas, A., Bikiaris, D., Zorba, T., Paraskevopoulos, K. M., & Jannakoudakis, A. (2008). Photochromic behavior of spiropyran in polystyrene and polycaprolactone thin films - Effect of UV absorber and antioxidant compound. *Dyes and Pigments*, *76*(2), 386–393. doi:10.1016/j.dyepig.2006.09.012

Scarmagnani, S., Walsh, Z., Benito-Lopez, F., Macka, M., Paull, B., & Diamond, D. (2010). Incorporation of acrylate based spiropyran monoliths in micro-fluidic devices for photo-controlled electroosmotic flow. *Advances in Science and Technology*, *76*, 100–105. doi:10.4028/www.scientific.net/AST.76.100

Schild, H. G. (1992). Poly (n-isopropylacrylamide) - experiment, theory and application. *Progress in Polymer Science*, *17*(2), 163–249. doi:10.1016/0079-6700(92)90023-R

Sershen, S. R., Mensing, G. A., Ng, M., Halas, N. J., Beebe, D. J., & West, J. L. (2005). Independent optical control of microfluidic valves formed from optomechanically responsive nanocomposite hydrogels. *Advanced Materials, 17*(11), 1366. doi:10.1002/adma.200401239

Singleton, T. A., Ramsay, K. S., Barsan, M. M., Butler, I. S., & Barrett, C. J. (2012). Azobenzene Photoisomerization under High External Pressures: Testing the Strength of a Light-Activated Molecular Muscle. *The Journal of Physical Chemistry B*, *116*(32), 9860–9865. doi:10.1021/jp3060872 PMID:22799454

Stitzel, S., Byrne, R., & Diamond, D. (2006). LED switching of spiropyran-doped polymer films. *Journal of Materials Science*, *41*(18), 5841–5844. doi:10.1007/s10853-006-0289-z

Such, G. K., Evans, R. A., & Davis, T. P. (2006). Rapid photochromic switching in a rigid polymer matrix using living radical polymerization. *Macromolecules*, *39*(4), 1391–1396. doi:10.1021/ma052002f

Sugiura, S., Sumaru, K., Ohi, K., Hiroki, K., Takagi, T., & Kanamori, T. (2007). Photoresponsive polymer gel microvalves controlled by local light irradiation. *Sensors and Actuators. A, Physical*, *140*(2), 176–184. doi:10.1016/j.sna.2007.06.024

Sugiura, S., Sumaru, K., Ohi, K., Hiroki, K., Takagi, T., & Kanamori, T. (2007). Photoresponsive polymer gel microvalves controlled by local light irradiation. *Sensors and Actuators. A, Physical*, *140*(2), 176–184. doi:10.1016/j.sna.2007.06.024

Sugiura, S., Szilagyi, A., Sumaru, K., Hattori, K., Takagi, T., Filipcsei, G., & Kanamori, T. et al. (2009). On-demand microfluidic control by micropatterned light irradiation of a photoresponsive hydrogel sheet. *Lab on a Chip*, *9*(2), 196–198. doi:10.1039/B810717C PMID:19107273

Sugiura, S., Szilágyi, A., Sumaru, K., Takagi, T., Zrínyi, M., & Kanamori, T. (2008). On-demand fluid control on microchip by micro-patterned light irradiation using photo-responsive hydrogels. *Paper presented at the Twelfth International Conference on Miniaturized Systems for Chemistry and Life Sciences*, San Diego, California, USA.

Sumaru, K., Kameda, M., Kanamori, T., & Shinbo, T. (2004). Reversible and Efficient Proton Dissociation of Spirobenzopyran-Functionalized Poly(N-isopropylacrylamide) in Aqueous Solution Triggered by Light Irradiation and Temporary Temperature Rise. *Macromolecules*, *37*(21), 7854–7856. doi:10.1021/ma0487249

Takagi, K., Kurematsu, T., & Sawaki, Y. (1995). Photochromic behavior of surfactant spiro 2h-1-benzopyran-2,2' 2,3 -dihydroindole s (spiropyrans) adsorbed into clay interlayers. *Journal of the Chemical Society-Perkin Transactions*, *2*(8), 1667–1671. doi:10.1039/p29950001667

Takumi, K., Sakamoto, H., Uda, R. M., Sakurai, Y., Kume, H., & Kimura, K. (2007). Photocontrol of anionic micelles containing lipophilic crowned spirobenzopyran. *Colloids and Surfaces a-Physicochemical and Engineering Aspects, 301*(1-3), 100-105. doi:10.1016/j.colsurfa.2006.12.029

Teh, S.-Y., Lin, R., Hung, L.-H., & Lee, A. P. (2008). Droplet microfluidics. *Lab on a Chip*, *8*(2), 198–220. doi:10.1039/b715524g PMID:18231657

Wang, D., Jiao, P., Wang, J., Zhang, Q., Feng, L., & Yang, Z. (2012). Fast photo-switched wettability and color of surfaces coated with polymer brushes containing spiropyran. *Journal of Applied Polymer Science*, *125*(2), 870–875. doi:10.1002/app.36254

Xu, X.-D., Wei, H., Zhang, X.-Z., Cheng, S.-X., & Zhuo, R.-X. (2007). Fabrication and characterization of a novel composite PNIPAAm hydrogel for controlled drug release. *Journal of Biomedical Materials Research. Part A*, *81A*(2), 418–426. doi:10.1002/jbm.a.31063 PMID:17117471

Yang, D., Piech, M., Bell, N. S., Gust, D., Vail, S., Garcia, A. A., & Picraux, S. T. et al. (2007). Photon control of liquid motion on reversibly photoresponsive surfaces. *Langmuir*, *23*(21), 10864–10872. doi:10.1021/la701507r PMID:17803327

Yang, L., Takisawa, N., Hayashita, T., & Shirahama, K. (1995). Colloid-chemical characterization of the photosurfactant 4-ethylazobenzene 4'-(oxyethyl)trimethylammonium bromide. *Journal of Physical Chemistry*, *99*(21), 8799–8803. doi:10.1021/j100021a054

Yoshida, M., Yang, J. S., Kumakura, M., Hagiwara, M., & Katakai, R. (1991). Artificially intelligent hydrogels responding to external stimuli such as temperature and ph. *European Polymer Journal*, *27*(9), 997–1005. doi:10.1016/0014-3057(91)90048-S

Zhang, J., Wang, S.-C., & Lee, C. T. Jr. (2009). Photoreversible Conformational Changes in Membrane Proteins Using Light-Responsive Surfactants. *The Journal of Physical Chemistry B*, *113*(25), 8569–8580. doi:10.1021/jp807875u PMID:19485396

Ziółkowski, B., Czugala, M., & Diamond, D. (2012). Integrating stimulus responsive materials and microfluidics – The key to next generation chemical sensors. *Journal of Intelligent Material Systems and Structures*.

Ziółkowski, B., Florea, L., Theobald, J., Benito-Lopez, F., & Diamond, D. (2013). Self-protonating spiropyran-co-NIPAM-co-acrylic acid hydrogel photoactuators. *Soft Matter*, *9*(36), 8754–8760. doi:10.1039/c3sm51386f

Chapter 11
High-Tech Applications of Functional Coatings:
Functional Coatings and Photovoltaic

Susana María Fernández
Ciemat, Spain

Antonio Garrido Marijuan
Austrian Institute of Technology, Austria

ABSTRACT

In the proposed chapter, the authors will present an overview about the role of the main transparent conductive oxides as functional coatings and its incorporation into emerging large area flexible photovoltaic technologies. In particular, the authors will describe the state of the art in this area, with the highlight of establishing the importance of using transparent conductive oxides as an essential part of an industrial chain. In this sense, it will be emphasized the main requirements of transparent conductive oxides, the main deposition techniques employed for its fabrication, the different material alternatives that emerge as substitutes of conventional ones and, finally, the evaluation of the main risks found when the coatings will be incorporated in an industrial production chain of flexible devices.

SCOPE OF THE CHAPTER

The present chapter focuses on the use of Transparent Conductive Oxides (TCOs) as functional coatings on the photovoltaic (PV) industry. First of all, it is explained the importance of these functional coatings in the near future and the role they play in the PV industry to reach device with higher nominal efficiencies and lower cost. The use of TCOs allows widening the applications of PV and producing new services industrially, such as large areas or flexible devices that can be easily folded, light and easy to carry, contributing to the cost reduction in the production process and being compatible with the use of roll-to-roll (R2R) deposition systems. The above applications are emerging as a potential choice being used in new products like wearable devices. These flexible devices aim to remain functional after repeated flexing with strong efforts, in order to guarantee its mechanical stability.

DOI: 10.4018/978 1 5225 0066 7.ch011

The second part is devoted to understand the use of TCOs as coating films and also to describe the main deposition techniques used for its fabrication. The focus of this part will be on the study of the TCO´s properties deposited onto different substrates used in industry. The outstanding role of ZnO:Al (AZO) will be remarked as one of the more important materials used nowadays for TCO applications due to the abundant of the raw materials, its beneficial properties as a representative of TCOs, comprising structural, electrical and optical properties, and its potential uses as integral part of PV devices. The interest emerging of the combination of TCOs with polymer substrates to flexible PV applications makes it into a hot topic. Polyethylene terephthalate (PET) and polyethylene naphthalate (PEN) are examples of proper cheap substrates used in this field, based on the fact that these polyesters are cheap and maintain its physical, mechanical, electrical and chemical properties for working temperatures up to 150ºC. It will be detailed how the use of polymer substrates affect over the functional, and the need of lower deposition temperatures for the deposition of TCOs on polymer substrates. Finally, in this second part, and taking into account the importance of mechanical stability for a flexible device, it will be discussed different approaches that could be applied, and easily incorporate in a chain production, in order to improve the adhesion of the TCO layer.

It is also important for a TCO to have a proper surface topography, with an efficient light-trapping scheme to enhance the intrinsically low absorbance of the thin-film absorbers at the long wavelength range. On the third part, different strategies employed via using post-treatments to obtain appropriated TCO surface texture will be discussed, as well and the main features and its effect on real devices, specifically those treatments especially suitable to be performed on AZO surface deposited at low temperatures for flexible applications. The authors will remark the advantage of optimizing light scattering behaviour separately from TCO optoelectronic properties.

Part 4 gathers the future of TCO's applied into solar devices, providing insight about the objectives of the present research and development activities. These objectives include developing viable PV products with better properties and lower cost with emerging materials such as AZO. Part 4 will be devoted to provide the fundamental information about the new families of materials that are proposed to substitute ITO, especially the emerging nanostructured TCO materials and their incorporation to the industry. The authors will also detail the main researches about different alternative schemes to enhance light trapping.

Finally, the authors will end with a conclusion to the exposed above.

PART 1: INTRODUCTION: PHOTOVOLTAIC'S IMPORTANCE IN THE FUTURE AND THE ROLE OF THE TCO

PV industry is called to be a major player in European electricity production in the medium (2030) and longer term (2050). It is a fairly common practice to assume an enormous potential for PV since it can also be installed on fields in large scale applications, such as roofs, building surfaces, fallow land, deserts, etc. (Defaix, van Sark, Worrell, & de Visser, 2012; Philbin et al., 2008). The recently published Solar Europe Industry Initiative (SEII) implementation plan 2010–2012 assumes a potential of 700 GWp installed capacity for the EU-27 plus Croatia, Norway and Turkey.

Additionally, favourable feed-in tariffs and a rising price for electricity at the consumer point are making PV installation as a real option for self-consumption. However, PV potential has a limit of energy generated per square meter, depending mainly on the PV efficiency. To make the above targets realistic it is important to increase not only the PV efficiency, but also to reduce costs and to spread the application fields. In this sense, nowadays, the PV industry is being focused on research and development (R&D) of PV flexible modules to reduce costs and create new applications.

Cost and Industrial Production: Towards Roll to Roll

The solar cells based on thin film of silicon or other materials can be made by batch process or low cost continuous process. The use of flexible substrates allows continuous R2R production, on a steady state processing that enhances significantly the reproducibility and stability of the manufacturing conditions and PV features. Based on these facts, it is possible to achieve higher growth rates without compromising film properties and also reducing production costs. In addition, R2R becomes much more environmentally benign and energy-efficient process than wafer-based or vacuum-based manufacturing processes. This new open field of low cost, light weight, flexible and unbreakable large area PV devices allows the use of large area plastic substrates as an available cheap solution (Waugh, 2011).

As an example, with R2R processes, the manufacturing of polymer organic solar cells is approaching the so called 10-10 target (10% efficiency and 10 years of stability) with an acceptable power conversion performance. For a high throughput production of flexible solar devices, spin coating and high vacuum thermal evaporation have to be neglected. A more compatible possibility is shown in figure 1, where it can be observed a typical solution processing with adequate features of speed, temperature and multilayer processing. If PV modules can be manufactured not only in a continuous R2R process without involving vacuum processes, but also with thin and cheap substrates, solar cells would reduce significantly its cost per unit area (Søndergaard, Hösel, Angmo, Larsen-Olsen, & Krebs, 2012).

The different deposition techniques used in the PV industry, mainly evaporation, sputtering, chemical vapour deposition, are compatible with R2R industrial processes. It is possible to load an entire roll into a vacuum system that facilitates the sputtering or evaporation processes. In this processes, thickness is closely related to rotation speed, initial position and orientation of the substrate (Watts, 2007).

New Applications

Finding new light-weight and flexible display technologies that can be manufactured in a continuous R2R process will open a wide list of multiple applications of PV modules and, hence, the square meters devoted to sun reception on PV displays. The new PV modules will suppose highly demanding products not only in efficient portable devices such as pagers, personal digital assistants (PDA), smart cards and the future "electronic paper" (Banerjee et al., 2006), but also integrated applications into buildings.

New applications include:

- **Building Integrated PV:** With flexible PV modules, the square meters of PV modules installed could be increased since the applications would not only be reduced to a roof installation. Hereby, the energy output would be also increased on a high extent. Their suitability for integration into building elements offers a clear opportunity for further reduction in PV system costs, (Gabriel et al., 2011).
- To produce energy in an *autonomous way in moving applications* where it's not normally available, such as backpacks, vehicles etc. These applications would reduce the weight of batteries and can be used for the Army, tourism etc.

Figure 1. R2R processing diagrams for organic electronics/thin films
Source: Hösel, Søndergaard, Angmo, & Krebs, 2013.

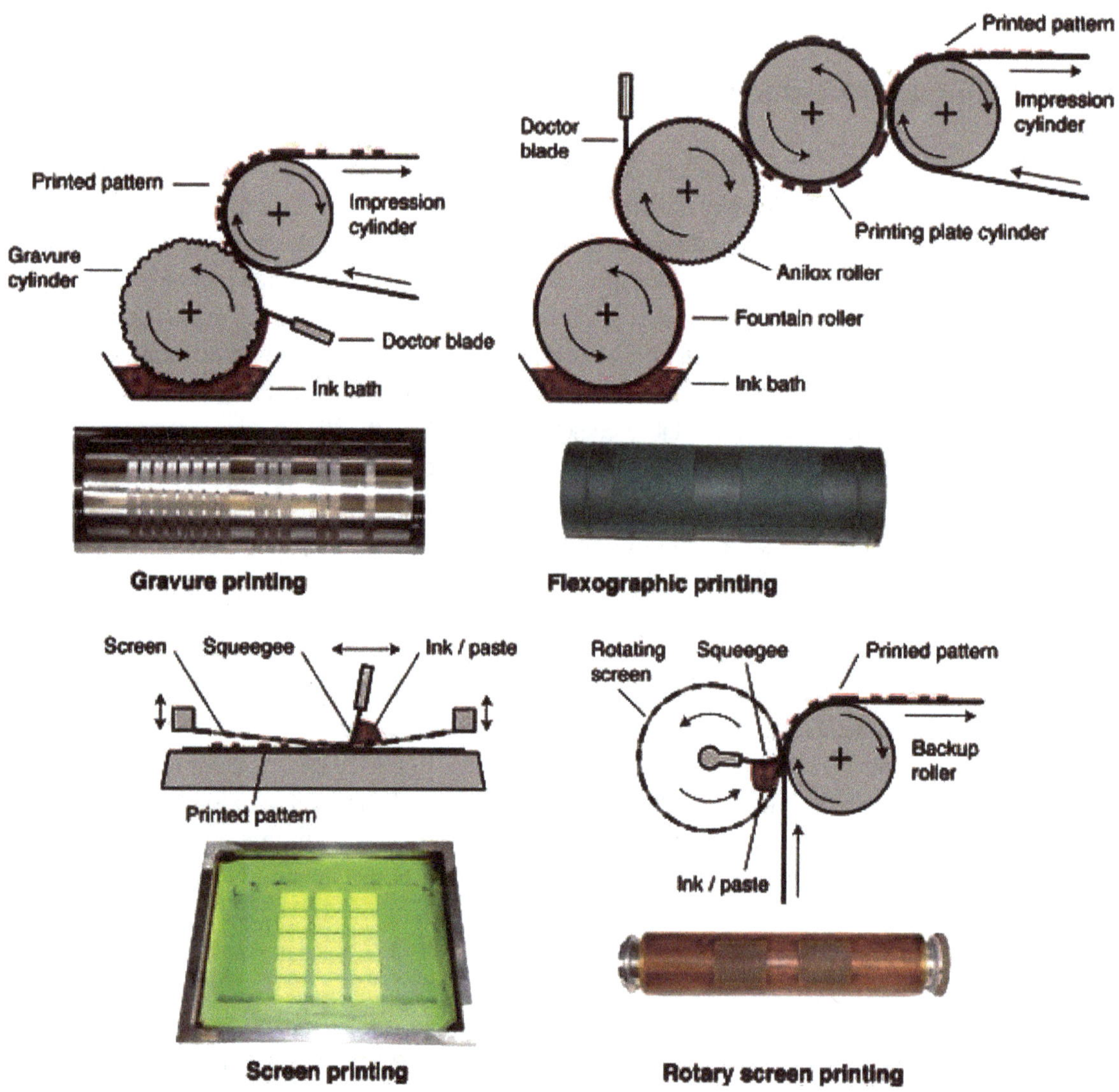

- **Printed PV:** With a set of printing methods used to create common printing equipment with a low-cost volume fabrication, (Lee, Lee, Park, & Choi, 2014).

R&D on Production of Flexible Devices

In order to achieve these goals, it is essential to find flexible, easy to produce, lightweight, thin but also cheap substrates, such as plastic materials. Polymers should have a plane surface, mechanical stability, being able to bend but not change its dimensions or deform, and permeability. SiO_2-coated glasses gather these properties, having been largely used in industry. However, the production of very large sheets of this kind of glasses that are thin enough is not easy. This material only becomes flexible when is thinner than 200 microns but a traditional display glass is typically only as thin as 500 microns.

Polymers also have to be permeable. In order to achieve this characteristic, functional coatings can be applied acting as barrier layers and allowing polymers to be usable in different applications integrated in our everyday life such as flexible displays, flexible organic photovoltaic cells (OPVs) or flexible electronic devices (FEDs) (Plichta et al., 2005). In all later devices, TCOs are considered as essential part, being responsible of collecting the generated current, increasing the light scattering or decreasing the surface defect density, essential parameter to maximize performance of the solar cells.

The deposition of TCO onto bendable and flexible polymer substrates in large areas opens up the possibility of using new ad intelligent production techniques in high volumes. However, it is well known that almost all plastic substrates used in those applications are polymeric materials with glass transition temperatures not high enough (<200ºC). Thus, how decreasing the deposition temperature of TCO without losing coating film qualities when they are deposited onto a polymer substrate becomes a key issue.

In the process of TCO deposition on flexible substrates, it is important to take into account that a high temperature process reaching the surface of the substrate deforms it due to the poor heat endurance that this substrate shows.

In the flexible technology industry, the research on the adhesion of coating films to the substrate is consider a relevant issue due to the low surface free energy that most of polymers presents. This low surface free energy results in poor adhesion and technical problems on the deposition of coatings. In this sense, the surface characteristics of polymers are one of the most important factors that should be evaluated in order to determine the interfacial properties substrate/subsequent films and with the objective to achieve mechanically stable coating films.

Among the main plastic substrates used for these last purposes are found **PET** and **PEN.**

PET is an excellent commercial thermoplastic polymer resin of the polyester family. PET can be semi-rigid to rigid, depending on its thickness, and is very lightweight. This polyester contains a benzene ring in a repeat unit and it can maintain its physical, mechanical, electrical and chemical properties for greater working temperatures than others plastic. PET is able to reach work temperatures up to 170ºC, although it begins to shrink at 150ºC. The most important reason to use PET in PV applications is that its transmission is around 90% in the wavelength range 0.4–2.4 µm, with a refractive index of 1.65.

PEN is a similar polyester to PET but permits its use at slight high temperatures. In this case, the repeated monomer is a naphthalene ring. It is semicrystalline and colourless slightly opaque. The optical, electrical and mechanical properties are maintained up to 155ºC, beginning to shrink at 190 ºC. The main drawback that this polyester presents is that it is more expensive that PET and less mechanical strength. Both PET and PEN can be easily manipulated and extruded on amorphous films (Mandal, Singha, Dhar, & Ray, 2007).

R&D on TCO Materials for Flexible Photovoltaic Applications

There is a real need to develop robust and low temperature deposition process for different solar cells components onto polymer substrates in order to facilitate the R2R manufacturing processes. The approach which promises a significant cost reduction in the solar devices is the use of TCOs as functional coatings, being one of the main areas of research to reach higher nominal efficiency and lower costs. TCOs requires low substrate temperatures to be deposited onto flexible substrates compared to the optimal temperatures, around 400-500 ºC, used in the process of manufacturing this material onto rigid substrates. These temperatures are normally used when deposited on conventional glass substrates, and it is not critical in the production of glass modules.

Since the cost of crystalline silicon is out of reach for the average consumer, there has been a push toward technology that can be readily available. With this approach, a new technology consisting on highly absorbent materials that can operate effectively at a thickness than varies from a few nanometers to tens of microns is being gaining interest as alternative. As thin film technology reduces the amount of active material in the solar cell, they are less expensive that thick crystalline silicon. Besides, it offers several advantages over traditional silicon panels. Manufacturing components with this built-in technology could cut down on cost, eliminating the need to spend money on a separate solar device and installation.

TCO are very important for thin film silicon solar cell structure, influencing decisively on the efficiencies of amorphous, microcrystalline or ''micromorph'' based solar cells. The optimization of TCO for solar cells constitutes a very complex problem especially on the up-scaling of the research results to industrially relevant substrate sizes (Müller, Rech, Springer, & Vanecek, 2004).

For a proper use in PV applications, TCOs should be highly transparent in the visible wavelength range to minimize optical losses and ensure that most of photons are captured in the device absorber, have high conductivity to reduce the resistive losses and low carrier concentration to avoid absorption losses in the red and near-infrared wavelength ranges (Exarhos & Zhou, 2007). Apart from these properties, TCOs surface should have a proper topography, with an efficient light-trapping scheme to enhance the intrinsically low absorbance of the thin-film absorbers in the long wavelength range (Krč, Zeman, Kluth, Smole, & Topič, 2003). For all these reasons, it is very important to study the application of these films on the performance of the solar cell.

PART 2: TCOS AS COATING FILMS: THE EMERGING ROLE OF AZO MAIN DEPOSITION TECHNIQUES

As seen, TCOs are one of the most important components for solar cells applications, being extensively use in flat panel displays, optical sensors or touch screens (Berginski, Hüpkes, Reetz, Rech, & Wuttig, 2008). For the described applications, TCO must have a remarkable combination of conductivity and transparency, with low resistivity and low optical absorption as significant pre-requisites.

TCOs are typically use on solar devices as front electrode and back reflector, this last application in the superstrate and substrate device configuration. It cannot be forgotten the potential use of TCOs as antireflection coating (AR) in silicon solar cells. Regardless the application, the main properties that a TCO should show are a resistivity less than 10^{-4} Ω cm and an average transparency in the visible (VIS) range higher than 85%. TCOs are generally n-type wide bandgap semiconductors (although p-type materials are being developed) with a relatively high concentration of free electrons in the conduction band. The wide bandgap is responsible for high optical transmittance and free electrons increase electrical conductivity.

The main requirements to be met by a TCO are:

Morphology: Thickness Uniformity and Topography

In the case of incorporating to a production large chain, the geometry of a TCO should be taken into account in terms of demonstrating the level of uniformity in both thickness and optoelectronic properties. TCO thickness should be of a certain minimum dimension to avoid light absorption in it and to develop the desired features to improve the light trapping; and it should be also as uniform as possible.

A way to estimate the uniformity of a TCO is to calculate its thickness from reflection/transmission spectra in several zones of the material deposition. The following expression is used to express global uniformity:

$$\Delta t = (t_{max} - t_{min})/t_{mid}$$

where "t" is the film thickness, "t_{max}", "t_{min}" and "t_{mid}" are the maximum, minimum and average value of the thicknesses measured in large areas, respectively. In order to give an idea about the margins for this last parameter, there are different studies on the literature about cell thickness where values of uniformity around 10% in areas of 10x10 cm^2 could be remarked as acceptable ones (Franken, 2006).

Regarding the TCO topography, one of the main characteristics is its surface roughness. Surface roughness has a great interest on light scattering, enhancing a better performance of the solar cell. In this sense, an efficient surface texture allows the reflections at rough internal interfaces of the device, leading to a more efficient light trapping and subsequently, increasing the light absorption into the cell. Most of TCOs present a smooth surface as-deposited. For this reason, it is important to achieve the desired roughness in terms of appropriated features on its surface, using TCO thicknesses as narrow as possible, in order not to lead an important increase in absorption, and different post-deposition treatments.

Different approaches are followed by the industry with the objective of obtaining an optimized surface texture, mainly the growth of smooth TCO films on surface textured substrates and the post-deposition surface treatment. Part 3 summarizes the main post-treatment techniques available in order to obtain the desired morphology, taking leaving absorption and surface morphology largely independent from each other.

Optical Properties

TCOs are wide band gap semiconductors with band gaps Eg>3 eV, whose properties are strongly dependent on the oxygen deficiency and on the impurity doping in film. Solar cell designs require reflective losses air-glass, glass-TCO and TCO-absorber interface less than 15-20% of incident light at normal incidence. At higher oblique angles, this amount of loss light increases. A way to reduce the TCO-absorber reflective losses and enhance the absorption in the active layer of the device is to have a rough surface for the TCO. This kind of surfaces is highly desired because this morphology increases the effective optical path length of the light within the absorber.

The requirements for the optical properties of the TCO depend on its structure and the absorber material of solar cell. In any case, TCO absorption should be as low as possible to avoid loss of the sunlight, and to allow a high percentage of light going through. Regarding the requirements for the average values of transmittance in the VIS region, a TCO should show values higher than 85%. Ideally, a wide band gap TCO should not absorb photons in the transmission "window" in the UV-VIS-NIR region.

Absorption is closely related to carrier density and its mobility. When the free carrier is increased, so does absorption. At the same time, the increase of absorption implies a decrease of transmission in the infra-red (IR) region (Kluth et al., 2003). Other parameters affecting absorption are structural properties such as defects, grain boundaries and film crystallinity.

As it can be described before, an ideal TCO should be as thin as possible with a low free carrier concentration in order to reduce absorption in the VIS/NIR region. Figure 2 illustrates how the light would through a thin TCO (around 100 nanometer of thickness) over the whole wavelength region measured, independently the absorbent use on the device.

Figure 2. AM 1.5 solar spectral irradiance, based on data from ASTMG-173-03 std, direct normal and hemispherical on 37º tilted surface
Source: ASTM G173-03(2012), Standard Tables for Reference Solar Spectral Irradiances: Direct Normal and Hemispherical on 37° Tilted Surface, ASTM International, West Conshohocken, PA, 2012, www.astm.org

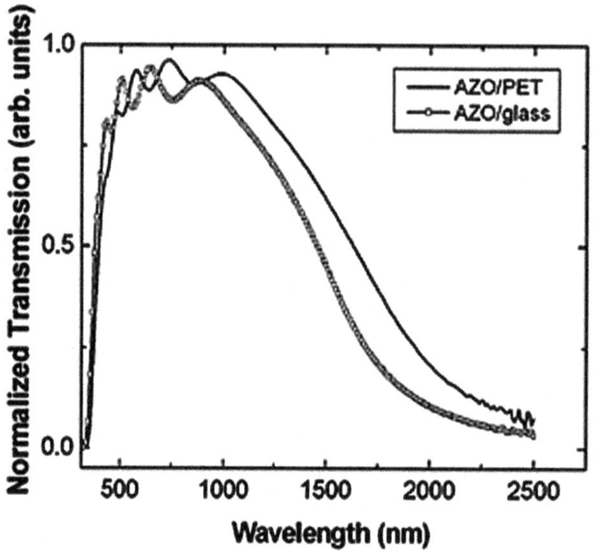

But at the same time, the previous considerations of using thin TCOs would lead to an increase of the sheet resistance Rsq. For this reason, it is necessary to look for a compromise between the optical and electrical TCO properties, as it will be shown later. The main objective is to find a proper TCO with low absorption by free-electrons and high conductivity both at the same time. Taking this into account, the average absorption in the wavelength range of 400-1100 nm of a TCO used as front electrode and deposited onto a highly transparent glass should be below 6–7%. For amorphous silicon absorber layers, a high transparency is required in the wavelength range of 400–750 nm. In the case of solar cells incorporating μc-Si:H, the TCO has to be highly transparent up to the near infrared (NIR) region (400–1100 nm). Figure 3 shows the strong effect of increasing TCO thickness on the optical properties.

Another possibility that permits the decrease of TCO thickness is to enhance the light trapping by textured coatings. The post-treatments use to enhance light trapping would be detailed in part 3.

Electrical properties

The electrical potential of TCOs has a large dependence on the device performance and for this reason, it has been considerably studied. As example, TCOs are used as front and back electrons on silicon based thin film solar cells, so they would show a high electrical conductivity in order to avoid resistance losses into devices.

It is accepted that sheet resistance, R_{sq} of a high-quality TCO should be around 10 Ω/sqr. However, several dependencies should be taken into account in order to incorporate a TCO into the device. Among them, as TCO thickness increases, resistance decreases in an exponential way. To measure the influence of sheet thickness and resistance, resistivity is defined as:

Figure 3. Transmittance spectra of ITO thin films of different thicknesses deposited by DC magnetron sputtering onto glass substrates at 350ºC

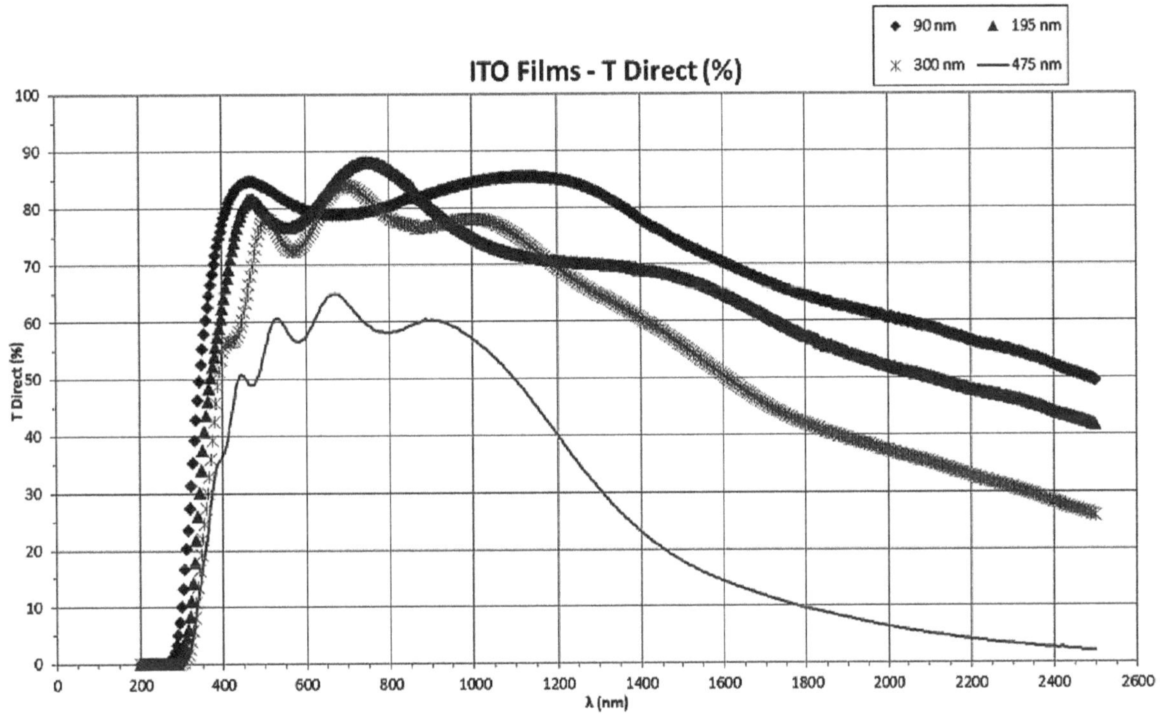

Resistivity= R_{sq}**thickness.*

Figure 4 shows the variation of the resistivity with the thickness of AZO thin films deposited onto glass by Radio-Frequency (RF) magnetron sputtering under the same sputtering conditions. This figure gives an idea about the importance of using appropriate TCO thickness on its electrical performance. As it can be observed, thin layers lead to high resistivity values. On contrary, according to the optical performance, films as thin as possible would be required.

On the other hand, deposition conditions may also affect on TCO electrical properties. As it can be appreciated, Figure 5 shows the strong influence of substrate temperature on resistivity of AZO thin films. In this particular case, the electrical improvement is basically attributed to a better quality crystalline of films with deposition temperature due to the enhancement of mobility of the species when reach the surface substrate.

Finally, another parameter that should be considered is the residual stress. Stress is an intrinsic parameter of thin films, so its quantification is an important matter. This parameter results from the addition of thermal stress, due to the difference between the thermal coefficients of film and substrate, and intrinsic stress due to itself growth. It also influences largely on the electrical properties since it can cause appearance of defects into the deposited films that could deteriorate resistivity.

From all these considerations, it can be deduced that it is necessary to look for a compromise between the structural, the optical and electrical properties of the TCO.

Figure 4. Variation of resistivity with thickness of the AZO thin films deposited onto glass by RF magnetron sputtering under the same sputtering conditions

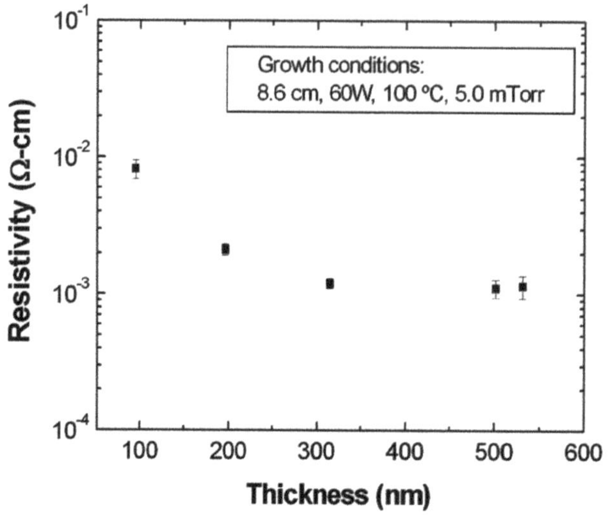

Main Materials Used as TCO: Emerging Role of AZO

Transparent conductive materials show the characteristics of coexistence of electrical conductivity and optical transparency. To meet these requirements, the main strategy have consisted in the doping the known binary and ternary TCOs with other elements, which can increase the density of conducting electrons. Among the main TCO proved to be adequate for the described applications are found pure

Figure 5. Evolution of resistivity with substrate temperature for 500 nm-thick AZO films deposited onto glass by RF magnetron sputtering under the same sputtering conditions

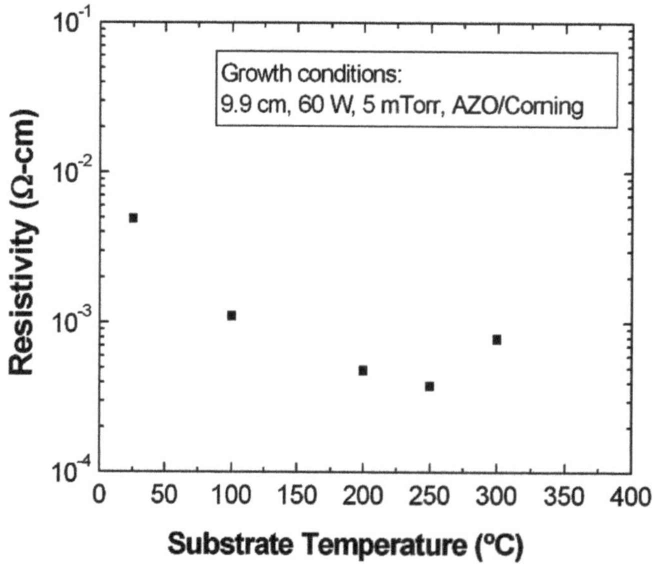

zinc oxide (ZnO) and its alternative n-type doping with Gallium (Ga), Aluminium (Al), Cooper (Cu), Silver (Ag), Indium (In) or even Boron (B)-doped ZnO, In2O3, SnO2 and CdO (Stadler, 2012). Among them, the most studied are Al-doped ZnO (AZO) and Ga-doped ZnO (Liu, Li, & Zeng, 2013). In the last years, p-doping of ZnO is being developed, and Nitrogen (N) and Phosphorus (P) seem to be also adequate dopants.

Of all the above, Sn2O:In2O3 (ITO) and F-doped SnO2 (FTO) are widely used for most solar applications because of its excellent electrical and optical characteristics (Agura, Suzuki, Matsushita, Aoki, & Okuda, 2003). However, these materials present some industrial limitations, such as the ability to deposit on large areas, the toxicology of the material, and the ability to introduce effective light scattering within the active layers of solar cells. In the particular case of FTO, its cost is comparable with ITO and this material is typically prepared by spray pyrolysis requiring high processing temperatures of around 500ºC. This fact together with the poor chemical stability against hydrogen-rich plasmas, leads to the necessity of searching alternative TCO materials.

In the case of ITO, this material has also several crucial drawbacks (Terasako et al., 2007), here described:

1. ITO films are inevitably quite brittle basically due to its nature of fabrication. The thermal expansion coefficient mismatch between ITO film and flexible substrates creates cyclic internal stress on the films, being more evident during the device fabrication steps where temperature is fluctuating. This fact could generate micro-cracks, damaging the deposited film, propagating throughout the device and producing short circuits in it.
2. The cost for its production becomes very high due to the limited supply of Indium, representing up to 30% of the overall module cost.
3. To maintain their favourable properties and to be incorporated in flexible industry, an additional annealing treatment is required to transform the ITO film structure from amorphous to polycrystalline. This last process is really difficult to apply in some flexible plastic substrates used in industry such as polyethylene terephthalate, polyimide, polyethylene naphthalene and polycarbonate. All these polymer substrates have low glass transition temperatures completely incompatible with an annealing process.
4. In the fabrication process of amorphous silicon solar cells in configuration of superstrate, the TCO is normally exposed to hydrogen plasma. In the case of ITO, this material is very sensitive to that environment. Under it, the ITO surface is reduced became very rough and causing a decrease of the visible transmittance due to the increased absorbance and reflectance.

Thus, it is essential to find some alternatives to both ITO and FTO for the low cost and low-temperature deposition of good quality TCO films onto plastic substrates. In this search of adequate TCO materials to be use in large area PV applications, Zinc oxide is being considered increasingly as a very promising candidate and reliable alternative, coming close to ITO in its electrical and optical performance. This material is a well-know semiconductor with a large exciton binding energy (60 meV) (Look & Claflin, 2004), and a hexagonal wurtzite structure. Un-doped ZnO usually shows n-type conductivity due to the oxygen vacancies and zinc in interstitial position. It should be pointed out that this material offers a series of advantages. Apart from ZnO-based films are cheap, nontoxic and compound of abundant materials, they are readily produced by sputtering technology, technique highly preferred for large-scale architectural coatings. ZnO also allows tailoring of the ultraviolet absorption which is possible because

Figure 6. Spectral solar irradiance and standard luminous efficiency function for photonic vision
Source: Jin, Hamberg, & Granqvist, 1988.

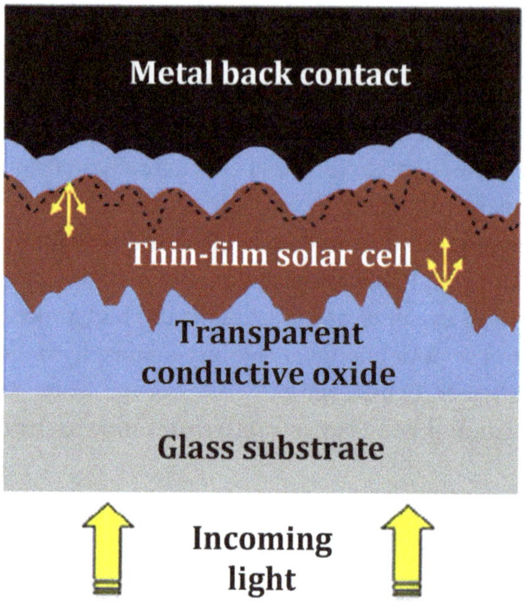

its fundamental band gap lies put out the end of the luminous spectrum. All these highlights permit that ZnO-based films can be used as window coatings, specifically, as ultraviolet protection, whereas other materials such as In_2O_3 and SnO_2 are not useful for this purpose. To illustrate this, Figure 6 shows the spectral solar irradiance at ground level for one air mass and the standard luminous efficiency of the light adapted human eye (Shan & Yu, 2003). Arrows indicate the band gap energy found in the literature for the main un-doped oxide semiconductors used as coatings, from 3.75 eV for In_2O_3, 3.7 eV for SnO_2, 3.4 eV ZnO to 2.3 eV for CdO (Mandal, Singha, Dhar, & Ray, 2007).

Highly conductive ZnO thin films with high transmittance in the VIS wavelength region has been achieved by doping with metals such as Al, B, Ga, and In. Among them, Al-doped ZnO, AZO, presents adequate structural, electrical and optical properties, as well as low cost, suitability for low-temperature deposition, wide availability of its constituent raw materials and high stability against the hydrogen plasmas. AZO is also easily etched with diluted acid to obtain a useful post-deposition texturing in order to improve the light trapping within the solar cell (Olvera, Maldonado, Asomoza, & Meléndez-Lira, 2000).

Taking into account the importance of replacing rigid substrates for flexible substrates in flexible solar technologies, the deposition of TCO onto polymer substrate becomes a great merit. This is basically due to the low substrate temperatures required for the material fabrication. Among deposition techniques used in coating fabrication, magnetron sputtering emerges as available choice. This practical method is commercially used and is suitable for large area deposition and mass production. An optimisation of deposition parameters such as sputtering power density, working gas pressure, substrate temperature, distance between target and substrate, and film thickness are required to achieve high quality of AZO films. On the other hand, the high energy of the sputtered particles can seriously affect the deformation of the substrate and, therefore, the film crystallinity, its residual stress, its adhesion to the substrate and its microstructure would be also affected (Fernández, Martínez-Steele, Gandía, & Naranjo, 2009; Fernández, & Naranjo, 2010).

Figure 7. Comparison of normalized transmission spectra of AZO films deposited on: glass substrate (white square) vs. PET substrate (black line)

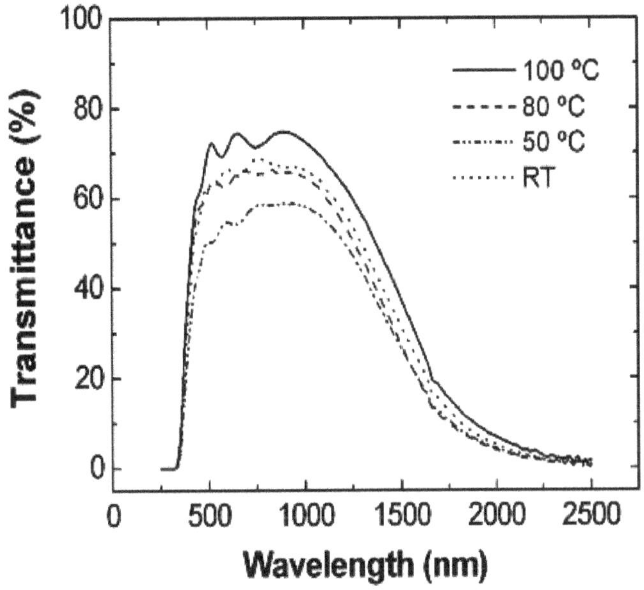

To analyse the suitability of AZO films deposited on PET, their properties are compared with those presented by films with similar thickness deposited on glass substrates. This comparison reveals several remarks:

- PET substrates have standard thicknesses of up to 0.25 mm, and 0.95 mm for the glass. This fact would contribute to the cost reduction in the production process.
- AZO thin films deposited on both glass and PET substrates can show similar optical properties by optimizing the sputtering conditions (see Figure 7). In this figure, to isolate the effect of the substrate, a normalization of transmission spectra to transmission of substrate has been performed, considering the following expression:

$$T_{normalized} = T_{film}/T_{substrate}$$

With last definition, 500 nm-thick AZO thin films presenting an average normalized transmission value in the VIS wavelength range of around 0.95, on glass substrate, and 0.9, on PET substrate, are achieved.

On the other hand, Figure 8 illustrates the importance of optimizing the deposition conditions. The strong dependence of the substrate temperature on the transmission spectra is appreciable.

- From Figure 7, it can be observed that the absorption edge of AZO/glass has a slightly shift to the short wavelength when compares with AZO/PET. This result is coherent with the electrical properties of both films, obtaining a widening of the band gap for the sample with the lowest resistivity (highest carrier concentration), according to the Burstein–Moss effect (Burstein, 1954).

Figure 8. Effect of substrate temperature on optical properties

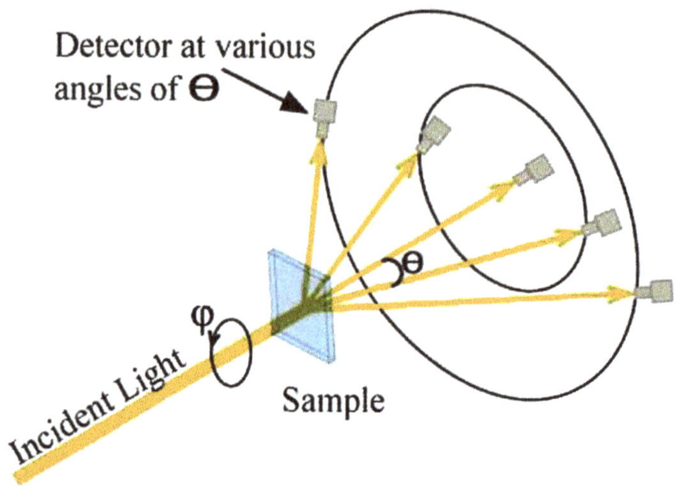

- To evaluate the performance of conductive films, a new parameter namely Figure of Merit (FOM) is defined as

$$FOM = -(1/\rho)^* \ln (T_r)$$

where ρ is the resistivity of the film and Tr is the average transmission in visible region (interval from 400 to 800 nm). The resistivity of AZO films deposited on PET is slightly worse than those on glass, basically attributed to the inferior structural properties obtained when deposition is at low temperatures (Fernández, Martínez-Steele, Gandía, & Naranjo, 2008). Under this low temperature regimen, it is necessary to look for a compromise between the minimum sputtered particle energy required to reach the flexible substrate with enough mobility to enhance the structural film properties and to avoid the damage on the surface substrate induced by too much energetic particles. Taking in mind all these considerations when optimizing sputtering deposition parameters, very close FOM values can be achieved regardless the substrate (Fernández et al., 2009).

- Thin films deposited on plastic substrates showed at least two times lower stress values than for samples deposited on glass, reaching a minimum of 0.05 GPa,. The polymeric substrate is capable to release the strain accumulated during the growth process, and it seems as a suitable reason for this advantageous feature, especially when thin films are incorporated to final devices (Fernández et al., 2010).

From all these remarks, it can be concluded that AZO films on flexible substrates showing appropriate quality can be obtained, even at very low substrate temperatures.

Other disadvantage to overcome when using flexible substrates is the existing difference in thermal expansion coefficients and lattice mismatch between coating films and substrates. This fact results in brittle functional layers affecting negatively to the device performance. Besides, the low surface free energy that most of polymers present results also in poor adhesion of additional coating, leading to technical challenges (Vijayalakshmi, Mekala, Yoganand, & Navaneetha Pandiyaraj, 2011). As possible solution, plasma surface treatments can be applied to the polymer surface to improve the adhesion of coatings. Specifically, Plasma-surface modification (PSM) is an effective and cheap surface treatment technique used widely in industry that is performed in regimen of low temperature and low gas pressure. This technique may cause physico/chemical changes on the polymer film structure and even its electrical properties. The plasma exposure can also cause effects on polymer surface such as surface cleaning, surface ablation or etching, cross-linking and modification of chemical properties. It is important to optimize the plasma conditions to reach the desired results. Among the main parameters to optimize, it can be founded the type of plasma (DC, RF or microwave), the discharge power density, the process pressure, the gas flow rate and the exposure time. Several studies have demonstrated the feasibility of PSM process using Ar plasma treatments carried out at room temperature to modify appropriately the surface of flexible substrates. Under optimized Ar plasma conditions, the polymer surface makes rougher, enhancing its crystallinity and increasing the polar functional groups, and consequently, its surface energy. All these modifications lead to improve the coating adhesion. In this sense, suitable AZO thin films mechanically stable and with good adhesion are obtained when flexible substrate is Ar-modified. The potential viability of this plasma treatment has been also demonstrated by fabricating flexible a-Si:H p-i-n-based solar devices that remain functional after several bending only when the optimized PSM is applied previously on the surface polymer (Fernández, Santos, Munuera, García-Hernández, & Naranjo, 2015).

Deposition Techniques

The morphology, electrical and optical properties of coating films depend on the fabrication techniques and parameters used in the deposition. In the case of AZO, there are several deposition techniques that can be used for its fabrication. Here, it is described most of them and their particularities.

- **Chemical Vapor Deposition (CVD):** This method permits good quality film deposition and industrial production, with a high deposition rate of more than 50 nm/min. AZO films can be deposited on glass substrates from diethyl zinc, triethyl aluminium, and ethanol by atmospheric pressure CVD in the temperature range between 367 – 444 °C, which is very high for polymers substrates. Using the plasma-assisted CVD, AZO deposited at lower temperatures (in the range of 200 to 300ºC) can be achieved (Seeber et al., 1999; Tomita, 2008).
- **Spray Pyrolysis:** Involves spraying of a solution containing a mixture of zinc acetate and aluminium chloride onto heated substrates (Kaid, & Ashour, 2007).
- **Pulsed Laser Deposition (PLD):** For this technique, a high power laser pulses are used for the growth of TCO films with low resistivity and a wide ranges of temperatures. The optical properties of TCO obtained by PLD remain similar in the deposition temperature range between 25 and 400 °C where the average transmittance in the visible wavelength range varied between 86 to 91%. The main disadvantage of PLD is its difficulty to be utilised for large applications (Shan, & Yu, 2003).
- **Sol-Gel:** In the sol-gel technique, a solution of ZnO precursor is obtained by dissolving zinc acetate ($Zn(CH_3COO)_2{\cdot}2H_2O$) in anhydrous ethanol or methanol. Al dope is introduced by the addition of $AlCl_3{\cdot}2H_2O$ and $Al(NO_3)_3{\cdot}9H_2O$ in the methanol solution, (Nunes et al., 2002).

- **Magnetron Sputtering:** Both Direct Current (DC) and Radio-Frequency (RF): This technique has the special advantage of a great scalability to large areas and the potential to synthesize materials outside thermodynamic equilibrium. This fact is very important to deposit high-quality coating on those material very sensitive to high temperatures (Constantin & Miremad, 2011)
    - The use of magnetron sputtering is based on the acceleration of gas ions towards the material to be deposited. The sputtered material is then deposited on the desired substrate contained in a closed recipient where the process takes places. The dc-sputtering is restricted to conducting materials like metals or doped semiconductors.
    - When using a metallic Zinc target for sputtering of ZnO layers, oxygen as reactive gas is inserted into the sputtering chamber in order to react with the sputtered Zn particles. This method is also called reactive sputtering. With increasing oxygen pressure a complete transition from metallic to highly oxidized state can be obtained. According to the literature, a very small amount of oxygen in the gas atmosphere drives to a stable reproducible target surface during sputtering over a long period of time.

In comparison with last deposition techniques such as spray pyrolysis and sol-gel, magnetron sputtering could be considered an expensive solution. However, most of later techniques require moderate substrate temperatures to achieve AZO films with low values of resistivity. For this reason, they would not be considered as appropriate to be used into thin film technologies on flexible substrates. Under this requirement, RF magnetron sputtering can be considered one of the most favourable technique because allows down to room substrate temperature (RT) maintaining high deposition rates, excellent reproducibility and good adhesion to the substrate. In this sense, several researches (Mandal, Singha, Dhar, & Ray, 2007; Fernández et al., 2010), have demonstrated the suitability of RF magnetron sputtering for low temperatures deposition, studying how this parameter effects on the opto-electrical properties of TCOs. As summary, the achievements of large area coatings at high deposition rates and low deposition temperatures as room temperature make it into a very attractive choice to be incorporated in the industrial production.

PART 3. IMPROVEMENT OF LIGHT TRAPPING IN SOLAR DEVICES

Trapping light in silicon solar cells is essential because it allows increasing the absorption of incident sunlight in optically thin silicon absorber layers. In this sense, the costs of solar cells can be reduced by lowering the material consumption and decreasing the physical constraints on the material quality. In the particular case of hydrogenated amorphous and microcrystalline silicon (a-Si:H and μc-Si:H, respectively) solar cells, the semiconductor structure (doped and intrinsic materials) are placed between a TCO that acts as front contact and a highly reflective back contact based on TCO and a metal layer. Both parts of devices result crucial for the performance of thin-film silicon-based solar cells. As it has been pointed out previously, the TCO acting as front contact must meet a series of requirements: highly transparency in the wavelength of the absorber layer (400-800 nm for a-Si:H and 400-1100 nm for μc-Si:H), highly conductive to transport carriers and optimized rough surfaces to scatter the incident light and increase the absorption in the active layers. Hence, the development of a suitable surface texture is essential to scatters the light efficiently and extends the effective path length within the active silicon layers. As example, Figure 9 shows a typical design desired for an efficient thin film solar cell.

Figure 9. Interface scheme for a thin film solar cell to achieve efficient light trapping
Source: Dewan, Jovanov, Hamraz, & Knipp, 2014.

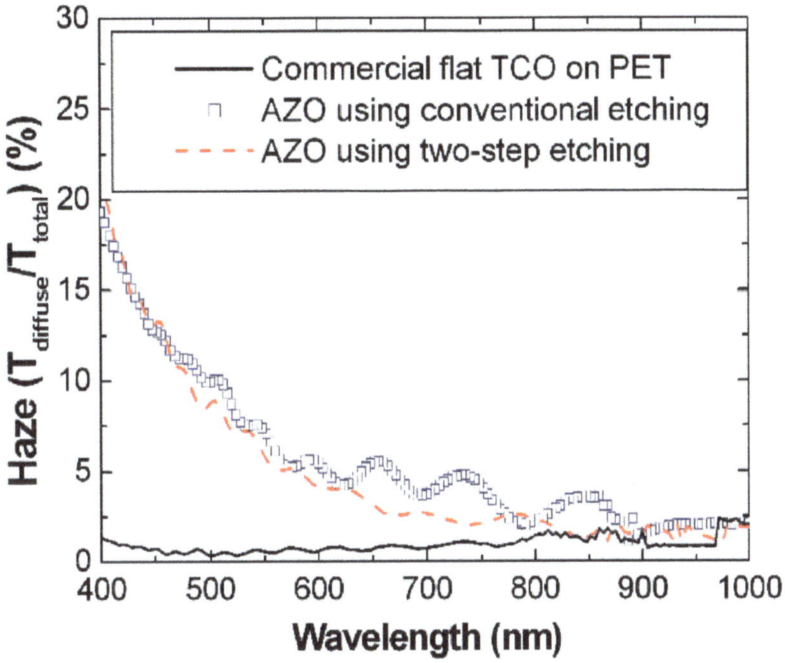

One way to achieve the desired morphology is using post-deposition chemical treatments. The main advantage of these etching processes is the possibility to separate the light scattering behaviour of the etched films from their optoelectronic properties. When using a wet etching to texture and before incorporating the rough TCO as front contact into device, several parameters should be analysed to evaluate its performance as light scattered element: the *root mean square* (rms) roughness and the height of the features, used to describe the surface morphology, the *haze parameter* H defined as the percentage of diffused light over the total transmitted light that can be referred for both transmitted and reflected lights, and finally, the *angular intensity distribution* that gives an idea about the intensity distribution of scattered light as function of the angle at which the scattered light propagates. Figure 10 illustrates how this last parameter can be easily measured.

Light scattering properties depend on the shape and size of the features obtained after the etching process.

To achieve appropriate features it has to be considered both the strong influence of the deposition technique and growth conditions, and etching mechanism/behavior of the etchant on the surfaces (Fernández, de Abril, Naranjo, & Gandía, 2012). The application of these etching methods implies in some cases a fast decreasing of layer thickness, and consequently, a rapid increase of its sheet resistance, and therefore it is crucial to control it. In the particular case of magnetron sputtering as deposition technique, low deposition pressure regimes lead to crater-like surfaces where the features are steep and sharp. On contrary, high pressure regimes lead to the appearance of granular-like features with bad scattering properties in spite of high rms roughness values (Kluth et al., 2003; Faÿ et al., 2006). The chemical etchant can also modify the shape and density with a less or more homogeneous distribution, being more or less effective to the scatter of light. In any case, it is really important to avoid deep holes spread on the surface of solar cells for risk of short-circuit in the devices.

Figure 10. Experimental system to measure the light scattering at different detector angles
Source: Farmer, Nichol, Seto, McSporran, & Deslandes, 2011.

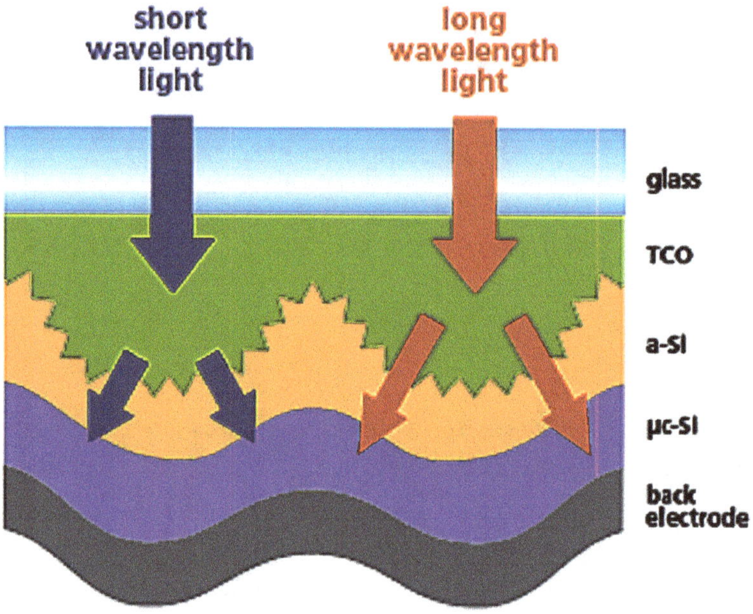

In the market, a solution for effective light trapping consists in a textured thick TCO layer, namely Asahi-U type SnO_2:F (FTO), deposited by chemical vapour deposition (CVD) on float glass. This rough TCO shows special features on its surface obtained during growth, without post-etching treatment. The concept of scattering all the wavelength of sunlight is based on combination of big and small features. Through this mechanism, all wavelengths are effectively utilized in the device. This behaviour combined with the excellent durability of glass make this product as a good choice for thin film PV technologies.

In the last years, RF magnetron sputtered AZO has emerged as a promising material to be used in thin film solar cells. Sputtered AZO films are initially smooth in most cases, but the achievement of different surface morphologies (using a wet chemical etching method with diluted acids makes this material more attractive. The typical acid used for the AZO surface texture process is the hydrochloric acid (HCl) at a concentration of 0.5% (v/v) (Fernández, de Abril, Naranjo, & Gandía, 2011)). Other acids used for the same purpose can be hydrofluoric acid (HF), ammonium chloride (NH_4Cl), nitric acid (HNO_3) and phosphoric acid (H_3PO_4).

According to the Thornton model, high deposition temperatures and low sputtering pressures are required to obtain AZO films showing these characteristics. However, as it can be described previously, in the flexible technology low substrate temperatures are required. Under this condition, the a-Si:H material exhibit worse performance than when it is deposited at high temperatures, and for this reason, an efficient light trapping is more necessary. The lack of a commercial substrate as Asahi-U for these applications makes the search of new strategies of texturing. However, the necessity of better scattering schemes performed on the AZO surfaces deposited at low substrate temperatures becomes a challenge to date. This is because the morphology of low temperature deposited AZO films is not so compact than those deposited at optimum high temperatures. And hence, etching rates considerably increase when using conventional acids, making more difficult the control of the etching process. As possible solution,

Figure 11. Comparison of spectral haze of a commercial flat TCO deposited on PET, an AZO surface etched with NH4Cl (10%wt) for 270 s, and an AZO surface etched with two step method (NH4Cl (10%wt) for 120s as first step and NH4Cl (2%) for 1500s as second step

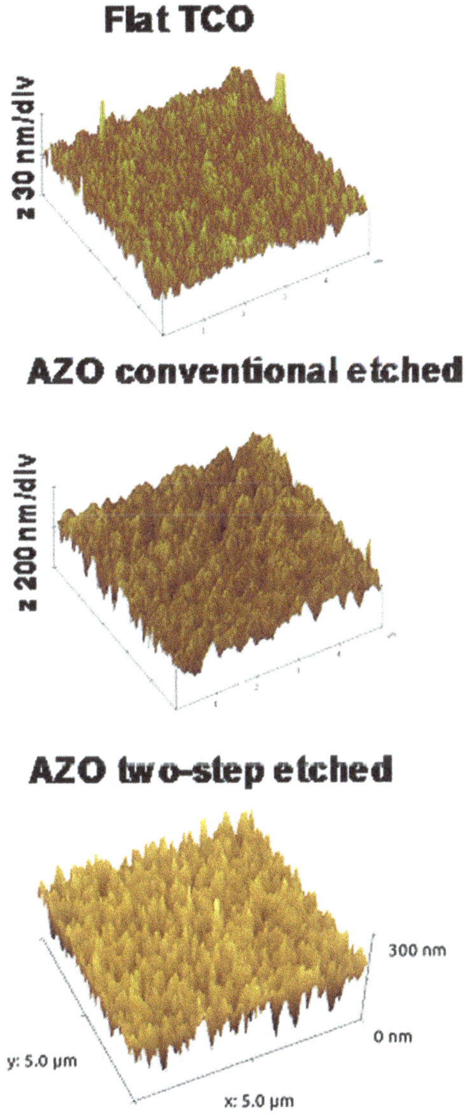

a combination of new strategies of texturing and more controlled etchant acids should be investigated. Following this idea, (Fernández, Pust, Hüpkes, & Naranjo, 2012) have developed a new wet-chemical etching method based on a two steps process using ammonium chloride, NH_4Cl, a salt of a weak base and a strong acid. This compound emerges as alternative candidate because of its relatively low cost and its controllable etch rate based on the slow etching effect.

In the case of ZnO, the reactive mechanism of NH_4Cl aqueous on film surface can be described by the following equations:

$NH_4Cl \rightarrow NH_4^+ + Cl^-$
$NH_4^+ + H_2O \leftrightarrow NH_3 + H_3O^+$
$ZnO + 2H_3O^+ \rightarrow Zn^{2+} + 3H_2O$

The etching mechanism of this salt is provided by the controlled amount of hydronium ion, H_3O^+ within the aqueous solution and its reaction with the oxygen in the ZnO lattice. The low etch rate is due to basically the low pH values of the solution in comparison to use other strong acids such as HCl. This fact results favourable in order to perform more controllable and reproducible etching process. As example, when solution concentration of this salt increased from 2 to 15 wt %, the estimated etch rate on AZO surfaces can vary from 3 to 50 nm/min, respectively.

Regarding the two step etching method, the sample is dipped shortly in highly concentrated solution with the aim to induce strong etches pit formation. Next, the second immersion is on a solution with a much lower concentration for a longer period of time with the objective of letting the pits generated in the previous step grow to optimize the roughness. This process allows controlling better film thickness and hence, shows a lower impact on the optoelectronic properties of the rough film.

To evaluate the suitability of both etching method and chemical etchant to achieve appropriate features on surfaces of AZO films deposited onto flexible substrate to scatter the light, Haze parameter and how the light is dispersive in whole angle are evaluated in a concrete case.

As example, Figure 11 shows the spectral haze in transmission of a commercial flat TCO on PET, and two AZO films with similar roughness (~35 nm) and optoelectronic properties (25 Ω/sqr of sheet resistance and around 10% of haze at 500 nm) etched using a conventional and two step etching method using NH_4Cl as etchant, respectively. The images of the surfaces of these three samples have been also included, where it can be observed features with hill-like shapes (Figure 12) (Fernández et al., 2012).

Figure 12. The images obtained with an atomic force microscope of previous three surfaces

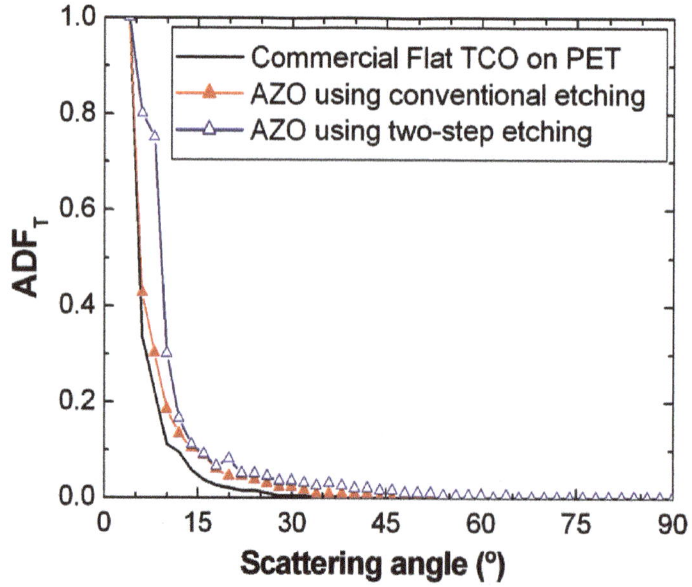

Figure 13. Measured angular distribution of a commercial flat TCO deposited on PET, an AZO surface etched with NH_4Cl (10%wt) for 270 s, and an AZO surface etched with two step method (NH_4Cl (10%wt) for 120s as first step and NH_4Cl (2%) for 1500s as second step
Source: Fernández et al., 2012.

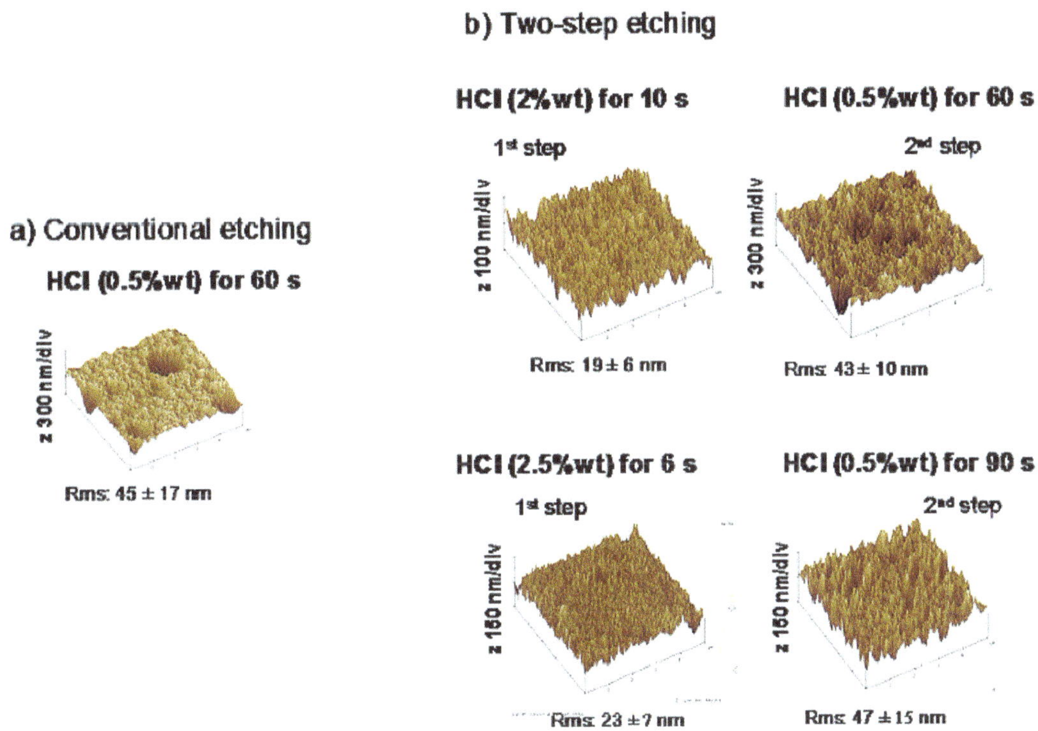

From this image, it can be appreciate that features achieved using NH_4Cl, present good haze factors at short wavelength, regardless the etching method used. This result could be useful in the case of using a-Si:H as absorber layer because that is the wavelength interval where it responses.

However, to discriminate the real scattering capability, Figure 13 illustrates the measured angular distribution of diffused transmitted light for the three surface morphologies. This figure establishes the better performance to scatter light when the surface is etched with in two steps, reaching scattering angles of almost 50º.

This two-step etching method has been also evaluated on optimized AZO thin films deposited onto glass substrates and using the common acid, HCl. Its effectiveness to achieve texture surfaces with appropriate morphology for light scattering maintaining good electrical and optical properties to thin film silicon solar cells applications can be observed in Figure 14 and 15. From the morphological point of view, this etching method clearly provides a better control rate and leads to hole-free surfaces, with homogeneous roughness values of around 50 nm, and sheet resistances less than 15 Ω/sqr after etching, (Fernández et al., 2011).

Finally, this method and also the NH_4Cl etchant have been already applied in a-Si:H/μc-Si:H tandem solar cells, (Fernández et al., 2012).

Figure 14. Images of AZO thin films deposited onto glass and etched using: A) conventional HCl (0.5%wt) for 60 s; B) two-steps etching using different acid concentrations and etching times

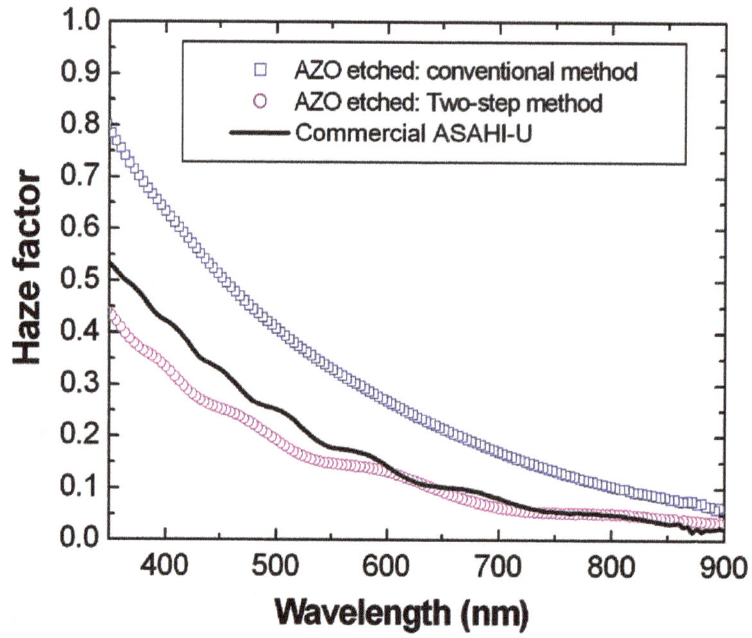

Figure 15. Comparison of spectral haze between AZO films etched using conventional and two steps method

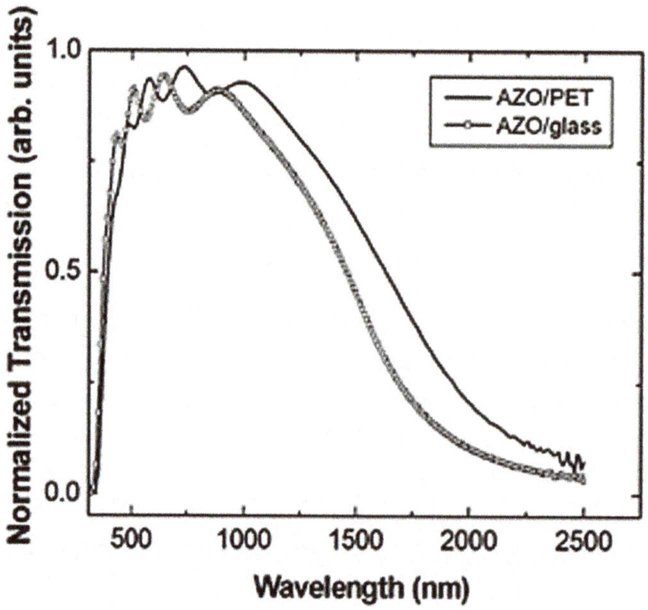

FUTURE RESEARCH DIRECTIONS

Main efforts are focused on increasing efficiency, durability and reducing the cost of the final PV product. For achieving this purpose, the replacement of some materials turns to be necessary because of their high cost and adverse effects in the environment. In the special case of amorphous silicon solar cells, the control of hydrogen content and morphology during the manufacturing process results essential in terms of avoiding the efficiency degradation of the cells when ageing.

Regarding TCOs materials, they are extensively incorporated in industry. To have an idea, in 2012, a 93% of the market used Indium Tin Oxide (ITO) as main TCO. However, this last material is expensive, because it is very dependent on the current price of indium, brittle, and barely flexible. All these drawbacks together with the increasing interest in flexible PV technologies have created the need of researching different TCOs that present additional properties for this application.

Key TCO technology must overcome certain challenges such as adequate conductivity and high transparency in films deposited at temperatures lower than 100ºC, improvement of the humidity reaching a good temperature resistance and thermal stability, easiness of tunable work function to improve the current extraction and finally, a better interfacial chemistry for optimum contact stability.

New families of materials compatible with R2R manufacturing processes and diverse chemical properties are challenging the supremacy of ITO. Some traditional poly-crystalline TCO materials can suffer easily film fracture, delaminating, or high residual film stress values that can adversely affect its optoelectronic properties and the flexible device performance. In this sense, amorphous TCO (a-TCO) materials are emerging as potential candidates to replace conventional TCOs. They are robust, easily chemically etched to achieve desired surface morphologies, and besides, they require lower deposition/processing substrate temperatures, while maintaining the optoelectronic properties showed by the conventional ones. The amorphous nature of these TCOs ensures very smooth surfaces, very useful for applications such as electrodes in polymers and organic light-emitting devices (OLEDs). Besides, amorphization process of oxides usually does not create deep trap levels, and for this reason, they can exhibit comparable transport properties relative to their crystalline counterparts.

As example of a-TCOs that are being researched, it can be found amorphous zinc tin oxide (a-ZTO), compound of zinc (Zn) and tin (Sn), which is emerging as a good candidate. It is cheap, abundant, non-toxic, with good thermal and chemical stability, and an excellent tolerance to the most of acid solutions commonly used for pattering in industrial processing. The main disadvantage is its low conductivity (Morales-Masis et al., 2014). Other possibilities that are being studied are ternary and quaternary systems, such as In_2O_3 $(ZnO)_n$ (IZO) and In_2O_3/SnO_2 $(ZnO)_n$ (ZITO). They possess comparable electrical properties, with an amorphous structure at temperatures above the crystallization temperature of ITO films (Kang, Lee, Lee, & Song, 2009; Walsh, Da Silva, & Wie, 2011).

In the search of ITO substitutes, nanostructured transparent conductive oxide materials are also emerging because of their potential applications in numerous fields. The main motivation for using these materials comes from their physical and chemical properties. The conventional and dense TCOs present fixed refractive index characteristics, limiting its performance as optical coating. Recently, oblique angle deposition techniques are opening the possibility of generating and controlling the shape of nanostructures that show modified and tunable optical properties. These columnar structure films also exhibit high surface roughness characteristics, which would enhance light trapping effects. The described deposition technique uses a physical-vapor process where the substrate is placed at non-normal angle relative to the incident atom flux. Under this geometrical condition, the substrate is not uniformly coated and a shadow region forms behind each self-assembled island where newly arriving atom flux cannot reach. So, the nano-columnar layers with a degree of porosity depending on the substrate tilt are fabricated. As the porosity increases, the refractive index decreases, becoming a tunable parameter. Step-graded antireflection coating based on ITO has been already tested. Excellent results have been obtained, improving the transmittance into PV devices (Welser et al., 2013). Graded AZO structures have been also recently demonstrated (Gondoni, 2014). In these structures, the bottom layer consists on a porous film that acts as a transparent light scatter mechanism, and the top layer is compact and acts as a good transparent electrical conductor. To evaluate the benefits deriving from this kind of mesoporous forest-like structures, a mayor effort to apply them in real devices should be carried out, being still a technology in study.

Following the development of the nanostructures materials, new concepts related with the further improvement of light trapping in thin-film silicon solar cells are being also developed. An appropriate optimization of the light trapping is required because optically, the absorption of the incident light is higher than the absorption of a single light crossing the absorber material. And besides, these schemes allow the reduction of the absorber material thickness and in consequence, the efficiency of solar cells increase and the cost of electricity generation is reduced. Among the novel concepts for light trapping, the use of plasmonic metal nanostructures is emerging with strength. The strong interaction of these nanostructures with light allows control over the propagation of light at nanoscale. Light can couple efficiently to localized surface Plasmon. Based on this approach, two different schemes to integrate in solar cells are being investigated to date. The first one based on the enhancement of light using an array of metal nanoparticles placed on top on the cell showing both antireflection effect and angular redistribution of the scattered light. The second one, using regular and random arrays of metal nanostructures designed to couple light in waveguide modes of thin semiconductor layers. The emerging investigation in this area can open new paths to improve the efficiency of thin solar cells and reduce the material costs.

Another hot topic could be the incorporation of these nanotechnologies to the industry. Currently, manufacturers are looking for new innovative continuous-feed processes for printing materials and structures onto roll-based flexible substrates. There is a huge interest in adapting R2R technologies for the extreme miniaturization of critical feature sizes to the nanoscale. Nanofabrication for R2R process platforms represents a disruptive manufacturing technology involving solution-processed, multi-layer precision coatings with functionalized nanostructures, materials, and patterning capability to realize unprecedented properties and functionalities for next generation consumer products.

The incorporation of emerging nanofabrication methods within R2R manufacturing processes will make possible to economically generate high value-added technology products at meters-per-minute rates on plastic film, paper, or foil, achieving feature dimensions as small as ten nanometers over areas. Nanofabrication processes should incorporate a range of advanced processes, tools, and materials. Future industrialization of this technology will require scaling system properties while retaining the collective properties of the nanoscopic elements over macroscale dimensions. Meeting this challenge is crucial for high-rate manufacturing of nano-enabled products and for establishing viable industrial-scale manufacturing platforms for continuous large-area R2R processing.

Nowadays, all these trends must be exploited and their availability of application in industry must be evaluated. In the particular case of flexible PV technologies, the main tendency that will mark a milestone will be the availability and production cost of developing low processing temperature ITO alternatives without losing performance.

CONCLUSION

In the present chapter, the role of TCO as functional coatings in the PV industry is analysed. The main objective consists on explaining the importance of searching adequate TCO to develop flexible lightweight devices, easy to produce in a continuous R2R process. Thanks to this development, the number of PV applications could increase at a lower cost and, therefore, more squares meters could be devoted to convert solar energy into electricity.

In the combination of TCO with flexible PV devices, the key issue is that common plastic substrates used in those applications are polymeric materials with glass transition temperatures lower than the optimal deposition temperatures characteristics of the main TCOs. The main purpose of the R&D activities carried out on this field is to achieve highly transparent and conductive thin films that can be successfully deposited on those flexible substrates at so low substrate temperature required.

In this search, ITO is the most common material used for most solar applications because of its excellent electrical and optical properties. However, ITO presents some important drawbacks and industrial limitations, mainly its high cost and its high sensibility to hydrogen plasma treatments. On the search of alternative materials, AZO outstands as a potential alternative, showing adequate properties, even deposited at low temperature, with low cost as well as being suitable to main industrial treatments. The literature concludes that under all these considerations, AZO emerges as possible and reliable candidate for PV applications.

Among the deposition techniques, RF magnetron sputtering outstands as a favourable technique to incorporate easily in industrial flexible production, basically because of allowing down to room substrate deposition temperature with high deposition rates and reproducibility at large areas.

Regarding the post-treatments to assure the desired TCO morphology to enhance light trapping into the solar devices, a new wet chemical etching method combined with new chemical etchants emerge as potential alternatives to achieve appropriated morphologies on sputtered flat AZO films to scatter light.

In the near future, the new families of materials compatible with continuous manufacturing processes will challenge the supremacy of ITO. Development of nanostructured and amorphous TCO materials open a new hot topic in the future trends, drawing the main R&D tendencies thanks to its adequate properties and its prospective employment into the industry of nanotechnologies.

REFERENCES

Agura, H., Suzuki, A., Matsushita, T., Aoki, T., & Okuda, M. (2003). Low resistivity transparent conducting Al-doped ZnO films prepared by pulsed laser deposition. *Thin Solid Films*, *445*(2), 263–267. doi:10.1016/S0040-6090(03)01158-1

Apreutesei, M., Arvinte, R., Marin, A., Andrei, O. C., & Munteanu, D. (2011). Magnetron sputtering technique used for coatings; technologies and applications. *Paper presented at 7th International Conference on Materials Science and Engineering BRAMAT '11*, Romania.

ASTM G173-03 (2012), Standard Tables for Reference Solar Spectral Irradiances: Direct Normal and Hemispherical on 37° Tilted Surface. *ASTM International.* Retrieved from http://www.astm.org/Standards/G173.htm

Banerjee, A. N., Ghosh, C. K., Chattopadhyay, K. K., Minoura, H., Sarkar, A. K., Akiba, A., & Endo, T. et al. (2006). Low-temperature deposition of ZnO thin films on PET and glass substrates by DC-sputtering technique. *Thin Solid Films*, *496*(1), 112–116. doi:10.1016/j.tsf.2005.08.258

Berginski, M., Hüpkes, J., Reetz, W., Rech, B., & Wuttig, M. (2008). Recent development on surface-textured ZnO:Al films prepared by sputtering for thin-film solar cell application. *Thin Solid Films*, *516*(17), 5836–5841. doi:10.1016/j.tsf.2007.10.029

Burstein, E. (1954). Anomalous Optical Absorption Limit in InSb. *Physical Review*, *93*(3), 632–633. doi:10.1103/PhysRev.93.632

Constantin, R., & Miremad, B. (1999). Performance of hard coatings, made by balanced and unbalanced magnetron sputtering, for decorative applications. *Surface and Coatings Technology*, *120–121*, 728–733. doi:10.1016/S0257-8972(99)00366-7

Defaix, P. R., van Sark, W. G. J. H. M., Worrell, E., & de Visser, E. (2012). Technical potential for photovoltaics on buildings in the EU-27. *Solar Energy*, *86*(9), 2644–2653. doi:10.1016/j.solener.2012.06.007

Dewan, R., Jovanov, V., Hamraz, S., & Knipp, D. (2014). Analyzing periodic and random textured silicon thin film solar cells by Rigorous Coupled Wave Analysis. *Scientific Reports*, *4*, 6029. doi:10.1038/srep06029 PMID:25112301

Exarhos, G. J., & Zhou, X. D. (2007). Discovery-based design of transparent conducting oxide films. *Thin Solid Films*, *515*(18), 7025–7052. doi:10.1016/j.tsf.2007.03.014

Farmer, J., Nichol, G., Seto, Y., McSporran, N., & Deslandes, L. M. (2011). Optical property measurements of high haze TCO coatings. *Proceedings of 26th European Photovoltaic Solar Energy Conference and Exhibition*, Hamburg: European Commission.

Faÿ, S., Feitknecht, L., Schlüchter, R., Kroll, U., Vallat-Sauvain, E., & Shah, A. (2006). Rough ZnO layers by LP-CVD process and their effect in improving performances of amorphous and microcrystalline silicon solar cells. *Solar Energy Materials and Solar Cells*, *90*(18-19), 2960–2967. doi:10.1016/j.solmat.2006.06.003

Fernández, S., de Abril, O., Naranjo, F. B., & Gandía, J. J. (2011). High quality textured ZnO:Al surfaces obtained by a two-step wet-chemical etching method for applications in thin film silicon solar cells. *Solar Energy Materials and Solar Cells*, *95*(8), 2281–2286. doi:10.1016/j.solmat.2011.03.042

Fernández, S., de Abril, O., Naranjo, F. B., & Gandía, J. J. (2012). Etching process optimization using NH4Cl aqueous solution to texture ZnO:Al films for efficient light trapping in flexible thin film solar cells. *Thin Solid Films*, *520*(12), 4144–4149. doi:10.1016/j.tsf.2011.04.207

Fernández, S., Martínez-Steele, A., Gandía, J. J., & Naranjo, F. B. (2008). Al-doped ZnO films deposited on flexible substrates by rf sputtering for thin film solar cells. *Proceedings of the 23rd European Photovoltaic Solar Energy Conference* (pp. 2347 – 2351). Valencia: EU PVSEC.

Fernández, S., Martínez-Steele, A., Gandía, J. J., & Naranjo, F. B. (2009). Radio frequency sputter deposition of high-quality conductive and transparent ZnO:Al films on polymer substrates for thin film solar cells applications. *Thin Solid Films*, *517*(10), 3152–3156. doi:10.1016/j.tsf.2008.11.097

Fernández, S., & Naranjo, F. B. (2010). Optimization of aluminium-doped zinc oxide films deposited at low temperature by radio-frequency sputtering on flexible substrates for solar cell applications. *Solar Energy Materials and Solar Cells*, *94*(2), 157–163. doi:10.1016/j.solmat.2009.08.012

Fernández, S., Pust, S. E., Hüpkes, J., & Naranjo, F. B. (2012). Development of two-step etching approach for aluminium doped zinc oxide using a combination of standard HCl and NH_4Cl etch steps. *Thin Solid Films*, *520*(14), 4678–4684. doi:10.1016/j.tsf.2011.10.190

Fernández, S., Santos, J. D., Munuera, C., García-Hernández, M., & Naranjo, B. (2015). Effect of argon plasma-treated polyethylene terephthalate on ZnO:Al properties for flexible thin film silicon solar cells applications. *Solar Energy Materials and Solar Cells*, *133*, 170–179. doi:10.1016/j.solmat.2014.10.030

Franken, R. H. (2006). *Transparent conducting oxide contacts and textured metal back reflectors for thin film silicon solar cells* [Doctoral dissertation].

Gabriel, O., Rohde, M., Wördenweber, J., Zelt, M., Neubert, S., & Kirner, S., … Schlatmann, R. (2011). Enhancement of a-Si/μc-Si solar cell performance made in an industrial PECVD reactor: The role of chamber history and in situ diagnostics. *Proceedings of PVSEC-21, Fukuoka, Japan.*

Gondoni, P. (2014). *Nanostructured Transparent Conducting Oxides for Advanced Photovoltaic Applications* [Doctoral dissertation].

Hösel, M., Søndergaard, R. R., Angmo, D., & Krebs, F. C. (2013). Comparison of fast roll-to-roll flexographic, inkjet, flatbed, and rotary screen printing of metal back electrodes for polymer solar cells. *Advanced Engineering Materials Communication*, *15*, 995–1001.

Jin, Z. C., Hamberg, I., & Granqvist, C. G. (1988). Optical properties of sputter-deposited ZnO:Al thin films. *Journal of Applied Physics*, *64*(10), 5117. doi:10.1063/1.342419

Kaid, M. A., & Ashour, A. (2007). Preparation of ZnO-doped Al films by spray pyrolysis technique. *Applied Surface Science*, *253*(6), 3029–3033. doi:10.1016/j.apsusc.2006.06.045

Kang, Y. M., Lee, D. Y., Lee, J. R., & Song, P. K. (2009). Transparent Conductive ITZO Films Deposited by Using a Magnetron Co-Sputtering System Using Two Cathodes. *Journal of the Korean Physical Society*, *55*(5), 1901–1905.

Kluth, O., Schöpe, G., Hüpkes, J., Agashe, C., Müller, J., & Rech, B. (2003). Modified Thornton model for magnetron sputtered zinc oxide: Film structure and etching behaviour. *Thin Solid Films*, *442*(1-2), 80–85. doi:10.1016/S0040-6090(03)00949-0

Krč, M., Zeman, M., Kluth, O., Smole, F., & Topič, M. (2003). Effect of surface roughness of ZnO:Al films on light scattering in hydrogenated amorphous silicon solar cells. *Thin Solid Films*, *426*(1-2), 296–304. doi:10.1016/S0040-6090(03)00006-3

Krebs, F. C. (2009). All solution roll-to-roll processed polymer solar cells free from indium-tin-oxide and vacuum coating steps. *Organic Electronics*, *10*(5), 761–768. doi:10.1016/j.orgel.2009.03.009

Lee, S., Lee, Y., Park, J., & Choi, D. (2014). Stitchable organic photovoltaic cells with textile electrodes. *Nano Energy*, *9*, 88–93.

Liu, Y., Li, Y., & Zeng, H. (2013). ZnO-Based Transparent Conductive Thin Films: Doping, Performance, and Processing. *Journal of Nanomaterials*, 2013.

Look, D. C., & Claflin, B. (2004). P-type doping and devices based on ZnO. [b]. *Physica Status Solidi*, *241*(3), 624–630. doi:10.1002/pssb.200304271

Mandal, S., Singha, R. K., Dhar, A., & Ray, S. K. (2007). Optical and structural characteristics of ZnO thin films grown by rf magnetron sputtering. *Materials Research Bulletin*, *43*(2), 244–250. doi:10.1016/j.materresbull.2007.05.006

Morales-Masis, M., Ding, L., Dauzou, F., Jeangros, Q., Hessler-Wyser, A., Nicolay, S., & Ballif, C. et al. (2014). Hydrogen plasma treatment for improved conductivity in amorphous aluminum doped zinc tin oxide thin films. *APL Materials*, *2*(9), 096113. doi:10.1063/1.4896051

Müller, J., Rech, R., Springer, J., & Vanecek, M. (2004). TCO and light trapping in silicon thin film solar cells. *Solar Energy*, *77*(6), 917–930. doi:10.1016/j.solener.2004.03.015

Nunes, P., Fortunato, E., Tonello, P., Braz Fernandes, F., Vilarinho, P., & Martins, R. (2002). Effect of different dopant elements on the properties of ZnO thin films. *Vacuum*, *64*(3-4), 281–285. doi:10.1016/S0042-207X(01)00322-0

Olvera, M. L., Maldonado, A., Asomoza, R., & Meléndez-Lira, M. (2000). Chemical stability of doped ZnO thin Films. *Journal of Materials Science Materials in Electronics*, *11*(1), 1–5. doi:10.1023/A:1008973800554

Philbin, P., Bitter, M., Fraile, D., Latour, M., Teske, S., & Wolfsegger, C. (2008). Solar Electricity for over one billion people and two million jobs by 2010. *Solar Generation, V*, 2008.

Plichta A., Habeck, A., Knoche, S., Kruse, A., Weber, A., & Hildebrand, N. (2005). Flexible glass substrates, *Flexible Flat Panel Displays*, 36-55.

Seeber, W. T., Abou-Helal, M. O., Barth, S., Beil, D., Höche, T., Afify, H. H., & Demian, S. E. et al. (1999). Transparent semiconducting ZnO:Al thin films prepared by spray pyrolysis. *Materials Science in Semiconductor Processing*, *2*(1), 45–55. doi:10.1016/S1369-8001(99)00007-4

Shan, F. K., & Yu, Y. S. (2003). Optical properties of pure and Al doped ZnO thin films fabricated with plasma produced by excimer laser. *Thin Solid Films*, *435*(1-2), 174–178. doi:10.1016/S0040-6090(03)00343-2

Søndergaard, R., Hösel, M., Angmo, D., Larsen-Olsen, T. T., & Krebs, F. C. (2012). Roll-to-roll fabrication of polymer solar cells. *Materials Today*, *15*(1-2), 36–49. doi:10.1016/S1369-7021(12)70019-6

Stadler, A. (2012). Transparent Conducting Oxides—An Up-To-Date Overview. *Materials*, *5*(12), 661–683. doi:10.3390/ma5040661

Terasako, T., Yagi, M., Ishizaki, M., Senda, Y., Matsuura, H., & Shirakata, S. (2007). Growth of zinc oxide films and nanowires by atmospheric-pressure chemical vapor deposition using zinc powder and water as source materials. *Surface and Coatings Technology*, *201*(22-23), 8924–8930. doi:10.1016/j.surfcoat.2007.04.017

Tomita, Y. (2008). *Alternative transparent electrodes for organic light emitting diodes* [Doctoral dissertation].

Vijayalakshmi, K. A., Mekala, M., Yoganand, C. P., & Navaneetha Pandiyaraj, K. (2011). Studies on modification of surface properties in polycarbonate (PC) film induced by DC glow discharge plasma. *International Journal of Polymer Science*, 2011.

Walsh, A., Da Silva, J. L. F., & Wie, S. (2011). Multi-component transparent conducting oxides: Progress in materials modelling. *Phys.: Condens. Matter*, *23*, 334210. PMID:21813942

Watts, M. (2007). Advances in roll to roll processing, *Impattern Solutions*. Retrieved from http://www.impattern.com/Download/RollToRollProcessing.pdf

Waugh, M. R. (2011). *The Synthesis, Characterisation and Application of Transparent Conducting Thin Films* [Doctoral dissertation].

Welser, R. E., Sood, A. W., Cho, J., Schubert, E. F., Harvey, J. L., & Dhar, N. L. … Sood, A. K. (2013). Nanostructured Transparent Conductive Oxides for Photovoltaic Applications. Mater. Res. Soc. Symp. Proc. 1493 (pp. 23-28). Materials Research Society.

KEY TERMS AND DEFINITIONS

Coating: A thin layer that covers a surface to protect, decorate and/or other specific function.

Etching Process: To wear away the surface of a layer by chemical action using a dilution of a single or a mix of acids.

Magnetron Sputtering: It is a physical vapor deposition method to fabricate thin-film materials onto a substrate in a high vacuum environment.

Plasma Surface Treatment: It is a process where a highly ionized gas with positive and negative charged ions impact on the surface of materials to modify or functionalizing it without changing its bulk properties and/or appearance.

Polymer: Large number of small molecules that link by chemical bonding to form a chain of many repeating units. It can be naturally occurring or synthetic.

Roll to Roll: It is an industrial vetted way to handle solution-based processes and coatings for high volume manufacturing.

Transparent Conductive Oxide: Doped metal oxide that is optically transparent and electrically conductive.

Chapter 12
Advances in Functional Nanocoatings Applied in the Aerospace Industry

Rafael Vargas-Bernal
Instituto Tecnológico Superior de Irapuato, Mexico

ABSTRACT

Surfaces of aerospace components are subjected to heat, water, ice, electromagnetic waves, corrosion, abrasion, wear, thermal shock and/or fire during their lifetime. Materials science and engineering have introduced the use of coatings to protect the surface of the materials under these operating conditions. Materials such as ceramics, polymers, and metals and their combinations either as composites or hybrid materials, as well as functionalized versions of them, have been proposed as different design alternatives. Six different types of coatings can be identified: anti-corrosion, abrasion and wear-resistant, thermal barrier and flame-retardant, conductive, anti-icing, and superhydrophobic. This chapter will review the most important advances in functional nanocoatings, which are being researched and those already used in aerospace applications, as well as the profits that these materials can provide to the surfaces of the components where them are used.

INTRODUCTION

Military and commercial aviation is suffering high maintenance cost and poor availability of assets (Del. 2013). Premature damages are found in engines, due to deterioration produced by hot erosion, distortion, clearance, and corrosion, as well as, by impact failure during operating cycles. Therefore, the use of coatings for aerospace industry must offer attractive design solutions, in terms of resistance at climate, durability and reliability during their operation. The use of coatings is imperative in harsh climatic conditions such as varying temperatures and humidity ratios, as well as when drag is presented. Researchers, around of the world, have been designing coatings with the aim of offering materials with low densi-

DOI: 10.4018/978-1-5225-0066-7.ch012

ties, anti-corrosive properties, reinforced-dragging, supporting to failure mechanisms both dynamic and static type, abrupt changes in harsh conditions, and good tribological properties. For example, a coating must be erosion resistant and must have low thermal conductivity, when it is applied as thermal barrier.

In aerospace industry, substrates based on aluminum (Al) and its alloys (Al); magnesium (Mg) and its alloys, as well as, titanium (Ti) and its alloys have been used to the manufacture of aerospace components (Wang, 2013). Moreover, conventional coatings based on chromium (Cr(VI)), phosphate and permanganate have been used in aerospace applications such as fasteners, adhesive joints, and corrosion protection. With the advent of nanomaterials, novel nanocoatings have been introduced in different technological applications. Nanostructured materials have the capacity of developing intelligent, functional or smart coatings such that they provide multiple or specific physicochemical properties exploited to protect surfaces of the substrates. These coatings are manufactured with composite materials and functional materials based on combinations of polymeric, ceramic and metallic materials. Composite materials are defined as a combination of a material called matrix and at least a material used as reinforcement, with the aim of providing better physicochemical properties that properties of the individual components. Functional materials are materials with designed physicochemical properties that serve for specified application needs and whose properties can be controlled. A composite material is considered as a particular case of the functional materials.

A coating is a covering applied to the surface of an object called substrate. The coatings have purpose decorative, functional, or both (Makhlouf, 2011). The coating itself may be an all-over coating on the substrate, or it only cover parts of the substrate. Functional coatings have the purpose of changing the surface properties of the substrate to offer adhesion, wettability, electrical conductivity or electrical insulation, thermal insulation or thermal conductivity, corrosion resistance, or abrasion and/or wear resistance. Coatings can be used in a variety of industries and applications and these can be applied to the substrate as liquids, gases or solids. A coating is formed by a system of materials involving the bulk material (substrate), interface (interaction between substrate and covering), and a modified surface layer. In practice, the interface region between the coating and the substrate material can represent the weakest link. A mechanical interlocking is favored if the substrate has a rough surface and contains large pores. In addition, a surface with high reactivity or with wetting conditions is preferred, since between the coating and substrate material will form a stronger interface. In particular, polymeric coatings are composed by binders, fillers, solvents, pigments, and additives (Wu, 2015). Coating processes may be classified as: vapor deposition either chemical (CVD) or physical (PVD), chemical and electrochemical techniques, spraying, and roll-to-roll processes.

Nanocoatings based on functional materials represent the technological proposal for replacing to conventional coatings, since these materials possess unique physicochemical properties and also to satisfy the environmental fulfillment (Makhlouf, 2014). A coating based on functional material must be lightweight and thin can be very attractive for practical applications in areas of aircraft and aerospace. Nanocoatings have the following functions in aerospace and aviation: 1) facilitate crack-healing by means of high-temperature strength and creep resistance, 2) increase service life of turboengines through of the application of multilayer structures to provide temperature resistance, thermal shock, as well as, anti-corrosion and erosive wear-resistance, 3) reduce high maintenance costs through of the use of anti-corrosion coatings in parts subjected to harsh environments, 4) save fuel-burn by means of the drag reduction using blended winglets to improve fuel efficiency and reduction of emissions into atmosphere, and 5) decrease ice accretion through of the control of climatic conditions of the surfaces of airfoils. The application of these coatings will be increased in next decade, until them can be manufactured in hundreds of thousands of tons, with the aim of being applied in areas such as consumer electronics, automobiles, aerospace, sporting goods, house cleaning, construction, and medicine.

Functional materials can overcome numerous technical challenges required by aerospace and defense industries, whose performance is almost an ideal solution for all requirements implicated in the different applications of the coatings (Wu, 2015). High performance coatings must protect the structures and surfaces of the aircrafts of harsh environments such as resistance to extreme temperatures, extreme climates, corrosion, abrasion and wear of engine parts, self-cleaning, improved hardness, and improved thermal performance and flame retardancy. Aerospace applications make use of nanomaterials in the materials found in the airframe and their components, as well as, in lubricants, fuel, adhesives, etc. The main materials that can be used in functional nanocoatings for the aerospace and aviation industry are carbon nanotubes, graphene, silicon dioxide (SiO_2) nanoparticles, aluminium oxide (Al_2O_3) nanoparticles, antimony tin oxide (SnO_2-Sb_2O_3) nanoparticles, and zirconium dioxide (ZrO_2) nanoparticles. Metal oxide nanoparticles, nanopowders and/or nanocrystals exhibit enhanced electrical, optical, magnetic and chemical properties with respect to the bulk ceramic material of which them are made. These nanomaterials are available in varying sizes, shapes and whose main use is associated to nanocoatings used in surfaces of aerospace components. Functional nanocoatings have found applications in industries such as aerospace, electronics, automotive, chemical, construction, ship-making, security, steel, and many other industries. Unfortunately, they present certain drawbacks which can be overcome, through optimization of the design parameters and experimental complexity, using a multidisciplinary approach, where the manipulation of nano-surfaces must integrating chemical and/or physical properties.

The research and development of functional nanocoatings has become an emerging technological area, thanks to improved physicochemical properties of such materials, which can be exploited in diverse applications associated with the military, industrial, and commercial sectors (Wu, 2015). These materials show significantly better mechanical properties and of barrier, flame retardancy, thermal resistance, electrical and magnetic properties, and environmental friendliness can be predicted for them. The most obvious benefit of using functional coatings versus conventional coatings, resides in that better properties are achieved with a very small loading (% weight) of fillers which are embedded in the coating, when this value is compared with quantity required by conventional coatings.

Functional nanocoatings have as aim serving as thermal barrier, icephobic or anti-icing protection, and protective coatings in aerospace materials. Six different types of nanocoatings can be distinguished: anti-corrosion coatings, abrasion-resistant coatings and wear-resistant coatings, thermal barriers and flame-retardant coatings, conductive coatings, anti-icing coatings, and superhydrophobic coatings, as shown in Figure 1. The main technological trend consists in the combination of more than one function in a single nanocoating. Nanocoatings are safer than conventional coatings, since these are manufactured with techniques that allow fulfilling global environmental regulatory factors and standards employed in the aviation, defense, and space domains. The chapter is divided into six sections, where each of the types of coatings is defined, analyzed and exemplified with technical innovations published in the last five years.

Figure 1. Different types of nanocoatings used in the aerospace industry

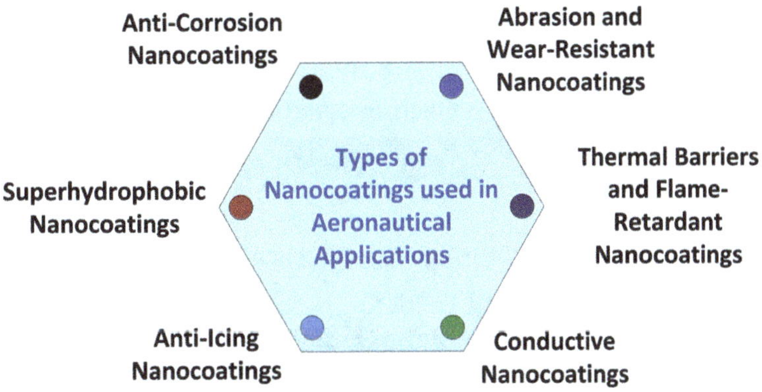

CONDUCTIVE NANOCOATINGS

Electrical conductive materials can be classified as conductive, static conductive, or static dissipative (Grady, 2011). The bulk electrical conductivity ranges for these materials are > 0.1 (useful for electromagnetic interference (EMI) shielding), 10^{-3} to 10^{-1}, and 10^{-3} to 10^{-9} S/m (useful for electrostatic dissipation (ESD)), respectively. In the following paragraphs, the effects of electrostatic discharge and electromagnetic interference that are used to protect the materials used in aerospace applications are defined.

Electrostatic discharge (ESD) can be defined as the sudden flow of electrical current between two electrically charged objects caused by contact, an electrical short or dielectric breakdown and often creating a visible spark. This effect is caused by tribocharging or by electrostatic induction. In the aerospace industry, electrostatic discharge (ESD), electrical overstress (EOS), latchup, electromagnetic interference (EMI) and electromagnetic compatibility (EMC) are important for airplanes, spacecraft, and satellites. Airplanes are concerned with electrical overstress (EOS) from lightning and partial discharge (Voldman, 2012). Satellites can exhibit problems due to electrostatic charging by a single event upset (SEU). Aerospace components such as those used in landing vehicles, rovers, and space station systems are vulnerable to charging phenomena, due to electrostatic charging effects; therefore, coatings must be applied to solve these effects. Heat-sealable electrostatic dissipative (ESD) films are used in aerospace applications, such as heat gun for cleanliness, electrostatic discharge and non-volatile residues and manufactured by Dun-Shield (News, 2011).

Electromagnetic interference (EMI) is defined as a disturbance that affects an electrical circuit, either due to electromagnetic induction or electromagnetic radiation emitted from an external source (Tong, 2009). The disturbance may interrupt, obstruct, or otherwise degrade or limit the effective performance of electrical circuits used in aerospace applications. These effects can range from a simple degradation of data to a total loss of data. This effect can be controlled by applying coatings with low electrical conductance or high electrical resistance.

A special emphasis must be placed in the development of nanocoatings used to protect aerospace materials, with the aim of enduring the harsh conditions of the space, such as cosmic rays (these can be presented as electromagnetic waves of very high energy) and thermal changes by means of conductive coatings and thermal barriers, respectively. In addition, functional nanocoatings based on carbon

Figure 2. Conductive effect found in aerospace surfaces and solved by nanocoatings

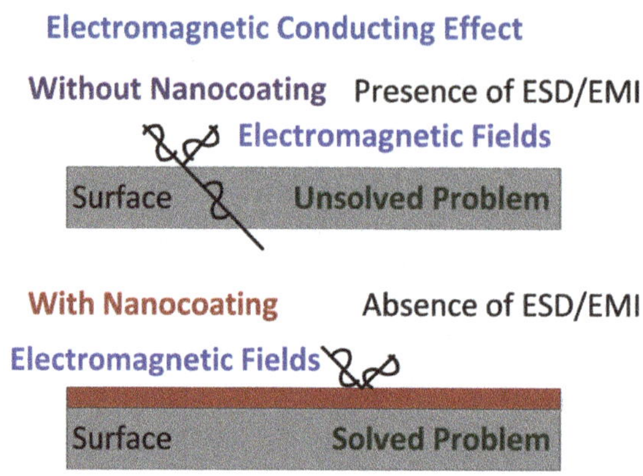

nanotubes and graphene with electrical and magnetic properties is possible. These materials can be applied for overcoming electromagnetic interference (EMI) shielding and for providing electrostatic dissipation (ESD).

Conductive nanocoatings are nanomaterials used to reduce or isolate electromagnetic or radiofrequency interference (EMI/RFI) applied by an external electromagnetic source to the surface, as is depicted in Figure 2. The electromagnetic shielding consists in blocking the electromagnetic waves with barrier materials made of conductive and/or magnetic materials. The use of polymers minimizes the loss of metallization through rubbing, and the metal or electrical conductive material by its part slows down conductivity degradation with age or working period of the metal. The flake shape on surface helps to ensure that a maximum number of points of contact improves the electrical conductivity, and therefore, a maximum electromagnetic shielding is achieved. In the following paragraphs, a set of technical solutions to solve the problem of electrostatic discharge and electromagnetic interference in aerospace materials is discussed.

Specific aerospace coatings applications such as electrostatic dissipation can be manufactured using polyurethane (PU) and carbon nanotubes (Zhao, 2010). These materials are durable, light color, and flexible nanocomposites with sufficiently low surface resistivity meet the requirements for aerospace coating applications.

Textile surfaces can be metallized, with the aim of offering multi-functional high-value-added products where electrical properties are exploited in smart textile applications, such as wearable sensors, antennas, electrostatic discharge, military camouflage, and electromagnetic shielding (Ersoy, 2014). In particular, some military and aerospace applications require materials with values of shielding electromagnetic of more of 80 dB to blind electromagnetic interferences of the electronic devices used in these applications. Coatings based on electroless silver coating on glass stitched fabrics can be used to develop electromagnetic shielding applications thanks to the high electrical conductivity achieved in the range of 79.5 S/m (> 0.1 S/m to be considered apt to electromagnetic shielding, Grady 2011), where ranges of filler between 40% and 80% in mass of silver nanoparticles are used.

Carbon nanotubes and graphene nanoplatelets (GNPs) are nanoscale additives that can work as the electrical conductive component of coatings or adhesives very useful for the development of innovative functional nanocoatings in aerospace industry. Carbon nanomaterials are being used in the development of conductive nanocoatings (Vargas-Bernal, R. 2015), thanks to that their electrical properties can be controlled in accordance with the volume fraction of the carbon nanotubes and/or graphene incorporated to the polymeric matrices based in materials such as polystyrene (PC), polycarbonate (PC), etc.

In the search of a synergistic interaction among materials with different sizes, composites based on fiber-polymer laminates and electrical conductive materials, are being proposed (Kandare, 2015). The fillers have two-dimensions (carbon nanotubes and/or graphene), one-dimension (silver nanowires), and zero-dimensions (silver nanoparticles in the nanoscale). The coatings, containing these materials possess good thermal and electrical conductivities through-thickness, and these are improved without adversely affecting mechanical properties.

Hybrid materials for coatings, based on graphene nanoribbon (GNR), polyaniline (PANI), and epoxic resin (ER), have been proposed with the aim of working as an shielding material for electromagnetic interference (EMI) in the X-band frequency of 8.2 to 12.4 GHz (Joshi, 2015). They found that shielding effectiveness depends of the percentage of GNR and the thickness of the film used as coating on surfaces under electromagnetic radiation.

Nanocomposites based on nylons reinforced with glass fiber have high rigidity, heat deflection temperature and creep resistance, low coefficient of friction, good low-frequency electric insulation properties and nonmagnetic properties to be used in aerospace applications (Kent, 2012). In general, nanofibers based on carbon and/or glass, can be used for reinforcing of polymeric materials such as epoxic or phenolic resins and metals, or aluminum or magnesium to provide also aerospace structural materials, with high strength, high modulus, and lightweight.

Polymer nanocomposites for structural and non-structural aerospace applications are incorporating nanomaterials such as nanoclays (such as montmorillonite), carbon nanotubes, nanofibers, and graphene to their polymeric matrices as nanofillers, for improving their physical properties. The main contribution of these materials is found in the replacing of metals, where high relative resistance, resilience to vibration and at fire, stiffness, toughness, and electrical properties are required. Epoxic resins can be applied as polymeric matrix in the development of composite materials for aerospace industry, but they require an improvement of their thermal and flame-resistance properties, which can be achieved through of nanofillers based on carbon nanotubes. Silsesquioxanes added to epoxic resins have led to better thermomechanical properties and processability of epoxic resin, which generates the adequate storage modulus for aerospace applications. Carbon nanotubes and graphene can be employed in aeronautics and astronautics to develop parts used in rotorcraft, unmanned aerial vehicles, satellites, commercial aircraft, military aircraft, and space launch vehicles. These allotropes of the carbon have high strength-to-weight ratio, and excellent mechanical, electrical, and thermal properties. The main functions of such materials, in these applications, consist in protection for aircraft, icing mitigation, reduced weight of airframes and/or satellites, and protection against cosmic-rays and thermal changes. Laminates based on carbon nanotube and epoxic resin, with aerospace-grade, have high interlaminar fracture toughness for Mode-I and/or Mode-II with respect to traditional materials. Moreover, the wettability of polymeric materials reinforced with carbon nanotubes can be adjusted to provide hydrophobic and hydrophilic surfaces, in accordance with the requirements established by the function of this surface. The use of electrical nanocoatings suggests the possibility of passing an electrical current through the coating to generate heat on the surface, and thus, the ice deposits localized on surface of the aircrafts can be subjected to vibration with the aim of being loosened and removed of it. Additionally, indirect advantages derived of the use of these materials in aerospace applications are the weight saving and cost saving.

ABRASION AND WEAR-RESISTANT NANOCOATINGS

Abrasion is defined as the process of scuffing, scratching, wearing down, marring, or rubbing away of materials (Aliofkhazraei, 2015). Wear can be defined as the erosion or sideways displacement of material from its primary position on a surface performed by the action of another surface due to contact between both surfaces (Mellor, 2006). There are many types of wear mechanisms including adhesive, abrasive, erosive, cavitation, and fretting. The wear and abrasion are two undesirable effects and considered as the most frequent reasons for equipment maintenance and downtime in aerospace industry.

Nanocoatings represent a modern solution for increasing the reliability and lifetime of mechanical components used in aerospace applications through of their ability to boost performance while the weight and cost of materials is reduced. In addition, they offer an environmentally friendly alternative to other coating technologies such as electroplating, and these offer an indirect impact on efficiency of manufacturing technology such as cutting, forming and casting of materials. Jet engine uses nanocoatings to provide protection against erosion, high-temperature oxidation and hot corrosion, fretting, galling/ seizing, surface fatigue, sticking, thermal impact, and general corrosion (Mellor, 2006).

Abrasion resistant nanocoating is a nanomaterial used to increase abrasion resistance or protection from galling, seizing and other physical damage. Wear-resistant nanocoatings are robust, ultra-hard nanomaterials used to reduce wear produced in material surfaces, as shown in Figure 3. Understanding the variables that cause wear and abrasion is very important for designing the proper wear resistant coatings. In the following paragraphs, a set of technical solutions to solve the problem of abrasion and wear in aerospace materials is discussed.

Airframe components, subjected to mechanical vibrations, are found in spherical bearings, mounts, and knuckle joints, and they require protection against fretting through the use of nitrides and carbon-based nanocoatings in fretting-prone materials based on titanium (Ti), aluminum (Al), and stainless steel (SS). During the coating deposition process is possible the formation of nodules or surface inhomogeneities containing diamond-like carbon structures, which offer high hardness and strong wear-resistance and are widely used for aerospace applications (Wang, 2013). In addition, erosion is found in aerospace components such as valves, missile domes, turbine blades, propulsors, pipelines and fluidized-bed combustion systems (Mellor, 2006). High-temperature aerospace applications require lightweight struc-

Figure 3. Wear-resistant effect found in aerospace surfaces and solved by nanocoatings

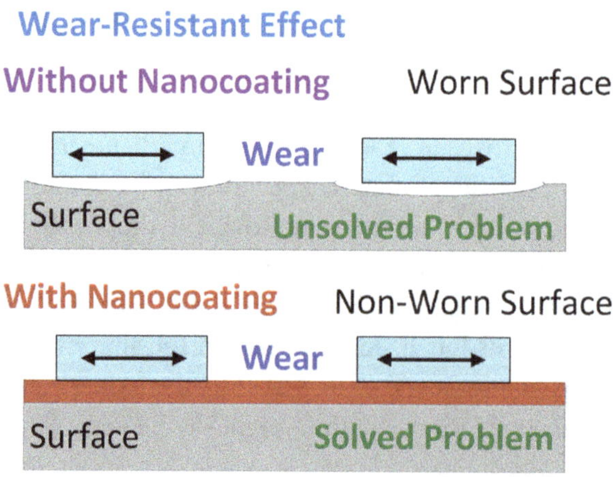

tural materials such as γ-TiAl-based alloy coated with nickel. Aircraft and aerospace applications use Ni-P conventional coatings with thickness in the order of 10-50 μm, lesser thickness can be used when nanomaterials. Yttria-stabilized ZrO_2 (YSZ) is used under operating conditions such as air humidity, air pressure and temperature and offering low friction and good wear resistance (Mellor, 2006). Tungsten Carbide/Diamond Like Carbon/Tungsten Disulfide (WC/DLC/WS_2) nanocomposite coatings present a desirable combination of hardness, good wear resistance and low friction in humid air, dry nitrogen and vacuum (Voevodin, 1999).

Bulk metals with nanoscaled crystalline structure and nanostructured metals are widely used in aircraft applications where properties such as yield strength, tensile strength, self-cleaning, self-repairing, self-healing, and corrosion resistance, as well as, a low density are required (Aliofkhazraei, 2015). Nanocoatings can improve durability of the metals used in aerospace applications, by improving their anti-corrosive properties, reducing the coefficient of friction and increasing the wear resistance. Magnesium alloys (such as ZE41, which contains rare-earth elements) are very useful due to their unique and strategic physical properties in the aerospace industry. Plasma sprayed coatings based on Al_2O_3-TiO_2 or TiO_2-ZrO_2 have been widely used as coatings for wear, corrosion, and thermal resistant in aerospace applications. Oxides based on silicon or boron, and other materials with morphology of nanocrystals can be used in anti-corrosive nanocoatings. Metallic and ceramic materials such as carbides, nitrides, etc., can act as friction-modifying agents, since these materials are used as nanostructured coating materials on aerospace materials. All these contributions allow that aerospace materials can support high temperatures, and the aircrafts can increase the efficiency of the engines.

A new generation of high-performance coatings provides wear-resistant effects, and is applied using physical vapor deposition (PVD) has been proposed (Fox-Rabinovich, 2012), these coatings make use of nanostructured materials, where extreme operating conditions of high performance machining are presented by materials that are hard of be cut. Two attractive aspects from a structural point-of-view can be identified: (1) the generation of tribofilms during friction, which develop functions of lubricating and of thermal barrier making use of surface layers in nanoscale, and (2) the presence of elemental nano/microscaled layers with complex structure and highly nonequilibrium state. Al-rich TiAlN-based coatings and Al-rich AlCrN-based coatings are widely used for machining of aerospace Ni-based alloys, which are hard to cut.

The addition of graphene to polytetrafluoroetylene (PTFE or Teflon) is used as solid lubricant and allows reduce the wear in aerospace parts (Kandanur, 2012). The incorporation of 10 wt % of graphene platelets has diminished until in four orders of magnitude the wear rate of unfilled PTFE. Also, it is possible maintaining low friction in dry sliding conditions in the composite, whose characteristic is offered by unfilled PTFE.

ANTI-CORROSION NANOCOATINGS

Anti-corrosion nanocoating is a corrosion resistant nanomaterial used to protect material surface against degradation due to moisture, salt spray, oxidation or exposure to a variety of environments or chemicals (Landolt, 2011). In addition, these materials allow provide the protection of metallic surfaces and operate as a barrier to inhibit or reduce the contact between corrosive chemical compounds and material surfaces used as substrate, as depicted in Figure 4. As a direct consequence of its use, the surfaces have the maximum possible lifespan. Regularly, these coatings are formed as thin films using alkoxides with

Figure 4. Anti-corrosion effect found in aerospace surfaces and solved by nanocoatings

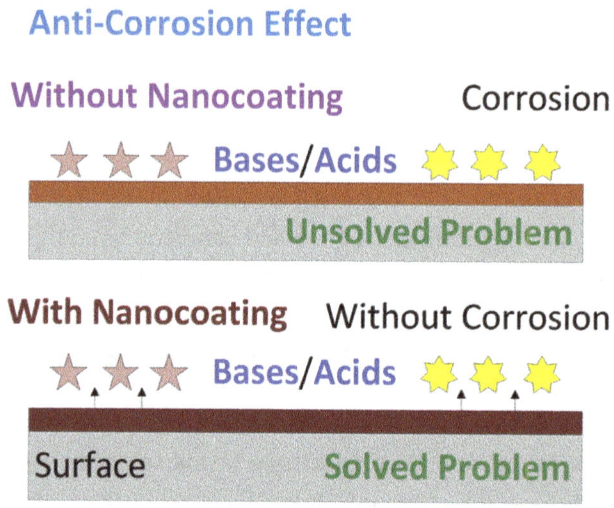

chemically attached ceramic nanoparticles (She, 2013). Effects of oxidation and corrosion to high-temperature found in metal surfaces can be reduced through of the use of thin layers of oxides that present chemical resistance and have a high melting point (Borisova, 2013). The main advantages of these coatings are their high purity, low-temperature processing and control of the composition (capacity of tuning of properties), as well as durability.

Tribocorrosion is defined as degradation of materials by fretting, sliding, rolling, or erosion conditions in presence of a corrosive environment through a combined action of chemical-electrochemical-mechanical processes (Wang, 2013). In aerospace industry, anti-tribocorrosion coatings have been used in aircraft turbine blades and turbine engine components, in inner surface of hydraulic actuating cylinders for the Joint Strike Fighter, landing gear, and helicopter components with the aim of increasing wear resistance and the reduction of the coefficient of friction. Nanocomposites based on boron-nitride (BN)-niquel (Ni), silicon carbide (SiC)-niquel (Ni), alumina Al_2O_3-niquel (Ni), chromia (Cr_2O_3)-niquel (Ni), and cobalt-phosphorus-silicon carbide (Co-P-SiC)-based niquel alloy (Ni) have been used in nanocoatings. In the following paragraphs, a set of technical solutions to solve the problem of corrosion of aerospace materials is discussed.

Magnesium alloys used in aerospace components require anticorrosive superhydrophobic surfaces with behavior robust and stable to increase its mechanical stability and durability, as well as, its chemical stability. It can be achieved making use of a process that combines the electrodeposition and the chemical modification, since contact angle and sliding angle must be carefully designed to avoid moisture conditions with potential trend to present corrosion (She, 2013).

Anticorrosive coatings have as main function protects surfaces of metals from corrosive attack, which can be classified as passive and active (Borisova, 2013). When a coating is damaged, it loses its protective function. An ideal coating must actively protect the surface for achieving a large adhesion, by means of a barrier function, and a high mechanical strength. In last decade, the use of coatings have been applied in inhibitor loaded containers, which respond to the changes in the coating integrity (cracks) or local environment (pH shift) that are originated by the corrosion attack. These inhibitors can operate as self-healing systems for the surface under a corrosive attack of chemicals or harsh environmental conditions. Inhibitors can be classified by size (nano and/or microcontainers), chemical nature (organic and/or inorganic), shape (spheres, tubes, or platelets) and structure (core-shell, porous, or by layered). Materials possess self-healing characteristic when they are capable of heal damage automatically wherever and whenever that it occurs in the material (Sun, 2015). A great quantity of different anticorrosive coating designs can be feasible using previous characteristics where can be included different stimuli for release of inhibitor such as pH shift, ion exchange or mechanical breakdown. Until now, it has been demonstrated that smaller containers exhibit better self-healing performance (Borisova, 2013). With the aim of releasing the inhibitor, capsules have been designed with protective shell or stoppers. However, novel designs are required to increase the efficiency of the releasing of the inhibitor from capsules. Common inhibitors are 2-mercaptobenzothiazole, 2-methylbenzothiazole, and 8-hydroxyquinoline. Containers are based on mesoporous silica (SiO_2) with diameters in the range of 80 to 700 nm.

Microcapsules based on double-layered polyuria and containing hexamethylene diisocyanate (HDI) were prepared to offer anti-corrosion effect and self-healing effects for surfaces under environments with non-polar organic solvents and thermal changes (Sun, 2015). These capsules undergo the anti-corrosion effect in scratched coatings via self-healing functionality directed to industrial coatings and paint systems of aerospace components.

Magnesium alloys as lightweight materials are replacing to the heavy alloys in aerospace industry to produce savings in fuel consumption and a reduction in CO_2 emissions. Magnesium alloys have high strength-to-weight ratio, low density, dimensional stability and castability. Nanostructured coatings have the potential for enhancing the functionality of the coating for improving environmental anti-corrosion coatings for aerospace industry, which represents a significant improvement in the corrosion resistance with respect to those materials processed using conventional coating methods (Makhlouf, 2011).

THERMAL BARRIERS AND FLAME-RETARDANT NANOCOATINGS

A thermal barrier coating (TBC) is an advanced materials system usually applied to metallic surfaces of gas turbines or aero-engine parts, operating at elevated temperatures, as a form of exhaust of heat or thermal insulation (Xu, 2011). These surfaces are subjected to large and prolonged heat loads and for establishing an appreciable temperature difference between the load-bearing alloys and the coating surface, which limits the thermal exposure of structural components, increases lifespan of the aerospace parts through the reduction of oxidation and thermal fatigue and also increases the melting point of the materials. Thermal barriers coatings improves the performance, in operating conditions such as high-temperature and/or stress on the friction surface due to the adhesive interaction with the workpiece, and harsh oxidative attack from the environment that are commonly found in aerospace components fabricated with nickel-based superalloys (Wang, 2013).

Thermal barrier nanocoatings are nanomaterials used to protect material surfaces of heat and thermal degradation, as is depicted in Figure 5. The use of thermal barrier coatings allows that metal alloys can be applied at higher temperatures with the aim of reducing the temperature to which the material surface is exposed. The most commonly used thermal barrier coating materials are ceramic oxides such as yttria-stabilized zirconia (YSZ), and metal alloys. Materials used for thermal insulation include YSZ, alumina compounds, refractory metals, and stainless steels. Multilayer TBCs are preferred for specialist applications based on nanostructured materials such as La_2O_3-YSZ (yttria-stabilized zirconia), ZrO_2-

Figure 5. Thermal barriers found in aerospace surfaces and solved by nanocoatings

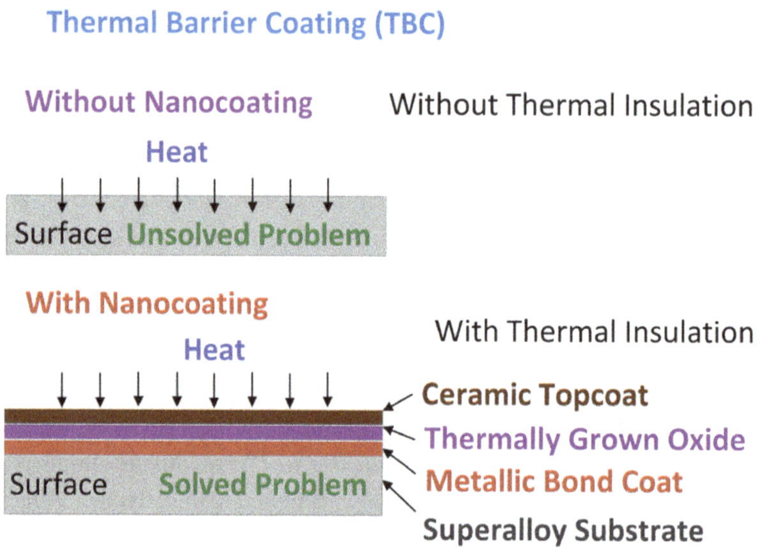

Y_2O_3-Gd_2O_3-Al_2O_3, YAG ($Y_3Al_5O_{12}$), CeO_2, *M*-Cr-Al-Y-based alloys, *M*-Cr-Al-based alloys, where *M* is a metal such as Ni, Fe, or Ti (Xu, 2011). In the following paragraphs, a set of technical solutions for developing thermal insulation in aerospace materials is discussed.

Thermal barrier coatings (TBCs) are fulfilling requirements on surfaces of the high-temperature materials such as for protecting to the material used as substrate at reducing its working temperature and in increasing its working efficiency (Wang, 2011), in high-performance applications such as aerospace, aircraft, marine, automotive, nuclear fusion reactors and heavy-duty utilities. The reliability of these materials is based on the role of the microstructure, pore content and residual stress to avoid the oxidation and to produce thermal insulation in the surface of the superalloys, where these nanomaterials are deposited to develop functions of nanocoatings. Regularly, a nanostructured ceramic is deposited using atmospheric plasma spray (APS), where powder nanoparticles are injected into a plasma flame in a melt state, and generated droplets are condensed in a substrate to be flattened and that is solidified forming a film. Particular attention must be put in the technique used in deposition of the coating, since it determines the stresses generated, porosity, and microstructure obtained; and therefore, their thermophysical properties such as thermal expansion coefficient, thermal conductivity, and thermal shock.

The use of nanostructured powders allow to obtain the same structure in the coating, for example, conventional coatings based on zirconia (ZrO_2) and yttia (Y_2O_3) contain *t*-ZrO_2 and *m*-ZrO_2 in place of *c*-ZrO_2 structure found in nanostructured coatings, which reduces the size of the average grain size of ZrO_2 to the range of 49.3 to 59.5 nm (Wang, 2011). Other characteristics are also achieved such as low elastic modulus, and monomodal distribution of residual stresses in the surface whose value is proportional to the thickness of the coating.

Ceramic composites based on zirconia (ZrO_2) and hafnia (HfO_2) have been used in mechanical applications, aerospace applications, and energy applications, as coatings against wear, heat, and harsh environments. As an interesting alternative, composite materials based on gadolinia (Gd_2O_3) and hafnia were developed with the aim of reducing thermal conductivity (Ramana, 2012). They found that at varying Gd_2O_3 content (0 to 38 mol %) it is possible incrementing the formation of cubic phase to guarantee a reduction of thermophysical properties, which is desirable for thermal barrier coatings.

Other type of coatings that are very important from the thermal point-of-view is called the fire-retardant property or low flammability. Aerospace applications make use of special coatings for this purpose. Fire-safe materials are materials that are resistant to degradation at high temperatures produced by its flammability. Flammability is the ability of a substance to burn or ignite, causing fire or combustion. The degree of flammability may be assessed in terms of the required time for that a certain degree of burned achieves the established effects for standards in aerospace applications. During the test of flammability is necessary quantifying the amount of smoke and glowing char or other by-products produced during the burning (Horrocks, 2008). These materials are used in the construction of small enclosed spaces such as airplane cabins or as adhesives in aerospace materials. Materials used as flame-retardant polymeric materials are silicones, poly(vinyl chloride) (PVC), polyamide-imide (PA-PI), polytetrafluoroethylene(PTFE, Teflon), polyethylene (PE)-PTFE, polytrifluoroethylene (PF3E), poly(vinylidene fluoride) (PVDF), polyvinylidenefluoride (PVDF)/hexafluoropropylene (C3F6) copolymer, polybenzimidazole (PBI), polyetheretherketone (PEEK), polyetherimide (PEI), polyimide (PI), poly(phenyl sulfone) (PPSF), poly("m"-phenylene isophthalamide), poly("p"-phenylene isophtalamide) (Kevlar), polyphosphazenes, siloxanes, and silphenylenes (Papaspyrides, 2014). Fire retardant structural materials are used in aircraft structures such as floor panels, keel beams, luggage bins, and galleys, found in aircraft interiors. In aerospace applications, composites based on carbon nanotubes and/or epoxic resins are used in the fuselage, wing and tail fin components, control surfaces and doors (Horrocks, 2008). A large percentage of the cabin interior of passenger aircraft is manufactured of polymer matrix composite, mostly made of nanocomposite based on glass and/or phenolic resin. Finally, composite based on polyaramid fibre and resins has been used in ceiling panels, interior wall panels, partitions, cabinet walls, structural flooring and overhead stowage bins using a honeycomb core as structural support (Horrocks, 2008). Coatings and/or finishing treatments may be applied to shield fabrics from heat sources and prevent volatilization of flammable materials.

Flame-retardant nanocoatings are nanomaterials used to reduce the spread of flames on material surfaces either combustible or non-combustible, as shown in Figure 6. In addition, flame-retardant coatings can be placed on substrates such as wood, sheetrock, metals, polymers, fiberglass, composites, concrete and open and closed cell foams. These coatings can be applied to surfaces through nanoparticle adsorption, layer by layer assembly, sol-gel and dual-cure processes, as well as, plasma deposition (Alongi, 2014). These novel coatings make use of the hybrid materials through of the organic-inorganic composition between matrices and fibers. In the following paragraphs, a set of technical solutions for developing flame retardant effect in aerospace materials is discussed.

Nanoparticle adsorption consists in the immersion of the material into an aqueous suspension of inorganic nanoparticles with the aim of promoting adsorption of them into fiber surface in the manner of a finishing treatment in a manner either natural or synthetic (Alongi, 2014). These nanoparticles protect to the material from heat, oxygen and flame, and therefore, these nanocoatings act as a thermal insulator, at absorbing the heat and oxygen found around of the polymer, and at entrapping the volatile species that can fuel the combustion. This effect is produced by any technique of treatment for obtaining flame-retardant nanocoatings.

Figure 6. Flame-retardant effect found in aerospace surfaces and solved by nanocoatings

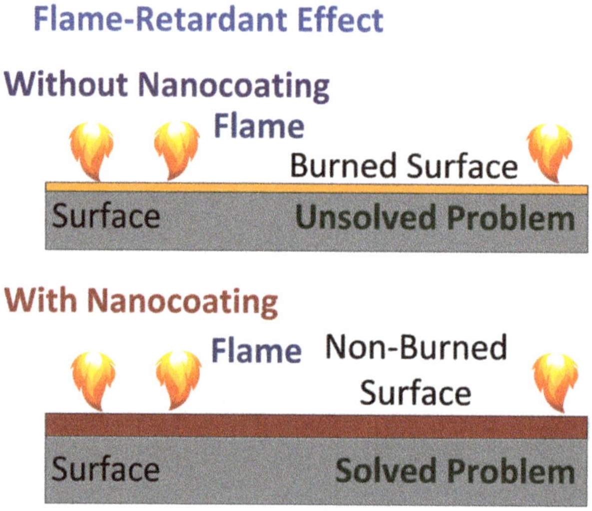

Layer-by-layer assembly consists in the build-up of films step-by-step through electrostatic interactions to form polyelectrolyte multilayers, and thus, inorganic nanoparticles are dispersed on a surface with the aim of exploiting different interactions (Alongi, 2014). It is achieved through of the alternate immersion of the substrate into an oppositely charged polyelectrolyte containing the inorganic nanoparticles that will realize the function of anti-flame protection.

The combination of sol-gel processes and dual-cure processes is based on a two-step reaction namely, hydrolysis (interaction between nanoparticles and substrate in aqueous solution) and condensation (deposition of the nanoparticles on surfaces), which are realized at or near of room temperature (Alongi, 2014). These reactions starting from (semi) metal alkoxides (they contain inorganic nanoparticles), result in hybrid materials.

The plasma technique allows that small functional groups and macromolecular compounds can be grafted to different substrates without to lead to modification or alteration of the bulky properties of the treated material (Alongi, 2014). Four different strategies can be used for this purpose: (1) the surface is etched to achieve a functionalization in non-polymerizable gases, (2) a thin film of thick polymer or inorganic layer is deposited on the surface of the material through compounds either volatile organic, organometallic or organosilicons, (3) a coating is placed on a preformed-polymer using two polymerization processes, and (4) a polimerization by graft is used to achieve the surface activation, and the grafting and polymerization of a non-volatile monomer on a surface.

Nanocoatings based on these techniques offer eco-sustainability and low toxicity, properties that cannot be obtained under conventional flame retardant systems based on halogenated products (Alongi, 2014). In addition, the use of proteins and nucleic acids in novel coatings leads to nanocoatings with characteristics sustainable and green.

A large area of opportunity of development is associated with the design of flame-retardant nanocomposites based on polymer-layered double hydroxide (LDH) (Gao, 2014). These materials operate under a combination of four functions that work in a synergistic manner: (1) heat absorption, (2) gaseous dilution, (3) char formation, and (4) dispersion. Polymers such as polyethylene (PE), polypropylene (PP), polystyrene (PS), polyvinyl chloride (PVC), polyamide 6 (PA 6), etc. as well as, hydroxides of zinc (Zn), magnesium (Mg), aluminum (Al), iron (Fe), and chromium (Cr) have been used for this purpose.

SUPERHYDROPHOBIC NANOCOATINGS

Nature is an unlimited source of functional surfaces, and the human being need mimetic substitutes, which can develop the same functions that artificial materials based on nanomaterials (Celia, 2013). The surface wettability represents the value of the contact angle that a water droplet forms with the surface. Self-cleaning surfaces have contact angles greater than 150°, and facility of sliding of water droplets, that is, the water droplets do not penetrate into the asperities of surface, and possible interactions among water and material surface are lowered or eliminated. Parameters such as surface free energy and surface geometrical structure can modify the wettability of the surface (Valipor, 2014).

Superhydrophobic nanocoatings are non-wettable nanomaterials with large water contact angles and with facility of sliding of drops (Ma, 2006), and therefore, it can be considered as a nanoscopic surface layer that repels water and possesses a self-cleaning effect (i.e., removes contaminants or powders simultaneously on material surface), as depicted in Figure 7. These coatings are also called ultrahydrophobic and they have other properties such as transparency and color, anisotropy, reversibility, flexibility and breathability (Ma, 2006). A superhydrophobic surface combines low surface energy materials and surface textures as the lotus leaf (Dai, 2011), which represents roughness in the range of micrometers and/or nanometers. These coatings are used in ultra-dry surface applications, and they can be considered as thin layers almost imperceptible on top of a surface. Thus, when coatings are sprayed onto materials, then their surface are make to waterproof. Moreover, the spray provides properties such as anti-corrosion, anti-fogging, anti-icing, anti-bacteria, anti-fouling, drag reduction, stain-resistant, and self-cleaning (Valipor, 2014). The control of the surface wettability is used in applications such as thin film technology, lubrication, antifouling paints, self-cleaning windows, microfluids, textiles and anti-snow-sticking surfaces (Contreras, 2014). Superhydrophobic coatings can be used in windshields and windows of aircrafts to prevent that rain droplets from clinging to the glass. Even, these coatings allow the removal of salt deposits in the surfaces without the necessity of using fresh water. These properties are achieved through of the micro and nanostructures that have qualities of repellence. Unfortunately, surfaces containing these coatings can be damaged by abrasion or cleaning, and therefore, these coatings do not support

Figure 7. Superhydrophobic effect found in aerospace surfaces and solved by nanocoatings

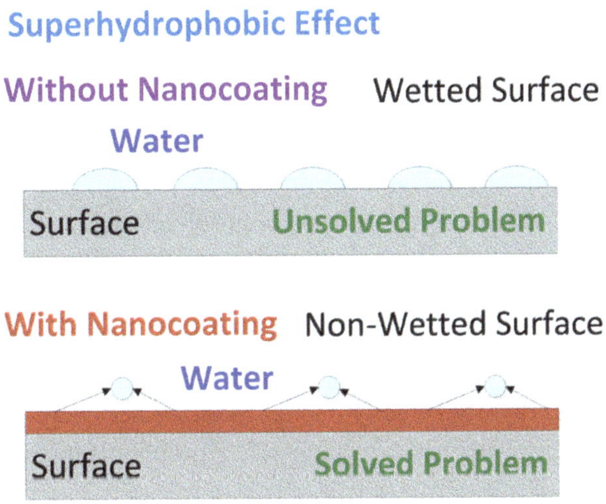

a continuum wear. Surfaces that support constant friction require a constant re-application of the coating to maintain a high degree of performance during operation of the part that has been covered. The main problems found in artificial and superhydrophobic surfaces are mechanical instability, long-term durability, diversity of pollutants and thermal instability (Zhang, 2013). In the following paragraphs, a set of technical solutions to solve the problem of hydrophobicity in aerospace materials is discussed.

Conditions such as high mechanical abrasion and high salinity change the hydrophobic state of the surface of materials based on magnesium alloys, that leading to the short-term corrosion of the aerospace components. In aerospace industry, it is highly desired that large-area superhydrophobic surfaces can be mechanically robust, and with properties such as anticorrosion, anti-icing, and self-cleaning can be deposited as nanocoatings on various substrate materials.

With the aim of mimicking superhydrophobic properties that are found in nature, many artificial superhydrophobic surfaces have been proposed. The use of nanomaterials is completely indispensable to increase surface roughness. Different materials have been used to develop superhydrophobic nanocoatings: zinc oxide-polystyrene (ZnO/PS) nanocomposite (Qing, 2013), titanium dioxide- trimethoxypropyl silane-polypropylene ($TiO_2/C_6H_{16}O_3Si/PP$) nanocomposite (Contreras, 2014), silica-polystyrene (SiO_2/PS) nanocomposite (Zhang, 2013), precipitated calcium carbonate ($CaCO_3$) (Hu, 2009), carbon nanotube structures (Hsieh, 2010), and silica (SiO_2) nanocoatings (Hsieh, 2009; Lin, 2013). The silica-based coatings are obtained by means of sol-gel processing, and these materials are applied either by dipping the surface into the gel or via aerosol spray, and moreover, this technique is considered the simplest of using and the most economically viable option at present.

Main precursors used in the fabrication of transparent superhydrophobic coatings based on silica nanocoatings are tetraethoxysilane (TEOS) and polymethylhydrosiloxane (PMHS) (Lin, 2013), cetyltrimethoxysilane (CTMS) and methyltrimethoxysilane (MTMS) (Fei, 2014). In this type of coatings, aspects such as low density, porous structure with high volume fraction and chemical groups on the surface, are the cause of their superhydrophobicity. In contrast, the zinc oxide-polystyrene composites require processes much more involved and costly, although, these materials are more durable than the gel-based coatings. Carbon nanostructures possess mechanical strength, thermal conductivity, pore structure, pore size distribution, surface acidity, and hydrophobicity/hydrophilicity, which can be exploited in these nanocoatings (Hsieh, 2010). In addition, carbon nanostructures (carbon nanotubes and graphene) facilitate the sliding of water drops through of their surface texture. Finally, coatings based on carbon nanotubes are difficult of producing and they are very expensive.

Two main approaches are used to obtain artificial, hierarchical superhydrophobic surfaces (Guo, 2011): (i) surfaces using micro- and nanostructures on hydrophobic substrates, and (ii) surfaces including a chemical modification with low surface free energy. Roughness surfaces can be prepared by means of numerous methods involving top-down or bottom-up processes (Qing, 2013) such as wet chemical reaction, hydrothermal reaction, electrochemical deposition, lithography (Dai, 2011), self-assembly layer-to-layer, plasma etching, chemical vapor deposition, sol-gel and polymerization reactions (Celia, 2013). Moreover, these methods allow obtaining the same nanocoatings with other functions such as electrowetting, anti-icing, drag-reduction, Leidenfrost drops, and reversible surfaces (Guo, 2011). The use of functional nanomaterials increases the possibility of separating water from surfaces with different surface tension (Zhang, 2013).

It can be concluded that superhydrophobicity implies a low surface energy and/or low adhesion, high-contact angle, low-contact angle hysteresis, bouncing-off, hydrophobic interaction, and minimization of the Gibbs energy (Hejazi, 2013).

ANTI-ICING OR ICEPHOBIC NANOCOATINGS

Although ice formation is very attractive by nature, it is disastrous when is accumulated in applications such as aviation, shipping, communications, and electrical power generation and transmission (Frankenstein, 2002). In cold-climate regions, adhesion and excessive accumulation of ice and wet-snow are causing serious problems in structures and equipment (Farhadi, 2011). The presence of ice has produced problems of communication among antennas, has destroyed satellite and microwave dishes, causing accidents with falling ice on infrastructure and injuries to people walking under cables with ice (Boinovich, 2013). These damages produce socioeconomic problems in diverse sectors of the economy such as power transmission and distribution, telecommunication networks, aircraft, boats, etc. An effective medium of melting ice from limited and critical areas, is the use of electrical resistance heating (Frankenstein, 2002). In the case of aerospace applications, the formation of ice on the aircraft's wings or blockade of its instruments will reduce the handling and stability performance, which may cause aviation accident (Ge, 2013). According to the Environmental Protection Agency (EPA), an estimated 25 million gallons of de-icing agents are applied to aircraft taking off from US commercial airports each year.

Three main types of methodologies to lower ice's adhesive strength have been proposed: electrical, chemical and mechanical. Mechanical methods have been the least studied and experimented, since these require a lot of time and they are costly. The electrical approach is based on the use of heaters with the aim of creating an electrical pulse that mechanically breaks the ice through a direct current bias to reduce ice's adhesion to the surfaces. The use of heaters is only effective in reduced areas, but impractical for removing ice from extended areas. The application of liquids with much lower crystallization temperatures than that of water is the most commonly used chemical approach in airplanes, helicopters and runways (Boinovich, 2013). Regularly, these commercial non-freezable liquids contain salts initiating the corrosion processes, and/or others that shift the salt balance in the environment or glycols. No doubt, the search of a coating with reduced adhesion to the ice is receiving a major attention since until now numerous proposals have been designed and experimented around of the world. Materials that have been proposed to be used as coatings are: Teflon, silicone, deicing fluids, rubbers, elastomers, oils and greases, urethane and polyurethane, polyethylene, polystyrene, acrylics, plasti-dip, Icex, polysiloxanes, polycarbonates, and neoprene (Frankenstein, 2002). Unfortunately, none of these materials has found fulfill the adhesive strength required in practical situations. However, these studies have allowed estimating that properties such as inert materials, hydrophobic materials, a high contact angle, low surface energy and low surface tension are required in the materials used as anti-icing coatings.

Anti-icing nanocoating is a nanomaterial used as coating that prevents the build-up of ice on a material surface, as shown in Figure 8. Anti-ice coatings form hydrophobic surfaces with characteristics slick, non-stick, when these are applied to substrates. The desired properties of these coatings are transparency, hardness, density, smooth finish, durability, and corrosion protection. These coatings must be used in a variety of surfaces such as plastics, metals, glass, ceramics, and over most painted surfaces which are applied by means of dip, spray (using low-pressure and high-volume) or brush (using a roller or brush with a cross pattern). Coatings are cured at room temperature for six to eight hours or by one hour or less at temperatures in the range of 100 to 150 °C. Due to the anti-stick property by nature of the anti-icing coatings, the large contact angle of the water removes the bead of ice formed away from the coating and reduced surface area to attach ice to the material surface.

Figure 8. Anti-icing effect found in aerospace surfaces and solved by nanocoatings

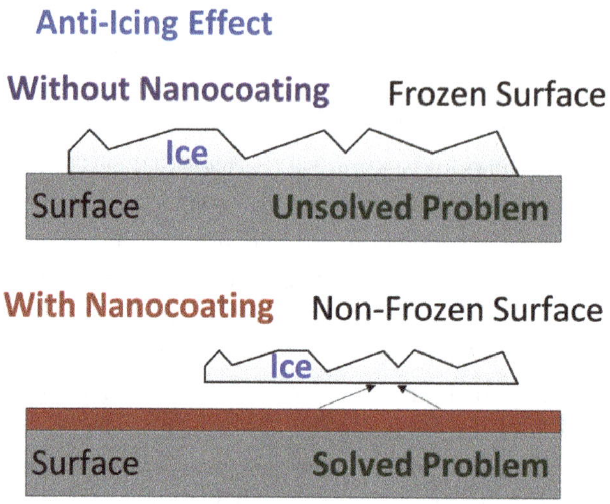

Typical applications of the anti-icing coatings are: cables, satellite dishes, microwave domes, communication towers, antennas, guy wires, fire watch stations, bridges (cables, railings), wind mills, overpasses, elevated walkways, roof edges, ships, train cars, tractor trailers, and heavy equipment exposed to ice/snow. Benefits of the anti-icing coatings are: easy clean, stain resistant, mar resistant, weather resistant, and chemical resistant to road salt, durable, reduces ice and snow accumulation, hydrophobic properties, icephobic properties, non-stick, non-toxic, waterborne, low volatile organic compounds (VOC), etc.

Five mechanisms are implied in the icephobicity that present superhydrophobic coatings used as anti-icing coatings (Boinovich, 2013): (i) removal of water drops on material surface with negative temperature before of their crystallization, (ii) decreasing heat transfer between drops and surface to reduce their solidification, (iii) delaying the crystallization of drops in a surface, (iv) shifting the triple point in water toward negative temperatures, and (v) lowering adhesion of water either liquid or crystalline on the surface. The knowledge of the mechanisms of anti-icing allows estimating the operation conditions of the coatings, and also allows knowing that wetting-regime is implied. In the following paragraphs, a set of technical solutions to solve the problem of icing in aerospace materials is discussed.

In the search of anti-icing coatings based in the use of superhydrophobic coatings (Farhadi, 2011), researchers have developed coatings based on CeO_2-Zonyl 8740 (perfluoroalkyl methacylic copolymer soluble in water, Dupont), FAS-13 (1H, 1 H, 2H, 2H-perfluoro-octyltrithoxysilane, Aldrich), Ag-Zonyl 8740, or RTV SR (Room-Temperature-Vulcanized Silicone Rubber, Dow Corning). They found that, after of eleven icing/de-icing cycles, results showed that the anti-ice performance of the samples was significantly deteriorated. A critical aspect is that superhydrophobicity of the coating can be retained at different contact angles and roll-off angles (Ge, 2013). De-icing coatings can be obtained by means of nanocomposites based on carbon nanotubes and polyurethane (PU) exploiting electrical properties such as electrostatic dissipation (Zhao, 2010).

Ge and collaborators (Ge, 2013) proposed the use of octadecyltrichlorosilane (OTS) films treated with dichloromethane using a simple phase-separation method. Their results showed that coating retained its superhydrophobicity even in subzero environment, and they attributed this behavior to the disordered reticular structure in submicron and/or nanometer scale leading to an excellent anti-icing property. In addition, it was found that the icing progress of the water droplets was markedly retarded due to the lower adhesion strength.

Different coupling agents can be applied to nano-/micro-aluminium surfaces for modifying these coatings, such as PTES (Perfluorodecyltriethoxysilane), TTPS (triethoxysilylethyl terminated polydimethylsiloxane) or PA (Palmitic acid) (Wang, 2015). These materials are used in anti-icing applications based on syperhydrophobic properties. These three coatings, partially retained their hydrophobicity, and in the case of coating with PTES, even it maintained the sliding and rebounding behavior under -10 °C and RH 90%. Nanocoatings with PTES were considered as the best option for anti-icing property, since this material presented adhesion strength of only about 20%. Moreover, surface with PTES has the better qualities necessary to support mechanical vibrations, which are directly related with hydrophobic property under the condition of condensed ice (Wang, 2015).

Chemical etching with 1:1 volume mixture of 40 wt % $FeCl_3$, 6 wt % HCl aqueous solution, and 10 wt % lauric acid ethanol solution was applied to aluminum alloy substrates (Ruan, 2013). This technique presented a contact angle of 159,1 ° and a small contact angle hysteresis of 4,0 °. These values allow for affirming that it is possible achieving an anti-icing effect with the use of nanocoatings. However, to date, the mechanism responsible of anti-icing continues unclear.

Superhydrophobic nanocoatings are not always ice-repellent since they are deteriorated during icing/deicing cycles due to that surface asperities appear gradually during operating time (Kulinich, 2011), that is, they present relatively low abrasive resistance and inability for deicing in all climatic conditions.

The general consensus affirms that there is no direct correlation between icephobicity and superhydrophobicity (Hejazi, 2013). It can be concluded that icephobicity implies a low adhesion, low normal strength (maximum stress), low shear strength, bouncing-off supercooled droplets, crystallization of ice and snowflakes, and surface roughening transition of the Gibbs energy (Hejazi, 2013). Furthermore, superhydrophobicity involves the repulsion of biological entities and minimizing surface contact with water.

CONCLUSION

The use of nanocoatings in aerospace industry is increasing each day more-and-more, however, these coatings until now provide a relatively exotic solution to protect the surface of the materials against different harsh conditions through of the use of functional materials. Six different types of nanocoatings can be found in aerospace industry: anti-corrosion coatings, abrasion and wear-resistant coatings, thermal barrier and flame-retardant coatings, conductive coatings, anti-icing coatings, and superhydrophobic nanocoatings. An exhaustive use of the nanomaterials has been identified for all types of nanocoatings where mechanical, physical and chemical properties are exploited. Despite the successes achieved so far, there is a tremendous need to develop new techniques and materials that can fully satisfy environmental and operating conditions presented in the aerospace industry.

ACKNOWLEDGMENT

This work is part of a project that is being financed by CONCYT-Mexico (152524).

REFERENCES

Aliofkhazraei, M. (2015). *Anti-Abrasive Nanocoatings: Current and Future Applications*. Cambridge, UK: Woodhead Publishing.

Alongi, J., Carosio, F., & Malucelli, G. (2014). Current Emerging Techniques to Impart Flame Retardancy to Fabrics: An Overview. *Polymer Degradation & Stability*, *106*, 138–149. doi:10.1016/j.polymdegradstab.2013.07.012

Boinovich, L. B., & Emelyanenko, A. M. (2013). Anti-Icing Potential of Superhydrophobic Coatings. *Mendeleev Communications*, *23*(1), 3–10. doi:10.1016/j.mencom.2013.01.002

Borisova, D., Akçakayiran, D., Schenderlein, M., Möhwald, H., & Shchukin, D. G. (2013). Nanocontainer-based Anticorrosive Coatings: Effect of the Container Size on the Self-Healing Performance. *Advanced Functional Materials*, *23*(30), 3799–3812. doi:10.1002/adfm.201203715

Celia, E., Darmanin, T., Taffin de Givenchy, E., Amigoni, S., & Guittard, F. (2013). Recent Advances in Designing Superhydrophobic Surfaces. *Journal of Colloid and Interface Science*, *402*, 1–18. doi:10.1016/j.jcis.2013.03.041 PMID:23647693

Contreras, C. B., Chagas, G., Strumia, M. C., & Weibel, D. E. (2014). Permanent Superhydrophobic Polypropylene Nanocomposite Coatings by a Simple One-Step Dipping Process. *Applied Surface Science*, *307*, 234–240. doi:10.1016/j.apsusc.2014.04.019

Dai, S., Ding, W., Wang, Y., Zhang, D., & Du, Z. (2011). Fabrication of Hydrophobic Inorganic Coatings on Natural Lotus Leaves for Nanoimprint Stamps. *Thin Solid Films*, *519*(16), 5523–5527. doi:10.1016/j.tsf.2011.03.118

Del, F., Kroes, M., Sterkenburg, R., & Watkins, W. (2013). *Aircraft Maintenance and Repair* (7th ed.). United States of America: McGraw-Hill.

Ersoy, M. S., & Onder, E. (2014). Electroless Silver Coating on Glass Stitched Fabrics for Electromagnetic Shielding Applications. *Textile Research Journal*, *84*(19), 2103–2114. doi:10.1177/0040517514530025

Farhadi, S., Farzaneh, M., & Kulinich, S. A. (2011). Anti-Icing Performance of Superhydrophobic Surfaces. *Applied Surface Science*, *257*(14), 6264–6269. doi:10.1016/j.apsusc.2011.02.057

Fei, T., Chen, H., & Lin, J. (2014). Transparent Superhydrophobic Films Possessing High Thermal Stability and Improved Moisture Resistance from the Deposition of MTMS-based Aerogels. *Colloids and Surfaces. A, Physicochemical and Engineering Aspects*, *443*, 255–264. doi:10.1016/j.colsurfa.2013.11.027

Fox-Rabinovich, G.S., Yamamoto, K., Beake, B.D., Gershman, I.S., Kovalev, A.I., Veldhuis, S.C., Aguirre, M.H., Dosbaeva, G. & Endrino, J.L. (2012). Hierarchical Adaptive Nanostructured PVD Coatings for Extreme Tribological Applications: The Quest for Nonequilibrium States and Emergent Behavior. *Science and Technology of Advanced Materials*, 13(4): 043001. doi:10.1088/1468-6996/13/4/043001

Frankenstein, S. & Tuthill., A.M. (2002). Ice Adhesion to Locks and Dams: Past Work; Future Directions?. *Journal of Cold Regions Engineering*, *16*(2), 83-96. doi:10.1061/(ASCE)0887-381X

Gao, Y., Wu, J., Wang, Q., Wilkie, C. A., & O'Hare, D. (2014). Flame Retardant Polymer/Layered Double Hydroxide Nanocomposites. *Journal of Materials Chemistry A*, *2*(29), 10996–11016. doi:10.1039/c4ta01030b

Ge, L., Ding, G., Wang, H., Yao, J., Cheng, P., & Wang, Y. (2013). Anti-Icing Property of Superhydrophobic Octadecyltrichlorosilane Film and Its Ice Adhesion Strength. *Journal of Nanomaterials*. 2013, 278936 doi:10.1155/2013/278936

Grady, B. P. (2011). *Carbon Nanotube-Polymer Composites: Manufacture, Properties, and Applications*. Hoboken, NJ: John Wiley & Sons; doi:10.1002/9781118084380

Guo, Z., Liu, W., & Su, B.-L. (2011). Superhydrophobic Surfaces: From Natural to Biomimetic to Functional. *Journal of Colloid and Interface Science*, *353*(2), 335–355. doi:10.1016/j.jcis.2010.08.047 PMID:20846662

Hejazi, V., Sobolev, K., & Nosonovsky, M. (2013). From Superhydrophobicity to Icephobicity: Forces and Interaction Analysis. *Scientific Reports*, *3*, 2194. doi:10.1038/srep02194 PMID:23846773

Horrocks, A. R., & Price, D. (2008). *Advances in Fire Retardant Materials*. Boca Raton, FL: Woodhead Publishing. doi:10.1533/9781845694701

Hsieh, C. T., & Chen, W. Y. (2010). Water/Oil Repellency and Drop Sliding Behavior on Carbon Nanotubes/Carbon Paper Composite Surfaces. *Carbon*, *48*(3), 612–619. doi:10.1016/j.carbon.2009.09.076

Hsieh, C. T., Wu, F.-L., & Chen, W.-Y. (2009). Super Water- and Oil-Repellencies from Silica-based Nanocoatings. *Surface and Coatings Technology*, *203*(22), 3377–3384. doi:10.1016/j.surfcoat.2009.04.025

Hu, Z., Zen, X., Gong, J., & Deng, Y. (2009). Water Resistance Improvement of Paper by Superhydrophobic modification with Microsized $CaCO_3$ and Fatty Acid Coating. *Colloids and Surfaces. A, Physicochemical and Engineering Aspects*, *351*(1-3), 65–70. doi:10.1016/j.colsurfa.2009.09.036

Joshi, A., Bajaj, A., Singh, R., Anand, A., Alegaonkar, P. S., & Datar, S. (2015). Processing of Graphene Nanoribbon based Hybrid Composite for Electromagnetic Shielding. *Composites. Part B, Engineering*, *69*, 472–477. doi:10.1016/j.compositesb.2014.09.014

Kandanur, S. S., Rafiee, M. A., Yavari, F., Schrameyer, M., Yu, Z.-Z., Blanchet, T. A., & Koratkar, N. (2012). Suppression of Wear in Graphene Polymer Composites. *Carbon*, *50*(9), 3178–3183. doi:10.1016/j.carbon.2011.10.038

Kandare, E., Khatibi, A. A., Yoo, S., Wang, R., Ma, J., Olivier, P., & Wang, C. H. et al. (2015). Improving the through-thickness Thermal and Electrical Conductivity of Carbon Fibre/Epoxy Laminates by exploiting Synergy between Graphene and Silver Nano-Inclusions. *Composites. Part A, Applied Science and Manufacturing*, *69*, 72–82. doi:10.1016/j.compositesa.2014.10.024

Kent, J. A. (2012). *Handbook of Industrial Chemistry and Biotechnology*. New York, NY: Springer. doi:10.1007/978-1-4614-4259-2

Kulinich, S. A., Farhadi, S., Nose, K., & Du, X. W. (2011). Superhydrophobic Surfaces: Are They Really Ice-Repellent? *Langmuir*, *27*(1), 25–29. doi:10.1021/la104277q PMID:21141839

Landolt, D., & Mischler, S. (2011). *Tribocorrosion of Passive Metals and Coatings*. Cambridge, UK: Woodhead Publishing. doi:10.1533/9780857093738

Lin, J., Chen, H., Fei, T., Liu, C., & Zhang, J. (2013). Highly Transparent and Thermally Stable Superhydrophobic Coatings from the Deposition of Silica Aerogels. *Applied Surface Science*, *273*, 776–786. doi:10.1016/j.apsusc.2013.02.134

Ma, M., & Hill, R. M. (2006). Superhydrophobic Surfaces. *Current Opinion in Colloid & Interface Science*, *11*(4), 193–202. doi:10.1016/j.cocis.2006.06.002

Makhlouf, A. S. H. (2014). Handbook of Smart Coatings for Materials Protection. Cambridge, UK: Woodhead Publishing. doi:10.1533/9780857096883

Makhlouf, A. S. H., & Tiginyanu, I. (2011). *Nanocoatings and Ultra-Thin Films: Technologies and Applications*. Cambridge, UK: Woodhead Publishing. doi:10.1533/9780857094902

Mellor, B. G. (2006). *Surface Coatings for Protection against Wear*. Boca Raton, FL: Woodhead Publishing.

News (2011). Heat-Sealable ESD Film Protects Sensitive Spacecraft Components during Construction. *Sealing Technology*, *2011*(11), 2-3. doi:10.1016/S1350-4789(12)70296-9

Papaspyrides, C. D., & Kiliaris, P. (2014). *Polymer Green Flame Retardants*. Oxford, UK: Elsevier.

Qing, Y., Zheng, Y., Hu, C., Wang, Y., He, Y., Gong, Y. & Mo, Q. (2013). Facile Approach in Fabricating Superhydrophobic ZnO/Polystyrene Nanocomposite Coating. *Applied Surface Science*, *285*(Part B), 583-587. doi:10.1016/j.apsusc.2013.08.097

Ramana, C. V., Gengler, J. J., & Jones, J. G. (2012). Crystal Structure, Phase Analysis, and Thermal Conductivity of Nanocrystalline GdO_3-HfO_2 Coatings. *The Journal of Physical Chemistry C*, *116*(48), 25178–25183. doi:10.1021/jp301008t

Ruan, M., Li, W., Wang, B., Deng, B., Ma, F., & Yu, Z. (2013). Preparation and Anti-Icing Behavior of Superhydrophobic Surfaces on Aluminum Alloy Substrates. *Langmuir*, *29*(27), 8482–8491. doi:10.1021/la400979d PMID:23718719

She, Z., Li, Q., Wang, Z., Li, L., Chen, F., & Zhou, J. (2013). Researching the Fabrication of Anticorrosion Superhydrophobic Surface on Magnesium Alloy and Its Mechanical Stability and Durability. *Chemical Engineering Journal*, *228*, 415–424. doi:10.1016/j.cej.2013.05.017

Sun, D., An, J., Wu, G., & Yang, J. (2015). Double-Layered Reactive Microcapsules with Excellent Thermal and Non-Polar Solvent Resistance for Self-Healing Coatings. *Journal of Materials Chemistry A*, *3*(8), 4435–4444. doi:10.1039/C4TA05339G

Tong, X. C. (2009). Advanced Materials and Design for Electromagnetic Interference Shielding. Boca Raton, FL: CRC Press. doi: 10.1201/9781420073591

Valipor, N. (2014). Super-Non-Wettable Surfaces: A Review. *Colloids and Surfaces. A, Physicochemical and Engineering Aspects*, *448*, 93–106. doi:10.1016/j.colsurfa.2014.02.016

Vargas-Bernal, R. (2015). Performance Analysis of Electromagnetic Interference Shielding based on Carbon Nanomaterials used in AMS/RF IC Design. In M. Fakhfakh, E. Tlelo-Cuatle, & M. H. Fino (Eds.), *Performance Optimization Techniques in Analog, Mixed-Signal, and Radio-Frequency Circuit Design* (pp. 268–294). Hershey, PA, USA: IGI Global. doi:10.4018/978-1-4666-6627-6.ch011

Voevodin, A. A., O'Neill, J. P., & Zabinski, J. S. (1999). Nanocomposite Tribological Coatings for Aerospace Applications. *Surface and Coatings Technology*, *116-119*, 36–45. doi:10.1016/S0257-8972(99)00228-5

Voldman, S. H. (2012). *ESD Basics: From Semiconductor Manufacturing to Product Use*. Chichester, United Kingdom: Wiley. doi:10.1002/9781118443323

Wang, L., Wang, Y., Sun, X. G., Pan, Z. Y., He, J. Q., Zhou, Y., & Wu, P. L. (2011). Microstructure and Surface Residual Stress of Plasma Sprayed Nanostructured and Conventional ZrO_2-8wt%Y_2O_3 Thermal Barrier Coatings. *Surface and Interface Analysis*, *43*(5), 869–880. doi:10.1002/sia.3648

Wang, Q. J., & Chung, Y.-W. (2013). *Encyclopedia of Tribiology*. New York, NY: Springer. doi:10.1007/978-0-387-92897-5

Wang, Y., Li, M., Lv, T., Wang, Q., Chen, Q., & Ding, J. (2015). Influence of Different Chemical Modifications on the Icephobic Properties of Superhydrophobic Surfaces in a Condensate Environment. *Journal of Materials Chemistry A*, *3*(9), 4967–49775. doi:10.1039/C4TA07077A

Wu, L., & Baghdachi, J. (2015). *Functional Polymer Coatings: Principles, Methods, and Applications*. Hoboken, New Jersey: Wiley. doi:10.1002/9781118883051

Xu, H., & Guo, H. (2011). *Thermal Barrier Coatings*. Cambridge, UK: Woodhead Publishing. doi:10.1533/9780857090829

Zhang, X., Geng, T., Guo, Y., Zhang, Z., & Zhang, P. (2013). Facile Fabrication of Stable Superhydrophobic SiO_2/Polystyrene Coating and Separation of Liquids with Different Surface Tension. *Chemical Engineering Journal*, *231*, 414–419. doi:10.1016/j.cej.2013.07.046

Zhao, W., Li, M., & Peng, H.-X. (2010). Functionalized MWNT-doped Thermoplastic Polyurethane Nanocomposites for Aerospace Coating Applications. *Macromolecular Materials and Engineering*, *295*(9), 838–845. doi:10.1002/mame.201000080

ADDITIONAL READING

Boinovich, L. B., Domantovskii, A. G., Emelyanenko, A. M., Miller, A. B., Potapov, Y. F., & Khodan, A. N. (2013). Antiicing Performance of Superhydrophobic Coatings on Aluminum and Stainless Steel, Russian Chemical Bulletin, International Edition, 62(2), 380-387. doi:10.1007/s11172-013-0049-6

Vargas-Bernal, R., Arizmendi-Morquecho, A. M., Bermudez-Reyes, B., & Herrera-Ramirez, J. M. (2013). Study of the Electromagnetic Shielding Effectiveness for Different Materials used in Aircraft and Spacecraft Materials. In A. P. Melendez (Ed.), *Congreso Internacional en Ciencia y Tecnologia Aeroespacial* (pp. 61–67). Zapopan, Jalisco, Mexico: CICYTA.

Wang, Y., Orol, D., Owens, J., Simpson, K., & Lee, H. J. (2013). Design and Development of Anti-Icing Aluminum Surface. *Materials Sciences and Applications*, *4*(6), 347–356. doi:10.4236/msa.2013.46045

Zhou, X., Xu, D., Lu, J., & Zhang, K. (2015). CuO/Mg/Fluorocarbon Sandwich-Structure Superhydrophobic Nanoenergetic Composite with Anti-Humidity Property. *Chemical Engineering Journal*, *266*, 163–170. doi:10.1016/j.cej.2014.12.087

KEY TERMS AND DEFINITIONS

Abrasion Resistant Nanocoating: Nanomaterial used to increase abrasion resistance or protection from galling, seizing and other physical damage.

Anti-Corrosion Nanocoating: Corrosion resistant nanomaterial used to protect material surfaces against degradation due to moisture, salt spray, oxidation or exposure to a variety of environments or chemicals.

Anti-Icing Nanocoating: Nanomaterial used as coating that prevents the build-up of ice on a material surface.

Conductive Nanocoatings: Nanomaterials used to reduce or isolate electromagnetic or radio frequency interference (EMI/RFI) provided by an electromagnetic source.

Flame-Retardant Nanocoatings: Nanomaterials used to reduce the spread of flames on combustible and non-combustible material surfaces.

Functional Nanomaterials: Nanomaterials that possess peculiar natural properties and functions of their own, which coupling multiple physical variables such as ferroelectricity, magnetism, etc.

Nanocoating: Thin film based on nanomaterials that is used as protective coating of material surface against harsh environments.

Superhydrophobic Nanocoatings: Non-wettable nanomaterials with water contact angles larger than 150° and with sliding angles lower than 18°.

Thermal Barrier Nanocoatings: Nanomaterials used to protect material surfaces from heat and thermal degradation.

Wear Resistant Nanocoatings: Robust, ultra-hard nanomaterials used to reduce wear produced in material surfaces.

Chapter 13
Fundamental, Fabrication and Applications of Superhydrophobic Surfaces

Adel M Mohamed
Qatar University, Qatar & Suez University, Egypt

Mariam Al-Maadeed
Qatar University, Qatar

Aboubakr Moustafa Abdullah
Qatar University, Qatar

Ahmed Bahgat
Qatar University, Qatar

ABSTRACT

The need to create new enhanced corrosion resistant coatings has grown because of the financial burden corrosion places on certain industries such as oil industry. Organic inhibitors have been extensively reported to protect metals and metallic alloys from corrosion although they have a negative impact on environment. Development of better corrosion resistant coatings such as metal alloys, metal-ceramics, polymers, and hydrophobic films are examples of corrosion resistant coatings for metals and alloys. Recently, superhydrophobic coatings have been widely implemented into many fields including anti-fogging transparent materials, self-cleaning surfaces, biomedical and corrosion applications. In the proposed chapter, a comprehensive review will be dedicated for the fundamentals and developments of superhydrophobic materials including theoretical background; superhydrophobicity in nature; preparation techniques; and recent attempts to develop superhydrophobic surfaces.

INTRODUCTION

The morphology and structure of the solid surfaces play a paramount function on the corrosion resistance.

Superhydrophobic surfaces are characterized by their extreme water-repellent properties. These surfaces have recently received much attention because of their potential implementations in many applications such as weather resistant garments, fluidic systems with reduced drag, self-cleaning windows, roof tiles, solar panels, and, as we shall see, anti-icing coatings (Carré & Mittal, 2009). A fact that demonstrates the

DOI: 10.4018/978-1-5225-0066-7.ch013

enormous growth of interest is the number of articles on superhydrophobic surfaces published between 2006 and 2007, which were more than double the total number of articles published before that date on this topic (Roach, Shirtcliffe, & Newton, 2008).

The right combination of surface properties can dramatically decrease the ability of water drops to stick and hence enhance the corrosion resistance of these surfaces. On the basis of the water contact angle (WCA) and the contact angle hysteresis (CAH), Quéré (Quéré, 2005), reported the requirements for a water drop to slither on a tilted surface. The CAH is known as the variance between the advancing and receding CAs of water acquired on a surface through the motion of droplet in one direction. Quéré (Quéré, 2005), postulates that a water droplet can sideslip off a surface, if CA is greater than 160° even when the CAH is higher, owing to the lower contact zone among the liquid and the solid. He concluded that "nonsticking" drops must therefore be observed on surfaces with CA > 160° and such surfaces are termed "superhydrophobic". Superhydrophobicity is noticed on definite natural tissues such as lotus leaves. The so-called "Lotus effect" is owing to the entity of a rough micro-nanostructure wrapped with waxy materials which are poorly wetting, resulting in a water contact angle greater than 150° (Roach, et al., 2008) . The rough micro-nanostructure allows the entrapment of air beneath the droplet in the pores (hollows) of the rough structure and the waxy coating, which has a low surface energy.

In this chapter, the basic theory behind superhydrophobicity is outlined including theoretical background; superhydrophobicity in nature; preparation techniques of superhydrophobic surfaces; applications of superhydrophobicity phenomenon and corrosion performance of superhydrophobic coatings.

BACKGROUND

Many researches have studied the role of superhydrophobicity on corrosion prevention (Ejenstam et al., 2013; Grignard et al., 2011; Jin, Li, Han, Wang, & Sun, 2014; Su & Yao, 2014; Zhou, Lu, Xin, & Liu, 2013). Learning from nature, a superhydrophobic surface can be produced by manufacturing an optimum surface roughness pursued by passivation with a low surface energy coating leading to a very low area of contact between the solid surface and the liquid, causing a water drop to roll off the surface. As the contact area of water on a superhydrophobic solid surface is omitted, such surfaces would effectively decrease the contact area of water. Therefore, the superhydrophobic film can resist corrosion and protect the metals. It is well known that, copper and its alloys have been one of the important materials in industry owing to its thermal and high electrical conductivities, mechanical workability and its comparatively noble properties. It is vastly utilized in various implementation in communications and electronic industries as a conductor in electrical power lines, pipelines for territorial and processing water utilities involving heat exchangers, heat conductors and seawater (Kim, 2008). Hence, corrosion of copper and its resistance in a wide variety of media, especially when they have chloride ions, have aggravated concern of a number of researchers (Barthlott & Neinhius, 1997; Bhushan, 2011b; Gennes, Wyart, & Quere, 2004). In addition to copper, aluminum and its alloy are also vastly utilized in marine, aerospace, and automotive applications due to their low density, convenient mechanical properties and comparatively good corrosion protection. Aluminum and its alloy possess a naturalistic corrosion resistance from its oxide layer, but it may be corroded if subject to aggressive environments (Traverso & Canepa, 2014). The appropriate coatings on copper, aluminum and their alloys may be reliable and add longer service life.

The natural defense of the metal against corrosion is its oxide layer (or rust, in the case of iron). The oxide layer is formed when the metal reacts with the aggressive species from the electrolyte solution. However, most of the time, the oxide layer is not dense enough to prevent further diffusion of aggressive species into the metal surface. Some corrosion protection methods focus on improving the property of the oxide layer. It has been shown that alloying can drastically increase the density and reduce the porosity of the oxide layer (Dewidar, Adamek, Jakubowicz, & AbdelRazik, 2014; Jakubowicz, 2008; Ro & Klar, 1981). However, alloying is expensive and can significantly affect the mechanical properties of the metal. Another approach is to use a coating to protect and isolate the metals from the corrosive environment media.

BASIC PRINCIPLES OF SUPERHYDROPHOBICITY

Surface Tension and Contact Angle

The surface of any material is associated with an excess energy which stems from the much higher fraction of broken bonds among the surface atoms than among the atoms in the bulk. This excess energy is commonly referred to as the surface tension, which is the work required per unit area in increasing the surface area at constant temperature and pressure (Sharma, 2007). In SI units, the surface tension is measured either in energy per unit area or force per unit length, i.e., J/m^2 or N/m. The surface tension of a given material can be chemically modified, for example by fluorination (He, Patankar, & Lee, 2003).

A droplet suspended in a gas phase is subjected to two forces which determine its shape: The surface tension, which acts to minimize the surface area, i.e., making the droplet spherical; and the gravitational force, which acts to flatten it. The effect of gravity can be neglected as long as the size of the water droplet is smaller than the so-called capillary length (Kim, 2008)

$$k_c = \sqrt{\frac{\gamma LA}{\rho g}} \tag{1}$$

Here, γLA is the surface tension of the liquid-air interface, ρ is the liquid density, and g is the acceleration of gravity. At ambient conditions, the capillary length of clean water is 2.7 mm (Kim, 2008). When a water droplet is brought in contact with a perfectly smooth and chemically homogeneous surface, a part of the solid-air interface is substituted with a solid-liquid interface of the same area. The energy of the solid-liquid air system is now reduced by an energy given by the Dupre equation (Bhushan, 2011b),

$$W_{SI} = \gamma SA + \gamma LA - \gamma SL \tag{2}$$

where W_{SL} is the work of adhesion per unit area, and γSA, γLA and γSL are the solid-air, liquid-air and solid-liquid interface energies, respectively. The shape of the water droplet changes due to the introduction of the new interface. In the case where the solid-liquid interface energy is lower than the solid vapor interface energy, the droplet will tend to spread spontaneously on the surface. If the solid-liquid interface energy is higher than the solid-vapor interface energy, on the other hand, then spontane-

Figure 1. The contact angle θ_Y and all the surface tensions in a system consisting of a water droplet resting on a smooth surface

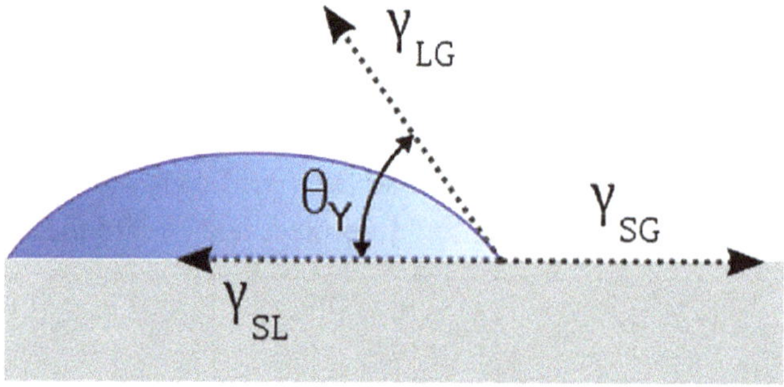

ous spreading will not occur. Instead, the droplet will bead up and assume a more spherical shape. The angle that the outline tangent of the droplet makes with the surface at the three-phase contact line is known as the contact angle (CA) of the system (see Figure 1).

On a smooth and chemically homogeneous surface, this contact angle is called Young's contact angle, usually denoted θ_Y or θ_0. When the size of the water droplet is less than the capillary length and when the volume and pressure are constant, then θ_Y can be pinpoint by diminishing the net surface tension of the three interfaces of the system (Bhushan, 2011b),

$$E_{tot} = \gamma LA\left(A_{LA} + A_{SL}\right) - W_{Sl}A_{Sl} \tag{3}$$

where A_{LA} and A_{SL} are the liquid-air and solid-liquid interface areas, respectively. In equilibrium

$$dE_{tot} = \gamma LA\left(dA_{LA} + dA_{SL}\right) - W_{Sl}dA_{Sl} = 0 \tag{4}$$

Using the geometrical relationship $dA_{LA} / dA_{Sl} = \cos\theta_Y$, (Bhushan, 2011b) and combining Equation 2 and 4, one obtains the well-known Young equation for the contact angle on a flat and chemically homogeneous surface (Bhushan, 2011b),

$$\cos\theta_Y = \frac{\gamma SA - \gamma SL}{\gamma lA} \tag{5}$$

This equation explains that a liquid droplet will spread on a surface when $\gamma SA - \gamma SL \geq \gamma lA$ $(\cos\theta_Y \geq 1,\ \theta_Y = 0)$, form a contact angle of 90° when $= \gamma SL(\cos\theta_Y = 0)$, and form a contact angle higher than 90° when $\gamma SA < \gamma SL(\cos\theta_Y < 0)$.

Surfaces that exhibit low SA are usually characterized by possessing few (or even no) polar surface groups (Barthlott & Neinhius, 1997). The CA is often used as a relative measurement of the hydrophobicity of a surface, i.e., of its tendency to repel water. A CA of 0° means that the surface is totally wetted, while a CA of 180° corresponds to complete non-wetting. Generally, the lower the surface tension

Figure 2. Schematic illustration of a droplet in the Wenzel state (Zheng et al., 2005)

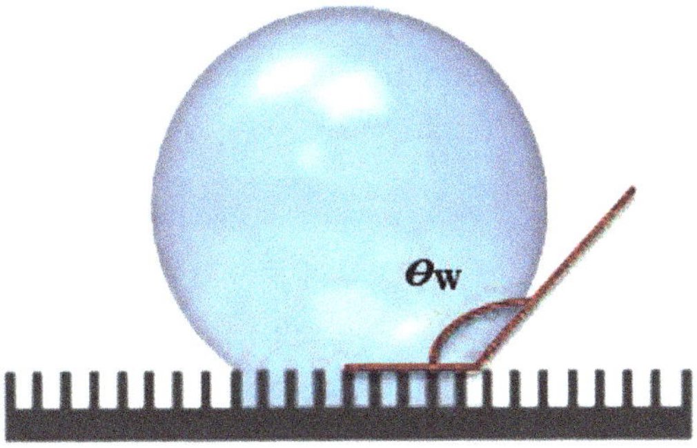

γSA, the higher the contact angle θ_Y of a surface. Surfaces with CA greater than 90° are classified as hydrophobic, whereas surfaces with CA less than 90° are referred to as hydrophilic (Gennes, et al., 2004). Subdividing even further: When the water CA is higher than 150°, the surface is said to be superhydrophobic. The highest water CA that can be obtained for poly(tetrafluoroethylene) or fluorine-terminated organic surfaces is ranged between 115 to 120° (S. H. Kim, 2008). In other words, surface chemistry alone is not sufficient to create superhydrophobic surfaces. Superhydrophobicity can be achieved, however, by combining an "intrinsically" hydrophobic material with an appropriate surface texture or roughness.

The Wenzel State

Young's equation cannot be applied when the surface is rough. In this case, the surface tension is not the only factor affecting the wettability of the water droplet; the surface roughness also enters the picture. On a rough surface, the deposited water droplet faces the alternatives of two main configurations, corresponding to two different equilibrium states of the droplet (Hirvi & Pakkanen, 2007). The first possibility that will be discussed is for the droplet to conform to the contours of the solid surface, i.e., "wetting" the grooves of the surface. This is called the Wenzel state, named after Robert N. Wenzel who outlined the principles of this model (Wenzel, 1936). The model is schematically depicted in two dimensions in Figure 2, where the surface roughness consists of regularly arranged square pillars.

Wenzel argued that, since the actual solid-liquid contact area is greater under the droplet if the surface is rough than if it is smooth, then a surface that is spontaneously wetted when it is smooth will be wetted more rapidly when the surface is rough, as there is a greater net energy decrease to induce wetting. In the same way, surfaces for which the solid-liquid interface energy is higher than the solid-air interface energy i.e., surfaces that are intrinsically water-repellent will be even harder to wet when the surface is rough. The equilibrium condition for a water droplet in the Wenzel state is given by (Wenzel, 1936),

$$\cos\theta_W = R_f \cos\theta_Y \tag{6}$$

where, θ_W is the apparent contact angle in the Wenzel theory, θ_Y is the Young CA for the surface of the same material, and R_f is the roughness factor which is defined as the ratio of the actual solid-liquid area A_{SL} to its projection on a flat plane A_F along the surface. The value of R_f on a rough surface must hence be larger than one.

Considering a surface with a roughness made up of a regular array of square cross-sectional pillars of length a, periodic inter-pillar spacing b, and height H. For such surface, the roughness is given by $R_f = 1 + (4aH) / -(a+b)^2$. The Wenzel equation can then be written as shown in equation (7) (Jiang & Feng, 2010),

$$\cos\theta_W = \left(1 + \frac{4A}{a / H}\right)\cos\theta_Y \tag{7}$$

where the parameter A is given by $R_f = 1 + ((b / a) + 1)^2$

It may be observed that water droplets in the Wenzel state will tend to amplify the "intrinsic" behavior of the surface. Whereas, hydrophilic surfaces ($\theta_Y > 90°$) will become more hydrophilic, and hydrophobic surfaces ($\theta_Y > 90°$) will become more hydrophobic. The Wenzel equation yields the Young equation (5) when $R_f = 1$. The dependence of the CA on the roughness factor R_f is schematically illustrated in Figure 3.

It may be concluded that, the measured CA decreases with increasing roughness factor R_f for the hydrophilic surfaces and increases for the hydrophobic surfaces, which is exactly what the Wenzel model predicts. The measured contact angles of water droplets of 5 μL large deposited on both hydrophilic and hydrophobic surfaces made from poly(methyl methacrylate) (PMMA), are shown in Figure 4 [24].

Figure 3. CA values for different Young contact angles θ_0 as a function of the roughness factor R_f according to Wenzel's model (Grewal et al., 2014)

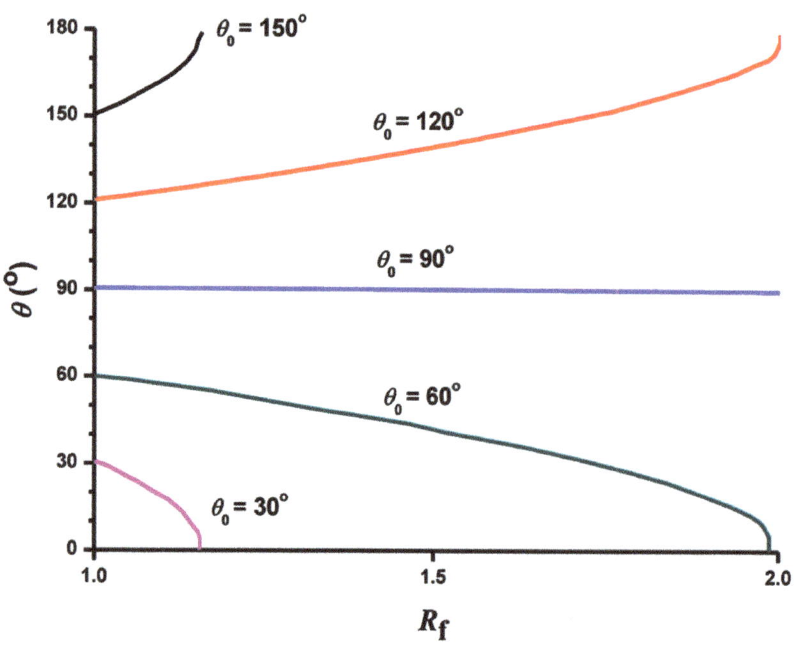

Figure 4. The measured contact angles on hydrophilic and hydrophobic surface (Grewal et al., 2014)

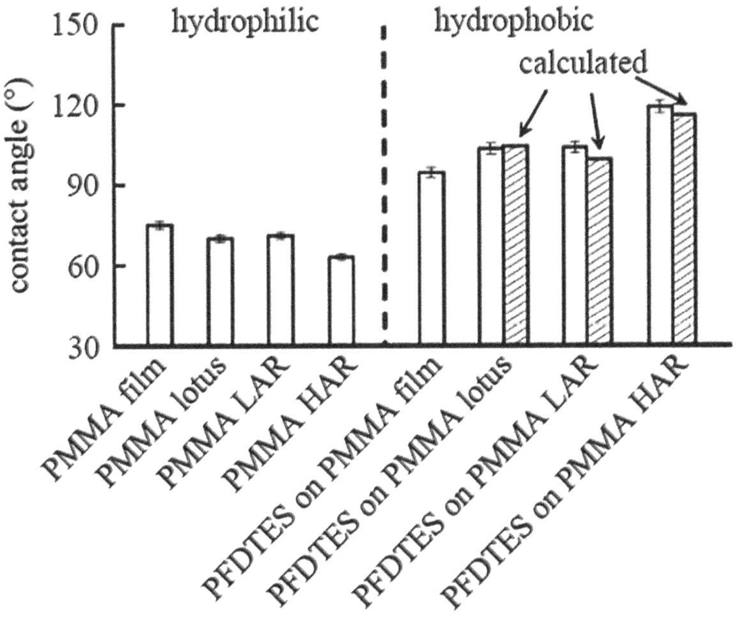

PMMA is a polymer that has polar (hydrophilic) groups with high surface energy, but the surfaces were made hydrophobic by coating them with a hydrophobic self-assembled monolayer (SAM) of peruorodecyltriethoxysilane (PFDTES). In addition to flat PMMA films, ordered nanostructured surfaces with asperities of both low and high height-to-diameter aspect ratio (LAR and HAR) were processed using soft lithography, as well as a replica of the lotus leaf microstructure with randomly distributed asperities (see Figure 5). These surfaces had roughness factors R_f of 2.1, 5.6 and 3.2, respectively.

The Cassie-Baxter State

The second possibility on a rough surface is for the water droplet to be suspended on summit of the surface asperities, trapping air in isolated cavities below itself. This condition is known as the Cassie-Baxter state (Cassie & Baxter, 1944). It is sometimes also known as the "fakir state" because the drop sits on a bed of micro-nails, as shown in Figure 6.

When a droplet comes to rest at a surface in the Cassie-Baxter state, a solid-air area is replaced by a solid-liquid area. Cassie and Baxter denoted f_1 and f_2 for the liquid-solid and the liquid-air fractional area under the droplet, respectively, where $(f_1 + f_2 = 1)$. They observed that in the process of spreading a water droplet on rough surface, an energy $f_{1\gamma SA}$ is gained in destroying the solid-air interface, while the energies f_1LS and f_2LA are expended in forming of solid-liquid and liquid-air interfaces, respectively (Cassie & Baxter, 1944). The net energy can be expressed by Eq. 8.

$$E_{net} = f_1\left(\gamma LS - \gamma SA\right) + f_2\gamma_{LA} \tag{8}$$

Figure 5. SEM micrographs of (a) two nano-patterned surfaces shown at two magnifications; (i)(ii) PMMA low-aspect-ratio (LAR) and (iii)(iv) PMMA high-aspect-ratio (HAR). (b) PMMA surface with a replica of the microstructure of a lotus leaf (Grewal et al., 2014)

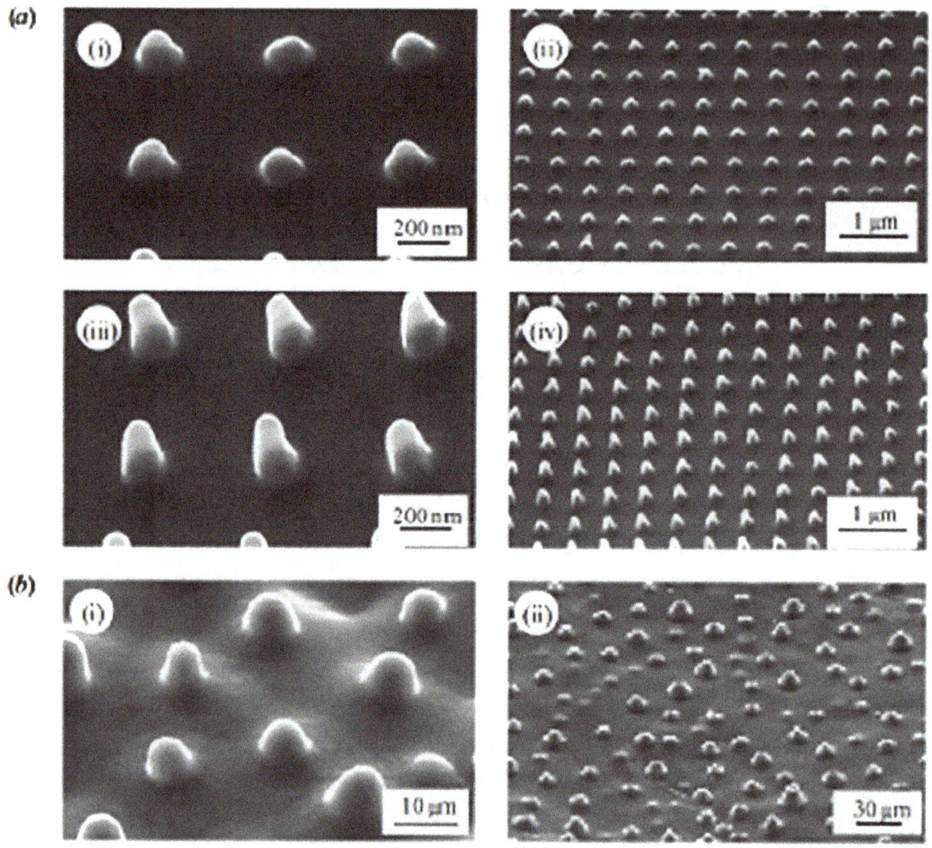

Figure 6. Schematic illustration of a droplet in the Cassie-Baxter state (Carré and Mittal, 2009)

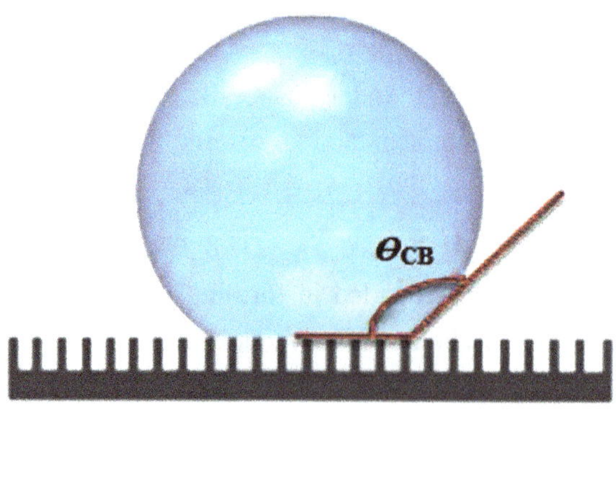

By substituting Equation 5 into Equation 8, the net energy is shown in Equation 9.

$$E_{net} = \gamma LA\left(f_2 - f_1 cos\theta_Y\right) \tag{9}$$

Since $\gamma LS - \gamma SA$ is the expended energy E in forming the new solid-liquid interface between a droplet and at surface, equation 5 can also be written

$$cos\theta_Y = \frac{-E}{\gamma_{LA}} \tag{10}$$

For a droplet in the Cassie-Baxter theory on a rough surface, one can exploit the following analogy (Cassie & Baxter, 1944),

$$cos\theta_{CB} = \frac{-E}{\gamma_{LA}} = f_1 cos\theta_Y - f_2 \tag{11}$$

where, θ_{CB} is the Cassie-Baxter contact angle, and θ_Y, is the "intrinsic" contact angle on a flat surface with the same chemistry, i.e., Young's CA. Equation 11 is known as the Cassie-Baxter equation. When air is not trapped in the valleys between the surface asperities, i.e., when $f_2 = 0$. A slightly more complex version of the Cassie-Baxter equation is given by Equation 12 (Roach, et al., 2008),

$$cos\theta_{CB} = r_f f_1 cos\theta_Y - f_2 \tag{12}$$

This equation applies when the tops of the surface asperities are rounded. The new factor r_f denotes the roughness value of the wet area.

When a droplet rests on summit of the surface asperities, the surface tension is reduced, and this leads to an increase in the CA. The higher the liquid-air fraction f_1 of the area under the droplet, the higher the CA according to the Cassie-Baxter model. This is true also for surfaces that are intrinsically hydrophilic. Like the Wenzel model, the Cassie-Baxter model can be modified to describe the three-dimensional pillared structure, with $f_1 = a^2 / (a+b)^2$ and $A = 1/((b/a)+1)^2$ the modified equation can be expressed as below:

$$cos\theta_{CB} = \mathrm{A}\left(1 + cos\theta_Y\right) - 1 \tag{13}$$

Equation 13 shows that highest value of aspect ratio pillars (i.e., high values of b/a for pillars of constant height H and intrinsic contact angle θ_Y) are beneficial for attaining large contact angles.

Mixed and Meta Stable States

To explain water contact angles that correspond neither with those predicted by Wenzel nor with those predicted by Cassie and Baxter, Zheng et al. (Zheng, Yu, & Zhao, 2005), proposed a model for a mixed wetting mode, where only a part of the water air interface has been able to reach down to the bottom of the valleys between the surface asperities:

$$cos\theta_M = xcos\theta_W + \left(1 - x\right)cos\theta_{CB} \tag{14}$$

where, x is the fraction of the area that has transitioned to the Wenzel state.

Intensive work has been conducted over the last years to identify the different wetting-mode transition mechanisms (Afferrante & Carbone, 2014; Grewal, Cho, Oh, & Yoon, 2014; Liu et al., 2014; Lv, Xue, Shi, Lin, & Duan, 2014). In general, both the Cassie-Baxter and the Wenzel states represent energy equilibrium positions. The Wenzel state represents a universal energy minimum, while Cassie-Baxter state represents a local minimum. Therefore Cassie-Baxter state is only a meta-stable position (He, et al., 2003). In all transitions between different energy minima, the interface between the droplet and the surface must overcome an energy barrier. The meta stable, intermediate, state depends on a number of factors such as pressure, vibration, humidity, texture, droplet size, surface roughness and how the water droplet is transferred to the surface (Kim, 2008). For example, when water droplets are gently deposited on the pillar tops of a superhydrophobic surface, the energy barrier is often too large to be overcome, and the surface will remain in the Cassie-Baxter state (the local energy minimum). On the other hand, the Wenzel state is more likely to materialize if the water droplet is produced by vapor condensation, or is impinged on the structured surface with kinetic energy and/or is subjected to vibration or electrical energy (Roach, et al., 2008).

Although the Wenzel state is generally expected to occur on surfaces with low to moderate roughness and the Cassie-Baxter state is more likely to appear on highly rough surfaces, there is no definite limit by means of which to separate the two states. The two states have indeed been observed to coexist on hydrophobic surfaces of moderate roughness $\left(R \approx 2\right)$ (Carré & Mittal, 2009).

Contact Angle Hysteresis and Critical Sliding Angle

When the surface on which a water droplet sits is tilted, its CA will increase and decrease on the downhill side and the uphill side, respectively. The CA on the downhill side is called the advancing CA $\left(\theta_A\right)$, while the CA on the uphill side is known as the receding CA (θ_R) (see Figure 7). When the tilt angle reaches a critical angle, the droplet will lose its grip and start moving down the plane. The difference between θ_A and θ_B just before the water droplet begins moving is termed the contact angle hysteresis (CAH) (Jiang & Feng, 2010). This value gives a measure of how "sticky" the surface.

The sticky of water droplet to the surface will increase as the CAH values increases, due to the presence of high energy barrier which resists the follow of droplet on the surface (Roach, et al., 2008). The tilt angle at which the droplet starts moving is called the critical angle. Equation 15, represents the relationship between tilt angle and CAH, which was proposed by Furmidge (Furmidge, 1962).

Figure 7. Advancing (θ_A) *and receding* (θ_R) *contact angles obtained by tilting the surface at the critical angle* α

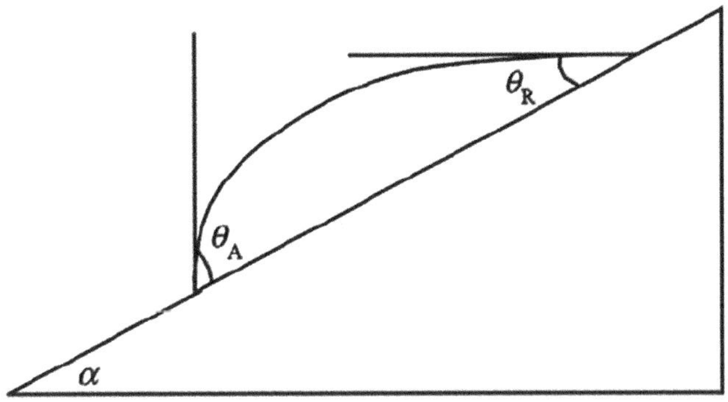

$$mg\frac{\sin\alpha}{\omega}=\gamma LV\left(\cos\theta_A-cos\theta_A\right) \tag{15}$$

where, m is the mass of the droplet, and ω is the width of the droplet perpendicular to the direction of motion.

Water droplets can move on a surface either by sliding, rolling, or a mixed propagation (Gao & McCarthy, 2009). The mechanism of propagation, which is consistent with the no-slip boundary condition of fluid mechanics (Öner & McCarthy, 2000), means that the energy barrier for advancing is much lesser than that for receding. To advance, the front molecules of the droplet simply need to descend onto the immediate surface ahead of it. To recede, on the other hand, the molecules at the back must disjoin from the surface, something which requires a lot more energy. This energy imbalance is believed to explain the CAH of droplets in the Cassie-Baxter theory. It also explains why droplets in the Cassie-Baxter state on rough surfaces have a smaller contact angle hysteresis than do droplets on smooth surfaces, because in the Cassie-Baxter state the solid-liquid contact area is smaller and therefore the energy barrier for receding is lower. Water droplets in the Wenzel state usually have a high water CAH, which can be explained by water droplets getting trapped in the surface texture. The droplets thus stick much better to the substrate than do droplets in the Cassie-Baxter state (Kim, 2008).

Generally, a decrease in surface roughness R_f will decrease the CAH for droplets in the Wenzel state, while a smaller CAH value can be achieved for droplets in the Cassie-Baxter State by minimizing the liquid-solid fractional interface (Bhushan, Jung, & Koch, 2009).

For water-repellent surfaces, the tilt angle should be as low as possible. Theoretically, if the CAH is equal to zero, the droplet will just slid without dissipating energy as soon as the surface is tilted just a little. In practice, however, there will always be some hysteresis due to friction caused by roughness and heterogeneity of the surface, but by carefully controlling the roughness on the micro- and nanoscale it is possible to achieve CAHs as low as 1° (Bhushan, et al., 2009).

Figure 8. (a) SEM micrographs of the surface of a lotus leaf at three magnifications. (b) A water droplet resting on the surface of a lotus leaf (Grewal et al., 2014)

SUPERHYDROPHOBICITY CONDITIONS

Smooth surfaces of many materials are known to exhibit "intrinsic" hydrophobicity with high CAs, e.g. fluorinated polymers and silicon resins (Kim, 2008). Thankfully, nature has already paved the way and provided us with many materials with superhydrophobic properties that we can learn from. Among the creatures in nature that exhibit extreme water-repellent properties, the self-cleaning lotus plant (Nelumbo nucifera) is probably the most famous example. Neinhuis and Barthlott (Barthlott & Neinhius, 1997) reported how the lotus leaf owes its superhydrophobicity to a surface with dual-scale roughness. The surface consists of micro-asperities formed by so-called papillose epidermal cells in addition to nano-protrusions introduced by a low-energy (hydrophobic) tubular epicuticular wax (see Figure 8).

Water droplets on lotus leaves are in the Cassie-Baxter state, which is clearly demonstrated by measured static CAs as high as 164° (Bhushan, 2011a). Water droplets on lotus leaves bead up and form spherical droplets, and roll easily off leaves that are slightly inclined. As water droplets roll down the leaves, they tend to drag dust, insects and contamination particles along on their way, leaving behind a clean and dry surface (see Figure 9) (Jiang & Feng, 2010). This unique property is referred to as self-cleaning. Due to its ability to remain clean in murky ponds, many cultures revere the lotus plant as a symbol of cleanliness and purity.

Pond skaters (Gerris remigis) are another species that have remarkable superhydrophobic properties. They are able to stand and walk stably on water without breaking through the surface, even in heavy rain and rushing currents. Careful investigation of these freshwater bugs has revealed a micronano hierarchical structure on their legs, consisting of nano-grooves superimposed on tiny microsetae (hairs) (Feng, Gao, Wu, Jiang, & Zheng, 2007). Figure 10 (b and c) show the microstructure and nanostructure of pond skater leg, respectively. This topography in combination with a secreted hydrophobic cuticle wax is the secret behind the pond skater's ability to resist water.

Figure 9. The movement of a water droplet down a tilted smooth (a) and a rough (b) surface. On the smooth surface, the droplet redistributes the particles. On the rough surface, the particles adhere to the surface of the water droplet as it rolls down, and they are entirely removed from the surface (Dewidar et al., 2008)

The same conclusion can be drawn after investigating several superhydrophobic materials in nature, including plants (Karamana, Çabuk, Özyurt, & Köysürena, 2012), insects (Hasan, Raj, Yadav, & Chatterjee, 2015), and animals (Darmanin & Guittard, 2014). The best water-repellency is obtained with a combination of a hydrophobic chemistry and a hierarchical topography in which a nanostructured texture/roughness is superimposed on a microstructure. Contact angles larger than 150° can be induced on surfaces comprising a microscale only or a nanoscale only texture, but the CAH (and the critical sliding angle) will be quite large, often higher than 30° (Jiang & Feng, 2010). To obtain a surface with both θ_A and θ_B greater than 150° in addition to low CAH, a complex hierarchical structure, as shown in Figure 11.

Such surface can reduce the critical sliding angle of a same-sized drop by a factor of ten (Jiang & Feng, 2010). This hypothesis was confirmed by Bhushan et al. (Bhushan, et al., 2009), who fabricated flat, nanostructured, microstructured and hierarchically structured surfaces covered with n-hexatriacontane, and studied the influence of the different structures on the static CA and CAH. SEM micrographs of the surfaces are shown in Figure 12.

Figure 10. (a) A pond skater resting on water. (b) SEM image of a pond skater leg covered with micro-sized hairs. (c) SEM image of nano-grooved structure on the hairs (Zheng et al., 2005)

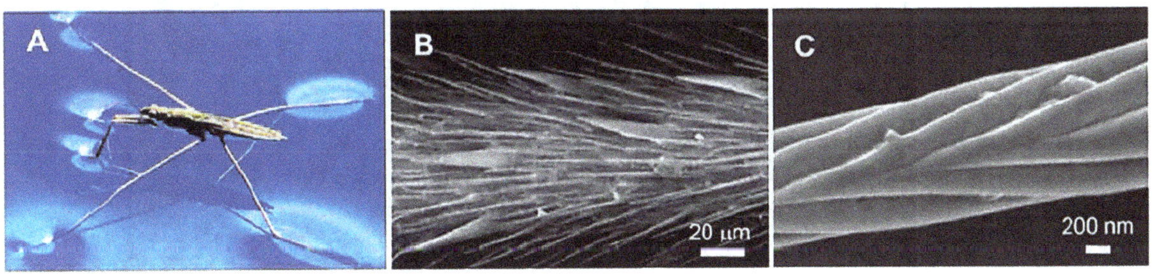

Figure 11. a) A simple microtextured surface. b) A complex nano-micro hierarchically textured surface (Afferrante and Carbone, 2014)

Figure 12. SEM micrographs in two different length scales of a flat (a), a nanostructured (b), a micro-structured (c) and a hierarchically structured (d) surfaces covered with 0.2 µg mm-2 of n- hexatriacontane (Grewal et al., 2014)

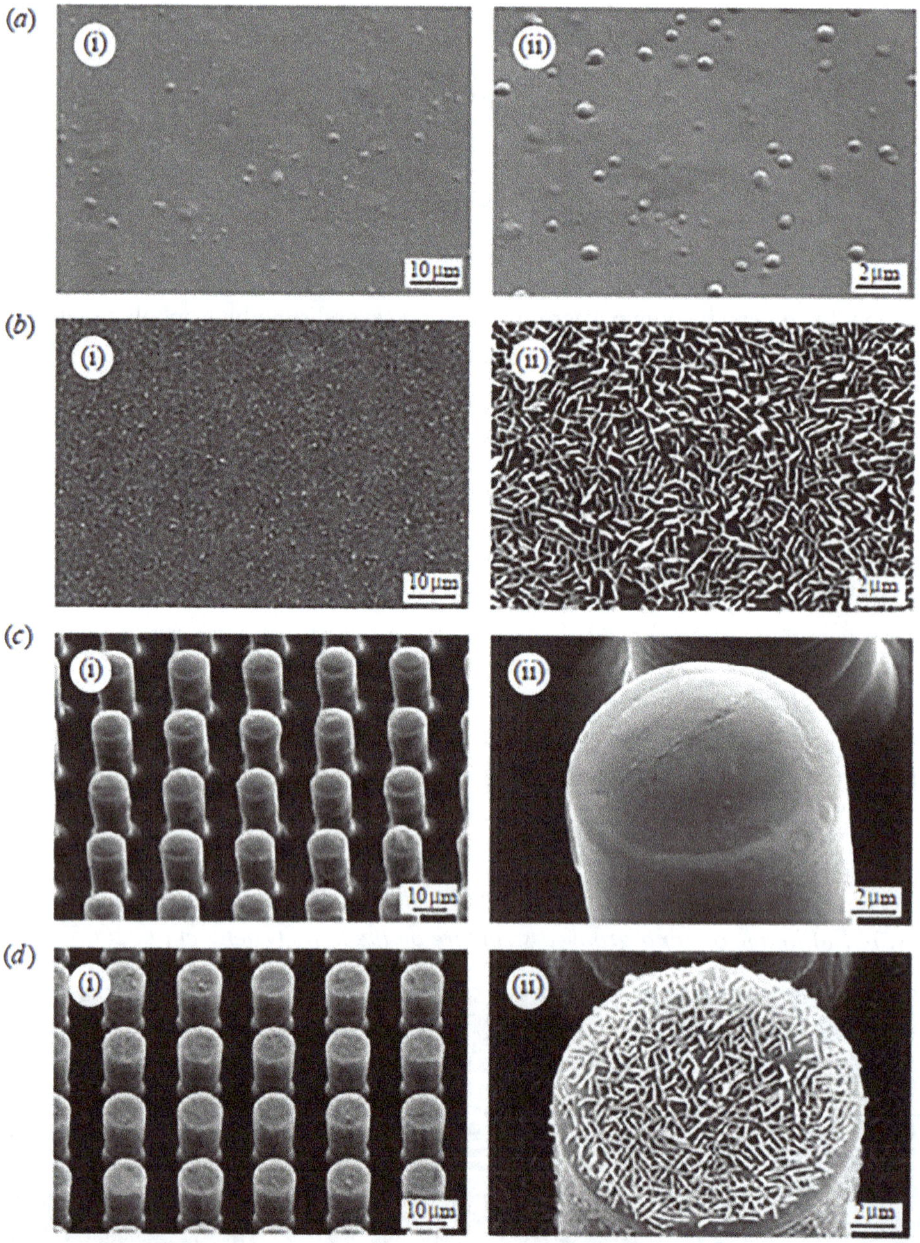

Figure 13. Bar charts showing the measured values of CA and CAH on four different surfaces, in addition to calculated values using both the Wenzel and the Cassie-Baxter equation (Grewal et al., 2014)

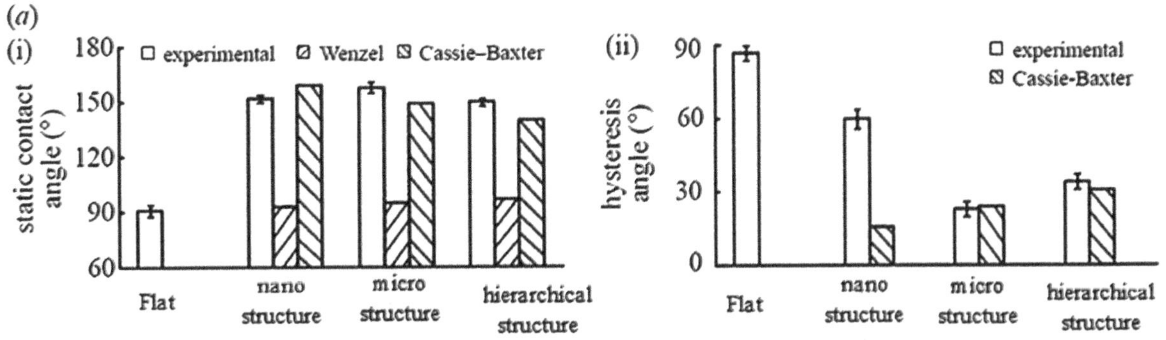

The roughness factor R_f was calculated based on geometrical considerations (for the microstructure) and measured using an AFM (for the nanostructure). For the nano, micro and hierarchical structure the roughnesses were 4.9, 3.5 and 8.4, respectively. The corresponding values of f were 0.85, 0.71 and 0.96.

The measured values of the static CA and the CAH are shown graphically in Figure 13, together with calculated values according to the Wenzel and Cassie-Baxter theories.

Even if the results suggest that all the three structured surfaces were in the Cassie-Baxter state, the hierarchical structure performed the best by far, having both the largest CA and the lowest CAH. With a measured CA of 169° and a CAH of 2°, it was the only surface to satisfy the requirements for superhydrophobicity. These values are actually superior to CA and CAH values measured on the lotus leaf. The runner-up in this "race" was the nanostructure, followed by the microstructure. The flat surface finished last. The tilt angle and the adhesion force also proved to be significantly lower on the hierarchically structured surface. A microstructure with a superimposed nanostructure ensures that the mechanical structure and the superhydrophobicity are stable. According to Nosonovsky and Bhushan (Bhushan, et al., 2009), the function of the micro-roughness in a dual-scale texture is to resist capillary waves by providing mechanical stabilization. The microstructure allows the nano-scale pillars to have lower height and lower density than if the nanopillars were to yield the same apparent contact angle alone. The nano-roughness functions to prevent liquid from entering the gaps between the surface asperities by pinning the liquid droplets, in addition to supporting nano droplets which may otherwise condense between the high asperities.

PREPARATION TECHNIQUES OF SUPERHYDROPHOBIC MATERIALS

Published research on microstructured hydrophobic surfaces has focused almost exclusively on flat surfaces, but many surfaces that could benefit from superhydrophobicity are curved such as an airplane wing or ship hull. Hydrophobic materials have low surface energy; hence a combination of a rough surface applied to a hydrophobic material can be used to produce a superhydrophobic material. Generally methods to prepare structures for superhydrophobic surfaces include: (i) preparation of a rough surface by etching of hydrophobic substrates or hydrophilic substrates followed by a surface hydrophobization treatment, or (ii) preparation of a rough surface by coating techniques using either hydrophobic materials or hydrophilic materials followed by hydrophobic treatment/coating. There are several ways of generating superhydrophobic materials based on low surface energy and roughness. Some of these methods will be elaborated in the following subsections:

Lithography Technique

Lithography is a method to transfer patterns from a mask onto a substrate surface or film. Photolithography is useful for generating well-defined patterns in a surface, e.g. pillars. Therefore, model surfaces can be created for the investigation of fundamental effects, such as the relationship between surface structure shape, geometries and superhydrophobicity (high contact angle and low sliding angle/roll-off angle/hysteresis). The idea is to expose a photoresist (a photoactive polymer) to radiation through a photomask, and then remove the exposed or the unexposed parts, depending on the nature of the polymer used. The patterned polymer can then either be used directly, or be utilized like a mask for deposition or etching. Different radiation sources can be used, such as UV, X-rays, electrons, and ions. Often it is necessary to apply a hydrophobic coating to generate superhydrophobicity. A promising candidate for the fabrication of superhydrophobic surfaces is nanoimprint lithography (NIL). NIL has several major advantages such as high throughput, high resolution, and low cost relative to other lithography techniques (photolithography, electron beam lithography, scanning probe lithography, etc.). Among the most interesting approaches is to directly use the dual-scale roughness of surfaces in nature such as plant leaves, insects, lichen, mould and fungus as stamps. For this purpose, Zhang et al. (Zhang, Zhang, Xie, Liu, & Shao, 2006), used the wings of the insect cicada. In addition to being abundant in nature, these insects have wings with a low surface tension which renders the use of an anti-adhesive layer unnecessary and the material can have a Young's modulus as high as 7-9 GPa. This is sufficient for imprinting polymers like PMMA (Zhang, et al., 2006). The first step is to form a nanowell array in the polymer by pressing the pillared cicada wing surface into a PMMA support at an elevated temperature. Subsequently, the pattern can either be transferred to an underlying silicon substrate using reactive ion etching (RIE), or it can be used as a mold to fabricate a patterned structure of another material.

Sol Gel Technique

For nanoparticle films, the surface wetting stability can depend on the specific chemistry of the sol-gel used as the source material. Sol-gel processes can produce rough surfaces on a variety of oxides such as silica, alumina, and titania. For example, superhydrophobic silica aerogels can be prepared by surface modification of standard silica gels with a heavily fluorinated silyl chloride followed by supercritical fluid solvent removal (Roig et al., 2004). Tadanaga et al. reported that the roughness of porous alumina gel films can be varied between 20 to 50 nm. As a result, high optical transparency can be achieved. After a surface hydrophobic treatment, a superhydrophobic surface with a contact angle of 165° can be achieved (Tadanaga, Katata, & Minami, 1997).

Electrospinning Technique

Electrospinning has become a popular method to generate continuous ultrathin fibers with micrometer and sub-micrometer diameters from a variety of polymeric materials. Electrospinning involves the use of an electrostatic force to draw thin (from nanoscale to microscale) fibers with a viscous polymer liquid on a substrate. Water CAs as high as 172° and very low CAH have been reported for such surfaces when a highly hydrophobic polymer is used (Roach, et al., 2008). By changing the process parameters such as applied voltage, distance between needle tip and collector, viscosity of polymer, the needle diameter, flow rate and temperature, fibers can adapt shapes ranging from uniform cylinders to globules, and this makes it possible to fabricate surfaces with a dual-scale roughness. Although some polymers are hydrophobic enough to form superhydrophobic surfaces, additional coatings are often necessary. For this purpose, fluorine-based materials are frequently used. The advantage of this process is the uniform distribution of nano-particles during process. The increase the concentration of nano particles lead to plug the need tip which considered a disadvantage during electrospun of polymer / nano-particle coating.

Plasma Techniques

Plasma techniques are a relatively simple method used for preparing superhydrophobic surfaces with high roughness and low surface energy. Plasma polymer vapor polymerization process is used to create superhydrophobic surfaces of PTFE coatings. Under appropriate plasma processing conditions (plasma etching for a sufficient time to generate the appropriate surface roughness), superhydrophobic and simultaneously transparent surfaces can be produced, which is of great practical importance and have several applications in food packaging (e.g. PE films) and biomaterials.

Electrochemical Deposition Technique

Generally, superhydrophobic coatings can be fabricated by electrochemical deposition through two steps; first, through preparing a surface with a micro/nanostructured roughness and then followed by modifying that surface using low-surface-energy materials (Wang, Li, She, Chen, & Li, 2012). However, Liu et al.

(Liu, Chen, & Kang, 2015) prepared a superhydrophobic coating on a magnesium alloy using a one-step cathodic electrochemical deposition. Zinc oxide conductive surface structures were also prepared by cathodic electrochemical deposition from aqueous solutions of $ZnCl_2$/KCl (Li et al., 2003). Wang et al. (Wang, Guo, & Dong, 2008) reported that using $Cu(NO_3)_2$ aqueous solutions leads to a formation of hierarchical spherical cupreous microstructures on an indium tin oxide substrate.

Phase Separation Technique

If a stable multi-component liquid mixture becomes unstable due to cooling or a change in pressure, then the phases will begin to separate. A bicontinuous network is formed, and if one of the phases becomes solid in the process (e.g. block copolymer films), then the remaining liquid phase can be removed. If the resulting wetting properties are not good enough, superhydrophobicity can sometimes be induced by applying an additional coating. Phase separation is a cheap and simple way of producing superhydrophobic materials (Roach, et al., 2008).

Layer-by-Layer Assembly Technique

This technique is based on the alternating physisorption of oppositely charged building blocks. By utilizing the attraction between electrostatically charged particles, one can form a multilayer structure by sequentially dipping a substrate in a positive and a negative polyelectrolyte. For example, using the polymers polyallylamine hydrochloride (PAH) and poly(acrylic acid) (PAA) will generate a micro-porous surface that can be coated with silica nanoparticles to make it superhydrophobic (Roach, et al., 2008). Zhai et al. prepared a polyelectrolyte multilayer surface by Layer-by-Layer (LbL) assembly and then overcoated the surface with silica nanoparticles to mimick the hierarchical scale present on lotus leaf surfaces. Superhydrophobicity was achieved after surface fluorination with a fluoroalkyl silane (Zhai, Cebeci, Cohen, & Rubner, 2004).

Etching and Anodic Oxidation Technique

Narita et al. (Narita, Kasuga, & Kiyotani, 2000) fabricated superhydrophobic surfaces by using electrolytic etching to form micro-sized pits in an Al substrate, subsequently using anodic oxidation to generate nano-sized pits superimposed on the microstructure, and finally functionalizing the surface with a fluoroalkylsilane (Roach, et al., 2008). The same method can also be used on other metals, such as zinc, copper, and polycrystalline metals.

Candle Soot Technique

Deng et al. (Deng, Mammen, Hans, & Vollmer, 2012), developed a simple way to make robust, transparent and superhydrophobic surfaces. They first held a glass slide above the flame of a paraffin candle, so that it was covered by a layer of black candle soot. The SEM micrographs reveal that the soot layer consists of a fractal-like network made up of carbon particles with a typical diameter of 30-40 nm (Deng, et al., 2012). To make the structure more robust, CVD was used to deposit a 205 nm thick shell of silica on top of the soot layer. The next step consisted in calcining the carbon/silica network at 600º C for 2 h in air. This caused the carbon core to combust, thereby reducing the shell thickness, however the coat roughness and texture was retained. Lastly, to decrease the surface energy, the hydrophilic silica layer was covered with a layer of semi-fluorinated silane, again using CVD. The resulting surface demonstrated a water CA of 165°±1, and a tilt angle smaller than 1°. Since this type of superhydrophobic coating is transparent, it may be a promising candidate for implementations on glass surfaces for example windows, goggles and touch screens.

Metal Microcasting Technique

Embossing is a low-cost high throughput technique that can produce superhydrophobic microstructures in polymer, and microstructured metal can be used as an embossing master (Chou, Krauss, & Renstrom, 1996a, 1996b; Cross, 2006). Heating polymer above the glass transition temperature (T_g) softens the polymer, and applying force to the master allows the softened polymer to flow into the cavities of the master. Cooling the polymer below the T_g then hardens the polymer, and when the master releases the polymer it is molded to the shape of the master. Key difference between embossing and casting methods is that casting methods do not use applied force to the master. Instead, the material to be microstructured is poured onto the microstructured master. Unlike paint, spray, or plasma coatings that can render surfaces more hydrophobic by creating disordered structures, embossing processes can fabricate geometrically anisotropic ordered three dimensional micro and nanostructures capable of directing the motion of liquids that interact with the ordered structures.

Laser Micro/Nano-Processing

Fabricating superhydrophobic surfaces by laser can be achieved by laser machining or lithography. These laser micro/nano-processing techniques can increase the surface roughness and enhance the hydrophobicity of many surfaces (Chen, Tsai, Liu, & Horng, 2015). Different kinds of laser beams have been used. For example;

1. **Laser Ablation:** A single beam process for eliminating a substance from a solid surface. The interaction mechanisms between the laser light and matter relies on the properties of the laser beam in addition to the chemical and physical properties of the material. In laser ablation, the heat affected zone causes modifications of the surface morphology and/or the material's properties which might affect the subsequent laser ablation performance. Therefore, using a picosecond or femtosecond laser can greatly reduce the area and the temperature of the heat affected zone (Hirayama & Obara, 2005).
2. **Laser Multi-Beam Processing:** The main disadvantage of the laser ablation technique was its low throughput. A multi-beam processing was developed to solve this problem. Steger et al fabricated hierarchical structure using direct laser writing to achieve microstructures and multi-beam interference to generate nanostructure on 3D surfaces (Steger, Hartmann, Beckemper, Holtkamp, & Gillner, 2013).

Injection Molding

Injection Molding technique is suitable for surface nanostructured plastic products. This technique involves a replicating process where identical parts are produced with mold and a master pattern engraved on the surface of the mold. The injection molding process can be done through four steps: filling, packing, cooling and releasing. In the filling and packing steps, the thermoplastic melt is injected and pressured into the mold cavity, while in the releasing step, the solidified nanostructured substrate is mechanically or chemically removed after a period of cooling (Giboz, Copponnex, & Mélé, 2007). Hopmann et al. (Hopmann, Behmenburg, Recht, & Zeuner, 2014) used design of experiment (DOE) to investigate the effect of different parameters of the injection molding technique, such as melt and mold temperatures and injection speed, on the possibility of producing superhydrophobic liquid silicone rubber surfaces. They concluded that the injection molding process can offer the merits of being able to produce complex microstructured parts in a single-step process.

APPLICATIONS OF SUPERHYDROPHOBIC MATERIALS

There are several applications for superhydrophobic materials in science and engineering. Most of these applications use the high water repellent, especially of architecture, vehicle windows, eyeglasses, biomedical, corrosion resistance, anti-biofouling, and optical windows for electronics (Gau, Herminghaus, Lenz, & Lipowsky, 1999; Liu, Yin, Chen, Chang, & Cheng, 2007; Samuel et al., 2005). Self-cleaning behavior can also be realized with superhydrophilic surfaces, though these approaches often attempt to utilize the photocatalytic properties of TiO_2 (Ferrara, Pilloni, Mazzarelli, & Tapfer, 2010). Liu *et al.* developed SiO_2/TiO_2 bilayer films that were both self-cleaning and antireflective (Liu, Zhang, Murakami,

& Fujishima, 2008). One of the more unique applications for modified surface wetting is in the area of prosthetic dentistry (Rupp et al., 2006). Oil or surfactant repellent or superoleophobic surfaces may also be desirable which add more constraints to superhydrophobicity design. As coatings improve and prices fall, use of superhydrophobic surfaces is likely to increase, particularly as understanding of many of the limitations is developed and perhaps mitigated by careful selection of coatings.

Anti-Adhesion and Self-Cleaning

Superhydrophobic surfaces with contact angles higher than 150° and small contact angle hysteresis have self-cleaning and low adhesion properties that can be exploited in many applications e.g. self-cleaning glass windows, textile industry and optics. The self-cleaning and anti-adhesion properties of a superhydrophobic film were studied by Nimittrakoolchai and Supothina using dust particles and red powder (Nimittrakoolchai & Supothina, 2008). In this study, the superhydrophobic coating was prepared on a glass substrate upon which a polyelectrolyte film was deposited followed by etching in $HClSiO_2$. Both of the red powder and the dust particles were removed in a quiet easier way without leaving remnants after dropping water on its surface in case of the superhydrophobic surface. In textile industry, stain-proof fabrics were produced on large scale (Xue, Jia, Zhang, & Ma, 2010). For optical applications, the self-cleaning properties were used in lenses, mirrors and solar panels that have to stay clean all the time to keep its efficiency as stable as possible (Nosonovsky & Bhushan, 2009).

Anti-Biofouling Applications

Biofouling is a critical issue in marine and many other industries. The growth of biological organisms to form colonies has many detrimental effects on the appearance, integrity and performance of many materials. Gudipati et al. and Townsin reported that biofouling of underwater structures and ships' hulls, in particular, increases the maintenance and operational costs (Gudipati, Finlay, Callow, Callow, & Wooley, 2005; Townsin, 2003). Biofouling can be resisted significantly using underwater superhydrophobic rough surfaces that can support an air film between itself and water (Marmur, 2006).The reduction in biofouling in this case is attributed to the reduction of the wetted area which minimizes the chances that microorganisms can stick to the solid surface. It is worthy to mention that the mechanical properties should be considered when these hydrophobic surfaces are designed. Zhang et al. studied the anti-fouling properties of superhydrophobic coatings (Zhang, Lamb, & Lewis, 2005). They revealed that while a normal surface was fouled within a day, superhydrophobic surfaces did not suffer from fouling after the first week of immersion in water.

Drag Reduction

Frictional drag (fluid resistance) is a macroscopic phenomenon that takes place when a liquid flows along a surface in a turbulent mode. It affects many applications e.g. the speed and efficiency of marine vessels, the cost of oil transportation through a pipeline, and many other engineering processes. The drag (resistance) is attributed to shear stress, the rate per unit area of momentum transfer from the flow to the surface. Superhydrophobic surfaces can be used to reduce the drag because of its high water repellent property. It can form a thin air film above an underwater surface which prevents the surface wetting. This air film formed at the interface between the fluid and the superhydrophobic surface has the capability to take in air supplied from outside because of the surface tension of the water. This application found its way to the ship building industry where superhydrophobic surfaces has much lower frictional drag compared to the conventional surfaces which act as air lubricant effect.

Corrosion Resistant Coatings

Corrosion is a spontaneous oxidation process of metals and/or alloys to form ionic species e.g. oxides and hydroxides. In order to take place, four conditions must coexist: (i) a metal or an alloy, (ii) an oxidant e.g. oxygen and/or protons, (iii) ionic conductor (electrolyte) and (iv) electronic conductor (the same metal or alloy). Absence of any of the aforementioned four elements will prevent corrosion from starting. One of the most common techniques to prevent corrosion is using organic coatings e.g. polyethylene that act as barriers to prevent the electrolyte to come in contact with the target metal or alloy. Superhydrophobic coatings have a more advantage over the conventional organic coatings because of its water repellence properties that increase the resistance for an electrolyte and hydrated aggressive ions to penetrate through the coating and come in contact with the metallic surface. Two mechanisms are explained to find out the reason behind the highly corrosion resistance properties of the super hydrophobic surfaces is ; First, the air retained on such superhydrophobic surfaces which can prevent aggressive ions, e.g., chloride ions in seawater from attacking the metal surface (Jeong, 2013; Li et al., 2015; Liu et al., 2007; Wang et al., 2015). Figure 14 summarizes this mechanism (Jeong, 2013). In addition, another mechanism was proposed in which seawater can be pushed out of porous superhydrophobic surfaces by the Laplace pressure as water transportation against gravity is easy in the porous structures (Liu, Chen, et al., 2007).

As a result, the coated substrate will be protected against corrosion even in an aggressive medium e.g. seawater. Mohamed et al. reviewed the use of superhydrophobic surfaces in the field of corrosion (Mohamed, Abdullah, & Younan, 2014). As examples, Kim et al., Barkhudarov et al., Li et al. and Vengatesh et al. studied different surface treatments to create superhydrophobic surfaces on Al and its alloys (Barkhudarov et al., 2008; Kim, Lee, Kim, & Kim, 2015; Vengatesh & Kulandainathan, 2015). Besides Barkhudarov et al. who used neutron reflectivity technique to evaluate their organosilica aerogel superhydrophobic coating, electrochemical impedance spectroscopy (EIS) was the most common used technique by the rest researchers in addition to open circuit potential (OCP) and linear polarization resistance. The latter technique can only be used when the coating is conductive or porous, so the electrolyte can find its way to the metallic surface. Liu et al. (2007 and 2009) for Cu and Zn metals (Li, et al., 2015; Liu, Szunerits, Xu, & Boukherroub, 2009; Liu, Chen, et al., 2007), Ishizaki et al. for Mg and its alloys (Ishizaki, Hieda, Saito, Saito, & Takai, 2010; Ishizaki, Masuda, & Sakamoto, 2011;

Figure 14. Basic concept of anti-corrosion using a superhydrophobic surface (Jeong, 2013)

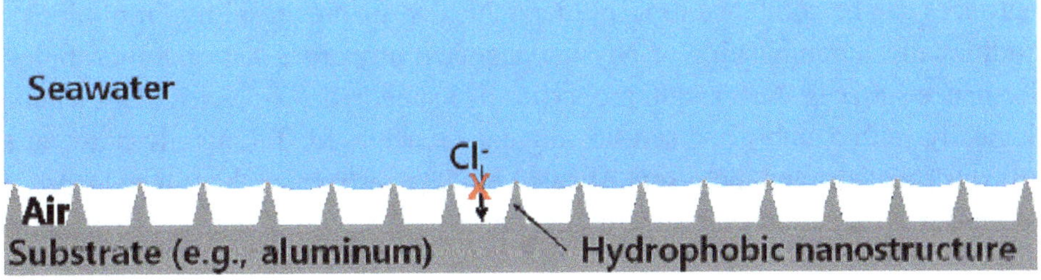

Ishizaki & Sakamoto, 2011) and de Leon et al. for C-steel (Leon, Pernites, & Advincula, 2012) used different coatings e.g. Myristic acid, perfluorodecyltrichlorosilane (PFTS), fluoroalkylsilane (FAS) and polythiophene respectively and evaluated them also using the polarization and EIS techniques. One of the critical drawbacks for the superhydrophobic coatings (sometimes besides their mechanical strengths) is the uncontrollable surface roughness, structural dimensions and shapes. Research work needs to be focused on these parameters in future.

Factors Affecting the Durability of Superhydrophobic Coatings

Fabrication of superhydrophobic coatings which retain their hydrophobicity and resist harsh operating conditions is a big challenge for successful applications. There are different factors that influence the durability of superhydrophobic coatings and transfer them from Cassie-Baxter regime (CB) to Wenzel state. For example, exposure to UV for certain time will result in photocatalytic decomposition of the superhydrophobic surface layers and turn them into hydrophobic ones. Lee et al. (Lee, Park, Kim, Jeong, & Kim, 2016) reported that the prepared PDMS-coated SiO_2 and N-TiO_2 superhydrophobic coatings with different ratios lose their hydrophobicity due to photooxidation processes. However, films that are composed of 60%PDMS-coated SiO_2 kept superhydrophobicity after exposure to UV for 48 h (Lee, et al., 2016).

In addition to UV exposure tests, superhydrophobic coatings should be evaluated and pass other tests in order to verify its chemical stability and mechanical durability such as immersion in saline water, effect of different pH (1-12), adhesion test, abrasion test, etc. Furthermore, other researchers studied the effect of the morphology and surface chemistry on the formation of durable superhydrophobic surface (Grewal, et al., 2014; Y. Liu & Li, 2012).

CONCLUSION

In this chapter, the fundamental theories of superhydrophobicity including Young model, Wenzel model, and Cassie–Baxter model were explained in details. The superhydrophobic property can be obtained by increasing the CA with more than 150° and decreasing the CAH with less than 10°. The superhydrophobic regime can present only when the substrate is hydrophobic. The formation of air pocket between surface grooves will decrease the water droplet surface tension and this will lead to increase the WCA. The increment in surface roughness leads to an increase in the hydrophobicity of the surfaces. Superhydrophobicity is critical to the survival of many insects. Inspired by superhydrophobic surfaces in nature, such as the lotus leaf, much effort have been devoted to developing superhydrophobic surfaces that are structured both at the micro- and nanoscale. The advantage of mimicking the biomaterials is that they can be highly organized from the molecular to the nano-, micro- and macro-scales. Furthermore, such surfaces can be combined to form hierarchical structures resulting in millions of different elements. Additionally, a combination of physico-chemical properties with the morphology can lead to high performance surfaces. As a result, properties like superhydrophobicity, self-cleaning, high adhesion, biologically self-assembly and self-healing can be achieved. Taking advantage of the Wenzel and Cassie-Baxter equations and more sophisticated models, researchers have developed a plethora of fabrication routes to create dual-scale superhydrophobic surfaces on a wide range of substrates, e.g., polymers, metals, wood, glass and textiles. Both bottom-up and top-down methods are used, including plasma and/or chemical etching, photolithography, lithography, phase separation and self-assembly of nanoparticles. Superhydrophobic protective coatings against corrosion are very promising for metals and alloys and their future is expected to flourish in the coming years.

REFERENCES

Afferrante, L., & Carbone, G. (2014). The effect of drop volume and micropillar shape on the apparent contact angle of ordered microstructured surfaces. *Soft Matter*, *10*(22), 3906–3914. doi:10.1039/c3sm53104j PMID:24643633

Barkhudarov, P. M., Shah, P. B., Watkins, E. B., Doshi, D. A., Brinker, C. J., & Majewski, J. (2008). Corrosion inhibition using superhydrophobic films. *Corrosion Science*, *50*(3), 897–902. doi:10.1016/j.corsci.2007.10.005

Barthlott, W., & Neinhuis, C. (1997). Purity of the sacred lotus, or escape from contamination in biological surface. *Planta*, *202*(1), 1–8. doi:10.1007/s004250050096

Bhushan, B. (2011a). Biomimetics inspired surfaces for drag reduction and oleo-phobicity/philicity. *Beilstein J. Nanotechnol.*, *2*, 66–84. doi:10.3762/bjnano.2.9 PMID:21977417

Bhushan, B. (2011b). *Nanotribology and Nanomechanics II*. Springer. doi:10.1007/978-3-642-15263-4

Bhushan, B., Jung, Y. C., & Koch, D. K. (2009). Micro-, nano- and hierarchical structures for superhydrophobicity, self-cleaning and low adhesion. *Phil. Trans. R. Soc. A*, *367*, 1631–1672. PMID:19376764

Carré, A., & Mittal, K. L. (2009). *Superhydrophobic Surfaces*. VSP. doi:10.1163/ej.9789004165939.i-496

Cassie, A. B. D., & Baxter, S. (1944). Wettability of Porous Surfaces. *Transactions of the Faraday Society*, *40*, 546–551. doi:10.1039/tf9444000546

Chen, Y. D., Tsai, W. J., Liu, S. H., & Horng, J. B. (2015). Hierarchical micro/nanostructures generated by femtosecond laser beams modified through a liquid-crystal spatial light modulator. *Applied Surface Science*, *353*, 568–573. doi:10.1016/j.apsusc.2015.06.152

Chou, S. Y., Krauss, P. R., & Renstrom, P. J. (1996a). Imprint lithography with 25-nanometer resolution. *Science,* 272, 85-87. doi:10.1126/science.272.5258.85

Chou, S. Y., Krauss, P. R., & Renstrom, P. J. (1996b). Nanoimprint lithography. *J. Vac. Sci. Technol., B,* 14, 4129-4133. doi:10.1116/1.588605

Cross, G. L. W. (2006). The production of nanostructures by mechanical forming. *J. Phys. D: Appl. Phys., 39*, R363-R386. doi:10.1088/0022-3727/39/20/r01

Darmanin, T., & Guittard, F. (2014). Recent advances in the potential applications of bioinspired superhydrophobic materials. *J. Mater. Chem. A*, *2*(39), 16319–16359. doi:10.1039/C4TA02071E

Deng, X., Mammen, L., Hans, J. B., & Vollmer, D. (2012). Candle soot as a template for transparent robust superamphiphobic coating. *Science*, *335*(6064), 67–70. doi:10.1126/science.1207115 PMID:22144464

Dewidar, M., Adamek, G., Jakubowicz, J., & AbdelRazik, K. K. (2014). Fabrication, Microstructure and Properties of Mechanically Alloyed and Electrochemically Etched Porous Ti-Y2O3. *International Journal of Electrochemical Science*, *9*, 7773–7783.

Ejenstam, L., Ovaskainen, L., Meizoso, I. R., Wågberg, L., Pan, J., Swerin, A., & Claesson, P. M. (2013). The effect of superhydrophobic wetting state on corrosion protection - The AKD example. *Journal of Colloid and Interface Science*, *412*, 56–64. PMID:24144374

Feng, X. Q., Gao, X., Wu, Z., Jiang, L., & Zheng, Q. S. (2007). Superior Water Repellency of Water Strider Legs with Hierarchical Structures: Experiments and Analysis. *Langmuir*, *23*(9), 4892–4896. doi:10.1021/la063039b PMID:17385899

Ferrara, M. C., Pilloni, L., Mazzarelli, S., & Tapfer, L. (2010). Hydrophilic and optical properties of nanostructured titania prepared by sol-gel dip coating. *J. Phys. D: Appl. Phys.*, *43*, 95301-95309. doi:10.1088/0022-3727/43/9/095301

Furmidge, C. G. L. (1962). Studies at phase interfaces. I. The sliding of liquid drops on solid surfaces and a theory for spray retention. *Journal of Colloid Science*, *17*(4), 309–324. doi:10.1016/0095-8522(62)90011-9

Gao, L., & McCarthy, T. J. (2009). Wetting 101. *Langmuir*, *25*(24), 14105–14115. doi:10.1021/la902206c PMID:19627073

Gau, H., Herminghaus, S., Lenz, P., & Lipowsky, R. (1999). Liquid morphologies on structured surfaces: from microchannels to microchips. *Science*, *283*, 46-49. doi:10.1126/science.283.5398.46

Gennes, P. G., Wyart, F. B., & Quere, D. (2004). *Capillarity and wetting phenomena: drops, bubbles, pearls, waves.*

Giboz, J., Copponnex, T., & Mélé, P. (2007). Microinjection molding of thermoplastic polymers: A review. *Journal of Micromechanics and Microengineering*, *17*(6), 96–109. doi:10.1088/0960-1317/17/6/R02

Grewal, H. S., Cho, I. J., Oh, J. E., & Yoon, E. S. (2014). Effect of topography on the wetting of nanoscale patterns: Experimental and modeling studies. *Nanoscale*, *6*(24), 15321–15332. doi:10.1039/C4NR04069D PMID:25387063

Grignard, B., Vaillant, A., Coninck, J., Piens, M., Jonas, A. M., Detrembleur, C., & Jerome, C. (2011). Electrospinning of a Functional Perfluorinated Block Copolymer as a Powerful Route for Imparting Superhydrophobicity and Corrosion Resistance to Aluminum Substrates. *Langmuir*, *27*(1), 335–342. doi:10.1021/la102808w PMID:21141949

Gudipati, C. S., Finlay, J. A., Callow, J. A., Callow, M. E., & Wooley, K. L. (2005). The Antifouling and Fouling-Release Performance of Hyperbranched Fluoropolymer (HBFP)-Poly(ethylene glycol) (PEG) Composite Coatings Evaluated by Adsorption of Biomacromolecules and the Green Fouling Alga Ulva. *Langmuir*, *21*, 3044-3053. doi:10.1021/la048015o

Hasan, J., Raj, S., Yadav, L., & Chatterjee, K. (2015). Engineering a nanostructured "super surface" with superhydrophobic and superkilling properties. *RSC Advances*, *5*(56), 44953–44959. doi:10.1039/C5RA05206H

He, B., Patankar, N. A., & Lee, J. (2003). Multiple Equilibrium Droplet Shapes and Design Criterion for Rough Hydrophobic Surfaces. *Langmuir*, *19*(12), 4999–5003. doi:10.1021/la0268348

Hirayama, Y., & Obara, M. (2005). Heat-affected zone and ablation rate of copper ablated with femtosecond laser. *Journal of Applied Physics*, *97*(6), 064903. doi:10.1063/1.1852692

Hirvi, J. T., & Pakkanen, T. A. (2007). Wetting of Nanogrooved Polymer Surfaces. *Langmuir*, *23*(14), 7724–7729. doi:10.1021/la700558v PMID:17559245

Hopmann, C., Behmenburg, C., Recht, U., & Zeuner, K. (2014). Injection Molding of Superhydrophobic Liquid Silicone Rubber Surfaces. *Silicon*, *6*(1), 35–43. doi:10.1007/s12633-013-9164-0

Ishizaki, T., Hieda, J., Saito, N., Saito, N., & Takai, O. (2010). Corrosion resistance and chemical stability of super-hydrophobic film deposited on magnesium alloy AZ31 by microwave plasma-enhanced chemical vapor deposition. *Electrochimica Acta*, *55*(23), 7094–7101. doi:10.1016/j.electacta.2010.06.064

Ishizaki, T., Masuda, Y., & Sakamoto, M. (2011). Corrosion Resistance and Durability of Superhydrophobic Surface Formed on Magnesium Alloy Coated with Nanostructured Cerium Oxide Film and Fluoroalkylsilane Molecules in Corrosive NaCl Aqueous Solution. *Langmuir*, *27*(8), 4780–4788. doi:10.1021/la2002783 PMID:21417352

Ishizaki, T., & Sakamoto, M. (2011). Facile Formation of Biomimetic Color-Tuned Superhydrophobic Magnesium Alloy with Corrosion Resistance. *Langmuir*, *27*(6), 2375–2381. doi:10.1021/la1051029 PMID:21319782

Jakubowicz, J. (2008). Formation of porous TiOx biomaterials in H3PO4 electrolytes. *Electrochemistry Communications*, *10*(5), 735–739. doi:10.1016/j.elecom.2008.02.027

Jeong, C. (2013). *Nano-Engineering of superhydrophobic aluminum surfaces for anti-corrosion [Ph.D Thesis].*. Stevens Institute of Technology.

Jiang, L., & Feng, L. (2010). *Bioinspired intelligent nanostructured interfacial materials*. Chemical Industry Press. doi:10.1142/7380

Jin, C., Li, J., Han, S., Wang, J., & Sun, Q. (2014). A durable, superhydrophobic, superoleophobic and corrosion-resistant coating with rose-like ZnO nanoflowers on a bamboo surface. *Applied Surface Science*, *320*, 322–327. doi:10.1016/j.apsusc.2014.09.065

Karamana, M. (2012). Self-supporting superhydrophobic thin polymer sheets that mimic the nature's petal effect. *Applied Surface Science*, *259*, 542–546. doi:10.1016/j.apsusc.2012.07.079

Kim, A., Lee, C., Kim, H., & Kim, J. (2015). Simple Approach to Superhydrophobic Nanostructured Al for Practical Antifrosting Application Based on Enhanced Self-propelled Jumping Droplets. *ACS Applied Materials & Interfaces*, *7*(13), 7206–7213. doi:10.1021/acsami.5b00292 PMID:25782028

Kim, S. H. (2008). Fabrication of Superhydrophobic Surfaces. *Journal of Adhesion Science and Technology*, *22*(3-4), 235–250. doi:10.1163/156856108X305156

Lee, J. H., Park, E. J., Kim, D. H., Jeong, M. G., & Kim, Y. D. (2016). Superhydrophobic surfaces with photocatalytic activity under UV and visible light irradiation. *Catalysis Today*, *260*, 32–38. doi:10.1016/j.cattod.2015.05.020

Leon, A. C. C., Pernites, R. B., & Advincula, R. C. (2012). Superhydrophobic Colloidally Textured Polythiophene Film as Superior Anticorrosion Coating. *ACS Applied Materials & Interfaces*, *4*(6), 3169–3176. doi:10.1021/am300513e PMID:22663552

Li, L., Huang, T., Lei, J., He, J., Qu, L., Huang, P., & Pan, F. et al. (2015). Robust biomimetic-structural superhydrophobic surface on aluminum alloy. *ACS Applied Materials & Interfaces*, *7*(3), 1449–1457. doi:10.1021/am505582j PMID:25545550

Li, M., Zhai, J., Liu, H., Song, Y., Jiang, L., & Zhu, D. (2003). Electrochemical Deposition of Conductive Superhydrophobic Zinc Oxide Thin Films. *J. Phys. Chem. B*, *107*, 9954-9957. doi:10.1021/jp035562u

Liu, H. Q., Szunerits, S., Xu, W. G., & Boukherroub, R. (2009). Preparation of Superhydrophobic Coatings on Zinc as Effective Corrosion Barriers. *ACS Applied Materials & Interfaces*, *1*(6), 1150–1153. doi:10.1021/am900100q PMID:20355904

Liu, Q., Chen, D., & Kang, Z. (2015). One-Step Electrodeposition Process to Fabricate Corrosion-Resistant Superhydrophobic Surface on Magnesium Alloy. *ACS Applied Materials & Interfaces*, *7*(3), 1859–1867. doi:10.1021/am507586u PMID:25559356

Liu, T., Chen, S. G., Cheng, S., Tian, J. T., Chang, X. T., & Yin, Y. S. (2007). Corrosion behavior of super-hydrophobic surface on copper in seawater. *Electrochimica Acta*, *52*(28), 8003–8007. doi:10.1016/j.electacta.2007.06.072

Liu, T., Yin, Y., Chen, S., Chang, X., & Cheng, S. (2007). Super-hydrophobic surfaces improve corrosion resistance of copper in seawater. *Electrochimica Acta*, *52*(11), 3709–3713. doi:10.1016/j.electacta.2006.10.059

Liu, Y., & Li, G. (2012). A new method for producing "Lotus Effect" on a biomimetic shark skin. *Journal of Colloid and Interface Science*, *388*(1), 235–242. doi:10.1016/j.jcis.2012.08.033 PMID:22995249

Liu, Y., Moevius, L., Xu, X., Qian, T., Yeomans, J. M., & Wang, Z. (2014). Pancake bouncing on superhydrophobic surfaces. *Nature Physics*, *10*(7), 515–519. doi:10.1038/nphys2980

Liu, Z., Zhang, X., Murakami, T., & Fujishima, A. (2008). Sol-gel SiO2/TiO2 bilayer films with self-cleaning and antireflection properties. *Sol. Energy Mater. Sol. Cells*, 92, 1434-1438. doi:10.1016/j.solmat.2008.06.005

Lv, P., Xue, Y., Shi, Y., Lin, H., & Duan, H. (2014). Metastable States and Wetting Transition of Submerged Superhydrophobic Structures. *PRL*, *112*(19), 196101. doi:10.1103/PhysRevLett.112.196101 PMID:24877948

Marmur, A. (2006). Super-hydrophobicity fundamentals: Implications to biofouling prevention. *Biofouling*, *22*, 107-115. doi:10.1080/08927010600562328

Mohamed, A. M. A., Abdullah, A. M., & Younan, N. A. (2014). Corrosion behavior of superhydrophobic surfaces: A review. *Arabian Journal of Chemistry*. doi:10.1016/j.arabjc.2014.03.006

Narita, M., Kasuga, T., & Kiyotani, A. (2000). Techniques for Surface Treatment of Light Metals. Super water-repellent aluminum by electrolytic etching and chemical adsorption. *Keikinzoku*, *50*(11), 594–597. doi:10.2464/jilm.50.594

Nimittrakoolchai, O. U., & Supothina, S. (2008). Deposition of organic-based superhydrophobic films for anti-adhesion and self-cleaning applications. *J. Eur. Ceram. Soc.*, *28*, 947-952. doi:10.1016/j.jeurceramsoc.2007.09.025

Nosonovsky, M., & Bhushan, B. (2009). Superhydrophobic surfaces and emerging applications: Non-adhesion, energy, green engineering. *Current Opinion in Colloid & Interface Science*, *14*(4), 270–280. doi:10.1016/j.cocis.2009.05.004

Öner, D., & McCarthy, T. J. (2000). Ultrahydrophobic Surfaces. Effects of Topography Length Scales on Wettability. *Langmuir*, *16*(20), 7777–7782. doi:10.1021/la000598o

Quéré, D. (2005). Non-sticking drops. *Reports on Progress in Physics*, *68*(11), 2495–2532. doi:10.1088/0034-4885/68/11/R01

Ro, D. H., & Klar, E. (1981). Corrosion Behavior of P/M Austenitic Stainless Steels. *Modern Developments in Powder Metallurgy*, *13*, 247–287.

Roach, P., Shirtcliffe, N. J., & Newton, M. I. (2008). Progress in superhydrophobic surface development. *Soft Matter*, *4*(2), 224–240. doi:10.1039/B712575P

Roig, A., Molins, E., Rodriguez, E., Martinez, S., Moreno, M. M., & Vallribera, A. (2004). Superhydrophobic silica aerogels by fluorination at the gel stage. *Chem. Commun.*, 2316-2317. doi:10.1039/b407987f

Rupp, F., Scheideler, L., Olshanska, N., Wild, M., Wieland, M., & Geis, G. J. (2006). Enhancing surface free energy and hydrophilicity through chemical modification of microstructured titanium implant surfaces, reply. *J. Biomed. Mater. Res. Part A*, 79A, 755-757. doi:10.1002/jbm.a.31045

Samuel, J. D. J. S., Ruther, P., Frerichs, H. P., Lehmann, M., Paul, O., & Ruehe, J. (2005). A simple route towards the reduction of surface conductivity in gas sensor devices. *Sens. Actuators, B, 110*, 218-224. doi:10.1016/j.snb.2005.01.032

Sharma, R. K. (2007). *Elementary Chemistry*. Discovery Publishing House.

Steger, M., Hartmann, C., Beckemper, S., Holtkamp, J., & Gillner, A. (2013). Fabrication of Hierarchical Structures by Direct Laser Writing and Multi-Beam-Interference. *JLMN-Journal of Laser Micro/Nanoengineering, 8*, 210-215.

Su, F., & Yao, K. (2014). Facile Fabrication of Superhydrophobic Surface with Excellent Mechanical Abrasion and Corrosion Resistance on Copper Substrate by a Novel Method. *ACS Applied Materials & Interfaces*, *6*(11), 8762–8770. doi:10.1021/am501539b PMID:24796223

Tadanaga, K., Katata, N., & Minami, T. (1997). Formation process of super-water-repellent Al_2O_3 coating films with high transparency by the sol-gel method. *J. Am. Ceram. Soc.*, *80*, 3213-3216.

Townsin, R. L. (2003). The ship hull fouling penalty. *Biofouling*, 19(Suppl.), 9-15.

Traverso, P., & Canepa, E. (2014). A review of studies on corrosion of metals and alloys in deep-sea environment. *Engineering*, *87*, 10–15.

Vengatesh, P., & Kulandainathan, M. A. (2015). Hierarchically Ordered Self-Lubricating Superhydrophobic Anodized Aluminum Surfaces with Enhanced Corrosion Resistance. *ACS Applied Materials & Interfaces*, *7*(3), 1516–1526. doi:10.1021/am506568v PMID:25529561

Wang, H., Liu, Z., Wang, E., Yuan, R., Gao, D., Zhang, X., & Zhu, Y. (2015). A robust superhydrophobic PVDF composite coating with wear/corrosion-resistance properties. *Applied Surface Science*, *332*, 518–524. doi:10.1016/j.apsusc.2015.01.213

Wang, L., Guo, S. L., & Dong, S. J. (2008). Facile electrochemical route to directly fabricate hierarchical spherical cupreous micro structures: Toward superhydrophobic surface. *Electrochemistry Communications*, *10*(4), 655–658. doi:10.1016/j.elecom.2008.01.034

Wang, Z., Li, Q., She, Z., Chen, F., & Li, L. (2012). Low-Cost and Large- Scale Fabrication Method for an Environmentally-Friendly Superhydrophobic Coating on Magnesium Alloy. *J. Mater. Chem. A*, *22*(9), 4097–4105. doi:10.1039/c2jm14475a

Wenzel, R. N. (1936). Resistance of Solid Surfaces to Wetting by Water. *Industrial & Engineering Chemistry*, *28*(8), 988–994. doi:10.1021/ie50320a024

Xue, C. H., Jia, S. T., Zhang, J., & Ma, J. Z. (2010). Large-area fabrication of superhydrophobic surfaces for practical applications: An overview. *Science and Technology of Advanced Materials, 11*(3). doi:10.1088/1468-6996/11/3/033002

Zhai, L., Cebeci, F. C., Cohen, R. E., & Rubner, M. F. (2004). Stable Superhydrophobic Coatings from Polyelectrolyte Multilayers. *Nano Lett., 4*, 1349-1353. doi:10.1021/nl049463j

Zhang, G., Zhang, J., Xie, G., Liu, H., & Shao, H. (2006). Cicada Wings: A Stamp from Nature for Nanoimprint Lithography. *Small*, *2*(12), 1440–1443. doi:10.1002/smll.200600255 PMID:17193002

Zhang, H., Lamb, R., & Lewis, J. (2005). Engineering nanoscale roughness on hydrophobic surface - preliminary assessment of fouling behaviour. *Science and Technology of Advanced Materials*, *6*(3-4), 236–239. doi:10.1016/j.stam.2005.03.003

Zheng, Q. S., Yu, Y., & Zhao, Z. H. (2005). Effects of Hydraulic Pressure on the Stability and Transition of Wetting Modes of Superhydrophobic Surfaces. *Langmuir*, *21*(26), 12207–12212. doi:10.1021/la052054y PMID:16342993

Zhou, C., Lu, X., Xin, Z., & Liu, H. (2013). Corrosion resistance of novel silane-functional polybenzoxazine coating on steel. *Corrosion Science*, *70*, 145–151. doi:10.1016/j.corsci.2013.01.023

Chapter 14
Electrodeposition of Nickel–Molybdenum (Ni–Mo) Alloys for Corrosion Protection in Harsh Environments

Teresa D. Golden
University of North Texas, USA

Jeerapan Tientong
Nakhon Sawan Rajabhat University, Thailand

Adel M.A. Mohamed
Qatar University, Qatar

ABSTRACT

Electrodeposition of only molybdenum onto substrates is difficult, therefore molybdenum is typically deposited with iron-based alloys such as nickel. The deposition of such alloys is known as an induced codeposition mechanism. The electrodeposition of nickel-molybdenum alloys using alkaline plating solutions is covered in this chapter. The mechanism for deposition of nickel-molybdenum is reviewed, as well as the influence of the plating parameters on the coatings. Characterization of the coatings by scanning electron microscopy and x-ray diffraction is discussed and how deposition parameters affect morphology, composition, and crystallite size. Nickel-molybdenum alloys offer enhanced corrosion protection and mechanical properties as coatings onto various substrates. A survey of the resulting hardness and Young's modulus is presented for several research studies. Corrosion parameters for several studies are also compared and show the percentage of molybdenum in the coatings affects these values.

INTRODUCTION

Stainless steels are corrosion resistant in a wide range of environments due to formation of a passivating film on their surfaces. Despite passivity, extensive localized corrosion commonly occurs on stainless steels when exposed to harsh environments containing chloride ions, such as seawater. Chloride ions

DOI: 10.4018/978-1-5225-0066-7.ch014

dislocate the passivating species at various sites within the passive film resulting in the initiation of pitting nuclei. Once the passivity has been destroyed, the reactions inside the growing pits propagate through an anodic process leading to a loss of strength in the overall structure (Caceres, 2009; Cvijović, 2006; Jung, 2012; Lu, 1993; Sedriks, 1996). Molybdenum, having been used as one of the fundamental metals incorporated in stainless steels, is of interest to alloy with nickel to enhance the repassivation behavior and deactivate pit growth in aggressive chloride media (Ahn, 1998; Bastidas, 2002; Ilevbare, 2001; Mishra, 2013; Tobler, 2006). Additionally, the presence of molybdenum improves the mechanical properties and temperature resistance of the alloys (Gomez, 2006). Molybdenum, in the same group as chromium and tungsten, has a melting point of 2620 °C. Molybdenum offers similar heat stability as tungsten and therefore is used to alloy with nickel leading to improved corrosion resistance, mechanical properties and heat stability. In addition, molybdenum is another alternative component to alloy with nickel to replace toxic components such as chromium which have been previously used for improved mechanical properties.

Due to the coating stability, nickel is commonly used to alloy with other metals to protect substrate materials under severe operating conditions such as corrosive environments, high temperatures and high stresses. Nickel can form intermetallic phases with some metals commonly used in nickel alloys leading to high strength for both low and high-temperature applications (Friend, 1980; Haubold, 1992). Nickel- molybdenum alloy coatings are of interest as a replacement for chromium due to environmental concerns (Brooman, 2000). Ni-Mo also exhibits high hardness and strength as well as good wear resistance, which makes it attractive as an anti-corrosion coating. Nickel-based alloys containing 9-16 weight percent molybdenum offer high corrosion resistance against chloride attack as well as protection from high temperatures (Chassaing, 1995; Friend, 1980; Sugimoto, 1977). Ni-Mo alloys have good resistance to corrosion in hydrochloric, phosphoric, and sulfuric acid solutions (Donten, 2005; Raj, 1988). Ni-Mo has been added to ferritic stainless steel to produce better interconnects in solid oxide fuel cells. Addition of 3 wt% Mo and 2 wt% Ni reduced the oxidation rate of interconnects (Safikhani, 2014). Thus Ni-Mo alloys are good choices for harsher environmental conditions such as high temperature, corrosive chemical attack, or high wear.

Electrodeposition techniques have been used to synthesize several materials including metals, ceramics, polymers, or composite coatings (Ahmad, 2014a; Conrad, 2012; Kahl, 2014; Karolus, 2006; Wang 2013). The coatings fabricated by electrodeposition techniques can provide a superior nanocrystalline structure improving corrosion protection and hardness properties of the materials (Koch, 2007; Mishra, 2004a; Mishra, 2004b; Wang, 2006; Xuetao, 2008; Youssef, 2004). Electrodeposition of only Mo metal onto substrates has proved elusive. An early attempt to deposit metallic Mo from aqueous solutions had a maximum current efficiency of only ~2% (Ksycki, 1949). It was postulated that with the large current density needed to obtain deposits, large amounts of hydrogen evolution occurred and instead of direct electroreduction of the Mo metal ion, more likely a reduction to oxides occurred through a multistep mechanism. Since it is not feasible to electrodeposit molybdenum alone, a more productive route was discovered in which the codeposition of Mo occurred with iron-based alloys. This pathway came to be known as an induced codeposition mechanism described by Brenner (Brenner, 1963). Some metal-Mo binary alloys such as Ni-Mo, Fe-Mo (Elezović, 2005; Kuznetsov, 2013; Niu, 1998; Sanches, 2007), Co-Mo (Gómez, 2003), and Zn-Mo (Kazimierczak, 2013a; Kazimierczak, 2013b), have been codeposited from aqueous solutions with much higher current efficiencies. The most common of these alloys has been Ni-Mo. While Ni-Mo alloys can be prepared by several different methods, Ni-Mo coatings are difficult to prepare by techniques other than electrodeposition. The differences in melting points between nickel

Figure 1. Simplified electrochemical cell setup

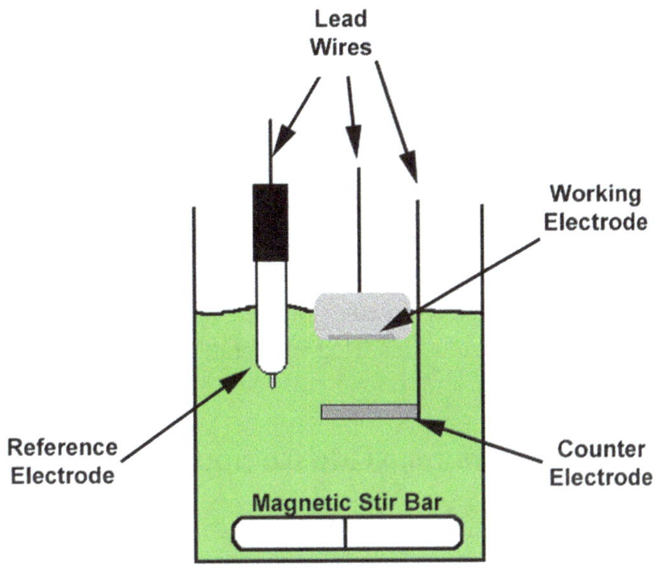

(1455 °C) and molybdenum (2620 °C) and limited mutual solubility make preparation by thermal, chemical vapor, sputter deposition or pulse laser techniques less studied (Gao, 2015; Kawashima, 1997). Using electrodeposition offers many advantages, e.g. relatively low cost, control of grain size, deposition onto various shapes, control of composition, low processing temperature, and ease of scale-up (Eliaz, 2008).

This chapter will cover the electrodeposition of nickel-molybdenum alloys onto steels and other substrates to improve corrosion resistance and mechanical properties. The influence of the deposition parameters will be covered as well as the electrodeposition mechanism. The resulting mechanical and corrosion properties will be also be discussed in the chapter.

ELECTRODEPOSTION OF NICKEL-MOLYBDENUM ALLOYS

Electrodeposition tends to be an easy and low cost technique for producing metallic coatings. The deposition cells can be simple setups comprised of a two or three electrode system. Figure 1 is a typical three-electrode arrangement of an electrochemical cell.

The deposition cell includes a counter electrode (CE), working electrode (WE) and reference electrode (RE). The electrodes are immersed in a conducting solution (plating bath) containing dissolved metal salts as metal ions and other additives that enhance coating properties. Deposition of the metal ions out of solution onto the working electrode as a coating can be done by applying a current (galvanostatic control) or a voltage (potentiostatic control).

For the induced co-deposition of nickel-molybdenum, the role of each component in the plating solution and the deposition mechanism is not immediately clear. However, over the years, a common plating solution has emerged for the deposition of the Ni-Mo coatings. While a pH range from acidic to alkaline has been used, the alkaline baths tend to have higher current efficiency and produce more dense uniform coatings. Therefore, this chapter will mainly focus on the research involving alkaline baths for the Ni-Mo alloys.

Plating Solutions (Bath Conditions) for Ni-Mo Coatings

The plating solutions that have been used for Ni-Mo electrodeposition are very similar. Deposition from citrate or pyrophosphate solutions give the best results, although other solutions using acetate (Singh, 1975), tartrate (Fukushima, 1978), and glycinate (Ahmad, 2014b) have been attempted. The majority of the research has used an alkaline based solution containing nickel sulfate, sodium molybdate dihydrate, and sodium citrate. The pH of the solution is adjusted to higher pH using ammonia. Earlier research studied the mechanism of deposition for the coatings and therefore copper was mostly used as the substrate. After some good deposition conditions were established, research on corrosion resistance of the coatings began and studies looked at coatings on steel substrates. Tables 1 and 2 lists the major deposition conditions and the plating solutions used in literature for alkaline based depositions under galvanostatic or potentiostatic control.

Influence of Plating Parameters on Ni-Mo Composition

The electrodeposition parameters such as pH, bath composition, and current density have a profound effect on the composition as well as the grain size of the Ni-Mo coatings. It should be noted that any change in experimental parameters will change the resulting composition or ratio of Ni to Mo in the coatings. Most researchers have observed that coatings with Mo content above 30 wt% tend to crack (are brittle) and contain oxide impurities within the coating. Thus higher Mo content can lead to undesirable properties for the coatings. Several researchers have shown that the Mo content in the coating decreases as the plating current density increases. Figure 2 shows a plot of %Mo versus j compiled from several studies.

Figure 2. Plot of %Mo versus current density showing a general trend of a decrease in Mo content in the Ni-Mo coatings as the applied deposition current density increased. Data compiled from (a) (Chassaing, 1978), b (Podlaha, 1996a), c (Halim, 2012), and d (Beltowska-Lehman, 2012b)

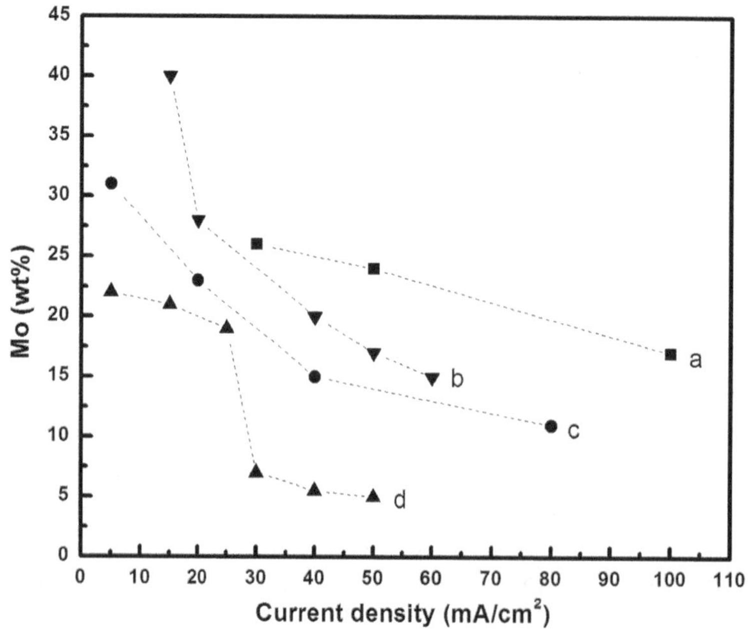

Table 1. Plating conditions for Ni-Mo electrodeposition under galvanostatic control

Ni/Mo Plating Salts	Substrate	pH	Applied j (mA/cm^2)	Temp. (°C)	Current eff. (%)	Mo (%wt)	Ref.
Composition #1[a]	Pt	10.5	10-100	25	80-87	25	(Ernst, 1958)
Composition #2[b]	-	9.5	30-100	40	-	15-26	(Chassaing, 1978)
Composition #1	Cu	10.5	2.5-200 pulse	60	-	6-25	(Nee, 1988)
Composition #1	Cu	9.5	45-150	25	46-67	45-50	(Podlaha, 1993)
Composition #1	Cu	9.2-10	50	25-40	50-60	20-50	(Podlaha, 1996a)
Composition #1	Cu	9.5	100	45	35-45	39-44	(Stepanova, 1998)
Composition #1	Cu	8.0-11	50-200	20-50	-	26-41	(Hu, 2000)
Composition #1	Cu	8.5	50 pulse	25	50-70	15-52	(Marlot, 2002)
Composition #1	Cu, Fe	8.5-9.5	10-150	30	30-70	5-40	(Chassaing, 2004)
Composition #3[c, d]	Cu	8.5-9.75	30	20	10-57	11-47	(Donten, 2005)
Composition #1	GC	10.5	70	55	-	22	(Martinez, 2006)
Composition #2 [e]	Ni	7.5-9.0	20-100	60	-	39-53	(Krastajić, 2008)
Composition #3	Cu, steel	8.0	20-40 pulse	35	-	30-38	(Han, 2010)
Composition #1	Cu	9.0	50	27	52-66	6-35	(Lima-Neto, 2010)
Composition #3	GC, Cu	7.0-10.0	50-150	25-65	-	15-45	(Prioteasa, 2010)
Composition #4[d, g]	steel	8.5-10.5	80 pulse	30-60	48-87	15-48	(Allahyarzadeh, 2011; Allahyarzadeh, 2012a,b,c)
Composition #1	steel	9.5	5-80	25	-	11-31	(Halim, 2012)
Composition #1 (no NH_3)	Cu	9.7	100-800	25	1	62-82	(Sun, 2012)
Composition #1	steel	8.0	5-50	25	60-80	5-22	(Beltowska-Lehman, 2012b)
Composition #3[h]	steel	9.3	50	20	-	17-26	(Ahmad, 2014b)
Composition #1	steel	9.5	100 pulse	30	29-63	26-30	(Huang, 2014, 2015)

[a]Ni/Mo plating composition #1: $NiSO_4$ $6H_2O$, Na_2MoO_4 $2H_2O$, [b]Ni/Mo plating composition #2: $NiCl_2$ $6H_2O$, Na_2MoO_4 $2H_2O$, [c]Ni/Mo plating composition #3: $NiSO_4$ $7H_2O$, Na_2MoO_4 $2H_2O$, [d]Ni/Mo plating composition #4: $NiSO_4$ $6H_2O$, $NiCl_2$ $6H_2O$ Na_2MoO_4 $2H_2O$, [e]Composition #5: $Na_4P_2O_7$ 10 H_2O, 2-butyne-1,4 diol, rokafenol N-10, [f]Composition #6: $K_4P_2O_7$ + $NaHCO_3$ and NH_4Cl or NaCl or $Na_2B_4O_7$, [g]$Na_3C_6H_5O_7$ $2H_2O$ + ionic liquids, ILs: 1-ethyl-3-methyl-imidazolium, 1-methyl-imidazolium, [h]Composition #7: Sodium glycinate, $NaC_2H_4NO_2$ + saccharin, $C_7H_5NO_3S$. (Unless indicated, the complexing agent used in the bath was trisodium citrate, $Na_3C_6H_5O_7$ $2H_2O$).

Chassing (1978) showed that an increase in current density from 10 to 150 mA/cm^2 led to a decrease in %Mo from 30 to 10 wt%. Other researchers showed the same trend (Beltowska-Lehman, 2012b; Halim, 2012; Podlaha, 1996a). However, the crystallite size, as measured by x-ray diffraction, decreases with increasing Mo content in the coating. Figure 3 is data compiled from literature of a plot of the crystallite size versus the Mo content in the Ni-Mo coatings.

Table 2. Plating conditions for Ni-Mo electrodeposition under potentiostatic control

Ni/Mo Plating Salts	Substrate	pH	Applied Potential (V)	Temperature (°C)	Current Efficiency (%)	Mo in Coating (%wt)	Reference
Composition #1[a]	Ni, Mo	9.5	-1.4 to -1.7 (SSE)	37	-	35-38[d]	(Chassaing, 1989)
Composition #1	C, Pd, Ni	10	-1.25 (SCE)	25	-	-	(Crousier, 1992)
Composition #1	Ni, steel, Pt, GC	9.0	-1.70 (SCE)	25	45	15-40	(Chassaing, 1995)
Composition #2[b]	steel	8.0-11.0	-1.15 to -1.35 (SCE)	-	70	14-30[d]	(Beltowska-Lehman,2012a)
Composition #2[c]	Cu	9.0	-1.5 (SCE)	-	22	24-32	(Ohgai, 2013)
Composition #2	steel	9.5	-1.35 to -0.7 (SCE)	25	-	15-29	(Tientong, 2014)

[a]Ni/Mo plating composition #1: $NiSO_4$ $7H_2O$, Na_2MoO_4 $2H_2O$, [b]Ni/Mo plating composition #2: $NiSO_4$ $6H_2O$, Na_2MoO_4 $2H_2O$, [c]$C_3H_4(OH)(COOH)_3$, [d]significant oxygen present in coatings as Mooxides.
(Unless indicated, the complexing agent used in the bath was trisodium citrate, $Na_3C_6H_5O_7$ $2H_2O$).

Figure 3. Plot of %Mo versus crystallite size for the Ni-Mo coatings. Data compiled from (a) (Chassaing, 2004 at pH 9.5), b (Donten, 2005), c (Halim, 2012), d (Chassaing, 1995), e (Chassaing, 2004 at pH 8.5)

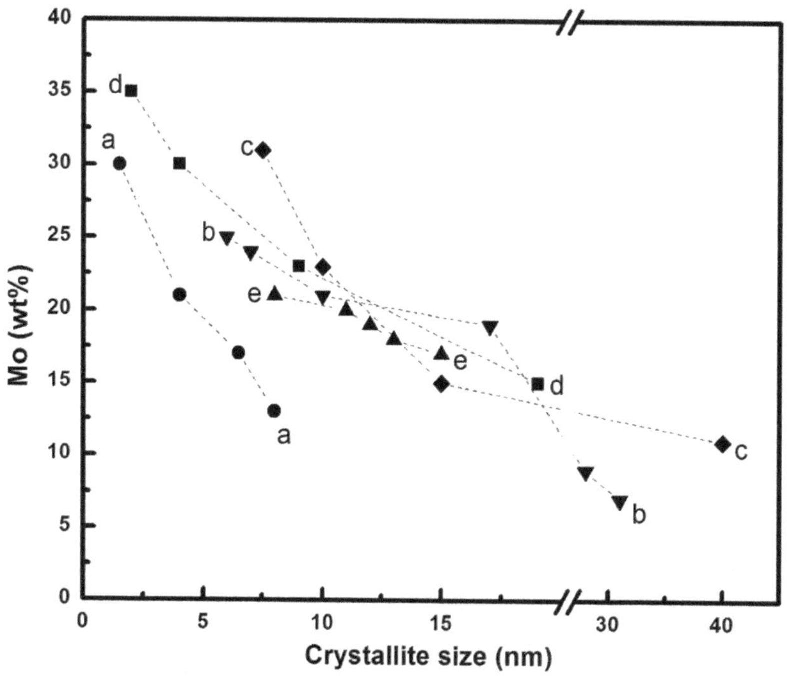

Table 3. The elemental composition of the alloy films, deposited from different $[MoO_4^{2-}]/[Ni^{2+}]$ molar ratios of 0.1 and 0.2 in the plating solution at pH 9.5 as measured by AAS and EDAX. Plating solution sodium molybdate; sodium citrate; and boric acid

$[MoO_4^{2-}]/[Ni^{2+}]$ Molar Ratio	Weight (%) by AAS (Avg. ± SD, n=3)		Weight (%) by EDX (Avg. ± SD, n=3)		Atomic (%) by EDX (Avg. ± SD, n=3)	
	Ni	Mo	Ni	Mo	Ni	Mo
Pure Ni	100	–	100	–	100	–
0.1	80.5 ± 1.2	19.5 ± 1.1	85 ± 5	15 ± 3	90 ± 3	10 ± 2
0.2	65.9 ± 1.4	34.1 ± 0.9	72 ± 3	29 ± 2	80 ± 3	20 ± 3

Since the solubility limit for Mo into the Ni crystal structure is ~10-13%, the typical fcc pattern seen for nickel begins to transform and become amorphous as the percentage of Mo in the coating approaches 20%. In fact, most studies show that the coatings almost always crack when the Mo content is over 25-30% or if the oxygen content in the films is higher than a few percent (codeposition of Mo oxides). Also, the Mo content in the coating varied with changes in pH (Allahyarzadeh, 2012b; Ohgai, 2013) and concentration of the complexing ligand in solution (Allahyarzadeh, 2012a; Hu, 2000). Inclusion of hydrogen in the coating is seen for some studies.

Another solution parameter that affects the composition of the Ni-Mo coatings is the $[MoO_4^{2-}]/[Ni^{2+}]$ molar ratio in solution. As an example, a correlation between the molybdate-to-nickel molar ratio in the plating solution and chemical composition of the nickel-molybdenum film can be determined by EDX and AAS (Table 3) (Tientong, 2014). When increasing the molybdate/nickel molar ratio from 0.1 to 0.2 in the plating solution, the molybdenum content incorporated in the nickel film increases from 15.5 to 28.5%. The composition of the films measured by AAS are slightly higher than that by EDX, which are 19.5% and 34.1% for molybdenum in the films deposited from molybdate/nickel ratio of 0.1 and 0.2, respectively. AAS data results from dissolving the entire film in acid and EDX data only measures from the top surface of the film, so some differences will be seen in the percentages.

This effect that the ratio of the $[MoO_4^{2-}]/[Ni^{2+}]$ has on the composition has been noted in several studies (Beltowska-Lehman, 2012a; Chassaing, 2004; Han, 2010; Ohgai, 2013). Chassaing (2004) showed that at a current density of 10 mA/cm^2, the Mo content in the coating increase from 25 to 30% for a 0.02 and 0.03 M Na_2MoO_4 solution, respectively. Ohgai (2013) also showed an effect between the MoO_4^{2-} concentration in the bath and the Mo content in the Ni-Mo alloy. However, these researchers discovered that the effect was greater at pH 9.0 compared to pH 2.0. They attributed this effect to the different Mo solution species that exist at the two different pH values. At pH 2, the predominant Mo species is $[(MoO_4)(Cit)H_2]^{3-}$ and at pH 9 is $[(MoO_4)(Cit)]^{5-}$. They surmised that the latter species is more effectively reduced at the substrate surface, showing a preferential electrodeposition at alkaline pHs. The same effect for the $[MoO_4^{2-}]/[Ni^{2+}]$ ratio in solution has been observed for complexes besides citrate (Ahmad, 2014b; Donten, 2005). Ahmad (2014b) showed that in an alkaline glycinate plating solution, the %Mo in the coatings was 17 and 26 wt% for a $[MoO_4^{2-}]/[Ni^{2+}]$ of 0.1 and 0.2, respectively.

Deposition Mechanism for Ni-Mo Alloy Coatings

While Ni-Mo has been electrodeposited from solutions having a pH of 2 to 11, the most common deposition has been out of alkaline baths. Specifically, from pH 8.5 to 10 in a nickel sulfate, sodium molybdate, and citrate solution, with added ammonia for pH adjustment. The combination of having many possible species in solution as well as at the surface of the electrode and an induced codeposition reaction have made it difficult to pin down an exact deposition mechanism. Typical species in the solution include citrate, which for the pH range 8 to 10 is predominately the triple charged anion, $C_6H_5O_7^{3-}$ (cit^{3-}), which can form complexes with Ni^{2+} and MoO_4^{2-} in solution (Green, 1998; Li, 2007; Rode, 2005; Zelenin, 2007). To complicate matters, for the alkaline plating baths, ammonia is added to adjust the pH, which can also form complexes. The initial Ni species for the citrate and ammonia bath at pH 8 were calculated as 50% $NiH_2Cit_2^{4-}$, 20% $Ni_2Cit_2^{4-}$, 22% $NiHCit^-$, and 7% $NiOH^+$; and for pH 11 as 1% $NiH_2Cit_2^{4-}$, 7% $Ni_2Cit_2^{4-}$, trace% $NiOH^+$, and 92% $Ni(NH_3)_n^{2+}$ (Beltowska-Lehman, 2012a). The deposition mechanism would be expected to be different for acidic and alkaline pHs.

In several studies by Podlaha and Landolt (1996a 1996b, 1997), using a rotating cell setup, the rate of Mo deposition was controlled by the flux of the nickel ions, pointing to an induced codeposition. Other factors that lead to an increase in Mo content in the coatings included; an increase in citrate concentration, an increase in MoO_4^{2-} concentration, and an increase in temperature. From their results they proposed an intermediate which would adsorb onto the substrate surface, $[NiHCit\text{-}MoO_2]_{ads}$. With an excess of the nickel ion to the molybdate species, a complex can form, $[NiHCit]^-$, which can catalyze the reduction of the intermediate. A mathematical model was developed to explain the proposed mechanism and data results (Podlaha, 1996b). However, this mechanism does not take into account the many other species that may be present and contribute to the mechanism, e.g. $[(MoO_4)(Cit)(H)]^{4-}$, or complexes of the metals with NH_3 (Eliaz, 2008).

Other researchers have suggested the formation of an intermediate Ni_4MoO_2 through the reduction of Mo (VI) to Mo(IV) species. Then the mixed oxide is further reduced in the presence of hydrogen to form the Ni-Mo coating. (Chassaing, 1989; Chassaing, 2004). There have been hydrogen impurities measured within the coatings (Chassaing, 1978), however, corresponding oxygen content is also detected. The difficulty in much of the literature is that the presence or amount of hydrogen and/or oxygen is many times not measured for the electrodeposited amorphous Ni-Mo coatings.

An interesting study by Crousier (1992) investigated the electrodeposition of Ni-Mo onto different substrates. The coatings would form on glassy carbon (GC) substrates but not on palladium (Pd). Deposition onto high purity nickel was more difficult than on GC, however if the nickel was annealed (forming larger grains), the deposits were much thicker. For glassy carbon, hydrogen diffusion does not occur but is rapid for Pd. And for nickel, thin films are formed on the non-annealed substrate (where hydrogen diffusion is faster), while thicker deposits occur on the annealed substrates (increasing grain size decreases hydrogen diffusion). Thus, instead of diffusion into the substrate, for GC and annealed Ni, the hydrogen (adsorbed on the surface) is available to participate in the Ni-Mo deposition and trapped in the alloy. The authors concluded that hydrogen plays some role in the codeposition of Ni-Mo. However, there is still no conclusive consensus on the exact mechanism for the alloy deposition.

Figure 4. SEM images of nickel-molybdenum alloys at different magnifications (5 and 10 μm) deposited from citrate-borate electrolyte (pH 9.5) plating solution for the $[MoO_4^{2-}]/[Ni^{2+}]$ molar ratio of 0.1

CHARACTERIZATION OF THE NICKEL-MOLYBDENUM COATINGS

Microscopy

For electrodeposited Ni-Mo coatings, typical coatings are light grey in color and display a nodular compact type texture of uniform coverage when imaged with scanning electron microscopy (SEM). Cracks in the coatings appear in the SEM images usually when the Mo content is close to or higher than ~30 wt % (Chassaing, 2004; Halim, 2012). Figure 4 is a typical SEM image for Ni-Mo films electrodeposited from an alkaline plating solution.

The globular structure is evident in the micrographs for the Ni-Mo coatings and typical for the electrodeposited coatings (Chassaing, 2004; Halim, 2012; Marlot, 2002). The morphology can change with different electrochemical parameters. Chassaing et. al showed that the nodular size became smaller at higher pH (9.5) compared to lower pH (8.5) (Chassaing, 2004). The size of the nodules also become courser as the applied current density is increased between 10 to 80 mA/cm^2 (Halim 2012). There is also smoothing of the morphology, when the applied deposition current is changed from DC to pulse, grain refinement occurs as seen in the SEM (Han, 2010, Huang, 2014). Solution composition also has an effect, addition of organic additives to the solution create a much smoother surface (Donten, 2005). Some researchers have been able to obtain a flattening or smoother texture for the coatings by adding imidazolium-based ionic liquids to the plating baths (Allahyarzadeh, 2012a).

In general, the electrodeposited Ni-Mo coatings are fairly smooth. Atomic force microscopy (AFM) can be used to measure the surface roughness of the electrodeposited films. Figure 5 shows the AFM image of an electrodeposited film from an alkaline glycinate plating bath (Ahmad, 2014b).

The surface roughness was expressed by an average deviation parameter (R_a). This value gives the average height of irregularities in the perpendicular direction to the surface. The surface of pure Ni-Mo was relatively smooth which is evident from its low value of R_a (7.83 nm). Ni-Mo alloy coatings deposited from pyrophosphate baths had even lower surface roughness values of 1-3 nm as measured by scanning tunneling microscopy (STM) (Donten, 2005).

Figure 5. AFM micrographs of (a) Ni-Mo prepared from plating solutions containing [MoO_4^{2-}]/[Ni^{2+}] ratio of 0.1

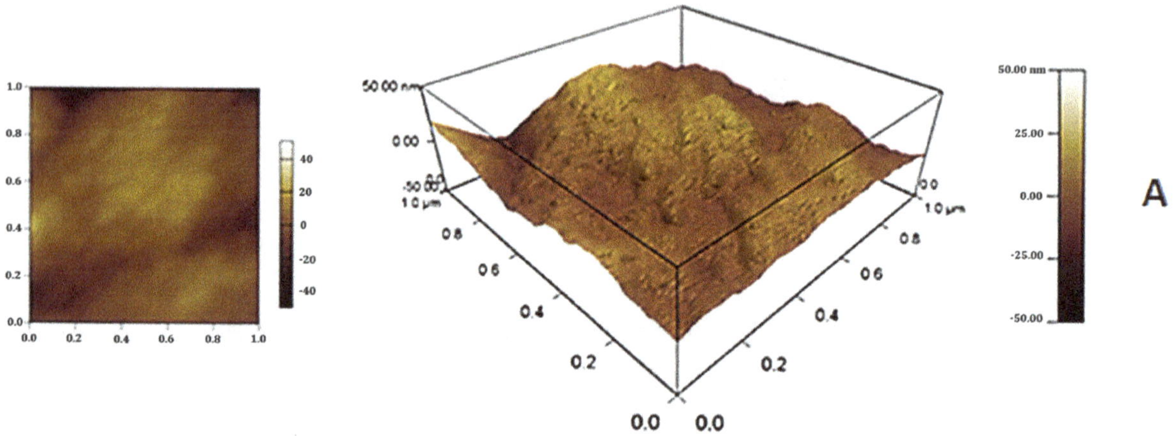

X-Ray Diffraction (XRD)

X-ray diffraction can be used to examine the structure and crystallite size of the electrodeposited Ni-Mo alloy coatings. Pure nickel has a face-centered cubic (fcc) crystalline structure. While molybdenum has a body-centered cubic (bcc) crystalline structure, when electrodeposited with nickel, the XRD pattern will show a single fcc phase. The codeposition of the Mo with the Ni forms a saturated Ni-Mo solid solution phase, where there is an expansion of the nickel lattice with the introduction of molybdenum. The Ni-Mo alloy will exhibit broaden reflections in the XRD pattern that correspond to the position of the nickel peaks. Figure 6a and b shows a Ni-Mo alloy electrodeposited from a citrate-borate plating solution at different [MoO_4^{2-}]/[Ni^{2+}] molar ratios of (a) 0.1 and (b) 0.2 (Tientong, 2014). Figure 6c shows an x-ray diffraction pattern of an electrodeposited pure nickel film for comparison. The peaks corresponding to the fcc structure are the (111), (200), (220), (311), and (222) reflections. The experimental pattern shown in figure 6c matches that for a randomly oriented nickel in the JCPDS database (PDF #04-0850).

The electrodeposited Ni-Mo alloy (Figure 6a and b) exhibits broaden reflections of 43–44°, 51°, and 74–75° (two theta) corresponding to the (111), (200), and (220) planes respectively (Halim, 2012). At Mo content up to ~15 wt%, there is noticeable broadening in all the peaks (Chassaing, 2004). As the Mo content increases (20-30wt%), a depressed and broaden (111) reflection is the main feature in the pattern. The (111) reflection at around 43-44° is the only or predominate peak left in the pattern, indicating the alloy becoming more nanocrystalline and amorphous (Donten, 2005; Ohgai, 2013). If the amorphous coating is annealed above 680 °C for 1 hour, the alloy structure is converted to a purely crystalline phase which is confirmed by the return of very sharp peaks in the XRD pattern (Allahyarzadeh, 2011). These peaks are associated with the formation of a mixture of Ni, Ni_4Mo, and NiMo phases (Lima-Neto, 2010).

When the Mo content is above 30%, the films become very brittle and cracking is seen in the SEM. Deposition of Mo oxides can occur at different conditions. Figure 7 shows the x-ray diffraction pattern of a Ni-Mo film which also contains measurable oxygen content in the coating (Ahmad, 2014b). The XRD shows the formation of a sharper peak on top of the broaden (111) Ni-Mo peak (Figure 7). This is probably due to the formation of Mo oxides along with the Ni-Mo saturated phase within the electrodeposited coating.

Figure 6. XRD patterns of nickel-molybdenum alloys electrodeposited from citrate-borate plating solution at different [MoO_4^{2-}]/[Ni^{2+}] molar ratios of (a) 0.1 (b) 0.2, and (c) pure nickel deposit

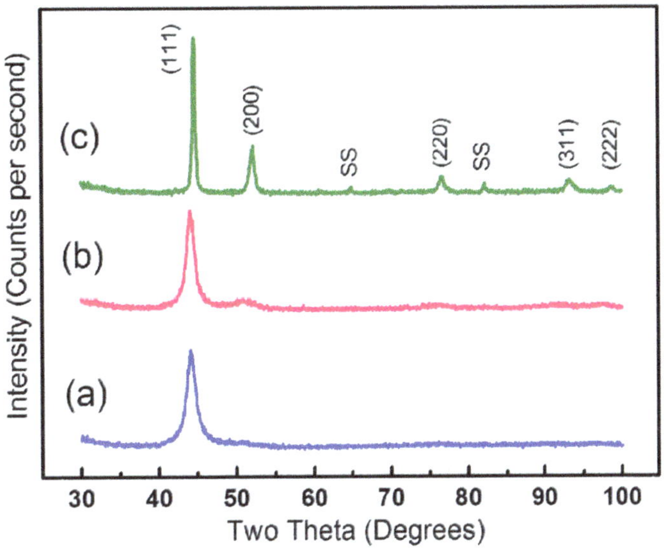

Figure 7. XRD patterns of Ni-Mo coating electrodeposited from a 0.2 [MoO_4^{2-}]/[Ni^{2+}] molar ratios solution in an alkaline glycinate bath. Elemental composition of the coating is Ni:74%, Mo:18%, and O:8% (Ahmad, 2014b)

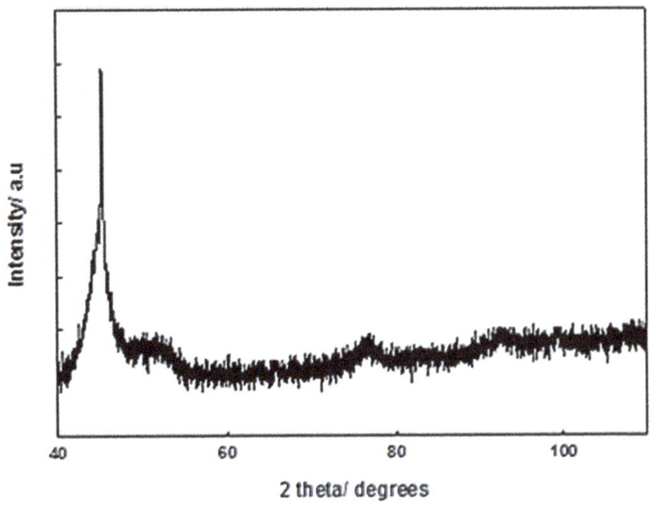

X-ray diffraction can also be used to calculate crystallite size of the Ni-Mo phase in the coatings. Since the XRD patterns for Ni-Mo usually just have the amorphous (111) reflection, the crystallite size can be estimated using the Scherrer equation (Patterson, 1939):

$$L = \frac{k\lambda}{\beta_r \cos\theta} \tag{1}$$

where λ is the wavelength of the x-rays, θ is the diffraction angle, L is the crystallite size, k is a constant (0.94 for Gaussian line profiles and small cubic crystals of uniform size). β_r (in radians) is the corrected full width at half maximum of the peak given by:

$$\beta_r^2 = \beta_m^2 - \beta_s^2 \tag{2}$$

where β_m is the experimental measured width at half maximum and β_s is the corresponding value of a standard with peaks corresponding to the same two theta region. This equation is used to correct for instrumental broadening, when the observed peaks have a near-Gaussian shape. As mentioned earlier, the crystallite size decreases with increasing Mo content in the coating (Figure 3) and can be calculated from XRD. Crystalline sizes from a few nanometers to ~100 nm can be obtained for the Ni-Mo coatings (Karolus, 2004).

MECHANICAL PROPERTIES

The mechanical properties of the electrodeposited coatings are influenced by the microstructure of the alloy matrix and the interaction between nucleation and crystal growth. This microstructure is influenced by parameters of the electrodeposition process, including bath composition, pH, and current density. This effect can be seen even for pure nickel coatings, where measured hardness values can range from ~1.4-5.8 GPa, depending on the experimental parameters.

There have been a few articles which measure the mechanical properties of the electrodeposited Ni-Mo coatings (Beltowska-Lehman, 2012b; Chassaing, 2004; Halim, 2012; Lima-Neto, 2010; Nee, 1988; Stepanova, 1998). The hardness values reported, however, have a large range and there is no consensus among the studies. Halim's (2012) microhardness values tended to be on the lower side, however, they did show a relationship between the % Mo in the coating and measured microhardness. At 11% Mo in the coating, the microhardness was 1.9 GPa, and at 23% Mo, the value was 2.8 GPa. This corresponded to a decrease in crystallite size. However, at the smallest crystallite size, 7 nm, the microhardness decreased to 1.7 GPa, a reversal of the trend. But the 7 nm crystallite size coating had Mo value above 30%, the level in which the Ni-Mo alloy tends to show cracking and non-uniform coatings in the SEM. Another study using pulse deposition showed best hardness values (8.32 GPa) at smaller grain sizes (5 nm) with Mo content around 30% (Huang, 2015). These films may exhibit better properties at the high Mo content due to using a pulse technique for deposition, which helps relieve hydrogen gas build up at the electrode. Chassaing (2004) also showed in their work that as the Mo content increased in the films, the microhardness values also increased up to a point. The best value was 7.9 GPa for coatings with 17-19% Mo and 10-15 nm crystallite size. An effect from annealing was also noticed for the Ni-Mo coatings (Lima-Neto, 2010). As deposited Ni-Mo with 13% Mo content, had a measured hardness value of 2.81 GPa. However when the coatings were annealed up to 600°C, the hardness values increased to

9.80 GPa. Another study reports values between 6.6-7.8 GPa for microhardness and 182-258 GPa for Young's modulus, with the best values for coatings containing 19% Mo (Beltowska-Lehman, 2012b). These values are close to those obtained by the authors using nanoindentation (Table 4).

The nanoindentation technique can be used for determining both modulus and hardness corresponding to the microstructure and mechanical properties of the materials. The indentation results were obtained using a continuous loading indenter. The hardness-depth relationship was obtained using continuous stiffness measurement (CSM) with a dynamic technique. The tests were conducted at a constant strain of 0.05 to a maximum depth of 400 nm. The load-displacement data obtained during the final unloading were analyzed to establish the hardness and elastic modulus using the method of Oliver and Pharr (Oliver, 1992). The depth of displacement together with the known geometry of the indenter offers an indirect measuring area of contact at full load, from which the mean contact pressure, and thus hardness is estimated. The hardness (H) is calculated from the indentation load divided by contact area from the formula:

$$H = \frac{P}{A_c} \tag{3}$$

where A_c is the actual contact area and P is a maximum load. When load is removed from the indenter, an analysis of the initial portion of this unloading response gives an estimate of the elastic modulus. The reduced modulus, E_r, attributed to an elastic displacement of the samples and indenter is represented as:

$$E_r = \frac{1}{\beta}\frac{\sqrt{\pi}}{2}\frac{S}{\sqrt{A_c}} \tag{4}$$

where β is a constant value of 1.034 for a Berkovich tip, S is a contact stiffness characterized by curve fitting the upper portion of the unloading curve and measuring its slope at peak load. A_c is contact area and can be deduced from an empirically determined shape function. The elastic modulus of the specimen (E) is calculated from the effective modulus as the following form:

$$\frac{1}{E_r} = \frac{1-\nu_i^2}{E_i} + \frac{1-\nu_s^2}{E_s} \tag{5}$$

where E_i is modulus of indenter, and ν_s and ν_i are Poisson's ratio for the specimen and the indenter, respectively. The values $E_i = 1140$ GPa, $\nu_i = 0.07$ and $\nu_s = 0.29$ are used in the calculation. The load-displacement data obtained during the final unloading were analyzed to establish the modulus (E_r) and hardness (H) using the method of Oliver and Pharr. The measured E_r was converted to elastic modulus using CSM method with the equation. The average data of elastic modulus (E) and hardness (H) are from 25 different indentations.

The averaged Young's modulus and hardness data with standard deviation for each sample are represented in Table 4. The modulus of the nickel-molybdenum alloys are 194 ± 35 and 189 ± 21 GPa for the molybdenum content of 15 and 29 weight percent, respectively. Obviously, the electrosynthesized

Table 4. The influence of [MoO_4^{2-}]/[Ni^{2+}] molar ratio in the solution on the hardness and Young's modulus of the Ni-Mo alloys

[MoO_4^{2-}]/[Ni^{2+}] Molar Ratio	Hardness (GPa ± SD, n=16)	Young's Modulus (GPa ± SD, n=19)
0 (Ni^{2+} only)	4.3 ± 1.7	143 ± 18
0.1	8.6 ± 1.5	194 ± 35
0.2	8.8 ± 1.8	189 ± 21

alloys exhibited higher modulus values than the deposited nickel films (143 ± 18 GPa). The hardness values of nickel-molybdenum alloy for the molybdenum content of 15 and 29 weight percent are 8.6 ± 1.5 and 8.8 ± 1.8, respectively, both of which are higher than that of the nickel film at 4.3 ± 1.7 GPa. Pure forged Hastelloy B, which contains 28% Mo, has a hardness value of 2.8 GPa. The hardness of nickel-molybdenum alloys increased over that of the nickel film due to the incorporation of high stiffness molybdenum metal into the nickel structure. These nickel-molybdenum alloys were synthesized through a modified pulsed potential electrochemical deposition with grain sizes of 5.6-6.4 nm for the Ni-Mo coatings and 19.8 nm for the nickel coatings.

CORROSION BEHAVIOR OF THE NICKEL-MOLYBDENUM COATINGS

Recently, much of the research for electrodeposited Ni-Mo on different substrates has been in studying either the corrosion behavior and/or the catalytic properties of the coatings. Studies for producing highly active and stable electrocatalysts for the hydrogen evolution reaction (HER) have focused on using nickel-based alloys such as Ni-Mo. (Halim, 2012; Han, 2010; Hu, 2000; Krastajić, 2008; Martinez, 2006; Wang, 2015).

Corrosion protection is another area of interest for Ni-Mo coatings for substrates such as steels. Stainless steels are iron based alloys and most extensively studied in corrosive environments (Sedriks, 1996). Steels are produced with varying compositions depending on the desired resulting property and use for the steel. Stainless steel type 304 is a general purpose grade and comprised of 18-20% chromium, 8-10% nickel, 2% manganese, 1% silicon, 0.080% carbon, 0.045% phosphorus, 0.030% sulfur and balance % iron. Steels containing a minimum of 11% chromium are good for increasing corrosion resistance. Additionally, many other elements are incorporated into steel to provide specific desired properties or ease production of the material. For instance, nickel, molybdenum and nitrogen are introduced to enhance corrosion resistance. Carbon, molybdenum, titanium, aluminum, copper and nitrogen increase mechanical strength of steel. Nickel improves formability and toughness while sulfur and selenium are used to improve machinability. Type 316 is a nickel based stainless steel and one of the largest produced steels in the United States (Frankel, 1998). Type 316, which includes 2-3% molybdenum, has increased resistance to pitting in seawater and chemical industry environments as compared to type 304. Inconel alloys are a type 625 steel, containing 22% chromium, 9% molybdenum, 5% iron and balance % of nickel.

Steels forms a thin metal oxide layer known as the passive film which helps protect the underlying layer from corrosion. Despite this passivity, extensive localized corrosion commonly occurs on stainless steels when exposed to harsh environments containing chloride ions, such as seawater. Consequently, it has been reported that many corrosion failures can occur in service due to localized attack such as pitting, crevice corrosion, intergranular corrosion and stress-corrosion cracking as a result of the breakdown of the passive film. This section will focus on the corrosion protection offered by the Ni-Mo coatings under various conditions. Since Ni-Mo has good corrosion resistance in aggressive environments, there is interest in using it as coatings for steels.

Nickel for Corrosion Protection

Nickel has been used for corrosion protection purposes since the beginning of commercial production in the twentieth century (Friend, 1980). Pure nickel offers high ductility and toughness due to its capacity to hold a face-centered cubic crystalline structure until its melting point (1455°C) is reached. Therefore, nickel provides an important basis for developing specialized alloys. Nickel can form intermetallic phases when alloyed with specific materials, resulting in enhanced strength for both low and high temperature systems (Haubold, 1992). Due to the stable nature of nickel, it is often used to alloy with other metals to protect materials operating under severe conditions such as corrosive environments, high temperature or high stress conditions (Revie, 2008). Nickel is tolerant of nonaerated reducing acids such as hydrochloric, sulfuric and phosphoric acids, but can be rapidly corroded by oxidizing acids such as nitric acid and other mineral acids containing significant amounts of oxidizing salts like ferric or cupric salts, nitrates or peroxides (Friend, 1980). Nickel is able to moderately withstand high-heated air and oxygen rich environments and has particular resistance at elevated temperatures to strong corrosive alkaline solutions such as caustic soda. Due to the formation of a protective nickel halide film, nickel is also able to offer protection at elevated temperatures in the presence of halogen gases, e.g. chlorine and fluorine, as opposed to the majority of other metals. Nickel is resistant to most deaerated organic acids and other organic compounds. In most natural fresh waters, including ground, distilled, deionized, and high-purity waters, nickel has a high degree of corrosion resistance. It is resistant in rapidly moving seawater systems; however severe pitting attacks arise under fouling organisms or other precipitates which can deposit on the surface of the metal in stagnant or low-velocity seawater (Friend, 1980).

As a general rule, nickel favors corrosion under oxidizing conditions, e.g. in nitric acid, whereas under reducing conditions, such as in hydrochloric acid and alkaline solutions, corrosion is retarded. However, nickel has the ability to protect itself from specific types of attack through formation of corrosion-resistant or passive oxide film formation allowing nickel to behave as a more noble material than copper. As a result of passive metal formation, the reducing condition does not typically accelerate corrosion. However, when an oxide film is locally destroyed such as in hot chloride solutions, it may cause nickel to undergo pitting in that area (Friend, 1980).

Molybdenum for Enhancement of Corrosion Resistance

Molybdenum, in the same periodic group as chromium and tungsten, has a melting point of 2620 °C. Molybdenum has similar heat stability as compared to tungsten (Friend, 1980). It is currently the most common material used in Cu(In, Ga)Se_2 (CIGS) solar cell back contacts due to its chemical and mechanical inertness, high conductivity, low resistance contact with the CIGS layer, and commensurate coefficient of thermal expansion (Khatri, 2008; Orgassa, 2003). Molybdenum metal offers high resistance to corrosive nonoxidizing solutions including hydrofluoric, hydrochloric and phosphoric acids under a wide range of concentrations and temperatures. It is particularly resistant to boiling sulfuric acid at concentrations up to 60%. When added to nickel and nickel alloys, molybdenum is particularly resistant to corrosion and pitting attack by chlorides (Friend, 1980; Revie, 2008).

Molybdenum, which improves pitting resistance, has been alloyed in stainless steel used in high chloride containing media (Lu, 1993). The incorporation of molybdenum in stainless steel improves corrosion resistance by decreasing current density. Molybdenum added to stainless steel increased pitting corrosion protection by formation of molybdenum oxide passive films which is more resistant to pitting corrosion (Lu, 1993). The addition of molybdenum to nickel increases resistance to reducing environments and pitting attack from chloride containing solutions, while the addition of sufficient chromium to nickel increases resistance to oxidizing environments such as nitric acid and high-temperature oxidation. Thus, molybdenum and chromium are suitable options for nickel alloy components to provide a broad range of corrosion protection for systems encountered in commercial practice under both oxidizing and reducing conditions.

As an example of the enhanced corrosion protection of molybdenum addition to an alloy, in 1986, a North America nuclear power plant equipped the service water piping system with INCO alloy 25-6MO stainless steel (6% molybdenum) (Maby, 2000). The Mo alloy replaced the 304 and 316 alloys previously used. It was found that the 6% molybdenum added to the stainless steel piping used in the nuclear power plant, where the water supply contained high chloride water fed as an open loop, resulted in an extended service life, twice that of the normal time frame previously reported by inspection cycles (Maby, 2000).

The typical chemical composition of stainless steel grade 304 (UNS No. S30400) is 18% Cr, 9% Ni; type 316 (UNS No. S31600) is 2% Mo, 17% Cr, 11% Ni; INCO alloy 25-6MO (UNS No. N08926) is 6% Mo, 20% Cr, 24% Ni, 0.2% N and 0.75% Cu, balance % of all alloys being iron (Maby, 2000). The service water temperature ranged from 32°F (0°C) to 85°F (28°C) with high chloride content. Type 316 was previously used as it could commonly resist water containing chloride ions up to 1000 ppm at a flow rate of 3ft (1m)/sec until the next maintenance cycle. By actual applications, it was estimated that 25-6MO stainless steel could potentially extend to three or four times the normal operational time frame as compared to Type 316. The plants equipped with 25-6MO stainless steel could save over $1.2 million in maintenance and inspection costs and over 1200 staff hours every eighteen months. Ultimately, they could omit every other inspection cycle, performing the inspection only every 36 months (Maby, 2000).

R. C. Newman (1985a) reported that 2.7% molybdenum in the steel FeCrNi alloy lowered the current density as compared to the alloy without molybdenum when polarized in a chloride media. Molybdenum was found to influence the corrosion behavior of the alloy through ennoblement of the anodic dissolution reaction in the pit area (Newman, 1985a; Newman, 1985b). The pitting dissolution was exhibited as the anodic polarized current. Additionally, molybdenum is able to recover from pitting or decrease the dissolution rate by forming and retaining molybdenum oxyhydroxide or molybdate (MoO_2) at the active surface sites (Ilevbare, 2001; Lu, 1993). The increase in molybdenum addition to stainless steel impressively reduced pitting corrosion in the presence of chromium (Bond, 1968; Cieslak, 1985; Clark, 1990; Frankel, 1998; Lu, 1993).

Molybdenum is one of the most promising elements included in alloys leading to improvement against pitting corrosion. In order to enhance the properties of the passive film, the incorporation of molybdenum is proposed to (1) increase the thickness of the passive film, (2) allow stronger bonding within the metal oxide film, (3) reduce the vacancy concentration of the passive film by forming vacancy-solute complexes and (4) increase the rate of passive film formation growth (Newman, 1985b). According to the pH-potential diagram by Pourbaix (1974), at applicable pH, the oxide films can approximate the metal in equilibrium with the electrolyte. Molybdenum is most stable in the form of a metal oxide with various oxidation states and under a range of pH values. When molybdenum alloys are exposed to water and aqueous solutions, they form several oxidation states of molybdenum oxide creating a passive film. Molybdenum in a hexavalent state can exist in both cationic and anionic states, known as molybdenum trioxide and ferrous molybdate, respectively. Molybdenum can also form in a quadrivalent state as molybdenum dioxide and molybdenum oxyhydroxide. Most studies agree that Mo(IV) exists in the inner region of the passive film, whereas Mo(VI) is present in the outer layer (Newman, 1985a; Newman, 1985b). The outer layer is defined as the salt precipitated layer or an extension of the barrier layer. However, molybdenum in the passive oxide layers may dissolve at a high rate when exposed to strongly acidic electrolytes, resulting in the dissolution of the free metal in the form of Mo(III).

Ni-Mo Alloy Coatings for Corrosion Protection

Nickel-molybdenum alloys comprising ~10 to 29% molybdenum exhibit useful resistance to corrosion under all concentrations and temperatures in hydrochloric acid (Friend, 1980). The alloys are resistant to boiling acids, up to 60% sulfuric acid and pure 85% phosphoric acid. However, the alloy does not offer resistance under strong oxidizing conditions (Friend, 1980). The performance of electrodeposited Ni-Mo coatings, especially, in chloride solutions show that they possess good corrosion protection properties.

Electrochemical techniques, such as poteniodynamic polarization and electrochemical impedance spectroscopy (EIS) can be used to study corrosion phenomenon and reactions. The measurements of the current in relation to the electrode potential can give information to predict a corrosion rate of the coatings within a specific environment. The coated electrode can be polarized by applying an external potential away from equilibrium to monitor corrosion potential and current. At the corrosion potential, the rate of metal dissolution is identical to the rate of hydrogen evolution. The deviation from an equilibrium potential is called polarization. The polarization resistance (R_p) is experimentally observed between the electrochemical current density and applied potential for the corroding electrode within a few millivolts of the polarization from the open circuit potential (E_{ocp}). The Tafel constants, β_a and β_c are obtained from the anodic and cathodic slope, respectively. The E_{corr} and i_{corr} can be determined graphically from the intercepts by an extrapolation method as shown in Figure 8.

The Stern-Geary equation can also be used to obtain an estimated i_{corr}. R_p values are obtained from linear polarization experiments over a small potential range (±20 mV of E_{ocp}). The slopes of the anodic and cathodic branches, β_a, β_c are obtained from the Tafel analysis.

$$i_{corr} = \frac{(\beta_a \beta_c)}{2.303 R_p (\beta_a + \beta_c)} \tag{6}$$

where R_p is the polarization resistance, i_{corr} is the corrosion current, and β_a and β_c are the anodic and cathodic Tafel slopes, respectively. The corrosion rate is inversely proportional to the polarization resistance.

Measuring the corrosion resistance of the coatings is important in assessing the use of electro-deposited Ni-Mo films for different industries, e.g. oil and gas, automobile, transport, military, etc. Bulk Ni-Mo alloys have been studied in several different media and show excellent corrosion resistance in aqueous solutions containing chloride ion. However, there have not been many studies on the corrosion resistance of electrodeposited Ni-Mo coatings. Table 5 shows the measured corrosion parameters from several studies in the literature.

Figure 8. Tafel extrapolation plots

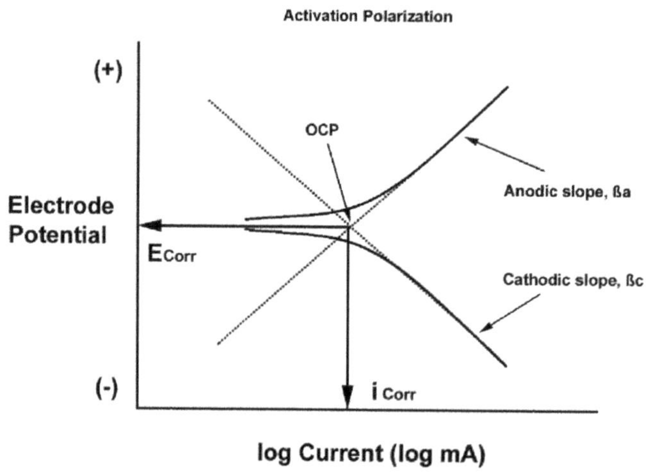

Table 5. Corrosion parameters for electrodeposited Ni-Mo coatings from listed references.

Solution	$E_{corr, vs SCE}$ (V)	i_{corr} (μA cm^{-2})	R_p or R_{coat}	r_{corr}	Reference
1 M HCl	-0.260 to -0.300	55 to 1000	-	-	(Chassaing, 2004)
0.5 M NaCl	-0.749 initial, 0.930 (after 96 hrs) (vs Ag/AgCl)	14 (initial), 20 (after 96 hrs)	3500 Ω (initial), 3752 Ω (after 96 hrs)	-	(Prioteasa, 2010)
3.5% NaCl	-0.459 to -0.517	11.7 to 14.6	1.77 to 2.87 Ω/cm^2	0.14 to 0.18 mpy	(Allahyarzadeh, 2012c)
0.5 M NaOH	-0.429 to -0.473	493 to 701	-	5.69 to 8.27 mm/yr	(Halim, 2012)
3.5% NaCl	-0.374 to -0.384	0.451 to 1.52	22 to 66 kΩ/cm^2	-	(Ahmad, 2014b)
3.5% NaCl	-0.500 to -0.530	0.47 to 1.72	-	-	(Huang, 2015)

Several researchers saw a correlation between the measured corrosion parameter values and the percentage of Mo in the coatings (Ahmad, 2014b; Chassaing, 2004; Halim, 2012; Prioteasa, 2010). The coatings with the best corrosion protection had values between 15 – 27 wt% Mo. All studies contributed the enhanced corrosion protection of the coatings to the formation of a passivating film when exposed to the corroding environment. In general, if the Mo content in the coating was too low (< 10%), then an incomplete or thin passivating layer would form. For a good passive layer to be formed, the Mo content needed to be between 15-25 wt%. Almost all of the literature for electrodeposited Ni-Mo coatings indicated a loss of corrosion protection when the Mo content increased above 30%. This loss of corrosion protection was attributed to an increase in the brittleness of the coatings as the formation of cracks was seen in the SEM images after deposition and also increase in oxygen content (Stepanova, 1998). One other parameter which affected the corrosion protection of the coatings was seen by a group who used ionic liquids in the electrodeposition of the Ni-Mo coatings. The corrosion protection was worse for coatings deposited in the presence of ionic liquids (Allahyarzadeh, 2012c).

Figure 9. Potentiodynamic polarization curves of nickel-molybdenum alloy in 3.5% NaCl solution. The alloys were electrodeposited from citrate-borate plating solutions at different $[MoO_4^{2-}]/[Ni^{2+}]$ molar ratios of 0.1, 0.15, and 0.2 compared to pure nickel deposit

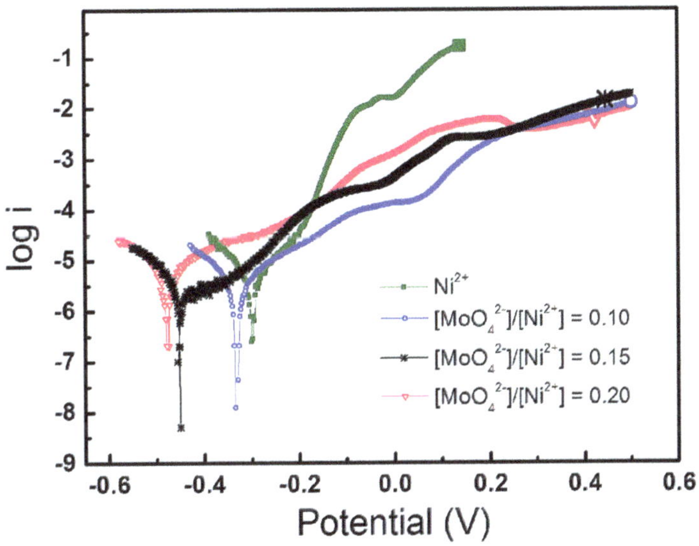

Examples of potentiodynamic polarization curves for electrodeposited nickel and nickel-molybdenum alloys in 3.5% NaCl solution are illustrated in Figure 9 (Tientong, 2014). The electrochemical parameters including corrosion potential, corrosion current, and the polarization resistance obtained from the Tafel analysis are summarized in Table 6.

The molybdenum influence on the passive behavior of the nickel-molybdenum alloy (Figure 9) indicates that a wider passive region (-0.15 to 0.1 V/SCE) was exhibited by the alloy film as compared to the nickel film (-0.1 to 0.0 V/SCE). The alloy containing molybdenum tends to retard corrosion better than the plain nickel film evidenced by the lower current density in the passive region. These results show that the beneficial effect of molybdenum incorporated into the nickel film is a lowering of the corrosion current density. Moreover, molybdenum in the alloy influences the anodic dissolution reaction by enhancing the growth of a passive oxide film, leading to a wider passivation region with higher molybdenum content. The corrosion potential (E_{corr}) of the nickel-molybdenum alloy electrodeposited from the $[MoO_4^{2-}]/[Ni^{2+}]$ molar ratio of 0.1 exhibited a more noble potential, i.e. shifted toward more positive values, compared to that of 0.2 molar ratio. Even though nickel films exhibited an E_{corr} of -0.29 V, the electrodeposited Ni showed the highest corrosion current (i_{corr}) value of 0.316 μA cm^{-2}, while nickel-molybdenum alloys give a corrosion current range from 0.039 to 0.100 μA cm^{-2}. Corrosion current

Table 6. E_{corr}, i_{corr} and R_p of the alloy film electrodeposited from the $[MoO_4^{2-}]/[Ni^{2+}]$ molar ratios of 0.1, and 0.2 and pure nickel film in the citrate-borate plating bath.

$[MoO_4^{2-}]/[Ni^{2+}]$ Molar Ratio	E_{corr} (V)	i_{corr} (μA cm^{-2})	R_p (kΩ cm^2)	β_c (V dec^{-1})	β_a (V dec^{-1})
Pure Ni	-0.29	0.316	38.89	-18	11
0.1	-0.33	0.039	1092	-25	20
0.2	-0.47	0.100	486.9	-16	14

is an important parameter for evaluating the kinetics of the corrosion reaction. Lower corrosion currents lead to better corrosion protection. The corrosion current of the synthesized alloy was low which shows better corrosion resistance. The alloy fabricated from the $[MoO_4^{2-}]/[Ni^{2+}]$ molar ratio of 0.1 provided the highest polarization resistance (R_p) value of 1092 kΩ cm², implying better corrosion resistance. In addition, the anodic polarization for the alloys fabricated from the 0.1 $[MoO_4^{2-}]/[Ni^{2+}]$ molar ratio plating bath revealed a longer passivation region with a lower current plateau than that obtained from the molar ratio of 0.2. The 0.1 molar ratio has between 16-20 wt% Mo in the coatings, while the 0.2 molar ratio is 29-34 wt%. The corrosion results match those in literature which shows the best corrosion resistance between 15-25 wt% Mo and loss of protection as the Mo content increases above 30%. In summary, the nickel-molybdenum alloy obtained from the ratio of 0.1 offers a better corrosion resistance shown by its lower current density and longer passivation region.

FUTURE RESEARCH DIRECTIONS

Using electrodeposition to obtain Ni-Mo coatings shows promise, however there are some areas of research that are still needed in order to understand how to produce optimal coatings.

- Elucidation of the exact deposition mechanism. While there have been several in depth studies, there is not a consensus on the exact mechanism. Also, the mechanism is most likely different for acidic versus alkaline solutions and this difference will have to be addressed.
- Comparison of the acidic and alkaline plating solutions under similar conditions. While the majority of the literature involving HER studies has used acidic deposition conditions, other studies for the elucidation of the mechanism and corrosion properties have deposited the coatings out of alkaline solutions. There are different research motivations when depositing films for properties such as corrosion protection or enhanced mechanical performance as opposed to an electrode for the hydrogen evolution reaction. An overall review on both the acidic and alkaline electrodeposition of Ni-Mo alloys would contribute to understanding the literature.
- An in-depth study of the separate influences which ammonia, citrate, and other complexing ligands have on the attainment and properties of the Ni-Mo coatings. Studies have noted that the number and type of species in the plating solution can be quite complex.

CONCLUSION

Electrodeposition of nickel-molybdenum is a promising pathway to produce technological important coatings for many industries. This chapter has explored the various plating solutions that have been used for electrodeposition of the coatings. Characterization of Ni-Mo by SEM, AFM, and XRD of the resulting coatings has been covered and showed that morphology, composition, and crystallite size are all influenced by the plating parameters. Mechanical and corrosion properties of the electrodeposited Ni-Mo alloys from alkaline baths have been reviewed and compared across several different studies.

REFERENCES

Ahmad, Y. H., Tientong, J., Nar, M., D'Souza, N., Mohamed, A. M. A., & Golden, T. D. (2014a). Characterization and corrosion resistance of electrodeposited Ni-Mo-silicate platelet nanocomposite coatings. *Surface and Coatings Technology*, *259*(Part C), 517–525. doi:10.1016/j.surfcoat.2014.10.036

Ahmad, Y. H., Mohamed, A. M. A., Golden, T. D., & D'Souza, N. A. (2014b). Electrodeposition of nanocrystalline Ni-Mo alloys from alkaline glycinate solutions. *International Journal of Electrochemical Science*, *9*(11), 6438–6450.

Ahn, M. K., Kwon, H. S., & Lee, H. M. (1998). Quantitative comparison of the influences of tungsten and molybdenum on the passivity of Fe-29Cr ferritic stainless steels. *Corrosion Science*, *40*(2-3), 307–322. doi:10.1016/S0010-938X(97)00138-8

Allahyarzadeh, M. H., Roozbehani, B., & Ashrafi, A. (2011). Electrodeposition of high Mo content amorphous/nanocrystalline Ni-Mo alloys using 1-ethyl-3-methyl-imidazolium chloride ionic liquid as an additive. *Electrochimica Acta*, *56*(27), 10210–10216. doi:10.1016/j.electacta.2011.09.011

Allahyarzadeh, M. H., Roozbehani, B., Ashrafi, A., Shadizadeh, S. R., Seddighian, A., & Kheradmand, E. (2012a). A novel aspect of two imidazolium-based ionic liquids in electrodeposition of amorphous/nanocrystalline Ni-Mo. *Journal of the Electrochemical Society*, *159*(8), D473–D478. doi:10.1149/2.065208jes

Allahyarzadeh, M. H., Roozbehani, B., Ashrafi, A., Shadizadeh, S. R., & Kheradmand, E. (2012b). Electrodeposition of high Mo content amorphous/nanocrystalline Ni-Mo using ionic liquids as additive. *ECS Transactions*, *41*(44), 11–28. doi:10.1149/1.4718388

Allahyarzadeh, M. H., Roozbehani, B., Ashrafi, A., & Shadizadeh, S. R. (2012c). Effect of surface preparation on corrosion resistance of electrochemically deposited Ni-Mo alloys in the presence 1-methyl-imidazolium chloride ionic liquid. *ECS Transactions*, *41*(44), 29–46. doi:10.1149/1.4718389

Bastidas, J. M., Torres, C. L., Cano, E., & Polo, J. L. (2002). Influence of molybdenum on passivation of polarised stainless steels in a chloride environment. *Corrosion Science*, *44*(3), 625–633. doi:10.1016/S0010-938X(01)00072-5

Beltowska-Lehman, E., & Indyka, P. (2012a). Kinetics of Ni-Mo electrodeposition from Ni-rich citrate baths. *Thin Solid Films*, *520*(6), 2046–2051. doi:10.1016/j.tsf.2011.10.024

Beltowska-Lehman, E., Bigos, A., Indyka, P., & Kot, M. (2012b). Electrodeposition and characterization of nanocrystalline Ni-Mo coatings. *Surface and Coatings Technology*, *211*, 67–71. doi:10.1016/j.surfcoat.2011.10.011

Bond, A. P., & Lizlovs, E. A. (1968). Anodic polarization of austenitic stainless steels in chloride media. *Journal of the Electrochemical Society*, *115*(11), 1130–1135. doi:10.1149/1.2410923

Brenner, A. (1963). Electrodeposition of alloys containing molybdenum and nickel, cobalt, or iron. Electrodeposition of Alloys: Principles and Practice (p. 413). Academic Press.

Brooman, E. W. (2000). Corrosion performance of environmentally acceptable alternatives to cadmium and chromium coatings: Chromium-Part I. *Metal Finishing*, *98*(7), 38–43. doi:10.1016/S0026-0576(00)82334-X

Caceres, L., Vargas, T., & Herrera, L. (2009). Influence of pitting and iron oxide formation during corrosion of carbon steel in unbuffered NaCl solutions. *Corrosion Science*, *51*(5), 971–978. doi:10.1016/j.corsci.2009.02.021

Chassaing, E., Cornet, M., & Kihn, V. Q. (1978). Influence of codeposited impurities on hydrogen permeation and properties of Ni-Mo and Co-Mo electrodeposited alloys. *Surface Technology*, *7*(2), 145–150. doi:10.1016/0376-4583(78)90007-9

Chassaing, E., Quang, K. V., & Wiart, R. (1989). Mechanism of nickel-molybdenum alloy electrodeposition in citrate electrolytes. *Journal of Applied Electrochemistry*, *19*(6), 839–843. doi:10.1007/BF01007931

Chassaing, E., Roumegas, M. P., & Trichet, M. F. (1995). Electrodeposition of Ni-Mo alloys with pulse reverse potentials. *Journal of Applied Electrochemistry*, *25*(7), 667–670. doi:10.1007/BF00241928

Chassaing, E., Portail, N., Levy, A.-F., & Wang, G. (2004). Characterisation of electrodeposited nanocrystalline Ni-Mo alloys. *Journal of Applied Electrochemistry*, *34*(11), 1085–1091. doi:10.1007/s10800-004-2460-z

Cieslak, W. R., & Duquette, D. J. (1985). An electrochemical study of the pit initiation resistance of ferritic stainless steels. *Journal of the Electrochemical Society*, *132*(3), 533–537. doi:10.1149/1.2113882

Clark, B. E., Thorpe, S. J., & Aust, K. T. (1990). The pitting behaviour of FeNiCrMo alloys. *Corrosion Science*, *31*, 551–556. doi:10.1016/0010-938X(90)90161-W

Conrad, H. C., Corbett, J. R., & Golden, T. D. (2012). Electrochemical deposition of γ-phase Zinc-Nickel alloys from alkaline solution. *Journal of the Electrochemical Society*, *159*(1), C29–C32. doi:10.1149/2.027201jes

Crousier, J., Eyraud, M., Crousier, J. P., & Roman, J.-M. (1992). Influence of substrate on the electrodeposition of nickel-molybdenum alloys. *Journal of Applied Electrochemistry*, *22*(8), 749–755. doi:10.1007/BF01027505

Cvijović, Z., & Radenković, G. (2006). Microstructure and pitting corrosion resistance of annealed duplex stainless steel. *Corrosion Science*, *48*(12), 3887–3906. doi:10.1016/j.corsci.2006.04.003

Donten, M., Cesiulis, H., & Stojek, Z. (2005). Electrodeposition of amorphous/nanocrystalline and polycrystalline Ni-Mo alloys from pyrophosphate baths. *Electrochimica Acta*, *50*(6), 1405–1412. doi:10.1016/j.electacta.2004.08.028

Eliaz, N., & Gileadi, E. (2008). Induced codeposition of alloys of tungsten, molybdenum and rhenium with transition metals. In Modern Aspects of Electrochemistry (Vol. 42, pp. 192-193). doi:10.1007/978-0-387-49489-0_4

Elezović, N. R., Jović, V. D., & Krstajić, N. V. (2005). Kinetics of the hydrogen evolution reaction on Fe-Mo film deposited on mild steel support in alkaline solution. *Electrochimica Acta*, *50*(28), 5594–5601. doi:10.1016/j.electacta.2005.03.037

Ernst, D. W., & Holt, M. L. (1958). Cathode potentials during the electrodeposition of molybdenum alloys from aqueous solutions. *Journal of the Electrochemical Society*, *105*(11), 686–692. doi:10.1149/1.2428691

Frankel, G. S. (1998). Pitting corrosion of metals. A review of the critical factors. *Journal of the Electrochemical Society*, *145*(6), 2186–2198. doi:10.1149/1.1838615

Friend, W. Z. (1980). *Corrosion of nickel and nickel-base alloys*. New York: Wiley-Interscience.

Fukushima, H., & Higashi, K. (1978). A study of the electrodeposition of Ni-Mo alloys from ammoniacal tartrate bath. *Journal of the Metal Finishing Society of Japan*, *29*(12), 627–635. doi:10.4139/sfj1950.29.627

Gao, Y., Xiong, J., Gong, D., Li, J., & Ding, M. (2015). Improvement of solar absorbing property of Ni-Mo based thermal spray coatings by laser surface treatment. *Vacuum*, *121*, 64–69. doi:10.1016/j.vacuum.2015.07.018

Gómez, E., Pellicer, E., & Vallés, E. (2003). Influence of the bath composition and the pH on the induced cobalt-molybdenum electrodeposition. *Journal of Electroanalytical Chemistry*, *556*, 137–145. doi:10.1016/S0022-0728(03)00339-5

Gomez, E., Pellicer, E., Duch, M., Esteve, J., & Valles, E. (2006). Molybdenum alloy electrodeposits for magnetic actuation. *Electrochimica Acta*, *51*(16), 3214–3222. doi:10.1016/j.electacta.2005.09.010

Green, T. A., Russell, A. E., & Roy, S. (1998). The development of a stable citrate electrolyte for the electrodeposition of copper-nickel alloys. *Journal of the Electrochemical Society*, *145*(3), 875–881. doi:10.1149/1.1838360

Halim, J., Abdel-Karim, R., El-Raghy, S., Nabil, M., & Waheed, A. (2012). Electrodeposition and characterization of nanocrystalline Ni-Mo catalysts for hydrogen production. *Journal of Nanomaterials*, *845673*, 1–9. doi:10.1155/2012/845673

Han, Q., Cui, S., Pu, N., Chen, J., Liu, K., & Wei, X. (2010). A study on pulse plating amorphous Ni-Mo alloy coating used as HER cathode in alkaline medium. *International Journal of Hydrogen Energy*, *35*(11), 5194–5201. doi:10.1016/j.ijhydene.2010.03.093

Haubold, T., Bohnn, R., Birringer, R., & Gleiter, H. (1992). Nanocrystalline intermetallic compounds—structure and mechanical properties. *Materials Science and Engineering A*, *153*(1-2), 679–683. doi:10.1016/0921-5093(92)90270-B

Hu, C. C., & Weng, C. Y. (2000). Hydrogen evolving activity on nickel-molybdenum deposits using experimental strategies. *Journal of Applied Electrochemistry*, *30*(4), 499–506. doi:10.1023/A:1003964728030

Huang, P. C., Hou, K. H., Sheu, H. H., Ger, M. D., & Wang, G. L. (2014). Wear properties of Ni–Mo coatings produced by pulse electroforming. *Surface and Coatings Technology*, *258*, 639–645. doi:10.1016/j.surfcoat.2014.08.024

Huang, P. C., Hou, K. H., Wang, G. L., Chen, M. L., & Wang, J. R. (2015). Corrosion resistance of the Ni-Mo alloy coatings related to coating's electroplating parameters. *International Journal of Electrochemical Science*, *10*(6), 4972–4984.

Ilevbare, G. O., & Burstein, G. T. (2001). The role of alloyed molybdenum in the inhibition of pitting corrosion in stainless steels. *Corrosion Science*, *43*(3), 485–513. doi:10.1016/S0010-938X(00)00086-X

Jung, R. H., Tsuchiya, H., & Fujimoto, S. (2012). XPS characterization of passive films formed on Type 304 stainless steel in humid atmosphere. *Corrosion Science*, *58*, 62–68. doi:10.1016/j.corsci.2012.01.006

Kahl, M., & Golden, T. D. (2014). Electrochemical determination of phenolic acids at a Zn/Al layered double hydroxide film modified glassy carbon electrode. *Electroanalysis*, *26*(8), 1664–1670. doi:10.1002/elan.201400156

Karolus, M., Rówiński, E., & Łągiewka, E. (2006). Structure and surface studies on Ni-Mo alloys with polymers. *Journal of Achievements in Materials and Manufacturing Engineering*, *18*(1-2), 119-122.

Kawashima, A., Akiyama, E., Habazaki, H., & Hashimoto, K. (1997). Characterization of sputter-deposited Ni-Mo and Ni-W alloy electrocatalysts for hydrogen evolution in alkaline solution. *Materials Science and Engineering A*, *226–228*, 905–909. doi:10.1016/S0921-5093(97)80095-0

Kazimierczak, H., Ozga, P., Świątek, Z., & Bielańska, E. (2013a). Characterisation of Zn-Mo alloy layers electrodeposited from aqueous citrate solution. *Journal of Alloys and Compounds*, *578*, 82–89. doi:10.1016/j.jallcom.2013.04.205

Kazimierczak, H., Ozga, P., & Socha, R. P. (2013b). Investigation of electrochemical co-deposition of zinc and molybdenum from citrate solutions. *Electrochimica Acta*, *104*, 378–390. doi:10.1016/j.electacta.2012.12.140

Khatri, H., & Marsillac, S. (2008). The effect of deposition parameters on radiofrequency sputtered molybdenum thin films. *Journal of Physics: Condensed Matter*, 20(5). doi:.10.1088/0953-8984/20/05/055206

Koch, C. C. (2007). Structural nanocrystalline materials: An overview. *Journal of Materials Science*, *42*(5), 1403–1414. doi:10.1007/s10853-006-0609-3

Krastajić, N. V., Jović, V. D., Gajic-Krstajić, Lj., Jović, B. M., Antozzi, A. L., & Martelli, G. N. (2008). Electrodeposition of Ni-Mo alloy coatings and their characterization as cathodes for hydrogen evolution in sodium hydroxide solution. *International Journal of Hydrogen Energy*, *35*(14), 3676–3687. doi:10.1016/j.ijhydene.2008.04.039

Ksycki, S. M. J., & Yntema, L. F. (1949). The electrodeposition of molybdenum from aqueous solutions. *Journal of the Electrochemical Society*, *96*(1), 48–56. doi:10.1149/1.2776770

Kuznetsov, V. V., Golyanin, K. E., Pshenichkina, T. V., Lyakhov, B. F., & Lyashenko, S. E. (2013). Chemical composition of Fe-Mo alloys obtained by electrodeposition. *Mendeleev Communications*, *23*(6), 331–333. doi:10.1016/j.mencom.2013.11.009

Li, C.-Q., Li, X.-H., Wang, Z.-X., & Guo, H.-J. (2007). Nickel electrodeposition from novel citrate bath. *Transactions of Nonferrous Metals Society of China*, *17*(6), 1300–1306. doi:10.1016/S1003-6326(07)60266-0

Lima-Neto, P. D., Correia, A. N., Vaz, G. L., & Casciano, P. N. (2010). Morphological, structural, microhardness and corrosion characterisations of electrodeposited Ni-Mo and Cr coatings. *Journal of the Brazilian Chemical Society*, *21*(10), 1968–1976. doi:10.1590/S0103-50532010001000024

Lu, Y. C., Ives, M. B., & Clayton, C. R. (1993). Synergism of alloying elements and pitting corrosion resistance of stainless steels. *Corrosion Science*, *35*(1-4), 89–96. doi:10.1016/0010-938X(93)90137-6

Maby, M. (2000, July). Moly does the job in nuclear power applications. *Newsletter, International Molybdenum Association*. London, UK.

Marlot, A., Kern, P., & Landolt, D. (2002). Pulse plating of Ni-Mo alloys from Ni-rich electrolytes. *Electrochimica Acta*, *48*(1), 29–36. doi:10.1016/S0013-4686(02)00544-3

Martinez, S., Metikoš-Huković, M., & Valek, L. (2006). Electrocatalytic properties of electrodeposited Ni-15Mo cathodes for the HER in acid solutions: Synergistic electronic effect. *Journal of Molecular Catalysis A Chemical*, *245*(1-2), 114–121. doi:10.1016/j.molcata.2005.09.040

Mishra, R., & Balasubramaniam, R. (2004a). Effect of nanocrystalline grain size on the electrochemical and corrosion behavior of nickel. *Corrosion Science*, *46*(12), 3019–3029. doi:10.1016/j.corsci.2004.04.007

Mishra, R., Basu, B., & Balasubramaniam, R. (2004b). Effect of grain size on the tribological behavior of nanocrystalline nickel. *Materials Science and Engineering A*, *373*(1-2), 370–373. doi:10.1016/j.msea.2003.09.107

Mishra, A. K., & Shoesmith, D. W. (2013). The activation/depassivation of nickel–chromium–molybdenum alloys: An oxyanion or a pH effect—Part II. *Electrochimica Acta*, *102*, 328–335. doi:10.1016/j.electacta.2013.03.177

Nee, C. C., Kim, W., & Weil, R. (1988). Pulsed electrodeposition of Ni-Mo alloys. *Journal of the Electrochemical Society*, *135*(5), 1100–1103. doi:10.1149/1.2095883

Newman, R. C. (1985a). The dissolution and passivation kinetics of stainless alloys containing molybdenum—1. Coulometric studies of FeCr and FeCrMo alloys. *Corrosion Science*, *25*(5), 331–339. doi:10.1016/0010-938X(85)90111-8

Newman, R. C. (1985b). The dissolution and passivation kinetics of stainless alloys containing molybdenum—II. Dissolution kinetics in artificial pits. *Corrosion Science*, *25*(5), 341–350. doi:10.1016/0010-938X(85)90112-X

Niu, Z.-J., Yao, S.-B., & Zhou, S.-M. (1998). In situ surface Raman investigation on induced-codeposition of an Fe-Mo alloy. *Journal of Electroanalytical Chemistry*, *455*(1-2), 205–207. doi:10.1016/S0022-0728(98)00148-X

Ohgai, T., Tanaka, Y., & Washio, R. (2013). Nanocrystalline structure and soft magnetic properties of nickel-molybdenum alloy thin films electrodeposited from acidic and alkaline aqueous solutions. *Journal of Solid State Electrochemistry*, *17*(3), 743–750. doi:10.1007/s10008-012-1924-z

Oliver, W. C., & Pharr, G. M. (1992). An improved technique for determining hardness and elastic modulus using load and displacement sensing indentation experiments. *Journal of Materials Research*, *7*(6), 1564–1583. doi:10.1557/JMR.1992.1564

Orgassa, K., Schock, H. W., & Werner, J. H. (2003). Alternative back contact materials for thin film Cu(In,Ga)Se_2 solar cells. *Thin Solid Films*, *431-432*, 387–391. doi:10.1016/S0040-6090(03)00257-8

Patterson, A. L. (1939). The Scherrer formula for X-ray particle size determination. *Physical Review*, *56*(10), 978–982. doi:10.1103/PhysRev.56.978

Podlaha, E. J., Matloez, M., & Landolt, D. (1993). Electrodeposition of high Mo content Ni-Mo alloys under forced convection. *Journal of the Electrochemical Society*, *140*(10), L149–L151. doi:10.1149/1.2220956

Podlaha, E. J., & Landolt, D. (1996a). Induced codeposition I. An experimental investigation of Ni-Mo alloys. *Journal of the Electrochemical Society*, *143*(3), 885–892. doi:10.1149/1.1836553

Podlaha, E. J., & Landolt, D. (1996b). Induced codeposition II. A mathematical model describing the electrodeposition of Ni-Mo alloys. *Journal of the Electrochemical Society*, *143*(3), 893–899. doi:10.1149/1.1836554

Podlaha, E. J., & Landolt, D. (1997). Induced codeposition III. Molybdenum alloys with nickel, cobalt, and iron. *Journal of the Electrochemical Society*, *144*(5), 1672–1680. doi:10.1149/1.1837658

Pourbaix, M. (1974). *Atlas of Electrochemical Equilibria in Aqueous Solutions, NACE International Cebelcor* (2nd ed.). Houston, Texas: National Association of Corrosion Engineers.

Prioteasa, P., Anicăi, L., & Vişan, T. (2010). Synthesis and corrosion characterization of electrodeposited Ni-Mo alloys obtained from aqueous solutions. *U.P.B. Science Bulletin, Series B.*, *72*(4), 11–24.

Raj, I. A., & Vekatesan, V. K. (1998). Characterization of nickel-molybdenum and nickel-molybdenum-iron alloy coatings as cathodes for alkaline water electrolysers. *International Journal of Hydrogen Energy*, *13*(4), 215–223. doi:10.1016/0360-3199(88)90088-2

Revie, R. W., & Uhlig, H. H. (2008). *Corrosion and corrosion control. An introduction to corrosion science and engineering* (4th ed.). New York, NY: John Wiley & Sons, Inc.

Rode, S., Henninot, C., & Matlosz, M. (2005). Complexation chemistry in nickel and copper-nickel alloy plating from citrate baths. *Journal of the Electrochemical Society*, *152*(4), C248–C254. doi:10.1149/1.1869980

Safikhani, A., Esmailian, M., Salmani, M. R., & Aminfard, M. (2014). Effect of Ni-Mo addition on cyclic and isothermal oxidation resistance and electrical behavior of ferritic stainless steel for SOFCs interconnect. *International Journal of Hydrogen Energy*, *39*(21), 11210–11223. doi:10.1016/j.ijhydene.2014.05.122

Sanches, L. S., Marino, C. B., & Mascaro, L. H. (2007). Investigation of the codeposition of Fe and Mo from sulphate-citrate acid solutions. *Journal of Alloys and Compounds*, *439*(1-2), 342–345. doi:10.1016/j.jallcom.2006.08.231

Scully, J. C. (1975). *The Fundamentals of Corrosion* (2nd ed.). Oxford, New York: Pergamon Press.

Sedriks, A. J. (1996). *Corrosion of Stainless Steel* (2nd ed.). New York, NY: John Wiley & Sons, Inc.

Singh, V. B., & Singh, D. (1975). Electrodeposition of Ni-Mo alloys from acetate bath. *Transactions of the Society for Advancement of Electrochemical Science and Technology*, *10*(2), 91-97.

Stepanova, L. I., & Purovskaya, O. G. (1998). Electrodeposition of nickel-based alloys with tungsten and molybdenum. *Metal Finishing*, *96*(11), 50–53. doi:10.1016/S0026-0576(98)80871-4

Sugimoto, K., & Sawada, Y. (1977). The role of molybdenum additions to austenitic stainless steels in the inhibition of pitting in acid chloride solutions. *Corrosion Science*, *17*(5), 425–445. doi:10.1016/0010-938X(77)90032-4

Sun, S., & Podlaha, E. J. (2012). Electrodeposition of Mo-rich, MoNi alloys from an aqueous electrolyte. *Journal of the Electrochemical Society*, *159*(2), D97–D102. doi:10.1149/2.091202jes

Tientong, J. (2014). *Electrodeposition of nickel and nickel alloy coatings with layered silicates for enhanced corrosion resistance and mechanical properties* [Doctoral dissertation].

Tobler, W. J., & Virtanen, S. (2006). Effect of Mo species on metastable pitting of Fe18Cr alloys-A current transient analysis. *Corrosion Science*, *48*(7), 1585–1607. doi:10.1016/j.corsci.2005.05.049

Wang, A. Q., & Golden, T. D. (2013). Electrodeposition of oriented cerium oxide films. *International Journal of Electrochemistry*, *2013*, 1–10. doi:10.1155/2013/482187

Wang, L., Zhang, J., Gao, Y., Xue, Q., Hu, L., & Xu, T. (2006). Grain size effect in corrosion behavior of electrodeposited nanocrystalline Ni coatings in alkaline solution. *Scripta Materialia*, *55*(7), 657–660. doi:10.1016/j.scriptamat.2006.04.009

Wang, M., Wang, Z., Yu, X., & Guo, Z. (2015). Facile one-step electrodeposition preparation of porous NiMo film as electrocatalyst for hydrogen evolution reaction. *International Journal of Hydrogen Energy*, *40*, 2173–2181. doi:10.1016/j.ijhydene.2014.12.022

Xuetao, Y., Yu, W., Dongbai, S., & Hongying, Y. (2008). Influence of pulse parameters on the microstructure and microhardness of nickel electrodeposits. *Surface and Coatings Technology*, *202*(9), 1895–1903. doi:10.1016/j.surfcoat.2007.08.023

Youssef, K. M. S., Koch, C. C., & Fedkiw, P. S. (2004). Improved corrosion behavior of nanocrystalline zinc produced by pulse-current electrodeposition. *Corrosion Science*, *46*(1), 51–64. doi:10.1016/S0010-938X(03)00142-2

Zelenin, O. Y. (2007). Interaction of the Ni^{2+} ion with citric acid in an aqueous solution. *Russian Journal of Coordination Chemistry*, *33*(5), 346–350. doi:10.1134/S1070328407050065

Chapter 15
Concept of Advanced Thermal Barrier Functional Coatings in High Temperature Engineering Components

A.H. Pakseresht
Materials and Energy Research Center, Iran

M. Alizadeh
Materials and Energy Research Center, Iran

M.R. Rahimipour
Materials and Energy Research Center, Iran

S.M.M. Hadavi
Materials and Energy Research Center, Iran

A. Shahbazkhan
Sharif University, Iran

ABSTRACT

In conventional thermal barrier coatings (TBCs) the major problem is the spalling of a ceramic coating. This could be due to the large thermal stresses which are generated during thermal cycling on the oxidizing environment. One of the ways to improve the life span and overcome mentioned problems is introducing the concept of functionally graded materials (FGM) into the TBC. Functionally graded materials are referred to as a class of advanced materials that are distinguished by variation in their properties with varying their dimensions. Through employing a functionally graded thermal barrier coating (FG-TBC), an intermediate layer with a gradual compositional variation is embedded between the top and the bond coats. This layer(s), composed of ceramic and metal in various ratios, can achieve a gradual composition variation, thereby leading to gradual changes in microstructures and better mechanical and physical properties.

INTRODUCTION AND BACKGROUND OF THERMAL BARRIER COATING

In the past few decades, several industries, e.g., gas turbines, have made considerable progress in the application of ceramic coatings that are used for the protection of the underlying base structures from oxidation, corrosion attack, and mechanical degradation upon exposure to high thermal stresses coupled

DOI: 10.4018/978-1-5225-0066-7.ch015

with fatigue and/or creep phenomena (Lee & Stinton, 1996). These ceramic coatings, generally known as TBCs, currently provide thermal insulation against hot gasses in turbines and engines (Rangaraj & Kokini, 2014).

It is known that TBCs have been employed in different applications for more than 40 years (Miller, R. A. 1997). In the mid-1970s, these coatings were first successfully explored and tested in the turbine engine. In the early 1980s, these were applied in the revenue service on the vane platforms of aircraft gas turbine engines (Widjaja, Limarga, & 2002; Miller, R. A. 1997). In this process, TBCs were applied to metallic surfaces operating at elevated temperature to increase their temperature.

Note that TBCs have already been widely applied to increase the efficiency of high-temperature components, such as aerospace, aircraft, marine automobiles, and heavy-duty utilities (Xiong, Kawasaki, Kang, & Watanabe, 2005). The reliability of turbine components can also be improved by applying the protective coating. The reliability of TBCs can be affected by various factors such as microstructure, porosity, and residual stress. Residual stress has received considerable attention because it can influence the effect of many factors, such as microcrack formation, adhesion strength, and creep behavior, on the performance of TBC (Widjaja, Limarga, & Yip, 2003). The mismatch stress of the coefficient of thermal expansion (CTE) between different layers and substrates should be minimized to increase the spallation resistance (Negahdari. Porda, & Scherm. 2010).

TBC systems that are applied in turbines and aircraft engines are composed of a top coat (TC) and a bond coat (BC). TC consists a 200–350 μm thermally insulating ceramic layer and BC a 50–150 μm metallic layer between the ceramic top coat and the metallic substrate. These TBC systems are generally known as the duplex or the conventional TBC (Kim, Kim, & Park, 2003). Ceramic layers have a low thermal resistance that can increase the inlet temperatures of the turbine from 100 °C to 300 °C, resulting in a decrease in the heating rate transferred to the component. Therefore, the service temperature and hence the productivity can be increased accordingly (Rangaraj, & Kokini, 2003b). It is noteworthy that the inlet temperatures of turbines have considerably been increased since 1940s. Nowadays, the inlet temperatures of some turbines are more than 1500–1600 °C. They are supposed to reach 1760 °C or more in the recent years. It is important for gas turbines to work at temperatures near the melting point of their substrates, implying that the new and modern thermal barrier coatings are required (Karaoglani, Ogawa, Turk, & Ozdemir, 2013). Therefore, the applications of the modern TBC have been increasing as a result of the requirement of higher engine efficiency, which can be obtained by increasing the operating temperatures (Widjaja, Limarga, & 2002). In addition, if the temperature of a component is reduced, its substrate material would show properties that are similar to those found at the room temperature (Rangaraj & Kokini, 2014; Viswanathan, Dwivedi, & Sampath, 2014). The modern TBCs have been used to improve the life of turbine components under critical environmental conditions such as corrosion, oxidation, and wear at high temperatures (Rangaraj, & Kokini, 2003 b; Wolfe, & Singh, 1998). Regardless of various advantages, the understanding the complexity of phenomena, such as thermal expansion, plastic and elastic deformation, fatigue, fracture, creep deformation, diffusion, phase transformation, oxidation, thermal conduction, radiation, and sintering, in the design and development of the modern TBCs, has become a challenging task (Goswami, Ray, & Sahay, 2004).

In recent years, several studies have focused on the following three areas to improve the modern TBCs: (1) improving the conventional powder coating (ZrO_2) by doping different metal oxide stabilizers such as ceria and yttria; (2) searching for new materials usable as TBCs at higher temperatures, such as oxides with pyrochlore, fluorite, and perovskite structures; and (3) developing architectural designs of TBCs based on the concepts of functionally graded TBCs (FG-TBCs) and double ceramic layered TBCs (Chen et al. 2011). TBC coatings with new materials, and structures known as new TBCs are described in the following sections. Recent studies have shown that the new TBCs, such as thermal-barrier-type FGM and multilayer TBC coatings, can exhibit better properties than those coatings that do not possess the graded structure (Xiong, Kawasaki, Kang, & Watanabe, 2005). Generally, the properties of FG-TBCs depend on various factors such as mechanical and thermal properties of the mixed materials and the processing techniques (Chen, Gu, Zou, Wang, & Cao, 2012). In this regard, FG-TBCs can improve the factors such as bonding strength, toughness, wear and corrosion resistance, and reduced residual and thermal stresses bestowed by the properties of the graded material.

The objective of this study was to introduce the concept of creating functionally graded microstructure for thermal barrier coatings and to describe some of the recent research on functionally graded TBCs, multilayer TBCs, new TBC fabrication methods, and new materials used for improving TBCs, applied in high-temperature applications.

CONVENTIONAL TBCs AND THEIR STRUCTURES

The application of zirconia-based coatings is appealing due to their low thermal conductivity, relatively high thermal expansion coefficient compared with ceramic materials, good microstructure stability, and good thermal shock resistance and oxidation resistance. They also possess good mechanical integrity when exposed to thermal shock and thermal cycling, high thermodynamic stability, and high melting temperature. These characteristics make ZrO_2 the leading material to be applied as a TBC. However, ZrO_2 undergoes polymorphic phase transformations because of the high quenching rate of the process during plasma spraying (Wolfe, & Singh, 1998; Khor, Dong & Gu, 1999).

The tetragonal phase transformation occurs around 1443 K with a 4% decrease in volume, and a tetragonal to cubic transformation occurs around 2643 K. This change in volume (during thermal cycling) can result in a spalling of the coating and eventually in the component failure.

Therefore, ZrO_2 can be mixed with 3–8 wt% phase-stabilizing agents, such as Al_2O_3, CaO, CeO_2, MgO, and Y_2O_3, in order to minimize the change in the volume due to phase transformations at higher temperatures (Wolfe, & Singh, 1998; Khor, Dong & Gu, 1999). Among them, Yttria-stabilized zirconia (YSZ) has been commonly used as the top coat due to its low thermal conductivity, relatively high thermal expansion coefficient, good fracture toughness and good chemical stability among all ceramic material. It is, therefore used as a commercial powder in many industries. A bond coat layer, which is typically MCrAlY (M= Ni, Co, Fe or a combination of these elements) produced from metal alloys with high thermal expansion coefficients, is often first deposited onto a metallic substrate (Widjaja, Limarga, & Yip 2002; Negahdari, Porda, & Scherm. 2010). The bonding between the ceramic top coat and substrate can be improved using BC. To improve the bonding, a material with a suitable thermal expansion ratio should be selected. In this process, the thermal expansion stress during heating and cooling, which is between the top coat and the substrate, could be remained at a minimum level. The protection of the substrate from hot corrosion and oxidation is the second function of BC. Note that YSZ has a porous structure, because it is a transparent material for oxygen, so it cannot be protected (Karaoglani, Ogawa, Turk, & Ozdemir, 2013). To achieve this objective, the bond coat is contacted with oxygen and an oxide layer between the top coat and the interface surface is created. The oxide layer with a thickness of about 2–8 μm is formed on the bond coat surface during the service. This layer is called the thermally grown aluminum-rich oxide (TGO) layer (Karaoglani, Ogawa, Turk, & Ozdemir, 2013). TGO is one of the critical mechanisms leading to the failure of TBCs at high temperatures. TGOs can provide high-temperature oxidation protection if they grow slowly and adhere to the substrate (Widjaja, Limarga, & Yip 2002).

Generally, Ni-based superalloys are applied as the substrate materials in gas turbines. The durability of these alloys at a high temperature of up to 1100 °C is the primary reason for selecting them as substrates (Karaoglani, Ogawa, Turk, & Ozdemir, 2013). To increase the creep resistance, the protection of substrates is carried out by directional grains or a single crystal structure. The application of coatings in the new generation of gas turbines increases the working temperature of the high-temperature components (Karaoglani, Ogawa, Turk, & Ozdemir, 2013).

The electron beam physical vapor deposition (EB-PVD) or the air plasma spraying (APS) techniques are used to fabricate TBCs. However, the plasma spray technique is more popular due to its high deposition efficiency and cost-effectiveness (Wolfe, & Singh, 1998; Nejati, Rahimipour, & Mobasherpour, 2014). These two methods are briefly described in the following sections.

Introduction of Air Plasma Spray Technique

The air plasma spray technique uses a plasma torch consisting of an anode in the form of a nozzle and a cathode. The ionization of a mixture of two or more gases, such as hydrogen and argon, produces an electric arc discharge in the plasma torch (Das, Saini, Pathak, & Kumar, 2012). In the thermal spraying process, particles are first fed into the plasma flame, and then they are melted and semi-melted due to the high temperature of the plasma flame. The plasma atmosphere with a temperature of 3000–15000 K can reach the melting point of any material. The feedstock is divided into millions of molten droplets because of the continuous stream of the pressurized gas and plasma formation (Wolfe, & Singh, 1998). The velocities of the droplets are in the range of 80-300 m s^{-1}. This depends on the properties of the plasma and the particle size (Das, Saini, Pathak, & Kumar, 2012). When the melted particles interact with the substrate, they cool rapidly and produce primary splats. This process continues alternatively until the final coating is produced.

When millions of particles are exposed simultaneously in flame or plasma, many of them will experience different levels of heating. Some of the particles may not melt at all, thus resulting in porosity or trapped unmelted particles in the coating structure (Wolfe, & Singh, 1998).

The high deposition rates (100–1000 μm min^{-1}) and various oxide and metallic coatings are among the important advantages of these processes. On the contrary, the inability to obtain homogeneous and dense microstructure is the disadvantage of the spray processes. In addition, the coating surface is relatively rough in the spray processes. Moreover, coatings contain defects such as non-bonded interfaces, micro-crack, and micro-porosity, thus resulting in lower mechanical properties such as ductility, resistance to oxidation, corrosion, impact, and wear (Wolfe, & Singh, 1998). Feedstock powders, disk-like splat, and lamella microstructure are illustrated in Figure 1.

In the applications of TBCs, porosity is required for high strain compliance to reduce the thermal conductivity during heating and to improve the thermal shock resistance. In general, the TBC porosity is ~15% (Wolfe, & Singh, 1998).

Figure 1. (a) feedstock powder, (b) the disk-like individual splat, and (c) lamella microstructure in the fracture cross-section micrograph of the deposit(marked splats by (S), voids by (V), micro-cracks by (C) and the non-bounded interfaces between splats by (N)

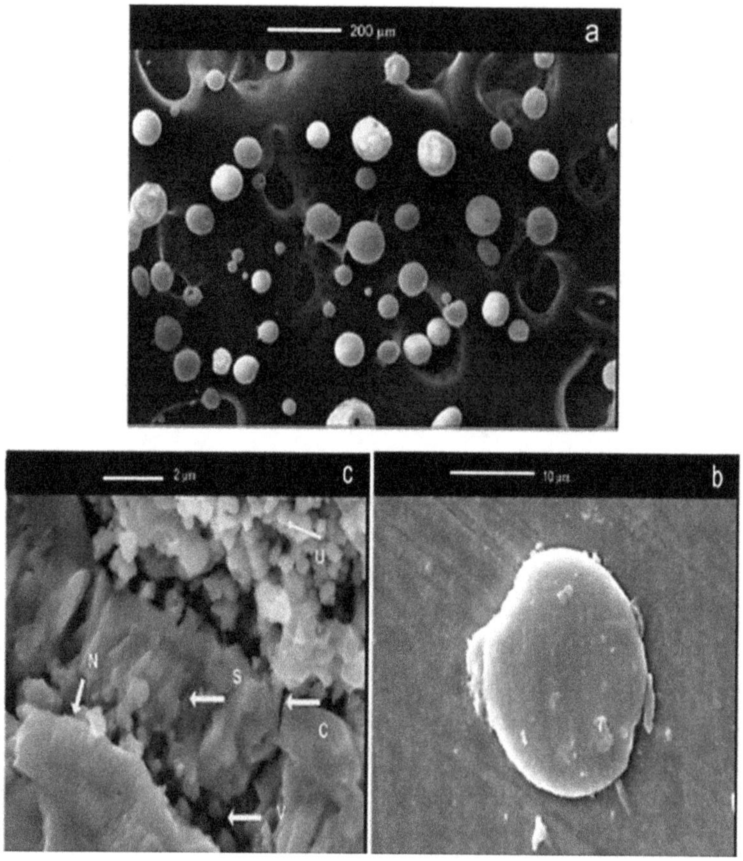

It is known that the properties of the coating depend on different factors, including the type of thermal spray processes, the operating conditions selected such as input power, primary gas flow rate, stand-off distance, powder feed rate, and carrier gas flow rate, and the feedstock being sprayed (Pakseresht et al., 2015) Thermal spray processes can be further classified based on the plasma formation and heat energy, i.e. plasma spray, high-velocity oxyfuel (HVOF), detonation gun (D-gun), low-pressure vacuum plasma spray processes (Wolfe, & Singh, 1998).

Introduction of Electron Beam-Physical Vapor Deposition Technique

Due to the inhomogeneous microstructure, poor bonding between splats, non-bonded interfaces, voids and unmelted particles in the APS process, electron beam-physical vapor deposition (EB-PVD) has been experimented to improve the life of turbine components and overcome some of the disadvantages in the thermal spray processes. EB-PVD has been shown to increase the life of turbine components by a factor of two due to the columnar structure in TBC (Wolfe, & Singh, 1998). In the EB-PVD process, electron beams are produced by an electron gun, which is directed to melt and evaporate ingots, and preheat the substrate in the vacuum chamber. The coating is formed by the condensation of vapor clouds on the surface of the substrate surface (Wolfe, & Singh, 1998).

The EB-PVD method can produce high purity coatings because the evaporated material does not come into contact with the crucible. Some of the advantages of this process are strong metallurgical bonding, columnar and polycrystalline microstructure, high deposition rates, precise composition control, dense coatings, high thermal efficiency, and low contamination. On the other hand, by changing the EB-PVD processing, multilayered ceramic and metallic coatings can be deposited on large components at relatively low temperatures. In addition, the coating surface is relatively smooth in the EB-PVD process (Schulz, Peters, Bach, & Tegeder, 2003; Wolfe, & Singh, 1998).

INVESTIGATION OF THE FAILURE MECHANISM OF TBCs

The failure mechanism for thermal barrier coatings is a complex multifaceted process with several intrinsic and extrinsic mechanisms. Intrinsic failure mechanisms occur due to the growth of a TGO layer at the BC/TC interface and the changes associated with thermal strains, which are involved in the kinetically driven processes. In addition, particularly in the case of plasma sprayed TBCs, sintering of the porous ceramic over the sustained thermal exposure causes the loss of compliance (Viswanathan, Dwivedi, & Sampath, 2014; Lee & Stinton, 1996). In addition, TC, BC, and substrate possess different thermomechanical properties. Generally, thermal stresses on the coating surface and at the interfaces are caused due to the thermal expansion mismatch and the applied temperature gradients. These are adequate to introduce surface and interface cracks (Kokini, K., Dejonge, Rangaraj, & Beardsley, 2002). The repeated application of thermal stresses leads to the growth of micro-cracks of the coating. As the surface cracks grow further, due to a decrease in the area of the TBC that resists the bending moment, an increasing driving force is generated, which results in the interface crack propagation (Rangaraj, & Kokini, 2003 b). In addition, the TBC systems possess various inherent limitations, such as low toughness of the TC–BC interface, large mismatch between the properties of the TC and BC, and the low fracture toughness of top coat materials.

The intrinsic failure mechanisms vary for the APS and EB-PVD processes, partly due to differences in their final microstructures. In the latter case, the delamination failure occurs at the BC–substrate interface due to a strong adhesion between TC and BC. However, in the sprayed coating, the failure of the YSZ layer is due to the spallation of the TC near the interface between the BC and ceramic layer (Lee & Stinton, 1996). Moreover, the sintering of the interface–crack and a consequent increase in stiffness can result in the loss of compliance in the layered coating (Viswanathan, Dwivedi, & Sampath, 2014). Extrinsic failure mechanisms, which are less predictable, occur due to erosion, foreign object damage particulate, and chemical attack by ingested sand, ash, and molten salt. These factors are same for both the methods (Viswanathan, Dwivedi, & Sampath, 2014).

Generally, the application of high heat flux thermal stresses contributes to the growth of TBC–BC interface cracks leading to delamination and spalling near the TC–BC interface and loss of structural integrity of these coatings (Xiong, Kawasaki, Kang, & Watanabe, 2005; Rangaraj & Kokini, 2003a). Hence, the development of functionally graded and multilayer TBCs can be a promising process to decrease the driving force in the TC–BC interface fracture because the FGM structure can decrease the thermal stresses generated at the TC–BC interfaces. Therefore, multilayer and FG-TBCs are supposed to be resistant to delamination and hence considered more durable (Rangaraj, & Kokini, 2003 b; Kawasaki & Watanabe, 1996).

INTRODUCTION OF FUNCTIONALLY GRADED MATERIALS

FGM system has been used to reduce some of the problems involved in the development of a sharp interface at the joint of two or more dissimilar materials (Joshi & Ng, 2011). FGMs are composite materials composed of more than single continuous or discontinuous gradient in structure, composition, phase distribution, particle content or texture. They can be composed of gradients of metals, ceramics, polymers or just a kind of variation in density and porosity (Khor, Dong & Gu, 1999; Schulz, U., Peters, Bach, & Tegeder, 2003). These materials have some type of flexibility in terms of the behavior. It should be noted that the properties of FMGs are totally different from those of their individual components. This can explain why their mechanical and thermal properties in the FGM system are continuously varying such that one side of the FGMs structures could have higher thermal resistance and mechanical strength while the other side has a different worth in terms of the property.

Many theoretical studies have been conducted to design and analyze the performance of FGMs since 1970s. The creation of the gradient is the first step in the preparation of FGMs. This process is known as the gradation process. FGM can be produced by different techniques involving plasma spraying, arc spraying, chemical and physical vapor deposition, laser cladding, powder metallurgy, hot pressing, cold pressing, sintering process (spark plasma sintering, microwave), infiltration process, and graded casting process such as centrifugal casting, slip casting, tape casting and thermal diffusion processing; however, in all cases, the central problem has been the need to reduce the residual and thermal stresses in the FGM structures. (Nur, Jamaludin, Mustapha, & Nuruzzaman, 2013; Schulz, U., Peters, Bach, & Tegeder, 2003). The compositionally and/or microstructurally graded interfaces, with proper design and preparation, can be more resistant to crack initiation and propagation in structural applications. In addition, graded interfaces can be considered valuable in reducing peak stress and eliminating stress that occur due to gradual changes in the properties of materials (Joshi & Ng, 2011).

CONCEPT OF FUNCTIONALLY GRADED THERMAL BARRIER COATING

The concept of FGM was introduced to the coating design to improve the reliability and longevity. Recently, functionally graded coatings with varying compositions, with their properties changing smoothly from one layer to another, have been applied to develop advanced materials for bio-implants, optoelectronic devices, thermionic and thermoelectric energy conversion, thermal barrier, corrosion prevention, and wear-resistant coatings (Schulz, U., Peters, Bach, & Tegeder, 2003; Khor, Dong & Gu, 1999).

FG-TBCs are the most important graded coating designs that have been primarily used for the protection of metallic substrates due to hot corrosion, oxidation, and heat penetration. It is possible to ensure the realization of the best properties of metals and ceramics in FG-TBCs. Simultaneously, it can impart good impact resistance due to the hard ceramic part. Furthermore, thermal shock resistance has improved due to the smooth thermal coefficient transition from the ceramic to the metal (Xiong, Kawasaki, Kang, & Watanabe, 2005). The structure of FG-TBCs can effectively reduce the thermal stresses and then improve the resistance of an FG-TBC to thermal shock cracking in comparison with the regular duplex TBC, thus providing excellent coating properties (Xiong, Kawasaki, Kang, & Watanabe, 2005).

Such TBCs might have variable composition through their thickness. The upper layer is a ceramic-rich layer, which generally consists of 100% YSZ. The layer near the BC possesses the maximum proportion of the BC intermetallic alloy. The composition varies in each subsequent layer of these layers. The variation progresses from ceramic-rich composition near the upper layer to a BC-alloy-rich composition close to the TBC–BC interface. Thus, the layers of a graded TBC are composed of composites of ceramic–BC alloy particulate with different YSZ volume fractions. A graded composition using the coating thickness results in good modification in terms of the thermomechanical properties and temperature distribution. It also decreases the mismatch between the properties of the TC and BC. Therefore, the spallation lifetime can be improved (Rangaraj & Kokini, 2003a). In a graded TBC, the introduction of BC alloys into the coating layers may also improve the fracture toughness of the TBC. It can, therefore, be inferred that the layers of a graded TBC are expected to show a higher resistance (in comparison to a conventional TBC) to the propagation of surface and TC–BC interface cracks (Rangaraj & Kokini, 2003 a).

FGM coatings have the potential to simultaneously decrease the thermal expansion mismatch and increase the interface bonding strength (Bao & Cai, 1997). Furthermore, the thermal cycling resistance of FG-TBCs is superior to that of the conventional TBCs. When functionally graded concept is introduced to replace the conventional TBCs, the difference in the operating temperature between TC and substrate can be increased by ~35% (Kokini, K., Dejonge, Rangaraj, & Beardsley, 2002).

Figure 2 shows the polished cross-sectional microstructures of FG-TBCs with a map analysis, from BC to TC of different layers of elements. The graded structure is composed of three layers, where the YSZ content is changed stepwise from 0 vol% at metallic bond coat to 40 in intermediate layers and 100 vol% in the ceramic top layer. For the samples, the thickness of the BC is fixed at 200 μm. The two other layers have an equal thickness of 100 μm. Figure 2 shows that there are merely a few micro-cracks and no macrocracks in the as-sprayed coatings. The polishing process can create some new defects in the polished cross-sectional structure. The interfaces are smooth and almost indistinguishable between different layers. No delaminations appear at the interface in the cross sections of the sample.

The concept of FG-TBC seems to be rather simple and intuitive. As the two kinds of feedstock powders may be different in terms of morphology, size, flowability, melting point, and density, it is not easy to control the plasma spray parameters to form a uniform film (Kim, Kim, & Park, 2003). Therefore, creating a compositional gradient is completely difficult. In this regard, several methods have been explored, which are as follows:

- Applying co-injection of the ceramic and metallic powders in a single plasma torch (Khoddami, Sabour, & Hadavi, 2007).
- Applying prealloyed composite powders as feedstock. In this method, mechanical alloying and plasma powder spheroidization produce composite powders with different compositions. Then, the spheroidized powder can be used successively in a single plasma torch to form the functionally graded coating. The advantage of applying composite powders is to develop uniform density along the graded layers and to ensure the chemical homogeneity of coating (Khor, Dong & Gu, 1999). In addition, this method can ensure the flexibility of depositing different numbers of interlayers by varying the composition of the composite powders (Gu, Khor, Fu, & Wang, 1997).
- For combining the benefits of graded coating architecture with the columnar microstructure, the following conditions should be considered. It should be noted that the vaporization of materials is difficult during EB-PVD because it depends on the vapor pressure of each component (Widjaja, Limarga, & Yip 2002; Kokini, Takeuchi, & Choules, 1996; Schulz, U., Peters, Bach, & Tegeder, 2003). For the deposition of functionally graded materials, the composite ceramic ingot is the key element of EB-PVD technology. Composite ingots allow the deposition of gradient coatings onto a metal substrate with metal or metal–ceramic bond coat and the outer ceramic layer, including TBC, during one process in an EB-PVD coater (Movchan, 2002).

Figure 2. The polished cross-sectional microstructures of FG-TBCs with a map analysis of elements

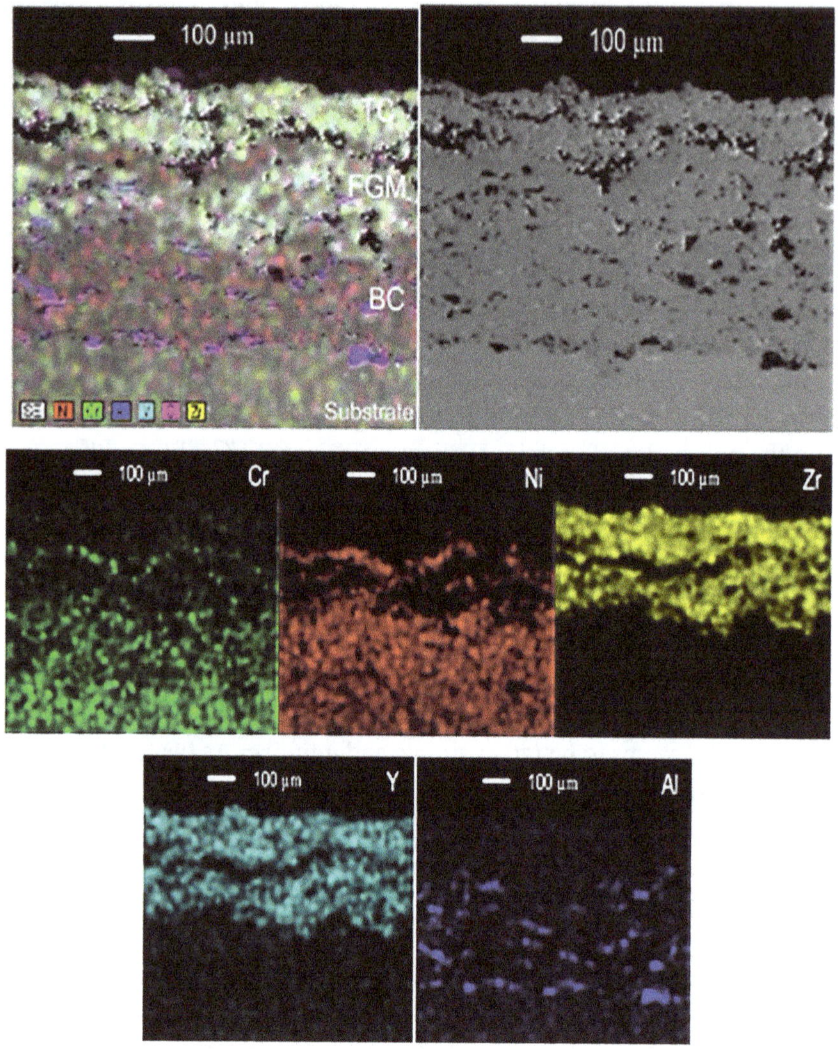

Comparison of the Properties of FGM Coatings with Those of Duplex Coatings

The advantages and disadvantages of functionally graded thermal barrier coating are discussed in this section. From physical and mechanical perspectives, functionally graded thermal barrier coatings provide the possibility of significantly improving the properties and lifespan in comparison with the conventional TBCs. There are many technical problems related to measuring and estimating the mechanical and physical properties of graded coatings though the concept of FGM is rather easy to understand.

Several experimental studies on FG-TBCs have been conducted by Khore et al. (1999; 2000 a), while preparing the functionally graded coatings, they first sprayed the NiCoCrAlY alloy onto grit-blasted inconel superalloy disc to form a bond coat that was approximately 200 µm thick. Then, the layers of 25% YSZ–75% NiCoCrAlY and 50% YSZ–50% NiCoCrAlY were formed. Similarly, 75% YSZ–25% NiCoCrAlY was used to form the fourth layer. The intermediate layers were formed with a thickness of 200 μm. Finally, YSZ was sprayed as the top layer with a thickness of 200 μm (Fu, Khor, Ng, & Teo, 2000). Khore et al. demonstrated that the mechanical properties, such as microhardness and elastic modulus, microstructure, density, thermal conductivity, coefficient of thermal expansion, and diffusivity, change gradually while using the five-layer functionally graded coatings, which was appropriate for improving the thermal and mechanical properties of the coatings (Khor & Gu, 2000 a).

The other advantage of this structure is that the BC part is mechanically very strong with a high thermal conductivity. On the other hand, the YSZ possesses good heat resistance, which is necessary for TBCs. The high thermal stress generated due to sharp differences in the modulus between the metallic and ceramic phases can be considerably reduced by the gradient distribution of the two phases in the coating (Khor, Gu, & Dong, 2001). In addition, the residual stress was the lowest for the five-layer functionally graded coating as compared with that of the conventional TBC (Khor & Gu, 2000 a). In the following sections, some of the properties of FG-TBCs are compared with those of duplex TBCs.

The FG-TBC shows the highest bond strength due to a significant decrease in the residual stresses, whereas the conventional TBC shows the lowest bond strength (Khor & Gu, 2000 a). The increase in the bond strength takes place due to the microstructure that has no sharp interfaces between layers (Khor, Gu, & Dong, 2000; Gu, Khor, Fu, & Wang, 1997). Furthermore, the layers are gradually changed from a ceramic to a metallic phase as they are deposited on the surface with different chemical compositions, and consequently the thermal stresses in the coating are decreased. Therefore, the bonding strength is increased. Khodami et al. (2007) showed that for the conventional TBC the mean strength was about 37.9 MPa, while for FG-TBC it was about 57.8 MPa.

Khore et al. (2000a) conducted a microhardness test on the FG-TBC and the conventional TBC. For the FG-TBC, the microhardness was changed gradually through the five-layer coating, whereas a significant microhardness difference of about 200 HV was observed between BC and YSZ layers for the duplex TBC. In addition, note that the large difference in the elastic modulus between the metallic BC and ceramic TC can be decreased by the gradient distribution of the microhardness. This is advantageous for the bond strength between the coating and the substrate using tensile or bending stress (Khor, Gu, & Dong, 2000).

The porosity and the thermal conductivity of five individual layers of FG-TBC are shown in Figure 3 YSZ possesses a higher melting point than the BC alloy (NiCoCrAlY). The particles of YSZ powder not only completely melt in the plasma atmosphere, but also produce larger pores. Although the porosity under a certain level is allowed in TBC applications, it needs to be retained at the optimum level by changing the spray parameters (Khor, Dong & Gu, 1999).

The thermal diffusivity and conductivity can be increased by increasing the heating temperature and BC material content (Khor & Gu, 2000 b). Figure 4(a) illustrates the thermal diffusivities of different layers of the FG-TBC at a temperature range of 25–1200 °C. The 100% YSZ layer showed the lowest thermal diffusivity of 0.00238 $cm^2 s^{-1}$ at an ambient temperature compared with other layers. This is most probably due to the sprayed YSZ layer that generally possesses a relatively high porosity and a lower thermal diffusivity in comparison with those of metals and graded layers. For the BC layer, the thermal diffusivity was much higher than that for the TC layer. As shown in Figure 4(a), the thermal diffusivity can be decreased gradually with an increase in the YSZ content of the individual layers. (Khor & Gu, 2000 a; Khor & Gu, 2000 b). Khodami et al. (2007) showed that the intensity of porosity in NiCrAlY was about 1% in the duplex layer, whereas it was about 13% in the YSZ layer. The high amount of porosity in the YSZ layer was produced because ceramic particles were not thoroughly melted during the coating process, thus resulting in less plastic deformation on collision with the substrate surface. In the layers

Figure 3. The relationship between porosity and thermal conductivity at room temperature (ELSEVIER,2000)

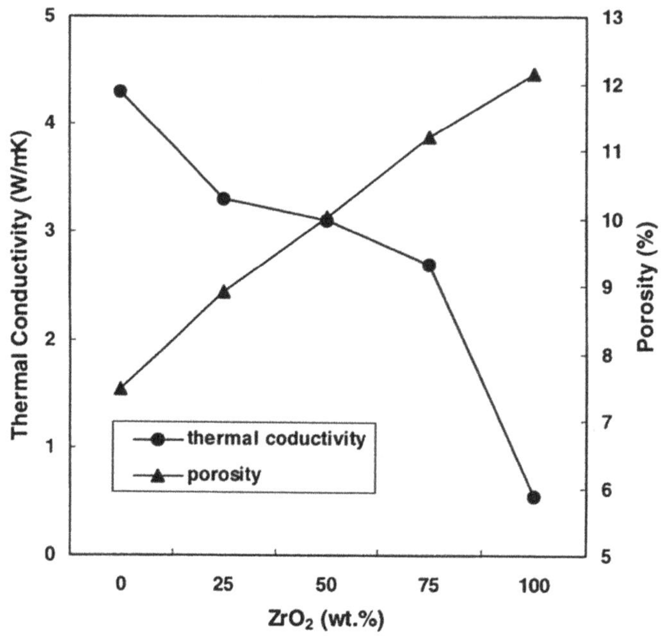

of FG-TBCs, the total pore and porosity increased with an increase in the YSZ content from about 4% in the 30% YSZ layer to about 12% in the 100% YSZ layer. This is one of the reasons that the thermal diffusivity is high in the bond coat, thus conforming to the results obtained by Khore. As shown in Figure 4(b), the change in the thermal conductivities of various layers at different temperatures follows a similar trend as that in the thermal diffusivities of various layers (Khor & Gu, 2000 b). In the FG-TBC, the thermal conductivity is gradually decreased with an increase in the YSZ content.

The results of Khore et al. (2000 b) showed a significant difference in the coefficients of the thermal expansion of CTE value between the YSZ and NiCoCrAlY layers. This difference can significantly be increased with an increase in temperature (Khor & Gu, 2000 a; Khor & Gu, 2000 b). However, the FGM coating has demonstrated some gradual variation in the coefficients of thermal expansion (Khor & Gu, 2000 b; Khor, Dong & Gu, 1999). Note that the mismatch in the thermal expansion between the BC and TC layers of the systems is the primary reason of the failure of the conventional TBC (Kim, Kim, & Park, 2003). A large thermal residual stress is caused due to this large thermal mismatch, thus resulting in the formation of spallation or cracks of the coating. For FG-TBCs composed of five graded layers, the thermal mismatch between the two adjacent layers is decreased due to the gradual change in the CTE of different layers (Khor & Gu, 2000 b) which is one of the most important factors of FG-TBCs.

In general, the thermal properties of the sprayed films are quite different and much lower than those of the bulk materials. This is due to the presence of pores, cracks, unmelted particles, oxides, and non-bounded interface between splats in the as-sprayed films. The common defects of the sprayed coating can be considered a critical factor that influences the thermal properties of FG-TBCs. Generally, the thermal properties are decreased by increasing the porosity (Khor & Gu, 2000 b). Figure 3, also, illustrates the amounts of porosity of layers and their corresponding thermal conductivity.

Figure 4. Thermal diffusivitiesrconductivities of individual layers of FGM coating (a) Thermal diffusivity; (b) thermal conductivity (ELSEVIER,2000)

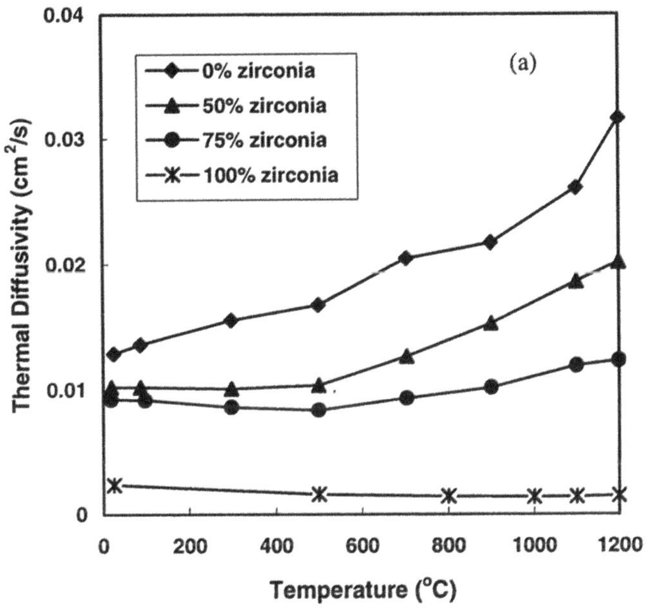

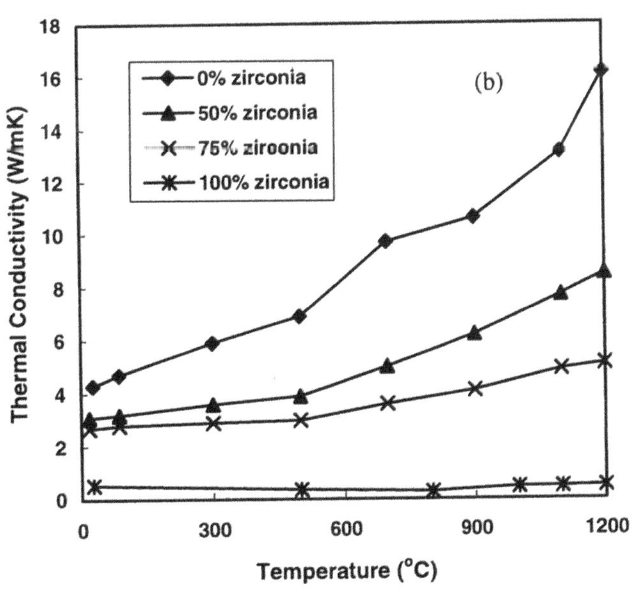

Khore et al. (2001) showed that the bending strength and elastic moduli of individual layers of FG-TBCs are decreased with an increase in the content of the YSZ layer caused by an increase in the porosities of the coating (Khor, Gu, & Dong, 2001). The bending strength of the layers strongly depends on the interlamellae contacts, splats boundaries, and the microstructure of coatings. When the YSZ content in the coating is increased, the porosity of the coating is increased and the bending strength is decreased (Khor, Gu, & Dong, 2001). In addition, the elastic modulus of BC is almost five times more than that of the ceramic layer. This is caused by the presence of the defects and pores in the plasma sprayed films. The porosity is increased and the density is decreased with an increase in the ceramic content in the as-sprayed layers, thus resulting in the decrease in the elastic modulus and bending strength (Khor, Gu, & Dong, 2001).

The FG-TBC shows better resistance to the thermal cycling than that of the conventional TBC. The values of the thermal shock tests are 15 cycles and 90 cycles for the duplex coating and FG-TBC, respectively. The improvement in the thermal cycling of FG-TBC is due to the gradual changes occurring in the material properties of different layers and a significant decrease in residual stresses. The thermal shock tests can be improved by reducing the difference in the thermal expansion coefficient between the ceramic and metallic layers and increasing the bond strength (Gu, Khor, Fu, & Wang, 1997; Fu, Khor, Ng, & Teo, 2000; Khor, Gu, & Dong, 2000).

Overall, a significant reduction in residual stresses and an increment in cohesive strength can be the reasons for the improvement of thermal cycling resistance of FG-TBCs (Khor & Gu, 2000 b).

The oxidation test conducted by Khore (2000b) also showed that for the long-term oxidation resistance the FG-TBC performs better than the conventional TBC. Note that in the beginning of the oxidation test, the conventional TBC showed more resistance to the oxidation than the FG-TBC. The NiCoCrAlY in the second layer (25% NiCoCrAl + 75% YSZ(can be easily oxidized due to the relatively small thickness of the top layer for FG-TBC. The rapid oxidation of the internal bond coat can be decreased with an increase in the oxidation time due to the gradual change in the composition of FG-TBCs. In the second layer, a layer of Al_2O_3 can be easily formed, and the combination of YSZ and Al_2O_3 can prevent the coating from further oxidation (Khor & Gu, 2000 b).

For the conventional TBC, the results of the fracture test showed that the fracture is in a complex adhesive/cohesive mode. In this type of coating, the adhesive fracture occurs at the BC–TC interface, whereas the cohesive fracture occurs within TC. For the FG-TBC, rupture can be seen within the glue, whereas the cohesive fracture can be seen within the YSZ layer (Khoddami, Sabour, & Hadavi, 2007). In general, the properties of the FG-TBCs are more directly related to the gradual variation of their microstructures and compositions, leading to the reduction in the residual stresses of the coating and stresses that are induced during heating and cooling cycles (Khor, Dong & Gu, 1999).

CONCEPT OF MULTILAYER COATING

In advanced technological applications and basic sciences, the researchers have shown much interest in multilayered ceramics and metallic coatings. Multilayered coatings can provide unique mechanical and physical properties due to their refined microstructures (Wolfe & Singh, 1998). Better thermal resistance and wear-resistant properties can be obtained with the proper selection of coating material, number of layers, and thickness of the layers (Wolfe & Singh, 1998). This kind of coating has more than two layers on the substrate surface with each layer exhibiting unique functions. The first layer is designed in order to prevent the interaction between the substrate and the coating materials. Crack arrestment, which reduces the thermal stresses in the thermal barrier ceramic layer, is incorporated in the second layer. This layer serves as a thermal stress controlling layer and arresting layer for crack progress initiated from this ceramic layer. Conversely, the third layer is a corrosion and oxidation resistant layer with improved adhesion. This layer prevents corrosion and oxidation of the substrate material apart from enhancing the adhesion between the thermal stress control layer and the following thermal barrier ceramic layer.

The fourth layer is composed of a ceramic material with low thermal conductivity and good chemical stability. This layer is a thermal barrier layer with such a thickness that the temperature of the substrate becomes less than a critical value. The fifth layer prevents the atmospheric impurities arising due to oxides, known as erosive material, which prevents the thickness loss of the thermal barrier ceramic layer (Tamura et al. 1999). It is possible to use two or more layers in a system depending on the application and the condition.

In recent years, many researchers have added an Al_2O_3 layer or the composite of Al_2O_3 and YSZ layers as an oxidation resistant layer to the duplex TBC (Saremi, Afrasiabi, & Kobayashi, 2008; Stover, Pracht, Lehmann, Dietrich, & Vaben, 2004). In this coating design, the low oxygen diffusivity of Al_2O_3 is combined with the high thermal insulation of YSZ. In these coatings, the infiltration of oxygen or other detrimental factors is prevented and zirconia phase transformation and formations of TGO layer are retarded due to the presence of the dense Al_2O_3 layer (Keyvani, Saremi, & Sohi, 2010; Negahdari. Porda, & Scherm. 2010; Schulz, Peters, Bach, & Tegeder, 2003). For example, Widjaja (2002) added an additional oxygen barrier layer in the TBC systems to suppress the growth of the TGO scale. The Al_2O_3 layer has been recommended as the interlayer barrier due to its low oxygen diffusivity. Hence, further oxygen penetration into the superalloy substrate can be reduced (Widjaja, Limarga, & Yip 2002). In addition, the oxidation of BC has been considered one of the important factors that can affect the durability of BC during the service. More residual stresses can be created at BC/TGO and TGO/TC interfaces by growing the TGO layer due to high-temperature exposure, thus leading to coating detachment. In order to preserve TC, the adhesion of TGO is essential because the spallation of the Al_2O_3 layer results in the loss of the overlying TC (Widjaja, Limarga, & Yip 2002).

Al_2O_3 is strongly believed to be an excellent option that can improve the performance and reliability of TBCs with further improvement in the TBC systems. To find the best architecture for this kind of coating, Keyvani et al. (2010 & 2011) have conducted many experimental studies. In this regard, they sprayed different samples with different architectures as follows.

First, the NiCrAlY powder was deposited on the substrates with a thickness of 150 μm to create BC. Then, the samples were sprayed with the following procedure.

- YSZ with a thickness of 350μm as a TC.
- A composite layer with 100μm Al_2O_3 on the bond coat and 250μm YSZ as a top layer.
- A composite layer of YSZ and Al_2O_3 mixed powder with 350μm thickness.
- A composite layer having 100 μm Al_2O_3 top coat and 250μm YSZ between TC and BC

In Figure 5, the schematic of the four types of TBCs are illustrated.

The Al_2O_3/YSZ composite and the YSZ/Al_2O_3 coating showed the most and the least resistance, respectively, to the hot corrosion test. The low resistance of the YSZ/Al_2O_3 coating could be caused by the mismatch of the thermal expansion of zirconia and alumina layers and also by Al_2O_3 with the BC layer. The composite coatings of YSZ/Al_2O_3 with alumina as a TC and a mixed YSZ/Al_2O_3 layer showed better hot corrosion and oxidation resistance than normal YSZ. Therefore, it was concluded that the deposition of Al_2O_3 on the YSZ layer improved the hot corrosion and oxidation resistance of the coatings by preventing the infiltration of molten salt and oxygen through the YSZ layer. (Keyvani, Saremi, & Sohi, 2010; Keyvani, Saremi, & Sohi, 2011).

Figure 5. Schematic of as-sprayed coatings: (a) YSZ/NiCrAlY; (b) YSZ/Al$_2$O$_3$/NiCrAlY; (c) YSZ+Al$_2$O$_3$/NiCrAlY; and (d) Al$_2$O$_3$/YSZ/NiCrAlY

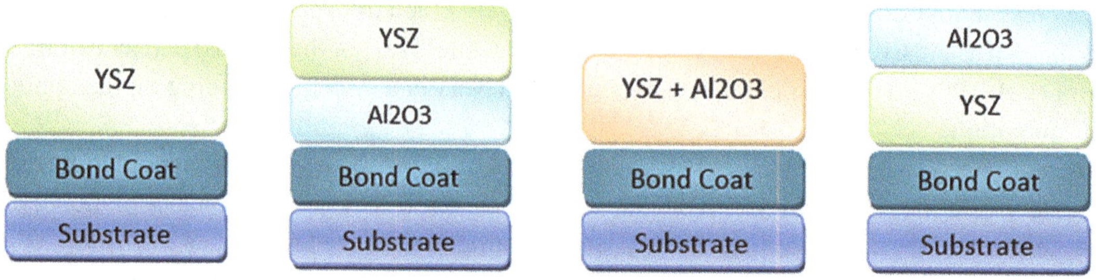

The embedding of Al_2O_3 hard particles into the YSZ matrix causes an increment in the TBCs wear resistance. Schulz (2003) studied many samples of Al_2O_3 particles in the YSZ matrix. The weight loss by erosion would become lower by increasing the Al_2O_3 content. In other words, Al_2O_3 contents exceeding an upper limit of 18% can cause poor thermal shock behavior, and therefore must be avoided. The possible delamination between a double-layer YSZ/Al_2O_3 coating can be avoided by the graded ceramic coatings produced by increasing the exchange of zirconia by alumina in a direction toward the coating surface, and thermal shock and wear resistance can also be improved (Schulz, Peters, Bach, & Tegeder, 2003). Schulz et al. (2003) showed that 25% Al_2O_3 content at the outer coating surface is the optimum with regard to wear resistance and good thermal shock.

FUTURE IN TBC AND FG-TBC

In this section, new materials and methods of producing TBC are analyzed.

New Materials

As discussed above, YSZ is considered as an important TBC material in industrial sector due to its low thermal conductivity (~2 W m^{-1} K^{-1}, bulk), high melting temperature (2973 K), and relatively high coefficient of thermal expansion (CTE, ~11 × 10^{-6} K^{-1}). However, the application of YSZ is limited due to strong sintering and phase transition in combination with the increase in Young's modulus as it limits the long-term service temperature of the YSZ layer above 1200 °C (Chen, Gu, Zou, Wang, & Cao, 2012; Keyvani, Saremi, & Sohi, 2011; Negahdari. Porda, & Scherm. 2010). For enhancing the stability of YSZ above 1200 °C, different attempts have been made to improve the TBC systems. In one of the cases, Y_2O_3 is substituted with more acidic stabilizing elements, such as Sc_2O_3, CeO_2, $YTaO_4$, and In_2O_3, for the partial stabilization of ZrO_2. (Nejati, Rahimipour, & Mobasherpour, 2014). Among these elements, ceria has been investigated intensively (Alfano, Di Girolamo, Pagnotta, & Sun, 2010; Park, Kim, Kim, Song, & Park, 2005). Some of the investigators have reported that the ceria-stabilized zirconia (CSZ) TBCs have a higher thermal expansion coefficient and a lower thermal conductivity than the conventional TBCs in addition to high-temperature stability, high fracture toughness, and good corrosion resistance (Nejati, Rahimipour, & Mobasherpour, 2014).

In another case, the ZrO_2 doping is applied with smaller or bigger cations of rare-earth elements to improve coating properties (Habibi, Wang, Liang, & Guo, 2013). However, the increasing demands for higher efficiency of gas turbines emphasize investigation of a new system and new materials. As a result, on the one hand, as mentioned earlier a ceramic–metal functionally graded TBC (FG-TBC) system has been widely investigated and, on the other hand, the applications of new materials have been increased (Chen, Gu, Zou, Wang, & Cao, 2012). In this regard, the new TBC materials that possess higher thermal insulation performance, lower sinterability, and better temperature capability are required. A lot of efforts have been put on developing the new TBC materials to replace the conventional YSZ in order to further increase the inlet temperature of the parts of the engine. The new materials are selected considering the factors such as excellent phase stability between room temperature and operating temperature, good resistance to sintering at high temperature, and low thermal conductivity (Guo, Wang, Wang, & Gong, 2009). For the new TBC materials, some candidates, such as $La_2Ce_2O_7$, Y_2O_3–HfO_2, $La_2Zr_2O_7$, $Gd_2Zr_2O_7$, $LaMgAl_{11}O_{17}$, and complex perovskites, which are applied at higher temperatures are being investigated by researchers (Guo, Wang, Wang, & Gong, 2009). Previous studies have reported that these candidates possess the thermal conductivity lower than or equal to YSZ and high thermal stability, which are the desirable physical properties of this type of new materials. However, the CTE of new TBC is typically lower than that of the conventional TBC, leading to higher thermal stresses in the new TBC systems (Vassen et al. 2009). Some of these materials are discussed below.

- **Pyrochlore Materials:** TBC materials with pyrochlore structure ($A_2B_2O_7$, where A= rare-earth element and B = Zr) have been applied at service temperatures above 1300 °C. These materials are among a number of more complex crystal structures that offer very attractive properties comparable to YSZ. (Vaßen, Jarligo, Steinke, Mack, & Stöver, 2010). The thermal conductivity of the new TBC can be reduced with the increase in the complexity of the crystal structure. This is caused by the increasing number of optical phonons that do not substantially contribute to the thermal conductivity. Pyrochlore compounds, for A = Gd, Nd, Sm, Eu, and La, possess extremely low thermal conductivity, ranging from ~1.1–1.7 W m^{-1} K^{-1} at temperatures between 700 °C and 1200 °C (Pan, Phillpot, Wan. Chernatynskiy, & Qu. 2012). Among these pyrochlores, some oxides, such as $Gd_2Z_{r2}O7$, $Sm_2Zr_2O_7$, $La_2Ce_2O_7$, $La_2Zr_2O_7$, and their composites are good candidates for new TBC materials, and provide better comprehensive thermal and mechanical performances (Pan, Phillpot, Wan. Chernatynskiy, & Qu. 2012). Two types of the combinations are introduced here. As a TBC substitute material, $La_2Ce_2O_7$ is considered a promising candidate due to its lower thermal conductivity, larger CTE, and higher phase stability up to 1400 °C than YSZ. However, the thermal cycling performance of the sprayed $La_2Ce_2O_7$ film requires improvement for the potential advantages of the TBC (Guo, Wang, Wang, & Gong, 2009). $La_2Zr_2O_7$ is another pyrochlore material, it has a relatively higher thermal expansion coefficients (TEC) and lower CTE than YSZ, lower sintering rate at high temperature, lower thermal conductivity, higher temperature phase stability and no phase transformation and better ability to accommodate defects in comparison to YSZ (Habibi, Wang, & Guo, 2012). A disadvantage of it, however, is the relatively low CTE of about 9×10^{-6} K^{-1} as compared to YSZ with 10–11×10^{-6} K^{-1} thereby leading to higher thermal stresses that resulted from more thermal expansion mismatch between TC and BC (Vaßen, Jarligo, Steinke, Mack, & Stöver, 2010).

- **Perovskite-Structured Oxides:** Another class of materials that could be considered as good candidates for use in TBC applications is the ABO_3-type perovskites. At elevated temperatures, most of these materials are stable, thus making them potential candidates for material improvement in TBC applications (Vaßen, Jarligo, Steinke, Mack, & Stöver, 2010). Pure ABO_3 perovskites including high atomic mass ions show low thermal conductivities. For example, thermal conductivities of $SrZrO_3$ and $BaZrO_3$ are 2.1 W m^{-1} K^{-1} and 3.4 W m^{-1} K^{-1}, respectively, due to the structural rigidity of corner-shared octahedral (Pan, Phillpot, Wan. Chernatynskiy, & Qu. 2012). Moreover, structures of the new complex perovskites are formed by increasing the doping level. For instance, $Ba(Mg_{1/3}Ta_{2/3})O_3$ is formed with a disordered cubic phase and an ordered trigonal structure which can be improved with an increase in the synthesis temperature up to 1650 °C (Vaßen, Jarligo, Steinke, Mack, & Stöver, 2010) Its thermal conductivity attains the value of 2.5 W $m^{-1}K^{-1}$ at 1100 °C. Similarly, the $La(Al_{1/4}Mg_{1/2}Ta1/_4)O_3$ complex perovskite system attains a low value for thermal conductivity (~2 W m^{-1} K^{-1}) and a high CTE (~11.9 × 10^{-6} K^{-1}), thus making it a promising combination for TBC systems. Nevertheless, TBCs composed of these complex perovskites exhibit a short thermal cycling lifespan considering their slight toughness and consequent weak thermal shock resistance (Karaoglani, Ogawa, Turk, & Ozdemir, 2013). However, the partial evaporation of ingredients, which are in their phase, is the primary problem with perovskite materials during plasma spraying, thus leading to impure phases in the coating and consequently to some detrimental effects on the performance of coating (Vassen et al. 2009 ; Pakseresht, Rahimipour, Vaezi, Salehi, 2015).
- **Lanthanide Orthophosphates And Silicates:** Another kind of material with low thermal conductivities is lanthanide orthophosphates ($LnPO_4$). The orthophosphates can have either a monazite or a xenotime crystal structure depending on the ionic size of the lanthanide ions. These materials possess higher thermal stability and lower thermal conductivity than YSZ. Lanthanum hexaaluminate, $LaAl_{11}O_{18}$ (LHA) with a magnetoplumbite structure and plate-like grain is a promising material used for replacing YSZ. This is due to low thermal conductivity and Young's modulus, lower sinter-ability, superior structural and thermo-chemical stability up to 1400 °C and the stable pore structure (Negahdari, Porda, & Scherm, 2010).
- **Defect Cluster TBCs:** Different types of rare-earth cations doped with ZrO_2 are considered as the new and promising TBC materials. As in the ZrO_2 system, these additions result in the formation of dopant clusters, leading to the reduction of the thermal conductivity by about 20–40% (Vaßen, Jarligo, Steinke, Mack, & Stöver, 2010).

However, when new materials with a lower thermal expansion coefficient are applied as lone ceramic top coats, they show shorter thermal cycling endurance due to the thermal mismatch in the TBC layers, as both BC and metallic substrates have relatively higher values of CTE than ceramic materials (Vassen et al. 2010). In the modern TBC system, this problem can be overcome by further dividing the ceramic TC into layers with different functions. A double-layer top coat system in which standard YSZ is applied as the bottom and new materials, such as perovskites or pyrochlores, are applied as the upper layers is an example of the modern TBC systems. These systems have demonstrated improved temperature capability compared with the conventional YSZ. The double-layer systems exhibit longer cycling lifetime than the single-layer systems (Vassen et al. 2010). These new materials can be used for producing FG-TBC or multilayered coating by using BC or other ceramic materials. Some of the examples are discussed as follows.

Negahdari et al. (2010) produced functionally graded alumina/lanthanum hexaaluminate (FGLHA) with a gradient in composition. They applied five different composition layers (alumina with *x* vol% LHA composite powder, where $x = 0$, 20, 40, 60, and 80). In their study, they compared the thermal conductivity and diffusivity of FGLHA with the monolithic composite ceramics. They found lower thermal diffusivity for FGLHA in comparison with alumina and alumina/lanthanum hexaaluminate (LHA). In conclusion, the CTE mismatch stress between TBC/TGO bond coat interfaces is minimized by the gradient in composition and porosity of FGLHA along the layers.

Chen et al. (2012) have designed a new FG-TBC based on $LaMgAl_{11}O_{19}$ (LaMA)/YSZ. LaMA with magnetoplumbite-type structure has exhibited excellent mechanical properties, such as high fracture toughness (3.7 MPa $m^{1/2}$), high thermochemical stability (up to 1400 °C), superior sintering resistance, etc. In addition, its coating has exhibited a thermal cycling lifetime similar to the conventional YSZ coating at an operating temperature of 1250 °C, thus obviously making it superior to the pyrochlore-type zirconate coatings (Chen et al. 2011). LaMA possesses the longest thermal cycling lifetime as compared with other MgO-doped rare-earth hexaaluminate counterparts. However, the relative lower CTE in combination with the re-crystallization behavior, which would decrease the bond strength between the BC and the ceramic TC, may make the single LaMA film less durable than the conventional YSZ at elevated temperatures. Therefore, they have explored the new LaMA/YSZ double ceramic and FG-TBC system in order to develop a new TBC with better performances than the conventional YSZ and LaMA at elevated temperatures (Chen, Gu, Zou, Wang, & Cao, 2012). They have produced a five-layered ceramic (LaMA)/ceramic (YSZ) FG-TBC with the weight ratios of 0%, 25%, 50%, 75%, and 100% on the surface substrate with MCrAlY, where M = Ni, Co, and Fe as a BC (BC thickness = 150 μm). The total thickness of the ceramic coating was about 360 μm, and it was about 60 μm for the YSZ coating applied as the bottom layer. Moreover, the thickness of each of the three intermediate composite coatings was about 60 μm. The thickness of the final LaMA top layer was about 120 μm. (Chen, Gu, Zou, Wang, & Cao, 2012).

CTE has found to reduce gradually from the bottom to the top layer, resulting in the reduction of the residual stress level as compared with the LaMA/YSZ double ceramic layered and the conventional TBC systems. The average CTE of YSZ is $\sim 10.2\times10^{-6}$ K^{-1} across the temperature range from ambient temperature to 877 °C, and the average CTE values measured for the pure LaMA coating and the LaMA-containing composite coatings with a LaMA content of 75 wt. %, 50 wt. % and 25 wt. % at the same temperature range are $\sim 5.13\times10^{-6} K^{-1}$, $\sim 7.33\times10^{-6} K^{-1}$, $\sim 8.14\times10^{-6} K^{-1}$ and $\sim 9.12\times10^{-6} K^{-1}$, respectively.

So, the specific graded architecture design led to a reduced residual stress level. Therefore, such a FG-TBC displays a great potential for advanced TBCs applications in the future. (Chen, Gu, Zou, Wang, & Cao, 2012).

New Production Methods

For the protection of numerous blades and vanes, the global market for gas turbines requires a new coating deposition technology with lower cost (a one-stage process). The fabrication process of a TBC system involves a large number of steps with two or multilayers (Boidot, Selezneff, Monceau, Oquab, & Estournès, 2010). In this regard, the spark plasma sintering method (SPS) has been introduced as a new technique to fabricate coating. SPS consists of heating by applying a pulsed direct current through the die with a uniaxial pressure. This technique is a novel sintering method with higher heating rate and shorter dwell times than the conventional sintering methods, thus enhancing the sintering activity and leading to limited grain growth. To build complex multilayered structures, SPS has the capability to fabricate coatings with a wide range of defined compositions (Song, Ma, Zhang, & Schoenung, 2010).

Figure 6. Schematic of factors that led to fracture in TBC and also, the solutions that can improve the efficiency and durability of TBC coatings

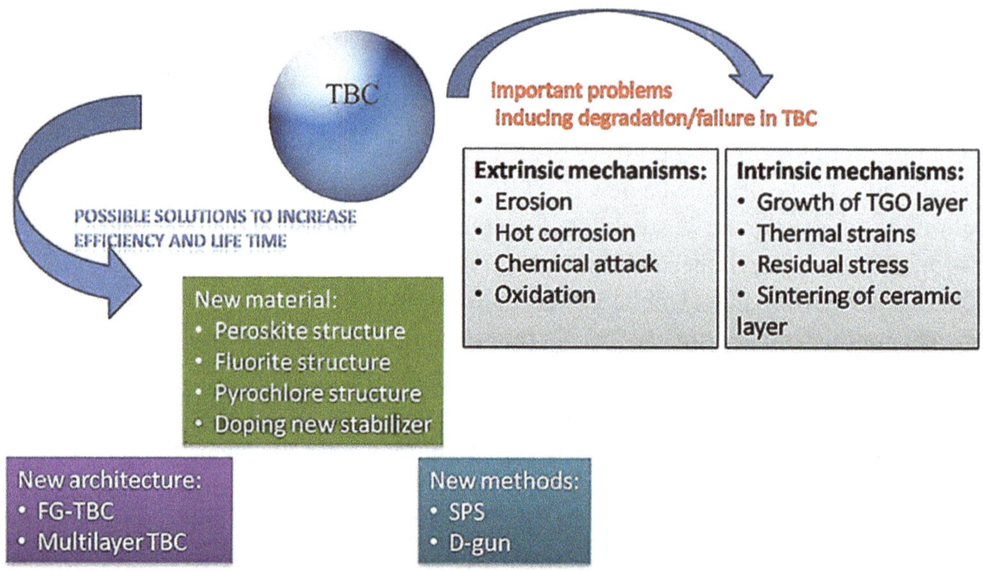

The SPS technique has been employed by a number of studies to develop the nanosized grain ceramics, functionally graded materials, fully dense materials, and difficult to sinter materials. However, there have been few studies that refer to the possibility of partial coating using the SPS technique (Boidot, Selezneff, Monceau, Oquab, & Estournès, 2010). Pakseresht et al. (2014) have produced FG-TBC in one step by using the SPS method. The SPS technique can eliminate the multiple steps in a process.

Beside the SPS method, Kim et al. (2003) fabricated FG-TBC by using a detonation gun spray process. In this method, the properties of each metal and ceramic are retained during the process. A new approach, known as a "shot control method," has also been proposed for the preparation of FG-TBC. This method alternately sprays the metal and ceramic powders using two powder feeders to deposit a mixture of metals and ceramics. In the area overlapped by the successive shots, an excellent mixture of ceramics and metals can be produced by this method at the level of each particle splat. By changing the ratio of the metal powder to the ceramic, it is possible to control the volume ratio of the coating components of metal and ceramic (Kim, Kim, & Park, 2003).

In conclusion, Figure 6 summarizes the causes of fracture and the problems with TBC coatings, as well as the solutions to improve the efficiency and durability of TBC coatings.

CONCLUSION

In this study, the gradient variations of the physical and mechanical properties, such as elastic moduli, thermal diffusivity/conductivity, coefficient of thermal expansion, of FG-TBC are discussed. Thus, the mismatch in residual stresses and thermal strains in the coatings can be reduced in FG-TBC as compared with the conventional one. In general, in the conventional TBC the thermal stresses produced after the plasma spraying process are extremely high. Consequently, these high stresses result in the cracking, delamination, and spallation of the conventional TBC. The residual stresses can be significantly reduced with the application of FG-TBC. In addition, the microstructure, chemical composition, porosity, and phase composition can be changed gradually in the FG-TBC. Generally, the application of FG-TBC has the following advantages: (1) it improves the mechanical properties of TBCs, (2) reduces the mismatch between the coefficients of the thermal expansion of the BC and TC, and (3) increases the high-temperature resistance. Moreover, interchanging a single-layer ceramic TBC with a YSZ–bond coat alloy functionally graded TBC can improve the effective fracture toughness and effectively prevent the growth of the delamination cracks along the TC–BC interface. Therefore, as a result, FGM coatings possess superior properties as compared with duplex coatings.

REFERENCES

Alfano, M., Di Girolamo, G., Pagnotta, L., & Sun, D. (2010). Processing, microstructure and mechanical properties of air plasma-sprayed ceria–yttria co-stabilized zirconia coatings. *Strain*, *46*(5), 409–418. doi:10.1111/j.1475-1305.2009.00659.x

Bao, G., & Cai, H. (1997). Delamination cracking in functionally graded coating/metal substrate systems. *Acta Materialia*, *45*(3), 1055–1066. doi:10.1016/S1359-6454(96)00232-7

Boidot, M., Selezneff, S., Monceau, D., Oquab, D., & Estournès, C. (2010). Proto-TGO formation in TBC systems fabricated by spark plasma sintering. *Surface and Coatings Technology*, *205*(5), 1245–1249. doi:10.1016/j.surfcoat.2010.09.042

Chen, X., Zhao, Y., Fan, X., Liu, Y., Zou, B., Wang, Y., & Ma, H. (2011). Thermal cycling failure of new $LaMgA_{11}O_{19}$ / YSZ double ceramic top coat thermal barrier coating systems. *Surface and Coatings Technology*, *205*, 3293–3300.

Chen, X., Gu, L., Zou, B., Wang, Y., & Cao, X. (2012). New functionally graded thermal barrier coating system based on $LaMgAl_{11}O_{19}$ / YSZ prepared by air plasma spraying. *Surface and Coatings Technology*, *206*(8-9), 2265–2274. doi:10.1016/j.surfcoat.2011.09.076

Das, D., Saini, A. K., Pathak, M. K., & Kumar, S. (2012). Application of functionally graded materials as thermal insulator in high temperature engineering. *International Journal of Research in Engineering & Applied Sciences*, *2*, 966–974.

Fu, L. U., Khor, K. A., Ng, H. W., & Teo, T. N. (2000). Non-destructive evaluation of plasma sprayed functionally graded thermal barrier coatings. *Surface and Coatings Technology*, *130*(2-3), 233–239. doi:10.1016/S0257-8972(00)00700-3

Goswami, B., Ray, A. K., & Sahay, S. K. (2004). Thermal barrier coatings system for gas turbine application- A review. *High Temperature Material Processes*, *23*, 73–92.

Guo, H., Wang, Y., Wang, L., & Gong, S. (2009). Thermo-physical properties and thermal shock resistance of segmented $La_2Ce_2O_7$ / YSZ thermal barrier coatings. *Journal of Thermal Spray Technology*, *18*(4), 665–671. doi:10.1007/s11666-009-9350-1

Gu, Y. W., Khor, K. A., Fu, Y. Q., & Wang, Y. (1997). Functionally graded ZrOz-NiCrA1Y coatings prepared by plasma spraying using pre-mixed, spheroidized powders. *Surface and Coatings Technology, 96*(2-3), 305–312. doi:10.1016/S0257-8972(97)00185-0

Habibi, M. H., Wang, L., & Guo, S. M. (2012). Evolution of hot corrosion resistance of YSZ, $Gd_2Zr_2O_7$, and $Gd_2Zr_2O_7$ + YSZ composite thermal barrier coatings in $Na_2SO_4 + V_2O_5$ at 1050 °C. *Journal of the European Ceramic Society, 32*(8), 1635–1642. doi:10.1016/j.jeurceramsoc.2012.01.006

Habibi, M. H., Wang, L., Liang, J., & Guo, S. M. (2013). An investigation on hot corrosion behavior of YSZ-Ta_2O_5 in $Na_2SO_4 + V_2O_5$ salt at 1100°C. *Corrosion Science, 75*, 409–414. doi:10.1016/j.corsci.2013.06.025

Joshi, S. C., & Ng, H. W. (2011). Optimizing functionally graded nickel – zirconia coating profiles for thermal stress relaxation. *Simulation Modelling Practice and Theory, 19*(1), 586–598. doi:10.1016/j.simpat.2010.08.013

Karaoglani, A., Ogawa, O., Turk, A., & Ozdemir, I. (2013). Thermal shock and cycling behavior of thermal barrier coating used in gas turbines. In E. Benin (Ed.), Progress in gas turbine performance (pp 237-260). InTech.

Kawasaki, A., Watanabe, R., Yuki, M., Nakanishi, Y., & Onabe, H. (1996). Effect of microstructure on the thermal shock cracking of functionally graded thermal barrier coatings studied by burner heating test. *Materials Transactions, 37*(4), 788–795. doi:10.2320/matertrans1989.37.788

Keyvani, A., Saremi, M., & Sohi, M. H. (2010). Microstructural stability of zirconia – alumina composite coatings during hot corrosion test at 1050°C. *Journal of Alloys and Compounds, 506*(1), 103–108. doi:10.1016/j.jallcom.2010.06.110

Keyvani, A., Saremi, M., & Sohi, M. H. (2011). Oxidation resistance of YSZ-alumina composites compared to normal YSZ TBC coatings at 1100°C. *Journal of Alloys and Compounds, 509*(33), 8370–8377. doi:10.1016/j.jallcom.2011.05.029

Khoddami, A. M., Sabour, A., & Hadavi, S. M. M. (2007). Microstructure formation in thermally-sprayed duplex and functionally graded NiCrAlY / Yttria-Stabilized Zirconia coatings. *Surface and Coatings Technology, 201*(12), 6019–6024. doi:10.1016/j.surfcoat.2006.11.020

Khor, K. A., Dong, Z. L., & Gu, Y. W. (1999). Plasma sprayed functionally graded thermal barrier coatings. *Materials Letters, 38*(6), 437–444. doi:10.1016/S0167-577X(98)00203-1

Khor, K. A., & Gu, Y. W. (2000 a). Effects of residual stress on the performance of plasma sprayed functionally graded ZrO_2 / NiCoCrAlY coatings. *Materials Science and Engineering A, 277*(1-2), 64–76. doi:10.1016/S0921-5093(99)00565-1

Khor, K. A., & Gu, Y. W. (2000 b). Thermal properties of plasma-sprayed functionally graded thermal barrier coatings. *Thin Solid Films, 372*(1-2), 104–113. doi:10.1016/S0040-6090(00)01024-5

Khor, K. A., Gu, Y. W., & Dong, Z. L. (2000). Plasma spraying of functionally graded yttria stabilized zirconia / NiCoCrAlY coating system using composite powders. *Journal of Journal of Thermal Spray Technology, 9*(2), 245–249. doi:10.1361/105996300770349999

Khor, K. A. U., Gu, Y. W., & Dong, Z. L. (2001). Mechanical behavior of plasma sprayed functionally graded YSZ/NiCoCrAlY composite coatings. *Surface and Coatings Technology*, *139*(2-3), 200–206. doi:10.1016/S0257-8972(01)00997-5

Kim, J. H., Kim, M. C., & Park, C. G. (2003). Evaluation of functionally graded thermal barrier coatings fabricated by detonation gun spray technique. *Surface and Coating Technology*, *168*, 275–280.

Kokini, K., Takeuchi, Y. R., & Choules, B. D. (1996). Thermal crack initiation mechanisms on the surface of functionally graded ceramic thermal barrier coatings. *Ceramics International*, *22*, 397–401.

Kokini, K., Dejonge, J., Rangaraj, S., & Beardsley, B. (2002). Thermal shock of functionally graded thermal barrier coatings with similar thermal resistance. *Surface and Coatings Technology*, *154*(2-3), 223–231. doi:10.1016/S0257-8972(02)00031-2

Lee, W., Stinton, D., Berndt, C. C., Erdogan, F., Lee, Y.-D., & Mutasim, Z. (1996). Concept of functionally graded materials for advanced thermal barrier coatings applications. *Journal of the American Ceramic Society*, *79*(12), 3002–3012. doi:10.1111/j.1151-2916.1996.tb08070.x

Miller, R. A. (1997). Thermal Barrier Coatings for Aircraft Engines: History and Directions. *Journal of Journal of Thermal Spray Technology*, *6*(1), 35–42. doi:10.1007/BF02646310

Movchan, B. A. (2002). Functionally graded EB PVD coatings. *Surface and Coatings Technology*, *149*(2-3), 252–262. doi:10.1016/S0257-8972(01)01439-6

Negahdari, Z., Porda, M., & Scherm, F. (2010). Development of Novel Functionally Graded Al_2O_3 -Lanthanum Hexaaluminate Ceramics for Thermal Barrier Coatings Zahra Negahdari. *Materials Science Forum*, *631-632*, 97–102. doi:10.4028/www.scientific.net/MSF.631-632.97

Nejati, M., Rahimipour, M. R., & Mobasherpour, I. (2014). Evaluation of hot corrosion behavior of CSZ, CSZ / micro Al_2O_3 and CSZ/nano Al_2O_3 plasma sprayed thermal barrier coatings. *Ceramics International*, *40*(3), 4579–4590. doi:10.1016/j.ceramint.2013.08.135

Nur, S., Jamaludin, S., Mustapha, F., & Nuruzzaman, D. M. (2013). A review on the fabrication techniques of functionally graded ceramic-metallic materials in advanced composites. *Scientific Research and Essays*, *8*, 828–840.

Pakseresht, A. H., Ghasali, E., Nejati, M., Shirvanimoghaddam, K., Javadi, A. H., & Teimouri, R. (2015). Development empirical-intelligent relationship between plasma spray parameters and coating performance of Yttria-Stabilized Zirconia. *International Journal of Advanced Manufacturing Technology*, *76*(5-8), 1031–1045. doi:10.1007/s00170-014-6212-x

Pakseresht, A. H., Rahimipour, M. R., Vaezi, M. R., & Salehi, M. (2015). Effect of splat morphology on the microstructure and dielectric properties of plasma sprayed barium titanate films. *Applied Surface Science*, *324*, 797–806. doi:10.1016/j.apsusc.2014.11.041

Pakseresht, A. H., Simchi, A., Bahrami, M., & Javadi, A. H. (2014). Application of spark plasma sintering for fabrication of functionally graded thermal barrier coating on a superalloy substrate. *Paper presented at Conference smart materials and surfaces*, Bangkok, Thailand.

Pan, W., Phillpot, S. R., Wan, C., Chernatynskiy, A., & Qu, Z. (2012). Low thermal conductivity oxides. *Materials Research Society Bulletin*, *37*(10), 917–921. doi:10.1557/mrs.2012.234

Park, S. Y., Kim, J. H., Kim, M. C., Song, H. S., & Park, C. G. (2005). Microscopic observation of degradation behavior in yttria and ceria stabilized zirconia thermal barrier coatings under hot corrosion. *Surface and Coatings Technology*, *190*(2-3), 357–365. doi:10.1016/j.surfcoat.2004.04.065

Rangaraj, S., & Kokini, K. (2003a). Estimating the fracture resistance of functionally graded thermal barrier coatings from thermal shock tests. *Surface and Coatings Technology*, *173*(2-3), 201–212. doi:10.1016/S0257-8972(03)00515-2

Rangaraj, S., & Kokini, K. (2003b). Interface thermal fracture in functionally graded zirconia – mullite – bond coat alloy thermal barrier coatings. *Acta Materialia*, *51*(1), 251–267. doi:10.1016/S1359-6454(02)00396-8

Rangaraj, S., & Kokini, K. (2004). A Study of thermal fracture in functionally graded thermal barrier coatings using a cohesive. *Journal of Engineering Materials and Technology*, *126*(1), 103–115. doi:10.1115/1.1631028

Saremi, M., Afrasiabi, A., & Kobayashi, A. (2008). Microstructural analysis of YSZ and YSZ / Al_2O_3 plasma sprayed thermal barrier coatings after high temperature oxidation. *Surface and Coatings Technology*, *202*(14), 3233–3238. doi:10.1016/j.surfcoat.2007.11.029

Schulz, U., Peters, M., Bach, F., & Tegeder, G. (2003). Graded coatings for thermal, wear and corrosion barriers. *Materials Science and Engineering A*, *362*(1-2), 61–80. doi:10.1016/S0921-5093(03)00579-3

Song, J., Ma, K., Zhang, L., & Schoenung, J. M. (2010). Technology Simultaneous synthesis by spark plasma sintering of a thermal barrier coating system with a NiCrAlY bond coat. *Surface and Coatings Technology*, *205*(5), 1241–1244. doi:10.1016/j.surfcoat.2010.08.064

Stover, D., Pracht, G., Lehmann, H. Dietrich, M. & Vaben, R. (2004) New material concepts for the next generation of plasma-sprayed thermal barrier coatings. *Journal ot Thermal Spray Technology*, 13, 76–83.

Tamura, M., Takahashi, M., Ishii, J., Suzuki, K., Sato, M., & Shimomura, K. (1999). Multilayered thermal barrier coating for land-based gas turbines. *Journal of Thermal Spray Technology*, *8*(1), 68–72. doi:10.1361/105996399770350584

Vaßen, R., Jarligo, M. O., Steinke, T., Mack, D. E., & Stöver, D. (2010). Overview on advanced thermal barrier coatings. *Surface and Coatings Technology*, *205*(4), 938–942. doi:10.1016/j.surfcoat.2010.08.151

Vassen, R., Stuke, A., Sto, D., Vassen, R., Stuke, A., Sto, D., & Ju, F. (2009). Recent developments in the field of thermal barrier coatings. *Journal of Thermal Spray Technology*, *18*(2), 181–186. doi:10.1007/s11666-009-9312-7

Vassen, R., Kassner, H., Stuke, A., Mack, D. E., Jarligo, M. O., & Stöver, D. (2010). Functionally Graded Thermal Barrier Coatings with Improved Reflectivity and High-Temperature Capability. *Materials Science Forum*, *631-632*, 73–78. doi:10.4028/www.scientific.net/MSF.631-632.73

Viswanathan, V., Dwivedi, D., & Sampath, S. (2014). Engineered Multilayer Thermal Barrier Coatings for Enhanced Durability and Functional Performance. *Journal of the American Ceramic Society*, *97*(9), 2770–2778. doi:10.1111/jace.13033

Widjaja, S., Limarga, A. M., & Yip, T. H. (2002). Oxidation behavior of a plasma-sprayed functionally graded ZrO_2 / Al_2O_3 thermal barrier coating. *Materials Letters*, *57*(3), 628–634. doi:10.1016/S0167-577X(02)00842-X

Widjaja, S., Limarga, A. M., & Yip, T. H. (2003). Modeling of residual stresses in a plasma-sprayed zirconiay alumina functionally graded-thermal barrier coating. *Thin Solid Films*, *434*(1-2), 216–227. doi:10.1016/S0040-6090(03)00427-9

Xiong, H., Kawasaki, A., Kang, Y., & Watanabe, R. (2005). Experimental study on heat insulation performance of functionally graded metal / ceramic coatings and their fracture behavior at high surface temperatures. *Surface and Coatings Technology*, *194*(2-3), 203–214. doi:10.1016/j.surfcoat.2004.07.069

Wolfe, D., & Singh, J. (1998). Functionally gradient ceramic/metallic coatings for gas turbine components by high-energy beams for high-temperature applications. *Journal of Materials Science*, *33*(14), 3677–3692. doi:10.1023/A:1004675900887

KEY TERMS AND DEFINITIONS

Bond Coat: It is a metallic coating which is inserted between the thermally insulating top layer and the substrate of a thermal barrier coating in order to increase the bonding strength and improve oxidation resistance.

Failure Mechanism: The mechanisms that cause the fracture in the coatings, the most important of which are the quick growth of TGO layer and creation of residual stress in the layers.

Functionally Graded Material: It is a compound obtained by mixing two or more materials; by deviating from part of the material, its composition and properties are varied.

Modern TBC: It is the new generation of TBC system using the new material and FGM structure in order to improve the efficiency and reduce the destructive effects of TBC systems.

New TBC Material: The new generation of thermal barrier coatings is made from some material with higher thermal stability, in comparison to YSZ, in order to increase the temperature of system.

TBC: These kinds of coatings are usually applied on superalloy materials in order to increase their working temperature and protect them against environmental factors.

YSZ: It is the most common ceramic oxide which is used as the upper layer of thermal barrier coatings due to its low thermal conductivity.

Compilation of References

Abgrall, P., & Gue, A. M. (2007). Lab-on-chip technologies: Making a microfluidic network and coupling it into a complete microsystem - a review. *Journal of Micromechanics and Microengineering*, *17*(5), R15–R49. doi:10.1088/0960-1317/17/5/R01

Abukawa, H., Papadaki, M., Abulikemu, M., Leaf, J., Vacanti, J. P., Kaban, L. B., & Troulis, M. J. (2006). The engineering of craniofacial tissues in the laboratory: A review of biomaterials for scaffolds and implant coatings. *Dental Clinics of North America*, *50*(2), 205–216. doi:10.1016/j.cden.2005.11.006 PMID:16530058

Abu-Thabit, N., & Makhlouf, A. S. (2015). Recent approaches for designing nanomaterials-based coatings for corrosion protection.Handbook of Nanoelectrochemistry, 1, 1–20. Switzerland, Springer International Publishing

Afferrante, L., & Carbone, G. (2014). The effect of drop volume and micropillar shape on the apparent contact angle of ordered microstructured surfaces. *Soft Matter*, *10*(22), 3906–3914. doi:10.1039/c3sm53104j PMID:24643633

Agudelo-Morales, C. E., Galian, R. E., & Pérez-Prieto, J. (2012). Pyrene-Functionalized Nanoparticles: Two Independent Sensors, the Excimer and the Monomer. *Analytical Chemistry*, *84*(18), 8083–8087. doi:10.1021/ac302276j PMID:22924743

Aguilera-Sigalat, J., Casas-Solvas, J. M., Morant-Minana, M. C., Vargas-Berenguel, A., Galian, R. E., & Perez-Prieto, J. (2012). Quantum dot/cyclodextrin supramolecular systems based on efficient molecular recognition and their use for sensing. *Chemical Communications*, *48*(20), 2573–2575. doi:10.1039/C1CC15312A PMID:22080219

Aguilera-Sigalat, J., Pais, V. F., Doménech-Carbó, A., Pischel, U., Galian, R. E., & Pérez-Prieto, J. (2013). Unconventional Fluorescence Quenching in Naphthalimide-Capped CdSe/ZnS Nanoparticles. *The Journal of Physical Chemistry C*, *117*(14), 7365–7375. doi:10.1021/jp3128252

Aguilera-Sigalat, J., Rocton, S., Galian, R. E., & Pérez-Prieto, J. (2011). Fluorescence enhancement of amine-capped CdSe/ZnS quantum dots by thiol addition. *Canadian Journal of Chemistry*, *89*(3), 359–363. doi:10.1139/V10-160

Aguilera-Sigalat, J., Rocton, S., Sanchez-Royo, J. F., Galian, R. E., & Perez-Prieto, J. (2012). Highly fluorescent and photostable organic- and water-soluble CdSe/ZnS core-shell quantum dots capped with thiols. *RSC Advances*, *2*(4), 1632–1638. doi:10.1039/C1RA01005K

Agura, H., Suzuki, A., Matsushita, T., Aoki, T., & Okuda, M. (2003). Low resistivity transparent conducting Al-doped ZnO films prepared by pulsed laser deposition. *Thin Solid Films*, *445*(2), 263–267. doi:10.1016/S0040-6090(03)01158-1

Ahmad, Y. H., Mohamed, A. M. A., Golden, T. D., & D'Souza, N. A. (2014b). Electrodeposition of nanocrystalline Ni-Mo alloys from alkaline glycinate solutions. *International Journal of Electrochemical Science*, *9*(11), 6438–6450.

Ahmad, Y. H., Tientong, J., Nar, M., D'Souza, N., Mohamed, A. M. A., & Golden, T. D. (2014a). Characterization and corrosion resistance of electrodeposited Ni-Mo-silicate platelet nanocomposite coatings. *Surface and Coatings Technology*, *259*(Part C), 517–525. doi:10.1016/j.surfcoat.2014.10.036

Ahn, M. K., Kwon, H. S., & Lee, H. M. (1998). Quantitative comparison of the influences of tungsten and molybdenum on the passivity of Fe-29Cr ferritic stainless steels. *Corrosion Science*, *40*(2-3), 307–322. doi:10.1016/S0010-938X(97)00138-8

Aissa, K. A., Achour, A., Elmazria, O., Simon, Q., Elhosni, M., Boulet, P., & Djouadi, M. et al. (2015). AlN films deposited by dc magnetron sputtering and high power impulse magnetron sputtering for SAW applications. *Journal of Physics. D, Applied Physics*, *48*(14), 145307. doi:10.1088/0022-3727/48/14/145307

Akazawa, H., & Ueno, Y. (2013). Growth of preferentially c-axis oriented hydroxyapatite thin films on Si (100) substrate by electron-cyclotron-resonance plasma sputtering. *Applied Surface Science*, *276*, 217–222. doi:10.1016/j.apsusc.2013.03.070

Akazawa, H., & Ueno, Y. (2014). Control of composition and crystallinity in hydroxyapatite films deposited by electron cyclotron resonance plasma sputtering. *Journal of Physics and Chemistry of Solids*, *75*(1), 94–99. doi:10.1016/j.jpcs.2013.08.014

Albella, P., Garcia-Cueto, B., Gonzalez, F., Moreno, F., Wu, P. C., Kim, T.-H., & Videen, G. et al. (2011). Shape matters: Plasmonic nanoparticle shape enhances interaction with dielectric substrate. *Nano Letters*, *11*(9), 3531–3537. doi:10.1021/nl201783v PMID:21848270

Aldana, J., Wang, Y. A., & Peng, X. (2001). Photochemical Instability of CdSe Nanocrystals Coated by Hydrophilic Thiols. *Journal of the American Chemical Society*, *123*(36), 8844–8850. doi:10.1021/ja016424q PMID:11535092

Alfano, M., Di Girolamo, G., Pagnotta, L., & Sun, D. (2010). Processing, microstructure and mechanical properties of air plasma-sprayed ceria–yttria co-stabilized zirconia coatings. *Strain*, *46*(5), 409–418. doi:10.1111/j.1475-1305.2009.00659.x

Aliofkhazraei, M. (2015). *Anti-Abrasive Nanocoatings: Current and Future Applications*. Cambridge, UK: Woodhead Publishing.

Alivisatos, A. P. (1996). Semiconductor Clusters, Nanocrystals, and Quantum Dots. *Science*, *271*(5251), 933–937. doi:10.1126/science.271.5251.933

Allahyarzadeh, M. H., Roozbehani, B., & Ashrafi, A. (2011). Electrodeposition of high Mo content amorphous/nanocrystalline Ni-Mo alloys using 1-ethyl-3-methyl-imidazolium chloride ionic liquid as an additive. *Electrochimica Acta*, *56*(27), 10210–10216. doi:10.1016/j.electacta.2011.09.011

Allahyarzadeh, M. H., Roozbehani, B., Ashrafi, A., & Shadizadeh, S. R. (2012c). Effect of surface preparation on corrosion resistance of electrochemically deposited Ni-Mo alloys in the presence 1-methyl-imidazolium chloride ionic liquid. *ECS Transactions*, *41*(44), 29–46. doi:10.1149/1.4718389

Allahyarzadeh, M. H., Roozbehani, B., Ashrafi, A., Shadizadeh, S. R., & Kheradmand, E. (2012b). Electrodeposition of high Mo content amorphous/nanocrystalline Ni-Mo using ionic liquids as additive. *ECS Transactions*, *41*(44), 11–28. doi:10.1149/1.4718388

Allahyarzadeh, M. H., Roozbehani, B., Ashrafi, A., Shadizadeh, S. R., Seddighian, A., & Kheradmand, E. (2012a). A novel aspect of two imidazolium-based ionic liquids in electrodeposition of amorphous/nanocrystalline Ni-Mo. *Journal of the Electrochemical Society*, *159*(8), D473–D478. doi:10.1149/2.065208jes

Allauddin, S., Narayan, R., & Raju, R. V. S. N. (2013). Synthesis and properties of alkoxysilane castor oil and their polyurethane/urea-silica hybrid coating films. *ACS Sustainable Chemistry & Engineering*, *1*(8), 910–918. doi:10.1021/sc3001756

Alongi, J., Carosio, F., & Malucelli, G. (2014). Current Emerging Techniques to Impart Flame Retardancy to Fabrics: An Overview. *Polymer Degradation & Stability*, *106*, 138–149. doi:10.1016/j.polymdegradstab.2013.07.012

Al-Taher, F., Telegdi, J., & Kalman, E. (2007). Effect of divalent cations in Langmuir-Blodgett films on the protection of copper against corrosion. *Materials Science Forum*, 537-538, 9-14.

Al-Telegdi, J., Rigo, T., & Kalman, E. (2005). Molecular layers of hydroxamic acids in copper corrosion inhibition. *Journal of Electroanalytical Chemistry*, *582*(1-2), 191–201. doi:10.1016/j.jelechem.2005.03.045

Alvarez-Lorenzo, C., & Concheiro, A. (2008). Intelligent Drug Delivery Systems: Polymeric Micelles and Hydrogels. *Mini Reviews in Medicinal Chemistry*, *8*(11), 1065–1074. doi:10.2174/138955708785909952 PMID:18855723

Amerio, E., Sangermano, M., Malucelli, G., Priola, A., & Voit, B. (2005). Preparation and characterization of hybrid nanocomposite coatings by photopolymerization and sol-gel process. *Polymer*, *46*(25), 11241–11246. doi:10.1016/j.polymer.2005.09.062

Anastasiadis, S. H. (2013). Development of functional polymer surfaces with controlled wettability. *Langmuir*, *29*(30), 9277–9290. doi:10.1021/la400533u PMID:23789943

Anastasiadis, S. H., Lygeraki, M. I., Athanassiou, A., Farsari, M., & Pisignano, D. (2008). Reversibly Photo-Responsive Polymer Surfaces for Controlled Wettability. *Journal of Adhesion Science and Technology*, *22*(15), 1853–1868. doi:10.1163/156856108X320014

Anderson, R. E., & Chan, W. C. W. (2008). Systematic Investigation of Preparing Biocompatible, Single, and Small ZnS-Capped CdSe Quantum Dots with Amphiphilic Polymers. *ACS Nano*, *2*(7), 1341–1352. doi:10.1021/nn700450g PMID:19206301

Angiolini, L., Benelli, T., Giorgini, L., & Raymo, F. M. (2009). Optical and chiroptical switches based on photoinduced photon and proton transfer in copolymers containing spiropyran and azopyridine chromophores in their side chains. *Polymer*, *50*(24), 5638–5646. doi:10.1016/j.polymer.2009.10.013

Antebi, B., Cheng, X., Harris, J.N., Gower, L.B., Chen, X.-D., & Ling, J. (2013). Biomimetic collagen-hydroxyapatite composite fabricated via a novel perfusion-flow mineralization technique. *Tissue Engineering Part C-Methods, 19*(7), 487-496. doi:10.1089/ten.tec.2012.0452

Anti-Fog PET Film Products. (n. d.). FSI Coating Technologies. Retrieved from http://www.fsicti.com/index.php?pid=19

An, Y. H., & Friedman, R. J. (1996). Prevention of sepsis in total joint arthroplasty. *The Journal of Hospital Infection*, *33*(2), 93–108. doi:10.1016/S0195-6701(96)90094-8 PMID:8808743

Apreutesei, M., Arvinte, R., Marin, A., Andrei, O. C., & Munteanu, D. (2011). Magnetron sputtering technique used for coatings; technologies and applications. *Paper presented at 7th International Conference on Materials Science and Engineering BRAMAT '11*, Romania.

Arafat, M.T., Savalani, M.M., Gibson, I., & Asme. (2009). *Improving the mechanical properties in tissue engineered scaffolds*.

Arafat, M. T., Gibson, I., & Li, X. (2014). State of the art and future direction of additive manufactured scaffolds-based bone tissue engineering. *Rapid Prototyping Journal*, *20*(1), 13–26. doi:10.1108/RPJ-03-2012-0023

Arafat, M. T., Lam, C. X. F., Ekaputra, A. K., Wong, S. Y., He, C., Hutmacher, D. W., & Gibson, I. et al. (2011). High performance additive manufactured scaffolds for bone tissue engineering application. *Soft Matter*, *7*(18), 8013–8022. doi:10.1039/c1sm05793f

Arafat, M. T., Lam, C. X. F., Ekaputra, A. K., Wong, S. Y., Li, X., & Gibson, I. (2011). Biomimetic composite coating on rapid prototyped scaffolds for bone tissue engineering. *Acta Biomaterialia*, *7*(2), 809–820. doi:10.1016/j.actbio.2010.09.010 PMID:20849985

Argentiere, S., Gigli, G., Mortato, M., Gerges, I., & Blasi, L. (2012). Advances in Microfluidics. In R. Kelly (Ed.), *Advances in Microfluidics*. InTech.

Arnebold, A., Schorsch, O., Beckmann, J., & Hartwig, A. (2013). Concomitant cationic polymerization of a hybrid monomer and an epoxy resin. *Reactive & Functional Polymers*, *73*(12), 1625–1631. doi:10.1016/j.reactfunctpolym.2013.09.004

Arun, M., Kantheti, S., Gaddam, R. R., Narayan, R., & Raju, K. V. S. N. (2014). Surface modification of TiO_2 nanoparticles with 1,3,5-triazine based silane coupling agent and its cumulative effect on the properties of polyurethane composite coating. *Journal of Polymer Research*, *21*(12), 600. doi:10.1007/s10965-014-0600-7

Arwin, H., Poksinski, M., & Johansen, K. (2004). Total internal reflection ellipsometry: Principles and applications. *Applied Optics*, *43*(15), 3028–3036. doi:10.1364/AO.43.003028 PMID:15176189

Arwin, H., Poksinski, M., & Johansen, K. (2008). Enhancement in ellipsometric thin film sensitivity near surface plasmon resonance conditions. *Physica Status Solidi*, *205*(4), 817–820. doi:10.1002/pssa.200777899

Asadi, K., de Leeuw, D. M., de Boer, B., & Blom, P. W. M. (2008). Organic non-volatile memories from ferroelectric phase-separated blends. *Nature Materials*, *7*(7), 547–550. doi:10.1038/nmat2207 PMID:18552851

Asghar, M., Shoaib, M., Placido, F., & Naseem, S. (2008). Wide bandpass optical filters with TiO2 and Ta2O5. *Central European Journal of Physics*.

Ashrafi, A., Golozar, M. A., & Mallakpour, S. (2003). Electropolymerization of pyrrole and N-methylpyrrole on mild steel. *Iranian Polymer Journal*, *12*(6), 485–490.

Ashurst, W. R., Carraro, C., & Maboudian, R. (2003). Vapor phase anti-stiction coatings for MEMS. *IEEE Transactions on Device and Materials Reliability*, *3*(4), 173–178. doi:10.1109/TDMR.2003.821540

Aspnes, D. E. (1975). Photometric ellipsometer for measuring partially polarized light. *Journal of the Optical Society of America*, *65*(11), 1274–1278. doi:10.1364/JOSA.65.001274

Aspnes, D. E. (2004). Expanding horizons: New developments in ellipsometry and polarimetry. *Thin Solid Films*, *455-456*, 3–13. doi:10.1016/j.tsf.2003.12.038

Astala, R., & Stott, M. (2008). First-principles study of hydroxyapatite surfaces and water adsorption. *Physical Review B: Condensed Matter and Materials Physics*, *78*(7), 075427. doi:10.1103/PhysRevB.78.075427

ASTM G173-03 (2012), Standard Tables for Reference Solar Spectral Irradiances: Direct Normal and Hemispherical on 37° Tilted Surface. *ASTM International.* Retrieved from http://www.astm.org/Standards/G173.htm

ASTM-F1185–03. Standard specification for composition of hydroxylapatite for surgical implants. (2009).

Attia, S., Wang, J., Wu, G., Shen, J., & Ma, J. (2002). Review on Sol—Gel Derived Coatings: Process, Techniques and Optical Applications. *Journal of Materials Science and Technology*, *18*(3), 211–218.

Audebert, P., & Miomandre, F. (2007). Electrochemistry of conducting polymers. In T. A. Skotheim and J. R. Reynolds (Eds.), Handbook of Conducting Polymers 3rd Edition Conjugated Polymers Theory, Synthesis, Properties and Characterization (pp. 18–1-18–40). Boca Raton, Florida: CRC Press.

Auernheimer, J., Zukowski, D., Dahmen, C., Kantlehner, M., Enderle, A., Goodman, S. L., & Kessler, H. (2005). Titanium implant materials with improved riocompatibility through coating with phosphonate-anchored cyclic RGD peptides. *ChemBioChem*, *6*(11), 2034–2040. doi:10.1002/cbic.200500031 PMID:16206226

Avelar-Batista Wilson, J. C., Wu, S., Gotman, I., Housden, J., & Gutmanas, E. Y. (2015). Duplex coatings with enhanced adhesion to Ti alloy substrate prepared by powder immersion nitriding and TiN/Ti multilayer deposition. *Materials Letters*, *157*, 45–49. doi:10.1016/j.matlet.2015.05.054

Aygüler, M. F., Weber, M. D., Puscher, B. M. D., Medina, D. D., Docampo, P., & Costa, R. D. (2015). Light-Emitting Electrochemical Cells Based on Hybrid Lead Halide Perovskite Nanoparticles. *The Journal of Physical Chemistry C*, *119*(21), 12047–12054. doi:10.1021/acs.jpcc.5b02959

Azzam, R. M. A., Bashara, N. M., & Burns, D. T. (1987). *Ellipsometry and polarized light*. Amsterdam: North Holland.

Bachon, T. (2005). Compositions based on silane-terminated polymers and hardened by radiation and humidity, production and use thereof. *Patent WO 2005/100482 A1*.

Bae, S. E., Choi, J., Joung, Y. K., Park, K., & Han, D. K. (2012). Controlled release of bone morphogenetic protein (BMP)-2 from nanocomplex incorporated on hydroxyapatite-formed titanium surface. *Journal of Controlled Release*, *160*(3), 676–684. doi:10.1016/j.jconrel.2012.04.021 PMID:22543042

Baghdachi, J., Perez, H., & Shah, A. (2010). Design and development of self-healing polymers and coatings. In J. Baghdachi and T. Provder (Eds.), Smart Coatings III, *ACS Symposium Series* 1050 (pp. 3-20). Washington, DC: American Chemical Society. doi:10.1021/bk-2010-1050.ch001

Baigl, D. (2012). Photo-actuation of liquids for light-driven microfluidics: State of the art and perspectives. *Lab on a Chip*, *12*(19), 3637–3653. doi:10.1039/c2lc40596b PMID:22864577

Balakrishnan, T., Sathiyanarayanan, S., & Mayavan, S. (2015). Advanced Anticorrosion Coating Materials Derived from Sunflower Oil with Bifunctional Properties. *ACS Applied Materials & Interfaces*, *150826103253006*. 7(35), 19781–19788 PMID:26292971

Balazs, A. C., Emrick, T., & Russell, T. P. (2006). Nanoparticle Polymer Composites: Where Two Small Worlds Meet. *Science*, *314*(5802), 1107–1110. doi:10.1126/science.1130557 PMID:17110567

Baleviciute, I., Balevicius, Z., Makaraviciute, A., Ramanaviciene, A., & Ramanavicius, A. (2013). Study of antibody/antigen binding kinetics by total internal reflection ellipsometry. *Biosensors & Bioelectronics*, *39*(1), 170–176. doi:10.1016/j.bios.2012.07.017 PMID:22938839

Balog, M., & Schieber, M., Patai, S., & Michman, M. (1972). Thin films of metal oxides on silicon by chemical vapor deposition with organometallic compounds. *Journal of Crystal Growth*, *17*, 298-301.

Banerjee, A. N., Ghosh, C. K., Chattopadhyay, K. K., Minoura, H., Sarkar, A. K., Akiba, A., & Endo, T. et al. (2006). Low-temperature deposition of ZnO thin films on PET and glass substrates by DC-sputtering technique. *Thin Solid Films*, *496*(1), 112–116. doi:10.1016/j.tsf.2005.08.258

Bao, G., & Cai, H. (1997). Delamination cracking in functionally graded coating/metal substrate systems. *Acta Materialia*, *45*(3), 1055–1066. doi:10.1016/S1359-6454(96)00232-7

Bao, X.-M., Cui, J.-F., Sun, X.-X., Liang, W.-D., Zhu, Z.-Q., An, J., & Li, A. et al. (2014). Facile preparation of superhydrophobic surfaces based on metal oxide nanoparticles. *Applied Surface Science*, *303*, 473–480. doi:10.1016/j.apsusc.2014.03.029

Barberis, F. (2013). Contact angle measurements over macro, micro and nano dimensional systems. *Progress in Colloid and Interface Science*, 3, 1-35.

Barkhudarov, P. M., Shah, P. B., Watkins, E. B., Doshi, D. A., Brinker, C. J., & Majewski, J. (2008). Corrosion inhibition using superhydrophobic films. *Corrosion Science*, *50*(3), 897–902. doi:10.1016/j.corsci.2007.10.005

Barletta, M., Rubino, G., Tagliaferri, V., & Vesco, S. (2014). Design and manufacture of photoluminescent coatings on stainless steel substrates. *Colloids and Surfaces. A, Physicochemical and Engineering Aspects*, *455*(1), 147–155. doi:10.1016/j.colsurfa.2014.04.044

Barthlott, W., & Neinhuis, C. (1997). Purity of the sacred lotus, or escape from contamination in biological surface. *Planta*, *202*(1), 1–8. doi:10.1007/s004250050096

Bartlett, P. N. (2004). *Electrodeposition of Nanostructured Films Using Self-Organizing Templates.* In Electrochemical Society Interface (pp. 28–33).

BASF iGloss: A breakthrough clearcoat for automotive applications. (2012). *Metalfinishing.com.* Retrieved from http://www.metalfinishing.com/view/25005/basf-igloss-a-breakthrough-clearcoat-for-automotive-applications/

Basharat, S., Betchley, W., Carmalt, C. J., Barnett, S., Tocher, D. A., & Davies, H. O. (2007). Synthesis of Group 13 Sesquialkoxides and Their Application as Precursors to Crystalline Oxide Films. *Organometallics*, *26*(2), 403–407. doi:10.1021/om0608657

Basharat, S., Carmalt, C. J., Barnett, S. A., Tocher, D. A., & Davies, H. O. (2007). Aerosol Assisted Chemical Vapor Deposition of In2O3 Films from Me3In and Donor Functionalized Alcohols. *Inorganic Chemistry*, *46*(22), 9473–9480. doi:10.1021/ic701372b PMID:17914814

Bassetti, M. J., Chatterjee, A. N., Aluru, N. R., & Beebe, D. J. (2005). Development and modeling of electrically triggered hydrogels for microfluidic applications. *Journal of Microelectromechanical Systems*, *14*(5), 1198–1207. doi:10.1109/JMEMS.2005.845407

Bastidas, J. M., Torres, C. L., Cano, E., & Polo, J. L. (2002). Influence of molybdenum on passivation of polarised stainless steels in a chloride environment. *Corrosion Science*, *44*(3), 625–633. doi:10.1016/S0010-938X(01)00072-5

Bauer, S., Schmuki, P., von der Mark, K., & Park, J. (2013). Engineering biocompatible implant surfaces: Part I: Materials and surfaces. *Progress in Materials Science*, *58*(3), 261–326. doi:10.1016/j.pmatsci.2012.09.001

Bäumle, M., Stamou, D., Segura, J.-M., Hovius, R., & Vogel, H. (2004). Highly Fluorescent Streptavidin-Coated CdSe Nanoparticles: Preparation in Water, Characterization, and Micropatterning. *Langmuir*, *20*(10), 3828–3831. doi:10.1021/la0363940 PMID:15969364

Bawendi, M. G., Steigerwald, M. L., & Brus, L. E. (1990). The Quantum Mechanics of Larger Semiconductor Clusters ("Quantum Dots"). *Annual Review of Physical Chemistry*, *41*(1), 477–496. doi:10.1146/annurev.pc.41.100190.002401

Beari, F., Brand, M., Jenkner, P., Lehnert, R., Metternich, H. J., Monkiewicz, J., & Siesler, H. W. (2001). Organofunctional alkoxysilanes in dilute aqueous solution: New accounts on the dynamic structural mutability. *Journal of Organometallic Chemistry*, *625*(2), 208–216. doi:10.1016/S0022-328X(01)00650-7

Bechet, D., Couleaud, P., Frochot, C., Viriot, M.-L., Guillemin, F., & Barberi-Heyob, M. (2008). Nanoparticles as vehicles for delivery of photodynamic therapy agents. *Trends in Biotechnology*, *26*(11), 612–621. doi:10.1016/j.tibtech.2008.07.007 PMID:18804298

Beebe, D. J., Moore, J. S., Bauer, J. M., Yu, Q., Liu, R. H., Devadoss, C., & Jo, B. H. (2000). Functional hydrogel structures for autonomous flow control inside microfluidic channels. *Nature*, *404*(6778), 588–590. doi:10.1038/35007047 PMID:10766238

Belabbes, A., de Carvalho, L., Schleife, A., & Bechstedt, F. (2011). Cubic inclusions in hexagonal AlN, GaN, and InN: Electronic states. *Physical Review B: Condensed Matter and Materials Physics*, *84*(12), 125108. doi:10.1103/PhysRevB.84.125108

Belon, C., Chemtob, A., Croutxé-Barghorn, C., Rigolet, S., Le Houérou, V., & Gauthier, C. (2011). A simple method for the reinforcement of UV-cured coatings via sol-gel photopolymerization. *Macromolecular Materials and Engineering*, *296*(6), 506–516. doi:10.1002/mame.201000350

Belon, C., Chemtob, A., Croutxé-Barghorn, C., Rigolet, S., Le Houréou, V., & Gauthier, C. (2010b). Combination of radical and cationic photoprocess for the single-step synthesis of organic-inorganic hybrid films. *Journal of Polymer Science*, *48*, 4150–4158.

Belon, C., Chemtob, A., Croutxé-Barghorn, C., Rigolet, S., Schmitt, M., Bistac, S., & Gauthier, C. et al. (2010a). Nanocomposite coatings via simultaneous organic-inorganic photo-induced polymerization: Synthesis, structural investigation and mechanical characterization. *Polymer International*, *59*, 1175–1186.

Beltowska-Lehman, E., Bigos, A., Indyka, P., & Kot, M. (2012b). Electrodeposition and characterization of nanocrystalline Ni-Mo coatings. *Surface and Coatings Technology*, *211*, 67–71. doi:10.1016/j.surfcoat.2011.10.011

Beltowska-Lehman, E., & Indyka, P. (2012a). Kinetics of Ni-Mo electrodeposition from Ni-rich citrate baths. *Thin Solid Films*, *520*(6), 2046–2051. doi:10.1016/j.tsf.2011.10.024

Benea, L. (2015). Nanocomposite layers obtained by electro-co-deposition: corrosion and tribocorrosion properties. *Solid State Phenomena,* 227, 1-5.

Benelmekki, M., Torrell, M., Xuriguera, E., Vaz, F., & Teixeira, V. (2012). Structure and Properties of Silver Clusters Implanted in PET by PVD Sputtering for Active Packaging Applications. *Journal of Nano Research*, *18-19*, 105–116. doi:10.4028/www.scientific.net/JNanoR.18-19.105

Benito-Lopez, F., Byrne, R., Raduta, A. M., Vrana, N. E., McGuinness, G., & Diamond, D. (2010). Ionogel-based light-actuated valves for controlling liquid flow in micro-fluidic manifolds. *Lab on a Chip*, *10*(2), 195–201. doi:10.1039/B914709H PMID:20066247

Benito-Lopez, F., Scarmagnani, S., Walsh, Z., Paull, B., Macka, M., & Diamond, D. (2009). Spiropyran modified micro-fluidic chip channels as photonically controlled self-indicating system for metal ion accumulation and release. *Sensors and Actuators. B, Chemical*, *140*(1), 295–303. doi:10.1016/j.snb.2009.03.080

Bera, D., Qian, L., Tseng, T.-K., & Holloway, P. H. (2010). Quantum Dots and Their Multimodal Applications: A Review. *Materials (Basel)*, *3*(4), 2260–2345. doi:10.3390/ma3042260

Berginski, M., Hüpkes, J., Reetz, W., Rech, B., & Wuttig, M. (2008). Recent development on surface-textured ZnO:Al films prepared by sputtering for thin-film solar cell application. *Thin Solid Films*, *516*(17), 5836–5841. doi:10.1016/j.tsf.2007.10.029

Berna, J., Leigh, D. A., Lubomska, M., Mendoza, S. M., Perez, E. M., Rudolf, P., & Zerbetto, F. et al. (2005). Macroscopic transport by synthetic molecular machines. *Nature Materials*, *4*(9), 704–710. doi:10.1038/nmat1455 PMID:16127455

Bhattacharyya, D., Howden, R. M., Borrelli, D. C., & Gleason, K. K. (2012). Vapor phase oxidative synthesis of conjugated polymers and applications. *Journal of Polymer Science. Part B, Polymer Physics*, *50*(19), 1329–1351. doi:10.1002/polb.23138

Bhushan, B. (2011a). Biomimetics inspired surfaces for drag reduction and oleo-phobicity/philicity. *Beilstein J. Nanotechnol.*, *2*, 66–84. doi:10.3762/bjnano.2.9 PMID:21977417

Bhushan, B. (2011b). *Nanotribology and Nanomechanics II.* Springer. doi:10.1007/978-3-642-15263-4

Bhushan, B., & Jung, Y. C. (2011). Natural and biomimetic artificial surfaces for superhydrophobicity, self-cleaning, low adhesion, and drag reduction. *Progress in Materials Science*, *56*(1), 1–108. doi:10.1016/j.pmatsci.2010.04.003

Bhushan, B., Jung, Y. C., & Koch, D. K. (2009). Micro-, nano- and hierarchical structures for superhydrophobicity, self-cleaning and low adhesion. *Phil. Trans. R. Soc. A*, *367*, 1631–1672. PMID:19376764

Bierbaum, S., Douglas, T., Hanke, T., Scharnweber, D., Tippelt, S., Monsees, T. K., & Worch, H. et al. (2006). Collageneous matrix coatings on titanium implants modified with decorin and chondroitin sulfate: Characterization and influence on osteoblastic cells. *Journal of Biomedical Materials Research. Part A*, *77A*(3), 551–562. doi:10.1002/jbm.a.30572 PMID:16498599

Bigi, A., Boanini, E., Capuccini, C., Fini, M., Mihailescu, I. N., Ristoscu, C., & Torricelli, P. et al. (2009). Biofunctional alendronate–Hydroxyapatite thin films deposited by Matrix Assisted Pulsed Laser Evaporation. *Biomaterials*, *30*(31), 6168–6177. doi:10.1016/j.biomaterials.2009.07.066 PMID:19692118

Bilan, R., Fleury, F., Nabiev, I., & Sukhanova, A. (2015). Quantum Dot Surface Chemistry and Functionalization for Cell Targeting and Imaging. *Bioconjugate Chemistry*, *26*(4), 609–624. doi:10.1021/acs.bioconjchem.5b00069 PMID:25710410

Billing, D. G., & Lemmerer, A. (2009). Inorganic-organic hybrid materials incorporating primary cyclic ammonium cations: The lead bromide and chloride series. *CrystEngComm*, *11*(8), 1549–1562. doi:10.1039/b819455f

Bischoff, C., & Holberg, S. (2012). Repellent coating composition and coating, method for making and uses thereof. *Patent WO 2012/083970 A1*.

Bishop, J. A., Palanca, A. A., Bellino, M. J., & Lowenberg, D. W. (2012). Assessment of compromised fracture healing. *The Journal of the American Academy of Orthopaedic Surgeons*, *20*(5), 273–282. doi:10.5435/JAAOS-20-05-273 PMID:22553099

Blanco-Canosa, J. B., Wu, M., Susumu, K., Petryayeva, E., Jennings, T. L., Dawson, P. E., & Medintz, I. L. et al. (2014). Recent progress in the bioconjugation of quantum dots. *Coordination Chemistry Reviews*, *263–264*, 101–137. doi:10.1016/j.ccr.2013.08.030

Bloor, L. G., Carmalt, C. J., & Pugh, D. (2011). Single-source precursors to gallium and indium oxide thin films. *Coordination Chemistry Reviews*, *255*(11-12), 1293–1318. doi:10.1016/j.ccr.2010.12.018

Boidot, M., Selezneff, S., Monceau, D., Oquab, D., & Estournès, C. (2010). Proto-TGO formation in TBC systems fabricated by spark plasma sintering. *Surface and Coatings Technology*, *205*(5), 1245–1249. doi:10.1016/j.surfcoat.2010.09.042

Boinovich, L. B., & Emelyanenko, A. M. (2013). Anti-Icing Potential of Superhydrophobic Coatings. *Mendeleev Communications*, *23*(1), 3–10. doi:10.1016/j.mencom.2013.01.002

Bologna, R. A., Tu, L. M., Polansky, M., Fraimow, H. D., Gordon, D. A., & Whitmore, K. E. (1999). Hydrogel/silver ion-coated urinary catheter reduces nosocomial urinary tract infection rates in intensive care unit patients: A multicenter study. *Urology*, *54*(6), 982–987. doi:10.1016/S0090-4295(99)00318-0 PMID:10604694

Bond, A. P., & Lizlovs, E. A. (1968). Anodic polarization of austenitic stainless steels in chloride media. *Journal of the Electrochemical Society*, *115*(11), 1130–1135. doi:10.1149/1.2410923

Bonini, M., Berti, D., Di Meglio, J. M., Almgren, M., Teixeira, J., & Baglioni, P. (2005). Surfactant aggregates hosting a photoresponsive amphiphile: Structure and photoinduced conformational changes. *Soft Matter*, *1*(6), 444–454. doi:10.1039/b511544b

Borisova, D., Akçakayiran, D., Schenderlein, M., Möhwald, H., & Shchukin, D. G. (2013). Nanocontainer-based Anticorrosive Coatings: Effect of the Container Size on the Self-Healing Performance. *Advanced Functional Materials*, *23*(30), 3799–3812. doi:10.1002/adfm.201203715

Bosetti, M., Masse, A., Tobin, E., & Cannas, M. (2002). Silver coated materials for external fixation devices: In vitro biocompatibility and genotoxicity. *Biomaterials*, *23*(3), 887–892. doi:10.1016/S0142-9612(01)00198-3 PMID:11771707

Boyd, A., Duffy, H., McCann, R., Cairns, M., & Meenan, B. (2007). The Influence of argon gas pressure on co-sputtered calcium phosphate thin films. *Nuclear Instruments & Methods in Physics Research. Section B, Beam Interactions with Materials and Atoms*, *258*(2), 421–428. doi:10.1016/j.nimb.2007.02.072

Boyd, A., Meenan, B., & Leyland, N. (2006). Surface characterisation of the evolving nature of radio frequency (RF) magnetron sputter deposited calcium phosphate thin films after exposure to physiological solution. *Surface and Coatings Technology*, *200*(20-21), 6002–6013. doi:10.1016/j.surfcoat.2005.09.032

Boyle, T. J., Hernandez-Sanchez, B. A., Baros, C. M., Brewer, L. N., & Rodriguez, M. A. (2007). Influencing the Morphology and Phase of Calcium Ceramic Nanoparticles Based on the Calcium Alkoxide Precursors' Characteristics. *Chemistry of Materials*, *19*(8), 2016–2026. doi:10.1021/cm0700091

Bradley, D. C., Charkravarti, B. N., Chatterjee, A. K., Wardlaw, W., & Whitley, A. (1958). Niobium and tantalum mixed alkoxides. *Journal of the Chemical Society*, 99. doi:10.1039/jr9580000099

Bradley, D. C., Mehrotra, R. C., & Gaur, D. P. (1978). Metal Alkoxides. *Zeitschrift für Chemie*, *20*(10), 396–396.

Brady, R. F., Jr. (2007). Twenty-first century materials: coatings that interact with their environment. In T. Provder, & J. Baghdachi (Eds.), *Smart Coatings ACS Symposium Series 957* (pp. 3-11). Washington, DC: American Chemical Society. doi:10.1021/bk-2007-0957.ch001

Brand, M., Frings, A., Jenker, P., Lehnert, R., Metternich, H. J., Monkiewicz, J., & Schramm, J. (1999). NMR-spectroscopic investigations on the hydrolysis of functional trialkoxysilanes. *Zeitschrift für Naturforschung*, *54b*, 155–164.

Branzoi, V., Golgovici, F., & Branzoi, F. (2008). Electronically conductive polymers type polypyrrole and polyqaniline obtained by electropolymerization onto aluminum substrate. *Revue Roumaine de Chemie*, *53*(7), 553–561.

Brenner, A. (1963). Electrodeposition of alloys containing molybdenum and nickel, cobalt, or iron. Electrodeposition of Alloys: Principles and Practice (p. 413). Academic Press.

Brinker, C. J. (1988). Hydrolysis and condensation of silicates: Effects on structure. *Journal of Non-Crystalline Solids*, *100*(1-3), 31–50. doi:10.1016/0022-3093(88)90005-1

Brinker, C. J., & Scherer, G. W. (1990). *Sol-gel science*. San Diego, CA: Academic Press.

Bronk, J. K., Russell, B. H., Rivera, J. J., Pasqualini, R., Arap, W., Hook, M., & Barbu, E. M. (2014). A multifunctional streptococcal collagen-mimetic protein coating prevents bacterial adhesion and promotes osteoid formation on titanium. *Acta Biomaterialia*, *10*(7), 3354–3362. doi:10.1016/j.actbio.2014.04.005 PMID:24732634

Brooman, E. W. (2000). Corrosion performance of environmentally acceptable alternatives to cadmium and chromium coatings: Chromium-Part I. *Metal Finishing*, *98*(7), 38–43. doi:10.1016/S0026-0576(00)82334-X

Brusciotti, F., Snihirova, D. V., Xue, H., Montemor, F., Lamaka, S. V., & Ferreira, M. G. S. (2013). Hybrid epoxy-silane coatings for improved corrosion protection of Mg alloy. *Corrosion Science*, *67*, 82–90. doi:10.1016/j.corsci.2012.10.013

Brzeziński, S., Malinowska, G., Kowalczyk, D., Kaleta, A., Borak, B., Jasiorski, M., & Tracz, A. et al. (2012). Antibacterial and fungicidal coating of textile-polymeric materials filled with bioactive nano-and submicro-particles. *Fibres and Textiles in Eastern Europe*, *90*(1), 70–77.

Bullen, C., & Mulvaney, P. (2006). The Effects of Chemisorption on the Luminescence of CdSe Quantum Dots. *Langmuir*, *22*(7), 3007–3013. doi:10.1021/la051898e PMID:16548550

Burstein, E. (1954). Anomalous Optical Absorption Limit in InSb. *Physical Review*, *93*(3), 632–633. doi:10.1103/PhysRev.93.632

Butt, H.-J., Vollmer, D., Deng, X., Klapper, M., Praven, M., Schuster, T., & Papadopoulos, P. (2014). *Process and device for particle synthesis on a supehydrophobic or superoleophobic surface*. European Patent Number 2,677,332.

Byrne, R. J., Stitzel, S. E., & Diamond, D. (2006). Photo-regenerable surface with potential for optical sensing. *Journal of Materials Chemistry*, *16*(14), 1332–1337. doi:10.1039/b516400a

Byrne, R., Benito-Lopez, F., & Diamond, D. (2010). Materials science and the sensor revolution. *Materials Today*, *13*(7-8), 9–16. doi:10.1016/S1369-7021(10)70124-3

Caceres, L., Vargas, T., & Herrera, L. (2009). Influence of pitting and iron oxide formation during corrosion of carbon steel in unbuffered NaCl solutions. *Corrosion Science*, *51*(5), 971–978. doi:10.1016/j.corsci.2009.02.021

Cadavid, D., Ibanez, M., Shavel, A., Dura, O. J., Lopez de la Torre, M. A., & Cabot, A. (2013). Organic ligand displacement by metal salts to enhance nanoparticle functionality: Thermoelectric properties of Ag2Te. *Journal of Materials Chemistry A*, *1*(15), 4864–4870. doi:10.1039/c3ta01455j

Caicedo, J., Pérez, J., Caicedo, H., & Riascos, H. (2013). Determination of physical response in (Mo/AlN) SAW devices. *Surface Review and Letters*, *20*(2), 1350017. doi:10.1142/S0218625X13500170

Caicedo, J., Zambrano, G., Aperador, W., Escobar-Alarcon, L., & Camps, E. (2011). Mechanical and electrochemical characterization of vanadium nitride (VN) thin films. *Applied Surface Science*, *258*(1), 312–320. doi:10.1016/j.apsusc.2011.08.057

Cai, H., An, X., Cui, J., Li, J., Wen, S., Li, K., & Shi, X. et al. (2013). Facile hydrothermal synthesis and surface functionalization of poyethyleneimine-coated iron oxide nanoparticles for biomedical applications. *ACS Applied Materials & Interfaces*, *5*(5), 1722–1731. doi:10.1021/am302883m PMID:23388099

Cai, X.-S., & Xiao, X.-Y. (2013). Progress in superhydrophobic surface coatings. *Xiandai Huagong*, *33*(1), 22–25.

Çakmakçı, E., Mülazim, Y., & Kahraman, M. V. (2013). UV-curable fluorine-containing hybrid coatings via thiol-ene “click” reaction and an in situ sol–gel method. *Polymer Bulletin*, *70*(3), 1037–1048. doi:10.1007/s00289-012-0871-2

Campbell, A. A., Song, L., Li, X. S., Nelson, B. J., Bottoni, C., Brooks, D. E., & DeJong, E. S. (2000). Development, characterization, and anti-microbial efficacy of hydroxyapatite-chlorhexidine coatings produced by surface-induced mineralization. *Journal of Biomedical Materials Research*, *53*(4), 400–407. doi:10.1002/1097-4636(2000)53:4<400::AID-JBM14>3.0.CO;2-Z PMID:10898881

Campbell, S. A. (2001). *The science and engineering of microelectronic fabrication*. Oxford, Oxford University Press.

Campos, R., Haddad, T. S., & Mabry, J. (2009). Toward mechanically-robust omniphobic composite coatings using fluoroalkyl-modified silica. *ACS Polymer Preprints*, *50*(2), 717–718.

Cao, M., Cao, C., Liu, M., Wang, P., & Zhu, C. (2009). Selective fluorometry of cytochrome c using glutathione-capped CdTe quantum dots in weakly basic medium. *Mikrochimica Acta*, *165*(3-4), 341–346. doi:10.1007/s00604-009-0140-8

Čapek, R. K., Lambert, K., Dorfs, D., Smet, P. F., Poelman, D., Eychmüller, A., & Hens, Z. (2009). Synthesis of Extremely Small CdSe and Bright Blue Luminescent CdSe/ZnS Nanoparticles by a Prefocused Hot-Injection Approach. *Chemistry of Materials*, *21*(8), 1743–1749. doi:10.1021/cm900248b

Carmalt, C. J., & King, S. J. (2006). Gallium(III) and indium(III) alkoxides and aryloxides. *Coordination Chemistry Reviews*, *250*(5-6), 682–709. doi:10.1016/j.ccr.2005.05.028

Carradò, A. (2011). Nano-crystalline pulsed laser deposition hydroxyapatite thin films on Ti substrate for biomedical application. *Journal of Coatings Technology and Research*, *8*(6), 749–755. doi:10.1007/s11998-011-9355-9

Carré, A., & Mittal, K. L. (2009). *Superhydrophobic Surfaces*. VSP. doi:10.1163/ej.9789004165939.i-496

Carrillo-Carrion, C., Cardenas, S., Simonet, B. M., & Valcarcel, M. (2009). Quantum dots luminescence enhancement due to illumination with UV/Vis light. *Chemical Communications*, (35): 5214–5226. doi:10.1039/b904381k PMID:19707627

Carter, V. E. (1977). Metallic Coatings for Corrosion Control. United Kingdom, Elsevier Ltd.

Carter, D., Sproul, W., & Christie, D. (n. d.). Effective Control for Reactive Sputtering Processes. *Advanced Energy Industries Inc.*

Cassagneau, T. P. (2004). Layer-by-layer self-assembly of metal nanoparticles on planar substrates. Fabrication and properties. *Colloids and Colloid Assemblies*, 398-436.

Cassie, A. B. D., & Baxter, S. (1944). Wettability of Porous Surfaces. *Transactions of the Faraday Society*, *40*, 546–551. doi:10.1039/tf9444000546

Celia, E., Darmanin, T., Taffin de Givenchy, E., Amigoni, S., & Guittard, F. (2013). Recent Advances in Designing Superhydrophobic Surfaces. *Journal of Colloid and Interface Science*, *402*, 1–18. doi:10.1016/j.jcis.2013.03.041 PMID:23647693

Cerkez, I., Kocer, H. B., Worley, S. D., Broughton, R. M., & Huang, T. S. (2011). N-halamine biocidal coatings via a layer-by-layer assembly technique. *Langmuir*, *27*(7), 4091–4097. doi:10.1021/la104923x PMID:21348478

Champagne, V. K. (2007). The Cold Spray Materials Deposition Process: Fundamentals and applications. United kingdom, Woodhead Publishing Books

Chang, W., Hartman, M. E., & Scriven, R. L. (1977). Ambient temperature, moisture-curable acrylic-silane coating compositions having improved potlife. *Patent US 4043953.*

Chang, C.-C., Chiu, N.-F., Lin, D. S., Chu-Su, Y., Liang, Y.-H., & Lin, C.-W. (2010). High-sensitivity detection of carbohydrate antigen 15-3 using a gold/zinc oxide thin film surface plasmon resonance-based biosensor. *Analytical Chemistry*, *82*(4), 1207–1212. doi:10.1021/ac901797j PMID:20102177

Chang, Y.-Y., Lai, C.-H., Hsu, J.-T., Tang, C.-H., Liao, W.-C., & Huang, H.-L. (2012). Antibacterial properties and human gingival fibroblast cell compatibility of TiO2/Ag compound coatings and ZnO films on titanium-based material. *Clinical Oral Investigations*, *16*(1), 95–100. doi:10.1007/s00784-010-0504-9 PMID:21234622

Charville, G. W., Hetrick, E. M., Geer, C. B., & Schoenfisch, M. H. (2008). Reduced bacterial adhesion to fibrinogen-coated substrates via nitric oxide release. *Biomaterials*, *29*(30), 4039–4044. doi:10.1016/j.biomaterials.2008.07.005 PMID:18657857

Chassaing, E., Cornet, M., & Kihn, V. Q. (1978). Influence of codeposited impurities on hydrogen permeation and properties of Ni-Mo and Co-Mo electrodeposited alloys. *Surface Technology*, *7*(2), 145–150. doi:10.1016/0376-4583(78)90007-9

Chassaing, E., Portail, N., Levy, A.-F., & Wang, G. (2004). Characterisation of electrodeposited nanocrystalline Ni-Mo alloys. *Journal of Applied Electrochemistry*, *34*(11), 1085–1091. doi:10.1007/s10800-004-2460-z

Chassaing, E., Quang, K. V., & Wiart, R. (1989). Mechanism of nickel-molybdenum alloy electrodeposition in citrate electrolytes. *Journal of Applied Electrochemistry*, *19*(6), 839–843. doi:10.1007/BF01007931

Chassaing, E., Roumegas, M. P., & Trichet, M. F. (1995). Electrodeposition of Ni-Mo alloys with pulse reverse potentials. *Journal of Applied Electrochemistry*, *25*(7), 667–670. doi:10.1007/BF00241928

Chattopadhyay, & D. K., Webster, D. C. (2009a). Hybrid coatings from novel silane-modified glycidyl carbamate resins and amine crosslinkers. *Progress in Organic Coatings, 66,* 73-85.

Chattopadhyay, D. K., Muehlberg, A. J., & Webster, D. C. (2008). Organic–inorganic hybrid coatings prepared from glycidyl carbamate resins and amino-functional silanes. *Progress in Organic Coatings, 63*(4), 405–415. doi:10.1016/j.porgcoat.2008.06.009

Chattopadhyay, D. K., Zakula, A. D., & Webster, D. C. (2009b). Organic-inorganic hybrid coatings prepared from glycidyl carbamate resin, 3-aminopropyl trimethoxy silane and tetraethoxyorthosilicate. *Progress in Organic Coatings, 64*(2-3), 128–137. doi:10.1016/j.porgcoat.2008.09.008

Chaudhury, M. K., & Whitesides, G. M. (1992). How to make water run uphill. *Science, 256*(5063), 1539–1541. doi:10.1126/science.256.5063.1539 PMID:17836321

Chemtob, A., Courtecuisse, F., Croutxé-Barghorn, C., & Rigolet, S. (2011). Simultaneous sol–gel and anionic photopolymerization of 3-(glycidyloxypropyl)trimethoxysilane via photobase catalysis. *New Journal of Chemistry, 35*(9), 1803–1808. doi:10.1039/c1nj20320g

Chemtob, A., De Paz-Simon, H., Croutxé-Barghorn, C., & Rigolet, S. (2014a). UV-activated silicone oligomer cross-linking through photoacid and photobase organocatalysts. *Journal of Applied Polymer Science, 131*(3), 39875. doi:10.1002/app.39875

Chemtob, A., De Paz-Simon, H., Dietlin, C., Croutxé-Barghorn, C., Salmi, H., Allonas, X., & Rigolet, S. et al. (2014b). A highly reactive photobase catalytic system for sol-gel polymerization. *Thin Solid Films, 550,* 177–183. doi:10.1016/j.tsf.2013.10.146

Chemtob, A., Ni, L., Dietlin, C., Croutxé-Barghorn, C., Kitzmann, P., Brogly, M., & Vidal, L. (2012). Spontaneous photoinduced formation of hybrid polymer films with functionalized macroporous patterns. *Surface and Coatings Technology, 209,* 64–72. doi:10.1016/j.surfcoat.2012.08.030

Chemtob, A., Versace, D.-L., Belon, C., Croutxé-Barghorn, C., & Rigolet, S. (2008). Concomitant organic-inorganic UV-curing catalyzed by photoacids. *Macromolecules, 41*(20), 7390–7398. doi:10.1021/ma801017k

Chen, C., Jing, M., Pi, Z., Zhu, S., & Shen, X. (2015). Preparation and Properties of Double-Sided AgNWs/PVC/AgNWs Flexible Transparent Conductive Film by Dip-Coating Process. *Nanoscale Research Letters, 10*(1), 315-321. doi:10.1186/s11671-015-1022-0 PMID:26245859

Chen, F. F., & Smith, M. D. (1984). *Plasma.* Wiley Online Library.

Chen, G., Svec, F., & Knapp, D. R. (2008). Light-actuated high pressure-resisting microvalve for on-chip flow control based on thermo-responsive nanostructured polymer. *Lab on a Chip, 8*(7), 1198–1204. doi:10.1039/b803293a PMID:18584098

Cheng, H., Li, Y., Huo, K., Gao, B., & Xiong, W. (2014). Long-lasting in vivo and in vitro antibacterial ability of nanostructured titania coating incorporated with silver nanoparticles. *Journal of Biomedical Materials Research. Part A, 102*(10), 3488–3499. doi:10.1002/jbm.a.35019 PMID:24178451

Chen, M., Huang, H., Zhu, Y., Liu, Z., Xing, X., Cheng, F., & Yu, Y. (2011). Photodeformable CLCP material: Study on photo-activated microvalve applications. *Applied Physics. A, Materials Science & Processing, 102*(3), 667–672. doi:10.1007/s00339-010-6103-4

Chen, N., He, Y., Su, Y., Li, X., Huang, Q., Wang, H., & Fan, C. et al. (2012). The cytotoxicity of cadmium-based quantum dots. *Biomaterials, 33*(5), 1238–1244. PMID:22078811

Chen, W., Liu, Y., Courtney, H. S., Bettenga, M., Agrawal, C. M., Bumgardner, J. D., & Ong, J. L. (2006). In vitro antibacterial and biological properties of magnetron co-sputtered silver-containing hydroxyapatite coating. *Biomaterials*, *27*(32), 5512–5517. doi:10.1016/j.biomaterials.2006.07.003 PMID:16872671

Chen, W., & McCarthy, T. J. (1997). Layer-by-Layer Deposition: A Tool for Polymer Surface Modification. *Macromolecules*, *30*(1), 78–86. doi:10.1021/ma961096d

Chen, X., Gong, Y., Suo, X., Huang, J., Liu, Y., & Li, H. (2015). Construction of mechanically durable superhydrophobic surfaces by thermal spray deposition and further surface modification. *Applied Surface Science*. 356(30), 639-644

Chen, X., Gu, L., Zou, B., Wang, Y., & Cao, X. (2012). New functionally graded thermal barrier coating system based on $LaMgAl_{11}O_{19}$ / YSZ prepared by air plasma spraying. *Surface and Coatings Technology*, *206*(8-9), 2265–2274. doi:10.1016/j.surfcoat.2011.09.076

Chen, X., Hutchison, J., Dobson, P., & Wakefield, G. (2009). Highly luminescent monodisperse CdSe nanoparticles synthesized in aqueous solution. *Journal of Materials Science*, *44*(1), 285–292. doi:10.1007/s10853-008-3055-6

Chen, X., & Mao, S. S. (2007). Titanium Dioxide Nanomaterials: Synthesis, Properties, Modifications, and Applications. *Chemical Reviews*, *107*(7), 289. doi:10.1021/cr0500535 PMID:17590053

Chen, X., Zhao, Y., Fan, X., Liu, Y., Zou, B., Wang, Y., & Ma, H. (2011). Thermal cycling failure of new $LaMgA_{11}O_{19}$ / YSZ double ceramic top coat thermal barrier coating systems. *Surface and Coatings Technology*, *205*, 3293–3300.

Chen, Y. D., Tsai, W. J., Liu, S. H., & Horng, J. B. (2015). Hierarchical micro/nanostructures generated by femtosecond laser beams modified through a liquid-crystal spatial light modulator. *Applied Surface Science*, *353*, 568–573. doi:10.1016/j.apsusc.2015.06.152

Chiefari, J., Chong, Y. K., Ercole, F., Kristina, J., Jeffrey, J., Le, T. P., & Thang, S. H. et al. (1998). Living free-radical polymerization by reversible addition-fragmentation chain transfer: The RAFT process. *Macromolecules*, *31*(16), 5559–5562. doi:10.1021/ma9804951

Chim, H., Hutmacher, D. W., Chou, A. M., Oliveira, A. L., Reis, R. L., Lim, T. C., & Schantz, J. T. (2006). A comparative analysis of scaffold material modifications for load-bearing applications in bone tissue engineering. *International Journal of Oral and Maxillofacial Surgery*, *35*(10), 928–934. doi:10.1016/j.ijom.2006.03.024 PMID:16762529

Chisholm, M. H., Folting, K., Hampden-Smith, M., & Smith, C. A. (1987). Preparation and characterization of hexacyclohexoxido-and trispinacolatoditungsten. *Polyhedron*, *6*(9), 1747.

Chisholm, M. H., Huffman, J. C., Kirkpatric, J., Leonelli, J., & Folting, K. (1981). Facile cleavage of the molybdenum. tplbond.molybdenum bond in reactions between Mo2(OBut)6 and aryl azides and molecular oxygen. *Journal of the American Chemical Society*, *103*(5), 1305–1306. doi:10.1021/ja00395a092

Cho, H., Zook, J., Banner, T., Park, S.-H., Min, B.-H., Hasty, K. A., & Lindner, E. et al. (2014). Immobilization of fibrinogen antibody on self-assembled gold monolayers for immunosensor applications. *Tissue Engineering and Regenerative Medicine*, *11*(1), 10–15. doi:10.1007/s13770-013-1119-4

Choi, H., Santra, P. K., & Kamat, P. V. (2012). Synchronized Energy and Electron Transfer Processes in Covalently Linked CdSe–Squaraine Dye–TiO2 Light Harvesting Assembly. *ACS Nano*, *6*(6), 5718–5726. doi:10.1021/nn301982e PMID:22658983

Choi, J.-M., Kim, H.-E., & Lee, I.-S. (2000). Ion-beam-assisted deposition (IBAD) of hydroxyapatite coating layer on Ti-based metal substrate. *Biomaterials*, *21*(5), 469–473. doi:10.1016/S0142-9612(99)00186-6 PMID:10674811

Chou, S. Y., Krauss, P. R., & Renstrom, P. J. (1996a). Imprint lithography with 25-nanometer resolution. *Science, 272,* 85-87. doi:10.1126/science.272.5258.85

Chou, S. Y., Krauss, P. R., & Renstrom, P. J. (1996b). Nanoimprint lithography. *J. Vac. Sci. Technol., B,* 14, 4129-4133. doi:10.1116/1.588605

Chung, D. J., Ito, Y., & Imanishi, Y. (1994). Preparation of porous membranes grafted with poly(spiropyran-containing methacrylate) and photocontrol of permeability. *Journal of Applied Polymer Science*, *51*(12), 2027–2033. doi:10.1002/app.1994.070511207

Chung, M.-H., Lin, J.-S., Hsieh, T.-E., Chen, N.-P., Juang, F.-S., Chen, C.-M., & Liu, L.-C. (2011). Preparation of organic/inorganic hybrid nanocomposite by ultraviolet irradiation and their packaging applications for organic optoelectronic devices. *Applied Surface Science*, *257*(21), 9142–9151. doi:10.1016/j.apsusc.2011.05.117

Chung, R. J., Hsieh, M. F., Huang, C. W., Perng, L. H., Wen, H. W., & Chin, T. S. (2006). Antimicrobial effects and human gingival biocompatibility of hydroxyapatite sol–gel coatings. *Journal of Biomedical Materials Research. Part B, Applied Biomaterials*, *76*(1), 169–178. doi:10.1002/jbm.b.30365 PMID:16080164

Chu, W., Mayer, J., & Nicolet, M. (1978). *Backscattering Spectrometry*. Academic Press.

Cieslak, W. R., & Duquette, D. J. (1985). An electrochemical study of the pit initiation resistance of ferritic stainless steels. *Journal of the Electrochemical Society*, *132*(3), 533–537. doi:10.1149/1.2113882

Clapp, A. R., Goldman, E. R., & Mattoussi, H. (2006). Capping of CdSe-ZnS quantum dots with DHLA and subsequent conjugation with proteins. *Nature Protocols*, *1*(3), 1258–1266. doi:10.1038/nprot.2006.184 PMID:17406409

Clark, B. E., Thorpe, S. J., & Aust, K. T. (1990). The pitting behaviour of FeNiCrMo alloys. *Corrosion Science*, *31*, 551–556. doi:10.1016/0010-938X(90)90161-W

Clarke, D. R. (1999). Varistor Ceramics. *Journal of the American Ceramic Society*, *82*(3), 485–502. doi:10.1111/j.1151-2916.1999.tb01793.x

Clarke, D. R., & Phillpot, S. R. (2005). Thermal barrier coating materials. *Materials Today*, *8*(6), 22–29. doi:10.1016/S1369-7021(05)70934-2

Clement, M., Vergara, L., Sangrador, J., Iborra, E., & Sanz-Hervás, A. (2004). SAW characteristics of AlN films sputtered on silicon substrates. *Ultrasonics,* 42(1–9), 403-407. doi:10.1016/j.ultras.2004.01.034

Combes, C., & Rey, C. (2002). Adsorption of proteins and calcium phosphate materials bioactivity. *Biomaterials*, *23*(13), 2817–2823. doi:10.1016/S0142-9612(02)00073-X PMID:12059033

Conrad, H. C., Corbett, J. R., & Golden, T. D. (2012). Electrochemical deposition of γ-phase Zinc-Nickel alloys from alkaline solution. *Journal of the Electrochemical Society*, *159*(1), C29–C32. doi:10.1149/2.027201jes

Constantin, R., & Miremad, B. (1999). Performance of hard coatings, made by balanced and unbalanced magnetron sputtering, for decorative applications. *Surface and Coatings Technology*, *120–121*, 728–733. doi:10.1016/S0257-8972(99)00366-7

Contreras, C. B., Chagas, G., Strumia, M. C., & Weibel, D. E. (2014). Permanent Superhydrophobic Polypropylene Nanocomposite Coatings by a Simple One-Step Dipping Process. *Applied Surface Science*, *307*, 234–240. doi:10.1016/j.apsusc.2014.04.019

Cooper, D. R., Dimitrijevic, N. M., & Nadeau, J. L. (2010). Photosensitization of CdSe/ZnS QDs and reliability of assays for reactive oxygen species production. *Nanoscale*, *2*(1), 114–121. doi:10.1039/B9NR00130A PMID:20648372

Cooper, J. K., Franco, A. M., Gul, S., Corrado, C., & Zhang, J. Z. (2011). Characterization of Primary Amine Capped CdSe, ZnSe, and ZnS Quantum Dots by FT-IR: Determination of Surface Bonding Interaction and Identification of Selective Desorption. *Langmuir, 27*(13), 8486–8493. doi:10.1021/la201273x PMID:21631120

Cressler, J. D. (2005). *Silicon Heterostructure Handbook: Materials, Fabrication, Devices, Circuits and Applications of SiGe and Si Strained-Layer Epitaxy.* CRC press. doi:10.1201/9781420026580

Crevillen, A., Hervas, M., Lopez, M., Gonzalez, M., & Escarpa, A. (2007). Real sample analysis on microfluidic devices. *Talanta, 74*(3), 342–357. doi:10.1016/j.talanta.2007.10.019 PMID:18371648

Cross, G. L. W. (2006). The production of nanostructures by mechanical forming. *J. Phys. D: Appl. Phys., 39*, R363-R386. doi:10.1088/0022-3727/39/20/r01

Crousier, J., Eyraud, M., Crousier, J. P., & Roman, J.-M. (1992). Influence of substrate on the electrodeposition of nickel-molybdenum alloys. *Journal of Applied Electrochemistry, 22*(8), 749–755. doi:10.1007/BF01027505

Croutxe-Barghorn, C., Chemtob, A., Moreau, N., & Bouder, T. (2013a). Radiation curable composition, and method for preparing a hybrid sol-gel layer on a surface of a substrate using said composition. *Patent WO 2013/171582 A1.*

Croutxé-Barghorn, C., Chemtob, A., Ni, L., De-Paz-Simon, H., & Belon, C. (2013c, June). Nanocomposite coatings by UV driven sol-gel process and photopolymerization. *Paper presented at Coating Science International*, Noordwijk, The Netherlands.

Croutxe-Barghorn, C., Chemtob, A., Ni, L., Moreau, N., & Bouder, T. (2013b). Radiation radically and cationically curable composition, and method for preparing a hybrid sol-gel layer on a surface of a substrate using said composition. *Patent WO 2013/171579 A1.*

Croutxé-Barghorn, C., De Brito, M., Allonas, X., Belon, C., Chemtob, A., Ni, L., & Dietlin, C. et al. (2012). Organic and hybrid interpenetrated polymer networks for advanced materials. *Journal of Photopolymer Science and Technology, 25*, 131–135. doi:10.2494/photopolymer.25.131

Cumurcu, A., Yazgan, P., & Eriyes, T. (2011). Synthesis and properties of alkoxysilane-functionalized styrenated oil based polyester. *Progress in Organic Coatings, 71*(2), 147–152. doi:10.1016/j.porgcoat.2011.01.007

Cvijović, Z., & Radenković, G. (2006). Microstructure and pitting corrosion resistance of annealed duplex stainless steel. *Corrosion Science, 48*(12), 3887–3906. doi:10.1016/j.corsci.2006.04.003

Cyracure cycloaliphatic epoxides - cationic UV cure [brochure]. (2001Dow Chemical Company. Midland, MI: Dow Chemical Company.

Czechowski, K., Królicka, B., Szlosek, P., Smuk, B., Wszołek, J., Kurleto, A, & Kasina, J. (2005). Effect of nitride nano-scale multilayer coatings on functional properties of composite ceramic cutting inserts. *Bulletin of the Polish Academy of Sciences, Technical Sciences, 53*(4),425-431.

Czugala, M., Ziolkowski, B., Byrne, R., Diamond, D., & Benito-Lopez, F. (2011). *Materials science: the key to revolutionary breakthroughs in micro-fluidic devices.*

Czugala, M., O'Connell, C., Blin, C., Fischer, P., Fraser, K. J., Benito-Lopez, F., & Diamond, D. (2014). Swelling and shrinking behaviour of photoresponsive phosphonium-based ionogel microstructures. *Sensors and Actuators. B, Chemical, 194*, 105–113. doi:10.1016/j.snb.2013.12.072

Dabbousi, B. O., Rodriguez-Viejo, J., Mikulec, F. V., Heine, J. R., Mattoussi, H., Ober, R., & Bawendi, M. G. et al. (1997). (CdSe)ZnS Core–Shell Quantum Dots: Synthesis and Characterization of a Size Series of Highly Luminescent Nanocrystallites. *The Journal of Physical Chemistry B, 101*(46), 9463–9475. doi:10.1021/jp971091y

Dai, S., Ding, W., Wang, Y., Zhang, D., & Du, Z. (2011). Fabrication of Hydrophobic Inorganic Coatings on Natural Lotus Leaves for Nanoimprint Stamps. *Thin Solid Films*, *519*(16), 5523–5527. doi:10.1016/j.tsf.2011.03.118

Darmanin, T., & Guittard, F. (2014). Recent advances in the potential applications of bioinspired superhydrophobic materials. *J. Mater. Chem. A*, *2*(39), 16319–16359. doi:10.1039/C4TA02071E

Darouiche, R. O. (2004). Current concepts - Treatment of infections associated with surgical implants. *The New England Journal of Medicine*, *350*(14), 1422–1429. doi:10.1056/NEJMra035415 PMID:15070792

Das, D., Saini, A. K., Pathak, M. K., & Kumar, S. (2012). Application of functionally graded materials as thermal insulator in high temperature engineering. *International Journal of Research in Engineering & Applied Sciences*, *2*, 966–974.

Das, S., Nayak, G. C., Sahu, S. K., Routray, P. C., Roy, K., & Baskey, H. (2015). Titania Coated Magnetite and Ni-Ferrite nano-composite based RADAR Absorbing Materials for Camouflaging Application. *Polymer-Plastics Technology and Engineering*, 54(14), 1483-1493

Dattilo, D., Armelao, L., Fois, G., Mistura, G., & Maggini, M. (2007). Wetting properties of flat and porous silicon surfaces coated with a spiropyran. *Langmuir*, *23*(26), 12945–12950. doi:10.1021/la7024878 PMID:18027976

Davies, D. (2003). Understanding biofilm resistance to antibacterial agents. *Nature Reviews. Drug Discovery*, *2*(2), 114–122. doi:10.1038/nrd1008 PMID:12563302

De Groot, K., Geesink, R., Klein, C. P. A. T., & Serekian, P. (1987). Plasma sprayed coatings of hydroxylapatite. *Journal of Biomedical Materials Research*, *21*(12), 1375–1381. doi:10.1002/jbm.820211203 PMID:3429472

de Jonge, L. T., Leeuwenburgh, S. C. G., van den Beucken, J. J. J. P., te Riet, J., Daamen, W. F., Wolke, J. G. C., & Jansen, J. A. et al. (2010). The osteogenic effect of electrosprayed nanoscale collagen/calcium phosphate coatings on titanium. *Biomaterials*, *31*(9), 2461–2469. doi:10.1016/j.biomaterials.2009.11.114 PMID:20022365

de Jonge, L. T., Leeuwenburgh, S. C. G., Wolke, J. G. C., & Jansen, J. A. (2008). Organic inorganic surface modifications for titanium implant surfaces. *Pharmaceutical Research*, *25*(10), 2357–2369. doi:10.1007/s11095-008-9617-0 PMID:18509601

De Keukeleere, K., Pollefeyt, G., Feys, J., De Roo, J., Rijckaert, H., Lommens, P., & Van Driessche, I. (2015). Chemical solution deposition of functional ceramic coatings using ink-jet printing. *Pure and Applied Chemistry*, *87*(3), 231–238. doi:10.1515/pac-2014-0938

De Paz, H., Chemtob, A., Croutxé-Barghorn, C., Le Nouen, D., & Rigolet, S. (2012). Insights into photoinduced sol–gel polymerization: An in situ infrared spectroscopy study. *The Journal of Physical Chemistry B*, *116*(17), 5260–5268. doi:10.1021/jp212386e PMID:22486534

Decher, G., Hong, J. D., & Schmitt, J. (1992). Buildup of ultrathin multilayer films by a self-assembly process: III. Consecutively alternating adsorption of anionic and cationic polyelectrolytes on charged surfaces. *Thin Solid Films*, 210-211(Part 2), 831-835 doi:10.1016/0040-6090(92)90417-A

Defaix, P. R., van Sark, W. G. J. H. M., Worrell, E., & de Visser, E. (2012). Technical potential for photovoltaics on buildings in the EU-27. *Solar Energy*, *86*(9), 2644–2653. doi:10.1016/j.solener.2012.06.007

DeJong, E. S., DeBerardino, T. M., Brooks, D. E., Nelson, B. J., Campbell, A. A., Bottoni, C. R., & McManus, A. T. et al. (2001). Antimicrobial efficacy of external fixator pins coated with a lipid stabilized hydroxyapatite/chlorhexidine complex to prevent pin tract infection in a goat model. *The Journal of Trauma Injury Infection and Critical Care*, *50*(6), 1008–1013. doi:10.1097/00005373-200106000-00006 PMID:11426113

Del, F., Kroes, M., Sterkenburg, R., & Watkins, W. (2013). *Aircraft Maintenance and Repair* (7th ed.). United States of America: McGraw-Hill.

Delgado-Pérez, T., Bouchet, L. M., de la Guardia, M., Galian, R. E., & Pérez-Prieto, J. (2013). Sensing Chiral Drugs by Using CdSe/ZnS Nanoparticles Capped with N-Acetyl-L-Cysteine Methyl Ester. *Chemistry (Weinheim an der Bergstrasse, Germany), 19*(33), 11068–11076. doi:10.1002/chem.201300875 PMID:23813622

Deng, X., Mammen, L., Hans, J. B., & Vollmer, D. (2012). Candle soot as a template for transparent robust superamphiphobic coating. *Science, 335*(6064), 67–70. doi:10.1126/science.1207115 PMID:22144464

Deng, Z., Cao, L., Tang, F., & Zou, B. (2005). A New Route to Zinc-Blende CdSe Nanocrystals: Mechanism and Synthesis. *The Journal of Physical Chemistry B, 109*(35), 16671–16675. doi:10.1021/jp052484x PMID:16853121

Devi, G. S., Manorama, S. V., & Rao, V. J. (1999). SnO2:Bi2O3 based sensor: Laser-Raman, temperature programmed desorption and X-ray photoelectron spectroscopic studies. *Sensors and Actuators. B, Chemical*, 1999, 98. doi:10.1016/S0925-4005(99)00164-1

Dewan, R., Jovanov, V., Hamraz, S., & Knipp, D. (2014). Analyzing periodic and random textured silicon thin film solar cells by Rigorous Coupled Wave Analysis. *Scientific Reports, 4*, 6029. doi:10.1038/srep06029 PMID:25112301

Dewidar, M., Adamek, G., Jakubowicz, J., & AbdelRazik, K. K. (2014). Fabrication, Microstructure and Properties of Mechanically Alloyed and Electrochemically Etched Porous Ti-Y2O3. *International Journal of Electrochemical Science, 9*, 7773–7783.

Diguet, A., Guillermic, R. M., Magome, N., Saint-Jalmes, A., Chen, Y., Yoshikawa, K., & Baigl, D. (2009). Photomanipulation of a Droplet by the Chromocapillary Effect. *Angewandte Chemie International Edition, 48*(49), 9281–9284. doi:10.1002/anie.200904868 PMID:19890927

Diguet, A., Li, H., Queyriaux, N., Chen, Y., & Baigl, D. (2011). Photoreversible fragmentation of a liquid interface for micro-droplet generation by light actuation. *Lab on a Chip, 11*(16), 2666–2669. doi:10.1039/c1lc20328b PMID:21727984

Dilger, S., Lizandara-Pueyo, C., Krumm, M., & Polarz, S. (2012). Hierarchical zinc oxide materials with multiple porosity prepared by ultrafast temperature gradient chemical gas-phase synthesis. *Advanced Materials, 24*(4), 543–548. doi:10.1002/adma.201103557 PMID:22213088

Dirin, D. N., Dreyfuss, S., Bodnarchuk, M. I., Nedelcu, G., Papagiorgis, P., Itskos, G., & Kovalenko, M. V. (2014). Lead Halide Perovskites and Other Metal Halide Complexes As Inorganic Capping Ligands for Colloidal Nanocrystals. *Journal of the American Chemical Society, 136*(18), 6550–6553. doi:10.1021/ja5006288 PMID:24746226

Dobrzański, L., & Mikuła, J. (2005). The structure and functional properties of PVD and CVD coated Al 2O3 + ZrO2 oxide tool ceramics. *Journal of Materials Processing Technology, 167*(2-3), 438–446. doi:10.1016/j.jmatprotec.2005.05.034

Dong, L., & Jiang, H. (2007). Autonomous microfluidics with stimuli-responsive hydrogels. *Soft Matter, 3*(10), 1223–1230. doi:10.1039/b706563a

Dong, Z. L., Khor, K. A., Quek, C., White, T. J., & Cheang, P. (2003). TEM and STEM analysis on heat-treated and in vitro plasma-sprayed hydroxyapatite/Ti-6Al-4V composite coatings. *Biomaterials, 24*(1), 97–105. doi:10.1016/S0142-9612(02)00267-3 PMID:12417183

Donten, M., Cesiulis, H., & Stojek, Z. (2005). Electrodeposition of amorphous/nanocrystalline and polycrystalline Ni-Mo alloys from pyrophosphate baths. *Electrochimica Acta, 50*(6), 1405–1412. doi:10.1016/j.electacta.2004.08.028

Doron, A., Katz, E., Tao, G. L., & Willner, I. (1997). Photochemically-, chemically-, and pH-controlled electrochemistry at functionalized spiropyran monolayer electrodes. *Langmuir, 13*(6), 1783–1790. doi:10.1021/la960729e

Dos Santos, E., Moldovan, M., Jacomine, L., Mateescu, M., Werckmann, J., Anselme, K., & Pelletier, H. et al. (2010). Oriented hydroxyapatite single crystals produced by the electrodeposition method. *Materials Science and Engineering B*, *169*(1), 138–144. doi:10.1016/j.mseb.2009.10.007

Dosta, S., Torrell, M., Cano, I. G., & Guilemany, J. M. (2012). Functional colored ceramic coatings obtained by thermal spray for decorative applications. *Journal of the European Ceramic Society*, *32*(14), 3685–3692. doi:10.1016/j.jeurceramsoc.2012.04.026

Dowbenko, R., Stewart, R. S., Hartman, M. E., Gorman, B., & Thomas, S. J. (1985). Colour plus clear coating method utilizing addition interpolymers containing alkoxy silane and/or alcoxy silane groups. *Patent US 4499150.*

Dowling, D. P., Betts, A. J., Pope, C., McConnell, M. L., Eloy, R., & Arnaud, M. N. (2003). Anti-bacterial silver coatings exhibiting enhanced activity through the addition of platinum. *Surface and Coatings Technology*, *163–164*(0), 637–640. doi:10.1016/S0257-8972(02)00689-8

Draft Guidance for Industry and FDA Staff - Premarket Notification [510(k)] Submissions for Medical Devices that Include Antimicrobial Agents. (2007).

Drafts, B. (2001). Acoustic wave technology sensors. *IEEE Transactions on Microwave Theory and Techniques*, *49*(4), 795–802. doi:10.1109/22.915466

Draine, B. T., & Flatau, P. J. (2013). User guide for the discrete dipole approximation code DDSCAT 7.3. Retrieved from http://arxiv.org/abs/1305.6497

Dreher, A. M., Krumm, M., Lizandara-Pueyo, C., & Polarz, S. (2010). Chromium containing zinc oxide materials from organobimetallic precursors. *Dalton Transactions (Cambridge, England)*, *39*(9), 2232. doi:10.1039/b916253b PMID:20162196

Drelich, J., & Chibowski, E. (2010). Superhydrophobic and superwetting surfaces: Definition and mechanism of control. *Langmuir*, *26*(24), 18621–18623. doi:10.1021/la1039893 PMID:21090661

Drelich, J., Chibowski, E., Meng, D. D., & Terpilowski, K. (2011). Hydrophilic and superhydrophilic surfaces and materials. *Soft Matter*, *7*(21), 9804. 9804-9828 doi:10.1039/c1sm05849e

Durr, H., & Bouas-Laurent, H. (2003). *Photochromism: Molecules and Systems*. Elsevier.

Dusek, K. (1993). Responsive gels, volume transitions I and II. In D. K. (Ed.), *Adv. Polym. Sci.* (Vol. 109 & 110).

Eastoe, J., Dominguez, M. S., Wyatt, P., Beeby, A., & Heenan, R. K. (2002). Properties of a stilbene-containing gemini photosurfactant: Light-triggered changes in surface tension and aggregation. *Langmuir*, *18*(21), 7837–7844. doi:10.1021/la0257384

Ebelmen. (1846). *Annal. Chem. Pharm.*, *57*, 334.

Ejenstam, L., Ovaskainen, L., Meizoso, I. R., Wågberg, L., Pan, J., Swerin, A., & Claesson, P. M. (2013). The effect of superhydrophobic wetting state on corrosion protection - The AKD example. *Journal of Colloid and Interface Science*, *412*, 56–64. PMID:24144374

Elezović, N. R., Jović, V. D., & Krstajić, N. V. (2005). Kinetics of the hydrogen evolution reaction on Fe-Mo film deposited on mild steel support in alkaline solution. *Electrochimica Acta*, *50*(28), 5594–5601. doi:10.1016/j.electacta.2005.03.037

Eliaz, N., & Gileadi, E. (2008). Induced codeposition of alloys of tungsten, molybdenum and rhenium with transition metals. In Modern Aspects of Electrochemistry (Vol. 42, pp. 192-193). doi:10.1007/978-0-387-49489-0_4

Ellies, L., Nelson, D., & Featherstone, J. (1992). Crystallographic changes in calcium phosphates during plasma-spraying. *Biomaterials*, *13*(5), 313–316. doi:10.1016/0142-9612(92)90055-S PMID:1318087

Ellsworth, M. W., & Novak, B. M. (1991). Mutually interpenetrating inorganic-organic networks. New routes into nonshrinking sol-gel composite materials. *Journal of the American Chemical Society*, *113*(7), 2756–2758. doi:10.1021/ja00007a062

Ellsworth, M. W., & Novak, B. M. (1993). "Inverse" organic-inorganic composite materials. 3. High glass content "nonshrinking" sol-gel composites via poly(silicic acid esters). *Chemistry of Materials*, *5*(6), 839–844. doi:10.1021/cm00030a020

Elmengaard, B., Bechtold, J. E., & Soballe, K. (2005). In vivo study of the effect of RGD treatment on bone ongrowth on press-fit titanium alloy implants. *Biomaterials*, *26*(17), 3521–3526. doi:10.1016/j.biomaterials.2004.09.039 PMID:15621242

Elsner, C., & Buchmeiser, M. R. (2006). Surface analysis on polymers. *Nachrichten aus der Chemie*, *54*(2), 141–144.

Emin, S., Singh, S. P., Han, L., Satoh, N., & Islam, A. (2011). Colloidal quantum dot solar cells. *Solar Energy*, *85*(6), 1264–1282. doi:10.1016/j.solener.2011.02.005

Ernst, D. W., & Holt, M. L. (1958). Cathode potentials during the electrodeposition of molybdenum alloys from aqueous solutions. *Journal of the Electrochemical Society*, *105*(11), 686–692. doi:10.1149/1.2428691

Ersoy, M. S., & Onder, E. (2014). Electroless Silver Coating on Glass Stitched Fabrics for Electromagnetic Shielding Applications. *Textile Research Journal*, *84*(19), 2103–2114. doi:10.1177/0040517514530025

Exarhos, G. J., & Zhou, X. D. (2007). Discovery-based design of transparent conducting oxide films. *Thin Solid Films*, *515*(18), 7025–7052. doi:10.1016/j.tsf.2007.03.014

Fan, J. A., Wu, C., Bao, K., Bao, J., Bardhan, R., Halas, N. J., & Capasso, F. et al. (2010). Self-assembled plasmonic nanoparticle clusters. *Science*, *328*(5982), 1135–1138. doi:10.1126/science.1187949 PMID:20508125

Fano, U. (1961). Effects of configuration interaction on intensities and phase shifts. *Physical Review*, *124*(6), 1866–1878. doi:10.1103/PhysRev.124.1866

Fan, Y. W., Duan, K., & Wang, R. Z. (2005). A composite coating by electrolysis-induced collagen self-assembly and calcium phosphate mineralization. *Biomaterials*, *26*(14), 1623–1632. doi:10.1016/j.biomaterials.2004.06.019 PMID:15576136

Farhadi, S., Farzaneh, M., & Kulinich, S. A. (2011). Anti-Icing Performance of Superhydrophobic Surfaces. *Applied Surface Science*, *257*(14), 6264–6269. doi:10.1016/j.apsusc.2011.02.057

Farmer, J., Nichol, G., Seto, Y., McSporran, N., & Deslandes, L. M. (2011). Optical property measurements of high haze TCO coatings.*Proceedings of 26th European Photovoltaic Solar Energy Conference and Exhibition*, Hamburg: European Commission.

Faÿ, S., Feitknecht, L., Schlüchter, R., Kroll, U., Vallat-Sauvain, E., & Shah, A. (2006). Rough ZnO layers by LP-CVD process and their effect in improving performances of amorphous and microcrystalline silicon solar cells. *Solar Energy Materials and Solar Cells*, *90*(18-19), 2960–2967. doi:10.1016/j.solmat.2006.06.003

Feddes, B., Wolke, J., Jansen, J., & Vredenberg, A. (2003). Radio frequency magnetron sputtering deposition of calcium phosphate coatings: The effect of resputtering on the coating composition. *Journal of Applied Physics*, *93*(12), 9503–9507. doi:10.1063/1.1576894

Fei, T., Chen, H., & Lin, J. (2014). Transparent Superhydrophobic Films Possessing High Thermal Stability and Improved Moisture Resistance from the Deposition of MTMS-based Aerogels. *Colloids and Surfaces. A, Physicochemical and Engineering Aspects*, *443*, 255–264. doi:10.1016/j.colsurfa.2013.11.027

Feng, Q. L., Wu, J., Chen, G. Q., Cui, F. Z., Kim, T. N., & Kim, J. O. (2000). A mechanistic study of the antibacterial effect of silver ions on Escherichia coli and Staphylococcus aureus. *Journal of Biomedical Materials Research*, *52*(4), 662–668. doi:10.1002/1097-4636(20001215)52:4<662::AID-JBM10>3.0.CO;2-3 PMID:11033548

Feng, X. Q., Gao, X., Wu, Z., Jiang, L., & Zheng, Q. S. (2007). Superior Water Repellency of Water Strider Legs with Hierarchical Structures: Experiments and Analysis. *Langmuir*, *23*(9), 4892–4896. doi:10.1021/la063039b PMID:17385899

Fenn, D. R., & Davies, S. P. (1993). Silane functional oligomer. *Patent EP 0571073 A2.*

Fernández, S., Martínez-Steele, A., Gandía, J. J., & Naranjo, F. B. (2008). Al-doped ZnO films deposited on flexible substrates by rf sputtering for thin film solar cells. *Proceedings of the23rd European Photovoltaic Solar Energy Conference* (pp. 2347 – 2351). Valencia: EU PVSEC.

Fernández, S., de Abril, O., Naranjo, F. B., & Gandía, J. J. (2011). High quality textured ZnO:Al surfaces obtained by a two-step wet-chemical etching method for applications in thin film silicon solar cells. *Solar Energy Materials and Solar Cells*, *95*(8), 2281–2286. doi:10.1016/j.solmat.2011.03.042

Fernández, S., de Abril, O., Naranjo, F. B., & Gandía, J. J. (2012). Etching process optimization using NH4Cl aqueous solution to texture ZnO:Al films for efficient light trapping in flexible thin film solar cells. *Thin Solid Films*, *520*(12), 4144–4149. doi:10.1016/j.tsf.2011.04.207

Fernández, S., Martínez-Steele, A., Gandía, J. J., & Naranjo, F. B. (2009). Radio frequency sputter deposition of high-quality conductive and transparent ZnO:Al films on polymer substrates for thin film solar cells applications. *Thin Solid Films*, *517*(10), 3152–3156. doi:10.1016/j.tsf.2008.11.097

Fernández, S., & Naranjo, F. B. (2010). Optimization of aluminium-doped zinc oxide films deposited at low temperature by radio-frequency sputtering on flexible substrates for solar cell applications. *Solar Energy Materials and Solar Cells*, *94*(2), 157–163. doi:10.1016/j.solmat.2009.08.012

Fernández, S., Pust, S. E., Hüpkes, J., & Naranjo, F. B. (2012). Development of two-step etching approach for aluminium doped zinc oxide using a combination of standard HCl and NH_4Cl etch steps. *Thin Solid Films*, *520*(14), 4678–4684. doi:10.1016/j.tsf.2011.10.190

Fernández, S., Santos, J. D., Munuera, C., García-Hernández, M., & Naranjo, B. (2015). Effect of argon plasma-treated polyethylene terephthalate on ZnO:Al properties for flexible thin film silicon solar cells applications. *Solar Energy Materials and Solar Cells*, *133*, 170–179. doi:10.1016/j.solmat.2014.10.030

Ferrara, M. C., Pilloni, L., Mazzarelli, S., & Tapfer, L. (2010). Hydrophilic and optical properties of nanostructured titania prepared by sol-gel dip coating. *J. Phys. D: Appl. Phys.*, *43*, 95301-95309. doi:10.1088/0022-3727/43/9/095301

Ferreira, A. M., Gentile, P., Chiono, V., & Ciardelli, G. (2012). Collagen for bone tissue regeneration. *Acta Biomaterialia*, *8*(9), 3191–3200. doi:10.1016/j.actbio.2012.06.014 PMID:22705634

Ferro, D., Barinov, S. M., Rau, J. V., Teghil, R., & Latini, A. (2005). Calcium phosphate and fluorinated calcium phosphate coatings on titanium deposited by Nd: YAG laser at a high fluence. *Biomaterials*, *26*(7), 805–812. doi:10.1016/j.biomaterials.2004.03.027 PMID:15350786

Fesik, E. V., Grebnev, V. V., Zarazhevskii, V. I., & Mal'chikov, G. D. (2014). Ruthenium coatings on zirconium dioxide ceramic: Physicochemical and functional properties. *Russian Journal of Applied Chemistry*, *87*(5), 591–595. doi:10.1134/S1070427214050097

Fischer, W., & Yuva, N. (2011). Self-healing UV-coatings. *Paper presented at RadTech Europe*, Basel, Switzerland. Retrieved from www.bayercoatings.com

Fish, C. (2014). High performance silicon based coating compositions. *Patent US 2014/0011009 A1*.

Florea, L., Diamond, D., & Benito-Lopez, F. Polyaniline Coated Micro-capillaries for Continuous Flow Analysis of Aqueous Solutions. *Analytica Chimica Acta*. doi:10.1016/j.aca.2012.11.027 PMID:23260671

Florea, L., Hennart, A., Diamond, D., & Benito-Lopez, F. (2012). Synthesis and characterisation of spiropyran-polymer brushes in micro-capillaries: Towards an integrated optical sensor for continuous flow analysis. *Sensors and Actuators. B, Chemical*, *175*, 92–99. doi:10.1016/j.snb.2011.12.055

Florea, L., Wagner, K., Wagner, P., Wallace, G. G., Benito-Lopez, F., Officer, D. L., & Diamond, D. (2014). Photo-Chemopropulsion–Light-Stimulated Movement of Microdroplets. *Advanced Materials*, *26*(43), 7339–7345. doi:10.1002/adma.201403007 PMID:25236879

For Powerful Connections [brochure]. (2012a). Wacker Chemie. Retrieved from http://www.wacker.com/cms/media/publications/downloads/6085_EN.pdf

Forster, Y., Rentsch, C., Schneiders, W., Bernhardt, R., Simon, J. C., Worch, H., & Rammelt, S. (2012). Surface modification of implants in long bone. *Biomatter*, *2*(3), 149–157. doi:10.4161/biom.21563 PMID:23507866

Foscante, R. E., Gysegem, A. P., Martinich, P. J., & Law, G. H. (1981). Interpenetrating polymer network comprising epoxy polymer and polysiloxane. *Patent US 4250074*.

Fox-Rabinovich, G.S., Yamamoto, K., Beake, B.D., Gershman, I.S., Kovalev, A.I., Veldhuis, S.C., Aguirre, M.H., Dosbaeva, G. & Endrino, J.L. (2012). Hierarchical Adaptive Nanostructured PVD Coatings for Extreme Tribological Applications: The Quest for Nonequilibrium States and Emergent Behavior. *Science and Technology of Advanced Materials*, 13(4): 043001. doi:10.1088/1468-6996/13/4/043001

Francés-Soriano, L., Pocoví-Martínez, S., González-Béjar, M., & Pérez-Prieto, J. (2014). Reversible phase transfer of quantum dots by gas bubbling. *Green Materials*, *2*(2), 62–68. doi:10.1680/gmat.13.00018

Francis, W., Fay, C., Florea, L., & Diamond, D. (2015). Self-propelled chemotactic ionic liquid droplets. *Chemical Communications*, *51*(12), 2342–2344. doi:10.1039/C4CC09214G PMID:25562698

Frankel, G. S. (1998). Pitting corrosion of metals. A review of the critical factors. *Journal of the Electrochemical Society*, *145*(6), 2186–2198. doi:10.1149/1.1838615

Franken, R. H. (2006). *Transparent conducting oxide contacts and textured metal back reflectors for thin film silicon solar cells* [Doctoral dissertation].

Frankenstein, S. & Tuthill., A.M. (2002). Ice Adhesion to Locks and Dams: Past Work; Future Directions?. *Journal of Cold Regions Engineering*, *16*(2), 83-96. doi:10.1061/(ASCE)0887-381X

Frederick, M. T., Amin, V. A., & Weiss, E. A. (2013). Optical Properties of Strongly Coupled Quantum Dot–Ligand Systems. *The Journal of Physical Chemistry Letters*, *4*(4), 634–640. doi:10.1021/jz301905n PMID:26281879

Friend, W. Z. (1980). *Corrosion of nickel and nickel-base alloys*. New York: Wiley-Interscience.

Fries, K., Samanta, S., Orski, S., & Locklin, J. (2008). Reversible colorimetric ion sensors based on surface initiated polymerization of photochromic polymers. *Chemical Communications*, 47, 6288–6290. doi:10.1039/b818042c PMID:19048131

Fujishima, A., & Zhang, X. T. (2006). Titanium dioxide photocatalysis: Present situation and future approaches. *Comptes Rendus. Chimie*, *9*(5-6), 750–760. doi:10.1016/j.crci.2005.02.055

Fukushima, H., & Higashi, K. (1978). A study of the electrodeposition of Ni-Mo alloys from ammoniacal tartrate bath. *Journal of the Metal Finishing Society of Japan*, *29*(12), 627–635. doi:10.4139/sfj1950.29.627

Fu, L. U., Khor, K. A., Ng, H. W., & Teo, T. N. (2000). Non-destructive evaluation of plasma sprayed functionally graded thermal barrier coatings. *Surface and Coatings Technology*, *130*(2-3), 233–239. doi:10.1016/S0257-8972(00)00700-3

Furmidge, C. G. L. (1962). Studies at phase interfaces. I. The sliding of liquid drops on solid surfaces and a theory for spray retention. *Journal of Colloid Science*, *17*(4), 309–324. doi:10.1016/0095-8522(62)90011-9

Future Markets. (2015). *The Global Nanocoatings Market.*

Gabriel, O., Rohde, M., Wördenweber, J., Zelt, M., Neubert, S., & Kirner, S., … Schlatmann, R. (2011). Enhancement of a-Si/μc-Si solar cell performance made in an industrial PECVD reactor: The role of chamber history and in situ diagnostics. *Proceedings of PVSEC-21, Fukuoka, Japan.*

Galian, R. E., & Guardia, M. (2009). The use of quantum dots in organic chemistry. *TrAC Trends in Analytical Chemistry*, *28*(3), 279–291. doi:10.1016/j.trac.2008.12.001

Galian, R. E., & Perez-Prieto, J. (2010). Catalytic processes activated by light. *Energy & Environmental Science*, *3*(10), 1488–1498. doi:10.1039/c0ee00003e

Galian, R. E., & Scaiano, J. C. (2009). Fluorescence quenching of CdSe quantum dots by tertiary amines and their surface binding effect. *Photochemical & Photobiological Sciences*, *8*(1), 70–74. doi:10.1039/B807580H PMID:19247532

Galian, R., Diaz, P., Ribera, A., Rincón-Bertolín, A., Agouram, S., & Pérez-Prieto, J. (2015). Controlled building of CdSe@ZnS/Au and CdSe@ZnS/Au2S/Au nanohybrids. *Nano Research*, *8*(7), 2271–2287. doi:10.1007/s12274-015-0738-6

Gallinet, B., & Martin, O. J. F. (2011). Influence of electromagnetic interactions on the line shape of plasmonic Fano resonances. *ACS Nano*, *5*(11), 8999–9008. doi:10.1021/nn203173r PMID:22026329

Gao, J., Chen, K., Xie, R., Xie, J., Lee, S., Cheng, Z., Peng, X., & Chen, X. (2010). Ultrasmall Near-infrared Non-cadmium Quantum Dots for in vivo Tumor Imaging. *Small*, 6(2), 256-261. Weinheim an der Bergstrasse, Germany.

Gao, J., Zhao, C., Zhou, J., Li, C., Shao, Y., Shi, C., & Zhu, Y. (2015). Plasma sprayed rutile titania-nanosilver antibacterial coatings. *Applied Surface Science*, *355*, 593–601.

Gao, L., & McCarthy, T. J. (2009). Wetting 101. *Langmuir*, *25*(24), 14105–14115. doi:10.1021/la902206c PMID:19627073

Gaonkar, S. R., Kumar, G. S., & Neckers, D. C. (1990). Photochromism of azobenzene-containing polymers. 4. effect of spacer groups. *Macromolecules*, *23*(24), 5146–5148. doi:10.1021/ma00226a018

Gao, Y., Wu, J., Wang, Q., Wilkie, C. A., & O'Hare, D. (2014). Flame Retardant Polymer/Layered Double Hydroxide Nanocomposites. *Journal of Materials Chemistry A*, *2*(29), 10996–11016. doi:10.1039/c4ta01030b

Gao, Y., Xiong, J., Gong, D., Li, J., & Ding, M. (2015). Improvement of solar absorbing property of Ni-Mo based thermal spray coatings by laser surface treatment. *Vacuum*, *121*, 64–69. doi:10.1016/j.vacuum.2015.07.018

Garcia, A., Marquez, M., Cai, T., Rosario, R., Hu, Z., Gust, D., & Park, C.-D. et al. (2007). Photo-, thermally, and pH-responsive microgels. *Langmuir*, *23*(1), 224–229. doi:10.1021/la061632n PMID:17190508

García-Marín, A., Abad, J. M., Ruiz, E., Lorenzo, E., Piqueras, J., & Pau, J. L. (2014). Glutathione immunosensing platform based on total internal reflection ellipsometry enhanced by functionalized gold nanoparticles. *Analytical Chemistry*, *86*(10), 4969–4976. doi:10.1021/ac5005212 PMID:24766219

García-Marín, A., Hernández, M. J., Ruiz, E., Abad, J. M., Lorenzo, E., Piqueras, J., & Pau, J. L. (2015). Immunosensing platform based on gallium nanoparticle arrays on silicon substrates. *Biosensors & Bioelectronics*, *74*, 1069–1075. doi:10.1016/j.bios.2015.08.002 PMID:26276543

Gau, H., Herminghaus, S., Lenz, P., & Lipowsky, R. (1999). Liquid morphologies on structured surfaces: from microchannels to microchips. *Science*, *283*, 46-49. doi:10.1126/science.283.5398.46

Geiger, E. J., Pisano, A. P., & Svec, F. (2010). A Polymer-Based Microfluidic Platform Featuring On-Chip Actuated Hydrogel Valves for Disposable Applications. *Journal of Microelectromechanical Systems*, *19*(4), 944–950. doi:10.1109/JMEMS.2010.2048702

Geissler, U., Hempel, U., Wolf, C., Scharnweber, D., Worch, H., & Wenzel, K. W. (2000). Collagen type I-coating of Ti6Al4V promotes adhesion of osteoblasts. *Journal of Biomedical Materials Research*, *51*(4), 752–760. doi:10.1002/1097-4636(20000915)51:4<752::AID-JBM25>3.0.CO;2-7 PMID:10880125

Ge, L., Ding, G., Wang, H., Yao, J., Cheng, P., & Wang, Y. (2013). Anti-Icing Property of Superhydrophobic Octadecyltrichlorosilane Film and Its Ice Adhesion Strength. *Journal of Nanomaterials*. 2013, 278936 doi:10.1155/2013/278936

Gennes, P. G., Wyart, F. B., & Quere, D. (2004). *Capillarity and wetting phenomena: drops, bubbles, pearls, waves.*

Ghoniem, M. A. (2014). A bronze Osiris statuette from the Egyptian museum in Cairo: Microstructural characterization and conservation. *Mediterranean Archaeology and Archaeometry*, *14*(1), 37–49.

Giboz, J., Copponnex, T., & Mélé, P. (2007). Microinjection molding of thermoplastic polymers: A review. *Journal of Micromechanics and Microengineering*, *17*(6), 96–109. doi:10.1088/0960-1317/17/6/R02

Gill, R., Zayats, M., & Willner, I. (2008). Semiconductor Quantum Dots for Bioanalysis. *Angewandte Chemie International Edition*, *47*(40), 7602–7625. doi:10.1002/anie.200800169 PMID:18810756

Gireesh, K. B., Jena, K. K., Allauddin, S., Radhika, K. R., Narayan, R., & Raju, K. V. S. N. (2010). Structure and thermomechanical properties study of polyurethane-urea/glycidoxypropyltrimethoxysilane hybrid coatings. *Progress in Organic Coatings*, *68*(3), 165–172. doi:10.1016/j.porgcoat.2010.02.008

Glavan, A. C., Martinez, R. V., Subramaniam, A. B., Yoon, H. J., Nunes, R. M. D., Lange, H., & Whitesides, G. M. et al. (2014). Omniphobic "rF paper" produced by silanization of paper with fluoroalkyltrichlorosilanes. *Advanced Functional Materials*, *24*(1), 60–70. doi:10.1002/adfm.201300780

Glinel, K., Thebault, P., Humblot, V., Pradier, C. M., & Jouenne, T. (2012). Antibacterial surfaces developed from bioinspired approaches. *Acta Biomaterialia*, *8*(5), 1670–1684. doi:10.1016/j.actbio.2012.01.011 PMID:22289644

Global Industry Trends. (2014). Nanocoatings Market (Anti-fingerprint, Anti-microbial, Anti-fouling & Easy-to-clean, Self-cleaning and Others) for Medical & Healthcare, Food Manufacturing, Packaging, Marine, Water Treatment, Electronics, Building & Construction, Automotive, Energy and Other End-user Industries: Global Industry Analysis, Size, Share, Growth, Trends and Forecast. And Segment Fore.

Gomez, E., Pellicer, E., Duch, M., Esteve, J., & Valles, E. (2006). Molybdenum alloy electrodeposits for magnetic actuation. *Electrochimica Acta*, *51*(16), 3214–3222. doi:10.1016/j.electacta.2005.09.010

Gómez, E., Pellicer, E., & Vallés, E. (2003). Influence of the bath composition and the pH on the induced cobalt-molybdenum electrodeposition. *Journal of Electroanalytical Chemistry*, *556*, 137–145. doi:10.1016/S0022-0728(03)00339-5

Gondoni, P. (2014). *Nanostructured Transparent Conducting Oxides for Advanced Photovoltaic Applications* [Doctoral dissertation].

Gonzalez-Carrero, S., Agudelo-Morales, C., Guardia, M., Galian, R. E., & Perez-Prieto, J. (2015). Three independent channel nanohybrids as fluorescent probes. *RSC Advances*, *5*(109), 90065–90070. doi:10.1039/C5RA18028G

González-Carrero, S., de la Guardia, M., Galian, R. E., & Pérez-Prieto, J. (2014). Pyrene-Capped CdSe@ZnS Nanoparticles as Sensitive Flexible Oxygen Sensors in Non-Aqueous Media. *ChemistryOpen*, *3*(5), 199–205. doi:10.1002/open.201402021 PMID:25478315

Gonzalez-Carrero, S., Espallargas, G. M., Galian, R. E., & Perez-Prieto, J. (2015). Blue-luminescent organic lead bromide perovskites: Highly dispersible and photostable materials. *Journal of Materials Chemistry A*, *3*(26), 14039–14045. doi:10.1039/C5TA01765C

Gonzalez-Carrero, S., Galian, R. E., & Perez-Prieto, J. (2015). Maximizing the emissive properties of CH3NH3PbBr3 perovskite nanoparticles. *Journal of Materials Chemistry A*, *3*(17), 9187–9193. doi:10.1039/C4TA05878J

González-Carrero, S., Galian, R. E., & Pérez-Prieto, J. (2015). Organometal Halide Perovskites: Bulk Low-Dimension Materials and Nanoparticles. *Particle & Particle Systems Characterization*, *32*(7), 709–720. doi:10.1002/ppsc.201400214

Gooch, J. W. (2002). Lead-Based Paint handbook. New York, Kluwer Academic Publishers.

Goodman, S. B., Yao, Z. Y., Keeney, M., & Yang, F. (2013). The future of biologic coatings for orthopaedic implants. *Biomaterials*, *34*(13), 3174–3183. doi:10.1016/j.biomaterials.2013.01.074 PMID:23391496

Goshtagore, R. N. (1970). Mechanism of heterogeneous deposition of thin film rutile. *Journal of the Electrochemical Society*, *117(4)*, 529. doi:10.1149/1.2407561

Goswami, B., Ray, A. K., & Sahay, S. K. (2004). Thermal barrier coatings system for gas turbine application- A review. *High Temperature Material Processes*, *23*, 73–92.

Grady, B. P. (2011). *Carbon Nanotube-Polymer Composites: Manufacture, Properties, and Applications*. Hoboken, NJ: John Wiley & Sons; doi:10.1002/9781118084380

Grasset, F., Starukh, G., Spanhel, L., Ababou-Girard, S., Su, D. S., & Klein, A. (2005). From ZnO colloids to nanocrystalline colored ZnxTi yOw-zNz spinel films. *Advanced Materials*, *17*(3), 294–297. doi:10.1002/adma.200400031

Grecu, M., Novac, G., Ionita, D., & Ungureanu, C. (2011). Incorporation of Tobramycin Biomimetic in Hydroxyapatite Coating on CoCrMo Alloy and its Antimicrobial Acitivity. *Revista De Chimie*, *62*(3), 352–356.

Green, T. A., Russell, A. E., & Roy, S. (1998). The development of a stable citrate electrolyte for the electrodeposition of copper-nickel alloys. *Journal of the Electrochemical Society*, *145*(3), 875–881. doi:10.1149/1.1838360

Gregorovich, B. V., Hazan, I., & Wilczek, L. (2001). Coatings that contain silicone oligomers. *Patent EP 1123951 A2.*

Grewal, H. S., Cho, I. J., Oh, J. E., & Yoon, E. S. (2014). Effect of topography on the wetting of nanoscale patterns: Experimental and modeling studies. *Nanoscale*, *6*(24), 15321–15332. doi:10.1039/C4NR04069D PMID:25387063

Grignard, B., Vaillant, A., Connick, J., & de, . (2011). Electrospinning of functional perfluorinated block copolymer as a powerful route for imparting superhydrophobbicity and corrosion resistance to aluminum substrates. *Langmuir*, *27*(1), 335–342. doi:10.1021/la102808w PMID:21141949

Grigoriev, D., Schenderlein, M., Akcakayiran, D., & Shchukin, D. (2014). Protective organic coatings with anticorrosive and other feedback active features: Micro- and nanocontainers based approach. *Corrosion*, *70*(5), 446–463. doi:10.5006/0976

Gristina, A. G. (1987). Biomaterial-centered infection-microbial adhesion versus tissue integration. *Science*, *237*(4822), 1588–1595. doi:10.1126/science.3629258 PMID:3629258

Groenewolt, M., & Müller, J. (2012a). Coating composition having a high solids content and good levelling and also multilayer surface coatings produced therefrom and their use. *Patent US 20140065313 A1.*

Groenewolt, M., Hesener, S., Stubbe, W., Niemeier, M., & Westhoff, E. (2009). Verwendung teilsilanisierter Verbindungen auf Polyisocyanatbasis als Vernetzungsmittel in Beschichtungszusammensetzungen und Beschichtungszusammensetzung enthaltend die Verbindungen. *Patent DE 102008030304 A1.* [Use of partially silanized polyisocyanate-based compounds as crosslinking-agents in coating compositions, and coating compositions comprising the compounds]. *Patent US 8658752 B2.*

Groenewolt, M., Klein, G., Hilge, O., Austrup, B., & Huebner, K. (2012b). Coating agent compositions, coatings made therefrom and exhibiting high scratch resistance and good polishability, and use thereof. *Patent WO 2012/168079 A1.*

Groenewolt, M., Poppe, A., Klein, G., Niemeier, M., Westhoff, E., Stübbe, W., & Hesener, S. (2008). Coating agents having high scratch resistance and weathering stability. *Patent US 8569438 B2.*

Grosso, D. (2011). How to exploit the full potential of the dip-coating process to better control film formation. *Journal of Materials Chemistry*, *21*(43), 17033-17038 doi:10.1039/c1jm12837j

Grundke, K., Poschel, K., Synytska, A., Frenzel, R., Drechsler, A., Nitschke, M., & Welzrl, P. B. et al. (2014). Experimental studies of contact angle hysteresis phenomena on polymer surfaces-toward the understanding and control of wettability for different applications. *Advances in Colloid and Interface Science*. PMID:25488284

Grund, S., Kempe, P., Baumann, G., Seifert, A., & Spange, S. (2007). Nanocomposites prepared by twin polymerization of a single-source monomer. *Angewandte Chemie International Edition*, *46*(4), 628–632. doi:10.1002/anie.200504327 PMID:17152104

Gudipati, C. S., Finlay, J. A., Callow, J. A., Callow, M. E., & Wooley, K. L. (2005). The Antifouling and Fouling-Release Performance of Hyperbranched Fluoropolymer (HBFP)-Poly(ethylene glycol) (PEG) Composite Coatings Evaluated by Adsorption of Biomacromolecules and the Green Fouling Alga Ulva. *Langmuir, 21*, 3044-3053. doi:10.1021/la048015o

Guidance for Industry and FDA Staff - Class II Special Controls Guidance Document: Root-form Endosseous Dental Implants and Endosseous Dental Abutments. (2004).

Guipont, V., Jeandin, M., Bansard, S., Khor, K. A., Nivard, M., Berthe, L., & Boustie, M. et al. (2010). Bond strength determination of hydroxyapatite coatings on Ti-6Al-4V substrates using the LAser Shock Adhesion Test (LASAT). *Journal of Biomedical Materials Research. Part A*, *95*(4), 1096–1104. doi:10.1002/jbm.a.32907 PMID:20878900

Guo, H., Wang, Y., Wang, L., & Gong, S. (2009). Thermo-physical properties and thermal shock resistance of segmented $La_2Ce_2O_7$ / YSZ thermal barrier coatings. *Journal of Thermal Spray Technology*, *18*(4), 665–671. doi:10.1007/s11666-009-9350-1

Guo, L., Yuan, W., Lu, Z., & Li, C. M. (2013). Polymer/nanosilver composite coatings for antibacterial applications. *Colloids and Surfaces. A, Physicochemical and Engineering Aspects*, *439*, 69–83. doi:10.1016/j.colsurfa.2012.12.029

Guo, Z. G., Fang, J., Wang, L., & Liu, W. (2007). Fabrication of superhydrophobic copper by wet chemical reaction. *Thin Solid Films*, *515*(18), 7190–7194. doi:10.1016/j.tsf.2007.02.100

Guo, Z., Liu, W., & Su, B.-L. (2011). Superhydrophobic surfaces: From natural to biomimetic to functional. *Journal of Colloid and Interfacial Science*, *353*(2), 335–355. doi:10.1016/j.jcis.2010.08.047 PMID:20846662

Gupta, G., & Satapathy, A. (2015). Processing and Characterization of Plasma Spray Coatings of Glass Microspheres Premixed with Al_2O_3 Particles. *Particulate Science and Technology*, *33*(2), 145–149. doi:10.1080/02726351.2014.947658

Gu, Y. W., Khor, K. A., Fu, Y. Q., & Wang, Y. (1997). Functionally graded ZrOz-NiCrA1Y coatings prepared by plasma spraying using pre-mixed, spheroidized powders. *Surface and Coatings Technology*, *96*(2-3), 305–312. doi:10.1016/S0257-8972(97)00185-0

Guyomard, A., De, E., Jouenne, T., Malandain, J. J., Muller, G., & Glinel, K. (2008). Incorporation of a hydrophobic antibacterial peptide into amphiphilic polyelectrolyte multilayers: A bioinspired approach to prepare biocidal thin coatings. *Advanced Functional Materials*, *18*(5), 758–765. doi:10.1002/adfm.200700793

Habibi, M. H., Wang, L., & Guo, S. M. (2012). Evolution of hot corrosion resistance of YSZ, $Gd_2Zr_2O_7$, and $Gd_2Zr_2O_7$ + YSZ composite thermal barrier coatings in Na_2SO_4 + V_2O_5 at 1050 °C. *Journal of the European Ceramic Society*, *32*(8), 1635–1642. doi:10.1016/j.jeurceramsoc.2012.01.006

Habibi, M. H., Wang, L., Liang, J., & Guo, S. M. (2013). An investigation on hot corrosion behavior of YSZ-Ta_2O_5 in Na_2SO_4 + V_2O_5 salt at 1100°C. *Corrosion Science*, *75*, 409–414. doi:10.1016/j.corsci.2013.06.025

Habibovic, P., Barrere, F., Blitterswijk, C. A., Groot, K., & Layrolle, P. (2002). Biomimetic hydroxyapatite coating on metal implants. *Journal of the American Ceramic Society*, *85*(3), 517–522. doi:10.1111/j.1151-2916.2002.tb00126.x

Hackford, J. E., Shaw, C., & Smith, W. E. (1946). Improvements in or relating to liquids comprising silicyl esters. *Patent GB 575752*.

Hadjoub, Z., Touati, I., Doghmane, M., & Doghmane, A. (2008). Application of negative velocity dispersion curves to the distinction between layer and substrate Rayleigh waves. *Comptes Rendus Physique*, *9*(8), 903–910. doi:10.1016/j.crhy.2008.08.001

Hahn, B.-D., Lee, J.-M., Park, D.-S., Choi, J.-J., Ryu, J., Yoon, W.-H., & Kim, H.-E. et al. (2011). Enhanced bioactivity and biocompatibility of nanostructured hydroxyapatite coating by hydrothermal annealing. *Thin Solid Films*, *519*(22), 8085–8090. doi:10.1016/j.tsf.2011.07.008

Halim, J., Abdel-Karim, R., El-Raghy, S., Nabil, M., & Waheed, A. (2012). Electrodeposition and characterization of nanocrystalline Ni-Mo catalysts for hydrogen production. *Journal of Nanomaterials*, *845673*, 1–9. doi:10.1155/2012/845673

Hammer, N. I., Emrick, T., & Barnes, M. D. (2007). Quantum dots coordinated with conjugated organic ligands: New nanomaterials with novel photophysics. *Nanoscale Research Letters*, *2*(6), 282–290. doi:10.1007/s11671-007-9062-8

Han, J. G. (2009). Recent progress in thin film processing by magnetron sputtering with plasma diagnostics. *Journal of Physics. D, Applied Physics*, *42*(4). doi:10.1088/0022-3727/42/4/043001

Han, M., Mang, J., & Seo, J. (2010). Syntheses of novel sol-gel precursor containing anti-corrosive functional group and their uses in organic-inorganic hybrid coatings. *Polymer Korea*, *34*(5), 405–409.

Han, Q., Cui, S., Pu, N., Chen, J., Liu, K., & Wei, X. (2010). A study on pulse plating amorphous Ni-Mo alloy coating used as HER cathode in alkaline medium. *International Journal of Hydrogen Energy*, *35*(11), 5194–5201. doi:10.1016/j.ijhydene.2010.03.093

Han, S. J., Chen, Y., & Sampath, S. (2014). Role of process conditions on the microstructure, stoichiometry and functional performance of atmospheric plasma sprayed La(Sr)MnO3 coatings. *Journal of Power Sources*, *259*, 245–254. doi:10.1016/j.jpowsour.2014.02.073

Hansen, R. G., Hays, D. S., Iyer, S., Frank, J. W., & Kumar, R. C. (2013). Moisture-curable siloxanes and siloxane polymers. *Patent US 2013/0011673 A1*.

Hansson, S., Trouillet, V., Tischer, T., Goldmann, A. S., Carlmark, A., Barner-Kowollik, C., & Malmstroem, E. (2013). Grafting efficiency of synthetic polymers onto biomaterials: A comparative study of grating-from versus grating-to. *Biomacromolecules*, *14*(1), 64–74. doi:10.1021/bm3013132 PMID:23043441

Hao, L., Sirong, Y., & Xiangxiang, H. (2015). Preparation of a biomimetic superhydrophobic ZnO coating on an X90 pipeline steel surface. *New Journal of Chemistry*, *39*(6), 4860–4868. doi:10.1039/C5NJ00576K

Harmon, M. E., Kuckling, D., & Frank, C.W. (2003). Photo-cross-linkable PNIPAAm copolymers. 5. Mechanical properties of hydrogel layers. *Langmuir*, *19*(26), 10660–10665. doi:10.1021/la030232m

Harris, C., & Kamat, P. V. (2009). Photocatalysis with CdSe Nanoparticles in Confined Media: Mapping Charge Transfer Events in the Subpicosecond to Second Timescales. *ACS Nano*, *3*(3), 682–690. doi:10.1021/nn800848y PMID:19226135

Hasan, J., Crawford, R. J., & Ivanova, E. P. (2013). Antibacterial surfaces: The quest for a new generation of biomaterials. *Trends in Biotechnology*, *31*(5), 295-304. doi:10.1016/j.tibtech.2013.01.017 PMID:23434154

Hasan, J., Raj, S., Yadav, L., & Chatterjee, K. (2015). Engineering a nanostructured "super surface" with superhydrophobic and superkilling properties. *RSC Advances*, *5*(56), 44953–44959. doi:10.1039/C5RA05206H

Hashimoto, K.-Y., & Yamaguchi, M. (2001). Excitation and propagation of shear-horizontal-type surface and bulk acoustic waves. *IEEE Transactions on* Ultrasonics, Ferroelectrics, and Frequency Control, *48*(5), 1181–1188.

Haubold, T., Bohnn, R., Birringer, R., & Gleiter, H. (1992). Nanocrystalline intermetallic compounds—structure and mechanical properties. *Materials Science and Engineering A*, *153*(1-2), 679–683. doi:10.1016/0921-5093(92)90270-B

Haynes, C. L., & Van Duyne, R. P. (2001). Nanosphere lithography: A versatile nanofabrication tool for studies of size-dependent nanoparticle optics. *The Journal of Physical Chemistry B*, *105*(24), 5599–5611. doi:10.1021/jp010657m

Heavend, O., Ring, J., & Smith, S. (1957). Interference filters for the infra-red. *Spectrochimica Acta.*

He, B., Patankar, N. A., & Lee, J. (2003). Multiple Equilibrium Droplet Shapes and Design Criterion for Rough Hydrophobic Surfaces. *Langmuir*, *19*(12), 4999–5003. doi:10.1021/la0268348

Heidong, H., Wei, Q., Xiaohong, W., Xianbo, D., & Long, C. (2007). The photocatalytic properties of bismuth oxide films prepared through the sol-gel method. *Thin Solid Films*, *515*(13), 5362–5365. doi:10.1016/j.tsf.2007.01.031

He, J., Huang, T., Gan, L., Zhou, Z. K., Jiang, B., Wu, Y., & Gu, Z. W. et al. (2012). Collagen-infiltrated porous hydroxyapatite coating and its osteogenic properties: In vitro and in vivo study. *Journal of Biomedical Materials Research. Part A*, *100*(7), 1706–1715. doi:10.1002/jbm.a.34121 PMID:22447757

Hejazi, V., Sobolev, K., & Nosonovsky, M. (2013). From Superhydrophobicity to Icephobicity: Forces and Interaction Analysis. *Scientific Reports*, *3*, 2194. doi:10.1038/srep02194 PMID:23846773

He, L., Zuo, Q., Xie, S., Huang, Y., & Xue, W. (2011). Intelligent hydrogels for drug delivery system. *Recent Patents on Drug Delivery & Formulation*, *5*(3), 265–274. doi:10.2174/187221111797200533 PMID:21834779

Helbig, R., Nickerl, J., Neinhuis, C., & Werner, C. (2011). Smart skin patterns protect springtails. *PLoS ONE*, *6*(9), e25105. doi:10.1371/journal.pone.0025105 PMID:21980383

Herrmann, C. F., Fabreguette, F. H., Finch, D. S., Geiss, R., & George, S. M. (2005). Multilayer and functional coatings on carbon nanotubes using atomic layer deposition. *Applied Physics Letters*, *87*(12), 1–3. doi:10.1063/1.2053358

Hetrick, E. M., & Schoenfisch, M. H. (2006). Reducing implant-related infections: Active release strategies. *Chemical Society Reviews*, *35*(9), 780–789. doi:10.1039/b515219b PMID:16936926

Higuchi, A., Hamamura, A., Shindo, Y., Kitamura, H., Yoon, B. O., Mori, T., & Umezawa, A. et al. (2004). Photon-modulated changes of cell attachments on poly(spiropyran-co-methyl methacrylate) membranes. *Biomacromolecules*, *5*(5), 1770–1774. doi:10.1021/bm049737x PMID:15360286

Hines, D. A., & Kamat, P. V. (2013). Quantum Dot Surface Chemistry: Ligand Effects and Electron Transfer Reactions. *The Journal of Physical Chemistry C*, *117*(27), 14418–14426. doi:10.1021/jp404031s

Hines, D. A., & Kamat, P. V. (2014). Recent Advances in Quantum Dot Surface Chemistry. *ACS Applied Materials & Interfaces*, *6*(5), 3041–3057. doi:10.1021/am405196u PMID:24506801

Hines, M. A., & Guyot-Sionnest, P. (1996). Synthesis and Characterization of Strongly Luminescing ZnS-Capped CdSe Nanocrystals. *Journal of Physical Chemistry*, *100*(2), 468–471. doi:10.1021/jp9530562

Hinzmann, D., Klotzbach, T., & Herrwerth, S. (2012). Ein starkes Netzwerk gegen Rost. *Farbe + Lack*, *118*(9), 22–26.

Hirayama, Y., & Obara, M. (2005). Heat-affected zone and ablation rate of copper ablated with femtosecond laser. *Journal of Applied Physics*, *97*(6), 064903. doi:10.1063/1.1852692

Hirshberg, Y. (1956). Reversible formation and eradication of colors by irradiation at low temperatures - a photochemical memory model. *Journal of the American Chemical Society*, *78*(10), 2304–2312. doi:10.1021/ja01591a075

Hirvi, J. T., & Pakkanen, T. A. (2007). Wetting of Nanogrooved Polymer Surfaces. *Langmuir*, *23*(14), 7724–7729. doi:10.1021/la700558v PMID:17559245

Ho-Baillie, A., & Green, M. (2010). Forty three per cent composite split-spectrum concentrator solar cell efficiency. *Progress in Photovoltaics: Research and Applications*, 2010, 42–47.

Hofacker, S., Mechtel, M., Hovestadt, W., & Kobusch, C. (2002). Two-component polyurethane binders as primers. *Patent US 2002/0142169 A1*.

Hoffmann, K., Resch-Genger, U., & Marlow, F. (2000). Photoinduced switching of nanocomposites consisting of azo-benzene and molecular sieves: Investigation of the switching states. *Microporous and Mesoporous Materials*, *41*(1-3), 99–106. doi:10.1016/S1387-1811(00)00277-8

Holberg, S., Bischoff, C., Poulsen, T., & Tabrizian, N. (2013, September). Versatile urea-siloxane hybrid coatings: Corrosion protection of aluminium and reduction of crude oil fouling. *Poster presented at Eurocorr 2013*, Estoril, Portugal.

Holberg, S., & Bischoff, C. (2014). Application of a repellent urea-siloxane hybrid coating in the oil industry. *Progress in Organic Coatings*, *77*(10), 1591–1595. doi:10.1016/j.porgcoat.2013.10.019

Hollister, S. J. (2005). Porous scaffold design for tissue engineering. *Nature Materials*, *4*(7), 518–524. doi:10.1038/nmat1421 PMID:16003400

Homola, J., Yee, S. S., & Gauglitz, G. (1999). Surface plasmon resonance sensors[review]. *Sensors and Actuators. B, Chemical*, *54*(1-2), 3–15. doi:10.1016/S0925-4005(98)00321-9

Hong, Z., Luan, L., Paik, S.-B., Deng, B., Ellis, D. E., Ketterson, J. B., & Rossi, A. et al. (2007). Crystalline hydroxy-apatite thin films produced at room temperature—an opposing radio frequency magnetron sputtering approach. *Thin Solid Films*, *515*(17), 6773–6780. doi:10.1016/j.tsf.2007.02.089

Hopmann, C., Behmenburg, C., Recht, U., & Zeuner, K. (2014). Injection Molding of Superhydrophobic Liquid Silicone Rubber Surfaces. *Silicon*, *6*(1), 35–43. doi:10.1007/s12633-013-9164-0

Horrocks, A. R., & Price, D. (2008). *Advances in Fire Retardant Materials*. Boca Raton, FL: Woodhead Publishing. doi:10.1533/9781845694701

Hösel, M., Søndergaard, R. R., Angmo, D., & Krebs, F. C. (2013). Comparison of fast roll-to-roll flexographic, inkjet, flatbed, and rotary screen printing of metal back electrodes for polymer solar cells. *Advanced Engineering Materials Communication*, *15*, 995–1001.

Howard, C., Hallack, M., Schlemann, U., & Bartek, A. (2008). Novel high-solids systems based on silicone epoxy resins. *Paint & Coating Industry*, *2008*(October), 42–48.

Hoyle, C., Lowe, A. B., & Bowman, C. N. (2010). Thiol-click chemistry: A multifaceted toolbox for small molecule and polymer synthesis. *Chemical Society Reviews*, *39*(4), 1355–1387. doi:10.1039/b901979k PMID:20309491

Hsieh, C. T., & Chen, W. Y. (2010). Water/Oil Repellency and Drop Sliding Behavior on Carbon Nanotubes/Carbon Paper Composite Surfaces. *Carbon*, *48*(3), 612–619. doi:10.1016/j.carbon.2009.09.076

Hsieh, C. T., Wu, F.-L., & Chen, W.-Y. (2009). Super Water- and Oil-Repellencies from Silica-based Nanocoatings. *Surface and Coatings Technology*, *203*(22), 3377–3384. doi:10.1016/j.surfcoat.2009.04.025

Huang, H., Susha, A. S., Kershaw, S. V., Hung, T. F., & Rogach, A. L. (2015). Control of Emission Color of High Quantum Yield CH3NH3PbBr3 Perovskite Quantum Dots by Precipitation Temperature. *Advanced Science*.

Huang, X., & Zacharia, N. S. (2015). Functional polyelectrolyte multilayer assemblies for surfaces with controlled wetting behavior. *Journal of Applied Polymer Science*. *132*(45), 42767-42778

Huang, Y., Chen, H., Roper, J., III, Tucker, C., Vandezande, G., Li, Y., & Zhang, Y. (2012a). Moisture curable anti-fouling coating compositions. *Patent US 2012/0269976 A1*.

Huang, Y., Chen, H., Roper, J., III, Tucker, C., Vandezande, G., Li, Y., & Zhang, Y. (2012b). Two-component moisture curable coating compositions. *Patent US 2012/0269977 A1*.

Huang, H.-C., Barua, S., Sharma, G., Dey, S. K., & Rege, K. (2011). Inorganic nanoparticles for cancer imaging and therapy. *Journal of Controlled Release*, *155*(3), 344–357. doi:10.1016/j.jconrel.2011.06.004 PMID:21723891

Huang, P. C., Hou, K. H., Sheu, H. H., Ger, M. D., & Wang, G. L. (2014). Wear properties of Ni–Mo coatings produced by pulse electroforming. *Surface and Coatings Technology*, *258*, 639–645. doi:10.1016/j.surfcoat.2014.08.024

Huang, P. C., Hou, K. H., Wang, G. L., Chen, M. L., & Wang, J. R. (2015). Corrosion resistance of the Ni-Mo alloy coatings related to coating's electroplating parameters. *International Journal of Electrochemical Science*, *10*(6), 4972–4984.

Huang, X. H., Gordon, M. J., & Zare, R. N. (1988). Current-monitoring method for measuring the electroosmotic flowrate in capillary zone electrophoresis. *Analytical Chemistry*, *60*(17), 1837–1838. doi:10.1021/ac00168a040

Huang, X., Lindgren, E., & Chelikowsky, J. R. (2005). Surface passivation method for semiconductor nanostructures. *Physical Review B: Condensed Matter and Materials Physics*, *71*(16), 165328. doi:10.1103/PhysRevB.71.165328

Hua, Z., Pal, R., Srivannavit, O., Burns, M. A., & Gulari, E. (2008). A light writable microfluidic "flash memory": Optically addressed actuator array with latched operation for microfluidic applications. *Lab on a Chip*, *8*(3), 488–491. doi:10.1039/b712983a PMID:18305870

Hu, C. C., & Weng, C. Y. (2000). Hydrogen evolving activity on nickel-molybdenum deposits using experimental strategies. *Journal of Applied Electrochemistry*, *30*(4), 499–506. doi:10.1023/A:1003964728030

Hu, C., Liu, S., Li, B., Yang, H., Fan, C., & Cui, W. (2013). Micro-/nanometer rough structure of a superhydrophobic biodegradable coating by electrospraying for initial anti-bioadhesion. *Advanced Healthcare Materials*, *2*(10), 1314–1321. doi:10.1002/adhm.201300021 PMID:23554405

Hu, H., Zhang, W., Qiao, Y., Jiang, X., Liu, X., & Ding, C. (2012). Antibacterial activity and increased bone marrow stem cell functions of Zn-incorporated TiO2 coatings on titanium. *Acta Biomaterialia*, *8*(2), 904–915. doi:10.1016/j.actbio.2011.09.031 PMID:22023752

Hu, L., Graaf, M. D., & Moeller, K. D. (2013). The use of UV-cross-linkable di-block copolymers as functional reaction surfaces for microelectrode arrays. *Journal of the Electrochemical Society*, *160*(7), G3020–G3029. doi:10.1149/2.004307jes

Hu, R.-G., Zhang, S., Bu, J.-F., Lin, C.-J., & Song, G.-L. (2012). Recent progress in corrosion protection of magnesium alloys by organic coatings. *Progress in Organic Coatings*, *73*(2-3), 129–141. doi:10.1016/j.porgcoat.2011.10.011

Hussain, M., Mehmooda, M., Ahmada, J., Tanvir, M., Khan, A., Ali, T., & Mahmood, A. (2013). RBS depth profiling and optical characterization of multilayers of TiO2 (20 nm) and Ge (15 nm). *Materials Chemistry and Physics*, *139*(1), 17–26. doi:10.1016/j.matchemphys.2012.10.037

Hu, Z., Zen, X., Gong, J., & Deng, Y. (2009). Water Resistance Improvement of Paper by Superhydrophobic modification with Microsized $CaCO_3$ and Fatty Acid Coating. *Colloids and Surfaces. A, Physicochemical and Engineering Aspects*, *351*(1-3), 65–70. doi:10.1016/j.colsurfa.2009.09.036

Ichimura, K., Oh, S. K., & Nakagawa, M. (2000). Light-driven motion of liquids on a photoresponsive surface. *Science*, *288*(5471), 1624–1626. doi:10.1126/science.288.5471.1624 PMID:10834837

Ide, T., Ozama, Y., & Matsui, K. (2011). Photochemistry of azobenzene in sol-gel systems. *Journal of Non-Crystalline Solids*, *357*(1), 100–104. doi:10.1016/j.jnoncrysol.2010.09.009

Igari, T., Oyamada, N., Nanao, H., & Mori, S. (2000). Relationship between the structure of organic thin films and their microtribological property. *Toraiborojisuto*, *45*(5), 414–420.

Ilevbare, G. O., & Burstein, G. T. (2001). The role of alloyed molybdenum in the inhibition of pitting corrosion in stainless steels. *Corrosion Science*, *43*(3), 485–513. doi:10.1016/S0010-938X(00)00086-X

Iliescu, M., Nelea, V., Werckmann, J., & Mihailescu, I. (2004). Transmission electron microscopy investigation of pulsed-laser deposited hydroxylapatite thin films prepared by tripod and focused ion beam techniques. *Surface and Coatings Technology*, *187*(1), 131–140. doi:10.1016/j.surfcoat.2004.01.022

Imbesi, P. M., Gohad, N. V., Eller, M. J., Orihuela, B., Rittschof, D., Schweikert, E. A., & Wooley, K. L. et al. (2012). Noradrenaline-functionalized hyperbranched fuoropolymer - poly(ethylene glycol) cross-linked networks as dual-mode, anti-biofouling coatings. *ACS Nano*, *6*(2), 1503–1512. doi:10.1021/nn204431m PMID:22276525

Imenes, A. G., & Mills, D. R. (2004). Spectral beam splitting technology for increased conversion efficiency in solar concentrating systems: A review. *Solar Energy Materials and Solar Cells*, *84*(1-4), 19–69. doi:10.1016/j.solmat.2004.01.038

Im, J.-H., Jang, I.-H., Pellet, N., Grätzel, M., & Park, N.-G. (2014). Growth of CH3NH3PbI3 cuboids with controlled size for high-efficiency perovskite solar cells. *Nat Nano*, *9*(11), 927–932. doi:10.1038/nnano.2014.181 PMID:25173829

Im, J.-H., Lee, C.-R., Lee, J.-W., Park, S.-W., & Park, N.-G. (2011). 6.5% efficient perovskite quantum-dot-sensitized solar cell. *Nanoscale*, *3*(10), 4088–4093. doi:10.1039/c1nr10867k PMID:21897986

Innovative sol-gel coatings with sivento silanes [brochure]. (n. d.). Degussa. Retrieved from www.dynasylan.com

Iqbal, N., Abdul Kadir, M. R., Nik Malek, N. A. N., Humaimi Mahmood, N., Raman Murali, M., & Kamarul, T. (2012). Rapid microwave assisted synthesis and characterization of nanosized silver-doped hydroxyapatite with antibacterial properties. *Materials Letters*, *89*, 118–122. doi:10.1016/j.matlet.2012.08.057

Irie, M., & Hosoda, M. (1985). Photoresponsive polymers—reversible solution viscosity change of poly(N,N-dimethylacrylamide) with pendant triphenyl- methane leucohydroxide residues in methanol. *Die Makromolekulare Chemie. Rapid Communications*, *6*(8), 533–536. doi:10.1002/marc.1985.030060803

Irie, M., & Kunwatchakun, D. (1986). Photoresponsivepolymers.8.Reversiblephoto- stimulated dilation of polyacrylamide gels having triphenylmethane leuco derivatives. *Macromolecules*, *19*(10), 2476–2480. doi:10.1021/ma00164a003

Irie, M., Menju, A., & Hayashi, K. (1979). Photoresponsive polymers. reversible solution viscosity change of poly (methyl methacrylate) having spiroben- zopyran side groups. *Macromolecules*, *12*(6), 1176–1180. doi:10.1021/ma60072a032

Ishizaki, T., Hieda, J., Saito, N., Saito, N., & Takai, O. (2010). Corrosion resistance and chemical stability of superhydrophobic film deposited on magnesium alloy AZ31 by microwave plasma-enhanced chemical vapor deposition. *Electrochimica Acta*, *55*(23), 7094–7101. doi:10.1016/j.electacta.2010.06.064

Ishizaki, T., Masuda, Y., & Sakamoto, M. (2011). Corrosion Resistance and Durability of Superhydrophobic Surface Formed on Magnesium Alloy Coated with Nanostructured Cerium Oxide Film and Fluoroalkylsilane Molecules in Corrosive NaCl Aqueous Solution. *Langmuir*, *27*(8), 4780–4788. doi:10.1021/la2002783 PMID:21417352

Ishizaki, T., & Sakamoto, M. (2011). Facile Formation of Biomimetic Color-Tuned Superhydrophobic Magnesium Alloy with Corrosion Resistance. *Langmuir*, *27*(6), 2375–2381. doi:10.1021/la1051029 PMID:21319782

ISO13779–2. Implants for surgery – hydroxyapatite – Part 2: Coatings of hydroxyapatite.(2008).

ISO13779–3. Implants for surgery – hydroxyapatite – Part 3: Chemical analysis and characterization of crystallinity and phase purity. (2008).

ISO13779–4. Implants for surgery – hydroxyapatite – Part 4: Determination of coating adhesion strength. (2002).

Ivanov, A. E., Eremeev, N. L., Wahlund, P. O., Galaev, I. Y., & Mattiasson, B. (2002). Photosensitive copolymer of N-isopropylacrylamide and methacryloyl derivative of spyrobenzopyran. *Polymer*, *43*(13), 3819-3823. doi:10.1016/s0032-3861(02)00191-x

Jagtap, R. N., & Ambre, A. H. (2005). Atomic Force microscopy (AFM): Basics and its important applications for polymer characterization: An overview. *Journal of Polymer Materials*, *22*(1), 1–26.

Jakubowicz, J. (2008). Formation of porous TiOx biomaterials in H3PO4 electrolytes. *Electrochemistry Communications*, *10*(5), 735–739. doi:10.1016/j.elecom.2008.02.027

Jallot, E., Nedelec, J.-M., Grimault, A. S., Chassot, E., Grandjean-Laquerriere, A., Laquerriere, P., & Laurent-Maquin, D. (2005). STEM and EDXS characterisation of physico-chemical reactions at the periphery of sol–gel derived Zn-substituted hydroxyapatites during interactions with biological fluids. *Colloids and Surfaces. B, Biointerfaces*, *42*(3), 205–210. doi:10.1016/j.colsurfb.2005.03.001 PMID:15893220

Jasieniak, J., & Mulvaney, P. (2007). From Cd-Rich to Se-Rich – the Manipulation of CdSe Nanocrystal Surface Stoichiometry. *Journal of the American Chemical Society*, *129*(10), 2841–2848. doi:10.1021/ja066205a PMID:17309253

Jedynski, M., Hoffman, J., Mroz, W., & Szymanski, Z. (2008). Plasma plume induced during ArF laser ablation of hydroxyapatite. *Applied Surface Science*, *255*(5), 2230–2236. doi:10.1016/j.apsusc.2008.07.070

Jeong, C. (2013). *Nano-Engineering of superhydrophobic aluminum surfaces for anti-corrosion [Ph.D Thesis]*.. Stevens Institute of Technology.

Jeong, Y.-H., Choe, H.-C., & Eun, S.-W. (2011). Hydroxyapatite coating on the Ti–35Nb–xZr alloy by electron beam-physical vapor deposition. *Thin Solid Films*, *519*(20), 7050–7056. doi:10.1016/j.tsf.2011.04.086

Jerzykiewicz, L. B., Utko, J., & Sobota, P. (2006). Structural Characterization of Manganese(II) Compounds with Methylalumonaxe (MAO) and AlMe2 Units. *Organometallics*, *25*(20), 4924–4926. doi:10.1021/om060207m

Jiang, L., & Feng, L. (2010). *Bioinspired intelligent nanostructured interfacial materials*. Chemical Industry Press. doi:10.1142/7380

Jiang, L., Shen, W., Ogawa, H., & Guo, Q. (2003). Temperature dependence of the optical properties in hexagonal AlN. *Journal of Applied Physics*, *94*(9), 5704–5709. doi:10.1063/1.1616988

Jiang, T., Guo, Z., & Liu, W. (2015). Biomimetic superoleophobic surfaces: Focusing on their fabrication and applications. *Journal of Materials Chemistry A. Materials for Energy and Sustainability*, *3*(5), 1811–1827.

Jin, C., Li, J., Han, S., Wang, J., & Sun, Q. (2014). A durable, superhydrophobic, superoleophobic and corrosion-resistant coating with rose-like ZnO nanoflowers on a bamboo surface. *Applied Surface Science*, *320*, 322–327. doi:10.1016/j.apsusc.2014.09.065

Jin, Z. C., Hamberg, I., & Granqvist, C. G. (1988). Optical properties of sputter-deposited ZnO:Al thin films. *Journal of Applied Physics*, *64*(10), 5117. doi:10.1063/1.342419

Ji, X., Copenhaver, D., Sichmeller, C., & Peng, X. (2008). Ligand Bonding and Dynamics on Colloidal Nanocrystals at Room Temperature: The Case of Alkylamines on CdSe Nanocrystals. *Journal of the American Chemical Society*, *130*(17), 5726–5735. doi:10.1021/ja710909f PMID:18396878

Joanni, E., Ferro, M. C., Mardare, C. C., Mardare, A. I., Fernandes, J. R. A., & Pina, S. C. A. (2004). Pulsed laser deposition of SiO2-P2O5-CaO-MgO glass coatings on titanium substrates. *Materials Research*, *7*(3), 431–436. doi:10.1590/S1516-14392004000300010

John, L., & Sobota, P. (2010). Alkoxide Molecular Precursors for Nanomaterials: A One Step Strategy for Oxide Ceramics. Intech Open Access.

John, Å., Utko, J., Szafert, S., Jerzykiewicz, L. B., Kępiński, L., & Sobota, P. (2008). Synthesis and Characterization of Mixed-Metal Aryloxo-Organometallic Precursors for Oxides Ceramic Materials. *Chemistry of Materials*, *20*(13), 4231–4239. doi:10.1021/cm702262n

Johnson, P. B., & Christy, R. W. (1972). Optical constants of the noble metals. *Physical Review B: Condensed Matter and Materials Physics*, *6*(12), 4370–4379. doi:10.1103/PhysRevB.6.4370

Johnson, R. E. Jr, & Dettre, R. H. (1964). Contact angle hysteresis. III. Study of an idealized heterogeneous surface. *Journal of Physical Chemistry*, *68*(7), 1744–1750. doi:10.1021/j100789a012

Jones, A. C. (2002). Molecular design of improved precursors for the MOCVD of electroceramic oxides. *Journal of Materials Chemistry*, *12*(9), 2576–2590. doi:10.1039/b202675a

Jones, A. C., Leedham, T. J., Wright, P. J., Crosbie, M. J., Fleeting, K. A., Otway, D. J., & Pemble, M. E. et al. (1998). Synthesis and characterisation of two novel titanium isopropoxides stabilised with a chelating alkoxide: Their use in the liquid injection MOCVD of titanium dioxide thin films. *Journal of Materials Chemistry*, *8*(8), 1773–1777. doi:10.1039/a802933d

Jonschker, G. (2014). *Sol-Gel-Technology in Praxis*. Hanover, Germany: Vincentz Network.

Joshi, A., Bajaj, A., Singh, R., Anand, A., Alegaonkar, P. S., & Datar, S. (2015). Processing of Graphene Nanoribbon based Hybrid Composite for Electromagnetic Shielding. *Composites. Part B, Engineering*, *69*, 472–477. doi:10.1016/j.compositesb.2014.09.014

Joshi, S. C., & Ng, H. W. (2011). Optimizing functionally graded nickel – zirconia coating profiles for thermal stress relaxation. *Simulation Modelling Practice and Theory*, *19*(1), 586–598. doi:10.1016/j.simpat.2010.08.013

Jubete, E., Loaiza, O. A., Ochoteco, E., Pomposo, J. A., Grande, H., & Rodríguez, J. (2009). Nanotechnology: A Tool for Improved Performance on Electrochemical Screen-Printed (Bio)Sensors. *Journal of Sensors*, 842575, 1–13. doi:10.1155/2009/842575

Judeinstein, P., & Sanchez, C. (1996). Hybrid organic-inorganic materials: A land of multidisciplinarity. *Journal of Materials Chemistry*, *6*(4), 511–525. doi:10.1039/jm9960600511

Jung, J.-P., Lee, J.-B., Kim, J.-S., & Park, J.-S. (2004). Fabrication and characterization of high frequency SAW device with IDT/ZnO/AlN/Si conFigureuration: Role of AlN buffer. *Thin Solid Films*, *447*, 605–609. doi:10.1016/j.tsf.2003.07.022

Jung, R. H., Tsuchiya, H., & Fujimoto, S. (2012). XPS characterization of passive films formed on Type 304 stainless steel in humid atmosphere. *Corrosion Science*, *58*, 62–68. doi:10.1016/j.corsci.2012.01.006

Jur, J. S., Sweet, W. J. III, Oldham, C. J., & Parsons, G. N. (2011). Atomic layer deposition of conductive coatings on cotton, paper, and synthetic fibers: Conductivity analysis and functional chemical sensing using "all-fiber" capacitors. *Advanced Functional Materials*, *21*(11), 1993–2002. doi:10.1002/adfm.201001756

Kahl, M., & Golden, T. D. (2014). Electrochemical determination of phenolic acids at a Zn/Al layered double hydroxide film modified glassy carbon electrode. *Electroanalysis*, *26*(8), 1664–1670. doi:10.1002/elan.201400156

Kaid, M. A., & Ashour, A. (2007). Preparation of ZnO-doped Al films by spray pyrolysis technique. *Applied Surface Science*, *253*(6), 3029–3033. doi:10.1016/j.apsusc.2006.06.045

Kamitakahara, M., Ohtsuki, C., & Miyazaki, T. (2007). Coating of bone-like apatite for development of bioactive materials for bone reconstruction. *Biomedical Materials (Bristol, England)*, *2*(4), R17–R23. doi:10.1088/1748-6041/2/4/R01 PMID:18458474

Kanaras, A. G., Sönnichsen, C., Liu, H., & Alivisatos, A. P. (2005). Controlled Synthesis of Hyperbranched Inorganic Nanocrystals with Rich Three-Dimensional Structures. *Nano Letters*, *5*(11), 2164–2167. doi:10.1021/nl0518728 PMID:16277446

Kandanur, S. S., Rafiee, M. A., Yavari, F., Schrameyer, M., Yu, Z.-Z., Blanchet, T. A., & Koratkar, N. (2012). Suppression of Wear in Graphene Polymer Composites. *Carbon*, *50*(9), 3178–3183. doi:10.1016/j.carbon.2011.10.038

Kandare, E., Khatibi, A. A., Yoo, S., Wang, R., Ma, J., Olivier, P., & Wang, C. H. et al. (2015). Improving the through-thickness Thermal and Electrical Conductivity of Carbon Fibre/Epoxy Laminates by exploiting Synergy between Graphene and Silver Nano-Inclusions. *Composites. Part A, Applied Science and Manufacturing*, *69*, 72–82. doi:10.1016/j.compositesa.2014.10.024

Kang, T., Niu, Y., & Jin, G. (2014). Visualization of the interaction between tris and lysozyme with a biosensor based on total internal reflection imaging ellipsometry. *Thin Solid Films*, *571*, 463–467. doi:10.1016/j.tsf.2014.02.098

Kang, Y. M., Lee, D. Y., Lee, J. R., & Song, P. K. (2009). Transparent Conductive ITZO Films Deposited by Using a Magnetron Co-Sputtering System Using Two Cathodes. *Journal of the Korean Physical Society*, *55*(5), 1901–1905.

Kantheti, S., Narayan, R., & Raju, K. V. S. N. (2015). The impact of 1,2,3-triazoles in the design of functional coatings. *RSC Advances*, *5*(5), 3687–3708. doi:10.1039/C4RA12739K

Kanzaki, N., Treboux, G., Onuma, K., Tsutsumi, S., & Ito, A. (2001). Calcium phosphate clusters. *Biomaterials*, *22*(21), 2921–2929. doi:10.1016/S0142-9612(01)00039-4 PMID:11561898

Kao, K.-S., Cheng, C.-C., & Chen, Y.-C. (2002). Synthesis and surface acoustic wave properties of AlN films deposited on LiNbO/sub 3/substrates. *Ultrasonics, Ferroelectrics, and Frequency Control. IEEE Transactions on*, *49*(3), 345–349.

Karamana, M. (2012). Self-supporting superhydrophobic thin polymer sheets that mimic the nature's petal effect. *Applied Surface Science*, *259*, 542–546. doi:10.1016/j.apsusc.2012.07.079

Karaoglani, A., Ogawa, O., Turk, A., & Ozdemir, I. (2013). Thermal shock and cycling behavior of thermal barrier coating used in gas turbines. In E. Benin (Ed.), Progress in gas turbine performance (pp 237-260). InTech.

Kar, J. P., & Bose, G. (2011). *Aluminum Nitride (AlN) film based acoustic devices: material synthesis and device fabrication*. INTECH Open Access Publisher.

Karolus, M., Rówiński, E., & Łągiewka, E. (2006). Structure and surface studies on Ni-Mo alloys with polymers. *Journal of Achievements in Materials and Manufacturing Engineering, 18*(1-2), 119-122.

Kato, M., Kamigaito, M., Sawamoto, M., & Higashimura, T. (1995). Polymerization of methyl methacrylate with the carbon tetrachloride/dichlorotris-(triphenylphosphine) ruthenium (II)/methylaluminum bis(2,6-di-tert-butylphenoxide) initiating system: Possibility of living radical polymerization. *Macromolecules*, *28*(5), 1721–1723. doi:10.1021/ma00109a056

Katsenelenbaum, B. Z. (2014). Antiradar protection of a metal body with a dielectric coating. *Journal of Communications Technology and Electronics*, *59*(11), 1095–1100. doi:10.1134/S1064226914110084

Kawasaki, A., Watanabe, R., Yuki, M., Nakanishi, Y., & Onabe, H. (1996). Effect of microstructure on the thermal shock cracking of functionally graded thermal barrier coatings studied by burner heating test. *Materials Transactions*, *37*(4), 788–795. doi:10.2320/matertrans1989.37.788

Kawashima, A., Akiyama, E., Habazaki, H., & Hashimoto, K. (1997). Characterization of sputter-deposited Ni-Mo and Ni-W alloy electrocatalysts for hydrogen evolution in alkaline solution. *Materials Science and Engineering A*, *226–228*, 905–909. doi:10.1016/S0921-5093(97)80095-0

Kaya, C., Kaya, F., Su, B., Thomas, B., & Boccaccini, R. (2005). Structural and functional thick ceramic coatings by electrophoretic deposition. *Surface and Coatings Technology*, *191*(2-3), 303–310. doi:10.1016/j.surfcoat.2004.03.042

Kazimierczak, H., Ozga, P., & Socha, R. P. (2013b). Investigation of electrochemical co-deposition of zinc and molybdenum from citrate solutions. *Electrochimica Acta*, *104*, 378–390. doi:10.1016/j.electacta.2012.12.140

Kazimierczak, H., Ozga, P., Świątek, Z., & Bielańska, E. (2013a). Characterisation of Zn-Mo alloy layers electrodeposited from aqueous citrate solution. *Journal of Alloys and Compounds*, *578*, 82–89. doi:10.1016/j.jallcom.2013.04.205

Keddie, D. J., Moad, G., Rizzardo, E., & Thang, S. H. (2012). RAFT agent design and synthesis. *Macromolecules*, 45(13), 5321-5342.

Kelly, S. A. M., Birkert, C. R., & Andrews, A. F. (1998). Coating compositions. *Patent EP 0941290 B1*.

Kempe, P., Löschner, T., Auer, A. A., Seifert, A., Cox, G., & Spange, S. (2014). Thermally induced twin polymerization of 4H-1,3,2-benzodioxasilines. *Chemistry (Weinheim an der Bergstrasse, Germany)*, *20*(26), 8040–8053. doi:10.1002/chem.201400038 PMID:24849004

Kent, J. A. (2012). *Handbook of Industrial Chemistry and Biotechnology*. New York, NY: Springer. doi:10.1007/978-1-4614-4259-2

Keske, A., Atar, A., Üstündağ, İ., & Çağlayan, M. O. (2014). Detection of Influenza A by Surface Plasmon Resonance Enhanced Total Internal Reflection Ellipsometry. *Journal of Computational and Theoretical Nanoscience*, *11*(4), 981–986. doi:10.1166/jctn.2014.3453

Kessler, G. V., Hubert-Pfalzgraf, G. L., Daniele, S., & Gleizes, A. (1994). Single-source precursors for BaTiO3: Synthesis and Characterization of b-diketonato alkoxides. *Chemistry of Materials, 6*, 2336–2342. doi:10.1021/cm00048a020

Kessler, V. (2004). Geometrical Molecular Structure Design Concept in Approach to Homo- and Heterometallic Precursors of Advanced Materials in Sol-Gel Technology. *Journal of Sol-Gel Science and Technology, 32*(1-3), 11–17. doi:10.1007/s10971-004-5757-9

Keyvani, A., Saremi, M., & Sohi, M. H. (2010). Microstructural stability of zirconia – alumina composite coatings during hot corrosion test at 1050°C. *Journal of Alloys and Compounds, 506*(1), 103–108. doi:10.1016/j.jallcom.2010.06.110

Keyvani, A., Saremi, M., & Sohi, M. H. (2011). Oxidation resistance of YSZ-alumina composites compared to normal YSZ TBC coatings at 1100°C. *Journal of Alloys and Compounds, 509*(33), 8370–8377. doi:10.1016/j.jallcom.2011.05.029

Khatri, H., & Marsillac, S. (2008). The effect of deposition parameters on radiofrequency sputtered molybdenum thin films. *Journal of Physics: Condensed Matter,* 20(5). doi:.10.1088/0953-8984/20/05/055206

Khoddami, A. M., Sabour, A., & Hadavi, S. M. M. (2007). Microstructure formation in thermally-sprayed duplex and functionally graded NiCrAlY / Yttria-Stabilized Zirconia coatings. *Surface and Coatings Technology, 201*(12), 6019–6024. doi:10.1016/j.surfcoat.2006.11.020

Khor, K. A. U., Gu, Y. W., & Dong, Z. L. (2001). Mechanical behavior of plasma sprayed functionally graded YSZ/NiCoCrAlY composite coatings. *Surface and Coatings Technology, 139*(2-3), 200–206. doi:10.1016/S0257-8972(01)00997-5

Khor, K. A., Dong, Z. L., & Gu, Y. W. (1999). Plasma sprayed functionally graded thermal barrier coatings. *Materials Letters, 38*(6), 437–444. doi:10.1016/S0167-577X(98)00203-1

Khor, K. A., & Gu, Y. W. (2000 a). Effects of residual stress on the performance of plasma sprayed functionally graded ZrO_2 / NiCoCrAlY coatings. *Materials Science and Engineering A, 277*(1-2), 64–76. doi:10.1016/S0921-5093(99)00565-1

Khor, K. A., & Gu, Y. W. (2000 b). Thermal properties of plasma-sprayed functionally graded thermal barrier coatings. *Thin Solid Films, 372*(1-2), 104–113. doi:10.1016/S0040-6090(00)01024-5

Khor, K. A., Gu, Y. W., & Dong, Z. L. (2000). Plasma spraying of functionally graded yttria stabilized zirconia / NiCoCrAlY coating system using composite powders. *Journal of Journal of Thermal Spray Technology, 9*(2), 245–249. doi:10.1361/105996300770349999

Khor, K., Dong, Z., Quek, C., & Cheang, P. (2000). Microstructure investigation of plasma sprayed HA/Ti6Al4V composites by TEM. *Materials Science and Engineering A, 281*(1), 221–228. doi:10.1016/S0921-5093(99)00717-0

Kim, S. J., Dang, V.-S., Xu, K., Barreca, D., Maccato, C., Carraro, G., et al. (2015). MOCVD of TiO2 thin films from a modified titanium alkoxide precursor. *Physica status solidi (a), 212*(7), 1563-1570.

Kim, A., Lee, C., Kim, H., & Kim, J. (2015). Simple Approach to Superhydrophobic Nanostructured Al for Practical Antifrosting Application Based on Enhanced Self-propelled Jumping Droplets. *ACS Applied Materials & Interfaces, 7*(13), 7206–7213. doi:10.1021/acsami.5b00292 PMID:25782028

Kim, B. S., Shin, S., Shin, S. J., Kim, K. M., & Cho, H. H. (2011). Control of superhydrophilicity/superhydrophobicity using silicon nanowires via electroless etching method and fluorine carbon coating. *Langmuir, 27*(16), 10148–10156. doi:10.1021/la200940j PMID:21728376

Kim, B., & Shin, Y. (2007). pH-sensitive swelling and release behaviors of anionic hydrogels for intelligent drug delivery system. *Journal of Applied Polymer Science, 105*(6), 3656–3661. doi:10.1002/app.26450

Kim, H.-S., Lee, C.-R., Im, J.-H., Lee, K.-B., Moehl, T., Marchioro, A., & Park, N.-G. et al. (2012). Lead Iodide Perovskite Sensitized All-Solid-State Submicron Thin Film Mesoscopic Solar Cell with Efficiency Exceeding 9%. *Scientific Reports*, *2*, 591. doi:10.1038/srep00591 PMID:22912919

Kim, J. H., Kim, M. C., & Park, C. G. (2003). Evaluation of functionally graded thermal barrier coatings fabricated by detonation gun spray technique. *Surface and Coating Technology*, *168*, 275–280.

Kim, S. H. (2008). Fabrication of Superhydrophobic Surfaces. *Journal of Adhesion Science and Technology*, *22*(3-4), 235–250. doi:10.1163/156856108X305156

Kimura, S.-I. (2006). Recent progress of surface analysis of polymers. 1. Introduction. *Nippon Gomu Kyokaishi*, *79*(6), 250–256. doi:10.2324/gomu.79.250

Kim, W.-G., & Choe, H.-C. (2011). Surface characteristics of hydroxyapatite/titanium composite layer on the Ti-35Ta-xZr surface by RF and DC sputtering. *Thin Solid Films*, *519*(20), 7045–7049. doi:10.1016/j.tsf.2011.04.090

Kirtay, S. (2014). Characterization of SiO2-TiO2 Hybrid Corrosion Protective Coatings on Mild Steel. *Journal of Materials Engineering and Performance*, *23*(12), 4309–4315.

Klajn, R., Wesson, P. J., Bishop, K. J. M., & Grzybowski, B. A. (2009). Writing Self-Erasing Images using Metastable Nanoparticle "Inks.". *Angewandte Chemie International Edition*, *48*(38), 7035–7039. doi:10.1002/anie.200901119 PMID:19533698

Kluth, O., Schöpe, G., Hüpkes, J., Agashe, C., Müller, J., & Rech, B. (2003). Modified Thornton model for magnetron sputtered zinc oxide: Film structure and etching behaviour. *Thin Solid Films*, *442*(1-2), 80–85. doi:10.1016/S0040-6090(03)00949-0

Knez, M. (2008). ALD- A versatile tool for nanostructuring. *Material Matters*, *3*(2), 28–33.

Knez, M. (2012). Diffusion phenomena in atomic layer deposition. *Semiconductor Science and Technology*, *27*(7), 074001. doi:10.1088/0268-1242/27/7/074001

Kobayashi, M., Terayama, Y., Yamaguchi, H., Tereda, M., Murakami, D., Ishihara, K., & Takahara, A. (2012). Wettability and Antifouling Behavior on the Surfaces of Superhydrophilic Polymer Brushes. *Langmuir*, *28*(18), 7212–7222. doi:10.1021/la301033h PMID:22500465

Koberstein, J. T. (2004). Molecular design of functional polymer surfaces. *Journal of Polymer Science. Part B, Polymer Physics*, *42*(16), 2942–2956. doi:10.1002/polb.20157

Koberstein, J. T., Hu, W., Duch, D., Bhatia, R., Lenk, T., Smith, D. A., & Gallot, Y. et al. (1997). Molecular design concepts for creating smart polymer surfaces.*Proceedings of the20thAnnual Meeting of the Adhesion Society* (pp. 283-285).

Koch, K., & Barthlott, W. (2009). Superhydrophobic and superhydrophilic plant surfaces: an inspiration for biomimetic materials. *Philosophical transactions. Series A, Mathematical, physical, and engineering sciences*, 367(1893), 1487-1509.

Koch, C. C. (2007). Structural nanocrystalline materials: An overview. *Journal of Materials Science*, *42*(5), 1403–1414. doi:10.1007/s10853-006-0609-3

Kofron, M. D., & Laurencin, C. T. (2004). Development of a calcium phosphate co-precipitate/poly(lactideco-glycolide) DNA delivery system: Release kinetics and cellular transfection studies. *Biomaterials*, *25*(13), 2637–2643. doi:10.1016/j.biomaterials.2003.09.042 PMID:14751750

Kojima, A., Ikegami, M., Teshima, K., & Miyasaka, T. (2012). Highly Luminescent Lead Bromide Perovskite Nanoparticles Synthesized with Porous Alumina Media. *Chemistry Letters*, *41*(4), 397–399. doi:10.1246/cl.2012.397

Kojima, A., Teshima, K., Shirai, Y., & Miyasaka, T. (2009). Organometal Halide Perovskites as Visible-Light Sensitizers for Photovoltaic Cells. *Journal of the American Chemical Society*, *131*(17), 6050–6051. doi:10.1021/ja809598r PMID:19366264

Kokini, K., Dejonge, J., Rangaraj, S., & Beardsley, B. (2002). Thermal shock of functionally graded thermal barrier coatings with similar thermal resistance. *Surface and Coatings Technology*, *154*(2-3), 223–231. doi:10.1016/S0257-8972(02)00031-2

Kokini, K., Takeuchi, Y. R., & Choules, B. D. (1996). Thermal crack initiation mechanisms on the surface of functionally graded ceramic thermal barrier coatings. *Ceramics International*, *22*, 397–401.

Kokubo, T., Ito, S., Huang, Z. T., Hayashi, T., Sakka, S., Kitsugi, T., & Yamamuro, T. (1990). Ca,P-rich layer formed on high-strength bioactive glass-ceramic A-W. *Journal of Biomedical Materials Research*, *24*(3), 331–343. doi:10.1002/jbm.820240306 PMID:2156869

Koleilat, G. I., Levina, L., Shukla, H., Myrskog, S. H., Hinds, S., Pattantyus-Abraham, A. G., & Sargent, E. H. (2008). Efficient, Stable Infrared Photovoltaics Based on Solution-Cast Colloidal Quantum Dots. *ACS Nano*, *2*(5), 833–840. doi:10.1021/nn800093v PMID:19206479

Kota, A. K., & Tuteja, A. (2012). Superoleophobic surfaces. In *Advances in fluorine-containing polymers, ACS Symposium Series* (*Vol. 1106*, pp. 171-185), American Chemical Society, Washington, DC.

Kovsman, E. P., Andruseva, S. I., Solovjeva, L. I., Fedyaev, V. I., Adamova, M. N., & Rogova, T. V. (1994). Electrochemical Synthesis of Metal Alkoxides. Prospects of commercial alkoxides production. *Journal of Sol-Gel Science and Technology*, *2*(1-3), 61–66. doi:10.1007/BF00486214

Krastajić, N. V., Jović, V. D., Gajic-Krstajić, Lj., Jović, B. M., Antozzi, A. L., & Martelli, G. N. (2008). Electrodeposition of Ni-Mo alloy coatings and their characterization as cathodes for hydrogen evolution in sodium hydroxide solution. *International Journal of Hydrogen Energy*, *35*(14), 3676–3687. doi:10.1016/j.ijhydene.2008.04.039

Krč, M., Zeman, M., Kluth, O., Smole, F., & Topič, M. (2003). Effect of surface roughness of ZnO:Al films on light scattering in hydrogenated amorphous silicon solar cells. *Thin Solid Films*, *426*(1-2), 296–304. doi:10.1016/S0040-6090(03)00006-3

Krebs, F. C. (2009). All solution roll-to-roll processed polymer solar cells free from indium-tin-oxide and vacuum coating steps. *Organic Electronics*, *10*(5), 761–768. doi:10.1016/j.orgel.2009.03.009

Krukiewicz, K., Herman, A. P., Turczyn, R., Szymanska, K., Koziol, K. K. K., Boncel, S., & Zak, J. K. (2014). A role of nanotube dangling pyrrole and oxygen functions in the electrochemical synthesis of polypyrrole/MWCNTs hybrid materials. *Applied Surface Science*, *317*, 794–802. doi:10.1016/j.apsusc.2014.08.165

Krumm, M., Pawlitzek, F., Weickert, J., Schmidt-Mende, L., & Polarz, S. (2012). Temperature-Stable and Optically Transparent Thin-Film Zinc Oxide Aerogel Electrodes As Model Systems for 3D Interpenetrating Organic-Inorganic Heterojunction Solar Cells. *ACS Applied Materials & Interfaces*, *4*(12), 6522–6529. doi:10.1021/am302458n PMID:23194020

Krumm, M., Pueyo, C. L., & Polarz, S. (2010). Monolithic Zinc Oxide Aerogels from Organometallic Sol-Gel Precursors. *Chemistry of Materials*, *22*(18), 5129–5136. doi:10.1021/cm1006907

Ksycki, S. M. J., & Yntema, L. F. (1949). The electrodeposition of molybdenum from aqueous solutions. *Journal of the Electrochemical Society*, *96*(1), 48–56. doi:10.1149/1.2776770

Kuckling, D., Harmon, M. E., & Frank, C. W. (2002). Photo-cross-linkable PNIPAAm copolymers. 1. Synthesis and characterization of constrained temperature-responsive hydrogel layers. *Macromolecules*, *35*(16), 6377–6383. doi:10.1021/ma0203041

Kulinich, S. A., Farhadi, S., Nose, K., & Du, X. W. (2011). Superhydrophobic Surfaces: Are They Really Ice-Repellent? *Langmuir*, *27*(1), 25–29. doi:10.1021/la104277q PMID:21141839

Kumar, D., Wu, X., Fu, Q., Ho, J. W. C., Kanhere, P. D., Li, L., & Chen, Z. (2015). Hydrophobic sol–gel coatings based on polydimethylsiloxane for self-cleaning applications. *Materials & Design*, *86*, 855–862. doi:10.1016/j.matdes.2015.07.174

Kumar, R., & Münstedt, H. (2005). Polyamide/silver antimicrobials: Effect of crystallinity on the silver ion release. *Polymer International*, *54*(8), 1180–1186. doi:10.1002/pi.1828

Kumar, R., & Munstedt, H. (2005). Silver ion release from antimicrobial polyamide/silver composites. *Biomaterials*, *26*(14), 2081–2088. doi:10.1016/j.biomaterials.2004.05.030 PMID:15576182

Kundu, D., Biswas, P., & Ganguli, D. (1989). Sol-gel preparation of wavelength-selective reflecting coatings in the system ZrO2-SiO2. *Journal of Non-Crystalline Solids*, *110*(1), 13–16. doi:10.1016/0022-3093(89)90176-2

Kungwatchakun, D., & Irie, M. (1988). Photoresponsive polymers—photocontrol of the phase-separation temperature of aqueous-solutions of poly N-isopropylacrylamide-co-N-(4-phenylazophenyl)acrylamide. *Die Makromolekulare Chemie. Rapid Communications*, *9*(4), 243–246. doi:10.1002/marc.1988.030090408

Kuznetsov, V. V., Golyanin, K. E., Pshenichkina, T. V., Lyakhov, B. F., & Lyashenko, S. E. (2013). Chemical composition of Fe-Mo alloys obtained by electrodeposition. *Mendeleev Communications*, *23*(6), 331–333. doi:10.1016/j.mencom.2013.11.009

Kwok, C. S., Horbett, T. A., & Ratner, B. D. (1999). Design of infection-resistant antibiotic-releasing polymers: II. Controlled release of antibiotics through a plasma-deposited thin film barrier. *Journal of Controlled Release*, *62*(3), 301–311. doi:10.1016/S0168-3659(99)00105-4 PMID:10528068

Kwok, C. S., Wan, C., Hendricks, S., Bryers, J. D., Horbett, T. A., & Ratner, B. D. (1999). Design of infection-resistant antibiotic-releasing polymers: I. Fabrication and formulation. *Journal of Controlled Release*, *62*(3), 289–299. doi:10.1016/S0168-3659(99)00106-6 PMID:10528067

Kwon, G. H., Jeong, G. S., Park, J. Y., Moon, J. H., & Lee, S.-H. (2011). A low-energy-consumption electroactive valveless hydrogel micropump for long-term biomedical applications. *Lab on a Chip*, *11*(17), 2910–2915. doi:10.1039/c1lc20288j PMID:21761057

Kwon, W., Seo, K. W., Yoon, S. H., Lee, S. S., & Shem, I.-W. (2005). Preparation of Silver Nanoparticles in Cellulose Acetate Polymer and the Reaction Chemistry of Sylver Complexes in the Polymer. *Bulletin of the Korean Chemical Society*, *26*(10).

Lackner, J. M., & Waldhauser, W. (2010). Inorganic PVD and CVD Coatings in Medicine — A Review of Protein and Cell Adhesion on Coated Surfaces. *Journal of Adhesion Science and Technology*, *24*(5), 925–961. doi:10.1163/016942409X12598231568023

Laferriere, M., Galian, R. E., Maurel, V., & Scaiano, J. C. (2006). Non-linear effects in the quenching of fluorescent quantum dots by nitroxyl free radicals. *Chemical Communications*, (3): 257–259. doi:10.1039/B511515A PMID:16391726

Lai, Y.-K., Chen, Z., & Lin, C.-J. (2011). Recent progress on the superhydrophobic surfaces with special adhesion: From natural to biomimetic to functional. *J. of Nanoengineering and Nanomanufacturing*, *1*(1), 1–17. doi:10.1166/jnan.2011.1007

Lamuta, C., Di Girolamo, G., & Pagnotta, L. (2015). Microstructural, mechanical and tribological properties of nano-structured YSZ coatings produced with different APS process parameters. *Ceramics International*, *41*(7), 8904–8914. doi:10.1016/j.ceramint.2015.03.148

Landolt, D., & Mischler, S. (2011). *Tribocorrosion of Passive Metals and Coatings*. Cambridge, UK: Woodhead Publishing. doi:10.1533/9780857093738

Langmuir, I., & Blodgett, K. B. (1935). A new method for investigating unimolecular films. *Kolloid-Zeitschrift*, *73*, 258–263.

Laryea, N., Thurn, C., & Sepeur, S. (2008b). Verfahren zur Herstellung eines Beschichtungsmaterials. *Patent DE 102007020404* A1. [Method for the production of a coating material. *Patent US 2010/0092686 A1]*.

Laryea, N., Thurn, C., Sepeur, S., & Schlick, G. (2008a). Silanbeschichtungsmaterial und Verfahren zur Herstellung eines Silanbeschichtungsmaterials. *Patent DE 102006044310 A1*. [Silane coating material and a process to produce silane coating. *Patent US 2009/0326146 A1]*.

Laser, D. J., & Santiago, J. G. (2004). A review of micropumps. *Journal of Micromechanics and Microengineering*, *14*(6), R35-R64. doi:10.1088/0960-1317/14/6/r01

Laubender, T., & Greene, J. (2006). Adding value to coatings by using unique organofunctional silicone hybrids. *Polimery*, *51*(2), 138–142.

Layrolle, P., Ito, A., & Tateishi, T. (1998). Sol-Gel Synthesis of Amorphous Calcium Phosphate and Sintering into Microporous Hydroxyapatite Bioceramics. *Journal of the American Ceramic Society*, *81*(6), 1421–1428. doi:10.1111/j.1151-2916.1998.tb02499.x

Lee, C. S., Kim, J., Son, J. Y., Choi, W., & Kim, H. (2009). Photocatalytic functional coatings of TiO2 thin films on polymer substrate by plasma enhanced atomic layer deposition. *Applied Catalysis B: Environmental*, *91*(3-4), 628–633. doi:10.1016/j.apcatb.2009.06.037

Lee, C. W., Song, Y. H., Lee, Y., Ryu, K. S., & Chi, K. W. (2009). Reversible photochromic switch ensemble and its photoimaging using H+ transfer between spiropyran and fluorescein in a polymer matrix. *Chemical Communications*, 41,6282–6284. doi:10.1039/b913096a PMID:19826695

Lee, D., Cohen, R. E., & Rubner, M. F. (2005). Antibacterial properties of Ag nanoparticle loaded multilayers and formation of magnetically directed antibacterial microparticles. *Langmuir*, *21*(21), 9651–9659. doi:10.1021/la0513306 PMID:16207049

Lee, H., Alcaraz, M. L., Rubner, M. F., & Cohen, R. E. (2013). Zwitter-wettability and antifogging coatings with frost-resistant capabilities. *ACS Nano*, *7*(3), 2172–2185. doi:10.1021/nn3057966 PMID:23360374

Lee, J. H., Park, E. J., Kim, D. H., Jeong, M. G., & Kim, Y. D. (2016). Superhydrophobic surfaces with photocatalytic activity under UV and visible light irradiation. *Catalysis Today*, *260*, 32–38. doi:10.1016/j.cattod.2015.05.020

Lee, J. Y., & Hong, S.-Y. (2005). Fabrication of an alternating multilayer film of poly(ethylene-alt-maleic anhydride) and poly(4-vinyl pyridine) by layer-by-layer self-assembly method. *Polymer(Korea)*, *29*(4), 392–398.

Lee, J.-M., & Lee, C.-W. (2007). Comparison of hydroxyapatite-coated non-hydroxyapatite-coated noncemented total hip arthroplasty in same patients. *The Journal of Arthroplasty*, *22*(7), 1019–1023. doi:10.1016/j.arth.2007.04.029 PMID:17920475

Lee, S., Lee, Y., Park, J., & Choi, D. (2014). Stitchable organic photovoltaic cells with textile electrodes. *Nano Energy*, *9*, 88–93.

Lee, T.-Z., Ma, C.-C. M., Hsu, C.-W., & Wu, H.-L. (2005). Effect of molecular structures and mobility on the thermal and dynamical mechanical properties of thermally cured epoxy-bridged polyorganosiloxanes. *Polymer, 46*(19), 8286–8296. doi:10.1016/j.polymer.2005.06.069

Lee, W., Ji, R., Gosele, U., & Nielsch, K. (2006). Fast fabrication of long-range ordered porous alumina membranes by hard anodization. *Nature Materials, 5*(9), 741–747. doi:10.1038/nmat1717 PMID:16921361

Lee, W., Stinton, D., Berndt, C. C., Erdogan, F., Lee, Y.-D., & Mutasim, Z. (1996). Concept of functionally graded materials for advanced thermal barrier coatings applications. *Journal of the American Ceramic Society, 79*(12), 3002–3012. doi:10.1111/j.1151-2916.1996.tb08070.x

Le, H. P. (1998). Progress and trends in ink-jet printing technology. *The Journal of Imaging Science and Technology, 42*(1), 49–62.

Lehmkhul, H., & Eisenbach, W. (1975). Elektrochemische Synthesen organischer Metallverbindungen, VII. Elektrosynthese von Alkoholaten und Acetylacetonaten des Eisens, Kobalts und Nickels. *Annual Chemistry, 4*, 672–691. doi:10.1002/jlac.197519750410

Lejars, M., Margaillan, A., & Bressy, C. (2012). Fouling release coatings: A nontoxic alternative to biocidal antifouling coatings. *Chemical Reviews, 112*(8), 4347–4390. doi:10.1021/cr200350v PMID:22578131

Leon, A. C. C., Pernites, R. B., & Advincula, R. C. (2012). Superhydrophobic Colloidally Textured Polythiophene Film as Superior Anticorrosion Coating. *ACS Applied Materials & Interfaces, 4*(6), 3169–3176. doi:10.1021/am300513e PMID:22663552

Levashov, E. A., Pogozhev, Y. S., & Kurbatkina, V. V. (2011). Advanced Ceramic Target Materials Produced by Self-Propagating High-Temperature Synthesis for Deposition of Functional Nanostructured Coatings - Part 1 : Four Elements and Less Systems. *Advances in Ceramics - Synthesis and Characterization- Processing and Specific Applications.*

Levinshtein, M. E., Rumyantsev, S. L., & Shur, M. S. (2001). *Properties of Advanced Semiconductor Materials: GaN, AIN, InN, BN, SiC, SiGe.* John Wiley & Sons.

Levy, D., DelMonte, F., Oton, J. M., Fiksman, G., Matias, I., Datta, P., & Lopez-Amo, M. (1997). Photochromic doped sol-gel materials for fiber-optic devices. *Journal of Sol-Gel Science and Technology, 8*(1-3), 931–935. doi:10.1007/BF02436963

Li, J., Zhang, Z., Xu, J., & Wong, C. P. (2006). Self-cleaning materials-lotus effect surfaces. In S. Arza (Ed.), Kirk-Othmer Encyclopedia of Chemical Technology (5th ed., Vol. 22, pp. 108-127). doi:10.1002/0471238961.selfwong.a01

Li, M., Zhai, J., Liu, H., Song, Y., Jiang, L., & Zhu, D. (2003). Electrochemical Deposition of Conductive Superhydrophobic Zinc Oxide Thin Films. *J. Phys. Chem. B, 107*, 9954-9957. doi:10.1021/jp035562u

Li, C., Hong, G., Yu, H., & Qi, L. (2010). Facile fabrication of honeycomb-patterned thin films of amorphous calcium carbonate and mosaic calcite. *Chemistry of Materials, 22*(10), 3206–3211. doi:10.1021/cm100363a

Li, C.-Q., Li, X.-H., Wang, Z.-X., & Guo, H.-J. (2007). Nickel electrodeposition from novel citrate bath. *Transactions of Nonferrous Metals Society of China, 17*(6), 1300–1306. doi:10.1016/S1003-6326(07)60266-0

Li, C., & Qi, L. (2008). Bioinspired fabrication of 3D ordered macroporous single crystals of calcite from a transient amorphous phase. *Angewandte Chemie International Edition, 47*(13), 2388–2393. doi:10.1002/anie.200705403 PMID:18300213

Li, D., Wan, D., Zhu, X., Wang, Y., Han, Z., Han, S., & Huang, F. et al. (2014). Broadband antireflection TiO2–SiO2 stack coatings with refractive-index-grade structure and their applications to Cu(In,Ga)Se2 solar cells. *Solar Energy Materials and Solar Cells.*

Lieberman, M. A., & Lichtenberg, A. J. (1994). Principles of plasma discharges and materials processing. *MRS Bulletin*, *30*, 899–901.

Lieberman, M., & Lichtenberg, A. (2005). *Principles of Plasma Discharges and Plasma Processing*. John Wiley and Sons.

Li, G., Zheng, H., Wang, Y., Wang, H., Dong, Q., & Bai, R. (2010). A facile strategy for the fabrication of highly stable superhydrophobic cotton fabric using amphiphilic fluorinated triblock azide copolymers. *Polymer*, *51*(9), 1940–1946. doi:10.1016/j.polymer.2010.03.002

Li, L., Huang, T., Lei, J., He, J., Qu, L., Huang, P., & Pan, F. et al. (2015). Robust biomimetic-structural superhydrophobic surface on aluminum alloy. *ACS Applied Materials & Interfaces*, *7*(3), 1449–1457. doi:10.1021/am505582j PMID:25545550

Lima-Neto, P. D., Correia, A. N., Vaz, G. L., & Casciano, P. N. (2010). Morphological, structural, microhardness and corrosion characterisations of electrodeposited Ni-Mo and Cr coatings. *Journal of the Brazilian Chemical Society*, *21*(10), 1968–1976. doi:10.1590/S0103-50532010001000024

Lim, P. N., Chang, L., Tay, B. Y., Guneta, V., Choong, C., Ho, B., & Thian, E. S. (2014). Proposed mechanism of antibacterial action of chemically modified apatite for reduced bone infection. *ACS Applied Materials & Interfaces*, *6*(19), 17082–17092. doi:10.1021/am504716g PMID:25198775

Lin, J. S., & Chiu, H. T. (2003). Photochromic behavior of spiropyran and fulgide in thin films of blends of PMMA and SBS. *Journal of Polymer Research-Taiwan*, *10*(2), 105–110. doi:10.1023/A:1024940921678

Lin, J., Chen, H., Fei, T., Liu, C., & Zhang, J. (2013). Highly Transparent and Thermally Stable Superhydrophobic Coatings from the Deposition of Silica Aerogels. *Applied Surface Science*, *273*, 776–786. doi:10.1016/j.apsusc.2013.02.134

Li, Q., Sun, B., Kinloch, I. A., Zhi, D., Sirringhaus, H., & Windle, A. H. (2006). Enhanced Self-Assembly of Pyridine-Capped CdSe Nanocrystals on Individual Single-Walled Carbon Nanotubes. *Chemistry of Materials*, *18*(1), 164–168. doi:10.1021/cm0512575

Liras, M., Peinado, E., Cañamero, P., Quijada-Garrido, I., & García, O. (2014). Smart photoluminescent nanohybrids based on CdSe quantum dots capped with multidentate thiolated pH-responsive and thermoresponsive polymers for nanosensing. *Journal of Polymer Science. Part A, Polymer Chemistry*, *52*(21), 3087–3095. doi:10.1002/pola.27358

Liu, Y., Li, J. P., Hunziker, E. B., & De Groot, K. (2006). Incorporation of growth factors into medical devices via biomimetic coatings. *Philosophical Transactions of the Royal Society a-Mathematical Physical and Engineering Sciences*, *364*(1838), 233-248. doi:10.1098/rsta.2005.1685

Liu, Y., Wang, X., Fei, B., Hu, H., Lai, C., & Xin, J. H. (2015). Bioinspired, Stimuli-Responsive, Multifunctional Superhydrophobic Surface with Directional Wetting, Adhesion, and Transport of Water. *Advanced Functional Materials*. *25*(31), 5047-5056

Liu, Z., Zhang, X., Murakami, T., & Fujishima, A. (2008). Sol-gel SiO2/TiO2 bilayer films with self-cleaning and antireflection properties. *Sol. Energy Mater. Sol. Cells*, 92, 1434-1438. doi:10.1016/j.solmat.2008.06.005

Liu, C.-H., Niu, P.-L., & Sung, C.-K. (2014). Integrating antireflection and superhydrophobicity of moth-eye-like surface morphology on large–area flexible substrate. *Journal of Physics. D, Applied Physics*, *47*(1), 015401/1–015401, 5. doi:10.1088/0022-3727/47/1/015401

Liu, D., Savino, K., & Yates, M. Z. (2011). Coating of hydroxyapatite films on metal substrates by seeded hydrothermal deposition. *Surface and Coatings Technology*, *205*(16), 3975–3986. doi:10.1016/j.surfcoat.2011.02.008

Liu, H. Q., Szunerits, S., Xu, W. G., & Boukherroub, R. (2009). Preparation of Superhydrophobic Coatings on Zinc as Effective Corrosion Barriers. *ACS Applied Materials & Interfaces*, *1*(6), 1150–1153. doi:10.1021/am900100q PMID:20355904

Liu, H., Szunerits, S., Pisarek, M., Xu, W., & Boukherroub, R. (2009). Preparation of superhydrophobic coatings on zinc, silicon, and steel by a solution-immersion technique. *ACS Applied Materials & Interfaces*, *1*(9), 2086–2091. doi:10.1021/am900488w PMID:20355837

Liu, I. S., Lo, H.-H., Chien, C.-T., Lin, Y.-Y., Chen, C.-W., Chen, Y.-F., & Liou, S.-C. et al. (2008). Enhancing photoluminescence quenching and photoelectric properties of CdSe quantum dots with hole accepting ligands. *Journal of Materials Chemistry*, *18*(6), 675–682. doi:10.1039/b715253a

Liu, L., Guo, X., Li, Y., & Zhong, X. (2010). Bifunctional Multidentate Ligand Modified Highly Stable Water-Soluble Quantum Dots. *Inorganic Chemistry*, *49*(8), 3768–3775. doi:10.1021/ic902469d PMID:20329710

Liu, M., Wang, S., & Jiang, L. (2013). Bioinspired multiscale surfaces with special wettability. *MRS Bulletin*, *38*(5), 375–382. doi:10.1557/mrs.2013.100

Liu, Q., Chen, D., & Kang, Z. (2015). One-Step Electrodeposition Process to Fabricate Corrosion-Resistant Superhydrophobic Surface on Magnesium Alloy. *ACS Applied Materials & Interfaces*, *7*(3), 1859–1867. doi:10.1021/am507586u PMID:25559356

Liu, S. L., Fujihira, M., & Saji, T. (1994). Formation of an organic thin-film by photochemical isomerization of a surfactant with a spiropyran moiety. *Journal of the Chemical Society. Chemical Communications*, 16,1855–1856. doi:10.1039/c39940001855

Liu, T., Chen, S. G., Cheng, S., Tian, J. T., Chang, X. T., & Yin, Y. S. (2007). Corrosion behavior of super-hydrophobic surface on copper in seawater. *Electrochimica Acta*, *52*(28), 8003–8007. doi:10.1016/j.electacta.2007.06.072

Liu, T., Yin, Y., Chen, S., Chang, X., & Cheng, S. (2007). Super-hydrophobic surfaces improve corrosion resistance of copper in seawater. *Electrochimica Acta*, *52*(11), 3709–3713. doi:10.1016/j.electacta.2006.10.059

Liu, W., X., Devaraju, M. K., Yin, S., & Sato, T. (2010). Calcium-doped ceria/titanate tabular functional nanocomposite by layer-by-layer coating method. *Journal of Solid State Chemistry*, *183*(7) 1545-1549

Liu, X., Chu, P. K., & Ding, C. (2010). Surface nano-functionalization of biomaterials. *Materials Science and Engineering R Reports*, *70*(3-6), 275–302. doi:10.1016/j.mser.2010.06.013

Liu, Y., Brandon, R., Cate, M., Peng, X., Stony, R., & Johnson, M. (2007). Detection of Pathogens Using Luminescent CdSe/ZnS Dendron Nanocrystals and a Porous Membrane Immunofilter. *Analytical Chemistry*, *79*(22), 8796–8802. doi:10.1021/ac0709605 PMID:17939743

Liu, Y., Fanguy, J. C., Bledsoe, J. M., & Henry, C. S. (2000). Dynamic coating using polyelectrolyte multilayers for chemical control of electroosmotic flow in capillary electrophoresis microchips. *Analytical Chemistry*, *72*(24), 5939–5944. doi:10.1021/ac000932l PMID:11140760

Liu, Y., Huse, R. O., de Groot, K., Buser, D., & Hunziker, E. B. (2007). Delivery mode and efficacy of BMP-2 in association with implants. *Journal of Dental Research*, *86*(1), 84–89. doi:10.1177/154405910708600114 PMID:17189469

Liu, Y., & Li, G. (2012). A new method for producing "Lotus Effect" on a biomimetic shark skin. *Journal of Colloid and Interface Science*, *388*(1), 235–242. doi:10.1016/j.jcis.2012.08.033 PMID:22995249

Liu, Y., Li, Y., & Zeng, H. (2013). ZnO-Based Transparent Conductive Thin Films: Doping, Performance, and Processing. *Journal of Nanomaterials*, 2013.

Liu, Y., Moevius, L., Xu, X., Qian, T., Yeomans, J. M., & Wang, Z. (2014). Pancake bouncing on superhydrophobic surfaces. *Nature Physics*, *10*(7), 515–519. doi:10.1038/nphys2980

Livage, J. (2000). The sol-gel process. *Verre (Versailles)*, *6*(5), 12–16.

Li, Y., Chen, S., Wu, M., & Sun, J. (2014). All spraying processes for the fabrication of robust, self-healing, superhydrophobic coatings. *Advanced Materials*, *26*, 3344–3348. PMID:24687272

Li, Y., Duan, G., & Cai, W. P. (2007). Controllable superhydrophobic and lipophobic properties of ordered pore indium oxide array films. *Journal of Colloid and Interface Science*, *314*(2), 615–620. doi:10.1016/j.jcis.2007.05.061 PMID:17583718

Li, Z., Han, W., Kozodaev, D., Brokken-Zijp, J. C. M., de With, G., & Thüne, P. C. (2006). Surface properties of poly(dimethylsiloxane)-based inorganic/organic hybrid materials. *Polymer*, *47*(4), 1150–1158. doi:10.1016/j.polymer.2005.12.057

Li, Z., Lee, D., Sheng, X., Cohen, R. E., & Rubner, M. F. (2006). Two-level antibacterial coating with both release-killing and contact-killing capabilities. *Langmuir*, *22*(24), 9820–9823. doi:10.1021/la0622166 PMID:17106967

Lizandara-Pueyo, C., van denBerg, M.W.E., DeToni, A., Goes, T., & Polarz, S. (2008). Nucleation and growth of ZnO in Organic Solvents- an in Situ Study. *Journal of the American Chemical Society*, *130*(49), 16601–16610. doi:10.1021/ja804071h PMID:19554730

Lizandara-Pueyo, C., Dilger, S., Wagner, M. R., Gerigk, M., Hoffmann, A., & Polarz, S. (2014). Li-doped ZnO nanorods with single-crystal quality - non-classical crystallization and self-assembly into mesoporous materials. *CrystEngComm*, *16*(8), 1525–1531. doi:10.1039/C3CE41670D

Lizandara-Pueyo, C., Morant-Miñana, M. C., Wessig, M., Krumm, M., Mecking, S., & Polarz, S. (2012). Biomomietic crystallization of anisotropic zinc oxide nanoparticles in the homogeneous phase: Shape control by surface additives applied under thermodynamic or kinetic control. *RSC Advances*, *2*(12), 5298. doi:10.1039/c2ra20343j

Lizandara-Pueyo, C., Siroky, S., Wagner, M. R., Hoffmann, A., Reparaz, J. S., Lehmann, M., & Polarz, S. (2011). Shape Anisotropy Influencing Functional Properties: Trigonal Prismatic ZnO Nanoparticles as an Example. *Advanced Functional Materials*, *21*(2), 295–304. doi:10.1002/adfm.201000997

Lo, C.-W., Zhu, D., & Jiang, H. (2011). An infrared-light responsive graphene-oxide incorporated poly(N-isopropylacrylamide) hydrogel nanocomposite. *Soft Matter*, *7*(12), 5604–5609. doi:10.1039/c1sm00011j

Lodewijks, K., Van Roy, W., Borghs, G., Lagae, L., & Van Dorpe, P. (2012). Boosting the figure of merit of LSPR-based refractive index sensing by phase-sensitive measurements. *Nano Letters*, *12*(3), 1655–1659. doi:10.1021/nl300044a PMID:22356465

Lombardi, A. V. Jr, Berend, K. R., & Mallory, T. H. (2006). Hydroxyapatite-coated titanium porous plasma spray tapered stem - Experience at 15 to 18 years. *Clinical Orthopaedics and Related Research*, *453*, 81–85. doi:10.1097/01.blo.0000238872.01767.09 PMID:17312587

Lomoelder, R., Görlitzer, H., Hallack, M., & Unkelhäußer, T. (2013). Beschichtungsmittel mit hoher Kratzbeständigkeit. *Patent EP 2676982 A1*.

Look, D. C., & Claflin, B. (2004). P-type doping and devices based on ZnO.[b]. *Physica Status Solidi*, *241*(3), 624–630. doi:10.1002/pssb.200304271

Look, D. C., Reynolds, D. C., Litton, C. W., Jones, R. L., Eason, D. B., & Cantwell, G. (2002). Characterization of homoepitaxial p-type ZnO grown by mononuclear beam epitaxy. *Applied Physics Letters*, *81*(10), 1830. doi:10.1063/1.1504875

López, E. O., Mello, A., Sendão, H., Costa, L. T., Rossi, A. L., Ospina, R. O., & Rossi, A. M. et al. (2013). Growth of crystalline hydroxyapatite thin films at room temperature by tuning the energy of the RF-magnetron sputtering plasma. *ACS Applied Materials & Interfaces*, *5*(19), 9435–9445. doi:10.1021/am4020007 PMID:24059686

López, E. O., Rossi, A. L., Archanjo, B. S., Ospina, R. O., Mello, A., & Rossi, A. M. (2015). Crystalline nano-coatings of fluorine-substituted hydroxyapatite produced by magnetron sputtering with high plasma confinement. *Surface and Coatings Technology*, *264*, 163–174.

Löschner, T., Mehner, A., Grund, S., Seifert, A., Pohlers, A., Lange, A. ... Spange, S. (2012). A modular approach for the synthesis of nanostructured hybrid materials with tailored properties: The simultaneous twin polymerization. Angewandte Chememie International Edition, 51, 3258–3261.

Louis-Achille, V., De Windt, L., & Defranceschi, M. (2000). Electronic structure of minerals: The apatite group as a relevant example. *International Journal of Quantum Chemistry*, *77*(6), 991–1006. doi:10.1002/(SICI)1097-461X(2000)77:6<991::AID-QUA7>3.0.CO;2-Z

Lou, L., MacDonough, G., Walde, H., Carter, D., Roche, G., & Scholl, R. (1999). *Closed loop controlled reactive dual magnetron sputtering*. Advanced Energy Insudries, Inc.

Love, J. C., Estroff, L. A., Kriebel, J. K., Nuzzo, R. G., & Whitesides, G. M. (2005). Self-assembled monolayers of thiolates on metals as a form of nanotechnology. *Chemical Reviews*, *105*(4), 1103–1169. doi:10.1021/cr0300789 PMID:15826011

Lu, G., Li, Y.-M., Lu, C.-H., & Xu, Z.-Z. (2010). Corrosion protection of iron surface modified by poly(methyl methacrylate) using surface-initiated atom transfer radical polymerization (SI-ATRP). *Colloid & Polymer Science*, *288*(14-15), 1445–1455. doi:10.1007/s00396-010-2283-x

Luk'yanchuk, B., Zheludev, N. I., Maier, S. A., Halas, N. J., Nordlander, P., Giessen, H., & Chong, C. T. (2010). The Fano resonance in plasmonic nanostructures and metamaterials. *Nature Materials*, *9*(9), 707–715. doi:10.1038/nmat2810 PMID:20733610

Luong, L. N., McFalls, K. M., & Kohn, D. H. (2009). Gene delivery via DNA incorporation within a biomimetic apatite coating. *Biomaterials*, *30*(36), 6996–7004. doi:10.1016/j.biomaterials.2009.09.001 PMID:19775750

Lu, X., Wang, Z., Yang, X., Xu, X., Zhang, L., Zhao, N., & Xu, J. (2011). Antifogging and antireflective silica film and its application on solar modules. *Surface and Coatings Technology*, *206*(6), 1490–1494. doi:10.1016/j.surfcoat.2011.09.031

Lu, Y. C., Ives, M. B., & Clayton, C. R. (1993). Synergism of alloying elements and pitting corrosion resistance of stainless steels. *Corrosion Science*, *35*(1-4), 89–96. doi:10.1016/0010-938X(93)90137-6

Lu, Y., Guan, S., Hao, L., & Yoshida, H. (2015). Review on the Photocatalyst Coatings of TiO2: Fabrication by Mechanical Coating Technique and Its Application. *Coatings*, *5*(3), 425–464. doi:10.3390/coatings5030425

Lu, Y., Sathasivan, S., Song, J., Crick, C. R., Carmalt, C. J., & Parklin, I. P. (2015). Robust self-cleaning surfaces that function when exposed to either air or oil. *Science*, *347*(6226), 1132–1135. doi:10.1126/science.aaa0946 PMID:25745169

Luzinov, I., Minko, S., & Tsukruk, V. V. (2008). Responsive brush layers: From tailored gradients to reversibly assembled nanoparticles. *Soft Matter*, *4*(4), 714–725. doi:10.1039/b718999k

Lv, P., Xue, Y., Shi, Y., Lin, H., & Duan, H. (2014). Metastable States and Wetting Transition of Submerged Superhydrophobic Structures. *PRL*, *112*(19), 196101. doi:10.1103/PhysRevLett.112.196101 PMID:24877948

Lygeraki, M. I., Tsiranidou, E., Anastasiadis, S. H., Fotakis, C., Pisignano, D., Cingolani, R., & Athanassiou, A. (2008). Controlling the reversible wetting capability of smart photochromic-polymer surfaces by micro patterning. *Applied Physics. A, Materials Science & Processing*, *91*(3), 397–401. doi:10.1007/s00339-008-4473-7

Maby, M. (2000, July). Moly does the job in nuclear power applications. *Newsletter, International Molybdenum Association*. London, UK.

Macaskie, L. E., Yong, P., Paterson-Beedle, M., Thackray, A. C., Marquis, P. M., Sammons, R. L., & Hall, L. D. et al. (2005). A novel non line-of-sight method for coating hydroxyapatite onto the surfaces of support materials by biomineralization. *Journal of Biotechnology*, *118*(2), 187–200. doi:10.1016/j.jbiotec.2005.03.006 PMID:15964651

Macleod, H. (1986). *Thin film optical filters*. New York: McGraw-Hill. doi:10.1887/0750306882

Madhankumar, A., Thangavel, E., Ramakrishna, S., Obot, I.B., Jung, H.C., Shin, K S., Kim, D.-E. (2014). Multi-functional ceramic hybrid coatings on biodegradable AZ31 Mg implants: electrochemical, tribological and quantum chemical aspects for orthopaedic applications. *RSC Advances, 4*(46), 24272.

Mahmood, A., Machorro, R., Muhl, S., Heiras, J., Castillon, F., Farıas, M., & Andrade, E. (2003). Optical and surface analysis of DC-reactive sputtered AlN films. *Diamond and Related Materials*, *12*(8), 1315–1321. doi:10.1016/S0925-9635(03)00076-1

Maidenberg, Y., Zhang, S., Luo, K., Akhavein, N., & Koberstein, J. T. (2013). Mixed silane monolayers for controlling the surface area density of click-reactive alkyne groups: A method to assess preferential surface adsorption on flat substrates and a method to verify compositional homogeneity on nanoparticle. *Langmuir*, *29*(38), 11959–11965. doi:10.1021/la402517m PMID:23985021

Maity, A. R., Palmal, S., Basiruddin, S. K., Karan, N. S., Sarkar, S., Pradhan, N., & Jana, N. R. (2013). Doped semiconductor nanocrystal based fluorescent cellular imaging probes. *Nanoscale*, *5*(12), 5506–5513. doi:10.1039/c3nr00549f PMID:23674276

Makhlouf, A. S. H. (2014). Handbook of Smart Coatings for Materials Protection. Cambridge, UK: Woodhead Publishing. doi:10.1533/9780857096883

Makhlouf, A. S. H., & Tiginyanu, I. (2011). *Nanocoatings and ultra-thin films*. Woodhead Publishing Limited doi:10.1533/9780857094902

Malik, M. A., Afzaal, M., & O'Brien, P. (2010). Precursor Chemistry for Main Group Elements in Semiconducting Materials. *Chemical Reviews*, *110*(7), 4417–4446. doi:10.1021/cr900406f PMID:20481563

Malinkiewicz, O., Roldán-Carmona, C., Soriano, A., Bandiello, E., Camacho, L., Nazeeruddin, M. K., & Bolink, H. J. (2014). Metal-Oxide-Free Methylammonium Lead Iodide Perovskite-Based Solar Cells: the Influence of Organic Charge Transport Layers. Advanced Energy Materials, 4(15).

Malinkiewicz, O., Yella, A., Lee, Y. H., Espallargas, G. M., Graetzel, M., Nazeeruddin, M. K., & Bolink, H. J. (2014). Perovskite solar cells employing organic charge-transport layers. *Nature Photonics*, *8*(2), 128–132. doi:10.1038/nphoton.2013.341

Ma, M., & Hill, R. M. (2006). Superhydrophobic Surfaces. *Current Opinion in Colloid & Interface Science*, *11*(4), 193–202. doi:10.1016/j.cocis.2006.06.002

Manara, S., Paolucci, F., Palazzo, B., Marcaccio, M., Foresti, E., Tosi, G., & Roveri, N. et al. (2008). Electrochemically-assisted deposition of biomimetic hydroxyapatite-collagen coatings on titanium plate. *Inorganica Chimica Acta*, *361*(6), 1634–1645. doi:10.1016/j.ica.2007.03.044

Mandal, S., Singha, R. K., Dhar, A., & Ray, S. K. (2007). Optical and structural characteristics of ZnO thin films grown by rf magnetron sputtering. *Materials Research Bulletin*, *43*(2), 244–250. doi:10.1016/j.materresbull.2007.05.006

Manickam, G., Gandhiraman, R., Vijayaraghavan, R. K., Kerr, L., Doyle, C., Williams, D. E., & Daniels, S. (2012). Protection and functionalisation of silver as an optical sensing platform for highly sensitive SPR based analysis. *Analyst (London)*, *137*(22), 5265–5271. doi:10.1039/c2an35826c PMID:23037590

Mann, S., Archibald, D. D., Didymus, J. M., Douglas, T., Heywood, B. R., Meldrum, F. C., & Reeves, N. J. (1993). Crystallization at inorganic-organic interfaces - biomaterials and biomimetic synthesis. *Science*, *261*(5126), 1286–1292. doi:10.1126/science.261.5126.1286 PMID:17731856

Marceaux, S., Bressy, C., Perrin, F.-X., Martin, C., & Margaillan, A. (2014). Development of polyorganosilazane-silicone marine coatings. *Progress in Organic Coatings*, *77*(11), 1919–1928. doi:10.1016/j.porgcoat.2014.06.020

Marchand, P., & Carmalt, C. J. (2013). Molecular precursor approach to metal oxide and pnictide thin films. *Coordination Chemistry Reviews*, *257*(23-24), 3202–3221. doi:10.1016/j.ccr.2013.01.030

Marlot, A., Kern, P., & Landolt, D. (2002). Pulse plating of Ni-Mo alloys from Ni-rich electrolytes. *Electrochimica Acta*, *48*(1), 29–36. doi:10.1016/S0013-4686(02)00544-3

Marmur, A. (2006). Super-hydrophobicity fundamentals: Implications to biofouling prevention. *Biofouling, 22*, 107-115. doi:10.1080/08927010600562328

Martin, T. P., Lau, K. K. S., Chan, K., Mao, Y., Gupta, M., Shannan O'Shaughnessy, W., & Gleason, K. K. (2007). Initiated chemical vapor deposition (iCVD) of polymeric nanocoatings. *Surface and Coatings Technology, 201*(22-23 SPEC. ISS.), 9400–9405.

Martinez, A. W., Phillips, S. T., Whitesides, G. M., & Carrilho, E. (2010). Diagnostics for the Developing World: Microfluidic Paper-Based Analytical Devices. *Analytical Chemistry*, *82*(1), 3–10. doi:10.1021/ac9013989 PMID:20000334

Martinez, S., Metikoš-Huković, M., & Valek, L. (2006). Electrocatalytic properties of electrodeposited Ni-15Mo cathodes for the HER in acid solutions: Synergistic electronic effect. *Journal of Molecular Catalysis A Chemical*, *245*(1-2), 114–121. doi:10.1016/j.molcata.2005.09.040

Maruska, H. P., & Tietjen, J. (1969). The preparation and properties of Vapor-Deposited single-crystal-line GaN. *Applied Physics Letters*, *15*(10), 327–329. doi:10.1063/1.1652845

Masse, A., Bruno, A., Bosetti, M., Biasibetti, A., Cannas, M., & Gallinaro, P. (2000). Prevention of pin track infection in external fixation with silver coated pins: Clinical and microbiological results. *Journal of Biomedical Materials Research*, *53*(5), 600–604. doi:10.1002/1097-4636(200009)53:5<600::AID-JBM21>3.0.CO;2-D PMID:10984710

Matchett, M. A., Chiang, M. Y., & Buhro, W. E. (1990). Soluble and volatile alkoxides of bismuth. The first structurally characterized bismuth trialkoxide: [Bi(.mu.-.eta.1-OCH2CH2OMe)2(.eta.1-OCH2OMe)].infin. *Inorganic Chemistry*, *29*(3), 358-360.

Mathews, T. C. S. N. (2014). Advancements in perovskite solar cells: Photophysics behind the photovoltaics. *Energy & Environmental Science*, *7*(8), 2518–2534. doi:10.1039/C4EE00673A

Mathiazhagan, A., & Joseph, R. (2011). Nanotechnology-A New Prospective in Organic Coating. *International Journal of Chemical Engineering and Applications*, *2*(4), 228–237.

Mathur, S., Shen, H., & Altmayer, J. (2007). Nanostructured Functional Ceramic Coatings By Molecule-Based Chemical Vapor Deposition. *Review of Advanced Materials Science*, *15*, 16–23.

Mattoussi, H., Cumming, A. W., Murray, C. B., Bawendi, M. G., & Ober, R. (1996). Characterization of CdSe nanocrystallite dispersions by small angle x-ray scattering. *The Journal of Chemical Physics*, *105*(22), 9890–9896. doi:10.1063/1.472854

Maurel, V., Laferrière, M., Billone, P., Godin, R., & Scaiano, J. C. (2006). Free Radical Sensor Based on CdSe Quantum Dots with Added 4-Amino-2,2,6,6-Tetramethylpiperidine Oxide Functionality. *The Journal of Physical Chemistry B*, *110*(33), 16353–16358. doi:10.1021/jp061115d PMID:16913763

Mayer, M. (2003). RBS – Rutherford Backscattering Spectrometry.*Workshop on nuclear data for science and technology: materials analysis.*

Mazuri, M., Woicieszak, D., Domaradk, J., Kaczmarek, D., Song, S., & Placido, F. (n. d.). TiO2/SiO2 multilayer as an antireflective and protective coating deposited by microwave assisted magnetron sputtering. *Versita Opto-Electronics review 21 (2)*, 233 - 238.

McDowell, M. T., Woo Lee, S., Wang, C., & Cui, Y. (2012). The effect of metallic coatings and crystallinity on the volume expansion of silicon during electrochemical lithiation/delithiation. *Nano Energy*, *1*(3), 401–410. doi:10.1016/j.nanoen.2012.03.004

McDowell, M., Wright, A. E., & Hammer, N. I. (2010). Semiconductor Nanocrystals Hybridized with Functional Ligands: New Composite Materials with Tunable Properties. *Materials (Basel)*, *3*(1), 614–637. doi:10.3390/ma3010614

Medintz, I. L., Clapp, A. R., Mattoussi, H., Goldman, E. R., Fisher, B., & Mauro, J. M. (2003). Self-assembled nanoscale biosensors based on quantum dot FRET donors. *Nature Materials*, *2*(9), 630–638. doi:10.1038/nmat961 PMID:12942071

Mehring, M. (2007). From molecules to bismuth oxide-based materials: Potential homo- and heterometallic precursors and model compounds. *Coordination Chemistry Reviews*, *251*(7-8), 974–1006. doi:10.1016/j.ccr.2006.06.005

Mehrotra, R. C. (1953). Aluminium alkoxides. *Journal of the Indian Chemical Society*, *30*, 585–591.

Mehrotra, R. C. (1954). *Journal of the Indian Chemical Society*, *31*, 904–910.

Melde, B. J., & Stein, A. (2002). Periodic macroporous hydroxyapatite-containing calcium phosphates. *Chemistry of Materials*, *14*(8), 3326–3331. doi:10.1021/cm020259b

Mello, A., Hong, Z., Rossi, A., Luan, L., Farina, M., Querido, W., & Webster, T. et al. (2007). Osteoblast proliferation on hydroxyapatite thin coatings produced by right angle magnetron sputtering. *Biomedical Materials (Bristol, England)*, *2*(2), 67–77. doi:10.1088/1748-6041/2/2/003 PMID:18458438

Mellor, B. G. (2006). *Surface Coatings for Protection against Wear*. Boca Raton, FL: Woodhead Publishing.

Menju, A., Hayashi, K., & Irie, M. (1981). Photoresponsive polymers. 3. Reversible solution viscosity change of poly(methacrylic acid) having spirobenzopy- ran pendant groups in methano. *Macromolecules*, *14*(3), 755–758. doi:10.1021/ma50004a055

Meyer, B. K., Alves, H., Hofmann, D. M., Kriegseis, W., Forster, D., & Bertram, F. et al.. (2004). Idntification of bound exciton complexes in ZnO. *Physica Status Solidi. B, Basic Research*, *241*, 231.

Miao, S., Cheng, K., Weng, W., Du, P., Shen, G., Han, G., & Zhang, S. et al. (2008). Fabrication and evaluation of Zn containing fluoridated hydroxyapatite layer with Zn release ability. *Acta Biomaterialia*, *4*(2), 441–446. doi:10.1016/j.actbio.2007.08.013 PMID:17942380

Miao, S., Weng, W., Li, Z., Cheng, K., Du, P., Shen, G., & Han, G. (2009). Electrolytic deposition of octacalcium phosphate/collagen composite coating on titanium alloy. *Journal of Materials Science. Materials in Medicine*, *20*(1), 131–134. doi:10.1007/s10856-008-3552-9 PMID:18704651

Michel, C., Hervieu, M., Borel, M. M., Grandin, A., Deslandes, F., Provost, J., & Raveau, B. (1987). Superconductivity in the Bi-Sr-Cu-O system. *Zeitschrift für Physik B, Condensed Matter*, *68*(4), 421–423. doi:10.1007/BF01471071

Michel, M., Toniazzo, V., Ruch, D., & Ball, V. (2012). Deposition Mechanisms in Layer-by-Layer or Step-by-Step Deposition Methods: From Elastic and Impermeable Films to Soft Membranes with Ion Exchange Properties. *ISRN Materials Science*, *2012*, 1–13. doi:10.5402/2012/701695

Miller, R. A. (1997). Thermal Barrier Coatings for Aircraft Engines: History and Directions. *Journal of Journal of Thermal Spray Technology*, *6*(1), 35–42. doi:10.1007/BF02646310

Minkin, V. I. (2004). Photo-, thermo-, solvato-, and electrochromic spiroheterocyclic compounds. *Chemical Reviews*, *104*(5), 2751–2776. doi:10.1021/cr020088u PMID:15137806

Miroiu, F. M., Socol, G., Visan, A., Stefan, N., Craciun, D., Craciun, V., & Ducu, C. et al. (2010). Composite biocompatible hydroxyapatite–silk fibroin coatings for medical implants obtained by Matrix Assisted Pulsed Laser Evaporation. *Materials Science and Engineering B*, *169*(1–3), 151–158. doi:10.1016/j.mseb.2009.10.004

Mishra, A. K., & Shoesmith, D. W. (2013). The activation/depassivation of nickel–chromium–molybdenum alloys: An oxyanion or a pH effect—Part II. *Electrochimica Acta*, *102*, 328–335. doi:10.1016/j.electacta.2013.03.177

Mishra, R., & Balasubramaniam, R. (2004a). Effect of nanocrystalline grain size on the electrochemical and corrosion behavior of nickel. *Corrosion Science*, *46*(12), 3019–3029. doi:10.1016/j.corsci.2004.04.007

Mishra, R., Basu, B., & Balasubramaniam, R. (2004b). Effect of grain size on the tribological behavior of nanocrystalline nickel. *Materials Science and Engineering A*, *373*(1-2), 370–373. doi:10.1016/j.msea.2003.09.107

Mitzi, D. B. (2004). Solution-processed inorganic semiconductors. *Journal of Materials Chemistry*, *14*(15), 2355–2365. doi:10.1039/b403482a

Mitzi, D. B., Chondroudis, K., & Kagan, C. R. (2001). Organic-inorganic electronics. *IBM Journal of Research and Development*, *45*(1), 29–45. doi:10.1147/rd.451.0029

Mitzi, D. B., Feild, C. A., Harrison, W. T. A., & Guloy, A. M. (1994). Conducting tin halides with a layered organic-based perovskite structure. *Nature*, *369*(6480), 467–469. doi:10.1038/369467a0

Mo, C., Zheng, Y., Wang, F., & Mo, Q. (2015). A Simple Process for Fabricating Organic / TiO 2 Super- Hydrophobic and Anti-Corrosion Coating, *Int. J. Electrochem. Sci.* 10, 7380–7391.

Mock, J. J., Hill, R. T., Tsai, Y. J., Chilkoti, A., & Smith, D. R. (2012). Probing dynamically tunable localized surface plasmon resonances of film-coupled nanoparticles by evanescent wave excitation. *Nano Letters*, *12*(4), 1757–1764. doi:10.1021/nl204596h PMID:22429053

Mohamed, A. M. A., Abdullah, A. M., & Younan, N. A. (2014). Corrosion behavior of superhydrophobic surfaces: A review. *Arabian Journal of Chemistry*. doi:10.1016/j.arabjc.2014.03.006

Monkiewicz, J., Edelmann, R., Edinger, A., & Borup, B. (2006). Air-drying silane coating compositions. *Patent US 7026398 B2*.

Monroy, E., Omnes, F., & Calle, F. (2003). Wide-bandgap semiconductor ultraviolet photodetectors. *Semiconductor Science and Technology*, *18*(4), R33–R51. doi:10.1088/0268-1242/18/4/201

Montazeri, L., Javadpour, J., Shokrgozar, M., Bonakdar, S., Moghaddam, M. K., & Asgary, V. (2013). The interaction of plasma proteins with nano-size fluoride-substituted apatite powders. *Ceramics International*, *39*(6), 6145–6152. doi:10.1016/j.ceramint.2013.01.033

Moorthy, J., Khoury, C., Moore, J. S., & Beebe, D. J. (2001). Active control of electroosmotic flow in microchannels using light. *Sensors and Actuators. B, Chemical*, *75*(3), 223–229. doi:10.1016/S0925-4005(01)00557-3

Morales-Masis, M., Ding, L., Dauzou, F., Jeangros, Q., Hessler-Wyser, A., Nicolay, S., & Ballif, C. et al. (2014). Hydrogen plasma treatment for improved conductivity in amorphous aluminum doped zinc tin oxide thin films. *APL Materials*, *2*(9), 096113. doi:10.1063/1.4896051

Morant-Miñana, M. C., & Elizalde, J. (2015). Microscale electrodes integrated on COP for real sample Campylobacter spp. detection. Biosensors & Bioelectronics, 70, 491–497. doi:10.1016/j.bios.2015.03.063 PubMed doi:10.1016/j.bios.2015.03.063 PMID:25889259

Morant-Miñana, M. C., & Korthals, B. (2011). Ultrathin films of variable polarity and crystallinity obtained from 1,2-polybutadiene nanoparticle dispersions. Langmuir, 27(12), 7516–7523. doi:10.1021/la200334m PMID:21591698

Morgan, H., Wilson, R., Elliott, J., Dowker, S., & Anderson, P. (2000). Preparation and characterisation of monoclinic hydroxyapatite and its precipitated carbonate apatite intermediate. *Biomaterials*, *21*(6), 617–627. doi:10.1016/S0142-9612(99)00225-2 PMID:10701462

Morkoc, H. (2013). *Nitride semiconductors and devices* (Vol. 32). Springer Science & Business Media. doi:10.1002/9783527649006

Morra, M., Cassinelli, C., Cascardo, G., Mazzucco, L., Borzini, P., Fini, M., & Giardino, R. et al. (2006). Collagen I-coated titanium surfaces: Mesenchymal cell adhesion and in vivo evaluation in trabecular bone implants. *Journal of Biomedical Materials Research. Part A*, *78A*(3), 449–458. doi:10.1002/jbm.a.30783 PMID:16721758

Morra, M., Cassinelli, C., Meda, L., Fini, M., Giavaresi, G., & Giardino, R. (2005). Surface analysis and effects on interfacial bone microhardness of collagen-coated titanium implants: A rabbit model. *The International Journal of Oral & Maxillofacial Implants*, *20*(1), 23–30. PMID:15747670

Movchan, B. A. (2002). Functionally graded EB PVD coatings. *Surface and Coatings Technology*, *149*(2-3), 252–262. doi:10.1016/S0257-8972(01)01439-6

Mróz, W., Jedyński, M., Prokopiuk, A., Ślósarczyk, A., & Paszkiewicz, Z. (2009). Characterization of calcium phosphate coatings doped with Mg, deposited by pulsed laser deposition technique using ArF excimer laser. *Micron (Oxford, England)*, *40*(1), 140–142. doi:10.1016/j.micron.2008.01.016 PMID:18407507

Müller, J., Rech, R., Springer, J., & Vanecek, M. (2004). TCO and light trapping in silicon thin film solar cells. *Solar Energy*, *77*(6), 917–930. doi:10.1016/j.solener.2004.03.015

Munikata, S., Yamaauchi, M., Hirono, T., Kodama, S.-I., & Matsuo, M. (2001). Towering developments. *Asia Pacific coatings Journal*, 14(6), pp. 8, 10, 14.

Murray, C. B., Norris, D. J., & Bawendi, M. G. (1993). Synthesis and characterization of nearly monodisperse CdE (E = sulfur, selenium, tellurium) semiconductor nanocrystallites. *Journal of the American Chemical Society*, *115*(19), 8706–8715. doi:10.1021/ja00072a025

Muthu, C., Nagamma, S. R., & Nair, V. C. (2014). Luminescent hybrid perovskite nanoparticles as a new platform for selective detection of 2,4,6-trinitrophenol. *RSC Advances*, *4*(99), 55908–55911. doi:10.1039/C4RA07884E

Nablo, B. J., Rothrock, A. R., & Schoenfisch, M. H. (2005). Nitric oxide-releasing sol–gels as antibacterial coatings for orthopedic implants. *Biomaterials*, *26*(8), 917–924. doi:10.1016/j.biomaterials.2004.03.031 PMID:15353203

Nanocoatings Market by Type, by Application, and by Region - Trends & Forecasts to 2019. (2015). Research and Markets.

Narayanan, R., Seshadri, S., Kwon, T., & Kim, K. (2008). Calcium phosphate-based coatings on titanium and its alloys. *Journal of Biomedical Materials Research. Part B, Applied Biomaterials*, *85*(1), 279–299. doi:10.1002/jbm.b.30932 PMID:17853421

Narita, M., Kasuga, T., & Kiyotani, A. (2000). Techniques for Surface Treatment of Light Metals. Super water-repellent aluminum by electrolytic etching and chemical adsorption. *Keikinzoku*, *50*(11), 594–597. doi:10.2464/jilm.50.594

Nauth, A., Ristevski, B., Li, R., & Schemitsch, E. H. (2011). Growth factors and bone regeneration: How much bone can we expect? *Injury-International Journal of the Care of the Injured*, *42*(6), 574–579. doi:10.1016/j.injury.2011.03.034 PMID:21489530

Near, R. D., Hayden, S. C., Hunter, R. E. Jr, Thackston, D., & El-Sayed, M. A. (2013). Rapid and efficient prediction of optical extinction coefficients for gold nanospheres and gold nanorods. *The Journal of Physical Chemistry C*, *117*(45), 23950–23955. doi:10.1021/jp4082596

Nee, C. C., Kim, W., & Weil, R. (1988). Pulsed electrodeposition of Ni-Mo alloys. *Journal of the Electrochemical Society*, *135*(5), 1100–1103. doi:10.1149/1.2095883

Negahdari, Z., Porda, M., & Scherm, F. (2010). Development of Novel Functionally Graded $Al_2 O_3$ -Lanthanum Hexaaluminate Ceramics for Thermal Barrier Coatings Zahra Negahdari. *Materials Science Forum*, *631-632*, 97–102. doi:10.4028/www.scientific.net/MSF.631-632.97

Nejati, M., Rahimipour, M. R., & Mobasherpour, I. (2014). Evaluation of hot corrosion behavior of CSZ, CSZ / micro Al_2O_3 and CSZ/nano Al_2O_3 plasma sprayed thermal barrier coatings. *Ceramics International*, *40*(3), 4579–4590. doi:10.1016/j.ceramint.2013.08.135

Nelea, V., Morosanu, C., Iliescu, M., & Mihailescu, I. (2003). Microstructure and mechanical properties of hydroxyapatite thin films grown by RF magnetron sputtering. *Surface and Coatings Technology*, *173*(2), 315–322. doi:10.1016/S0257-8972(03)00729-1

Neumüller, B. (2003). Organometallic sesquialkoxides of aluminium, gallium and indium. *Chemical Society Reviews*, *32*, 50–55. doi:10.1039/b205568f PMID:12596545

Newman, R. C. (1985a). The dissolution and passivation kinetics of stainless alloys containing molybdenum—1. Coulometric studies of FeCr and FeCrMo alloys. *Corrosion Science*, *25*(5), 331–339. doi:10.1016/0010-938X(85)90111-8

Newman, R. C. (1985b). The dissolution and passivation kinetics of stainless alloys containing molybdenum—II. Dissolution kinetics in artificial pits. *Corrosion Science*, *25*(5), 341–350. doi:10.1016/0010-938X(85)90112-X

News (2011). Heat-Sealable ESD Film Protects Sensitive Spacecraft Components during Construction. *Sealing Technology*, *2011*(11), 2-3. doi:10.1016/S1350-4789(12)70296-9

Nguyen, T. D. H. (2011). *Revêtements polysilazane a activités antibactériennes*. (PhD thesis). Université de Toulon, La Garde, France. Retrieved from tel.archives-ouvertes.fr/tel-00769942/document

Nicoletta, F. P., Cupelli, D., Formoso, P., De Filpo, G., Colella, V., & Gugliuzza, A. (2012). Light Responsive Polymer Membranes: A Review. *Membranes*, *2*(1), 134–197. doi:10.3390/membranes2010134 PMID:24957966

Niinistö, J., & Niinistö, L. (2012). Introduction: Basic Features and Historical Development of Atomic Layer Deposition. Atomic Layer Deposition of Nanostructured Materials. Wiley-VCH;

Nimittrakoolchai, O. U., & Supothina, S. (2008). Deposition of organic-based superhydrophobic films for anti-adhesion and self-cleaning applications. *J. Eur. Ceram. Soc.*, *28*, 947-952. doi:10.1016/j.jeurceramsoc.2007.09.025

Nishimoto, S., & Bhushan, B. (2013). Bioinspired self-cleaning surfaces with superhydrophobicity, superoleophobicity, and superhydrophilicity. *RSC Advances*, *3*(3), 671–690. doi:10.1039/C2RA21260A

Niu, G., Guo, X., & Wang, L. (2015). Review of recent progress in chemical stability of perovskite solar cells. *Journal of Materials Chemistry A*, *3*(17), 8970–8980. doi:10.1039/C4TA04994B

Niu, G., Wang, L., Gao, R., Li, W., Guo, X., Dong, H., & Qiu, Y. (2013). Inorganic halogen ligands in quantum dots: I-, Br-, Cl- and film fabrication through electrophoretic deposition. *Physical Chemistry Chemical Physics*, *15*(45), 19595–19600. doi:10.1039/c3cp52678j PMID:23958930

Niu, S., Zhang, L., Wang, N., Zhu, J., Zhang, W., Cheng, Z., & Zhu, X. (2013). Fabrication of magnetic nanofibers via a surface-initiated RAFT polymerization and coaxial electrospinning. *Reactive & Functional Polymers*, *73*(11), 1447–1454. doi:10.1016/j.reactfunctpolym.2013.07.011

Niu, Z.-J., Yao, S.-B., & Zhou, S.-M. (1998). In situ surface Raman investigation on induced-codeposition of an Fe-Mo alloy. *Journal of Electroanalytical Chemistry*, *455*(1-2), 205–207. doi:10.1016/S0022-0728(98)00148-X

Niu, Z., & Li, Y. (2014). Removal and Utilization of Capping Agents in Nanocatalysis. *Chemistry of Materials*, *26*(1), 72–83. doi:10.1021/cm4022479

Noguez, C. (2007). Surface plasmons on metal nanoparticles: The influence of shape and physical environment. *Journal of Physical Chemistry*, *111*(10), 3806–3819. doi:10.1021/jp066539m

Noomen, A. (1997). Applications of Michael addition chemistry in coatings technology. *Progress in Organic Coatings*, *32*(1-4), 137–142. doi:10.1016/S0300-9440(97)00070-2

Nordlander, P., Oubre, C., Prodan, E., Li, K., & Stockman, M. I. (2004). Plasmon hybridization in nanoparticle dimers. *Nano Letters*, *4*(5), 899–903. doi:10.1021/nl049681c

Nosonovsky, M., & Bhushan, B. (2009). Superhydrophobic surfaces and emerging applications: Non-adhesion, energy, green engineering. *Current Opinion in Colloid & Interface Science*, *14*(4), 270–280. doi:10.1016/j.cocis.2009.05.004

Novak, B. M., & Davies, C. (1991). "Inverse" organic-inorganic composite materials. 2. Free-radical routes into non-shrinking sol-gel composition. *Macromolecules*, *24*(19), 5481–5483. doi:10.1021/ma00019a044

Nuernberg, A., Stolle, R., Wahl, G., & Raic, K. T. (2003). Plasma activated CVD of Co-C-O-layers. deposition rate phenomena. *Advances in Science and Technology*, 32, 91-98.

Nunes, P., Fortunato, E., Tonello, P., Braz Fernandes, F., Vilarinho, P., & Martins, R. (2002). Effect of different dopant elements on the properties of ZnO thin films. *Vacuum*, *64*(3-4), 281–285. doi:10.1016/S0042-207X(01)00322-0

Nur, S., Jamaludin, S., Mustapha, F., & Nuruzzaman, D. M. (2013). A review on the fabrication techniques of functionally graded ceramic-metallic materials in advanced composites. *Scientific Research and Essays*, *8*, 828–840.

O'Rourke-Muisener, P. A. V., Koberstein, J. T., & Kumar, S. (2003). Optimal chain architectures for the molecular design of functional polymer surfaces. *Macromolecules*, *36*(3), 771–781. doi:10.1021/ma011969q

Oertli, M., & Jucker, B. (2013). Moisture-curing compositons comprising at least two polymers having silane groups. *Patent US 8372514 B2*.

Ohgai, T., Tanaka, Y., & Washio, R. (2013). Nanocrystalline structure and soft magnetic properties of nickel-molybdenum alloy thin films electrodeposited from acidic and alkaline aqueous solutions. *Journal of Solid State Electrochemistry*, *17*(3), 743–750. doi:10.1007/s10008-012-1924-z

Oh, K. W., & Ahn, C. H. (2006). A review of microvalves. *Journal of Micromechanics and Microengineering*, *16*(5), R13–R39. doi:10.1088/0960-1317/16/5/R01

Ohring, M. (1992). *The material science of thin film*. Academic Press.

Okyay, T. O., Bala, R. K., Nguyen, H. N., Atalay, R., Bayam, Y., & Rodrigues, D. F. (2015). Antibacterial properties and mechanisms of toxicity of sonochemically grown ZnO nanorods. RSC Advances, 5(4), 2568–2575. doi:10.1039/C4RA12539H

Olakanmi, E. O., & Doyoyo, M. (2014). Laser-assisted cold-sprayed corrosion- and wear-resistant coatings: A review. Journal of Thermal Spray Technology, 23(5), 765–785. doi:10.1007/s11666-014-0098-x

Oliver, W. C., & Pharr, G. M. (1992). An improved technique for determining hardness and elastic modulus using load and displacement sensing indentation experiments. *Journal of Materials Research, 7*(6), 1564–1583. doi:10.1557/JMR.1992.1564

Olvera, M. L., Maldonado, A., Asomoza, R., & Meléndez-Lira, M. (2000). Chemical stability of doped ZnO thin Films. *Journal of Materials Science Materials in Electronics, 11*(1), 1–5. doi:10.1023/A:1008973800554

Öner, D., & McCarthy, T. J. (2000). Ultrahydrophobic Surfaces. Effects of Topography Length Scales on Wettability. *Langmuir, 16*(20), 7777–7782. doi:10.1021/la000598o

Orgassa, K., Schock, H. W., & Werner, J. H. (2003). Alternative back contact materials for thin film Cu(In,Ga)Se_2 solar cells. *Thin Solid Films, 431-432*, 387–391. doi:10.1016/S0040-6090(03)00257-8

Oroszi, Laszlo, Der, Andras, Kirei, Huba, Ormos, Pal, & Rakovics, Vilmos. (2006). Control of electro-osmostic flow by light. *Applied Physics Letters, 89*(26). doi:10.1063/1.2424648

Oroszi, L., Dér, A., Kirei, H., Rakovics, V., & Ormos, P. (2009). Manipulation of microfluidic flow pattern by optically controlled electroosmosis. *Microfluidics and Nanofluidics, 6*(4), 565–569. doi:10.1007/s10404-008-0346-9

Owen, J. S., Park, J., Trudeau, P.-E., & Alivisatos, A. P. (2008). Reaction Chemistry and Ligand Exchange at Cadmium−Selenide Nanocrystal Surfaces. *Journal of the American Chemical Society, 130*(37), 12279–12281. doi:10.1021/ja804414f PMID:18722426

Ozeki, K., Aoki, H., & Masuzawa, T. (2010). TEM Observation of Nano-Interface between Hydroxyapatite Sputtered Film and Titanium Substrate. *Journal of the Australasian Ceramic Society, 46*(2), 59–62.

Padgett, M., & Di Leonardo, R. (2011). Holographic optical tweezers and their relevance to lab on chip devices. *Lab on a Chip, 11*(7), 1196–1205. doi:10.1039/c0lc00526f PMID:21327211

Padilla, S., Roman, J., Carenas, A., & Vallet-Regı, M. (2005). The influence of the phosphorus content on the bioactivity of sol–gel glass ceramics. *Biomaterials, 26*(5), 475–483. doi:10.1016/j.biomaterials.2004.02.054 PMID:15276355

Paduano, Q., Weyburne, D., Jasinski, J., & Liliental-Weber, Z. (2004). Effect of initial process conditions on the structural properties of AlN films. *Journal of Crystal Growth, 261*(2), 259–265. doi:10.1016/j.jcrysgro.2003.11.017

Pakseresht, A. H., Simchi, A., Bahrami, M., & Javadi, A. H. (2014). Application of spark plasma sintering for fabrication of functionally graded thermal barrier coating on a superalloy substrate. *Paper presented atConference smart materials and surfaces*, Bangkok, Thailand.

Pakseresht, A. H., Ghasali, E., Nejati, M., Shirvanimoghaddam, K., Javadi, A. H., & Teimouri, R. (2015). Development empirical-intelligent relationship between plasma spray parameters and coating performance of Yttria-Stabilized Zirconia. *International Journal of Advanced Manufacturing Technology, 76*(5-8), 1031–1045. doi:10.1007/s00170-014-6212-x

Pakseresht, A. H., Rahimipour, M. R., Vaezi, M. R., & Salehi, M. (2015). Effect of splat morphology on the microstructure and dielectric properties of plasma sprayed barium titanate films. *Applied Surface Science, 324*, 797–806. doi:10.1016/j.apsusc.2014.11.041

Palik, E. D. (1998). *Handbook of optical constants of solids*. Academic press.

Pan, W., Phillpot, S. R., Wan, C., Chernatynskiy, A., & Qu, Z. (2012). Low thermal conductivity oxides. *Materials Research Society Bulletin*, *37*(10), 917–921. doi:10.1557/mrs.2012.234

Papaspyrides, C. D., & Kiliaris, P. (2014). *Polymer Green Flame Retardants*. Oxford, UK: Elsevier.

Papavassiliou, G. C., Pagona, G., Karousis, N., Mousdis, G. A., Koutselas, I., & Vassilakopoulou, A. (2012). Nanocrystalline/microcrystalline materials based on lead-halide units. *Journal of Materials Chemistry*, *22*(17), 8271–8280. doi:10.1039/c2jm15783g

Papavassiliou, G. C., Pagona, G., Mousdis, G. A., & Karousis, N. (2013). Enhanced phosphorescence from nanocrystalline/microcrystalline materials based on (CH3NH3)(1-naphthylmethyl ammonium)2Pb2Cl7 and similar compounds. *Chemical Physics Letters*, *570*, 80–84. doi:10.1016/j.cplett.2013.03.041

Parashar, G., Srivastava, D., & Kumar, P. (2001). Ethyl silicate binders for high performance coatings. *Progress in Organic Coatings*, *42*(1-2), 1–14. doi:10.1016/S0300-9440(01)00128-X

Pardo, R., Zayat, M., & Levy, D. (2011). Photochromic organic-inorganic hybrid materials. Chemical Society Reviews, 40(2), 672–687. doi:10.1039/c0cs00065e PMID:21229130

Park, C. W., Kang, S.-K., Hernandez, H. L., Kaitz, J. A., Wie, D. S., Shin, J., . . . White, S. R. (2015). Thermally Triggered Degradation of Transient Electronic Devices. Advanced Materials, 27(25), 3783–3788. doi:10.1002/adma.201501180 PMID:25991389

Park, B. H., Kang, B. S., Bu, S. D., Noh, T. W., Lee, J., & Jo, W. (1999). Lanthanum-substituted bismuth titanate for use in non-volatile memories. *Nature*, *401*(6754), 682–684. doi:10.1038/44352

Park, I. S., Choi, U. J., Yi, H. K., Park, B. K., Lee, M. H., & Bae, T. S. (2008). Biomimetic apatite formation and biocompatibility on chemically treated Ti-6Al-7Nb alloy. *Surface and Interface Analysis*, *40*(1), 37–42. doi:10.1002/sia.2666

Park, J. P., Kim, S., Park, J.-Y., Ok, K. M., & Shim, I.-W. (2009). Preparation of ZnO thin films using Zn/O-containing single source precursor through MOCVD method. *Bulletin of the Korean Chemical Society*, *30*(1), 114–118. doi:10.5012/bkcs.2009.30.1.114

Park, J., Ries, J., Gelse, K., Kloss, F., von der Mark, K., Wiltfang, J., & Schneider, H. et al. (2003). Bone regeneration in critical size defects by cell-mediated BMP-2 gene transfer: A comparison of adenoviral vectors and liposomes. *Gene Therapy*, *10*(13), 1089–1098. doi:10.1038/sj.gt.3301960 PMID:12808439

Park, S. Y., Kim, J. H., Kim, M. C., Song, H. S., & Park, C. G. (2005). Microscopic observation of degradation behavior in yttria and ceria stabilized zirconia thermal barrier coatings under hot corrosion. *Surface and Coatings Technology*, *190*(2-3), 357–365. doi:10.1016/j.surfcoat.2004.04.065

Park, Y. S., Ito, Y., & Imanishi, Y. (1998). Photocontrolled gating by polymer brushes grafted on porous glass filter. *Macromolecules*, *31*(8), 2606–2610. doi:10.1021/ma9605199

Partridge, K. A., & Oreffo, R. O. C. (2004). Gene delivery in bone tissue engineering: Progress and prospects using viral and nonviral strategies. *Tissue Engineering*, *10*(1-2), 295–307. doi:10.1089/107632704322791934 PMID:15009954

Patterson, A. L. (1939). The Scherrer formula for X-ray particle size determination. *Physical Review*, *56*(10), 978–982. doi:10.1103/PhysRev.56.978

Paul, W., & Sharma, C. P. (2006). Nanoceramic matrices: Biomedical applications. *Am J Biochem Biotechnol*, *2*(2), 41–48. doi:10.3844/ajbbsp.2006.41.48

Pavithra, G. P., & Hegde, C. (2012). Magnetic property and corrosion resistance of electrodeposited nanocrystalline iron-nickel alloys. Applied Surface Science, 258(18), 6884–6890. doi:10.1016/j.apsusc.2012.03.124

Pelletier, H., Carradò, A., Faerber, J., & Mihailescu, I. N. (2011). Microstructure and mechanical characteristics of hydroxyapatite coatings on Ti/TiN/Si substrates synthesized by pulsed laser deposition. *Applied Physics. A, Materials Science & Processing*, *102*(3), 629–640. doi:10.1007/s00339-010-6144-8

Peng, L., Mendelsohn, A. D., LaTempa, T. J., Yoriya, S., Grimes, C. A., & Desai, T. A. (2009). Long-term small molecule and protein elution from TiO2 nanotubes. *Nano Letters*, *9*(5), 1932–1936. doi:10.1021/nl9001052 PMID:19323554

Peng, Z. A., & Peng, X. (2001a). Formation of High-Quality CdTe, CdSe, and CdS Nanocrystals Using CdO as Precursor. *Journal of the American Chemical Society*, *123*(1), 183–184. doi:10.1021/ja003633m PMID:11273619

Peng, Z. A., & Peng, X. (2001b). Mechanisms of the Shape Evolution of CdSe Nanocrystals. *Journal of the American Chemical Society*, *123*(7), 1389–1395. doi:10.1021/ja0027766

Peniredd, S. R., Janczewski, D., Go, D. P., Zhu, X., Guo, S., Teo, S. L. M., & Vancso, G. J. et al. (2015). Imprinting of metal receptors into multilayers polyelectrolyte films: Fabrication and applications in marine antifouling. *Chemical Sciences*, *6*(1), 372–383.

Pereiro, I., Rodríguez-Valencia, C., Serra, C., Solla, E., Serra, J., & González, P. (2012). Pulsed laser deposition of strontium-substituted hydroxyapatite coatings. *Applied Surface Science*, *258*(23), 9192–9197. doi:10.1016/j.apsusc.2012.04.063

Perez Taborda, J. A., Riascos, H., & Vera, L. (n. d.). Correlation between optical, morphological and compositional properties of Aluminum Nitride thin films by Pulsed Laser Deposition.

Pérez Taborda, J. A., Riascos Landázuri, H., Jiménez García, F. N., & Caicedo Angulo, J. C. (2010). Structural and morphological analysis of aluminum nitride films of obtained by pulsed laser deposition. *Ciencia e Ingeniería Neogranadina*, *20*(2), 107–115.

Pérez, J., Riascos, H., Caicedo, J., Cabrera, G., & Yate, L. (2011a). *Optical Properties Dependence with Gas Pressure in AlN Films Deposited by Pulsed Laser Ablation.Paper presented at the Journal of Physics:Conference Series.* doi:10.1088/1742-6596/274/1/012119

Perez-Jorge, C., Conde, A., Arenas, M. A., Perez-Tanoira, R., Matykina, E., de Damborenea, J. J., & Esteban, J. et al. (2012). In vitro assessment of Staphylococcus epidermidis and Staphylococcus aureus adhesion on TiO2 nanotubes on Ti-6Al-4V alloy. *Journal of Biomedical Materials Research. Part A*, *100A*(7), 1696–1705. doi:10.1002/jbm.a.34118 PMID:22447745

Pérez-Prieto, J. (2013). Symbiosis Between Photoactive Nanoparticles and their Organic Ligands. *Photochemistry and Photobiology*, *89*(6), 1291–1298. doi:10.1111/php.12109 PMID:23742191

Pernites, R. B., Ponnapati, R. R., & Advincula, R. C. (2011). Superhydrophobic-superoleophobic polythiophene films with tunable wetting and electrochromism. *Advanced Materials*, *23*(28), 3207–3213. doi:10.1002/adma.201100469 PMID:21638343

Petrie, T. A., Raynor, J. E., Reyes, C. D., Burns, K. L., Collard, D. M., & Garcia, A. J. (2008). The effect of integrin-specific bioactive coatings on tissue healing and implant osseointegration. *Biomaterials*, *29*(19), 2849–2857. doi:10.1016/j.biomaterials.2008.03.036 PMID:18406458

Philbin, P., Bitter, M., Fraile, D., Latour, M., Teske, S., & Wolfsegger, C. (2008). Solar Electricity for over one billion people and two million jobs by 2010. *Solar Generation*, *V*, 2008.

Pierre, A. C. (1998). Introduction sol-gel processing. Kluwer Academic Publishers.

Pierre, A. C. (1998). *Introduction to Sol-Gel Processing*. Norwell, MA: Kluwer Academic Publishers. doi:10.1007/978-1-4615-5659-6

Pin, L., Ansart, F., Fenech, J., Lours, P., Bonino, J. P., Sniezewski, J., & Le Maoult, Y. (2010). Advances in the Field of New Smart Thermal Barrier Coatings. Advances in Science and Technology, 66, 136–141. doi:10.4028/www.scientific.net/AST.66.136

Plassmann, W. (2001). Coating compositions having improved adhesion to aluminum substrates. *Patent WO 01/98393 A1.*

Plichta A., Habeck, A., Knoche, S., Kruse, A., Weber, A., & Hildebrand, N. (2005). Flexible glass substrates, *Flexible Flat Panel Displays*, 36-55.

Plueddemann, E. P. (1969). Room temperature curable acrylate rubbers. *Patent US 3453230.*

Podlaha, E. J., & Landolt, D. (1996a). Induced codeposition I. An experimental investigation of Ni-Mo alloys. *Journal of the Electrochemical Society, 143*(3), 885–892. doi:10.1149/1.1836553

Podlaha, E. J., & Landolt, D. (1996b). Induced codeposition II. A mathematical model describing the electrodeposition of Ni-Mo alloys. *Journal of the Electrochemical Society, 143*(3), 893–899. doi:10.1149/1.1836554

Podlaha, E. J., & Landolt, D. (1997). Induced codeposition III. Molybdenum alloys with nickel, cobalt, and iron. *Journal of the Electrochemical Society, 144*(5), 1672–1680. doi:10.1149/1.1837658

Podlaha, E. J., Matloez, M., & Landolt, D. (1993). Electrodeposition of high Mo content Ni-Mo alloys under forced convection. *Journal of the Electrochemical Society, 140*(10), L149–L151. doi:10.1149/1.2220956

Polarz, S., Pueyo, C. L., & Krumm, M. (2010). The molecular path to inorganic materials Zinc oxide and beyond. *Inorganica Chimica Acta, 363*(15), 4148–4157. doi:10.1016/j.ica.2010.06.053

Pong, B.-K., Trout, B. L., & Lee, J.-Y. (2008). Modified Ligand-Exchange for Efficient Solubilization of CdSe/ZnS Quantum Dots in Water: A Procedure Guided by Computational Studies. *Langmuir, 24*(10), 5270–5276. doi:10.1021/la703431j PMID:18412382

Poppe, A., Klein, G., Brosseit, A., Wermelt, K., & Möllers, M. (2006). Hochkratzfeste und hochelastische Beschichtungsmittel auf Basis von Alkoxysilan-funktionellen Komponenten. *Patent DE 102004050746 A1*. [Highly scratch-resistant and highly elastic coating agents based on alkoxysilane functional components. *Patent US 7772320 B2].*

Pourbaix, M. (1974). *Atlas of Electrochemical Equilibria in Aqueous Solutions, NACE International Cebelcor* (2nd ed.). Houston, Texas: National Association of Corrosion Engineers.

Prabhakaran, P., Kim, W. J., Lee, K.-S., & Prasad, P. N. (2012). Quantum dots (QDs) for photonic applications. *Optical Materials Express, 2*(5), 578–593. doi:10.1364/OME.2.000578

Prioteasa, P., Anicăi, L., & Vişan, T. (2010). Synthesis and corrosion characterization of electrodeposited Ni-Mo alloys obtained from aqueous solutions. *U.P.B. Science Bulletin, Series B., 72*(4), 11–24.

Psaltis, D., Quake, S. R., & Yang, C. (2006). Developing optofluidic technology through the fusion of microfluidics and optics. *Nature, 442*(7101), 381–386. doi:10.1038/nature05060 PMID:16871205

Pueyo, C. L., Siroky, S., Landsmann, S., Berg, M. W. E. d., Wagner, M. R., & Reparaz, J. S. et al.. (2010). Monolithic zinc Oxide Aerogels from Organometallic Sol-Gel Precursors. *Chemistry of Materials, 22*(14), 4263.

Puleo, D. A., & Nanci, A. (1999). Understanding and controlling the bone-implant interface. *Biomaterials, 20*(23-24), 2311–2321. doi:10.1016/S0142-9612(99)00160-X PMID:10614937

Pulker, H. K. (1984). *Coatings on Glass*. Elsevier Amsterdam.

Puzder, A., Williamson, A. J., Zaitseva, N., Galli, G., Manna, L., & Alivisatos, A. P. (2004). The Effect of Organic Ligand Binding on the Growth of CdSe Nanoparticles Probed by Ab Initio Calculations. *Nano Letters*, *4*(12), 2361–2365. doi:10.1021/nl0485861

Qing, Y., Zheng, Y., Hu, C., Wang, Y., He, Y., Gong, Y. & Mo, Q. (2013). Facile Approach in Fabricating Superhydrophobic ZnO/Polystyrene Nanocomposite Coating. *Applied Surface Science*, *285*(Part B), 583-587. doi:10.1016/j.apsusc.2013.08.097

Quéré, D. (2005). Non-sticking drops. *Reports on Progress in Physics*, *68*(11), 2495–2532. doi:10.1088/0034-4885/68/11/R01

Qu, L., & Peng, X. (2002). Control of Photoluminescence Properties of CdSe Nanocrystals in Growth. *Journal of the American Chemical Society*, *124*(9), 2049–2055. doi:10.1021/ja017002j PMID:11866620

Qu, M., Zhang, B., Song, S., Chen, L., Zhang, J., & Junyan, C. X. (2007). Fabrication of superhydrophobic surfaces on engineering materials via a solution-immersion process. *Advanced Functional Materials*, *217*(4), 593–596. doi:10.1002/adfm.200600472

Rabiei, A., Thomas, B., Jin, C., Narayan, R., Cuomo, J., Yang, Y., & Ong, J. (2006). A study on functionally graded HA coatings processed using ion beam assisted deposition with in situ heat treatment. *Surface and Coatings Technology*, *200*(20), 6111–6116. doi:10.1016/j.surfcoat.2005.09.027

Rabinovich, G. S., Yamamoto, K., Aguirre, M. H., Cahill, D. G., Veldhuis, S. C., Biksa, A., . . . Shuster, L. S. (2010). Multi-functional nano-multilayered AlTiN/Cu PVD coating for machining of Inconel 718 superalloy. Surface and Coatings Technology, 204(15), 2465–2471. doi:10.1016/j.surfcoat.2010.01.024 doi:10.1016/j.surfcoat.2010.01.024

Radhakrishnan, C., Lo, M. K. F., Knobler, C. M., Garcia-Garibay, M. A., & Monbouquette, H. G. (2011). Capping-Ligand Effect on the Stability of CdSe Quantum Dot Langmuir Monolayers. *Langmuir*, *27*(6), 2099–2103. doi:10.1021/la104244x PMID:21329375

Radin, S., Campbell, J. T., Ducheyne, P., & Cuckler, J. M. (1997). Calcium phosphate ceramic coatings as carriers of vancomycin. *Biomaterials*, *18*(11), 777–782. doi:10.1016/S0142-9612(96)00190-1 PMID:9177855

Raether, H. (2013). *Surface plasmons on smooth and rough surfaces and on gratings. Springer Tracts in Modern Physics* (Vol. 111). Springer Berlin Heidelberg; doi:10.1007/BFb0048317

Raj, I. A., & Vekatesan, V. K. (1998). Characterization of nickel-molybdenum and nickel-molybdenum-iron alloy coatings as cathodes for alkaline water electrolysers. *International Journal of Hydrogen Energy*, *13*(4), 215–223. doi:10.1016/0360-3199(88)90088-2

Ramana, C. V., Gengler, J. J., & Jones, J. G. (2012). Crystal Structure, Phase Analysis, and Thermal Conductivity of Nanocrystalline GdO_3-HfO_2 Coatings. *The Journal of Physical Chemistry C*, *116*(48), 25178–25183. doi:10.1021/jp301008t

Rammelt, S., Schulze, E., Bernhardt, R., Hanisch, U., Scharnweber, D., Worch, H., & Biewener, A. et al. (2004). Coating of titanium implants with type-I collagen. *Journal of Orthopaedic Research*, *22*(5), 1025–1034. doi:10.1016/j.orthres.2004.02.011 PMID:15304275

Rangaraj, S., & Kokini, K. (2003a). Estimating the fracture resistance of functionally graded thermal barrier coatings from thermal shock tests. *Surface and Coatings Technology*, *173*(2-3), 201–212. doi:10.1016/S0257-8972(03)00515-2

Rangaraj, S., & Kokini, K. (2003b). Interface thermal fracture in functionally graded zirconia – mullite – bond coat alloy thermal barrier coatings. *Acta Materialia*, *51*(1), 251–267. doi:10.1016/S1359-6454(02)00396-8

Rangaraj, S., & Kokini, K. (2004). A Study of thermal fracture in functionally graded thermal barrier coatings using a cohesive. *Journal of Engineering Materials and Technology*, *126*(1), 103–115. doi:10.1115/1.1631028

Raniero, W., Campostrini, M., Maggioni, G., & Quaranta, A. (2014). Physical vapour deposition reactive magnetron sputtering for the production and application of dichroics in photovoltaic system with solar spectral splitting. *Applied Surface Science*, *308*, 170–175. doi:10.1016/j.apsusc.2014.04.130

Raniero, W., Maggioni, G., Della Mea, G., Campostrini, M., Marigo, S., & Nardo, M. (2013). Rutherford Backscattering Spectrometry (RBS) analysis of dichroic systems for optical applications. *AIP Conference Proceedings*, *1530*, 133–139. doi:10.1063/1.4812915

Raole, P., Mukherjee, S., & John, P. (2005). X-ray photoelectron spectroscopic study of plasma source nitrogen ion implantation in single crystal natural diamond. *Diamond and Related Materials*, *14*(3), 482–485. doi:10.1016/j.diamond.2004.11.009

Rathore, A. S., & Horvath, C. (1997). Capillary electrochromatography: Theories on electroosmotic flow in porous media. *Journal of Chromatography. A*, *781*(1-2), 185–195. doi:10.1016/S0021-9673(97)00627-4 PMID:9368385

Rau, J., Smirnov, V., Laureti, S., Generosi, A., Varvaro, G., Fosca, M., & Barinov, S. et al. (2010). Properties of pulsed laser deposited fluorinated hydroxyapatite films on titanium. *Materials Research Bulletin*, *45*(9), 1304–1310. doi:10.1016/j.materresbull.2010.04.028

Raut, H. K., Ganesh, V. A., Nair, A. S., & Ramakrishna, S. (2011). Anti-reflective coatings: A critical, in-depth review. Energy & Environmental Science, 4(10), 3779-3804. doi:10.1039/c1ee01297e

Reactive silicones [brochure]. (2013). *Gelest, Inc.* Retrieved from http://www.gelest.com/goods/PDF/brochures/reactivesilicones.pdf

Rekondo, A., Fernándes-Berridi, M. J., & Irusta, L. (2006). Synthesis of silanized polyether urethane hybrid systems. Study of the curing process through hydrogen bonding interactions. *European Polymer Journal*, *42*(9), 2069–2080. doi:10.1016/j.eurpolymj.2006.03.024

Rentsch, B., Hofmann, A., Breier, A., Rentsch, C., & Scharnweber, D. (2009). Embroidered and surface modified polycaprolactone-co-lactide scaffolds as bone substitute: In vitro characterization. *Annals of Biomedical Engineering*, *37*(10), 2118–2128. doi:10.1007/s10439-009-9731-0 PMID:19626441

Revie, R. W., & Uhlig, H. H. (2008). *Corrosion and corrosion control. An introduction to corrosion science and engineering* (4th ed.). New York, NY: John Wiley & Sons, Inc.

Rezaei, S., Manoucheri, I., Moradian, R., & Pourabbas, B. (2014). One-step chemical vapor deposition and modification of silica nanoparticles at the lowest possible temperature and superhydrophobic surface fabrication. *Chemical Engineering Journal (Amsterdam, Netherlands)*, *252*, 11–16.

Rezania, A., & Healy, K. E. (1999). Integrin subunits responsible for adhesion of human osteoblast-like cells to biomimetic peptide surfaces. *Journal of Orthopaedic Research*, *17*(4), 615–623. doi:10.1002/jor.1100170423 PMID:10459771

Rhee, J. H., Chung, C.-C., & Diau, E. W.-G. (2013). A perspective of mesoscopic solar cells based on metal chalcogenide quantum dots and organometal-halide perovskites. *NPG Asia Mater*, *5*(10), e68. doi:10.1038/am.2013.53

Richter, A., Klatt, S., Paschew, G., & Klenke, C. (2009). Micropumps operated by swelling and shrinking of temperature-sensitive hydrogels. *Lab on a Chip*, *9*(4), 613–618. doi:10.1039/B810256B PMID:19190798

Rizk, S. D., & Hsieh, H. W. S. (1986). Moisture curable polyurethane polymers. *Patent US 46235012.*

Rizk, S. D., Hsieh, H. W. S., & Mazzeo, M. P. (1985). Acrylic resin having pendant silane groups thereon, and methods of making and using the same. *Patent US 4491650.*

Roach, P., Shirtcliffe, N. J., & Newton, M. I. (2008). Progress in superhydrophobic surface development. *Soft Matter*, *4*(2), 224–240. doi:10.1039/B712575P

Ro, D. H., & Klar, E. (1981). Corrosion Behavior of P/M Austenitic Stainless Steels. *Modern Developments in Powder Metallurgy*, *13*, 247–287.

Rode, S., Henninot, C., & Matlosz, M. (2005). Complexation chemistry in nickel and copper-nickel alloy plating from citrate baths. *Journal of the Electrochemical Society*, *152*(4), C248–C254. doi:10.1149/1.1869980

Rodríguez, A., Morant-Miñana, M. C., Dias-ponte, A., Martínez-calderón, M., Gómez-aranzadi, M., & Olaizola, S. M. (2015). Femtosecond laser-induced periodic surface nanostructuring of sputtered platinum thin films. Applied Surface Science, 351, 135–139.

Rogach, A. (Ed.). (2008). *Semiconductor Nanocrystal Quantum Dots* (1st ed.). Vienna: Springer-Verlag Wien. doi:10.1007/978-3-211-75237-1

Rogach, A. L., Kornowski, A., Gao, M., Eychmüller, A., & Weller, H. (1999). Synthesis and Characterization of a Size Series of Extremely Small Thiol-Stabilized CdSe Nanocrystals. *The Journal of Physical Chemistry B*, *103*(16), 3065–3069. doi:10.1021/jp984833b

Roig, A., Molins, E., Rodriguez, E., Martinez, S., Moreno, M. M., & Vallribera, A. (2004). Superhydrophobic silica aerogels by fluorination at the gel stage. *Chem. Commun.*, 2316-2317. doi:10.1039/b407987f

Rojaee, R., Fathi, M., & Raeissi, K. (2014). Comparing Nanostructured Hydroxyapatite Coating on AZ91 Alloy Samples via Sol-gel and Electrophoretic Deposition for Biomedical Applications. IEEE Transactions on Nanobioscience, 13(4), 409–414.

Roldan-Carmona, C., Malinkiewicz, O., Soriano, A., Minguez Espallargas, G., Garcia, A., Reinecke, P., & Bolink, H. J. et al. (2014). Flexible high efficiency perovskite solar cells. *Energy & Environmental Science*, *7*(3), 994–997. doi:10.1039/c3ee43619e

Rosenberger, L., Baird, R., McCullen, E., Auner, G., & Shreve, G. (2008). XPS analysis of aluminum nitride films deposited by plasma source molecular beam epitaxy. *Surface and Interface Analysis*, *40*(9), 1254–1261. doi:10.1002/sia.2874

Rosenthal, S. J., Chang, J. C., Kovtun, O., McBride, J. R., & Tomlinson, I. D. (2011). Biocompatible Quantum Dots for Biological Applications. *Chemistry & Biology*, *18*(1), 10–24. doi:10.1016/j.chembiol.2010.11.013 PMID:21276935

Ruan, M., Li, W., Wang, B., Deng, B., Ma, F., & Yu, Z. (2013). Preparation and Anti-Icing Behavior of Superhydrophobic Surfaces on Aluminum Alloy Substrates. *Langmuir*, *29*(27), 8482–8491. doi:10.1021/la400979d PMID:23718719

Rubner, M. F. (2003). PH-controlled fabrication of polyelectrolyte multilayers: assembly and applications. In G. Decher & J.B. Schlenoff (Eds.), Multilayer Thin Films (pp. 133-154). Weinheim, Germany: Wiley-VCH Verlag GmbH & Co. KGaA.

Ruoslahti, E., & Yamaguchi, Y. (1991). Proteoglycans as modulators of growth factor activities. *Cell*, *64*(5), 867–869. doi:10.1016/0092-8674(91)90308-L PMID:2001586

Rupp, F., Scheideler, L., Olshanska, N., Wild, M., Wieland, M., & Geis, G. J. (2006). Enhancing surface free energy and hydrophilicity through chemical modification of microstructured titanium implant surfaces, reply. *J. Biomed. Mater. Res. Part A*, 79A, 755-757. doi:10.1002/jbm.a.31045

Ruterana, P., Albrecht, M., & Neugebauer, J. (2006). *Nitride semiconductors: handbook on materials and devices*. John Wiley & Sons.

Rybiński, P., Pająk, A., Janowska, G., & Jóźwiak, M. (2015). Effect of hybrid filler (HNTs-phthalocyanine) on the thermal properties and flammability of diene rubber. *Journal of Applied Polymer Science, 132*(40).

Safety Data Sheet AGS 3550 (Trion Tensid AB). (2012). *Intersolia Sweden*. Retrieved from http://cloud.intersolia.com/pdf/610/%7B693D0CF5-5ED3-42DC-B2DE-2668D571D32E%7D/100/AGS_3550_EN

Safikhani, A., Esmailian, M., Salmani, M. R., & Aminfard, M. (2014). Effect of Ni-Mo addition on cyclic and isothermal oxidation resistance and electrical behavior of ferritic stainless steel for SOFCs interconnect. *International Journal of Hydrogen Energy, 39*(21), 11210–11223. doi:10.1016/j.ijhydene.2014.05.122

Saju, K., Reshmi, R., Jayadas, N., James, J., & Jayaraj, M. (2009). Polycrystalline coating of hydroxyapatite on TiAl6V4 implant material grown at lower substrate temperatures by hydrothermal annealing after pulsed laser deposition. *Proceedings of the Institution of Mechanical Engineers. Part H, Journal of Engineering in Medicine, 223*(8), 1049–1057. doi:10.1243/09544119JEIM568 PMID:20092101

Sakai, H., Ebana, H., Sakai, K., Tsuchiya, K., Ohkubo, T., & Abe, M. (2007). Photo-isomerization of spiropyran-modified cationic surfactants. *Journal of Colloid and Interface Science, 316*(2), 1027–1030. doi:10.1016/j.jcis.2007.08.042 PMID:17884070

Sakai, K., Imaizumi, Y., Oguchi, T., Sakai, H., & Abe, M. (2010). Adsorption Characteristics of Spiropyran-Modified Cationic Surfactants at the Silica/Aqueous Solution Interface. *Langmuir, 26*(12), 9283–9288. doi:10.1021/la100161a PMID:20225812

Sakiyama-Elbert, S. E. (2014). Incorporation of heparin into biomaterials. *Acta Biomaterialia, 10*(4), 1581–1587. doi:10.1016/j.actbio.2013.08.045 PMID:24021232

Sakka, S. (Ed.). (2005). *Applications of sol-gel technology*. In Handbook of sol-gel science and technology: processing, characterization, and application (Vol. 3). Norwell, MA: Kluwer Academic Publishers.

Samanta, S., & Locklin, J. (2008). Formation of photochromic spiropyran polymer brushes via surface-initiated, ring-opening metathesis polymerization: Reversible photocontrol of wetting behavior and solvent dependent morphology changes. *Langmuir, 24*(17), 9558–9565. doi:10.1021/la8017387 PMID:18642863

Samoladas, A., Bikiaris, D., Zorba, T., Paraskevopoulos, K. M., & Jannakoudakis, A. (2008). Photochromic behavior of spiropyran in polystyrene and polycaprolactone thin films - Effect of UV absorber and antioxidant compound. *Dyes and Pigments, 76*(2), 386–393. doi:10.1016/j.dyepig.2006.09.012

Samuel, J. D. J. S., Ruther, P., Frerichs, H. P., Lehmann, M., Paul, O., & Ruehe, J. (2005). A simple route towards the reduction of surface conductivity in gas sensor devices. *Sens. Actuators, B, 110*, 218-224. doi:10.1016/j.snb.2005.01.032

Sanches, L. S., Marino, C. B., & Mascaro, L. H. (2007). Investigation of the codeposition of Fe and Mo from sulphate-citrate acid solutions. *Journal of Alloys and Compounds, 439*(1-2), 342–345. doi:10.1016/j.jallcom.2006.08.231

Sangermano, M., Amerio, E., Priola, A., Di Gianni, A., & Voit, B. (2006). Preparation and characterization of acrylic resin/titania hybrid nanocomposite coatings by photopolymerization and sol-gel process. *Journal of Applied Polymer Science, 102*(5), 4659–5664. doi:10.1002/app.24707

Sangermano, M., Vitale, A., & Dietliker, K. (2014). Photolatent amines producing a strong base as photocatalyst for the in-situ preparation of organic-inorganic hybrid coatings. *Polymer, 55*(7), 1628–1635. doi:10.1016/j.polymer.2014.02.045

Santos, C. M., Mangadlao, J., Ahmed, F., Leon, A., Advincula, R. C., & Rodrigues, D. F. (2012). Graphene nanocomposite for biomedical applications: Fabrication, antimicrobial and cytotoxic investigations. *Nanotechnology*, *23*(39), 395101-395110. doi:10.1088/0957-4484/23/39/395101 PMID:22962260

Saran, N., Zhang, R. W., & Turcotte, R. E. (2011). Osteogenic protein-1 delivered by hydroxyapatite-coated implants improves bone ingrowth in extracortical bone bridging. *Clinical Orthopaedics and Related Research*, *469*(5), 1470–1478. doi:10.1007/s11999-010-1573-4 PMID:20878288

Saremi, M., Afrasiabi, A., & Kobayashi, A. (2008). Microstructural analysis of YSZ and YSZ / Al_2O_3 plasma sprayed thermal barrier coatings after high temperature oxidation. *Surface and Coatings Technology*, *202*(14), 3233–3238. doi:10.1016/j.surfcoat.2007.11.029

Satayaprassad, A., Jian, V., & Nema, S. K. (2007). Deposition of superhydrophobic nanostructured Teflon-like coating using expanding plasma arc. *Applied Surface Science*, *253*(12), 5462–5466. doi:10.1016/j.apsusc.2006.12.085

Sato, H., Morita, S., & Ozaki, Y. (2005). Polymer Surface explored by infrared spectroscopy and atomic force microscopy. *Nippon Setchaku Gakkaishi*, *41*(5), 183–195.

Scaife, D. E., Weller, P. F., & Fisher, W. G. (1974). Crystal preparation and properties of cesium tin(II) trihalides. *Journal of Solid State Chemistry*, *9*(3), 308–314. doi:10.1016/0022-4596(74)90088-7

Scarmagnani, S., Walsh, Z., Benito-Lopez, F., Macka, M., Paull, B., & Diamond, D. (2010). Incorporation of acrylate based spiropyran monoliths in micro-fluidic devices for photo-controlled electroosmotic flow. *Advances in Science and Technology*, *76*, 100–105. doi:10.4028/www.scientific.net/AST.76.100

Schaffner, P., & Dard, M. M. (2003). Structure and function of RGD peptides involved in bone biology. *Cellular and Molecular Life Sciences*, *60*(1), 119–132. doi:10.1007/s000180300008 PMID:12613662

Schierholz, J. M., Steinhauser, H., Rump, A. F. E., Berkels, R., & Pulverer, G. (1997). Controlled release of antibiotics from biomedical polyurethanes: Morphological and structural features. *Biomaterials*, *18*(12), 839–844. doi:10.1016/S0142-9612(96)00199-8 PMID:9184747

Schild, H. G. (1992). Poly (n-isopropylacrylamide) - experiment, theory and application. *Progress in Polymer Science*, *17*(2), 163–249. doi:10.1016/0079-6700(92)90023-R

Schilp, S., Rosenhahn, A., Pettitt, M. E., Bowen, J., Callow, M. E., Callow, J., & Grunze, M. (2009). Physicochemical properties of (ethylene glycol)-containing self-assembled monolayers relevant for protein and algal cell resistance. *Langmuir*, *25*(17), 10077–10082. doi:10.1021/la901038g PMID:19469528

Schindler, W., & Schwiebacher, E. (2009). Alkoxysilane cross-linked polymers having improved elastic recovery properties. *Patent US 20090131591 A1.*

Schmidt, H., Jonschker, G., Goedicke, S., & Mennig, M. (2000). The sol-gel process as a basic technology for nanoparticle-dispersed inorganic-organic composites. *Journal of Sol-Gel Science and Technology*, *19*(1/3), 39–51. doi:10.1023/A:1008706003996

Schmidt, L. C., Pertegás, A., González-Carrero, S., Malinkiewicz, O., Agouram, S., Mínguez Espallargas, G., & Pérez-Prieto, J. et al. (2014). Nontemplate Synthesis of CH3NH3PbBr3 Perovskite Nanoparticles. *Journal of the American Chemical Society*, *136*(3), 850–853. doi:10.1021/ja4109209 PMID:24387158

Schneiders, W., Reinstorf, A., Biewener, A., Serra, A., Grass, R., Kinscher, M., & Rammelt, S. et al. (2009). In vivo effects of modification of hydroxyapatite/collagen composites with and without chondroitin sulphate on bone remodeling in the sheep tibia. *Journal of Orthopaedic Research*, *27*(1), 15–21. doi:10.1002/jor.20719 PMID:18634066

Scholl, R. (1997). Power System for Reactive Sputtering of Insulating Films. *Surface and Coatings Technology*, 93(1), 7-13.

Schonherr, E., & Hausser, H. J. (2000). Extracellular matrix and cytokines: A functional unit. *Developmental Immunology*, *7*(2-4), 89–101. doi:10.1155/2000/31748 PMID:11097204

Schottner, G. (2001). Hybrid sol-gel-derived polymers: Applications of multifunctional materials. *Chemistry of Materials*, *13*(10), 3422–3435. doi:10.1021/cm011060m

Schreurs, W. J. A., & Rosenberg, H. (1982). Effect of silver ions on transport and retention of phosphate by escherichia coil. *Journal of Bacteriology*, *152*(1), 7–13. PMID:6749823

Schubert, M. (1996). Polarization-dependent optical parameters of arbitrarily anisotropic homogeneous layered systems. *Physical Review B: Condensed Matter and Materials Physics*, *53*(8), 4265–4274. doi:10.1103/PhysRevB.53.4265 PMID:9983979

Schubert, U., Huesing, N., & Lorenz, A. (1995). Hybrid inorganic-organic materials by sol-gel processing of organofunctional metal alkoxides. *Chemistry of Materials*, *7*(11), 2010–2027. doi:10.1021/cm00059a007

Schuler, M., Owen, G. R., Hamilton, D. W., de Wild, M., Textor, M., Brunette, D. M., & Tosatti, S. G. P. (2006). Biomimetic modification of titanium dental implant model surfaces using the RGDSP-peptide sequence: A cell morphology study. *Biomaterials*, *27*(21), 4003–4015. doi:10.1016/j.biomaterials.2006.03.009 PMID:16574219

Schulz, U., Peters, M., Bach, F., & Tegeder, G. (2003). Graded coatings for thermal, wear and corrosion barriers. *Materials Science and Engineering A*, *362*(1-2), 61–80. doi:10.1016/S0921-5093(03)00579-3

Schwarzacher, W. (2006). Electrodeposition: A Technology for the Future. *The Electrochemical Society Interface, 15*(1), 32–35.

Schweitzer, P. A. (Ed.). (2006). *Paint and coatings - application and corrosion resistance*. Boca Raton, FL: CRC Press.

Schwertfeger, F., Emmerling, A., Gross, J., Schubert, U., Fricke, J., & Attia, Y. (1994). *Sol-Gel Processing and Applications* (p. 343). New York. doi:10.1007/978-1-4615-2570-7_29

Scully, J. C. (1975). *The Fundamentals of Corrosion* (2nd ed.). Oxford, New York: Pergamon Press.

Sedriks, A. J. (1996). *Corrosion of Stainless Steel* (2nd ed.). New York, NY: John Wiley & Sons, Inc.

Seeber, W. T., Abou-Helal, M. O., Barth, S., Beil, D., Höche, T., Afify, H. H., & Demian, S. E. et al. (1999). Transparent semiconducting ZnO:Al thin films prepared by spray pyrolysis. *Materials Science in Semiconductor Processing*, *2*(1), 45–55. doi:10.1016/S1369-8001(99)00007-4

Seifert, A., & Spange, S. (2013). Aus eins mach zwei: Die Zwillingspolymerisation. *Nachrichten aus der Chemie*, *61*(10), 1028–1032. doi:10.1002/nadc.201390319

Seo, J. Y., & Han, M. (2011). Multi-functional hybrid coatings containing silica nanoparticles and anti-corrosive acrylate monomer for scratch and corrosion resistance. *Nanotechnology*, *22*(2), 25601-25609 doi:10.1088/0957-4484/22/2/025601 PMID:21135477

Sepeur, S., Laryea, N., Goedicke, S., & Groß, F. (2008). *Nanotechnology - Technical Basics and Applications*. Hannover, Germany: Vincentz Network.

Sershen, S. R., Mensing, G. A., Ng, M., Halas, N. J., Beebe, D. J., & West, J. L. (2005). Independent optical control of microfluidic valves formed from optomechanically responsive nanocomposite hydrogels. *Advanced Materials, 17*(11), 1366. doi:10.1002/adma.200401239

Seymour, R. B., & Mark, H. F. (1990). Organic coatings: their origin and development. *Proceedings of theInternational Symposium on the History of Organic Coatings*. Florida, Elsevier.

Shaginyan, L. R., Han, J. G., Shaginyan, V. R., & Musil, J. (2006). Evolution of film temperature during magnetron sputtering. *Journal of Vacuum Science & Technology A, 24*(4), 1083-1090. doi:10.1116/1.2210947

Shaik, A., Narayan, R., & Raju, K. V. S. N. (2014). Synthesis and properties of siloxane-crosslinked polyurethane-urea/silica hybrid films from castor oil. *Journal of Coatings Technology and Research, 11*(3), 397–407. doi:10.1007/s11998-013-9548-5

Shan, F. K., & Yu, Y. S. (2003). Optical properties of pure and Al doped ZnO thin films fabricated with plasma produced by excimer laser. *Thin Solid Films, 435*(1-2), 174–178. doi:10.1016/S0040-6090(03)00343-2

Shang, Z. B., Hu, S., Wang, Y., & Jin, W. J. (2011). Interaction of β-cyclodextrin-capped CdSe quantum dots with inorganic anions and cations. *Luminescence, 26*(6), 585–591. doi:10.1002/bio.1274 PMID:21268229

Sharifnabi, a., Fathi, M. H., Eftekhari Yekta, B., & Hossainalipour, M. (2014). The structural and bio-corrosion barrier performance of Mg-substituted fluorapatite coating on 316L stainless steel human body implant. *Applied Surface Science, 288*, 331–340.

Sharma, R. K. (2007). *Elementary Chemistry*. Discovery Publishing House.

Shchukin, D. G., & Moehwald, H. (2012). Multilayer polyelectrolyte assembly in feedback active coatings and films. In G. Decher & J. B. Schlenoff (Eds.), *Multilayer Thin Films, 2* (pp. 1039–1052). Weinheim, Germany: Wiley-VCH Verlag GmbH & Co. doi:10.1002/9783527646746.ch43

Shen, S., Zhang, Y., Peng, L., Xu, B., Du, Y., Deng, M., & Wang, Q. et al. (2011). Generalized synthesis of metal sulfide nanocrystals from single-source precursors: Size, shape and chemical composition control and their properties. *CrystEngComm, 13*(14), 4572–4579. doi:10.1039/c0ce00982b

Shen, X., Hu, Y., Xu, G., Chen, W., Xu, K., Ran, Q., & Cai, K. et al. (2014). Regulation of the biological functions of osteoblasts and bone formation by Zn-incorporated coating on microrough titanium. *ACS Applied Materials & Interfaces, 6*(18), 16426–16440. doi:10.1021/am5049338 PMID:25148131

She, Z., Li, Q., Wang, Z., Li, L., Chen, F., & Zhou, J. (2013). Researching the Fabrication of Anticorrosion Superhydrophobic Surface on Magnesium Alloy and Its Mechanical Stability and Durability. *Chemical Engineering Journal, 228*, 415–424. doi:10.1016/j.cej.2013.05.017

Shi, J., Chen, C., Yu, H., & Zhang, S. (2008). Application of magnetron sputtering for producing bioactive ceramic coatings on implant materials. *Bulletin of Materials Science, 31*(6), 877–884. doi:10.1007/s12034-008-0140-z

Shishkovsky, I., & Smurov, I. (2012). Titanium base functional graded coating via 3D laser cladding. *Materials Letters, 73*, 32–35. doi:10.1016/j.matlet.2011.12.099

Shi, Z., Neoh, K. G., & Kang, E. T. (2004). Surface-grafted viologen for precipitation of silver nanoparticles and their combined bactericidal activities. *Langmuir, 20*(16), 6847–6852. doi:10.1021/la049132m PMID:15274594

Shizuma, M., Shimomura, O., Moriwaki, K., Yamauchi, A., Sato, H., Kawano, S., & Ono, D. (2014). Evaluation of surface free energy by means of contact angle measurements. *Kagaku to Kogyo, 88*(5), 193–199.

Shukla, S., Gaur, P., & Rai, N. (2015). Complexes of tetraethylthiuram disulphide with group 12 metals: Single-source precursor in metal sulphide nanoparticles synthesis. *Applied Nanoscience, 5*(5), 583–593.

Shun'ko, E. V. (2009). *Langmuir probe in theory and practice*. Universal-Publishers.

Sichert, J. A., Tong, Y., Mutz, N., Vollmer, M., Fischer, S., Milowska, K. Z., & Feldmann, J. et al. (2015). Quantum Size Effect in Organometal Halide Perovskite Nanoplatelets. *Nano Letters*, *15*(10), 6521–6527. doi:10.1021/acs.nanolett.5b02985 PMID:26327242

Si, F., Zhang, L., Zhao, N., Chen, L., & Xu, J. (2011). Preparation method and application of superhydrophilic surfaces. *Huaxue Jinzhan*, *23*(9), 1831–1840.

Silres HP 2000 [brochure]. (2012b). Wacker Chemie. Retrieved from http://www.wacker.com/cms/media/publications/downloads/6870_EN.pdf

Silva, M., Rodrigues, J., Oliveira, M., Fernandes, A., Pereira, S., Costa, C., . . . Correia, J. (2014). Optical Filter for Providing the Required Illumination to Enable Narrow Band Imaging. *EUROSENSORS 2014, the XXVIII edition of the conference series*. Elsevier.

Simchi, A., Tamjid, E., Pishbin, F., & Boccaccini, A. R. (2011). Recent progress in inorganic and composite coatings with bactericidal capability for orthopaedic applications. *Nanomedicine; Nanotechnology, Biology, and Medicine*, *7*(1), 22–39. doi:10.1016/j.nano.2010.10.005 PMID:21050895

Singh, J., & Wolfe, D. E. (2005). Nano and macro-structured component fabrication by electron beam-physical vapor deposition (EB-PVD). *Journal of Materials Science, 40*, 1–26.

Singh, V. B., & Singh, D. (1975). Electrodeposition of Ni-Mo alloys from acetate bath. *Transactions of the Society for Advancement of Electrochemical Science and Technology, 10*(2), 91-97.

Singh, M. K., Yang, Y., & Takoudis, C. G. (2009). Synthesis of multifunctional multiferroic materials from organometallics. *Coordination Chemistry Reviews*, *253*(23-24), 2920–2934. doi:10.1016/j.ccr.2009.09.003

Singh, M., Haverinen, H. M., Dhagat, P., & Jabbour, G. E. (2010). Inkjet Printing-Process and Its Applications. *Advanced Materials*, *22*(6), 673–685. doi:10.1002/adma.200901141 PMID:20217769

Singleton, T. A., Ramsay, K. S., Barsan, M. M., Butler, I. S., & Barrett, C. J. (2012). Azobenzene Photoisomerization under High External Pressures: Testing the Strength of a Light-Activated Molecular Muscle. *The Journal of Physical Chemistry B*, *116*(32), 9860–9865. doi:10.1021/jp3060872 PMID:22799454

Snyders, R., Bousser, E., Music, D., Jensen, J., Hocquet, S., & Schneider, J. M. (2008). Influence of the Chemical Composition on the Phase Constitution and the Elastic Properties of RF-Sputtered Hydroxyapatite Coatings. *Plasma Processes and Polymers*, *5*(2), 168–174. doi:10.1002/ppap.200700112

Snyders, R., Music, D., Sigumonrong, D., Schelnberger, B., Jensen, J., & Schneider, J. (2007). Experimental and ab initio study of the mechanical properties of hydroxyapatite. *Applied Physics Letters*, *90*(19), 193902. doi:10.1063/1.2738386

Sobota, P., Utko, J., Sztajnowska, K., Ejfler, J., & Jerzykiewicz, L. B. (2000). Polynuclear magnesium and magnesium-titanium species. *Inorganic Chemistry*, *39*(2), 235–239. doi:10.1021/ic990542l PMID:11272530

Socol, G., Torricelli, P., Bracci, B., Iliescu, M., Miroiu, F., Bigi, A., & Mihailescu, I. et al. (2004). Biocompatible nanocrystalline octacalcium phosphate thin films obtained by pulsed laser deposition. *Biomaterials*, *25*(13), 2539–2545. doi:10.1016/j.biomaterials.2003.09.044 PMID:14751739

Søndergaard, R., Hösel, M., Angmo, D., Larsen-Olsen, T. T., & Krebs, F. C. (2012). Roll-to-roll fabrication of polymer solar cells. *Materials Today*, *15*(1-2), 36–49. doi:10.1016/S1369-7021(12)70019-6

Song, J., Ma, K., Zhang, L., & Schoenung, J. M. (2010). Technology Simultaneous synthesis by spark plasma sintering of a thermal barrier coating system with a NiCrAlY bond coat. *Surface and Coatings Technology*, *205*(5), 1241–1244. doi:10.1016/j.surfcoat.2010.08.064

Sorensen, J. H., Lilja, M., Astrand, M., Sorensen, T. C., Procter, P., Stromme, M., & Steckel, H. (2014). Co-precipitation of Tobramycin into Biomimetically Coated Orthopedic Fixation Pins Employing Submicron-Thin Seed Layers of Hydroxyapatite. *Current Drug Delivery*, *11*(4), 501–510. doi:10.2174/1567201811666140311154104 PMID:24611653

Spange, S., Kempe, P., & Seifert, A. Auer, A. A., Ecorchard, P., Lang, H., ... Kaskel, S. (2009b). Nanocomposites with structure domains of 0.5 to 3 nm by polymerization of silicon spiro compounds. Angewandte Chemie International Edition, 48, 8254–8258.

Spange, S., & Grund, S. (2009a). Nanostructured organic-inorganic composites by twin polymerisation of hybrid monomers. *Advanced Materials*, *21*(20), 2111–2116. doi:10.1002/adma.200802797

Spanhel, L. (2006). Colloidal ZnO nanostructures and functional coatings: A survey. *Journal of Sol-Gel Science and Technology*, *39*(1), 7–24. doi:10.1007/s10971-006-7302-5

Speier, P., Plogmann, F., Lomoelder, R., & Hoppe, D. (2003). Nonaqueous thermosetting two-component coating composition. *Patent US 2003/0027921 A1*.

Sperling, R.A., & Parak, W.J. (2010). Surface modification, functionalization and bioconjugation of colloidal inorganic nanoparticles. *Philosophical Transactions of the Royal Society of London A: Mathematical, Physical and Engineering Sciences*, 368(1915), 1333-1383.

Stadler, A. (2012). Transparent Conducting Oxides—An Up-To-Date Overview. *Materials*, *5*(12), 661–683. doi:10.3390/ma5040661

Stadlinger, B., Bierbaum, S., Grimmer, S., Schulz, M. C., Kuhlisch, E., Scharnweber, D., & Mai, R. et al. (2009). Increased bone formation around coated implants. *Journal of Clinical Periodontology*, *36*(8), 698–704. doi:10.1111/j.1600-051X.2009.01435.x PMID:19531092

Stanjek, V., Bachmeier, B.-J., & Zander, L. (2014). Cross-linkable materials based on organyl oxysilane-terminated polymers. *Patent US 8907027 B2*.

Stanjek, V., Jung, S., & Schauer, F. (2004). Silanvernetzbare Beschichtungsformulierungen. *Patent DE 10237270 A1*.

Steger, M., Hartmann, C., Beckemper, S., Holtkamp, J., & Gillner, A. (2013). Fabrication of Hierarchical Structures by Direct Laser Writing and Multi-Beam-Interference. *JLMN-Journal of Laser Micro/Nanoengineering, 8*, 210-215.

Stenzel, O. (2005). *The Physics of Thin Film Optical Spectra*. Springer.

Stenzel, V., & Rehfeld, N. (2011). *Functional Coatings*. Hannover, Germany: Vincentz Network.

Stepanova, L. I., & Purovskaya, O. G. (1998). Electrodeposition of nickel-based alloys with tungsten and molybdenum. *Metal Finishing*, *96*(11), 50–53. doi:10.1016/S0026-0576(98)80871-4

Stevens, M. M. (2008). Biomaterials for bone tissue engineering. *Materials Today*, *11*(5), 18–25. doi:10.1016/S1369-7021(08)70086-5

Stigter, M., Bezemer, J., de Groot, K., & Layrolle, P. (2004). Incorporation of different antibiotics into carbonated hydroxyapatite coatings on titanium implants, release and antibiotic efficacy. *Journal of Controlled Release*, *99*(1), 127–137. doi:10.1016/j.jconrel.2004.06.011 PMID:15342186

Stigter, M., de Groot, K., & Layrolle, P. (2002). Incorporation of tobramycin into biomimetic hydroxyapatite coating on titanium. *Biomaterials*, *23*(20), 4143–4153. doi:10.1016/S0142-9612(02)00157-6 PMID:12182316

Stillman, J. M. (1924). *The Story of Alchemy and Early Chemistry*. Dover Publications, Inc.

Stitzel, S., Byrne, R., & Diamond, D. (2006). LED switching of spiropyran-doped polymer films. *Journal of Materials Science*, *41*(18), 5841–5844. doi:10.1007/s10853-006-0289-z

Stobie, N., Duffy, B., Hinder, S. J., McHale, P., & McCormack, D. E. (2009). Silver doped perfluoropolyether-urethane coatings: Antibacterial activity and surface analysis. *Colloids and Surfaces. B, Biointerfaces*, *72*(1), 62–67. doi:10.1016/j.colsurfb.2009.03.014 PMID:19427177

Stover, D., Pracht, G., Lehmann, H. Dietrich, M. & Vaben, R. (2004) New material concepts for the next generation of plasma-sprayed thermal barrier coatings. *Journal ot Thermal Spray Technology*, 13, 76–83.

Strite, S., & Morkoç, H. (1992). GaN, AlN, and InN: A review. *Journal of Vacuum Science & Technology B Microelectronics and Nanometer Structures*, *10*(4), 1237–1266. doi:10.1116/1.585897

Stuart, M.A., C., , Huck, W.T.S., Genzer, J., Müller, M., Ober, C., Stamm, M., & Sukhorukov, G.B. (2010). Emerging applications of stimuli-responsive polymer materials. *Nature Materials*, *9*(2), 101–113. doi:10.1038/nmat2614 PMID:20094081

Suarez-Gonzalez, D., Lee, J. S., Diggs, A., Lu, Y., Nemke, B., Markel, M., & Murphy, W. L. et al. (2014). Controlled multiple growth factor delivery from bone tissue engineering scaffolds via designed affinity. *Tissue Engineering. Part A*, *20*(15-16), 2077–2087. doi:10.1089/ten.tea.2013.0358 PMID:24350567

Suchanek, W., & Yoshimura, M. (1998). Processing and properties of hydroxyapatite-based biomaterials for use as hard tissue replacement implants. *Journal of Materials Research*, *13*(01), 94–117. doi:10.1557/JMR.1998.0015

Such, G. K., Evans, R. A., & Davis, T. P. (2006). Rapid photochromic switching in a rigid polymer matrix using living radical polymerization. *Macromolecules*, *39*(4), 1391–1396. doi:10.1021/ma052002f

Su, F., & Yao, K. (2014). Facile Fabrication of Superhydrophobic Surface with Excellent Mechanical Abrasion and Corrosion Resistance on Copper Substrate by a Novel Method. *ACS Applied Materials & Interfaces*, *6*(11), 8762–8770. doi:10.1021/am501539b PMID:24796223

Sugimoto, K., & Sawada, Y. (1977). The role of molybdenum additions to austenitic stainless steels in the inhibition of pitting in acid chloride solutions. *Corrosion Science*, *17*(5), 425–445. doi:10.1016/0010-938X(77)90032-4

Sugiura, S., Szilágyi, A., Sumaru, K., Takagi, T., Zrínyi, M., & Kanamori, T. (2008). On-demand fluid control on microchip by micro-patterned light irradiation using photo-responsive hydrogels. *Paper presented at theTwelfth International Conference on Miniaturized Systems for Chemistry and Life Sciences*, San Diego, California, USA.

Sugiura, S., Sumaru, K., Ohi, K., Hiroki, K., Takagi, T., & Kanamori, T. (2007). Photoresponsive polymer gel microvalves controlled by local light irradiation. *Sensors and Actuators. A, Physical*, *140*(2), 176–184. doi:10.1016/j.sna.2007.06.024

Sugiura, S., Szilagyi, A., Sumaru, K., Hattori, K., Takagi, T., Filipcsei, G., & Kanamori, T. et al. (2009). On-demand microfluidic control by micropatterned light irradiation of a photoresponsive hydrogel sheet. *Lab on a Chip*, *9*(2), 196–198. doi:10.1039/B810717C PMID:19107273

Sumaru, K., Kameda, M., Kanamori, T., & Shinbo, T. (2004). Reversible and Efficient Proton Dissociation of Spirobenzopyran-Functionalized Poly(N-isopropylacrylamide) in Aqueous Solution Triggered by Light Irradiation and Temporary Temperature Rise. *Macromolecules*, *37*(21), 7854–7856. doi:10.1021/ma0487249

Sun, D., An, J., Wu, G., & Yang, J. (2015). Double-layered reactive microcapsules with excellent thermal and non-polar solvent resistance for self-healing coatings. *J. Mater. Chem. A*, *3*(8), 4435–4444. doi:10.1039/C4TA05339G

Sun, S., & Podlaha, E. J. (2012). Electrodeposition of Mo-rich, MoNi alloys from an aqueous electrolyte. *Journal of the Electrochemical Society*, *159*(2), D97–D102. doi:10.1149/2.091202jes

Sun, X.W., Chen, J., Song, J.L., Zhao, D.W., Deng, W.Q., & Lei, W. (2010). Ligand capping effect for dye solar cells with a CdSe quantum dot sensitized ZnO nanorod photoanode. *Optics Express*, *18*(2), 1296–1301. doi:10.1364/OE.18.001296 PMID:20173955

Surmeneva, M. A., Chaikina, M. V., Zaikovskiy, V. I., Pichugin, V. F., Buck, V., Prymak, O., & Surmenev, R. A. et al. (2013). The structure of an RF-magnetron sputter-deposited silicate-containing hydroxyapatite-based coating investigated by high-resolution techniques. *Surface and Coatings Technology*, *218*, 39–46. doi:10.1016/j.surfcoat.2012.12.023

Surmenev, R. A. (2012). A review of plasma-assisted methods for calcium phosphate-based coatings fabrication. *Surface and Coatings Technology*, *206*(8), 2035–2056. doi:10.1016/j.surfcoat.2011.11.002

Surmenev, R. A., Surmeneva, M. A., Evdokimov, K. E., Pichugin, V. F., Peitsch, T., & Epple, M. (2011). The influence of the deposition parameters on the properties of an rf-magnetron-deposited nanostructured calcium phosphate coating and a possible growth mechanism. *Surface and Coatings Technology*, *205*(12), 3600–3606. doi:10.1016/j.surfcoat.2010.12.039

Surmenev, R. A., Surmeneva, M. A., & Ivanova, A. A. (2014). Significance of calcium phosphate coatings for the enhancement of new bone osteogenesis - A review. *Acta Biomaterialia*, *10*(2), 557–579. doi:10.1016/j.actbio.2013.10.036 PMID:24211734

Syed, M., Hutt, N. J., Shah, N., & Edge, J. (2015). Hydroxyapatite ceramic-coated femoral components in young patients followed up for 17 to 25 years: An update of a previous report. *The Bone & Joint Journal*, *97-B*(6), 749–754. doi:10.1302/0301-620X.97B6.35278 PMID:26033053

Szafert, S., John, L., & Sobota, P. (2008). Well-defined polynuclear heterobimetallic complexes as precursors for ceramic and magnetic materials. *Dalton Transactions (Cambridge, England)*, (46): 6509–6520. doi:10.1039/b806167j PMID:19030610

Sze, S., & Ng, K. K. (2006). *Physics of semiconductor devices*. John Wiley & sons. Inc. doi:10.1002/0470068329

Szilard, B. (1906). *Electrochemistry*, *12*, 373–378.

Taborda, J. A. P., Caicedo, J., Grisales, M., Saldarriaga, W., & Riascos, H. (2015). Deposition pressure effect on chemical, morphological and optical properties of binary Al-nitrides. *Optics & Laser Technology*, *69*, 92–103. doi:10.1016/j.optlastec.2014.12.009

Tadanaga, K., Katata, N., & Minami, T. (1997). Formation process of super-water-repellent Al_2O_3 coating films with high transparency by the sol-gel method. *J. Am. Ceram. Soc.*, *80*, 3213-3216.

Tadmor, R. (2011). Approaches in wetting phenomena. *Soft Matter*, *7*(5), 1577–1580. doi:10.1039/C0SM00775G

Taft, E. A., & Philipp, H. R. (1961). Optical constants of silver. *Physical Review*, *121*(4), 1100–1103. doi:10.1103/PhysRev.121.1100

Takagi, K., Kurematsu, T., & Sawaki, Y. (1995). Photochromic behavior of surfactant spiro 2h-1-benzopyran-2,2' 2,3 -dihydroindole s (spiropyrans) adsorbed into clay interlayers. *Journal of the Chemical Society-Perkin Transactions*, *2*(8), 1667–1671. doi:10.1039/p29950001667

Takumi, K., Sakamoto, H., Uda, R. M., Sakurai, Y., Kume, H., & Kimura, K. (2007). Photocontrol of anionic micelles containing lipophilic crowned spirobenzopyran. *Colloids and Surfaces a-Physicochemical and Engineering Aspects*, *301*(1-3), 100-105. doi:10.1016/j.colsurfa.2006.12.029

Talapin, D. V., Rogach, A. L., Kornowski, A., Haase, M., & Weller, H. (2001). Highly Luminescent Monodisperse CdSe and CdSe/ZnS Nanocrystals Synthesized in a Hexadecylamine–Trioctylphosphine Oxide–Trioctylphospine Mixture. *Nano Letters*, *1*(4), 207–211. doi:10.1021/nl0155126

Tamura, M., Takahashi, M., Ishii, J., Suzuki, K., Sato, M., & Shimomura, K. (1999). Multilayered thermal barrier coating for land-based gas turbines. *Journal of Thermal Spray Technology*, *8*(1), 68–72. doi:10.1361/105996399770350584

Tanaka, K., Takahashi, T., Ban, T., Kondo, T., Uchida, K., & Miura, N. (2003). Comparative study on the excitons in lead-halide-based perovskite-type crystals CH3NH3PbBr3 CH3NH3PbI3. *Solid State Communications*, *127*(9–10), 619–623. doi:10.1016/S0038-1098(03)00566-0

Tang, J., Kemp, K. W., Hoogland, S., Jeong, K. S., Liu, H., Levina, L., & Sargent, E. H. et al. (2011). Colloidal-quantum-dot photovoltaics using atomic-ligand passivation. *Nature Materials*, *10*(10), 765–771. doi:10.1038/nmat3118 PMID:21927006

Tang, Y. Z., Chappell, H. F., Dove, M. T., Reeder, R. J., & Lee, Y. J. (2009). Zinc incorporation into hydroxylapatite. *Biomaterials*, *30*(15), 2864–2872. doi:10.1016/j.biomaterials.2009.01.043 PMID:19217156

Tanioka, A. (2010). Electrospray in the future. *Maku*, *35*(3), 100–104.

TB3: KiON Polysilazanes - Reactivity with Isocyanates. (n. d.). Kion Defense Technologies. Retrieved in 2014 from www.kiondefense.com/bulletins/TB3.pdf

TB4. KiON Polysilazanes - Reactivity with Phenolic Resins. (n. d.). Kion Defense Technologies. Retrieved from www.kiondefense.com/bulletins/TB4.pdf

TB5. KiON Polysilazanes - Reactivity with Epoxy Resins. (n. d.). Kion Defense Technologies. Retrieved from www.kiondefense.com/bulletins/TB5.pdf

Teh, S.-Y., Lin, R., Hung, L.-H., & Lee, A. P. (2008). Droplet microfluidics. *Lab on a Chip*, *8*(2), 198–220. doi:10.1039/b715524g PMID:18231657

Teng, S.-H., Lee, E.-J., Park, C.-S., Choi, W.-Y., Shin, D.-S., & Kim, H.-E. (2008). Bioactive nanocomposite coatings of collagen/hydroxyapatite on titanium substrates. *Journal of Materials Science. Materials in Medicine*, *19*(6), 2453–2461. doi:10.1007/s10856-008-3370-0 PMID:18219556

Terasako, T., Yagi, M., Ishizaki, M., Senda, Y., Matsuura, H., & Shirakata, S. (2007). Growth of zinc oxide films and nanowires by atmospheric-pressure chemical vapor deposition using zinc powder and water as source materials. *Surface and Coatings Technology*, *201*(22-23), 8924–8930. doi:10.1016/j.surfcoat.2007.04.017

Tesmer, J., Nastasi, M., Barbour, J., Maggiore, C., & Mayer, J. (1995). *Handbook of modern ion beam material analysis*. Material Research Society.

Thi, M. M., Suadicani, S. O., Schaffler, M. B., Weinbaum, S., & Spray, D. C. (2013). Mechanosensory responses of osteocytes to physiological forces occur along processes and not cell body and require alphaVbeta3 integrin. *Proceedings of the National Academy of Sciences of the United States of America*, *110*(52), 21012–21017. doi:10.1073/pnas.1321210110 PMID:24324138

Tientong, J. (2014). *Electrodeposition of nickel and nickel alloy coatings with layered silicates for enhanced corrosion resistance and mechanical properties* [Doctoral dissertation].

Tiwari, P. M., Vig, K., Dennis, V. A., & Singh, S. R. (2011). Functionalized gold nanoparticles and their biomedical applications. *Nanomaterials*, *1*(1), 31–63. doi:10.3390/nano1010031

Tobler, W. J., & Virtanen, S. (2006). Effect of Mo species on metastable pitting of Fe18Cr alloys-A current transient analysis. *Corrosion Science*, *48*(7), 1585–1607. doi:10.1016/j.corsci.2005.05.049

Tomasulo, M., Yildiz, I., Kaanumalle, S. L., & Raymo, F. M. (2006). pH-Sensitive Ligand for Luminescent Quantum Dots. *Langmuir*, *22*(24), 10284–10290. doi:10.1021/la0618014 PMID:17107034

Tomita, Y. (2008). *Alternative transparent electrodes for organic light emitting diodes* [Doctoral dissertation].

Tong, X. C. (2009). Advanced Materials and Design for Electromagnetic Interference Shielding. Boca Raton, FL: CRC Press. doi: 10.1201/9781420073591

Tonova, D., Patrini, M., Tognini, P., Stella, A., Cheyssac, P., & Kofman, R. (1999). Ellipsometric study of optical properties of liquid Ga Nanoparticles. *Journal of Physics Condensed Matter*, *11*(10), 2211–2222. doi:10.1088/0953-8984/11/10/008

Touati, I., Hadjoub, Z., & Doghmane, A. (2006). Investigations of elastic properties of Al2O3 layers on sic substrate via acoustic microscopie. *Physical & chemical news*, 27, 26-30.

Townsin, R. L. (2003). The ship hull fouling penalty. *Biofouling*, 19(Suppl.), 9-15.

Trang, H., Marie-Helene, V., Jean-Sebastien, D., Philippe, R., Philippe, B., & Pierre, L.-A. (2007). Temperature-compensated structure for SAW pressure sensor in very high temperature. *Paper presented at the Frequency Control Symposium, 2007 Joint with the 21st European Frequency and Time Forum.* IEEE International. doi:10.1109/FREQ.2007.4319026

Traverso, P., & Canepa, E. (2014). A review of studies on corrosion of metals and alloys in deep-sea environment. *Engineering*, *87*, 10–15.

Treboux, G., Layrolle, P., Kanzaki, N., Onuma, K., & Ito, A. (2000). Existence of Posner's cluster in vacuum. *The Journal of Physical Chemistry A*, *104*(21), 5111–5114. doi:10.1021/jp994399t

Tredwin, C. J., Young, A. M., Neel, E. A. A., Georgiou, G., & Knowles, J. C. (2014). Hydroxyapatite, fluor-hydroxyapatite and fluorapatite produced via the sol–gel method: Dissolution behaviour and biological properties after crystallisation. *Journal of Materials Science. Materials in Medicine*, *25*(1), 47–53. doi:10.1007/s10856-013-5050-y PMID:24052344

Tsai, P.-I., Lee, S.-S., Chou, S.-T., Chang, Y.-T., Lee, A. S.-Y., & Lee, C. K. (2014). Integrating SPR-ellipsometry and electrochemical measurements for performance evaluation of label-free thiophene-based biosensor. *Proc. SPIE 8933. Frontiers in Biological Detection: From Nanosensors to Systems*, *VI*, 89330H. doi:10.1117/12.2038955

Uchida, M., Oyane, A., Kim, H. M., Kokubo, T., & Ito, A. (2004). Biomimetic Coating of Laminin–Apatite Composite on Titanium Metal and Its Excellent Cell-Adhesive Properties. *Advanced Materials*, *16*(13), 1071–1074. doi:10.1002/adma.200400152

Ueda, K., Narushima, T., Goto, T., Taira, M., & Katsube, T. (2007). Fabrication of calcium phosphate films for coating on titanium substrates heated up to 773 K by RF magnetron sputtering and their evaluations. *Biomedical Materials*, *2*(3), S160–S166. doi:10.1088/1748-6041/2/3/S14 PMID:18458462

Uhlmann, P., Frenzel, R., Voit, B., Mock, U., Szyszka, B., Schmidt, B., & Roths, K. et al. (2007). Research agenda surface technology: Future demands for research in the field of coating materials. *Progress in Organic Coatings*, *58*(2-3), 122–126. doi:10.1016/j.porgcoat.2006.08.020

Ulrich, S., Holleck, H., Leiste, H., Niederberger, L., Nold, E., Sell, K., … Sattel, S. (2005). Nano-scale, multi-functional coatings in the material system B-C-N-H. *Surface and Coatings Technology*, *200*(1-4), 7–13.

Unkelhäußer, T., Görlitzer, H., Hallack, M., & Lomölder, R. (2014). The best of two worlds - Silane-urethane hybrid crosslinkers create scratch-resistant clearcoats. *European Coatings Journal*, *2014*(07/08), 21-25.

Uygun, M., Cook, W. D., Moorhoff, C., Chen, F., Vallo, C., Yagci, Y., & Sangermano, M. (2011). Photopolymerization kinetics and dynamic mechanical properties of silanes hydrolyzed without evolution of byproducts. Tetrakis (methacryloyloxy)ethoxy]silane–diethylene glycol dimethacrylate. *Macromolecules*, *44*(7), 1792–1800. doi:10.1021/ma102903q

Valipor, N. (2014). Super-Non-Wettable Surfaces: A Review. *Colloids and Surfaces. A, Physicochemical and Engineering Aspects*, *448*, 93–106. doi:10.1016/j.colsurfa.2014.02.016

Valizadeh, A., Mikaeili, H., Samiei, M., Farkhani, S., Zarghami, N., kouhi, M., & Davaran, S. et al. (2012). Quantum dots: Synthesis, bioapplications, and toxicity. *Nanoscale Research Letters*, *7*(1), 1–14. doi:10.1186/1556-276X-7-480 PMID:22929008

Vallet-Regí, M., & Ruiz-Hernández, E. (2011). Bioceramics: From bone regeneration to cancer nanomedicine. *Advanced Materials*, *23*(44), 5177–5218. doi:10.1002/adma.201101586 PMID:22009627

Van Der Wal, E., Oldenburg, S., Heij, T., Van Der Gon, A. D., Brongersma, H., Wolke, J., & Vredenberg, A. et al. (2006). Adsorption and desorption of Ca and PO 4 species from SBFs on RF-sputtered calcium phosphate thin films. *Applied Surface Science*, *252*(10), 3843–3854. doi:10.1016/j.apsusc.2005.06.006

Van der Wal, E., Wolke, J., Jansen, J., & Vredenberg, A. (2005). Initial reactivity of rf magnetron sputtered calcium phosphate thin films in simulated body fluids. *Applied Surface Science*, *246*(1), 183–192.

van Dijk, K., Schaeken, H. G., Wolke, J. G. C., & Jansen, J. A. (1996). Influence of annealing temperature on RF magnetron sputtered calcium phosphate coatings. *Biomaterials*, *17*(4), 405–410. doi:10.1016/0142-9612(96)89656-6 PMID:8938234

van Embden, J., Jasieniak, J., Gómez, D. E., Mulvaney, P., & Giersig, M. (2007). Review of the Synthetic Chemistry Involved in the Production of Core/Shell Semiconductor Nanocrystals. *Australian Journal of Chemistry*, *60*(7), 457–471. doi:10.1071/CH07046

Van Ooij, W. J., Suryanarayanan, K., & Metroke, T. L. (2007). Integral epoxy resin-silane primer system for hot-dip galvanized steel. In K. L. Mittal (Ed.), *Silanes and other coupling agents* (Vol. 4, pp. 275–286). Boca Raton, FL: CRC Press. doi:10.1163/ej.9789067644525.i-410.122

Vargas-Bernal, R. (2015). Performance Analysis of Electromagnetic Interference Shielding based on Carbon Nanomaterials used in AMS/RF IC Design. In M. Fakhfakh, E. Tlelo-Cuatle, & M. H. Fino (Eds.), *Performance Optimization Techniques in Analog, Mixed-Signal, and Radio-Frequency Circuit Design* (pp. 268–294). Hershey, PA, USA: IGI Global. doi:10.4018/978-1-4666-6627-6.ch011

Vargas-Reus, M. A., Memarzadeh, K., Huang, J., Ren, G. G., & Allaker, R. P. (2012). Antimicrobial activity of nanoparticulate metal oxides against peri-implantitis pathogens. *International Journal of Antimicrobial Agents*, *40*(2), 135–139. doi:10.1016/j.ijantimicag.2012.04.012 PMID:22727529

Vasiljevic, J., Tomsic, B., Jerman, I., Orel, B., Aska, G., Kovac, J., & Simoncic, B. (2014). Multifunctional superhydrophobic/oleophobic and flame-retardant cellulose fibres with improved ice-releasing properties and passive antibacterial activity prepared via the sol-gel method. *Journal of Sol-Gel Science and Technology*, *70*(3), 385–399. doi:10.1007/s10971-014-3294-8

Vaßen, R., Jarligo, M. O., Steinke, T., Mack, D. E., & Stöver, D. (2010). Overview on advanced thermal barrier coatings. *Surface and Coatings Technology*, *205*(4), 938–942. doi:10.1016/j.surfcoat.2010.08.151

Vassen, R., Kassner, H., Stuke, A., Mack, D. E., Jarligo, M. O., & Stöver, D. (2010). Functionally Graded Thermal Barrier Coatings with Improved Reflectivity and High-Temperature Capability. *Materials Science Forum*, *631-632*, 73–78. doi:10.4028/www.scientific.net/MSF.631-632.73

Vassen, R., Stuke, A., Sto, D., Vassen, R., Stuke, A., Sto, D., & Ju, F. (2009). Recent developments in the field of thermal barrier coatings. *Journal of Thermal Spray Technology*, *18*(2), 181–186. doi:10.1007/s11666-009-9312-7

Vengatesh, P., & Kulandainathan, M. A. (2015). Hierarchically Ordered Self-Lubricating Superhydrophobic Anodized Aluminum Surfaces with Enhanced Corrosion Resistance. *ACS Applied Materials & Interfaces*, *7*(3), 1516–1526. doi:10.1021/am506568v PMID:25529561

Verre, R., Fleischer, K., Smith, C., McAlinden, N., McGilp, J. F., & Shvets, V. (2011). Probing the out-of-plane optical response of plasmonic nanostructures using spectroscopic ellipsometry. *Physical Review B: Condensed Matter and Materials Physics*, *84*(8), 085440. doi:10.1103/PhysRevB.84.085440

Verre, R., Modreanu, M., Ualibek, O., Fox, D., Fleischer, K., Smith, C., & Shvets, I. V. et al. (2013). General approach to the analysis of plasmonic structures using spectroscopic ellipsometry. *Physical Review B: Condensed Matter and Materials Physics*, *87*(23), 235428. doi:10.1103/PhysRevB.87.235428

Vijayalakshmi, K. A., Mekala, M., Yoganand, C. P., & Navaneetha Pandiyaraj, K. (2011). Studies on modification of surface properties in polycarbonate (PC) film induced by DC glow discharge plasma. *International Journal of Polymer Science*, 2011.

Vijayalakshmi, U., Chellappa, M., Anjaneyulu, U., Manivasagam, G., & Sethu, S. (2016). Influence of Coating Parameter and Sintering Atmosphere on the Corrosion Resistance Behaviour of Electrophoretically Deposited Composite Coatings. *Materials and Manufacturing Processes*. 31(1), 95-106.

Vilardell Navarro, M. (2014). *Inkjet printing : A flexible manufacturing of functional ceramic coatings by Chemical Solution Deposition*. Universitat Autonoma de Barcelona.

Vilar, R. (1999). Laser cladding. *Journal of Laser Applications*, *11*(2), 64. doi:10.2351/1.521888

Viswanathan, V., Dwivedi, D., & Sampath, S. (2014). Engineered Multilayer Thermal Barrier Coatings for Enhanced Durability and Functional Performance. *Journal of the American Ceramic Society*, *97*(9), 2770–2778. doi:10.1111/jace.13033

Voevodin, A. A., O'Neill, J. P., & Zabinski, J. S. (1999). Nanocomposite Tribological Coatings for Aerospace Applications. *Surface and Coatings Technology*, *116-119*, 36–45. doi:10.1016/S0257-8972(99)00228-5

Voldman, S. H. (2012). *ESD Basics: From Semiconductor Manufacturing to Product Use*. Chichester, United Kingdom: Wiley. doi:10.1002/9781118443323

Vu, C., Osterod, F., Brand, S., & Ryan, K. (2008). Silicon and nitrogen. *European Coatings Journal, 2008*(02).

Wadhavane, P. D., Galian, R. E., Izquierdo, M. A., Aguilera-Sigalat, J., Galindo, F., Schmidt, L., & Luis, S. V. et al. (2012). Photoluminescence Enhancement of CdSe Quantum Dots: A Case of Organogel–Nanoparticle Symbiosis. *Journal of the American Chemical Society*, *134*(50), 20554–20563. doi:10.1021/ja310508r PMID:23214451

Walsh, A., Da Silva, J. L. F., & Wie, S. (2011). Multi-component transparent conducting oxides: Progress in materials modelling. *Phys.: Condens. Matter*, *23*, 334210. PMID:21813942

Wang, A. Q., & Golden, T. D. (2013). Electrodeposition of oriented cerium oxide films. *International Journal of Electrochemistry*, *2013*, 1–10. doi:10.1155/2013/482187

Wang, B., Lilja, M., Ma, T., Sörensen, J., Steckel, H., Ahuja, R., & Stromme, M. (2014). Theoretical and experimental study of the incorporation of tobramycin and strontium-ions into hydroxyapatite by means of co-precipitation. *Applied Surface Science*, *314*, 376–383. doi:10.1016/j.apsusc.2014.06.193

Wang, D., Jiao, P., Wang, J., Zhang, Q., Feng, L., & Yang, Z. (2012). Fast photo-switched wettability and color of surfaces coated with polymer brushes containing spiropyran. *Journal of Applied Polymer Science*, *125*(2), 870–875. doi:10.1002/app.36254

Wang, D., Zheng, Z., Hong, C., Liu, Y., & Pan, C. (2006). Michael addition polymerizations of difunctional amines (AA') and triacrylamides (B3). *Journal of Polymer Science. Part A, Polymer Chemistry*, *44*(21), 6226–6242. doi:10.1002/pola.21716

Wang, F., Qi, B., Wang, G., & Cui, W. (2014). Catalyst coating deposition behavior by cold spray for fuel reforming. *International Journal of Hydrogen Energy*, *39*(25), 13852–13858. doi:10.1016/j.ijhydene.2014.01.133

Wang, F., Shen, T., Li, C., Li, W., & Yan, G. (2014). Low temperature self-cleaning properties of superhydrophobic surfaces. *Applied Surface Science*, *317*, 1107–1112. doi:10.1016/j.apsusc.2014.08.200

Wang, F., Tang, R., & Buhro, W. E. (2008). The Trouble with TOPO; Identification of Adventitious Impurities Beneficial to the Growth of Cadmium Selenide Quantum Dots, Rods, and Wires. *Nano Letters*, *8*(10), 3521–3524. doi:10.1021/nl801692g PMID:18754691

Wang, G. C., & Zreiqat, H. (2010). Functional coatings or films for hard-tissue applications. *Materials (Basel)*, *3*(7), 3994–4050. doi:10.3390/ma3073994

Wang, H., Eliaz, N., & Hobbs, L. W. (2011). The nanostructure of an electrochemically deposited hydroxyapatite coating. *Materials Letters*, *65*(15), 2455–2457. doi:10.1016/j.matlet.2011.05.016

Wang, H., Eliaz, N., Xiang, Z., Hsu, H.-P., Spector, M., & Hobbs, L. W. (2006). Early bone apposition in vivo on plasma-sprayed and electrochemically deposited hydroxyapatite coatings on titanium alloy. *Biomaterials*, *27*(23), 4192–4203. doi:10.1016/j.biomaterials.2006.03.034 PMID:16618502

Wang, H., Liu, Z., Wang, E., Yuan, R., Gao, D., Zhang, X., & Zhu, Y. (2015). A robust superhydrophobic PVDF composite coating with wear/corrosion-resistance properties. *Applied Surface Science*, *332*, 518–524. doi:10.1016/j.apsusc.2015.01.213

Wang, H., Patil, A. J., Liu, K., Petrov, S., Mann, S., Winnik, M. A., & Manners, I. (2009). Fabrication of continuous and segmented polymer/metal oxide nanowires using cylindrical micelles and block comicelles as templates. *Advanced Materials*, *21*(18), 1805–1808. doi:10.1002/adma.200803015

Wang, H., Xie, G., Fang, M., Ying, Z., Tong, Y., & Zeng, Y. (2015). Electrical and mechanical properties of antistatic PVC films containing multi-layer graphene. *Composites. Part B, Engineering*, *79*, 444–450.

Wang, J., Guo, J., Liu, J., Wei, L., & Wu, G. (2014). BMP-functionalised coatings to promote osteogenesis for orthopaedic implants. *International Journal of Molecular Sciences*, *15*(6), 10150–10168. doi:10.3390/ijms150610150 PMID:24914764

Wang, J.-H., & Matyjaszewski, K. (1995). Controlled/"living" radical polymerization atom transfer radical polymerization in the presence of transition-metal complexes. *Journal of the American Chemical Society*, *117*(20), 5614–5615. doi:10.1021/ja00125a035

Wang, L., Guo, S. L., & Dong, S. J. (2008). Facile electrochemical route to directly fabricate hierarchical spherical cupreous micro structures: Toward superhydrophobic surface. *Electrochemistry Communications*, *10*(4), 655–658. doi:10.1016/j.elecom.2008.01.034

Wang, L., Wang, Y., Sun, X. G., Pan, Z. Y., He, J. Q., Zhou, Y., & Wu, P. L. (2011). Microstructure and Surface Residual Stress of Plasma Sprayed Nanostructured and Conventional ZrO_2-8wt%Y_2O_3 Thermal Barrier Coatings. *Surface and Interface Analysis*, *43*(5), 869–880. doi:10.1002/sia.3648

Wang, L., Zhang, J., Gao, Y., Xue, Q., Hu, L., & Xu, T. (2006). Grain size effect in corrosion behavior of electrodeposited nanocrystalline Ni coatings in alkaline solution. *Scripta Materialia*, *55*(7), 657–660. doi:10.1016/j.scriptamat.2006.04.009

Wang, M., Wang, Z., Yu, X., & Guo, Z. (2015). Facile one-step electrodeposition preparation of porous NiMo film as electrocatalyst for hydrogen evolution reaction. *International Journal of Hydrogen Energy*, *40*, 2173–2181. doi:10.1016/j.ijhydene.2014.12.022

Wang, P., Zhang, J., He, H., Xu, X., & Jin, Y. (2015). The important role of surface ligand on CdSe/CdS core/shell nanocrystals in affecting the efficiency of H2 photogeneration from water. *Nanoscale*, *7*(13), 5767–5775. PMID:25757912

Wang, Q. J., & Chung, Y.-W. (2013). *Encyclopedia of Tribiology*. New York, NY: Springer. doi:10.1007/978-0-387-92897-5

Wang, S., Feng, L., & Jiang, L. (2006). One-step solution-immersion process for the fabrication of stable bionic superhydrophobic surfaces. *Advanced Materials*, *18*(6), 767–700. doi:10.1002/adma.200501794

Wang, Y., Li, M., Lv, T., Wang, Q., Chen, Q., & Ding, J. (2015). Influence of Different Chemical Modifications on the Icephobic Properties of Superhydrophobic Surfaces in a Condensate Environment. *Journal of Materials Chemistry A*, *3*(9), 4967–49775. doi:10.1039/C4TA07077A

Wang, Z., Li, Q., She, Z., Chen, F., & Li, L. (2012). Low-Cost and Large- Scale Fabrication Method for an Environmentally-Friendly Superhydrophobic Coating on Magnesium Alloy. *J. Mater. Chem. A*, *22*(9), 4097–4105. doi:10.1039/c2jm14475a

Warrier, M., Lo, M. K. F., Monbouquette, H., & Garcia-Garibay, M. A. (2004). Photocatalytic reduction of aromatic azides to amines using CdS and CdSe nanoparticles. *Photochemical & Photobiological Sciences*, *3*(9), 859–863. doi:10.1039/b404268a PMID:15346187

Watin, M. (1773). *L´Art du peintre, doreur et vernisseur.*

Watson, D. F. (2010). Linker-Assisted Assembly and Interfacial Electron-Transfer Reactivity of Quantum Dot–Substrate Architectures. *The Journal of Physical Chemistry Letters*, *1*(15), 2299–2309. doi:10.1021/jz100571u

Watts, M. (2007). Advances in roll to roll processing, *Impattern Solutions*. Retrieved from http://www.impattern.com/Download/RollToRollProcessing.pdf

Waugh, M. R. (2011). *The Synthesis, Characterisation and Application of Transparent Conducting Thin Films* [Doctoral dissertation].

Weaver, J. H., Krafka, C., Lynch, D. W., & Koch, E. E. (1981). *Optical properties of metals Pt 2*. Germany: Fachinformationszentrum Energie, Physik, Mathematik.

Weeks, M. D., Subramanian, R., Vaidya, A., & Mumm, D. R. (2015). Surface & Coatings Technology Defining optimal morphology of the bond coat – thermal barrier coating interface of air-plasma sprayed thermal barrier coating systems. *Surface and Coatings Technology*, *273*, 50–59. doi:10.1016/j.surfcoat.2015.02.012

Wei, C., Li, J., Gao, F., Guo, S., Zhou, Y., & Zhao, D. (2015). One-Step Synthesis of High-Quality Water-Soluble CdSe Quantum Dots Capped by N-Acetyl-L-cysteine via Hydrothermal Method and Their Characterization. *Journal of Spectroscopy*, *2015*, 7. doi:10.1155/2015/369145

Welser, R. E., Sood, A. W., Cho, J., Schubert, E. F., Harvey, J. L., & Dhar, N. L. … Sood, A. K. (2013). Nanostructured Transparent Conductive Oxides for Photovoltaic Applications. Mater. Res. Soc. Symp. Proc. 1493 (pp. 23-28). Materials Research Society.

Wen, C., Xu, W., Hu, W., & Hodgson, P. (2007). Hydroxyapatite/titania sol–gel coatings on titanium–zirconium alloy for biomedical applications. *Acta Biomaterialia*, *3*(3), 403–410. doi:10.1016/j.actbio.2006.10.004 PMID:17204459

Weng, J., Liu, X., Zhang, X., Ma, Z., Ji, X., & Zyman, Z. (1993). Further studies on the plasma-sprayed amorphous phase in hydroxyapatite coatings and its deamorphization. *Biomaterials*, *14*(8), 578–582. doi:10.1016/0142-9612(93)90174-Z PMID:8399949

Wenzel, R. N. (1936). Resistance of Solid Surfaces to Wetting by Water. *Industrial & Engineering Chemistry*, *28*(8), 988–994. doi:10.1021/ie50320a024

Westin, G. (1997). A Study of the Hydrolysis-Condensation Pathways of a Hetero Bimetallic Alkoxide, Ni2Sb4(OEt)16. *Journal of Sol-Gel Science and Technology*, *8*(23-28).

Widjaja, S., Limarga, A. M., & Yip, T. H. (2002). Oxidation behavior of a plasma-sprayed functionally graded ZrO_2 / Al_2O_3 thermal barrier coating. *Materials Letters*, *57*(3), 628–634. doi:10.1016/S0167-577X(02)00842-X

Widjaja, S., Limarga, A. M., & Yip, T. H. (2003). Modeling of residual stresses in a plasma-sprayed zirconiay alumina functionally graded-thermal barrier coating. *Thin Solid Films*, *434*(1-2), 216–227. doi:10.1016/S0040-6090(03)00427-9

Wilhelm, O., Madler, L., & Pratsinis, S. E. (2003). Electrospray evaporation and deposition. *Journal of Aerosol Science*, *34*(7), 815–836. doi:10.1016/S0021-8502(03)00034-X

Willey, R. R. (2012). *Practical Production of Optical Thin Films* (2nd ed.). Charlevoix, USA: Willey Optical Consultants.

Williams, P. A., Jones, A. C., Crosbie, M. J., Wright, P. J., Bickley, J. F., Steiner, A., & Critchlow, G. W. et al. (2001). Crystal structure of Bi(OCMe2CH2OMe)3 and its use in the MOCVD of Bi2O3. *Chem. Vap. Deposition*, *7*(5), 205. doi:10.1002/1521-3862(200109)7:5<205::AID-CVDE205>3.0.CO;2-9

Witucki, G. L. (2006). Silicone organic hybrid coatings for the industrial maintenance market. Retrieved from http://www.dowcorning.com/content/publishedlit/26-1469-01.pdf

Witucki, G. L. (2013). *Practical applications of polysiloxane coatings* [PowerPoint slides]. Retrieved from http://www.paintsquare.com/education/branding_images/gwwebinar.ppt

Wojtowicz, A. M., Shekaran, A., Oest, M. E., Dupont, K. M., Templeman, K. L., Hutmacher, D. W., & Garcia, A. J. et al. (2010). Coating of biomaterial scaffolds with the collagen-mimetic peptide GFOGER for bone defect repair. *Biomaterials*, *31*(9), 2574–2582. doi:10.1016/j.biomaterials.2009.12.008 PMID:20056517

Wolfbeis, O. S. (2015). An overview of nanoparticles commonly used in fluorescent bioimaging. *Chemical Society Reviews*, *44*(14), 4743–4768. doi:10.1039/C4CS00392F PMID:25620543

Wolfe, D., & Singh, J. (1998). Functionally gradient ceramic/metallic coatings for gas turbine components by high-energy beams for high-temperature applications. *Journal of Materials Science*, *33*(14), 3677–3692. doi:10.1023/A:1004675900887

Woo Kim, J., Kim, B., Won Park, S., Kim, W., & Hyung Shim, J. (2014). Atomic layer deposition of ruthenium on plasma-treated vertically aligned carbon nanotubes for high-performance ultracapacitors. *Nanotechnology*, *25*(43), 435404. 435404-435412 doi:10.1088/0957-4484/25/43/435404 PMID:25299427

Wood, N., Schiemann, U., & Hallack, M. (2005). New Resin Hybrid. *Paint & Coating Industry*, *2005*(April), 70–74.

Wuister, S. F., de Mello Donegá, C., & Meijerink, A. (2004). Influence of Thiol Capping on the Exciton Luminescence and Decay Kinetics of CdTe and CdSe Quantum Dots. *The Journal of Physical Chemistry B*, *108*(45), 17393–17397. doi:10.1021/jp047078c

Wu, L., & Baghdachi, J. (2015). *Functional Polymer Coatings: Principles, Methods, and Applications*. Hoboken, New Jersey: John Wiley & Sons, Inc. doi:10.1002/9781118883051

Wulf, M., Wehling, A., & Reis, O. (2002). Coatings with Self-Cleaning Properties. *Macromolecular Symposia*, *187*(1), 459–467. doi:10.1002/1521-3900(200209)187:1<459::AID-MASY459>3.0.CO;2-Q

Wu, P. C., Kim, T.-H., Brown, A. S., Losurdo, M., Bruno, G., & Everitt, H. O. (2007). Real-time plasmon resonance tuning of liquid Ga nanoparticles by in situ spectroscopic ellipsometry. *Applied Physics Letters*, *90*(10), 103119. doi:10.1063/1.2712508

Wu, P., & Grainger, D. W. (2006). Drug/device combinations for local drug therapies and infection prophylaxis. *Biomaterials*, *27*(11), 2450–2467. doi:10.1016/j.biomaterials.2005.11.031 PMID:16337266

Wu, X. (1999). In situ formation by laser cladding of a TiC composite coating with a gradient distribution. *Surface and Coatings Technology*, *115*(2-3), 111–115. doi:10.1016/S0257-8972(99)00045-6

Wu, Y., Bekke, M., Inoue, Y., Sugimura, H., Kitaguchi, H., Liu, C., & Takai, O. (2004). Mechanical durability of ultra-water-repellent thin film by microwave plasma-enhanced CVD. *Thin Solid Films*, *457*(1), 122–127. doi:10.1016/j.tsf.2003.12.007

Wu, Y.-T., Ho, Y.-C., Lin, C.-C., & Gau, H.-M. (1996). Stepwise Reactions of TiCl4 and Ti(OiPr)Cl3 with 2-Propanol. Variable-Temperature NMR Studies and Crystal Structures of [TiCl2(OiPr)(HOiPr)(Î1/4-Cl)]2 and [TiCl2(OiPr)(HOiPr)(Î1/4-OiPr)]2. *Inorganic Chemistry*, *35*(20), 5948–5952. doi:10.1021/ic960147k

Xia, Z., Yu, X., & Wei, M. (2012). Biomimetic collagen/apatite coating formation on Ti6Al4V substrates. *Journal of Biomedical Materials Research. Part B, Applied Biomaterials*, *100*(3), 871–881. doi:10.1002/jbm.b.31970 PMID:22102365

Xiong, H., Kawasaki, A., Kang, Y., & Watanabe, R. (2005). Experimental study on heat insulation performance of functionally graded metal / ceramic coatings and their fracture behavior at high surface temperatures. *Surface and Coatings Technology*, *194*(2-3), 203–214. doi:10.1016/j.surfcoat.2004.07.069

Xiong, J., Li, Y., Wang, X., Hodgson, P., & Wen, C. (2008). Mechanical properties and bioactive surface modification via alkali-heat treatment of a porous Ti-18Nb-4Sn alloy for biomedical applications. *Acta Biomaterialia*, *4*(6), 1963–1968. doi:10.1016/j.actbio.2008.04.022 PMID:18524702

Xue, C. H., Jia, S. T., Zhang, J., & Ma, J. Z. (2010). Large-area fabrication of superhydrophobic surfaces for practical applications: An overview. *Science and Technology of Advanced Materials, 11*(3). doi:10.1088/1468-6996/11/3/033002

Xue, C.-H., Ji, X.-Q., Zhang, J., Ma, J.-Z., & Jia, S.-T. (2015). Biomimetic superhydrophobic surfaces by combining mussel-inspired adhesion with lotus-inspired coating. *Nanotechnology*, *26*(33), 335602. 335602-335611 doi:10.1088/0957-4484/26/33/335602 PMID:26222622

Xuetao, Y., Yu, W., Dongbai, S., & Hongying, Y. (2008). Influence of pulse parameters on the microstructure and microhardness of nickel electrodeposits. *Surface and Coatings Technology*, *202*(9), 1895–1903. doi:10.1016/j.surfcoat.2007.08.023

Xu, F., Neoh, K. G., & Kang, E. T. (2009). Bioactive surfaces and biomaterials via atom transfer radical polymerization. *Progress in Polymer Science*, *34*(68), 719–761. doi:10.1016/j.progpolymsci.2009.04.005

Xu, H., & Guo, H. (2011). *Thermal Barrier Coatings*. Cambridge, UK: Woodhead Publishing. doi:10.1533/9780857090829

Xu, S., Long, J., Sim, L., Diong, C. H., & Ostrikov, K. K. (2005). RF plasma sputtering deposition of hydroxyapatite bioceramics: Synthesis, performance, and biocompatibility. *Plasma Processes and Polymers*, *2*(5), 373–390. doi:10.1002/ppap.200400094

Xu, X.-D., Wei, H., Zhang, X.-Z., Cheng, S.-X., & Zhuo, R.-X. (2007). Fabrication and characterization of a novel composite PNIPAAm hydrogel for controlled drug release. *Journal of Biomedical Materials Research. Part A*, *81A*(2), 418–426. doi:10.1002/jbm.a.31063 PMID:17117471

Yakovlev, A., Bertrand, P., & Smurov, I. (2003). Laser cladding of wear resistant metal matrix composite coatings. *Thin Solid Films*,453-454, 133-138

Yamaguchi, T., Takahashi, H., & Sudoh, A. (1978). Optical behavior of a metal island film. *Journal of the Optical Society of America*, *68*(8), 1039–1044. doi:10.1364/JOSA.68.001039

Yamaguchi, T., Tanaka, Y., & Ide-Ektessabi, A. (2006). Fabrication of hydroxyapatite thin films for biomedical applications using RF magnetron sputtering. *Nuclear Instruments & Methods in Physics Research. Section B, Beam Interactions with Materials and Atoms*, *249*(1), 723–725. doi:10.1016/j.nimb.2006.03.126

Yamane, H., Kurosawa, H., & Hirai, T. (1988). Preparation of Bi-Sr-Ca-Cu-O films by chemical vapor deposition with metal chelate and alkoxide. *Chemistry Letters*, 9, 1515–1516. doi:10.1246/cl.1988.1515

Yamashita, H., Fukui, K., Misawa, S., & Yoshida, S. (1979). Optical properties of AlN epitaxial thin films in the vacuum ultraviolet region. *Journal of Applied Physics*, *50*(2), 896–898. doi:10.1063/1.326007

Yamashita, K., Arashi, T., Kitagaki, K., Yamada, S., Umegaki, T., & Ogawa, K. (1994). Preparation of Apatite Thin Films through rf-Sputtering from Calcium Phosphate Glasses. *Journal of the American Ceramic Society*, *77*(9), 2401–2407. doi:10.1111/j.1151-2916.1994.tb04611.x

Yang, W. J., Tao, X., Zhao, T-T., Weng, L., Kamg, E-T., & Wang, L. (2015). Antifouling.

Yang, Y., Semonick, M. A., Moore, G. G. I., Boardman, L. D., & Craton, M. A. (2012), Curable-on-demand polysiloxane coating composition. *Patent WO 2012/003153 A1*.

Yang, D., Piech, M., Bell, N. S., Gust, D., Vail, S., Garcia, A. A., & Picraux, S. T. et al. (2007). Photon control of liquid motion on reversibly photoresponsive surfaces. *Langmuir*, *23*(21), 10864–10872. doi:10.1021/la701507r PMID:17803327

Yang, H. S., La, W. G., Bhang, S. H., Lee, T. J., Lee, M., & Kim, B. S. (2011). Apatite-coated collagen scaffold for bone morphogenetic protein-2 Delivery. *Tissue Engineering. Part A*, *17*(17-18), 2153–2164. doi:10.1089/ten.tea.2010.0702 PMID:21529263

Yang, H. S., La, W.-G., Bhang, S. H., Jeon, J.-Y., Lee, J. H., & Kim, B.-S. (2010). Heparin-conjugated fibrin as an injectable system for sustained delivery of bone morphogenetic protein-2. *Tissue Engineering. Part A*, *16*(4), 1225–1233. doi:10.1089/ten.tea.2009.0390 PMID:19886733

Yang, H. S., La, W.-G., Cho, Y.-M., Shin, W., Yeo, G.-D., & Kim, B.-S. (2012). Comparison between heparin-conjugated fibrin and collagen sponge as bone morphogenetic protein-2 carriers for bone regeneration. *Experimental & Molecular Medicine*, *44*(5), 350–355. doi:10.3858/emm.2012.44.5.039 PMID:22322342

Yang, H., Dou, X., Fang, Y., & Jiang, P. (2013). Self-assembled biomimetic superhydrophobic hierarchical arrays. *Journal of Colloid and Interface Science*, *405*, 51–57. doi:10.1016/j.jcis.2013.05.040 PMID:23786830

Yang, J. Y., Neoh, K.-G., Kang, E.-T., Lee, S. S. C., Teo, S. L.-M., & Rittschof, D. (2012b). Functional polymer brushes via surface-initiated atom transfer radical graft polymerization for compacting marine biofouling. *Biofouling: The Journal of Bioadhesion and Biofilm Research*, *28*(9), 895–912. doi:10.1080/08927014.2012.719895 PMID:22963034

Yang, J., Zhang, Z., Xu, X., Zhu, X., Men, X., & Zhou, X. (2012a). Superhydrophilic-superoleophilic coatings. *Journal of Materials Chemistry*, *22*(7), 2834–2837. doi:10.1039/c2jm15987b

Yang, L., Takisawa, N., Hayashita, T., & Shirahama, K. (1995). Colloid-chemical characterization of the photosurfactant 4-ethylazobenzene 4'-(oxyethyl)trimethylammonium bromide. *Journal of Physical Chemistry*, *99*(21), 8799–8803. doi:10.1021/j100021a054

Yang, W. J., Neoh, K. G., Kang, E. T., Teo, S. L. M., & Rittschof, D. (2014). Polymer brush coatings for combating marine biofouling. *Progress in Polymer Science*, *39*(5), 1017–1042.

Yang, W. J., Pranantyo, D., Neoh, K. G., Kang, E. T., Teo, S. L. M., & Rittschof, D. (2012). Layer-by-layer click deposition of functional polymer coatings for combating marine biofouling. *Biomacromolecules*, *13*(9), 2769–2780. doi:10.1021/bm300757e PMID:22924814

Yan, Y., Zhang, X., Huang, Y., Ding, Q., & Pang, X. (2014). Antibacterial and bioactivity of silver substituted hydroxyapatite/TiO2 nanotube composite coatings on titanium. *Applied Surface Science*, *314*, 348–357. doi:10.1016/j.apsusc.2014.07.027

Yeatman, E. M. (1996). Resolution and sensitivity in surface plasmon microscopy and sensing. *Biosensors & Bioelectronics*, *11*(6-7), 635–649. doi:10.1016/0956-5663(96)83298-2

Yonggang, Y., Wolke, J. G., Yubao, L., & Jansen, J. A. (2007). In vitro evaluation of different heat-treated radio frequency magnetron sputtered calcium phosphate coatings. *Clinical Oral Implants Research*, *18*(3), 345–353. doi:10.1111/j.1600-0501.2006.01332.x PMID:17298487

Yoshida, H., Lu, Y., Nakayama, H., & Hirohashi, M. (2009). Fabrication of TiO2 film by mechanical coating technique and its photocatalytic activity. *Journal of Alloys and Compounds*, *475*(1-2), 383–386. doi:10.1016/j.jallcom.2008.07.059

Yoshida, M., Yang, J. S., Kumakura, M., Hagiwara, M., & Katakai, R. (1991). Artificially intelligent hydrogels responding to external stimuli such as temperature and ph. *European Polymer Journal*, *27*(9), 997–1005. doi:10.1016/0014-3057(91)90048-S

Yoshinari, M., Oda, Y., Kato, T., & Okuda, K. (2001). Influence of surface modifications to titanium on antibacterial activity in vitro. *Biomaterials*, *22*(14), 2043–2048. doi:10.1016/S0142-9612(00)00392-6 PMID:11426884

Youssef, K. M. S., Koch, C. C., & Fedkiw, P. S. (2004). Improved corrosion behavior of nanocrystalline zinc produced by pulse-current electrodeposition. *Corrosion Science*, *46*(1), 51–64. doi:10.1016/S0010-938X(03)00142-2

Yuan, Z., Wu, J., Hao, X., Wang, X., Xu, X., & Zhang, H. (2011). Preparation and characterization of superhydrophobic polypropylene surface with excellent self-cleaning property. *Advanced Materials Research*, 295-297, 1613-1616.

Yuan, J., Huang, X., Li, P., Li, L., & Shen, J. (2013). Surface-initiated RAFT polymerization of sulfobetaine from cellulose membranes to improve hemocompatibility and antibiofouling property. *Polymer Chemistry*, *4*(19), 5074–5085. doi:10.1039/c3py00565h

Yuan, S. J., Pehkonen, S. O., Liang, B., Ting, Y. P., Neoh, K. G., & Kang, E. T. (2011). Superhydrophobic fluoropolymer-modified copper surface via surface graft polymerisation for corrosion protection. *Corrosion Science*, *53*(9), 2738–2747. doi:10.1016/j.corsci.2011.05.008

Yuan, S. J., Pehkonen, S. O., Ting, Y. P., Neoh, K. G., & Kang, E. T. (2009). Inorganic-organic hybrid coatings on stainless steels by layer-by-layer deposition and surface- initiated atom-transfer-radical polymerization for combating biocorrosion. *ACS Applied Materials & Interfaces*, *1*(3), 640–652. doi:10.1021/am800182d PMID:20355986

Yuan, X., Li, C., Guan, G., Xiao, Y., & Zhang, D. (2009). Surface grafting modification of fibrous silicates with polyvinylpyrrolidone and its application in nanocomposites. *Journal of Applied Polymer Science*, *111*(1), 566–575. doi:10.1002/app.29109

Yurkin, M. A. (2014). Comment on “Rapid and efficient prediction of optical extinction coefficients for gold nanospheres and gold nanorods”. *Journal of Physical Chemistry*, *118*(37), 21738–21739. doi:10.1021/jp5054524

Zarras, P., & Stenger-Smith, J. D. (2014). Electro-active polymer (EAP) coatings for corrosion protection of metals. In H. M. Abdel Salam (Ed.), *Handbook of Smart Coatings for Materials Protection* (pp. 328–369). Kidlington, United Kingdom: Woodhead Publishing Limited. doi:10.1533/9780857096883.2.328

Zayats, A. V., Smolyaninov, I. I., & Maradudin, A. A. (2005). Nano-optics of surface plasmon polaritons. *Physics Reports*, *408*(3-4), 131–314. doi:10.1016/j.physrep.2004.11.001

Zayats, N., Boĭko, V., Gentsar, P., Litvin, O., Papusha, V., & Sopinskiĭ, N. (2008). Optical studies of AlN/n-Si (100) films obtained by the method of high-frequency magnetron sputtering. *Semiconductors*, *42*(2), 195–198. doi:10.1134/S1063782608020139

Zelenin, O. Y. (2007). Interaction of the Ni^{2+} ion with citric acid in an aqueous solution. *Russian Journal of Coordination Chemistry*, *33*(5), 346–350. doi:10.1134/S1070328407050065

Zelikin, A. N. (2010). Drug releasing polymer thin films: New era of surface-mediated drug delivery. *ACS Nano*, *4*(5), 2494–2509. doi:10.1021/nn100634r PMID:20423067

Zengin, A., & Caykara, T. (2012). RAFT-mediated synthesis of poly[oligoethylene glycol) methyl ether acrylate] brushes for biological functions. *Journal of Polymer Science. Part A, Polymer Chemistry*, *50*(21), 4443–4450. doi:10.1002/pola.26250

Zeuner, M., Pagano, S., & Schnick, W. (2011). Nitridosilicates and oxonitridosilicates: From ceramic materials to structural and functional diversity. *Angewandte Chemie International Edition*, *50*(34), 7754–7775. doi:10.1002/anie.201005755 PMID:21774043

Zhai, L., Cebeci, F. C., Cohen, R. E., & Rubner, M. F. (2004). Stable Superhydrophobic Coatings from Polyelectrolyte Multilayers. *Nano Lett.*, *4*, 1349-1353. doi:10.1021/nl049463j

Zhang, B. G. X., Myers, D. E., Wallace, G. G., Brandt, M., & Choong, P. F. M. (2014). Bioactive coatings for orthopaedic implants-recent trends in development of implant coatings. *International Journal of Molecular Sciences*, *15*(7), 11878–11921. doi:10.3390/ijms150711878 PMID:25000263

Zhang, D., Yu, W., Hao, D., Li, L., Liu, H., & Lu, Z. (2012). Functional nanostructured surfaces in hybrid sol–gel glass in large area for antireflective and super-hydrophobic purposes. *Journal of Materials Chemistry*, *22*(33), 17328. 17328-17331 doi:10.1039/c2jm33557c

Zhang, F., Zhong, H., Chen, C., Wu, X.-g, Hu, X., Huang, H., & Dong, Y. et al. (2015). Brightly Luminescent and Color-Tunable Colloidal CH3NH3PbX3 (X = Br, I, Cl) Quantum Dots: Potential Alternatives for Display Technology. *ACS Nano*, *9*(4), 4533–4542. doi:10.1021/acsnano.5b01154 PMID:25824283

Zhang, G., Zhang, J., Xie, G., Liu, H., & Shao, H. (2006). Cicada Wings: A Stamp from Nature for Nanoimprint Lithography. *Small*, *2*(12), 1440–1443. doi:10.1002/smll.200600255 PMID:17193002

Zhang, H., Jang, J., Liu, W., & Talapin, D. V. (2014). Colloidal Nanocrystals with Inorganic Halide, Pseudohalide, and Halometallate Ligands. *ACS Nano*, *8*(7), 7359–7369. doi:10.1021/nn502470v PMID:24988140

Zhang, H., Lamb, R., & Lewis, J. (2005). Engineering nanoscale roughness on hydrophobic surface - preliminary assessment of fouling behaviour. *Science and Technology of Advanced Materials*, *6*(3-4), 236–239. doi:10.1016/j.stam.2005.03.003

Zhang, J., Wang, S.-C., & Lee, C. T. Jr. (2009). Photoreversible Conformational Changes in Membrane Proteins Using Light-Responsive Surfactants. *The Journal of Physical Chemistry B*, *113*(25), 8569–8580. doi:10.1021/jp807875u PMID:19485396

Zhang, J.-Y., Wang, X.-Y., Xiao, M., Qu, L., & Peng, X. (2002). Lattice contraction in free-standing CdSe nanocrystals. *Applied Physics Letters*, *81*(11), 2076–2078. doi:10.1063/1.1507613

Zhang, K., Francis, L. F., Yan, H., & Stein, A. (2005). Apatite Converted from 3-D Ordered Macroporous Sol–Gel Bioactive Glass (3DOM-BG) Particles. *Journal of the American Ceramic Society*, *88*(3), 587–592. doi:10.1111/j.1551-2916.2005.00143.x

Zhang, L., Zeng, H., & Liu, Q. (2012). Probing molecular and surface interactions of comb-type polymer polystyrene-graft-poly(ethylene oxide) (PS-*g*-PEO) with an SFA. *The Journal of Physical Chemistry C*, *116*(33), 17554–17562. doi:10.1021/jp303561p

Zhang, S. (2013). *Hydroxyapatite Coatings for Biomedical Applications*. CRC Press.

Zhang, S., Wang, Y., Zeng, X., Cheng, K., Qian, M., Sun, D., & Chia, W. et al. (2007). Evaluation of interfacial shear strength and residual stress of sol–gel derived fluoridated hydroxyapatite coatings on Ti6Al4V substrates. *Engineering Fracture Mechanics*, *74*(12), 1884–1893. doi:10.1016/j.engfracmech.2006.04.021

Zhang, X., Geng, T., Guo, Y., Zhang, Z., & Zhang, P. (2013). Facile Fabrication of Stable Superhydrophobic SiO_2/Polystyrene Coating and Separation of Liquids with Different Surface Tension. *Chemical Engineering Journal*, *231*, 414–419. doi:10.1016/j.cej.2013.07.046

Zhang, X.-S., Zhu, F.-Y., Han, M.-D., Sun, X.-M., Peng, X.-H., & Zhang, H.-X. (2013). Self-cleaning poly(dimethylsiloxane) film with functional micro/nano hierarchical structures. *Langmuir*, *29*(34), 10769–10775. doi:10.1021/la4023745 PMID:23906343

Zhang, Y., & Clapp, A. (2011). Overview of Stabilizing Ligands for Biocompatible Quantum Dot Nanocrystals. *Sensors (Basel, Switzerland)*, *11*(12), 11036–11055. doi:10.3390/s111211036 PMID:22247651

Zhang, Z., Finlay, J. A., Wang, L., Gao, Y., Callow, J. A., Callow, M. E., & Jiang, S. (2009). Polysulfobetaine-grafted surfaces as environmentally benign ultralow fouling marine coatings. *Langmuir*, *25*(23), 13516–13521. doi:10.1021/la901957k PMID:19689148

Zhan, X., Yan, Y., Zhang, Q., & Chen, F. (2014). A novel superhydrophobic hybrid nanocomposite material prepared by surface-initiated AGET ATRP and its anti-icing properties. *Journal of Materials Chemistry A. Materials for Energy and Sustainability*, *2*(24), 9390–9399.

Zhao, L., Yan, H., Fang, Z., Wang, J., & Wang, H. (2015). On the flameproof treatment of ramie fabrics using a spray-assisted layer-by-layer technique. *Polymer Degradation & Stability*, *121*, 11–17. doi:10.1016/j.polymdegradstab.2015.08.007

Zhao, W., Li, M., & Peng, H.-X. (2010). Functionalized MWNT-doped Thermoplastic Polyurethane Nanocomposites for Aerospace Coating Applications. *Macromolecular Materials and Engineering*, *295*(9), 838–845. doi:10.1002/mame.201000080

Zheng, Q. S., Yu, Y., & Zhao, Z. H. (2005). Effects of Hydraulic Pressure on the Stability and Transition of Wetting Modes of Superhydrophobic Surfaces. *Langmuir*, *21*(26), 12207–12212. doi:10.1021/la052054y PMID:16342993

Zheng, Y., Gao, S., & Ying, J. Y. (2007). Synthesis and Cell-Imaging Applications of Glutathione-Capped CdTe Quantum Dots. *Advanced Materials*, *19*(3), 376–380. doi:10.1002/adma.200600342

Zhou, L., Li, X., Yu, H., & Li, W. (2007). Layer-by-layer assembled highly reflective optical coatings. *Proceedings of SPIE-The International Society for Optical Engineering* (Vol. 6722, pp.1-4).

Zhou, C., Lu, X., Xin, Z., & Liu, H. (2013). Corrosion resistance of novel silane-functional polybenzoxazine coating on steel. *Corrosion Science*, *70*, 145–151. doi:10.1016/j.corsci.2013.01.023

Zhou, F., & Huck, W. T. S. (2006). Surface grafted polymer brushes as ideal building blocks for "smart" surfaces. *Physical Chemistry Chemical Physics*, *8*(33), 3815–3823. doi:10.1039/b606415a PMID:19817040

Zhuang, C., Zhao, J., Jia, F., Guan, C., Wu, Z., Bai, Y., & Jiang, X. (2009). Tuning bond contents in B–C–N films via temperature and bias voltage within RF magnetron sputtering. *Surface and Coatings Technology*, *204*(5), 713–717. doi:10.1016/j.surfcoat.2009.09.031

Zhu, H., Hu, M. Z., Shao, L., Yu, K., Dabestani, R., Zaman, M. B., & Liao, S. (2014). Synthesis and Optical Properties of Thiol Functionalized CdSe/ZnS (Core/Shell) Quantum Dots by Ligand Exchange. *Journal of Nanomaterials*, *2014*, 14.

Zia, A. W., Wang, Y-Q., & Lee, S. (2015). Effects of physical and chemical plasma etching on surface wettability of carbon fiber-reinforced polymer composites for bone plate applications. *Advances in Polymer Technology*, 34(1), 1-4.

Ziółkowski, B., Czugala, M., & Diamond, D. (2012). Integrating stimulus responsive materials and microfluidics – The key to next generation chemical sensors. *Journal of Intelligent Material Systems and Structures*.

Ziółkowski, B., Florea, L., Theobald, J., Benito-Lopez, F., & Diamond, D. (2013). Self-protonating spiropyran-co-NIPAM-co-acrylic acid hydrogel photoactuators. *Soft Matter*, *9*(36), 8754–8760. doi:10.1039/c3sm51386f

Zou, Y., Yeh, P.-Y. J., Rossi, N. A. A., Brooks, D. E., & Kizhakkeedathu, J. N. (2010). Nonbiofouling polymer brushes with latent aldehyde functionality as a template for protein micropatterning. *Biomacromolecules*, *11*(1), 284–293. doi:10.1021/bm901159d PMID:20000794

Zuzuarregui, A., Coto, B., Rodríguez, J., Gregorczyk, K. E., Ruiz de Gopegui, U., Barriga, J., & Knez, M. (2015). Highly reflective polymeric substrates functionalized utilizing atomic layer deposition. *Applied Physics Letters*, *107*(6), 061602. 61602-61607. doi:10.1063/1.4928375

About the Contributors

Ana Zuzuarregui at present is a post-doctoral researcher of the Nanomaterials group at CIC Nanogune. She received her PhD in Applied Engineering from the University of Navarra in 2013, developed in the BioMEMS Research Line of the Microelectronics and Microsystems Area of CEIT working on the design and fabrication of biosensors for pathogen detection by means of electrochemical techniques. She received her Chemical Engineering degree from the Engineering School of the Polytechnic University of Madrid (Spain) in 2009. She is the referee of several international journals. At present, her research interest field is focused on the fabrication of thin-film coatings based on inorganic-organic materials and polymer modification for several applications.

Maria Carmen Morant-Miñana is a Postdoctoral Researcher in the Nanoengineering group at CIC nanoGUNE. She received her PhD degree in Pharmacy from the University of València in 2007 and for her postdoctoral fellowship she worked in the field of Material Science at the University of Konstanz. Before she assumed her current position in October 2015, she worked in the development of electrochemical sensors for bio and chemical applications based on thin gold films and new advanced materi als. She has authored several articles in recent years, mainly in chemistry and material science journals.

***

Aboubakr Moustafa Abdullah After graduation in 1993 from the Chemistry Department, Faculty of Science, Cairo University with high honors, Dr. Aboubakr was awarded his M.Sc. degree in 1997 in Physical Chemistry from the same school on the metal-complex modification of conducting polymers and their use as sensors for some neurotransmitters. In 1998, he received a Ph.D. scholarship from The Pennsylvania State University (PSU), USA to pursue his Ph.D. degree, which he was awarded in 2003 from PSU, from the Department of Materials Science and Engineering in the field of localized corrosion of aluminum and its alloys. From 2004 - 2006, he joined Kuwait University and continued his corrosion studies, but this time for corrosion of Cu and its alloys. Between October 2006 and January 2009, he joined the Tokyo Institute of Technology, Japan, as a Research Associate. His work there was focusing on the performance, catalyst poisoning, and the chemical degradation of the membrane electrode assembly of the polymer electrolyte membrane fuel cells. In January 2009, he received a third post doc fellowship from The University of Calgary, Alberta, Canada for one more year. During this year, he focused on studying the surface design and tailoring of carbon materials. In addition, as a member in the solid oxide fuel cells strategic research network in Canada (SOFC Canada), he worked on optimizing the loading of Ni nano and micro particles as an anode catalyst using different loading techniques within tubular porous YSZ electrolyte. Currently, Dr. Aboubakr is on leave from Cairo University to work as an associate professor at the Center for Advanced Materials of Qatar University, Qatar. Besides his gained experience over many years in the corrosion science and engineering field, Dr. Aboubakr is currently working on superhydrophobic materials. In addition to aforementioned research interest, he is also currently working on new carbonaceous materials for supercapacitors and wastewater treatment applications.

M. Tarik Arafat is a Research Fellow in the Clothworkers' Centre for Textile Materials Innovation for Healthcare & Biomaterials and Tissue Engineering Group, the School of Dentistry at the University of Leeds. He is working on the stratified design and manufacture of nonwoven collagen scaffolds, and the project is funded by MeDe Innovation of EPSRC centre for innovative manufacturing in medical devices. Prior to joining the University of Leeds as a Research fellow, he worked at Micron Technology Inc. as a Senior Process and Equipment Engineer (Photolithography) for 3 years. During this job, he worked in some world-wide collaborative projects, and was also involved in collaborative work with other companies. He joined Micron just after finishing his PhD. He completed his PhD (majoring in scaffold design, biomaterials and additive manufacturing) from National University of Singapore (NUS). His PhD work consists of designing, developing and fabricating scaffolds using a new generation of biomaterials which are osteoconductive and mechanically compatible to bone. During his PhD, he was also attached to Institute of Materials Research and Engineering (IMRE, A*STAR), Singapore.

Masoud Alizadeh received his PhD in Materials Engineering from Materials and Energy Research Center of karaj-Iran in 2006. He has been researched on thermal barrier coating and engineering ceramics. He has many articles in his research area and also he has been worked on many industry projects. At this time, He is a faculty member of materials and energy research center.

Ahmed Bahgat obtained his B.Sc. & M.Sc. degrees in Physical Chemistry from Zagazig University (Egypt) in 2004 and 2013. He has been working as a research assistant at the Centre for Advanced Materials at Qatar University. His research interests focus on materials science, physical chemistry and superhydrophobic coatings.

Rafael Vargas-Bernal was born in Irapuato, Guanajuato, México, September 30th, 1972. He received a B. E. degree in Electronics October 6th 1995, from the University of Guanajuato. After, a M. S. degree in Electronics from INAOE September 4th 1997. Lastly, he earned a Sc. D. degree in Electronics from the same institute July 19th, 2000. During October 2000 - September 2001, he worked as a professor at the University of Guanajuato in the Department of Communications and Electronics Engineering. Since January 2002 to December 2009, he has been with the Department of Electronics Engineering of ITESI as a Professor-Researcher. Since August 2005, he has been working in the Department of Materials Science in the area of Advanced Materials. He is a technical reviewer of standards on microtechnology and MEMS in SEMI since November 2006. Also, he is a reviewer of the magazine IEEE Latin America Transactions of the IEEE since August 2003; the Sensors and Actuators B: Chemical of Elsevier since February 2010; the IEEE Sensors Journal since October 2012; the Nanoscale of RSC since November 2012, and RSC Advances published by RSC since April 2013.

Manuel Cervera earned his Ph.D. in Physics at the Autonomous University of Madrid (UAM) in 1994 with a work on thermal treatments with electron beams. In 1996 he made a postdoctoral stay of one year in the Istituto Lamel Bologna (Italy), working with a linear accelerator for RBS and ERDA analysis and high energy ion implantation. Since 2003, he has been a tenured professor in the Department of Applied Physics at the UAM. He has worked on the synthesis and characterization of different compounds, mainly in thin films, and on the modification of the properties of material surfaces, by deposition techniques in a high vacuum (sputtering DC and RF, and CVD deposition assisted by electron cyclotron resonance plasma) and/or ion implantation. Materials studied were BCN, BN, and analysis techniques including electrical (IV, CV, DLTS), optical (spectroscopic ellipsometry) and structural (FTIR, X-rays, RBS, ERDA). His current work focuses on the growth of silicon nanowires by the VLS method in a CVD low pressure reactors and more recently, the growth of a new III-V semiconductors for application in solar cells.

Matteo Campostrini is a PhD student in the industrial engineering department at the University of Trento. The aim of his Ph.D. project is the development of a metallic multilayer for protection from intense electric fields, these kind of coatings are used to prevent the damage of X-band RF accelerating cavities. These devices are used to increase the energy of the particles in high energy nuclear accelerator. The coating is made by PVD (Physical vapour deposition) Magnetron sputtering using HiPIMS technique. HiPIMS (high power impulse magnetron sputtering) is a sputtering technique used to cover 3D shape with high quality and dense full coating. This project is in a collaboration with INFN (National institute of nuclear physic), an Italian research organisation, and usually he works at LNL (National Laboratory of Legnaro) in Padua. Matteo completed his master's degree in Engineering of Materials, at the University of Trento in October 2012. His thesis was on the design and development of optical filters applied in photovoltaic fields, always using PVD reactive magnetron sputtering techniques.

Dermot Diamond received his Ph.D. and D.Sc. from Queen's University Belfast (Chemical Sensors, 1987, Internet Scale Sensing, 2002), and was VP for Research at Dublin City University (2002–2004). He has published over 300 peer-reviewed papers in international journals, is a named inventor in 19 patents, and is co-author and editor of three books. He is currently the director of the National Centre for Sensor Research (www.ncsr.ie) and SFI INSIGHT Investigator (http://www.insight-centre.org/), a major research initiative focused on Big Data and Sensor Networks. In 2002, he was awarded the inaugural silver medal for Sensor Research by the RSC, London.

Susana María Fernández is a doctor in Science since October 2004. Her experience in the research field has been on different deposition techniques for the fabrication of semiconductors, from nitrides by molecular beam epitaxy to transparent conductive oxides by magnetron sputtering and also, evaporation of metals for contacts to apply in optoelectronic and solar devices.

Larisa Florea studied organic chemistry and chemical engineering at the University "Politehnica" in Timisoara, Romania (B.Sc. Hons 2009). In 2009, she joined the Adaptive Sensors Group at Dublin City University where she earned her Ph.D. degree under the supervision of Prof. Dermot Diamond and Dr. Fernando Benito-Lopez. Her research interests include the development of stimuli-responsive polymers as novel sensing materials in microfluidics.

Raquel Eugenia Galian has obtained her Ph.D. in Chemistry from the National University of Cordoba in Argentina (2001). She was a post-doc researcher at the Polytechnic University of Valencia (2002-

2003, Prof. Miranda), the University of Valencia (2004-2005, Prof. Pérez-Prieto) and the University of Ottawa (2005-2006, Prof. Scaiano). She has made contributions in different areas of research related to photosensitized process using aromatic ketones and functionalization of new materials such as photonic crystal fibers. In 2007, she has won a Ramón y Cajal Contract and is now a Researcher in the General Foundation of the University of Valencia since 2012. Her main interests include the design of hybrid nanomaterial and functionalization of nanoparticles (semiconductors, metallic, magnetic, mesoporous and perovskites) for different applications such as for sensors, carries of photosensitizers, photocatalysis and the preparation of luminescent devices.

Teresa D. Golden is a Full professor at the University of North Texas. Her research area covers the electrodeposition of nanomaterials of unique compositions and phases, including rare earth oxides, cermets, and metallic composites. Dr. Golden has published over 85 peer-reviewed papers in international journals and continues work on coatings for corrosion protection.

Paul A Goodman received his Ph. D. in chemistry from the University of Nebraska-Lincoln in 2013. He is currently a Research Chemist at NAWCWD China Lake, and his research interests include electrochemistry, energy storage/conversion, corrosion, and water purification.

S. M. Mehdi Hadavi obtained his Ph.D in Materials science and engineering- Corrosion and its protection from university of New South Wales, Sydney, Australia in 1996. He trained MSc and PhD students in Amir Kabir University of Technology as well as MA University in Tehran. He has published his findings in more than 150 papers. He is the sole author of 4 books and co-author of another 3 books. He is currently Associate Professor and serves as the president of the National Institute of Materials and Energy.

María Jesús Hernández got her PhD in Physics from Autonomous University of Madrid (UAM), Spain in 1998. Her PhD thesis was about CVD plasma growth of dielectrics for electronic applications at low temperatures. She performed a postdoctoral stage of fifteen months in Lyon, France at the laboratory of Multimatériaux et Interfaces of CNRS (UMR 5615), studying heteroepitaxial growth of Silicon Carbide on Silicon. Other research lines have been the synthesis of new wide band gap semiconductors, as Silicon Boron Carbonitrides obtained by plasma CVDs, as well as the modification of silicon substrates by high dose ionic implantation, or CVD deposited films of zinc oxide and zinc nitride. Throughout his research career, he has specialized in various techniques of optical, electrical and structural characterization, as spectroscopic ellipsometry, FTIR, C-V and I-V measurements or XRD, among others. In the last few years, she has focused her work on the study of VLS catalyzed growth of silicon nanowires by plasma processes. Actually, her work is focused on two fields, the growth of new III-V semiconductors for application in solar cells and the use of ellipsometry in biosensing platforms. Her research has been performed at the Electronics and Semiconductors Laboratory of the UAM, were she is a permanent professor in the Department of Applied Physics since June 2003.

Stefan Holberg received his Ph.D. in 2000 from the Institute of Organic Chemistry at Stuttgart University, Germany. In 2001, he joined Prof. H. Schmidt's group at the INM Leibniz-Institute for New Materials GmbH, Saarbrücken, Germany, as a Research Associate. Since 2006, he has been a Research

Associate and Project Manager at The Danish Technological Institute in Aarhus, Denmark. His research focus is on functional coatings (esp. repellent coatings) development projects in cooperation with industry.

Xu Li is a Senior Scientist at the Institute of Materials Research and Engineering (IMRE), Agency for Science, Technology and Research (A*STAR). He finished his PhD study in chemistry from the Department of Chemistry, at the National University of Singapore in 2001. He then joined IMRE as a Research Associate in 2001 and was promoted first to a Research Scientist, then to Senior Scientist in 2004 and 2011, respectively. He has been an adjunct Associate Professor of the Department of Chemistry at the National University of Singapore since 2012. As a principle investigator, he is now leading a research team on hybrid and nano-structured material development for various applications, especially nanostructured particles for controlled drug delivery and bioimaging.

Elvis O. López has a PhD. in Physics from the Centro Brasileiro de Pesquisas Físicas (2010 - 2014), working in the following areas: RF-magnetron sputtering (experimental prototype of right angle magnetron sputtering - RAMS), producing stoichiometric and crystalline hydroxyapatite thin films and crystalline hydroxyapatite thin films with ionic substitution by fluor, zinc, silicate and carbonate for biomedical applications. Moreover, he has experience in plasma processing and diagnostic of confined RF plasmas using Hall, Langmuir and Faraday probes and optical emission spectroscopy (EOS), as well as experience in structural characterization techniques of: grazing incidence X-ray diffraction by synchrotron radiation (GIXRD), Fourier transform infra-red spectroscopy (FTIR), atomic force microscopy (AFM), X-ray photoelectron spectroscopy (XPS), focused ion beam (FIB) and transmission electron microscopy (TEM).

Fernando Benito-Lopez studied chemistry at the Universidad Autonoma de Madridand and completed his masters studies in the Department of Inorganic Chemistry in 2002.He obtained his Ph.D. at the University of Twente, The Netherlands, under the super-vision of Prof. David N. Reinhoudt in 2007. He carried out his post-doctoral research in the group of Prof. Dermot Diamond at Dublin City University, Dublin, Ireland. In 2010, he was the team leader in polymer microfluidics at CLARITY: Centre for Sensor WebTechnology, National Centre for Sensor Research, Dublin City University. In 2012, he became a senior scientist at CIC microGUNE, Spain. In 2015, he moved to the University of the Basque Country UPV/EHU as Ramón y Cajal Fellow.

Carolina Egler Lucas received her Diploma in chemistry from the University of Barcelona in 2007. At the same year, she moved to the University of Zürich where she has worked in the group of Prof. H. Berke in the areas of 'organometallic chemistry'' and 'electronic materials.' After finishing her PhD, she moved to the University of Heidelberg, where she works as research associate together with Dr. C. Romero-Nieto in the field of organic materials.

Mariam Al-Maadeed is the director of the Center for Advanced Materials at Qatar University and the founder and coordinator of the Materials Science and Technology Master Program. She received her PhD in Materials Science from Alexandria University in Alexandria, Egypt in 2001 and joined the Physics Department at Qatar University in the same year. Mariam has had a long experience and a large body of internationally recognized work in the field of polymers characterization and structure, as well as in nanocomposites and nanotechnology techniques. Her work is featured in more than 70 journals and conference publications. Professor Mariam has many research grants as PI and Co-PI on different projects with total funds amounting to approximately $7,000,000 during the last five years. Professor Al-Maadeed has given many consultations and workshops to a diverse group of organizations such as industries and ministries, as well as the community. Professor Mariam designed and contributed to many workshops, symposiums and conferences in the field of materials science and in teaching and learning strategies. Among the accolades she has received is the Physics State Award on 2010/ 2012, Leadership Excellence for Women Award (Mentorship Award), Middle East Process Engineering Conference and Exhibition, 2015 and the Gulf Petrochemical Association (GPCA) Plastic Excllence Award 2014.

Antonio Garrido Marijuán (MSc.) works as a Junior Scientist in the Energy Department of the Austrian Institute of Technology (AIT). He graduated as Dipl.-Eng. in Chemical Engineering from the Universidad Politecnica de Madrid (Spain) and holds a M.Sc in Industrial Technologies, specialized in solar-thermal trigeneration technologies at a building level. He previously worked in the field of energy efficiency of buildings at the Spanish Research Center for Energy, Environment and Technology (CIEMAT).

Antonio García Marín is a Ph.D. student from the Applied Physics Department at Universidad Autónoma de Madrid (UAM). He received his BS degree from Universidad Complutense de Madrid (UCM) in 2010. He was granted a research fellowship in 2009 from Spanish Ministry of Science and Innovation (MCINN) to study the metaloproteomic of cisplatin, carboplatin and oxaliplatin and their interactions with several proteins by means of electrophoretic techniques and inductively coupled plasma mass spectrometry (ICP-MS). He received the MS degree from UCM in July of 2011. In October of 2011, he joined UAM as a researcher, where he was developing new biosensors based on the plasmonic resonance of gallium and gold using ellipsometry, and optoelectronic devices with metal oxide nanowires in order to get his PhD. He has written 12 publications about these topics in peer-reviewed journals.

Gianantonio Della Mea has been a professor of materials physics at the University of Trento since 1987. At present, his research activities are mainly devoted to the production and characterization of thin films for radiation detectors and sensor devices. For more than 25 years, he has been the scientific director of the Laboratory of Materials for Detectors at INFN-LNL. He is author of more than 250 publications on materials physics and on the interaction between ion beams and solid-state matter for materials analyses.

Adel M Mohamed received his B.Sc and M.Sc degrees in Metallurgical Engineering from the Suez Canal University (Egypt). In April 2008, obtained his Ph.D degree from the University of Quebec at Chicoutimi (Canada). From May 2008 to Jan 2011, he completed his post-doctorate at the same university. From Feb 2011 to Feb 2012, he worked as an assistant professor. He is the co-supervisor of PhD and Master's students. Currently, he is working as a researcher at the Advanced Materials Center, Qatar University. His research focuses on the advanced technology of light metals for automotive applications and nanotechnology. He has published about 40 research articles in various international journals and for conferences.

A.H. Pakseresht has worked as a researcher at the Materials and Energy Research Center (MERC) Karaj-Iran Iran, since 2010. Pakseresht completed his Ph.D. in materials science and engineering at Materials and Energy Research Center in 2015. He received his undergraduate study in materials science and engineering from Theran University. His research interests lie in the area of plasma spraying, with a focus on splat morphology and new thermal barrier coatings. He also, has collaborated actively with researchers in several other disciplines of materials science, particularly film, coating and metal matrix composite. Dr. Pakseresht has also published a book that has been published 5 times in 7 years. Totally, he has published more than 30 journal and conference paper (till 2015). He is an industry expert with more than 10 years of successful experience. He also, teaches physical chemistry of materials, thermodynamics of materials, non-destructive testing, and physical properties of materials.

Jose Luis Pau is an Associate Professor at Universidad Autónoma de Madrid (UAM). He received the BS degree from UAM in 1998 and a Ph.D. from the Universidad Politécnica de Madrid (UPM) in 2003. From 2003 to 2006, he was an Assistant Researcher and Project Manager at ISOM-UPM. In 2006, he received a Postdoctoral Fulbright Fellowship to join the Center for Quantum Devices at Northwestern University for two years. In 2008, he was awarded a Ramon y Cajal Grant within the Excellence R+D Program of the Spanish Ministry of Education and Science, becoming permanent member of the Applied Physics Department at UAM in 2013. He has been a visiting researcher at the University of California, Santa Barbara; Université Paris-Sud 11; the Massachusetts Institute of Technology; and The University of Alabama. His current research interests include the study of scalable tools for the integration of semiconductor nanowires and 2D materials in electronic devices and the development of plasmonic biosensors based on the ellipsometric properties. He has written over 80 publications in peer-reviewed journals, being cited more than 900 times, and reaching an H-index of 18.

Juan Piqueras is a Full Professor of Electronics at the Autonomous University of Madrid. He got his PhD in Physics at the same university in 1975. His research falls within the fields of semiconductor materials for electronic and optoelectronic applications. He is the Head of the Microelectronics Laboratory, a center devised for the training of undergraduate and graduate students in microfabrication technologies. Lately, his research has been focused on the optical characterization on plasmonic structures and the development of optical biosensors. He has been a member of the National Plan for Microelectronics, the Director of the Applied Physics Department and the Coordinator of the Electronics and Communications Areas of the National Agency. He has written more than 100 publications in international peer-reviewed journals and made over 100 contributions to international conferences.

Julia Pérez-Prieto is Full Professor of Organic Chemistry and the leader of the "Photochemistry Reactivity Group" at the Instituto de Ciencia Molecular at the Universitat de Valencia. She is the Coordinator of the Master and Doctorate of "Sustainable Chemistry" (sustainablechemistryval.com) at the UV since 2008. This is an official PhD Program and has an inter-universitary character. Julia Pérez-Prieto has acted as the supervisor for a total of 11 PhD theses. Currently, Prof. Pérez-Prieto supervises five PhD theses (1 in the writing phase). Furthermore, Prof. Pérez-Prieto has directed a considerable number of research works/dissertations and also master thesis. Currently, the Photochemistry Reactivity Group is working with semiconducting NPs (CdSe, CdSe/ZnS), plasmonic metal NPs (AuNP,AgNP), and upconversion NPs (UCNP, $NaYF_4$:Yb,Er). In addition, they are interested in the preparation of highly luminescent organolead halide perovskite NPs; and have learned how to prepare and stabilize this type of NPs as solid and colloidal solutions. They have gained experience in the preparation of functional nanosystems based on these types of nanparticles, semiconducting or metallic, and are now pursuing the preparation of more complex structures (multifunctional systems) based either on these materials or new type of photoactive NPs. Prof. Julia Pérez-Prieto has authored over 143 papers in the field of photochemistry, chemical reactivity, and nanomaterials.

Carles Lizandara Pueyo received his Diploma in chemistry from the University of Barcelona in 2007. At the same year he moved to the University of Konstanz where has worked in the group of Prof. S. Polarz in the areas of 'precursor chemistry' and 'metal oxide semiconductors.' After finishing his PhD, he worked at the Institute of Chemical Research of Catalonia together with Prof. M.A. Pericàs as a Post-Doctoral researcher in the areas of 'heterogenization of homogeneous catalysts' and 'asymmetric catalysis.'

Mohammad Reza Rahimipour born in 1961. He is presently professor of Materials and Energy Research Center (MERC). He obtained B.S and M.Sc. degrees in Materials Engineering in 1986 and 1992 , respectively, from Isfahan university of Technology and Ph.D. degree in material engineering in MERC in 1999. He has to his credit more than 62 Technical Reports and, as well as, 280 publications including papers in referred journals, book and conference papers. His research interests lie in the composite Materials and Coatings. Prof. M.R. Rahimipour is editor-in-chief of Advanced Ceramics and Progress, as well as, Nano Materials Journal.

Walter Raniero is a PhD in Material Sciences and Engineering in the Department of Industrial Engineering at the University of Trento. After his degree in materials engineering at the University of Trento in 2010 with a research on optical nanostructured multilayer for photovoltaic application, he worked for some Italian companies in the renewable energy sector involving the research of a new prototype concentrated photovoltaic system. His expertise as material engineer lies in the measurement and instrumentation techniques of optical and surface analysis he has conducted, helping him to develop his competences since 2012 attending a PhD Course in Material Sciences and Engineering (MSE). During his academic years he was the author of publications on the subject of dichroic optical multilayers design, production and characterization. His main research interests are related to the optical material design, with special attention to optical design matching between different optical components. He is presently preparing his dissertation on Optimization of a PVD deposition plant for the realization of dichroic filters used in CPV spectral separation systems for energy production, focusing on the oxide material analysis.

A.Shahbazkhan is a PhD student in materials science and engineering at the Sharif University, Tehran, Iran. She received her M. S. in materials science and engineering in 2011. She teaches in Islamic Azad University. In addition to teaching, she has conducted several research about nano-composite, plasma spraying of different materials and synthesis of Materials. She has published a several publication in her research area.

John D Stenger-Smith received a Bachelor of Chemical Engineering from the University of Delaware in 1984 and a Ph.D. in Chemical Engineering from the University of Massachusetts at Amherst in 1989. He has been employed at the NAWCWD, China Lake as a Chemical Engineer since 1990 and is working on non-linear optical polymers, electroactive polymers for charge storage devices and electroactive polymers for corrosion inhibition.

Jaime Andrés Pérez Taborda Engineering Physics from the Universidad Tecnológica de Pereira, Colombia, Colombia. He has over 25 publications in scientific journals, is currently vice president and founder of the Colombian Society of Physics Engineering (SCIF) (SCIF) and has participated in major international meetings and as an invited speaker. Master of synchrotron light radiation and particle accelerators at the Autonomous University of Barcelona - ALBA Synchrotron. He is a pre-doctoral researcher at the Institute of Microelectronics of Madrid - CSIC.

Jeerapan Tientong is a Lecturer in Faculty of Science and Technology (Chemistry) at Nakhon Sawan Rajabhat University.

Peter Zarras received his Ph.D. in polymer chemistry from the Polytechnic University, New York. Dr. Zarras is currently employed by the Department of the Navy and his research interests include polymer synthesis, conductive polymers, corrosion-inhibiting coatings and smart coatings.

Index

Q

R

S

T

U

www.ingramcontent.com/pod-product-compliance
Lightning Source LLC
LaVergne TN
LVHW060254011026
849161LV00011B/2714